Physics
for
Technology

Physics
for
Technology

Second Edition

John E. Betts

Camosun College
Victoria, British Columbia

Reston Publishing Company, Inc.
A Prentice-Hall Company
Reston, Virginia

Library of Congress Cataloging in Publication Data

Betts, John E
  Physics for technology.

  Includes index.
  1. Physics.   2. Engineering.   I. Title.
QC21.2.B49   1981   530'.02462   80-25329
ISBN 0-8359-5544-3
ISBN 0-8359-5545-1 (pbk.)

©1981 by
Reston Publishing Company, Inc.
A Prentice-Hall Company
Reston, Virginia 22090

10   9   8   7   6   5   4   3   2   1

Printed in the United States of America

Contents

Preface

Physics is the applied science from which most of the engineering technologies are derived. A knowledge of the basic principles of physics will help students to understand many aspects of technology.

This book is intended for use in a one-year course of applied physics at colleges, universities and technical institutes. No previous knowledge of physics is assumed, and only a basic understanding of elementary algebra and trigonometry is required. The relationships between physics and the engineering technologies are stressed. Many practical examples and recent developments in technology are included.

The International System of Units (SI) is introduced and discussed extensively. Only SI and engineering units and symbols are used. In general, students will not be expected to study the complete book, but there should be a sufficient number of examples and applications to make physics relevant to their areas of study.

I would like to express my appreciation to my colleagues E. Engel, L. Williams, A. Dixon, H. Stiebert, E. Tong, N. Ernst, R. Pierce, N. Preston, M. Bancroft, L. Waye and C. Wehrfritz at Mohawk College and Camosun College for their assistance. M. Bushner of the Milwaukee School of Engineering has also been very helpful.

My thanks to the staff of Reston Publishing Company, Inc., particularly Ben Wentzell, Barbara Gardetto and Meryl Goodman Thomas for their enthusiasm, cooperation and research.

Finally, I would like to thank my wife, Ursule, for her contributions to both the typing and proofreading.

John Betts

Man's natural curiosity and his abilities to think, to record information, and to construct tools have enabled him to be dominant in his environment. His quest for knowledge has produced a vast accumulation of facts and the discovery of general laws which govern many natural phenomena.

Knowledge is usually obtained through a self-correcting process of systematic investigation called the *scientific method*. The scientific method of investigation proceeds as follows:

1. *Observation:* Natural phenomena are observed and are recorded accurately in detail. Questions arise during this process because of these observations.

2. *Experiments:* The system under investigation is subjected to a series of different controlled conditions, and the responses are accurately recorded.

3. *Analysis:* All the data obtained from the controlled experiments are analyzed, and an imaginary model of the system is formed. This model gives a visual picture of the system. A suitable *theory* can then be developed to explain the observations.

4. *Hypothesis:* The model of the system is used to predict new results and characteristics.

5. *Verification:* More experiments are performed in order to test the predictions. If the predictions are inaccurate, the model is revised.

All models and laws obtained by the scientific method are always subject to change. Instruments are continually being developed and improved so that measurements can be made with more precision. This sometimes leads to the abandonment or modification of old theories.

The term *natural science* is used to denote a knowledge of facts and laws that has been obtained systematically by the scientific method. *Physics* is a branch of natural science that is concerned with the study of energy, forces, motion and matter. The extremely large scope

1

Introduction

of physics includes the fields of mechanics, fluids, waves, sound, light, heat, electricity, magnetism, solids, the atom and the atomic nucleus.

1-1 MATTER

All physical objects are composed of a substance called matter. A particle is a minute subdivision of matter.

The ultimate structure of matter is still a mystery and is an active area of research; but we do know that all matter is composed of small particles called atoms, which contain smaller particles called electrons, protons and neutrons. Many even smaller particles have also been discovered in atoms.

All matter possesses the basic property of *inertia*, which is a resistance to any change in motion and is described by the term mass. The *mass* of an object is an indication of the quantity of matter that it possesses; the more massive an object, the greater its inertia. An object has the same mass at any location, e.g. on the earth, on the moon and even in free space.

Any action, such as a push or a pull, that tends to make a stationary object move or changes the speed or direction of motion of a moving object is called a *force*. One type of force is due to the natural physical phenomenon that every particle of matter attracts all other particles of matter in the universe. This attraction between masses is called the *force of gravity*. The magnitude of the force of gravity between any two objects depends on their distance apart and their masses. The *weight* of an object is actually mainly due to its gravitational attraction towards some celestial body (usually the earth), therefore the weight of an object depends on its location. For example, the weight of an object on the earth's surface is not the same as its weight on the moon or in free space.

Matter may exist as a solid, liquid or gas (vapor) depending on its energy content. *Gases* and *vapors* are the highest energy states of matter; they flow to take the shape and to occupy the total volume of any container. When a gas or vapor loses energy (in the form of heat) to its surroundings, it condenses to a *liquid*. Liquids are also able to flow, but they are virtually incompressible. Matter that flows (i.e. a gas, vapor or liquid) is called a *fluid*. Solids constitute the lowest energy states of matter. *Solids* usually possess rigid structures, they do not flow and are relatively difficult to compress. While all matter is deformed by applications of loads, in many cases the deformations are relatively small and may be ignored. *Rigid bodies* are objects that undergo negligible deformations.

Note that all matter tends to exist in the configuration corresponding to the lowest possible energy state. Whenever possible, matter reduces its energy.

Problem

1. Can an object ever have no mass or no weight?

1-2 MEASUREMENT AND STANDARDS

Any scientific observation, and the use of the data taken, depends on accurate measurement and on a statement of the errors involved in taking that measurement. The statement of accuracy is extremely important because it specifies the limitations of the data. Scientists must never underestimate the errors in their measurements.

Measurement is the comparison of an unknown quantity with some well defined standard quantity. The measured value is expressed in terms of that standard. For example, before we can measure a distance, we must define a standard of length, such as a meter; this is called a *unit*. We then measure how many times (or what fraction of) that standard will equal the distance, and the result is expressed as a multiple or fraction of the meter. A measured value of 5.3 meters implies that the distance is 5.3 times the standard length, the meter.

It is important to remember that the units must be considered part of any statement defining a physical quantity that has been compared with a standard. *If the units are omitted, the statement is useless.* For example, if a length is expressed only as 5.0, it could mean 5.0 meters, 5.0 kilometers, 5.0 miles, etc. *The units are required to complete the statement.*

In order to define a complete, self-contained system of units, we must specify a complete set of *base units* for some chosen fundamental physical quantities. These fundamental physical quantities must be independent of each other, i.e. one base unit cannot be expressed in terms of other base units. All other physical quantities can be written in terms of a combination of two or more base units; this combination is called a *derived unit*. For convenience, some derived units may be given special names and symbols.

While there are many self-contained systems of units, we shall only consider the International Systems of Units (SI) and the British Engineering System because they are the most important in technology.

To make a measurement, we must first choose a suitable instrument, such as a meter stick, thermometer or chemical balance. These instruments are normally calibrated so that we can read the measured value directly. The accuracy of any measurement is limited by the *precision* of the instrument used and by our ability to take the readings. We normally indicate the accuracy in terms of the number of digits used to specify the multiple of the unit. For example, if we wish to measure the length of a table top, we could visually estimate it to be 2 meters. Using a stick graduated in tenths of a meter, we might estimate the length to

be 1.87 meters, i.e. between 1.8 meters and 1.9 meters. Finally, using a meter stick graduated in millimeters (one-thousandth of a meter), we could measure the length to be 1.8625 meters, i.e. between 1.862 and 1.863 meters. This can also be written as 1.8625 ± 0.0005 meters.

As the *precision* of the measuring instrument increases, more digits are retained in the stated measured value. In practice we usually retain only *one* doubtful digit (the last). The stated digits are then called *significant figures*. Thus in the measured value 1.8625 meters, there are five significant figures, and the last digit, 5, is uncertain since it was estimated.

In general a digit is significant when it is a part of a statement indicating accuracy. Zeros are not significant when they can be written as a power of 10, since they then represent "size" rather than accuracy. For example, in the numbers 93 000 000 and 0.000 583, the zeros are *not* significant since the numbers can be rewritten as 9.3×10^7 and 5.83×10^{-4}, respectively. However the zeros are significant in 4.5080 or 9.20×10^8 since they cannot be replaced by a power of 10.

1-3 INTERNATIONAL SYSTEM OF UNITS (SI)

The International System of Units (SI) was adopted by the General Conference of Weights and Measures in 1960 and is recommended for use in all areas of science and technology. While the British Engineering System is still used in North America, it is being replaced by SI.

There are actually three classes of SI units:

1. Seven base units represent the seven fundamental physical quantities that were chosen because they are independent of each other and can be measured very accurately. Some of the standard physical quantities and the corresponding base units are listed in Table 1-1.

Table 1-1
SI Base Units

Physical Quantity	Symbol	Unit	Unit Symbol	Dimension
Length	l	meter*	m	L
Mass	m	kilogram	kg	M
Time	t	second	s	T
Electric current	I	ampere	A	A
Quantity of a substance	n	mole	mol	n
Temperature	T	kelvin	K	θ
Luminous intensity	I_L	candela	cd	I
Supplementary Units				
Angle	θ	radian	rad	—
Solid angle	Ω	steradian	sr	—

*Also written as metre.

Some of the standards used to define the base SI units are:

Length | A *meter* (m) is 1 650 763.73 wavelengths in vacuum of the orange-red spectral line from the $2p_{10}$ and $5d_5$ levels of the krypton-86 atom.

Mass | A standard *kilogram* (kg) mass is a cylindrical platinum-iridium alloy kept at the International Bureau of Standards.

Time | A *second* (s) is the duration of 9 192 631 770 cycles of radiation from the transition between two ground-state energy levels in an atom of the isotope cesium-133.

Quantity of Substance | A *mole* (mol) is the amount of a substance which contains as many elementary entities as there are atoms in 0.012 kg of carbon-12, i.e. Avogadro's number—approximately 6.02×10^{23} elementary entities of that substance.

The other base units are the *ampere* (A) for electric current (Chapter 19), the *kelvin* (K) for temperature (Chapter 15) and the *candela* (cd) for luminous intensity (Chapter 27).

2. Two supplementary units are used to represent angles and solid angles. These units are also listed in Table 1-1.

3. There are three types of derived units. Some are written in terms of base units and supplementary units (Table 1-2). Others are given special names for convenience (Table 1-3). Some derived units are normally written in terms of the derived units with special names (Table 1-4).

Figure 1-1
Standard kilogram mass.
(National Research Council of Canada)

Figure 1-2
Atomic clock.
(National Research Council of Canada)

The use of a few other special units is also acceptable even though they are not really SI units; these units, their symbols and definitions are listed in Table 1-5.

A useful fact to remember is that a 1.0 kg mass of pure liquid water occupies a volume of approximately 1.0 L.

Table 1-2
Examples of SI Derived Units
Expressed in Terms of Base Units

Physical Quantity	Symbol	SI Unit Description	Symbol	Dimension
Area	A	square meter	m²	L²
Volume	V	cubic meter	m³	L³
Speed, velocity	v, $\mathbf{v}$	meter per second	m/s	L/T
Acceleration	$\mathbf{a}$	meter per second squared	m/s²	L/T²
Density	ρ	kilogram per cubic meter	kg/m³	M/L³
Luminance	L	candela per square meter	cd/m²	I/L²
Angular velocity	ω	radian per second	rad/s	—
Radiant intensity	I	watt per steradian	W/sr	—

Table 1-3
Derived SI Units with Special Names

Physical Quantity	Symbol	Unit	Unit Symbol	Definition in Base Units	Dimension
Force (weight)	$\mathbf{F}$ ($\mathbf{w}$)	newton	N	$kg \cdot m/s^2$	ML/T^2
Pressure	p	pascal	Pa	$kg/(m \cdot s^2)$	M/LT^2
Work (energy)	W (E)	joule	J	$kg \cdot m^2/s^2$	ML^2/T^2
Power	P	watt	W	$kg \cdot m^2/s^3$	ML^2/T^3
Frequency	f	hertz	Hz	$1/s$	$1/T$
Electric charge	Q	coulomb	C	$A \cdot s$	AT
Electric potential difference	U (V)	volt	V	$kg \cdot m^2/(A \cdot s^3)$	ML^2/AT^3
Electric resistance	R	ohm	Ω	$kg \cdot m^2/(A^2 \cdot s^3)$	ML^2/A^2T^3
Electric conductance	G	siemen	S	$A^2 \cdot s^3/(kg \cdot m^2)$	A^2T^3/ML^2
Electric capacitance	C	farad	F	$A^2 \cdot s^4/(kg \cdot m^2)$	A^2T^4/ML^2
Magnetic flux	ϕ_m	weber	Wb	$kg \cdot m^2/(s^2 \cdot A)$	ML^2/T^2A
Magnetic flux density (Magnetic induction)	$\mathbf{B}$	tesla	T	$kg/(A \cdot s^2)$	M/AT^2
Inductance	L	henry	H	$kg \cdot m^2/(s^2 \cdot A^2)$	ML^2/T^2A^2
Luminous flux	F_L	lumen	lm	cd/sr	I
Illuminance	E	lux	lx	$cd/(sr \cdot m^2)$	I/L^2

Table 1-4
Examples of Derived SI Units Expressed in Terms of Units with Special Names

Physical Quantity	Symbol	Unit	Unit Symbol	Expression in Base Units	Dimension
moment of force (torque)	M_0 (τ)	newton meter	N·m	$m^2 \cdot kg/s^2$	L^2M/T^2
specific heat capacity	c	joule per kilogram kelvin	J/(kg·K)	$m^2/(s^2 \cdot K)$	$L^2/T^2\theta$
thermal conductivity	K	watt per meter kelvin	W/(m·K)	$m \cdot kg/(s^3 \cdot K)$	$LM/T^3\theta$
electric field strength (intensity)	$\mathbf{E}$	volt per meter	V/m	$m \cdot kg/(s^3 \cdot A)$	LM/T^3A
magnetic field strength (intensity)	$\mathbf{H}$	ampere per meter	A/m	A/m	A/L

Table 1-5
Units Accepted for Use with SI

Physical Quantity	Unit	Symbol	Definition
Mass	tonne	t	1 t = 1 000 kg
Time	minute	min	1 min = 60 s
	hour	h	1 h = 3 600 s
	day	d	1 d = 86 400 s
	year	a	1 a ≈ 31.56 × 10⁶ s*
Area	hectare	ha	1 ha = 10⁴ m²
Volume	liter	L	1 L = 10⁻³ m³
Temperature	degree Celsius	°C	0°C = 273.15 K; in terms of intervals (or changes) in temperature 1°C = 1 K
Plane angle	degree	°	1° = (π/180) rad
	minute	′	1′ = (π/10 800) rad
	second	″	1″ = (π/648 000) rad

*This depends on the type of year, i.e. calendar, sidereal or tropical.

SI Rules

The following rules have been adopted for the use of SI units:

1. All units are written in the lower case, i.e. without a capital letter. However, if the unit was derived from the name of a person, the unit *symbol* is capitalized. For example, meter is abbreviated as m, kilogram is kg, newton is N, and hertz is Hz. The liter (symbol L) is the exception to the rule.

2. A space is left between the symbol and the numerical value. For example, two meters is written as 2 m. However, if the unit is not SI (Table 1-5), no space is left when the first character of the symbol is not a letter. For example, we write 23°C, *not* 23° C, and 49°12′18″, *not* 49 ° 12 ′ 18 ″.

3. Symbols are written without a period except at the end of a sentence, and they are not changed to indicate the plural. For example, twelve newtons is written as 12 N, *not* 12 N. or 12 Ns.

4. Symbols should be used with numerical values, but the full unit must be written if numbers are not involved. For example, we write 12 m², *not* 12 square meters, but the watt—*not* the W—is a unit of power.

5. Do not mix names and symbols. For example, we write kg·m or kilogram meter, *not* kg·meter.

6. Central dots are used to indicate the multiplication of units, and like units may be squared, cubed, etc.* A *space* is left between multiplied units in written form. For example, we write (2 kg) (5 m) = 10 kg·m, (2 m) (3 m) = 6 m² and twelve newton meters, *not* twelve newton-meters.

7. A stroke or negative power is used to indicate division, e.g.

*See also rules for units with prefixes.

$$\frac{m}{s^2} = m/s \quad \text{or} \quad m \cdot s^{-2}$$

The negative power represents division in exactly the same way that $10^{-2} = \frac{1}{10^2}$. The word "per" also means division, e.g. speed in meters per second. The central dot must be included when multiplication of different units occurs.

8. Long numbers are written with a *space* between each group of three digits starting at each side of the decimal point, e.g. 93 042.135 679 2.

9. A zero must be written before the decimal point if a number is less than one, e.g. 0.05, *not* .05.

Prefixes

Certain prefixes are used to designate multiples and submultiples of SI units. The prefix itself is equivalent to some power of ten (Table 1-6).

Rules for the Use of Prefixes

1. Prefixes are written *without* a space between the prefix symbol and the unit symbol, and the prefix name is combined with the unit name. For example, one megameter (1 Mm) is equal to one million meters (10^6 m), and one microsecond (1 μs) is equivalent to one one-millionth of a second (10^{-6} s).

2. Prefixes are considered to be combined directly with the unit for all algebraic manipulations. This is somewhat different from the rules for algebra! For example, in algebraic form, $x \cdot y^2 = x \cdot y \cdot y$. But with SI prefixes and units, $1 \text{ cm}^2 = (1 \text{ cm})(1 \text{ cm}) = (10^{-2} \text{ m})(10^{-2} \text{ m}) = 10^{-4} \text{ m}^2$. That is, the prefix "centi" is assumed to be a part of the total unit which is being squared.

Table 1-6
Numerical Prefixes for SI Units

Prefix	Multiplying Factor	Symbol
exa	10^{18}	E
peta	10^{15}	P
tera	10^{12}	T
giga	10^9	G
mega	10^6	M
kilo	10^3	k
hecto	10^2	h
deca	10^1	da
deci	10^{-1}	d
centi	10^{-2}	c
milli	10^{-3}	m
micro	10^{-6}	μ
nano	10^{-9}	n
pico	10^{-12}	p
femto	10^{-15}	f
atto	10^{-18}	a

3. Compound prefixes should not be used; this rule also applies to the base unit kilogram. For example, $\mu\mu$F should be written pF, and μkg should be written as mg.

4. When a prefix is used to replace scientific notation, the numerical value should be between 0.1 and 1 000. For example, 8.9×10^{-7} s, 0.89 μs and 890 ns are all acceptable; however, 0.02 μs or 1 200 μs must be rewritten.

5. Avoid the use of the prefixes *deci, centi, deca* and *hecto*. Exceptions are "centimeter" and "square hectometer," 1 hm^2 (abbreviated 1 ha = 10^4 m^2), which is used to measure areas and is called a *hectare*.

6. Only one prefix should be used in a compound unit, and it should appear in the numerator. The exception is the base unit kilogram. For example, we write Mg/m^3, *not* g/cm^3, and kJ/kg, *not* J/g.

Conversions of Units

In terms of SI prefixes, the conversion of one unit into another is relatively simple because prefixes represent powers of ten. The conversion factor is merely ten raised to a power equal to the original power *minus* the required power. We can use direct substitution by applying this conversion factor.

Example 1-1

How many nanometers are equivalent to 78 km?

Solution:

The prefixes "kilo" and "nano" represent 10^3 and 10^{-9} respectively. Since we are converting *from* kilo *to* nano, we subtract the exponent of nano from that of kilo in order to get the power of ten for the conversion factor: $3 - (-9) = 12$.
Thus

$$1 \text{ km} = 10^{12} \text{ nm}$$

and

$$78 \text{ km} = 78 \times 10^{12} \text{ nm} = 7.8 \times 10^{13} \text{ nm}.$$

Example 1-2

How many cubic millimeters are there in 0.8 km³?

Solution:

Milli represents 10^{-3}, and kilo represents 10^3. Therefore, since the power of ten of the conversion factor between milli and kilo is $3 - (-3) = 6$, $1 \text{ km} = 10^6 \text{ mm}$. But in this case the units are cubed; therefore we must cube both sides of the conversion factor, *numbers and units*.

$$(1 \text{ km})^3 = (10^6 \text{ mm})^3 = 10^{18} \text{ mm}^3$$

Thus

$$0.8 \text{ km}^3 = 0.8 \times 10^{18} \text{ mm}^3 = 8 \times 10^{17} \text{ mm}^3.$$

It is often necessary to convert units that are not related by prefixes. In such a case, the conversion factor is not merely a power of ten and must be found in a table of conversions. The conversion is again accomplished by substituting the equivalent term in the expression.

Example 1-3

Convert 54 km/h into meters per second.

Solution:

In this case, there are two conversion factors: in the numerator, $1 \text{ km} = 10^3 \text{ m}$, and in the denominator, $1 \text{ h} = 3\ 600 \text{ s}$. Thus by direct substitution,

$$54 \frac{\text{km}}{\text{h}} = \frac{54 \times 10^3 \text{ m}}{3\ 600 \text{ s}} = 15 \text{ m/s}.$$

Example 1-4

Convert 500 J/s into kilojoules per hour.

Solution:

The conversion factors are:

$1 \text{ kJ} = 10^3 \text{ J}$; therefore $1 \text{ J} = 10^{-3} \text{ kJ}$ (numerator)

$1 \text{ h} = 3\ 600 \text{ s}$; thus $1 \text{ s} = \dfrac{1}{3\ 600} \text{ h}$ (denominator)

Thus

$$500 \frac{\text{J}}{\text{s}} = \frac{500 \times 10^{-3} \text{ kJ}}{\dfrac{1}{3\ 600} \text{ h}} = 1\ 800 \frac{\text{kJ}}{\text{h}}.$$

Problems

2. What is the area of a rectangular field with a length of 200 m and a width of 120 m?

3. What is the volume of a room 3.0 m high, 8.0 m wide and 12 m long?

4. A circular swimming pool has a diameter of 5.50 m and is filled to an average depth of 1.50 m. Determine (a) the volume of

water and (b) the mass of water in the pool.

5. Write the following in terms of SI base units using scientific notation: (a) 36 cm, (b) 930 km, (c) 260 μs, (d) 130 ps, (e) 120 Gm, (f) 86 ms, (g) 300 μA, (h) 32 Mg, (i) 8.0 L, (j) 760 mm², (k) 0.94 cm³.

6. Write the following in terms of a unit with a prefix: (a) 4.6×10^3 m, (b) 2×10^{-6} s, (c) 8.3×10^{10} Hz, (d) 5.2×10^{-7} s, (e) 1.2×10^{-11} F, (f) 3.2×10^3 kg, (g) 4.2×10^7 N/C, (h) 8.6×10^2 V/mm, (i) 7.8 g/cm³.

7. How many kilograms are equivalent to: (a) 3 Gg? (b) 5.0 cg? (c) 420 μg?

8. How many centimeters are equivalent to: (a) 8.0 μm? (b) 5 mm? (c) 4 km?

9. Express the density 6.8 g/cm³ in correct SI units.

10. Convert 90 km/h into meters per second.

11. Convert 20 m/s into kilometers per hour.

12. A car uses 54.0 L of gasoline to travel 351 km. Determine the gasoline consumption (a) in kilometers per liter and (b) in liters per 100 km.

1-4 THE BRITISH ENGINEERING SYSTEM

The British Engineering System of units (Table 1-7) is still widely used in North America. In mechanics, this system is based on a set of standards for *length*, *time* and *weight* or *force* (rather than mass). As in SI, the base unit for time is the second (s). The other standards are:

1. *Length—the foot.* An *inch* (in) is defined as 2.54×10^{-2} m, and a foot (ft) is equal to 12 in or 0.3048 m. 1 *mile* (mi) = 1.609 km.

2. *Weight—the pound.* A pound (lb) is the weight of a 0.453 592 37 kg mass at sea level and a latitude of 45° on the earth's surface. Note that weight (or force) units are not equivalent to mass units; a non-relativistic object always has the same mass, but it may have different weights at different locations.

Some of the conversion factors between SI and British Engineering Systems of units are listed in Appendix 1.

Table 1-7
British Engineering System

Physical Quantity	Symbol	Unit	Unit Symbol (or Definition)
Base Units			
Length	L	foot	ft
Weight (force)	w (F)	pound	lb
Time	t	second	s
Derived Units			
Mass	m	slug	lb·s²/ft
Pressure	p	pound per square foot	lb/ft²
Work (energy)	W (E)	foot pound	ft·lb
Power	P	horsepower	$1 \text{ hp} = 550 \dfrac{\text{ft·lb}}{\text{s}}$

*Example 1-5*_____

Convert 2 700 kg/m³ into slug/ft³.

Solution:

The conversion factors are 1 kg = 0.0685 slug and 1 m = 3.281 ft. The last expression must be *cubed* (both the number and the units) because the original statement involves *cubic* meters. Thus

$$1 \text{ m}^3 = (3.281 \text{ ft})^3$$

and

$$2\,700\,\frac{\text{kg}}{\text{m}^3} = \frac{2\,700\,(0.0685 \text{ slug})}{(3.281 \text{ ft})^3} = 5.24 \text{ slug/ft}^3.$$

*Problems*_____

13. How many meters are there in one mile (5 280 ft)?

14. How many feet are there in one kilometer?

15. Convert (a) 1 mile (5 280 ft) into kilometers, (b) 35 kg into slugs, (c) 120 lb into newtons, (d) 30 ft/s into m/s, (e) 25 m/s² into ft/s², (f) 4.5 slug/ft³ into kg/m³, (g) 120 ft² into m², (h) 1.2 m³ into ft³, (i) 101 Pa (N/m²) into lb/ft² and lb/in², (j) 10 hp into watts (N·m/s).

16. Add the following. Express the answer in both SI and British Engineering System units: (a) 3.3 lb + 2.0 N, (b) 800 ft + 300 m, (c) 44 ft/s + 21 m/s, (d) 101 m² + 30 ft², (e) 0.80 km³ + 1 mile³, (f) 2.9 slug/ft³ + 1 400 kg/m³.

1-5 DIMENSIONAL ALGEBRA

Every physical quantity may be expressed in terms of a combination of base units; each base unit defines a physical dimension (Table 1-1). For example, an area is the product of two lengths, therefore it has dimensions of length squared (L^2).

Algebraic operations may be performed with units and dimensions in exactly the same way as with numerals. Dimensional algebra is extremely useful since it enables us:

1. To reduce a combination of units to the simplest form;

2. To check equations; and

3. To establish equations which have a dimensionless constant (no units).

*Example 1-6*_____

Reduce the units of work (joules) divided by pressure (pascals) to the simplest form.

Solution:

From Table 1-3

$$\frac{\text{Work}}{\text{Pressure}} = \frac{\text{J}}{\text{Pa}} = \frac{ML^2/T^2}{M/LT^2} = L^3.$$

L^3 represents the dimension for volume, which has SI units of m³.

*Example 1-7*_____

Is the following equation dimensionally correct?

$$\text{Time} = 2\pi\sqrt{\frac{\text{Force}}{\text{Mass} \times \text{Length}}}$$

Solution:

From Table 1-3, the right-hand side reduces to the dimensions of frequency.

$$\sqrt{\frac{ML/T^2}{ML}} = \sqrt{\frac{1}{T^2}} = \frac{1}{T}$$

Therefore the equation is incorrect because the left-hand side has dimensions of time (T). Note that the constant 2π is dimensionless and need not be considered in the dimensional analysis. The expression is dimensionally correct if it is changed to:

$$\text{Time} = 2\pi\sqrt{\frac{\text{Mass} \times \text{Length}}{\text{Force}}}.$$

*Example 1-8*_____

An object of mass m and speed v moves in a circular path of radius r about some fixed point, and a force (centripetal force) is required to maintain the motion. Derive an expression for centripetal force F.

Solution:

The centripetal force F must be some function of the mass m, speed v and radius r; thus

$$F = Km^\alpha v^\beta r^\gamma$$

where α, β and γ are constants representing powers and K is a proportionality constant. In terms of dimensions,

$$F = \frac{ML}{T^2} \quad \text{and} \quad Km^\alpha v^\beta r^\gamma = M^\alpha \left(\frac{L}{T}\right)^\beta L^\gamma.$$

If the equation is dimensionally correct,

$$\frac{ML}{T^2} = M^\alpha \left(\frac{L}{T}\right)^\beta L^\gamma.$$

Thus, equating the powers of the *independent* dimensions, we obtain

$$M = M^\alpha \quad \text{or} \quad \alpha = 1$$
$$1/T^2 = (1/T)^\beta \quad \text{or} \quad \beta = 2$$

and

$$L = L^\beta L^\gamma = L^2 L^\gamma = L^{2+\gamma} \quad \text{or} \quad \gamma = -1$$

Therefore

$$F = Km^\alpha v^\beta r^\gamma = Kmv^2/r.$$

Note that the magnitude of K cannot be determined because it is dimensionless, but it must be included.

Problems

17. Reduce the units of force (newtons) divided by acceleration (m/s²) to simplest form.

18. Reduce the units of electrical resistance (ohms) divided by magnetic flux (webers) to the simplest form in SI.

19. Reduce the units of inductance (henry) divided by electric potential difference (volts) to the simplest form.

20. Reduce the units of electric charge (coulombs) divided by capacitance (farads) to the simplest form.

21. Reduce the units of the product of electric charge and potential difference to the simplest form.

22. Check to see if the following equations are dimensionally correct:
    (a) Energy = ½Mass × (Speed)².
    (b) Work = Pressure × Volume.
    (c) Electric resistance = (Electric current) × (Electric potential difference).
    (d) Magnetic flux density = Magnetic flux/area.
    (e) Magnetic flux density = Inductance × Electric current.

23. Derive an expression for the magnetic force F on a charge Q which moves with a speed v through a region where the magnetic flux density is B. *Hint:* $x^0 = 1$.

24. Derive an expression for electric power P in terms of electric current I and electric resistance R.

25. Derive an expression for the frequency of an oscillatory circuit that has an inductance L and a capacitance C.

REVIEW PROBLEMS

1. (a) Determine the volume of a room that has dimensions of 2.4 m × 5.0 m × 8.0 m. (b) What is the volume in liters?

2. A rectangular swimming pool is 100 m long, 30 m wide and is filled to an average depth of 1.5 m. Determine (a) the volume of water and (b) the mass of water in the pool.

3. The gasoline tank of a car has a capacity of 55 L. How far can the car travel on a full tank if the average gasoline consumption is (a) 9.2 km/L? (b) 11 L/100 km?

4. Complete the following:
   (a) 36 m = ____ cm = ____ μm = ____ km

(b) $188 \text{ mL} = \underline{\quad} \text{ L} = \underline{\quad} \text{ m}^3$
$= \underline{\quad} \text{ cm}^3$
(c) $120 \text{ kg} = \underline{\quad} \text{ mg} = \underline{\quad} \text{ Mg}$
$= \underline{\quad} \mu\text{g}$
(d) $30 \text{ cm}^2 = \underline{\quad} \text{ mm}^2 = \underline{\quad} \text{ m}^2$
(e) $9 \times 10^4 \text{ mm}^3 = \underline{\quad} \text{ cm}^3 = \underline{\quad} \text{ m}^3$

5. Express the following in SI base units, using scientific notation where appropriate: (a) 36 μm, (b) 240 ns, (c) 800 μA, (d) 18 Gg, (e) 28 mL.

6. Express the following in prefix form: (a) 5.7×10^{-8} s, (b) 1.2×10^4 kg, (c) 3.4×10^{-4} m, (d) 7.2×10^{-2} L, (e) 9×10^{10} Hz, (f) 1.25×10^{-10} F.

7. How many 250 mL flasks of water are needed to fill completely an aquarium with inside dimensions of 1.2 m $\times$ 15 cm $\times$ 40 cm?

8. Convert 57.6 km/h into meters per second.

9. Convert 18.0 m/s into kilometers per hour.

10. Express the following in correct SI notation: (a) 2.4 g/cm^3, (b) 3.6 μW/mm^2, (c) 5×10^4 kV/cm, (d) 26 mJ/g, (e) 36 μA/mm^2, (f) 15 mA/mm^2.

11. A cardboard carton measures 60 mm $\times$ 20 cm $\times$ 30 cm. What is the maximum number of such cartons that could be packed into a freight car of inside dimensions 6 m $\times$ 3 m $\times$ 5 m?

12. Convert: (a) 1 mile into centimeters, (b) 12 slugs into kilograms, (c) 7 880 kg/m^3 into slug/ft^3.

13. Reduce the units of force (newtons) multiplied by distance (meters) to the simplest form.

14. Reduce the units of power (watts) multiplied by time (seconds) and divided by potential difference (volts) to the simplest form.

15. Check to see if the following equations are dimensionally correct:
(a) Energy $= \frac{1}{2}$(capacitance)$\times$(voltage)^2.
(b) Energy $= \frac{1}{2}$(inductance) $\times$ (current)^2.
(c) Power $=$ inductance $\times$ capacitance.

16. Derive an expression for energy E in terms of inductance L and electric current I.

17. Derive an expression for power P in terms of force F and velocity v.

18. Derive an expression for the period T (time required for one complete orbit) of a charged particle in a magnetic field in terms of its mass m, charge Q and the magnetic flux density B.

An adequate knowledge of vectors is a prerequisite to the study of physics because an essential part of physics is concerned with the manipulation of vector quantities.

2-1 VECTOR AND SCALAR QUANTITIES

Many physical quantities are completely specified in terms of a *magnitude* (size and units). These quantities are called *scalars*; they can usually be measured directly by means of a scale such as that on a ruler, thermometer or balance. Some examples of scalar quantities are mass, length, speed, work, power, temperature, density and volume.

Other physical quantities are only completely described when a direction and sometimes even a point of application are given in addition to the magnitude. These physical quantities are called *vectors*. Some common vector quantities are displacement, velocity, acceleration and force. For example, if a man leaves his home and drives a distance of 60 km, we cannot determine his final position because no direction is given. We can only say that the man is somewhere inside a circle that is centered at his home and has a radius of 60 km. However, if we know the direction in which the man drives in addition to the starting point (his home) and the distance (60 km), we can easily determine his final position.

2-2 REPRESENTATION OF VECTORS

In most textbooks the symbols representing vector quantities are written in boldface (heavy) type whereas the symbols for scalar quantities and magnitudes of vectors are written in italics. For example, $\mathbf{F}$ is the symbol used to represent a force vector which has a magnitude of F, and m is the symbol used to represent the scalar quantity mass.

2

Vectors

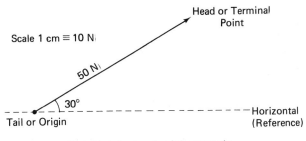

Scale 1 cm ≡ 10 N

50 N

30°

Tail or Origin

Head or Terminal Point

Horizontal (Reference)

Note that the length of the arrow is 5.0 cm which corresponds to 50 N in the chosen scale.

Figure 2-1
Scale diagram of a 50 N force directed 30° above the horizontal.

It is often convenient to represent vectors by arrows in scale diagrams. The "tail" of the arrow is called the *origin* of the vector, and the "head" of the arrow is called the *terminal point*. The length of the arrow represents the magnitude of the vector to scale in the appropriate units, and the direction of the arrow represents the direction of the vector. Usually the direction of a vector is specified by the angle it makes with a certain reference direction. Note that the scale and angles must be included in scale diagrams (Fig. 2-1).

2-3 VECTORS USED TO DESCRIBE MOTION

Both scalar and vector quantities are used extensively in the analysis of motion. The term *position* is used to specify a location relative to some reference point, e.g. 500 km north of New York, or in terms of a coordinate system, e.g. latitude and longitude. A moving object continually changes its position, and the line connecting all of the successive positions is called the *path*. The *distance d* that the object travels is the length measured along the path; it is a scalar quantity since no direction is specified.

The *change* in position of an object, called *displacement* **s**, is a vector quantity representing the length in the direction of the straight line *from* the initial position *to* the final position considered (Fig. 2-2). Note that the magnitude *s* of the displacement **s** can never exceed the distance *d* travelled.

Since both distances and displacements represent lengths, they are expressed in length units such as meters, kilometers, feet or miles.

Directions are often given in terms of degrees towards one direction measured from

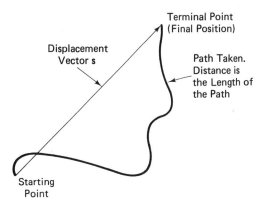

Displacement Vector **s**

Terminal Point (Final Position)

Path Taken. Distance is the Length of the Path

Starting Point

Figure 2-2
Illustration of the difference between distance and displacement.

some other direction. For example, 20° south of west is the direction measured *from* west and is 20° *towards* the south, i.e. 20° anticlockwise from west.

In navigation, directions are measured clockwise through 360° from north. These directions are normally expressed as a three digit angle with a "T" following to indicate that the angle is measured from "true north." For example, east, which is 90° clockwise from north, is represented as 090 T, south is 180 T, and west is 270 T.

Speed is a scalar quantity which is used to describe how fast an object is moving; the direction of motion is not specified. The *average speed* $\bar{v}$ of an object is defined as the ratio of the distance d travelled to the time t taken.

$$\bar{v} = \frac{d}{t} \qquad (2\text{-}1)$$

If the object travels equal distances in equal time intervals, it is said to have a *constant* (or *uniform*) *speed*. For example, a car has a constant speed of 60 km/h when it travels exactly one kilometer in each minute.

The *velocity* of a moving object is a vector quantity defined as the time rate of change of position; thus it represents a speed in a given direction. If an object moves from an initial position $\mathbf{r}_0$ to a final position $\mathbf{r}_1$, we define its *average velocity* $\bar{\mathbf{v}}$ as the ratio of the displacement $\mathbf{s}$ (or change in position $\mathbf{r}_1 - \mathbf{r}_0 = \Delta\mathbf{r}$)* to the time t taken.

$$\bar{\mathbf{v}} = \frac{\mathbf{s}}{t} = \frac{\mathbf{r}_1 - \mathbf{r}_0}{t} = \frac{\Delta\mathbf{r}}{t} \qquad (2\text{-}2)$$

If the object moves in a straight line at a constant speed, it is said to have a *constant velocity*. This is the simplest motion to analyze because changes in direction are more difficult to describe. Note from the definitions that speed and velocity are expressed in units of length/time, e.g. m/s, km/h, ft/s or mi/h.

_____

*The symbol Δ is used to indicate "a change in"; thus $\Delta\mathbf{r}$ represents a change in position.

*Example 2-1*_____

A car travels a total distance $d = 600$ km in 6.0 h along a typical highway. If the final position of the car is 360 km due east of its starting point, determine (a) the average speed v, and (b) the average velocity $\bar{\mathbf{v}}$ of the car.

Solution:

(a) $\qquad \bar{v} = \frac{d}{t} = \frac{600 \text{ km}}{6.0 \text{ h}} = 100$ km/h

(b) $\quad \bar{\mathbf{v}} = \frac{\mathbf{s}}{t} = \frac{360 \text{ km east}}{6.0 \text{ h}} = 60 \frac{\text{km}}{\text{h}}$ east

_____

Objects rarely travel in straight lines at constant speeds. They may change their speed, direction of motion, or both speed and direction of motion frequently. When they do, they are said to be accelerating. The *acceleration* $\mathbf{a}$ of an object is a vector quantity defined as the time rate of change of velocity. Therefore, if the velocity of an object changes from $\mathbf{v}_0$ to $\mathbf{v}_1$ in a time interval t, the *average acceleration* is

$$\bar{\mathbf{a}} = \frac{\mathbf{v}_1 - \mathbf{v}_0}{t} = \frac{\Delta\mathbf{v}}{t} \qquad (2\text{-}3)$$

The units of acceleration have dimensions of velocity/time² = length/time², e.g. m/s², km/h² and ft/s².

_____

*Example 2-2*_____

What is the average acceleration of an aircraft if it travels in a straight line due south and increases its speed from $v_0 = 84$ m/s to $v_1 = 120$ m/s in a time interval $t = 2.0$ min?

Solution:

$$\bar{\mathbf{a}} = \frac{\mathbf{v}_1 - \mathbf{v}_0}{t} = \frac{(120 \text{ m/s} - 84 \text{ m/s})}{120 \text{ s}} \text{ south}$$

$$= 0.30 \text{ m/s}^2 \text{ south}$$

_____

In Chapter 4 we shall see that accelerations are produced by actions called *forces*, which are also vector quantities. For now, we shall interpret a force as a "push" or a "pull."

Problems_____

1. A clock has a minute hand 2.0 ft long and an hour hand 1.5 ft long. During the time change from 12 o'clock to 6 o'clock, determine (a) the distances travelled by the tips of the minute and the hour hands, and (b) the displacements of the tips of the minute and hour hands.

2. Determine the displacement of an object if its initial position was 90 km due west and its final position was 60 km due east of some reference location.

3. What is the average speed of an aircraft that travels 480 km in exactly 50 min?

4. The star Sirius is 8.1×10^{13} km from the earth. If the speed of light is 3.0×10^8 m/s, how many years does it take the light emitted by Sirius to reach the earth?

5. A ship is initially 120 km north of a lighthouse. What is its average velocity if after 6.0 h it is 360 km north of the lighthouse?

6. A boy sees a flash of lightning and hears the thunder 8.0 s later. If the speed of sound in air is 330 m/s and the speed of light is 3.0×10^8 m/s, how far away is the storm?

7. A ship transmits a radar pulse and receives a reflection from an aircraft after 500 μs. If the speed of the radar pulse is 3.0×10^8 m/s, how far away is the target?

8. How far does a car go in 10 s if it travels at a constant speed of 60 mi/h?

9. A train starts from rest and accelerates to 110 km/h in 3.0 min while heading north along a straight track. Determine its average acceleration in SI base units.

10. Determine the final velocity of an object after 8.0 s if it is falling freely with an initial velocity of 15.0 m/s and its average acceleration is 9.8 m/s² downwards.

2-4 VECTOR ADDITION BY SCALE DIAGRAMS

Two or more vectors with the same units may be "added" to produce a single vector called the *resultant*, which represents their combined effect. However, the process of *vector addition* is more difficult than the addition of scalar quantities because the directions of the vectors must be considered. The rules for the addition of vectors can be determined directly from our knowledge of displacements, forces and geometry.

Collinear Vectors_____

Vectors are said to be *collinear* when they are located on the same straight line; they are either parallel (in the same direction) or anti-parallel (in opposite directions).

Suppose a man walks 4.0 km east and then an additional 3.0 km east. He has actually walked a total distance of 7.0 km, and his final position is 7.0 km east of his starting position (Fig. 2-3a). Thus the vector sum of the displacements 4.0 km east and 3.0 km east is 7.0 km east. In general, if two or more vectors are parallel, their vector sum is merely the sum of their magnitudes in the same direction. Note that the *commutative law* holds for vector addition, i.e. for any two vectors **A** and **B**, **A** + **B** = **B** + **A**: the order does not matter!

If two vectors are in opposite directions, the resultant representing their vector sum is equal to the difference in their magnitudes in the direction of the larger. For example, if a man walks 4.0 km east and then 3.0 km west, his final position is 1.0 km east of his starting point (Fig. 2-3b).

Negative Vectors_____

In order to simplify the addition of vectors, we often designate one direction as the posi-

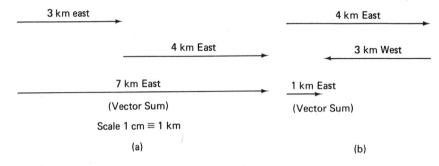

Figure 2-3

tive direction; vectors in the opposite direction are given a negative sign. Thus down is the negative of up, north is the negative of south, left is the negative of right, etc. Using this sign convention, we see that the vector sum of *collinear vectors* (with the same units) is their *algebraic* sum in the positive direction.

*Example 2-3*_____

A car travels 120 km east and then turns around and goes 80 km west. If the total driving time is 4.0 h, determine (a) the total distance travelled, (b) the total displacement, (c) the average speed and (d) the average velocity.

Solution:

(a) $d = 120 \text{ km} + 80 \text{ km} = 200 \text{ km}$
                         (i.e. scalar quantities)

(b) Since west is the negative of east,

$$\mathbf{s} = (120 \text{ km east}) + (-80 \text{ km east})$$
$$= 40 \text{ km east}$$

(c) $\bar{\mathbf{v}} = \dfrac{d}{t} = \dfrac{200 \text{ km}}{4.0 \text{ h}} = 50 \dfrac{\text{km}}{\text{h}}$

(d) $\bar{\mathbf{v}} = \dfrac{\mathbf{s}}{t} = \dfrac{40 \text{ km east}}{4.0 \text{ h}} = 10 \dfrac{\text{km}}{\text{h}} \text{ east}$

_____

If two or more similar vectors are not collinear, their vector sum is not equal to the algebraic sum of their magnitudes because the directions must be considered. For example, if a man walks 4.0 km east and then 3.0 km north, his final position is only 5.0 km from his starting point even though he walked a total distance of 7.0 km (Fig. 2-4).

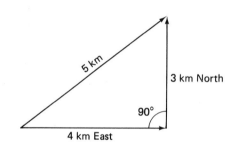

Figure 2-4

The magnitude and direction of a resultant vector may be determined by several different methods.

Polygon of Vectors_____

This method was actually used in Fig. 2-3 and Fig. 2-4. In general, to add two vectors **A** and **B** that have the same units, we draw one of the vectors (**A** for example) to scale, and from the terminal point of **A** we draw **B** to the same scale. The resultant vector **R** is drawn *from* the origin of vector **A** *to* the terminal point of vector **B**. The magnitude of **R** is the length of the arrow in the chosen scale, and its direction is measured directly from the diagram (Fig. 2-5).

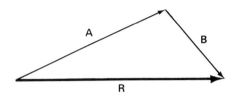

Figure 2-5
A + **B** = **R** from the polygon method.

To add more than two vectors, just follow a "head-to-tail" prodedure for all of the vectors in the system, using each vector *once and only once, but in any order*. The resultant vector is the line drawn from the origin of the first vector to the terminal point of the last vector.

*Example 2-4*_____

Determine the vector sum of the following displacements: **OA** = 80 km east, **OB** = 60 km 30° west of south, **OC** = 50 km 60° north of west and **OD** = 75 km north.

Solution:
The four displacements are illustrated in Fig. 2-6 using a scale of 1 cm ≡ 20 km. Choose any vector (**OC** for example) and draw it; then draw **OB** from the terminal point of **OC**, **OA** from the terminal point of **OB** and finally **OD** from the terminal point of **OA**. The resultant **R** is the vector from the origin O of **OC** *to* the terminal point A of **OA**. With the

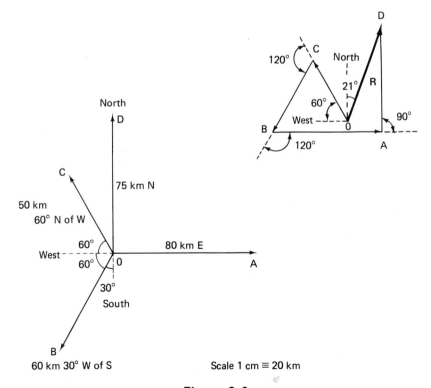

Scale 1 cm ≡ 20 km

Figure 2-6

chosen scale, since the length of **R** is approximately 3.5 cm, the resultant magnitude is 70 km, and the direction is 21° east of north (021 T).

We used the polygon method to add vectors, such as displacements, that occur in sequence, but it can also be used to add vectors that act simultaneously, such as velocities and forces.

*Example 2-5*_____

An aircraft is flying on a heading of 030 T at a constant airspeed, i.e. speed relative to the surrounding air, of 400 km/h. If the wind is blowing from the west, i.e. towards the east, at 50 km/h, what is the speed and direction of the aircraft relative to the ground?

Solution:

In this case the motion of the aircraft was given relative to the air, which in turn was moving relative to the ground because of the wind. Therefore the motion of the aircraft over the ground is the vector sum of its motion in air and the wind velocity. The addition of these vectors on a scale of 1 cm ≡ 50 km/h is illustrated in Fig. 2-7. The resultant speed of the aircraft over the ground is 427 km/h, and its direction is 036 T.

_____

Parallelogram Method_____

This method is often useful when we wish to add *two* vectors. It is actually quite similar to the polygon method, but it is easier to understand in terms of simultaneous vectors.

Suppose that we are given two vectors **A** and **B** with the same units. To find their vector sum, we draw **A** and **B** to scale from a common origin and complete a parallelogram using **A** and **B** as the sides (Fig. 2-8). The diagonal drawn from the common origin O represents the resultant **R** to scale.

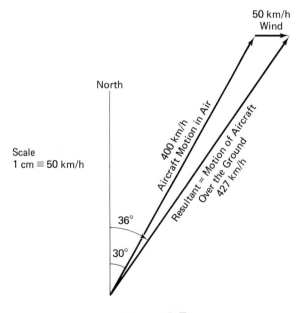

Figure 2-7

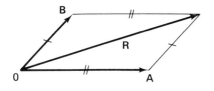

Figure 2-8

Example 2-6

Two forces with magnitudes of 80 N and 70 N act at a common point. If the angle between the forces is 60°, what are the magnitude and direction of their resultant?

Solution:

Choose a suitable scale (1 cm ≡ 20 N) and draw arrows **OA** and **OB** from a common origin 0 to represent the 80 N and 70 N forces respectively. From point A, draw a line parallel to **OB**, and from point B draw a line parallel to **OA**. The intersection of these lines is the terminal point of the resultant drawn from O (Fig. 2-9). The magnitude of the resultant is the length of this diagonal to scale, 6.5 cm ≡ 130 N, and the direction may be measured from either **OA** or **OB** (28° from **OA**).

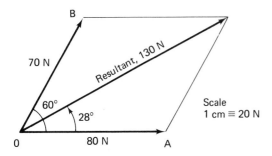

Figure 2-9

Note that one-half of the parallelogram is equivalent to the polygon method; therefore this method may also be used to add vectors occurring in sequence.

Problems

11. Two forces of magnitudes 36 N and 45 N are applied to a common point. If the angle between them is 48°, determine the magnitude and direction of the resultant.

12. Two forces are applied at a common point. If the forces have magnitudes of 60 N and 110 N and the angle between them is 36°, determine the magnitude and direction of the resultant.

13. A boat travelling at a speed of 12 km/h heads directly across a river whose current speed is 4.0 km/h. Determine the speed and direction of the boat's path with respect to the bank of the river.

14. What is the resultant of a 50.0 m northward displacement and a 85.0 m eastward displacement?

15. What is the resultant of a 7.0 lb horizontal force and a 10.0 lb vertical force?

16. An aircraft is flying on a heading of 45° south of east at a constant air speed (speed relative to air) of 330 km/h. If the wind is 50 km/h from 60° west of north, what are the speed and direction of the aircraft over the ground?

17. An aircraft flies a northwest (315 T) heading at a constant air speed of 400 km/h. If the wind is 30 km/h *from* the west, what are the speed and direction of the aircraft over the ground?

18. A man drives 110 km north, 60 km west and then 85 km 45° north of east in a total time of 5.0 h. (a) Determine his displacement from the starting point. (b) What distance did he travel? (c) What was his average speed? (d) What was his average velocity?

2-5 COMPONENTS AND COMPONENT METHOD OF VECTOR ADDITION

We have seen that two or more vectors may be added to produce a single vector, the resultant, which represents their combined effect. The reverse is also true. A single vector can be replaced by the sum of two or more other vectors called *components*. This process is called *vector resolution*. A vector may be *resolved* into a number of components, but it is convenient to choose components that are perpendicular to each other because they are independent, i.e. a vector in one direction has no effect in a perpendicular direction. For example, a displacement towards the east does not produce any change in the perpendicular directions north or south, but a northeast displacement produces changes towards both the north and east since it has components in those directions.

Consider any vector **OA** that makes an angle θ with a specified direction (the x-axis for example) and whose origin O is at the center of a coordinate system (Fig. 2-10). From the terminal point A of the vector **OA**,

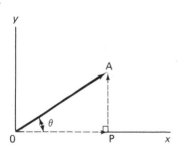

Figure 2-10
Vector resolution.

construct a line AP that is perpendicular to the x-axis at the point P. The vector **OA** is the resultant of the sum of the vectors **OP** and **PA**. Thus

$$\mathbf{OA} = \mathbf{OP} + \mathbf{PA}$$

But **OP** and **PA** are in the directions of the axes of the coordinate system, and their magnitudes are the lengths of their arrows. Therefore in the right triangle OPA:

$$OP = OA \cos \theta \qquad (2\text{-}4)$$
$$PA = OA \sin \theta \qquad (2\text{-}5)$$

Thus the vector **OA** may be written as the sum of a component vector of magnitude $OA \cos \theta$ along the x-axis and a component vector of magnitude $OA \sin \theta$ along the y-axis. Each component vector is the effective value of the original vector in the direction of the component. A component in the negative x or the negative y direction is given a negative value. Note that the coordinates of the terminal point of the vector are given by the components. In this example, the terminal point of the vector **OA** has coordinates (OP, PA) or ($OA \cos \theta$, $OA \sin \theta$).

We can use the Pythagorean Theorem to check answers since the perpendicular components (**OP** and **PA**) and the original vector **OA** form a right-angled triangle.

$$OP^2 + PA^2 = OA^2 \qquad (2\text{-}6)$$

Example 2-7

A boy pulls a wagon by exerting a 60 N force at an angle of 30° with the horizontal (Fig. 2-11). What are the vertical and horizontal forces exerted by the boy on the wagon?

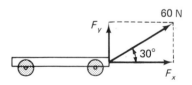

Figure 2-11

Solution:

Horizontal component: $F_x = (60 \text{ N}) \cos 30°$ $= 52 \text{ N}$.
Vertical component: $F_y = (60 \text{ N}) \sin 30° = 30 \text{ N}$.
As a check: $\sqrt{(30 \text{ N})^2 + (52 \text{ N})^2} = 60 \text{ N}$.

In general, vector quantities may have any orientation in a three dimensional space, and three perpendicular components are required to describe them.

Vector Addition by Component Method_____

Once vectors have been resolved into perpendicular components, they are easy to add because the components are independent. The algebraic sum of the x-components is the x-component of the resultant, and the algebraic sum of the y-components is the y-component of the resultant.

Example 2-8_____

When three displacements are drawn from the origin of a perpendicular coordinate system, their terminal points have the coordinates $\mathbf{A} = (25$ km E, 20 km N$)$, $\mathbf{B} = (15$ km W, 12 km N$)$ and $\mathbf{C} = (30$ km E, 18 km S$)$. What are the coordinates of the terminal point of their vector sum?

Solution:

$$\mathbf{A} + \mathbf{B} + \mathbf{C} = (25 \text{ km E, 20 km N})$$
$$+ (15 \text{ km W, 12 km N})$$
$$+ (30 \text{ km E, 18 km S})$$
$$= (25 \text{ km E} - 15 \text{ km E} + 30 \text{ km E,}$$
$$20 \text{ km N} + 12 \text{ km N} - 18 \text{ km N})$$

since west is the negative of east and south is the negative of north. Thus

$$\mathbf{A} + \mathbf{B} + \mathbf{C} = (40 \text{ km E, 14 km N})$$

___

The principles of components can also be used to add vectors that are in polar form, i.e. written in terms of a magnitude and a direction. The procedure is as follows:

1. Resolve each vector into its components along convenient perpendicular axes.

2. Algebraically sum the components in each direction to determine the resultant vector

$\mathbf{R}$ in terms of its x- and y-components R_x and R_y, respectively.

3. Use the Pythagorean Theorem to determine the magnitude R of the resultant (Fig. 2-12).

$$R = \sqrt{R_x^2 + R_y^2} \qquad (2\text{-}7)$$

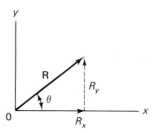

Figure 2-12
Component method.

4. The direction in space (θ with respect to the x-axis) of the *resultant vector* $\mathbf{R}$ is given by

$$\tan \theta = \frac{R_y}{R_x} = \frac{y\text{-component}}{x\text{-component}} \qquad (2\text{-}8)$$

Example 2-9_____

Determine the resultant of the vector system in Fig. 2-13.

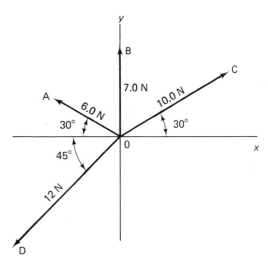

Figure 2-13

Solution:

Vector	x-component	y-component
OA	$-(6.0 \text{ N}) \cos 30 = -5.2 \text{ N}$	$(6.0 \text{ N}) \sin 30 = 3.0 \text{ N}$
OB	$(7.0 \text{ N}) \cos 90 = 0 \text{ N}$	$(7.0 \text{ N}) \sin 90 = 7.0 \text{ N}$
OC	$(10.0 \text{ N}) \cos 30 = 8.7 \text{ N}$	$(10.0 \text{ N}) \sin 30 = 5.0 \text{ N}$
OD	$-(12 \text{ N}) \cos 45 = -8.5 \text{ N}$	$-(12 \text{ N}) \sin 45 = -8.5 \text{ N}$

Determine the components of each vector of the system (above).

Note that some components have negative signs corresponding to the negative x- and y-directions. The components of the resultant **R** are:

$$R_x = -5.2 \text{ N} + 0 \text{ N} + 8.7 \text{ N} - 8.5 \text{ N}$$
$$= -5.0 \text{ N}$$

and

$$R_y = (3.0 \text{ N}) + (7.0 \text{ N}) + (5.0 \text{ N}) - (8.5 \text{ N})$$
$$= 6.5 \text{ N}$$

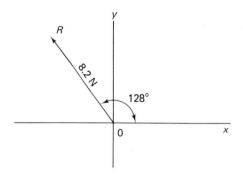

Figure 2-14

These are illustrated in Fig. 2-14. Thus, using the Pythagorean Theorem, we obtain the magnitude and direction of the resultant:

$$R = \sqrt{R_x^2 + R_y^2} = \sqrt{(-5.0 \text{ N})^2 + (6.5 \text{ N})^2}$$
$$= 8.2 \text{ N}$$
$$\tan \theta = \frac{R_y}{R_x} = \frac{-6.5 \text{ N}}{5.0 \text{ N}} = -1.3$$
$$\theta = 128°$$

Problems

19. The terminal points of velocity vectors, which are drawn from the origin of a coordinate system, are **A** = (80 km/h E, 35 km/h S), **B** = (65 km/h W, 80 km/h S), **C** = (120 km/h E, 50 km/h N) and **D** = (90 km/h E, 60 km/h S). Determine the vector sum **A** + **B** + **C** + **D** and express the answer (a) in terms of east and north components and (b) as a speed in a certain direction.

20. A boy pulls a wagon by exerting a 35.0 N force at an angle of 30° with the horizontal. What are the horizontal and vertical components of the force?

21. Resolve the following velocities into their east and north components: (a) 30 km/h at 045 T, (b) 20 km/h at 022 T, (c) 15 km/h at 340 T, (d) 25 km/h at 245 T, (e) 18 km/h at 160 T.

22. A weight of 50 lb rests on an inclined plane that makes an angle of 30° with the horizontal. Find the components of the weight parallel and perpendicular to the plane.

23. Resolve the following into horizontal and vertical components: (a) 72 N at 20° with the horizontal, (b) 28 N at 45° with the horizontal, (c) 35 N at 60° with the horizontal, (d) 85 N at 62° with the horizontal.

24. Resolve the following forces into their perpendicular x- and y-components: (a) 46 N at 35° with respect to the x-axis, (b) 38 N at 84° with the x-axis, (c) 55 N at

145° with the *x*-axis, (d) 63 N at 290° with the *x*-axis.

25. Add the vectors in Problem 23 using the component method and express the answer in polar form.

26. Determine the resultant of the sum of the following forces: 6.0 N at 30° to the *x*-axis, 10 N at 130° to the *x*-axis and 7.0 N at 255° to the *x*-axis.

27. Add the vectors in Problem 24 using the component method and express the answer in polar form.

2-6 VECTOR SUBTRACTION

It is sometimes necessary to subtract one vector quantity from another. This process is equivalent to vectorially adding the negative value (a vector of the same magnitude but opposite direction) of the vector to be subtracted. For example, to subtract vector **B** from vector **A**, form the negative of **B**, which is written as $-$**B**, and then add $-$**B** to **A** (Fig. 2-15). This is written as:

$$\mathbf{A} - \mathbf{B} = \mathbf{A} + (-\mathbf{B})$$

To subtract a vector that is written in component form, we must change the sign of *each* of its components and then add them algebraically.

Example 2-10

A car travelling at an initial velocity with components $\mathbf{v}_0 = (25 \text{ km/h E}, \quad 35 \text{ km/h N})$ accelerates to a final velocity with compo-

nents $\mathbf{v}_1 = (50 \text{ km/h E}, \quad 15 \text{ km/h S})$ in 50 s. Determine (a) the change in velocity and (b) the average acceleration of the car in component form.

Solution:

(a) $\Delta\mathbf{v} = \mathbf{v}_1 - \mathbf{v}_0 = (50 \text{ km/h E}, 15 \text{ km/h S})$
$\qquad - (25 \text{ km/h E}, -35 \text{ km/h S})$
$\qquad = (25 \text{ km/h E}, 50 \text{ km/h S})$

(b) $\mathbf{a} = \dfrac{\Delta\mathbf{v}}{t} = \dfrac{(25 \text{ km/h E}, 50 \text{ km/h S})}{50 \text{ s}}$
$\qquad = \dfrac{(7.0 \text{ m/s E}, 14 \text{ m/s S})}{50 \text{ s}}$
$\qquad = (0.14 \text{ m/s}^2 \text{ E}, 0.28 \text{ m/s}^2 \text{ S})$

We can now add these components to find the average acceleration of 0.31 m/s² 63° S of E in polar form.

Relative Motion

Occasionally we may want to determine the motion of one object relative to some other moving object. For example, if we are in a car or aircraft, our view of other moving objects is distorted by our own motion. In order to obtain the motion of one object relative to a moving observer, we must subtract the motion of the observer.

Suppose that we drive 50 km due north in one hour while a second car, which started at the same position, drives 60 km due south. After an hour, the second car is 110 km south of us, and therefore its velocity *relative to us* was 110 km/h south. This result can also be obtained by subtracting our velocity from the velocity of the other car, 60 km/h S $-$ ($-$50 km/h S) = 110 km/h S.

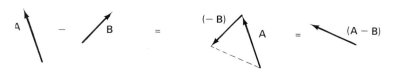

Figure 2-15
Vector subtraction.

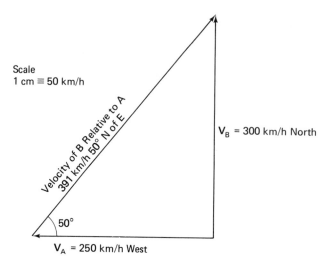

Scale
1 cm ≡ 50 km/h

Velocity of B Relative to A
391 km/h 50° N of E

V_B = 300 km/h North

50°

V_A = 250 km/h West

Figure 2-16

Example 2-11

Determine the velocity of one aircraft B relative to another aircraft A if B flies due north at 300 km/h while A flies due west at 250 km/h.

Solution:

Sketch the two velocities (Fig. 2-16). The velocity of B relative to A is represented by the arrow *from* the terminal point of the vector to be subtracted, v_A, *to* the terminal point of v_B. This is equivalent to completing the parallelogram with v_B and $-v_A$. Since the velocities are perpendicular, the relative velocity is $\sqrt{(300 \text{ km/h})^2 + (250 \text{ km/h})^2} = 391$ km/h in a direction given by

$$\tan \theta = \frac{300 \text{ km/h}}{250 \text{ km/h}} = 1.2$$
$$\theta = 50° \text{ N of E}$$

Problems

28. If $A = (25$ km north$)$ and $B = (35$ km west$)$, find (a) $A - B$ and (b) $B - A$.

29. If $A = 80$ km/h north and $B = 55$ km/h west, find the velocity A relative to B.

30. If $A = 15$ km $45°$ north of west, $B = 8.0$ km south and $C = 12$ km at $20°$ east of north, determine (a) $A + B - C$, (b) $A - B + C$, (c) $A - B - C$.

31. Two forces combine to produce a resultant of 150 N north. If one of the forces is 85 N at $30°$ east of north, what is the other force?

2-7 MULTIPLICATION OF VECTORS

The multiplication of any vector A by any scalar n is a vector nA with a magnitude equal to the product of the absolute value of the scalar n and the magnitude of the vector A. If n is positive, the resultant vector nA is in the same direction as the vector A; but if n is negative, the resultant vector is in the opposite direction to the vector A.

The multiplication of two or more vectors is not as simple however, because there is an added complication due to the directions of the vectors. It is convenient to define two different ways to multiply vectors, which do not necessarily have the same units.

Scalar (Dot) Product

The scalar product between two vectors **A** and **B** is defined as the product of the magnitudes of **A** and **B** and the cosine of the smallest angle θ between their directions (Fig. 2-17).

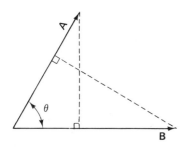

Figure 2-17
Scalar (dot) product.

This is written as:

$$\mathbf{A} \cdot \mathbf{B} = \mathbf{B} \cdot \mathbf{A} = AB \cos \theta, \quad 0 \leq \theta \leq \pi \quad (2\text{-}9)$$

Note that the magnitude of the component of **A** in the direction of **B** is $A \cos \theta$; similarly $B \cos \theta$ is the magnitude of the component of **B** in the direction of **A**. Therefore a scalar product is merely the product of the magnitudes of the components of both **A** and **B** in the direction of either **A** or **B**, and *the result is a scalar quantity.*

Example 2-12

Determine the scalar product between a force **F** = 50 N due east at 60° above the horizontal, and a horizontal displacement **s** = 20 m towards the east.

Solution:

$$\mathbf{F} \cdot \mathbf{s} = Fs \cos \theta = (50 \text{ N})(20 \text{ m}) \cos 60°$$
$$= 500 \text{ N} \cdot \text{m}$$

Vector (Cross) Product

Consider a vector **A** that is inclined at some angle θ with respect to some other vector **B**.

The vector product of two vectors **A** and **B** is defined as a *vector* **C**, whose magnitude is equal to the product of the magnitudes of **A** and **B** and the sine of the angle θ between them. The direction of **C** is perpendicular to both **A** and **B**, and it is determined by a right-hand rule. Align the fingers of the right hand in the direction of the first vector mentioned and rotate the hand (the natural direction in which the fingers curl) through the angle θ to the second vector. The straightened thumb gives the direction of the resultant vector (Fig. 2-18). A vector product is written as:

$$\mathbf{A} \times \mathbf{B} = -\mathbf{B} \times \mathbf{A} = AB \sin \theta \quad (2\text{-}10)$$
$$0 \leq \theta \leq \pi$$

in a direction determined by the right-hand rule. Note that the magnitude of the component of **A** in a direction perpendicular to

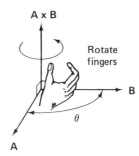

Figure 2-18
Vector **A** is directed out of the paper.

B is $A \sin \theta$. Therefore the magnitude of **A** × **B** is the product of the magnitude of **B** and the magnitude of the component of **A** that is perpendicular to **B**.

Example 2-13

Determine the vector product (**s** × **F**) between a 5.0 m horizontal displacement **s** towards 30° north of east and a 30 N horizontal easterly force **F**.

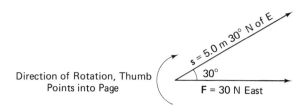

Direction of Rotation, Thumb
Points into Page

Figure 2-19

Solution:

The magnitude of the vector product (Fig. 2-19) is:

$$\mathbf{s} \times \mathbf{F} = sF \sin \theta = (5.0 \text{ m})(30 \text{ N}) \sin 30°$$
$$= 75 \text{ N·m}$$

Using the right-hand rule (rotating the fingers from $\mathbf{s}$ to $\mathbf{F}$), we can see that the direction of the vector product is down, i.e. into the earth!

---

*Problems*_____

32. What is the scalar product of two vectors of magnitudes 15 m and 9.0 N if the angle between their directions is 45°?

33. What is the magnitude of (a) the scalar product and (b) the vector product of two vectors of magnitudes 10 m and 6.0 N if the angle between their directions is 40°?

34. Determine the vector product $\mathbf{v} \times \mathbf{B}$ between a horizontal velocity $\mathbf{v} = 2.0 \times 10^4$ m/s towards the north and a magnetic flux density $\mathbf{B} = 0.250$ T towards the north but 30° below the horizontal.

REVIEW PROBLEMS

1. Show that 60 mi/h is equivalent to 88 ft/s.

2. What is the average speed of a car that travels 35 m in 2.1 s? Express the answer in kilometers per hour.

3. A ship transmits a radar pulse and receives a reflection from an aircraft after 200 μs. If the speed of the radar pulse is 1.86×10^5 mi/s, how far away is the aircraft?

4. How far does an aircraft travel in 25 s if it has a constant speed of 500 km/h?

5. What is the average acceleration of an object if it is travelling due east and changes its speed from 20 m/s to 35 m/s in 5.0 s?

6. An electron accelerates in a straight line at 3.0×10^{10} m/s² in a cathode ray tube. If its initial speed was 3.0×10^4 m/s, what is its speed after 3.0 μs?

7. What is the vector sum of a 12 N horizontal force and a 5.0 N vertical force?

8. The sum of two perpendicular forces has a magnitude of 380 N. If one of the forces has a magnitude of 230 N, what is the magnitude of the other?

9. A man drives 60 km east, 80 km in a direction 20° north of east and then 130 km south in a total time of 4.5 h. Determine (a) the distance travelled, (b) the displacement from the starting point, (c) his average speed and (d) his average velocity.

10. A man pushes a lawn mower by exerting a 150 N force along the handle, which is inclined at 60° with the horizontal. Find the horizontal and vertical components of the force.

11. (a) Resolve the following into their perpendicular x- and y-components:

 $A = 80$ N at $48°$ from the x-axis
 $B = 64$ N at $134°$ from the x-axis
 $C = 72$ N at $200°$ from the x-axis
 $D = 51$ N at $305°$ from the x-axis

(b) Find $A + B + C + D$. (c) Find $A - B - C + D$.

12. (a) Resolve the following vectors into their perpendicular east and north components:

 $A = 120$ km, $60°$ N of E
 $B = 100$ km, $20°$ E of S
 $C = 130$ km, $30°$ S of W
 $D = 80$ km N

(b) Find $A + B + C + D$. (c) Find $A - B + C + D$.

13. Boat A moves due north at 25 km/h and boat B moves $30°$ S of W at 30 km/h. What is the velocity of A relative to B?

14. Determine the scalar product between a force $F = 30$ lb at $48°$ above the horizontal towards the north and a horizontal displacement of 150 ft towards the north.

15. Determine the vector product $(s \times F)$ between an 18 m horizontal displacement s towards the north and a 42 N horizontal force towards $20°$ west of north.

Even though the causes of their motion may be different, the motion of most objects, such as celestial bodies, trains, cars, golf balls, falling objects, atoms and electrons, may all be described by similar equations.

The study of relative motion without reference to the cause of that motion is called **kinematics**. Since all bodies in the universe are in continuous motion, kinematics is a major topic in the science of physics.

3

Kinematics

3-1 GRAPHICAL ANALYSIS OF MOTION

We have seen that the average velocity $\bar{\mathbf{v}}$ of an object is the ratio of its displacement **s** (or change in position $\Delta\mathbf{r}$) to the corresponding elapsed time t:

$$\bar{\mathbf{v}} = \frac{\mathbf{s}}{t} = \frac{\Delta\mathbf{r}}{t} \qquad (3\text{-}1)$$

When an object is moving in a straight line and changes its position by equal amounts in equal line intervals, its velocity **v** is constant, i.e. it travels in a straight line at a constant speed. This is the simplest type of motion to analyze because the displacement **s** (or change in position $\Delta\mathbf{r}$) is directly proportional to the elapsed time t

$$\mathbf{s} = \Delta\mathbf{r} = \bar{\mathbf{v}}t \quad \text{and} \quad \mathbf{s} = \Delta\mathbf{r} \propto t$$

since **v** is constant. Therefore, if we plot a graph with position (or displacement) as the ordinate (y-axis) and elapsed time as the abscissa (x-axis), we obtain a straight line that has a slope equal to the velocity **v** (Fig. 3-1):

$$\text{slope} = \frac{\text{rise}}{\text{run}} = \frac{\Delta\mathbf{r}}{t} = \frac{\mathbf{s}}{t} = \mathbf{v}$$
$$\text{(position-time graph)} \qquad (3\text{-}2)$$

Example 3-1

A car travels in a straight line due west at a constant speed of 80 km/h. If the initial position of the car was 30 km west of some reference, plot the position-versus-elapsed time graph.

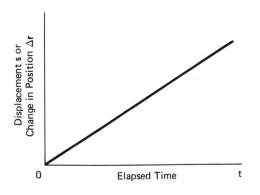

Figure 3-1
Displacement-versus-elapsed time graph for an object with a constant velocity.

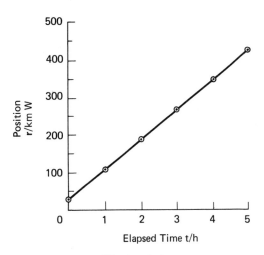

Figure 3-2
Position-versus-elapsed time graph.

Solution:

We can tabulate the positions of the car after certain elapsed times by merely adding 80 km for each hour that it travels (Table 3-1).

The corresponding graph is illustrated in Fig. 3-2. Note that the slope of the line is equal to the velocity of the car:

$$\text{slope} = \frac{\text{rise}}{\text{run}} = \frac{400 \text{ km W}}{5 \text{ h}} = 80 \frac{\text{km}}{\text{h}} \text{ W} = \mathbf{v}$$

Of course, objects do not always travel in straight lines at constant speeds, but *velocity* is always the time rate of change of position and corresponds to the slope of the position- (or displacement) versus-elapsed time graph. If the slope changes with time, the velocity also changes. In general, if we plot a graph with the displacement (or position) as the ordinate and the elapsed time as the abscissa, the *instantaneous velocity* $\mathbf{v}$ at some time t is the slope of the tangent* to the curve at that instant. Thus if $\mathbf{\Delta r}$ is the change in position during some small time interval $(t - t_0)$, the instantaneous velocity $\mathbf{v}$ at the time t_0 is obtained by making the increment in time $(t - t_0)$ infinitesimally small (by making t extremely close to t_0) (Fig. 3-3). This is often written as a limit:

_____

*The line just touching the curve.

Table 3-1

Elapsed time t/h	0	1	2	3	4	5
Position $\mathbf{r}$/km W	30	110	190	270	350	430
Displacement s/km W (change in position)	0	80	160	240	320	400

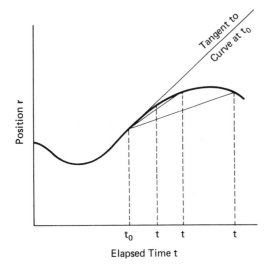

Position r

Tangent to
Curve at t_0

t_0 t t t

Elapsed Time t

Figure 3-3

Position-versus-elapsed time graph. As the time increment $t - t_0$ decreases, the slope of the curve segment approaches that of the tangent to the curve t_0.

$$\mathbf{v} = \underset{t \to t_0}{limit} \frac{\Delta\mathbf{r}}{t - t_0}$$

As the increment in time $(t - t_0)$ approaches zero, the corresponding change in position $\Delta\mathbf{r}$ becomes infinitesimally small, and therefore the *magnitude* of the velocity is equal to the instantaneous speed. In the future, we shall use the term velocity to mean instantaneous velocity, unless otherwise stated.

*Example 3-2*_____

If the positions of a car relative to some intersection are recorded at one-second time intervals (Table 3-2), (a) plot the correspond-

ing position-versus-elapsed time graph. From the graph, find (b) the average velocity of the car during the first ten seconds and (c) the instantaneous velocity at the sixth second.

Solution:

(a) See Fig. 3-4.

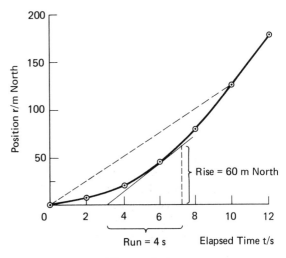

Rise = 60 m North

Run = 4 s   Elapsed Time t/s

Figure 3-4

Position-versus-elapsed time graph.

(b) $\bar{\mathbf{v}} = \frac{\mathbf{s}}{t} = \frac{125 \text{ m north}}{10 \text{ s}} = 12.5$ m/s north

This corresponds to the slope of the dotted line.

(c) Draw the tangent AB to the curve at the sixth second. The slope of AB is the instantaneous velocity

$$\mathbf{v} = \frac{\text{rise}}{\text{run}} = \frac{60 \text{ m}}{4 \text{ s}} \text{ north} = 15 \text{ m/s north}$$

Table 3-2

Elapsed time t/s	0	2	4	6	8	10	12
Position $\mathbf{r}$/(m north)	0	5	20	45	80	125	180

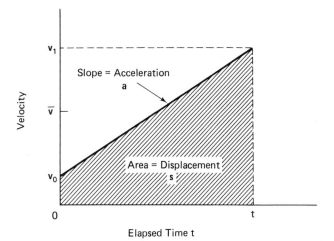

Figure 3-5
Velocity – versus – elapsed time graph for
uniformly accelerated motion.

If an object accelerates from an initial velocity v_0 to a final velocity v_1 in a time interval t, its average acceleration is:

$$\bar{\mathbf{a}} = \frac{\mathbf{v}_1 - \mathbf{v}_0}{t} = \frac{\Delta \mathbf{v}}{t} \qquad (3\text{-}3)$$

Frequently, objects travel in straight lines with constant accelerations, i.e. they change velocity by equal amounts in equal time intervals. Objects with this special type of motion are said to be *uniformly accelerated*. In this special case, Eq. 3-3 becomes:

$$\mathbf{v}_1 - \mathbf{v}_0 = \Delta \mathbf{v} = \mathbf{a}t \quad \text{and} \quad \Delta \mathbf{v} \propto t$$

since $\mathbf{a}$ is constant. Therefore if we plot a graph with velocity as the ordinate and elapsed time as the abscissa, we obtain a straight line that has a slope equal to the acceleration (Fig. 3-5):

$$\text{slope} = \frac{\text{rise}}{\text{run}} = \frac{\Delta \mathbf{v}}{t} = \mathbf{a}$$
$$\text{(velocity-elapsed time graph)} \quad (3\text{-}4)$$

Since the acceleration is constant, the average velocity equals:

$$\bar{\mathbf{v}} = \frac{\mathbf{v}_1 + \mathbf{v}_0}{2} \qquad (3\text{-}5)$$

where $\mathbf{v}_1$ and $\mathbf{v}_0$ are the final and initial veloc-

ities, respectively. From Eq. 3-1, we see that the area bounded by the curve and the elapsed time axis is equal to the displacement $\mathbf{s}$:*

$$\text{area} = \left(\frac{\mathbf{v}_1 + \mathbf{v}_0}{2}\right)t = \bar{\mathbf{v}}t = \mathbf{s}$$
$$\text{(velocity-elapsed time graph)} \quad (3\text{-}6)$$

Finally, if we plot a graph of acceleration-versus-elapsed time (Fig. 3-6), the area bounded by the curve and the elapsed time axis corresponds to the change in velocity $\Delta \mathbf{v}$:

$$\text{area} = \mathbf{a}t = \Delta \mathbf{v}$$
$$\text{(acceleration-elapsed time graph)} \quad (3\text{-}7)$$

*Example 3-3*_____

An aircraft travelling due north with an initial speed of 100 m/s accelerates uniformly at 5.0 m/s² for 8.0 s. Plot the corresponding velocity-elapsed time graph and find the displacement of the aircraft while it accelerates.

_____

*This is true for nonuniform motion also, but for nonuniform motion, the area must be determined by integral calculus as follows

$$s = \int_0^t v \, dt.$$

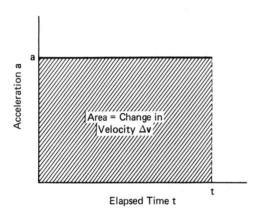

Figure 3-6
Acceleration-versus-elapsed time graph for uniformly accelerated motion.

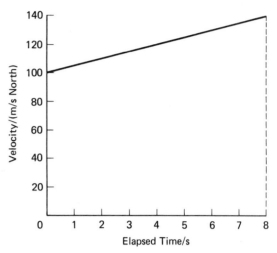

Figure 3-7
Velocity-versus-elapsed time graph for the aircraft.

Solution:

Since the acceleration is 5.0 m/s², the aircraft increases its velocity by 5.0 m/s each second (Table 3-3).

The displacement **s** is the area bounded by the curve and the elapsed time axis of the velocity-elapsed time graph (Fig. 3-7).

$$\mathbf{s} = \left(\frac{\mathbf{v}_1 + \mathbf{v}_0}{2}\right)t$$

$$= \frac{1}{2}\left(100\ \frac{m}{s}\ \text{north} + 140\ \frac{m}{s}\ \text{north}\right)(8.0\ \text{s})$$

$$= 960\ \text{m north}$$

Note that the slope of the line is the acceleration

$$= \frac{(140\ \text{m/s} - 100\ \text{m/s})}{8.0\ \text{s}}\ \text{north}$$

$$= 5.0\ \frac{m}{s^2}\ \text{north}$$

Even if an object does not have constant acceleration, we can find its instantaneous

acceleration from a velocity-elapsed time graph.

If we plot a graph with the velocity of the body as the ordinate and the elapsed time as the abscissa, the *instantaneous acceleration* **a** at a particular time t_0 is the slope of the curve at that instant (Fig. 3-8). Thus:

$$\mathbf{a} = \lim_{t \to t_0} \frac{\Delta \mathbf{v}}{t - t_0} \text{*} \qquad (3\text{-}8)$$

This is equivalent to defining the instantaneous acceleration at the time t_0 as the slope of the tangent to the velocity–time curve at that instant. As with velocity, the term accelera-

---

*This is also the definition of a derivative in calculus; it is written as $\mathbf{a} = d\mathbf{v}/dt$.

Table 3-3

Elapsed time t/s	0	1	2	3	4	5	6	7	8
Velocity v/(m/s) north	100	105	110	115	120	125	130	135	140

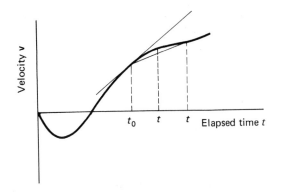

Figure 3-8

Velocity-versus-elapsed time graph. As t approaches t_0, the slope of the curve approaches that of the tangent to the curve at t_0.

tion will now refer to instantaneous acceleration, unless otherwise stated.

Problems

1. A particle starts from rest and accelerates uniformly at 8.0 m/s² for 12 s. Sketch the velocity-elapsed time graph, and find from the graph the distance that it travels while accelerating.

2. An electron in an oscilloscope accelerates uniformly from rest to a speed of 5.0×10^5 m/s in 2.0 μs. (a) Sketch the speed-elapsed time graph. (b) How far did the electron go in the 2.0 μs? (c) What was its acceleration?

3. A cyclist starts from rest and travels due east, accelerating uniformly to a speed of 12 m/s in 15 s. He then pedals steadily for 8.0 s. Finally he applies the brakes for 7.0 s, slowing uniformly to rest. (a) Plot the speed-versus-elapsed time graph, (b) Find the total displacement.

4. From the velocity—time curve in Fig. 3-9, determine (a) the time at which the velocity is a maximum, (b) the acceleration when the velocity is a maximum, (c) the time at which the acceleration is a maximum, (d) the maximum value of the acceleration, (e) over what period of time the acceleration is positive, (f) over what period of time the acceleration is negative, (g) over what period of time the acceleration is constant.

5. A car heading due west passes a starting point at a speed of 5.0 m/s and accelerates uniformly at 5.0 m/s² for 8.0 s. (a) Draw the velocity-versus-time graph. (b) Calculate the overall displacement. (c) What is the average velocity?

6. A car starts from rest and accelerates uniformly to a speed of 30 ft/s in 12 s. It travels at this speed for 6.0 s and then brakes

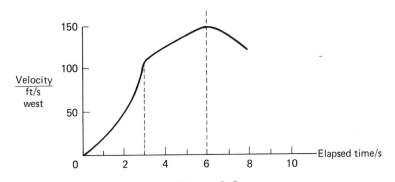

Figure 3-9

Velocity-versus-elapsed time graph.

uniformly to rest in a further 15 s. Sketch the graph of speed-versus-time. From the graph, determine (a) the acceleration during the first 12 s and (b) the total distance covered. (c) Sketch the distance–time graph.

7. An electron starts from rest and accelerates uniformly in an electric field to a speed of 10^6 m/s in 2×10^{-4} s. It travels at 10^6 m/s for 10^{-4} s before decelerating uniformly to rest in a further 3×10^{-4} s. (a) Sketch the speed-versus-elapsed time graph for the electron. (b) From the graph, determine the acceleration during the first 2×10^{-4} s. (c) Determine the total distance covered during the first 2×10^{-4} s. (d) Sketch the distance-versus-elapsed time curve for the electron's motion.

8. (a) Sketch the displacement-elapsed time and the acceleration-elapsed time graphs from the velocity-elapsed time graph in Fig. 3-10. Assume that the initial displacement is zero. (b) What is the total displacement after the 6.0 s? (c) What is the instantaneous velocity at 3.5 s? (d) What is the average acceleration during the 6.0 s?

9. Sketch the velocity-time curve and the displacement-time curve from the acceleration-time curve in Fig. 3-11.

3-2 EQUATIONS OF UNIFORMLY ACCELERATED MOTION

When an object has an average velocity $\bar{\mathbf{v}}$ for an elapsed time t, its displacement is:

$$\mathbf{s} = \Delta \mathbf{r} = \bar{\mathbf{v}}t \tag{3-1}$$

If its initial and final velocities were $\mathbf{v}_0$ and $\mathbf{v}_1$, its average acceleration was:

$$\bar{\mathbf{a}} = \frac{\mathbf{v}_1 - \mathbf{v}_0}{t} = \frac{\Delta \mathbf{v}}{t} \tag{3-3}$$

These two equations are valid for all types of motion!

In the special case when an object travels in a straight line and undergoes uniform acceleration, its acceleration is a constant and its average acceleration is equal to its instantaneous acceleration. If one direction along the line of motion is taken as positive and the opposite direction is taken as negative,

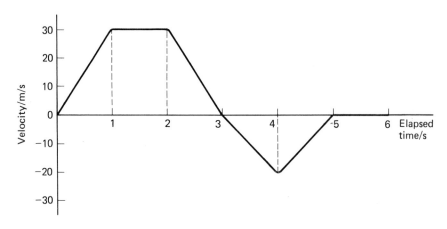

Figure 3-10
Velocity–versus–elapsed time graph.

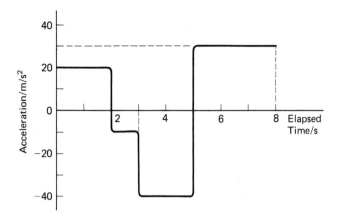

Figure 3-11
Acceleration-versus-elapsed time graph.

the magnitude of the acceleration is equal to the difference between the final speed v_1 and the initial speed v_0 divided by the elapsed time t.

$$a = \frac{v_1 - v_0}{t} \quad \text{or} \quad v_1 = v_0 + at * \quad (3\text{-}9)$$

The average speed $\bar{v}$ in the positive direction of the motion is equivalent to the average

---

*These equations involve vector quantities, but since the direction is constant we shall omit the vector notation for convenience.

between the initial and final values (Fig. 3-12). Thus

$$\bar{v} = \frac{v_1 + v_0}{2} * \quad (3\text{-}10)$$

From Eq. (3-1), the magnitude of the displacement in an elapsed time t, is

$$s = \bar{v}t = \left(\frac{v_1 + v_0}{2}\right)t * \quad (3\text{-}11)$$

Example 3-4

If a train accelerates uniformly from 15.0 mi/h to 60.0 mi/h in 50.0 s, how far does it travel while it is accelerating?

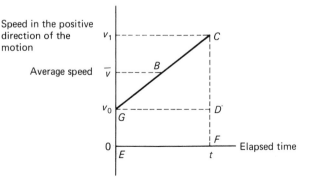

Figure 3-12
Velocity-versus-elapsed time graph for uniformly accelerated motion.

Solution:

$$v_1 = 60.0 \text{ mi/h} = 88.0 \text{ ft/s} \quad \text{and} \quad t = 50.0 \text{ s}$$

$$v_0 = 15.0 \text{ mi/h} = (15.0 \text{ mi/h})\left(\frac{88 \text{ ft/s}}{60 \text{ mi/h}}\right)$$

$$= 22.0 \text{ ft/s}$$

Thus

$$s = \left(\frac{v_0 + v_1}{2}\right)t$$

$$= \left[\frac{(88.0 \text{ ft/s}) + (22.0 \text{ ft/s})}{2}\right](50.0 \text{ s})$$

$$= 2\,750 \text{ ft}$$

Example 3-5

Figure 3-13 is the velocity-versus-elapsed time curve of a complete journey. Assume that the initial displacement was zero and determine (a) the instantaneous acceleration at the second hour, (b) the average acceleration for the whole trip, and (c) the total displacement after 10 h.

Solution:

(a) The instantaneous acceleration at the second hour is the slope of the tangent to the curve at that instant, i.e., the slope of the line AB.

$$\mathbf{a} = \frac{30 \text{ mi/h north}}{3 \text{ h}} = 10 \text{ mi/h}^2 \text{ north}$$

(b) The average acceleration for the whole trip is the slope of the line AF.

$$\bar{\mathbf{a}} = -\frac{15 \text{ mi/h north}}{10 \text{ h}} = -1.5 \text{ mi/h}^2 \text{ north}$$

$$= 1.5 \text{ mi/h}^2 \text{ south}$$

(c) The displacement after ten hours is the area beneath the curve from A to B to C to D to E to F.

$$s = [\tfrac{1}{2} (3 \text{ h})(30 \text{ mi/h}) + (2 \text{ h})(30 \text{ mi/h})$$
$$+ \tfrac{1}{2}(1 \text{ h})(30 \text{ mi/h})$$
$$+ 0 - \tfrac{1}{2}(3 \text{ h})(15 \text{ mi/h})] \text{ north}$$
$$= 97\tfrac{1}{2} \text{ mi north}$$

Two other useful equations may be derived by eliminating v_1 and t in Eqs. (3-9), (3-10), and (3-11). Substituting Eq. (3-9) in Eq. (3-11), we obtain:

$$s = \frac{v_1 t}{2} + \frac{v_0 t}{2} = (v_0 + at)\frac{t}{2} + \frac{v_0 t}{2}$$

$$= v_0 t + \frac{at^2}{2} \qquad (3\text{-}12)$$

Also, solving for t in Eqs. (3-9) and (3-11) and equating, we obtain:

$$t = \frac{2s}{v_1 + v_0} = \frac{(v_1 - v_0)}{a}$$

Thus

$$2as = (v_1 + v_0) \times (v_1 - v_0) = v_1^2 - v_0^2 \quad (3\text{-}13)$$

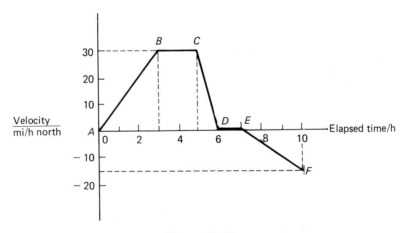

Figure 3-13
Velocity-versus-elapsed time graph.

To summarize, the five equations of *uniformly accelerated motion* are:

$$v_1 = v_0 + at \qquad (3\text{-}9)$$
$$\bar{v} = (v_1 + v_0)/2 \qquad (3\text{-}10)$$
$$s = \bar{v}t = \left(\frac{v_1 + v_0}{2}\right)t \qquad (3\text{-}11)$$
$$s = v_0 t + \tfrac{1}{2}at^2 \qquad (3\text{-}12)$$
$$v_1^2 = v_0^2 + 2as \qquad (3\text{-}13)$$

If an object accelerates uniformly and any three of the quantities s, v_0, v_1, a, or t are known, the two other quantities may be determined from these equations. Remember, these equations really involve vector quantities.

Example 3-6

A car is travelling at a speed of 30 m/s when the driver applies the brakes. If the car slows uniformly to a final speed of 2.0 m/s in a distance of 120 m, what is its acceleration?

Solution:

$v_0 = 30$ m/s, $v_1 = 2.0$ m/s, $s = 120$ m and $a = ?$ Thus from Eq. (3-13)

$$a = \frac{v_1^2 - v_0^2}{2s}$$
$$= \frac{(2.0 \text{ m/s})^2 - (30 \text{ m/s})^2}{2(120 \text{ m})} = -3.7 \text{ m/s}^2$$

The negative sign indicates a *deceleration*, which is an acceleration in the direction opposite to the motion.

Example 3-7

A body moving with an initial speed of 10 m/s is subjected to a uniform acceleration of 8.0 m/s² for 5.0 s. (a) What is the final speed of the body? (b) How far does the body travel while it is being accelerated?

Solution:

$v_0 = 10$ m/s, $\quad a = 8.0$ m/s², $\quad t = 5.0$ s,
(a) $v_1 = ?$ and (b) $s = ?$

(a) $v_1 = v_0 + at = 10$ m/s $+ (8.0 \text{ m/s}^2)(5.0 \text{ s})$
$\qquad\qquad = 50$ m/s

(b) $\quad s = v_0 t + \tfrac{1}{2}at^2$
$\qquad = (10 \text{ m/s})(5.0 \text{ s}) + \tfrac{1}{2}(8.0 \text{ m/s}^2)(5.0 \text{ s})^2$
$\qquad = 150$ m

Problems

10. A car accelerates from 30 mi/h to 80 mi/h in 20 s. (a) What is its acceleration in ft/s²? (b) How far does the car travel while accelerating?

11. A train starts from rest and accelerates uniformly to a speed of 20 mi/h in 35 s. (a) What is its acceleration in ft/s²? (b) How far does it travel while accelerating?

12. A car accelerates uniformly from rest to a speed of 80 mi/h in 13 s. (a) What is the acceleration? (b) How far does it travel while accelerating?

13. An electron accelerates uniformly at 5.0 × 10¹¹ m/s² for 2.0 μs. If the electron was initially at rest, calculate (a) its final speed and (b) the distance that it travels while accelerating.

14. An electron in a cathode-ray tube accelerates from rest to a speed of 400 km/s in an elapsed time of 1.5 μs. (a) What is its acceleration? (b) How far does it travel while accelerating?

15. An object accelerates uniformly at 10 m/s² for 10 s. If its initial speed was 30 m/s, (a) what is its final speed? (b) How far does it travel in the 10 s?

16. A car is travelling at 80 km/h when the driver applies the brakes, bringing it uniformly to rest in 4.0 s. (a) How far did the car travel after the brakes were applied? (b) What was the acceleration?

17. A driver exceeds an 80 km/h speed limit by an average of 5.0 km/h on a 272 km journey. How much time does he save?

18. An object accelerates uniformly at 9 m/s² to a final speed of 90 m/s in a distance of 80 m. (a) What was the initial speed? (b) How long does it accelerate?

19. A car starts from rest and accelerates uniformly. If it travels 36 m during the fifth second, find (a) its acceleration and (b)

the total distance that it travels from rest in 6.0 s.

20. A train travelling at a constant speed of 25 ft/s passes a stationary car. If the car starts at the instant that the train passes it and accelerates uniformly in the same direction at 5 ft/s², (a) how long does it take the car to overtake the train? (b) How far does the car go before it over-takes the train? (c) What is the speed of the car as it overtakes the train?

21. If the car in Problem 20 accelerates to 40 ft/s and then maintains that speed, (a) how long does it take the car to overtake the train? (b) How far does it travel before it overtakes the train?

3-3 FREELY FALLING BODIES

It is a well known fact that when an object is dropped near the surface of the earth, it increases its speed as it falls. Therefore freely falling objects must be accelerated towards the center of the earth. By rolling balls down inclined planes, Galileo discovered that this acceleration, which is called the *acceleration due to gravity*, is the same for all bodies, independent of their mass! This may be illustrated by simultaneously dropping a book of many pages and a single sheet of paper, made into a compact ball, from the same height; they both hit the ground at the same instant!

The magnitude g of the acceleration due to gravity is approximately 32 ft/s² or 9.8 m/s² at the surface of the earth. Even though the acceleration due to gravity changes with the distance from the surface of the earth (see Chapter 4), it is usually considered to be constant for small changes in height near the earth's surface. Therefore the equations of uniformly accelerated motion may be used for objects falling through distances (or thrown vertically upwards to heights), which are small compared with the radius of the

earth. In these problems the displacement **s** corresponds to the change in height h measured from the origin to the terminal point of the path, and the acceleration **a** is the acceleration due to gravity **g**, which always acts downwards. It is important to remember that these equations involve *vector* quantities.

Example 3-8

If a stone is dropped from a 100 ft high cliff into the sea and air resistance is neglected, (a) how long does the fall last? (b) What is the speed of the stone as it strikes the water?

Solution:

$s = h = 100$ ft, $v_0 = 0$ ft/s, $a = g = 32$ ft/s²
(a) $t = ?$ and (b) $v_1 = ?$

(a) $s = h = v_0 t + \frac{1}{2}at^2 = 0 + \frac{1}{2}gt^2$ (3-14)

Thus,

$$t = \sqrt{\frac{2s}{g}} = \sqrt{\frac{2(100 \text{ ft})}{32 \text{ ft/s}^2}} = 2.5 \text{ s}$$

(b) $v_1^2 = v_0^2 + 2as = 0 + 2gh$
or $v_1 = \sqrt{2gh}$ (3-15)

Thus

$$v_1 = \sqrt{2(32 \text{ ft/s}^2)(100 \text{ ft})} = 80 \text{ ft/s}$$

Example 3-9

A ball thrown vertically upwards with an initial speed of 20 m/s from the top of a bridge falls into the water 8.0 s later directly below the bridge. (a) How high did the ball go above the bridge? (b) How high is the bridge? Neglect air resistance.

Solution:

(a) If up is taken as the positive direction, $v_0 = 20$ m/s; $a = -g = -9.8$ m/s², i.e. **g** acts downwards; $v_1 = 0$ m/s, the ball comes momentarily to rest at the top of its flight; and $s = h = ?$ Substituting in Eq. (3-13), we obtain:

$$v_1^2 = v_0^2 + 2(-g)h$$

Therefore

$$h = \frac{v_1^2 - v_0^2}{-2g} = \frac{0 - (20 \text{ m/s})^2}{-2(9.8 \text{ m/s}^2)} = 20 \text{ m*}$$

(b) $v_0 = 20$ m/s, $a = -g = -9.8$ m/s², $t = 8.0$ s and $s = h = ?$ Note that the *displacement* (origin to terminal point of the path) is merely the height of the bridge even though the ball is thrown upwards to 20 m above the bridge (Fig. 3-14).

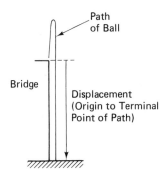

Figure 3-14

Substituting in Eq. (3-12) the height of the bridge,

$$h = v_0 t + \tfrac{1}{2}(-g)t^2$$
$$= (20 \text{ m/s})(8.0 \text{ s}) + \tfrac{1}{2}(-9.8 \text{ m/s}^2)(8.0 \text{ s})^2$$
$$= -150 \text{ m*}$$

The minus sign merely indicates that the displacement was downwards.

*Problems*_____

22. A stone is dropped from an altitude of 500 ft. (a) How long does it take to reach the ground? (b) What is its speed when it strikes the ground? Neglect air resistance.

23. A stone dropped from a tall building takes 3.6 s to reach the ground. How high is the building? Neglect air resistance.

_____

*To two significant figures.

24. While over enemy lines in his balloon, a soldier throws a bomb vertically downward with an initial speed of 12 m/s from a height of 1 000 m. (a) How long does it take the bomb to reach the ground? (b) What is the speed of the bomb as it strikes the ground? Neglect air resistance.

25. A stone is catapulted straight upwards from a high bridge with an initial speed of 25 m/s. On its return, it narrowly misses the bridge and plunges into the water below exactly 8.0 s after launch. How high is the bridge above the water?

26. A ball thrown straight downwards from a tower with an initial speed of 8.0 m/s hits the ground below exactly 2.5 s later. (a) How high is the tower? (b) What is the speed of the ball just before impact? Neglect air resistance.

27. A ball is thrown vertically upwards with an initial speed of 90 ft/s. (a) How high will it go? (b) How long will it take it to return to the ground? (c) What is its speed as it strikes the ground? Neglect air resistance.

28. A bomber in a vertical dive at 60 m/s releases a bomb that strikes the ground 18 s later. (a) What was the speed of the bomb just before impact? (b) From what height was the bomb released?

29. A bullet is fired vertically upward with an initial speed of 600 ft/s. (a) How high will it go? (b) How long will it take it to return to the earth?

30. With what initial velocity will a body moving vertically have to be thrown if, after 50 s, it is 125 ft below its starting position?

3-4 PROJECTILES

A *projectile* is an object that is launched into the atmosphere or into space without the ability to propel itself. The path of the projectile is called its *trajectory*, and the angle between the direction in which the projectile is launched and the horizontal is called the *angle of elevation*. Some examples of projectiles are rockets (after they have burned their fuel), balls, bullets and cannon shells in flight.

Consider two identical objects A and B at the same location at height h above the ground. If object A is dropped at the same time that object B is thrown horizontally, both objects reach the ground simultaneously. Therefore we may conclude that the action of gravity is independent of horizontal velocity.

Since displacement, velocity and acceleration are *vector* quantities, it is possible to separate the motion of a projectile into two simultaneous *independent* motions. The acceleration due to gravity acts vertically and has no horizontal component; therefore if air resistance is neglected, there is no change in the horizontal component of the velocity of a projectile. Consequently it is convenient to use the equations of uniformly accelerated motion *independently* in the horizontal and vertical directions in order to solve for the motion of a projectile.

Example 3-10

An aircraft, flying horizontally at 300 m/s, drops a bomb from an altitude of 490 m. (a)

How long will it take for the bomb to reach the ground? (b) What is the horizontal distance that the bomb will travel to the target? Neglect air resistance and the curvature of the earth.

Solution:

(a) Consider the *vertical* motion of the bomb. If down is taken as the positive direction, then $s = h = 490$ m, $v_0 = 0$ m/s, $a = g = 9.8$ m/s^2 and $t = ?$ Thus

$$s = v_0 t + \tfrac{1}{2}at^2 \quad \text{becomes} \quad h = \tfrac{1}{2}gt^2$$

or

$$t = \sqrt{\frac{2h}{g}} = \sqrt{\frac{2(490 \text{ m})}{9.8 \text{ m/s}^2}} = \sqrt{100 \text{ s}^2}$$
$$t = 10 \text{ s}.$$

(b) Consider the *horizontal* motion of the bomb. Since $t = 10$ s, as determined in (a), $v_0 = 300$ m/s and $a = 0$ m/s^2. Substituting in Eq. (3-12), we obtain:

$$s = v_0 t + \tfrac{1}{2}at^2 = v_0 t$$
$$= (300 \text{ m/s})(10 \text{ s}) = 3\,000 \text{ m}$$

In general, if a projectile is launched from the ground with an initial speed of v_0 at an angle of elevation α, it has initial speeds of $v_0 \sin \alpha$ and $v_0 \cos \alpha$ in the vertical and horizontal directions respectively (Fig. 3-15).

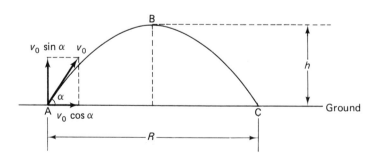

Figure 3-15
The trajectory of a projectile.

*Example 3-11*_____

A boy throws a ball with an initial speed $v_0 = 100$ m/s at an elevation $\alpha = 60°$. Find (a) the greatest height h attained, (b) the time T of the flight, and (c) the range R of the flight in the horizontal plane. Neglect air resistance.

Solution:

(a) At the instant the ball reaches its maximum height h (point B), it has no *vertical* velocity (Fig. 3-15); therefore substituting into Eq. (3-13) for the *vertical* motion, taking upwards as positive,

$$v_1^2 = v_0^2 + 2as$$

becomes:

$$0 = (v_0 \sin \alpha)^2 + 2(-g)h$$

Thus

$$h = \frac{v_0^2 \sin^2 \alpha}{2g} = \frac{(100 \text{ m/s})^2 (\sin 60°)^2}{2(9.8 \text{ m/s}^2)} \quad (3\text{-}16)$$
$$= 383 \text{ m}$$

(b) The ball returns to the same horizontal plane from which it is launched; therefore its total *vertical displacement* is zero. Using Eq. (3-12) for the vertical direction,

$$s = v_0 t + \tfrac{1}{2}at^2$$

becomes

$$0 = (v_0 \sin \alpha)T - \tfrac{1}{2}gT^2$$

Thus

$$T = \frac{2v_0 \sin \alpha}{g} = \frac{2(100 \text{ m/s})(\sin 60°)}{9.8 \text{ m/s}^2} \quad (3\text{-}17)$$
$$= 17.7 \text{ s}$$

(c) Using Eq. (3-12) for the *horizontal* component of the motion,

$$s = v_0 t + \tfrac{1}{2}at^2,$$

therefore

$$R = (v_0 \cos \alpha)T + 0$$

Thus

$$R = (v_0 \cos \alpha)T$$
$$= (100 \text{ m/s})(\cos 60°)(17.7 \text{ s}) \quad (3\text{-}18)$$
$$= 884 \text{ m}$$

Note that combining Eqs. (3-17) and (3-18), the horizontal range of the flight is

$$R = (v_0 \cos \alpha)T = (v_0 \cos \alpha)\left(\frac{2v_0 \sin \alpha}{g}\right)$$
$$= \frac{v_0^2 \sin 2\alpha}{g} \quad (3\text{-}19)$$

since $2 \sin \alpha \cos \alpha = \sin 2\alpha$. Since the maximum value of $\sin 2\alpha = 1$ occurs when $2\alpha = 90°$, neglecting air resistance, if a projectile returns to the same horizontal plane, its maximum horizontal range $R = v_0^2/g$ is attained when it is launched at an elevation $\alpha = 45°$.

*Example 3-12*_____

A projectile is launched with a speed of 78.4 m/s at an elevation of 30° from the top of a 98 m high building (Fig. 3-16). Find (a) the

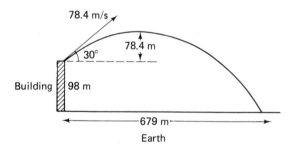

Figure 3-16
Trajectory of the projectile.

maximum height h attained, (b) the time T of the flight, and (c) the horizontal distance R from the bottom of the building to the landing place of the projectile. Neglect air resistance.

Solution:

(a) Consider the *vertical* component of the motion. Let s be the maximum height attained above the building. If we take upwards as the positive direction, $v_1 = 0$ (the projectile has no *vertical* velocity at the maximum altitude), $a = -g = -9.8$ m/s^2 and $v_0 = (78.4 \text{ m/s}) \sin 30° = 39.2$ m/s. Solving for s in Eq. (3-13), we obtain

$$s = \frac{v_1^2 - v_0^2}{2a} = \frac{0 - (39.2 \text{ m/s})^2}{2(-9.81 \text{ m/s}^2)} = 78.4 \text{ m}$$

Therefore $h = s + 98 \text{ m} = 176 \text{ m}$ above the ground.

(b) After the complete flight, the projectile has travelled a total vertical *displacement* equal to the height of the building, i.e. if up is taken as positive, in the vertical direction $s = -98 \text{ m}$, $a = -g = -9.8$ m/s² and $v_0 = 39.2$ m/s. Using Eq. (3-12),

$$s = v_0 t + \tfrac{1}{2} a t^2$$

becomes

$$-98 \text{ m} = (39.2 \text{ m/s})T + \tfrac{1}{2}(-9.8 \text{ m/s}^2)T^2$$

Dividing by 4.9 and rearranging (we shall temporarily drop the units for convenience), we obtain

$$T^2 - 8T - 20 = 0$$

and

$$(T - 10)(T + 2) = 0$$

Therefore $T = 10$ s is the time of the flight. The other solution $T = -2$ s corresponds to the time taken for the projectile to reach the height of the building if initially thrown from the ground.

(c) In the horizontal direction, $v_0 = (78.4$ m/s) cos 30° = 67.9 m/s, $a = 0$, $t = 10$ s and $s = R$. Therefore using Eq. (3-12), we obtain

$$R = v_0 t = (67.9 \text{ m/s})(10 \text{ s}) = 679 \text{ m}$$

Problems

Neglect air resistance and the curvature of the earth.

31. An aircraft in level flight at an altitude of 1 000 ft fires a rocket with an initial speed of 3 000 mi/h. (a) How long does it take for the rocket to reach the ground? (b) What horizontal distance from the firing point will the rocket travel before it strikes the ground?

32. An electron in a cathode ray tube travels 60 m horizontally with respect to the earth's surface at a speed of 3.0×10^6 m/s. If its initial vertical velocity was zero, what is its vertical deflection caused by the acceleration due to gravity?

33. A golfer drives a ball horizontally at 60.0 m/s from the top of a vertical cliff 44.1 m above the sea. (a) How long does it take the ball to hit the water? (b) How far from the foot of the cliff does the ball land? (c) What is the velocity of the ball just before it hits the water?

34. A missile is launched from the earth's surface with an initial speed of 98 m/s at an elevation of 30°. Find (a) the greatest height attained, (b) the time of flight and (c) the horizontal range of the flight.

35. What is the maximum range of an arrow if it can be launched with an initial speed of 100 m/s?

36. A cannon shell is fired with a speed of 1 500 ft/s at an elevation of 40°. Determine the horizontal distance it travels and the time of flight.

37. If the shell in Problem 36 strikes the side of a hill a horizontal distance of 1 000 ft from the firing point, how high above the level at which the shell was fired did it land?

38. An antiaircraft gun fires a shell from the ground with an initial speed of 600 m/s at an elevation of 60°, and an aircraft in level flight is flying at a constant speed of 200 m/s towards the gun at an altitude of 1 000 m. In order to hit the aircraft, at what horizontal range between the aircraft and the gun must the gun be fired?

REVIEW PROBLEMS

Neglect air resistance and the earth's curvature.

1. A car travelling at 25 m/s decelerates uniformly to 10 m/s in 6.0 s. It then travels

at 10 m/s for 12 s and finally accelerates uniformly to 30 m/s in another 5.0 s. (a) Draw the velocity-elapsed time graph. (b) How far did the car travel during the 23 s? (c) What were the initial and final accelerations?

2. A speeding car passes a motorcycle patrolman at a constant speed of 90 km/h. At the instant that the car passes, the patrolman accelerates from rest at 6.0 m/s² for 5.0 s and then travels at a constant speed until he overtakes the car. (a) Draw the velocity-elapsed time graphs for the car and the patrolman. (b) How long does it take the patrolman to overtake the car? (c) How far does the patrolman travel before he overtakes the car?

3. An electron in a television tube accelerates uniformly from rest to a final speed of 6.0×10^6 m/s in a distance of 5.0 cm. Find (a) the acceleration and (b) the time it takes the electron to travel the 5.0 cm.

4. A racing car starts from rest and accelerates uniformly to 216 km/h in 12 s. (a) What is its acceleration? (b) How far does it travel while accelerating?

5. A racehorse starts from rest with a uniform acceleration of 1.5 m/s² for 8.0 s. (a) What is its final speed? (b) How far does it travel in the 8.0 s?

6. A rocket accelerates uniformly from 2.0 km/s to 8.0 km/s in a distance of 100 km. (a) What is its acceleration? (b) How long does it take?

7. How long does it take a rock to fall 256 m?

8. A man fires at a duck that is 59.8 m directly above him. If the initial speed of the bullet is 294 m/s, (a) how long does it take the bullet to reach the duck? (b) With what speed does the bullet strike the duck?

9. A stone fired straight upwards from the top of a 50 m high bridge falls into the water below the bridge 12 s later. (a) What was the initial velocity of the stone? (b) How high above the bridge did it go? (c) What was the velocity of the stone just before it struck the water?

10. A bullet has a speed of 360 m/s as it leaves a rifle. If it is fired horizontally at a target 150 m away, how far below the center of the target does it strike?

11. An arrow is fired from the top of a building 78.4 m above the earth's surface with an initial speed of 58.8 m/s at an elevation of 30°. Find (a) the maximum height that it rises above the earth's surface, (b) the time of the flight if it lands on the earth's surface and (c) the horizontal distance that it travels.

12. A projectile is launched with a speed of 200 ft/s at an elevation of 60° from an altitude of 500 ft. Find (a) the maximum height attained by the projectile, (b) the time of the flight and (c) the horizontal distance travelled by the projectile before it strikes the ground.

13. A projectile is launched from an altitude of 80 ft above the earth's surface with a speed of 200 ft/s at an 18.7° angle of elevation. Find (a) the greatest height attained by the projectile above the earth's surface, (b) the time of the flight if it lands on the earth's surface and (c) the horizontal distance that it travels.

The study of *dynamics* is concerned with the relationships between changes in the motion of solid objects and the actions that caused these changes. These actions, called *forces*, tend to make stationary objects move or to change the motion of moving objects. Forces have many different origins; they may be mechanical, electrical, magnetic, gravitational, chemical or nuclear, but their results are identical.

The basic laws governing dynamics were first formulated by Sir Isaac Newton (1642–1727) in his *Principia Mathematica* (1686). Newton's conception of the laws of motion was one of the most profound developments in physics, and these laws constitute the basis of what is now called classical (or Newtonian) mechanics. It was not until the twentieth century and the development of quantum mechanics that Newton's laws proved to be inadequate because they could not fully describe the motion of extremely small atomic-sized particles or particles that travel at speeds close to the speed of light. However, because of their mathematical simplicity and accuracy, the laws of classical mechanics are still used to predict and describe the motions of many objects.

4

Dynamics

4-1 NEWTON'S FIRST LAW OF MOTION—THE LAW OF INERTIA

Newton's first law of motion merely describes the nature of a force. The law states that:

An object at rest will remain at rest, and an object in motion will continue that motion with a constant velocity unless it is acted upon by some external unbalanced force.

In other words, a force must be applied to a stationary object in order to make it change its speed, direction of motion, or both. The direction and magnitude of these changes in motion depends on the direction and magnitude of the applied force vectors. However, if an external force results in a "push" on an

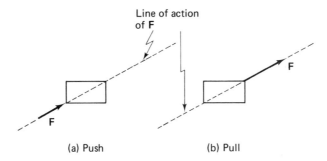

Figure 4-1
Lines of action.

object, the same effect could be obtained by "pulling" the object with an equal force. Consequently the direction of the force extended both ahead and behind the actual force vector has a special significance; it is called the *line of action* of that force (Fig. 4-1).

Since some kind of force is required to change the motion of a solid object, this law indicates that matter itself resists changes in motion. This inherent property of matter is called *inertia*.

4-2 NEWTON'S SECOND LAW OF MOTION—THE LAW OF ACCELERATION

We have previously stated that the *mass* of an object is an indication of the amount of matter that it contains compared with some standard, and that a *force* tends to produce a change in motion. Newton's second law gives more precise definitions and describes the changes in motion of an object when it is acted upon by some net external force. The second law states that:

When a solid object is acted upon by some net external unbalanced force **F**, *it accelerates in the direction of that force, the magnitude a of the acceleration* **a** *being directly proportional*

to the magnitude of the force and inversely proportional to the mass m *of the object.**

That is

$$\mathbf{a} \propto \text{net } \mathbf{F} \quad \text{and} \quad a \propto 1/m$$

Therefore

$$\mathbf{a} \propto \text{net } \frac{\mathbf{F}}{m}$$

By the appropriate definitions of the units, the constant of proportionality may be chosen as unity (one), and Newton's second law becomes:

$$\text{net } \mathbf{F} = m_i\mathbf{a} \qquad (4\text{-}1)$$

Therefore, in SI, force is expressed in terms of a derived unit called a *newton* (N), which is defined as the net force that accelerates a one kilogram mass at one meter per second squared.

$$1 \text{ N} = (1 \text{ kg})(1 \text{ m/s}^2) = 1 \text{ kg·m/s}^2$$

However, in the British Engineering System, force units are chosen as base units, and the inertial mass m_i, in *slugs*, of any object is defined as the constant of proportionality between the net force in pounds (lb) and the

---

*Notice that this assumes that the mass is constant. Newton actually stated the law as follows: The net force varies directly as the time rate of change of momentum (the product of the mass and the velocity).

resulting acceleration in feet per second squared.

$$1 \text{ slug} = \frac{1 \text{ lb}}{1 \text{ ft/s}^2} = \frac{1 \text{ lb} \cdot \text{s}^2}{\text{ft}}$$

The term inertial mass is used because more massive objects require larger forces to give them a particular acceleration. Thus mass itself is a measure of an object's *inertia*.

Note that acceleration is a consequence of a *net force*, which is the vector sum of all forces acting on the object. If several forces act simultaneously on the same object, each force independently accelerates that object according to Newton's second law. The *vector sum* of these accelerations is produced by the *vector sum* of the independent forces. If the net external force on an object is equal to zero, there will be no change in its velocity, and the object is said to be in a *state of equilibrium*.

*Example 4-1*_____

Determine the magnitude of the net force that causes an 1 800 kg car to accelerate at 2.5 m/s².

Solution:

net $F = ma = (1 \ 800 \text{ kg})(2.5 \text{ m/s}^2) = 4 \ 500$ N

*Example 4-2*_____

If a mass $m_1 = 18$ kg has an acceleration $a_1 = 8.0$ m/s² when acted upon by some net force, find the acceleration a_2 that the same net force would give a mass $m_2 = 24$ kg.

Solution:

$$\text{net } F = m_1 a_1 = m_2 a_2$$

Therefore

$$a_2 = a_1 \frac{m_1}{m_2} = \frac{(8.0 \text{ m/s}^2)(18 \text{ kg})}{24 \text{ kg}} = 6.0 \text{ m/s}^2$$

_____

Mass and Weight_____

When any object is allowed to fall freely to the earth from some height, it accelerates as it falls. This *acceleration due to gravity* varies with the distance from the earth, but it has the same magnitude for all objects if air resistance is neglected. At the surface of the earth, the acceleration due to gravity has the value 9.81 m/s² or 32.2 ft/s².

Since all freely falling objects experience this acceleration, Newton's second law implies the presence of some net external force. This force is due to the attraction between the object and the earth. In fact, in Chapter 10 we shall see that every object in the universe attracts every other object with a force which depends on the amount of matter in each object and the distance between their centers. This attractive force between objects is called the *force of gravity*. The *weight* of an object at the earth's surface is mainly due to the force of gravity between the earth and that object.* By analogy with Newton's second law, the proportionality constant between the weight **w** and the acceleration due to gravity **g** of the object is called its *gravitational mass* m_g. Thus

$$\mathbf{w} = m_g \mathbf{g} \tag{4-2}$$

If the units of weight are chosen to be the same as those for force, the gravitational mass of any object is found to be equal to its inertial mass ($m_i = m_g$). The term *mass m* is used to describe both. Therefore the *mass* of an object indicates the amount of matter that it contains and also describes its inertia. Mass is a fundamental quantity in SI; it is independent of the position of the object, but its measured value is affected by relativity.*

The *force of gravity* exerted by the earth

_____

*The measured value for weight is *slightly* different because of the effects of the earth's rotation and the buoyancy of air.

*According to Einstein the mass m of an object that moves with a speed v is given by $m = m_0/\sqrt{1 - v^2/c^2}$ where m_0 is the mass of the object when it is stationary and c is the speed of light.

on an object (the weight of the object) varies with its distance from the center of the earth. Thus the weight of the object is *not* the same at the earth's surface as it is on the moon or in free space. Astronauts can float in free space where there is approximately no net force of gravity and hence no weight.

*Example 4-3*_____

Determine the weight of a 5.00 kg mass (a) on the earth's surface and (b) on the moon where the acceleration due to gravity is 1.63 m/s².

Solution:

(a) $w = mg = (5.00 \text{ kg})(9.81 \text{ m/s}^2) = 49.1 \text{ N}$
(b) $w = mg = (5.00 \text{ kg})(1.63 \text{ m/s}^2) = 8.15 \text{ N}$

*Problems*_____

1. Can the mass of an object ever be zero? Can its weight ever be zero? Why?

2. What net force is required to give a 12.0 kg mass an acceleration of 5.0 m/s²?

3. What is the acceleration of a 16.0 lb object when it is acted upon by a net force of 64.0 lb?

4. Determine the acceleration of a 25 kg object if it experiences a net force of 70 N.

5. What is the mass of an object that is accelerated at 2.4 m/s² by a 72 N net force?

6. If a 45 N net force accelerates an object at 6.0 m/s², what net force would accelerate that object at 8.0 m/s²?

7. An object is accelerated at 3.0 m/s² by some net force. What is the acceleration if the net force is tripled?

8. A net force **F** acts on a 5 kg mass and accelerates it at 5 m/s². What is the acceleration of a 25 kg mass when it is acted upon by the same force?

9. A force of 12.0 lb gives a body an acceleration of 20 ft/s². (a) What force will accelerate the same body at 80 ft/s²? (b)

If the same body is acted upon by a 30 lb force, what is its acceleration?

10. If a force produces an acceleration of 35 m/s², what acceleration would it produce if the mass were half?

11. An electron, which has a mass of 9.1×10^{-31} kg, experiences an acceleration of 3×10^8 m/s² when it passes through an electric field. What net force is acting on it?

12. Determine the weight of a 3.0 kg body on earth.

13. Determine the mass of an object if it weighs 12 lb on the earth's surface.

14. Determine the mass of a body whose weight is 19.6 N on earth.

4-3 NEWTON'S THIRD LAW— ACTION AND REACTION

Newton's third law describes the effect of interactions between objects. The law states that:

Whenever one object exerts a force on a second object, the second object exerts a reactive force of equal magnitude and opposite direction on the first object.

For example, if a book is at rest on a flat horizontal table, the book exerts a force equal to its weight on the table, and the table in turn exerts an equal but opposite reactive force on the book. If the table did not exert this reactive force, the weight of the book would not be balanced, and, according to Newton's second law, the book would accelerate and fall through the table. Reaction forces must always be considered in calculations involving equilibrium (Chapter 6).

In mechanics, it is usually convenient to draw a diagram of the system under study by considering all items only in terms of their effect on the system. This is called a *free-body*

Reaction force of the
surface on the load

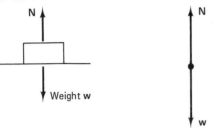

(a) Load **w** on a table   (b) Free body diagram

Figure 4-2

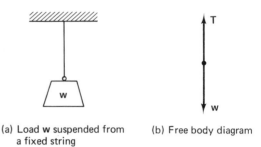

(a) Load **w** suspended from   (b) Free body diagram
a fixed string

Figure 4-3

diagram. A free-body diagram is a sketch of the forces present and their points of application. It is important to note that Newton's third law must be used in order to determine any reaction forces in the system. Consider a load **w** at rest on a horizontal surface (Fig. 4-2a). The free-body diagram must include the reaction force of the table on the load (Fig. 4-2b). This reaction force is called a *normal force* **N** and always acts perpendicularly to the surfaces in contact.

When a load **w** is suspended by a string, the string is in a state of *tension* because the load tends to elongate the string. If the string is in equilibrium and has negligible weight, the tensions at each end of the string must have equal magnitudes but opposite directions, otherwise the string would move. Since a tension force **T** will oppose the applied load, the free-body diagram must contain the tension force as well as the load (Fig. 4-3).

Newton's third law also describes the interaction of moving objects, such as collisions, the recoil of a gun as it fires, and even the propulsion of a rocket (Chapter 9).

4-4 THE SOLUTION OF PROBLEMS

Newton's laws of motion may be used to solve many types of problems. The procedure is as follows:

1. Draw a free-body diagram, if necessary. All reaction forces must be included and masses must be converted into weights (forces) according to Eq. 4-2.

2. List the quantities given and those to be found in terms of their symbols. It may be necessary to choose a positive direction.

3. Use Newton's second law and kinematic equations to find the unknown quantities.

*Example 4-4*_____

Determine the force that a 75.0 kg man exerts on the floor of an elevator (a) when it ascends with a constant speed of 1.20 m/s, (b) when it descends with a constant acceleration of 1.40 m/s² and (c) when it ascends with a constant acceleration of 1.40 m/s².

Solution:

Draw the free-body diagram (Fig. 4-4). The only forces present are the weight **w** of the man and the reaction force **R** of the elevator floor. Choose a positive direction, upwards for example. The weight of the man $w = mg$ = (75.0 kg)(9.81 m/s²)   downwards = 736 N downwards.

(a) Since the velocity 1.20 m/s upwards is constant, the acceleration **a** = 0 m/s². Thus from Newton's second law

$$\text{net } F_{up} = R - w = 0$$

or

$$R = w = 736 \text{ N}$$

(b) **a** = 1.40 m/s² downwards = −1.40 m/s² upwards. Thus from Newton's second law

Figure 4-4

$$\text{net } F_{\text{up}} = R - w = ma$$

or

$$R - 736 \text{ N} = (75.0 \text{ kg})(-1.40 \text{ m/s}^2)$$
$$R = 631 \text{ N}$$

(c) $a = 1.40 \text{ m/s}^2$ upwards. Therefore

$$\text{net } F_{\text{up}} = R - w = ma$$

or

$$R - 736 \text{ N} = (75.0 \text{ kg})(1.40 \text{ m/s}^2)$$
$$R = 841 \text{ N}.$$

Example 4-5

Masses $m_1 = 1.00 \text{ kg}$ and $m_2 = 3.00 \text{ kg}$ are joined by a light string, which passes over a frictionless pulley (Fig. 4-5a). If the masses are allowed to fall freely, what is the acceleration of the system and the tension in the string?

Solution:

The more massive 3.00 kg load will fall, accelerate downwards and pull (via the string) the lighter 1.00 kg load upwards. Since these loads constitute some of the forces involved, their weights must be determined. Also, since the string does not break or buckle, the tension **T** in the string must be equal in magnitude on both sides of the pulley, and the acceleration **a** of each load must be of the same magnitude a (but opposite direction).

Although Newton's second law may be applied to the system as a whole, it also describes the motion of each individual load. Consequently each load may be considered separately. The weight of the 1.00 kg load $\mathbf{w}_1 = m_1\mathbf{g} = (1.00 \text{ kg})(9.81 \text{ m/s}^2) = 9.81$ N. Draw the free-body diagram for the 1.00kg load (Fig. 4-5b).

$$\text{net } F_{\text{up}} = T - 9.81 \text{ N} = m_1 a = (1.00 \text{ kg})a$$

or

$$T = 9.81 \text{ N} + (1.00 \text{ kg})a \qquad (1)$$

The weight of the 3.00 kg load $w_2 = m_2 g = (3.00 \text{ kg})(9.81 \text{ m/s}^2) = 29.4$ N. Draw the free-body diagram for this load (Fig. 4-5c).

$$\text{net } F_{\text{down}} = 29.4 \text{ N} - T = m_2 a = (3.00 \text{ kg})a$$

or

$$29.4 \text{ N} - T = (3.00 \text{ kg})a \qquad (2)$$

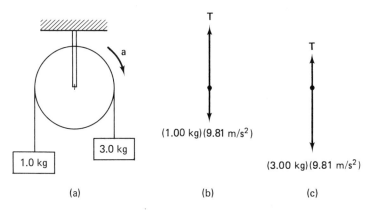

Figure 4-5
Atwood machine.

Solving for T in (1) and substituting this into (2), we obtain:

$$29.4 \text{ N} - [(1.00 \text{ kg})a + 9.81 \text{ N}] = (3.00 \text{ kg})a$$

$$a = \frac{19.6 \text{ N}}{4.00 \text{ kg}} = 4.90 \text{ m/s}^2$$

in the direction shown. Substituting the calculated value of a into (1), we obtain

$$T = (1.00 \text{ kg})(4.90 \text{ m/s}^2) + 9.81 \text{ N} = 14.7 \text{ N}$$

*Example 4-6*_____

A constant horizontal force of 20 N is exerted on a 10 kg block that is at rest on a horizontal, frictionless surface. (a) What is the acceleration of the block? (b) How far will the block move in 5.0 s? (c) What is the speed of the block after the 5.0 s?

Solution:

(a) Net $\mathbf{F} = 20$ N horizontally, $v_0 = 0$, $m = 10$ kg and $a = ?$
From Newton's second law:

$$\mathbf{a} = \frac{\text{net } \mathbf{F}}{m} = \frac{20 \text{ N}}{10 \text{ kg}} \text{ horizontally}$$

$$= 2.0 \frac{\text{m}}{\text{s}^2} \text{ horizontally}$$

(b) $t = 5.0$ s, $v_0 = 0$, $\mathbf{a} = 2.0$ m/s² horizontally and $s = ?$ Thus

$$s = v_0 t + \tfrac{1}{2}at^2 = 0 + \tfrac{1}{2}(2.0 \text{ m/s}^2)(5.0 \text{ s})^2$$
$$= 25 \text{ m}$$

(c) $t = 5.0$ s, $v_0 = 0$, $a = 2.0$ m/s² and $v_1 = ?$
Thus

$$v_1 = v_0 + at = 0 + (2.0 \text{ m/s}^2)(5.0 \text{ s})$$
$$= 10 \text{ m/s}$$

*Example 4-7*_____

What magnitude of force is required to accelerate a 1 500 kg car uniformly from rest to 90.0 km/h in 12.5 s on a horizontal road if 60% of this force must be used to overcome friction and other retarding effects?

Solution:

$$v_0 = 0, v_1 = 90.0 \frac{\text{km}}{\text{h}} = \frac{90.0 \times 10^3 \text{ m}}{3\ 600 \text{ s}}$$
$$= 25.0 \text{ m/s}, t = 12.5 \text{ s and}$$

the mass of the car $m = 1\ 500$ kg

$$a = \frac{v_1 - v_0}{t} = \frac{(25.0 \text{ m/s}) - (0)}{12.5 \text{ s}} = 2.00 \text{ m/s}^2$$

Since 60% of the required force $\mathbf{F}$ is used to overcome the retarding forces (Fig. 4-6), from

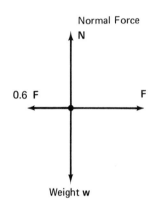

Normal Force

0.6 F   F

Weight w

Figure 4-6
Free body diagram.

Newton's second law in the direction of motion

$$\text{net } F = F - 0.6\ F = ma$$

or

$$0.4F = (1\ 500 \text{ kg})(2.00 \text{ m/s}^2)$$
$$F = \frac{(1\ 500 \text{ kg})(2.00 \text{ m/s}^2)}{0.4} = 7\ 500 \text{ N}$$

_____

*Problems*_____

15. Determine the tension in a rope when it is used to pull 30 lb of bricks to the top of a house with (a) a constant speed of 3.0 ft/s, (b) a constant acceleration of 3.0 ft/s² upwards.

16. What is the tension in the cable of a 1000 kg elevator: (a) If it ascends with an acceleration of 5.0 m/s²? (b) If it descends with an acceleration of 5.0 m/s²? (c) If it descends with an acceleration of 9.8 m/s²?

17. A loaded elevator has a weight of 2 000 lb. If the supporting cable pulls upwards with a 2 500 lb force, what is the upward acceleration?

18. A loaded 1 200 kg elevator is pulled upwards by a 12 000 N force on the cable. What is the upward acceleration?

19. Determine the total thrust developed by the five engines of a 2.82×10^6 kg Saturn rocket at the instant of blast off from the earth's surface if its acceleration is 2.60 m/s² upwards.

20. A 1.80×10^4 kg aircraft starts from rest and accelerates uniformly at 3.00 m/s² to its take-off speed of 198 km/h. Calculate (a) the length of runway used, (b) the time required to attain take-off speed, and (c) the net force on the aircraft.

21. A 100 kg pile driver strikes a pile while travelling at 30 m/s. If it drives the pile 3.0 cm into the ground, what is the average retarding force due to the pile?

22. A net force of 24 N acts on a 3.0 kg object for 8.0 s. If the object was initially at rest, (a) what is its acceleration? (b) What is its final speed? (c) How far does it travel in the 8.0 s?

23. A light string passes over a light frictionless pulley and has a 500 g load attached to one end and a 200 g load attached to the other. If the system is allowed to fall freely, what are (a) the acceleration and (b) the tension in the string?

24. Determine the minimun net average force required to increase the speed of a 1500 kg car from 30 km/h to 60 km/h in 6.0 s.

25. Calculate the displacement and final velocity of a 18.0 kg object after 12.0 s if its original speed was 4.00 m/s and it was acted upon by a 27.0 N net force (a) in the same direction as the motion and (b) in the opposite direction to the motion.

26. A constant 52 N net force acts on a 6.5 kg object for 12 s. If the object was initially at rest, (a) what is its acceleration? (b) What is its final speed? (c) How far does it travel in the 12 s?

27. A net force of 600 N acts on a 2.00 t object, accelerating it from rest to a final speed of 9.00 m/s. (a) How far does the object travel while accelerating? (b) For how long does it accelerate?

28. A 3 200 lb car is travelling at a speed of 60 mi/h when the driver applies the brakes. If the car comes to rest after it has travelled another 300 ft, determine the net average retarding force.

29. What is the *total* force exerted by an 1 800 kg car if it is accelerated uniformly from rest to a final speed of 30 m/s in 25 s and if 60% of this total force must be used merely to overcome friction?

30. While driving a car at 30 mi/h, a driver is suddenly confronted with a child playing in the path of the car. If the driver requires 0.50 s to react and then applies a braking force equal to one-half of the weight of the car, find (a) the acceleration of the car, (b) the distance that the car travels after the driver sights the child, (c) the time that it takes for the car to come to rest.

31. A catapult on an aircraft carrier accelerates a 12 000 kg aircraft uniformly from rest to 173 km/h in 2.40 s. Find (a) the acceleration of the aircraft and (b) the net thrust on the aircraft.

32. A cable can withstand a maximum tension of 750 N. If the cable is used to hoist a 50 kg load vertically from rest through 15 m, (a) what is the maximum acceleration of the load? (b) Calculate the minimum time required.

REVIEW PROBLEMS

1. What is the acceleration of a 15 kg mass if it is acted upon by a 12 N net force?

2. If a certain net force accelerates an object at 36 m/s², what acceleration would the same force produce if the object had four times the mass?

3. (a) Determine the mass of a man who weighs 180 lb on the earth. (b) What is the weight of the man on the moon, where the acceleration due to gravity is 5.37 ft/s²?

4. What would the weight of a 75.0 kg man be on the following planets with the given accelerations due to gravity? (a) Earth? (b) Venus, $g = 8.63$ m/s²? (c) Mars, $g = 3.83$ m/s²? (d) Jupiter, $g = 26$ m/s²? (e) Mercury, $g = 3.73$ m/s²? (f) Saturn, $g = 11.5$ m/s²?

5. If an electron is accelerated at 3.0×10^{10} m/s² by some force, what is its acceleration if the force is tripled?

6. If the rockets of a spaceship deliver a constant net thrust and it accelerates initially at 12.5 m/s², what is its acceleration after it has reduced its mass to one-third of its initial value by burning fuel?

7. Determine the minimum tensile force that the supporting cable of an elevator must withstand if the total mass of the fully loaded elevator is 2 700 kg and it is capable of a 1.20 m/s² acceleration.

8. A 40 000 lb aircraft accelerates uniformly from rest at 8.0 ft/s² to its take-off speed of 120 mi/h. Calculate (a) the length of runway used, (b) the time required, and (c) the net force on the aircraft.

9. Determine the average tension in the cable of a crane (a) when it lowers a 2 500 kg load to the earth at a constant speed of 1.20 m/s and (b) if the operator applies the brake when the load is 5.00 m above the earth so that the velocity of the load is zero just as it touches the earth.

10. A light string, which passes over a frictionless pulley, has a 7.5 lb load attached to one end and a 4.5 lb load on the other end. If the system is allowed to fall freely, what is (a) the acceleration and (b) the tension in the string?

11. A 7 200 N net force acts on a 4 800 kg truck for 15 s. If the truck was initially at rest, (a) what is its acceleration? (b) What is its final speed? (c) How far does it travel in the 15 s?

In mechanics, some objects are considered as rigid bodies, but, in reality, rigid bodies do not exist because all known materials are deformed to some extent by the application of force. To avoid structural failures, engineers must know the limitations of the materials that they use, and they must be able to calculate the magnitudes of any deformations.

A deformation depends on the type of material used and on the nature (magnitude and orientation) of the applied loads. *Isotropic materials* have the same properties in all directions from any given point in the material. Materials that have different properties in different directions are called *anisotropic*. For example, wood is anisotropic because it is easily split along its grain but is difficult to split in a direction perpendicular to its grain.

Materials are usually utilized according to their properties and cost. For example, steel girders are relatively strong; therefore they are often used as the main supports in large structures. Wooden beams are used in smaller structures because they have sufficient strength and are lighter and less expensive than steel girders.

In some areas of technology, materials must be able to withstand excessive temperatures or temperature variations, resist corrosion, conduct heat or resist heat flow, conduct electricity or resist the flow of electric charge. The choice of a suitable material is often quite critical.

5

Mechanical Properties of Materials

5-1 DENSITY

In order to indicate the characteristic property of "heaviness," we define the *density ρ* (or *mass density*) of a substance as its mass *m* per unit volume *V*.

$$\rho = \frac{\text{mass}}{\text{Volume}} = \frac{m}{V} \qquad (5\text{-}1)$$

Density is specified in units of kg/m^3 in SI and of $slug/ft^3$ in the British Engineering System. Under fixed conditions, the density of a material is constant and independent of the dimensions. Some typical values of density are listed in Table 5-1.

In the British Engineering System, since weight is used more frequently than mass, it is convenient to define a quantity called *weight density D* as the weight *w* per unit volume *V* of the substance.

$$D = \frac{\text{weight}}{\text{volume}} = \frac{w}{V} \qquad (5\text{-}2)$$

The corresponding units are lb/ft³ or N/m³.

Also, since $\mathbf{w} = m\mathbf{g}$,

$$D = \frac{\mathbf{w}}{V} = \frac{m\mathbf{g}}{V} = \rho\mathbf{g} \qquad (5\text{-}3)$$

where $\mathbf{g}$ is the acceleration due to gravity.

Example 5-1

Determine the volume of an irregularly shaped concrete block if its mass $m = 6\,900$ kg.

Table 5-1

Substance	Relative Density (specific gravity)ρ_r	Density ρ		Weight Density D
		kg/m³	slug/ft³	lb/ft³
Solids:	*(relative to water)*			
Aluminum	2.7	2 700	5.25	169
Brass	8.6	8 600	16.8	537
Concrete	2.3	2 300	4.5	144
Copper	8.89	8 890	17.3	555
Gold	19.3	19 300	37.7	1 204
Ice	0.92	920	1.8	57.5
Iron	7.85	7 850	15.3	490
Lead	11.3	11 300	22.0	705
Steel	7.8	7 800	15.2	486
Liquids at 20°C:				
Alcohol, ethyl	0.79	790	1.53	49.3
Benzene	0.88	880	1.71	54.7
Gasoline	0.68	680	1.32	42.4
Mercury	13.6	13 600	26.5	850
Water	1.0	1 000	1.95	62.4
Gases at 0°C and 760 mm Hg:	*(relative to air)*			
Air	1.0	1.29	2.52×10^{-3}	8.07×10^{-2}
Carbon dioxide	1.53	1.98	3.85×10^{-3}	1.24×10^{-1}
Helium	1.40	0.18	3.5×10^{-4}	1.12×10^{-2}
Hydrogen	0.07	0.09	1.75×10^{-4}	5.6×10^{-3}
Nitrogen	0.97	1.25	2.44×10^{-3}	7.8×10^{-2}
Oxygen	1.11	1.43	2.78×10^{-3}	8.9×10^{-2}

Solution:

The density ρ of concrete (Table 5-1) is 2 300 kg/m³. Thus, from Eq. 5-1, the volume:

$$V = \frac{m}{\rho} = \frac{6\ 900\ \text{kg}}{2\ 300\ \text{kg/m}^3} = 3.0\ \text{m}^3$$

Since solids and liquids are only slightly compressible, their densities are almost constant. Gases are easily compressed, however, so their densities will vary considerably with the conditions under which they are measured.

The *relative density* of a substance is the ratio of its density to that of some reference material at a certain temperature. Pure water at 4°C (or 39.2°F)* is usually taken as the reference for solids and liquids; air is usually used for gases unless otherwise stated. Thus the relative density of some substance, which has a density ρ and a weight density D, is given by

$$\rho_r = \frac{\rho}{\rho_{\text{reference}}} = \frac{D}{D_{\text{reference}}} \qquad (5\text{-}4)$$

where $\rho_{\text{reference}}$ and $D_{\text{reference}}$ are the density and weight density of the reference. Also, since $\rho = \frac{m}{V}$, the relative density

$$\rho_r = \frac{\text{mass of material}}{\text{mass of equal volume of reference material}}$$
$$(5\text{-}5)$$

$$\rho_r = \frac{\text{weight of material}}{\text{weight of equal volume of reference material}}$$

Note that the numerical value of density will vary with the system of units, but relative density is a dimensionless quantity that has the same value in all systems of units.

Example 5-2

An irregulary shaped block of metal has a mass of 12.0 kg. When it is lowered into a

---

*The relative density is often referred to as *specific gravity*, but in SI this term is no longer used.

---

container, which is completely full of water, 1.40 L of water overflows. Determine the density of the metal.

Solution:

The volume of the block must be equal to the volume of water displaced:

$$V = 1.40L = 1.40 \times 10^{-3}\ \text{m}^3$$

Thus

$$\rho = \frac{m}{V} = \frac{12.0\ \text{kg}}{1.40 \times 10^{-3}\ \text{m}^3} = 8\ 570\ \frac{\text{kg}}{\text{m}^3}$$

Problems

1. If the mass of the earth is 5.98×10^{24} kg and its radius is 6.38×10^6 m, what is its average density? *Hint*: Volume of a sphere $= (4/3)\pi r^3$.

2. If the sun has an average density of 1.4×10^3 kg/m³ and its radius is 6.97×10^8 m, what is its mass?

3. What is the volume of a pure copper block if its weight is 3.5 lb?

4. A man attempts to sell two 5 oz nuggets, which he claims are pure gold. To determine if the nuggets are gold, they are lowered into a container, which is completely full of water. The first nugget displaces 0.50 in³ of water, and the second displaces 0.259 oz of water. Check the values of the weight densities in each case. What is the relative density of each nugget?

5. An irregularly shaped metal block has a mass of 11.0 kg. When it is totally immersed in a full container of water, 1.25 kg of water overflows. Determine the density of the block.

6. What is the volume of a 300 kg concrete block?

7. What is the weight of an 8.0 m long steel pipe if its inner and outer diameters are 1.5 cm and 2.5 cm respectively?

5-2 ELASTICITY

A system of forces acting on a body will cause it to change its shape in the directions of the applied forces. The body is said to be *strained*, and it resists the forces that tend to deform it. The magnitude of the deformation of the body depends on the material of the body itself, and the magnitude, direction and area of application of the applied forces. When a force **F** is applied perpendicularly to the end of a rod, the smaller the cross-sectional area of the rod, the larger its deformation.

The *stress* (or pressure) is defined as the force F per unit of cross-sectional area A of the body.

$$\text{stress } \sigma = F/A \qquad (5\text{-}6)$$

The units of stress are N/m² or pascal (Pa) in SI, and lb/ft² in the British Engineering System although lb/in² (psi) are commonly used.

Strain ϵ is a dimensionless quantity that is defined as the fractional deformation of the body, and it is a direct consequence of the stress.

There are three basic kinds of stress and strain:

1. When equal and opposite forces are applied along the same line of action so that they tend to elongate the body, the body is said to be in a state of *tensile stress*. Tensile strain is therefore the fractional elongation of the body (Fig. 5-1a)

$$\epsilon = \frac{\Delta L}{L_0} = \frac{(L_1 - L_0)}{L_0} \qquad (5\text{-}7)$$

where L_1 is the final length, L_0 is the initial length and $\Delta L = (L_1 - L_0)$ is the elongation of the body.

2. A body is in a state of *compressive stress* when equal and opposite forces are applied along the same line of action and tend to shorten the body. The resulting strain is the fractional compression of the body (Fig. 5-1b)

$$\epsilon = \Delta L/L_0 \qquad (5\text{-}8)$$

where ΔL is the compression and L_0 is the initial length.

3. A *shearing stress* is applied to a body when two parallel, equal and opposite forces are applied along different lines of action. These forces tend to change the shape of the body, but not its volume (Fig. 5-1c). This kind of stress may be illustrated by pushing in opposite directions along the top and bottom covers of a book. The shape of the book changes, but the thickness and the volume of the book do not change since the number of pages and dimensions of the covers remain the same.

Consider a rectangular block of material of thickness y and surface area A. Shearing forces F will act *tangentially* to the surfaces and the larger the surface area A, the smaller the distortion of the body. Therefore shearing stress is also defined as:

$$\tau = F/A$$

However the plane of the surface area A is parallel to the applied force **F** in this case. If the shearing stresses displace these surfaces by a total relative distance x, the shearing strain

$$\epsilon = x/y = \tan \phi \qquad (5\text{-}9)$$

where ϕ, called the *angle of shear*, represents the total angular displacement of the surfaces. If ϕ is small, $\tan \phi \approx \phi$ (in radians) and the shearing strain $\epsilon \approx \phi$.

Example 5-3

What is the tensile stress in a nylon rope that has a diameter $d = 1.2$ cm when it is used to lift a 75 kg load?

Solution:

In this case the tensile force is the weight of the load, **w** = mg. Thus

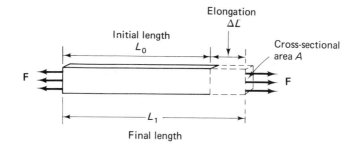

(a) Tensile stress and strain

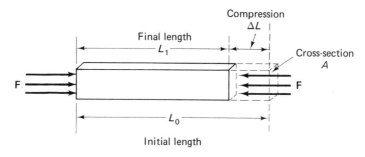

(b) Compressive stress and strain

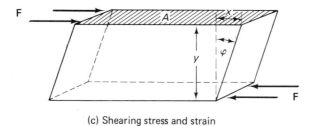

(c) Shearing stress and strain

Figure 5-1
Stress and strain.

$$\sigma = \frac{F}{A} = \frac{mg}{\pi d^2/4} = \frac{(75 \text{ kg})(9.8 \text{ m/s}^2)}{\pi(1.2 \times 10^{-2} \text{ m})^2/4}$$
$$= 6.5 \times 10^6 \text{ Pa.}$$

If one end of a uniform wire (or spring) is suspended from a rigid support and weights are added to the other end, the wire elongates. A typical graph of load-versus-elongation is illustrated in Fig. 5-2. Provided a certain limit (called the *elastic limit*) is not exceeded, many materials will return almost exactly to their original length or configuration after the loads are removed. These materials are said to have an elastic behavior.

In 1679, Robert Hooke determined experimentally that the elongation ΔL of an object is directly proportional to the applied

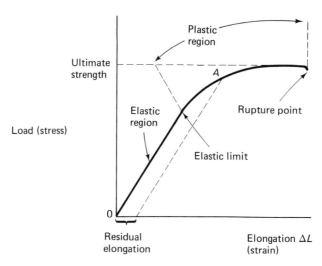

Figure 5-2
Load-versus-elongation graph.

load **F**, provided that the elastic limit of that object is not exceeded.

$$\mathbf{F} = k\,\Delta\mathbf{L} \qquad (5\text{-}10)$$

This is known as *Hooke's Law*. The proportionality constant k, called the *force constant* of the object, depends on the dimensions and the material structure of that object. It is expressed in units of N/m in SI and lb/ft in the British Engineering System.

The region beyond the elastic limit is known as the *plastic region*. For example, if the elastic limit of the material is exceeded and loading reaches point A unloading the material will result in the dotted-line return path (approximately parallel to the elastic region), leaving a permanent residual deformation in the material (Fig. 5-2). Any subsequent loading of this material will follow the dotted line path to point A, where it again resumes the original path. Many brittle materials break close to their elastic limit.

Most metals, however, may be deformed far beyond their elastic limits. This property, known as *ductility*, permits the material to be drawn into a wire. Heating a metal will often improve its ductility. Brittle materials, such as cast iron, rupture near their elastic limit, and they do not behave plastically. A few materials, such as lead, never behave elastically; when they are stressed, they always behave plastically.

In the plastic region, as the load is increased, increasingly larger deformations result, until the maximum stress that the material can withstand (its *ultimate strength*) is reached (Table 5-2). To ensure a safe design of a structure, engineers must allow for any unexpected overload and defects in the structure. This is accomplished by the use of a *factor of safety*, which is defined as the ratio of the ultimate strength to the maximum allowable stress for the material.

$$\text{Factor of safety} = \frac{\text{ultimate strength}}{\text{allowable stress}} \qquad (5\text{-}11)$$

Example 5-4

What minimum force is required to punch a hole of diameter $d = 1.4\,\text{cm}$ in a 5.0 mm thick steel plate?

Table 5-2
Typical Elastic Properties and Ultimate Strengths

Material	Young's Modulus 10^{10} Pa or 10^6 lb/in²		Bulk Modulus 10^{10} Pa or 10^6 lb/in²		Shear Modulus 10^{10} Pa or 10^6 lb/in²	
Aluminum	7.0	10	7.6	11	2.5	3.6
Brass	9.0	13	5.9	8.5	3.5	5.1
Copper	12.5	18	14	20	4.5	6.5
Glass	5.8	8.4	3.7	5.3	2.8	4.0
Iron, wrought	18	26	9.7	14	6.9	10
Lead, rolled	1.6	2.3	3.5	5.0	0.59	0.81
Nickel (cast)	21	30			7.6	11
Steel	20	29	17	25	8.4	12
Timber (parallel to grain)	8.3	12				

Material	Elastic Limit 10^5 Pa	10^4 lb/in²	Ultimate Tensile Strength 10^8 Pa	10^4 lb/in²	Ultimate Shear Strength 10^8 Pa	10^4 lb/in²
Aluminum	1.4	2.0	1.4	2.0	1.4	2.0
Brass	3.5	5.5	4.5	6.5		
Copper	1.6	2.3	4.1	6.0		
Glass						
Iron, wrought	1.7	2.5	3.2	4.7	2.7	3.9
Lead, rolled			0.2	0.3		
Nickel (cast)			4.2	5.5		
Steel	2.4	3.5	4.8	7.0	3.8	5.5
Timber (parallel to grain)	0.2	0.3	0.69	1	0.035	0.05

Solution:

The ultimate sheering strength for steel $\tau_m = 3.8 \times 10^8$ Pa (Table 5-2). Since the sheering force F must act tangentially to the surface area A around the circumference of the hole,

$$A = \pi \, dt = \pi (1.4 \times 10^{-2} \text{ m})(5.0 \times 10^{-3} \text{ m})$$
$$= 2.2 \times 10^{-4} \text{ m}$$

Thus

$$F = \tau_m A = (3.8 \times 10^8 \text{ Pa})(2.2 \times 10^{-4} \text{ m}^2)$$
$$= 8.4 \times 10^4 \text{ N}$$

Example 5-5

Determine the factor of safety of a steel girder that has a cross-sectional area of 50 in² if it is to be used to support a compressive load not exceeding 600 000 lb. The ultimate strength of structural steel is 70 000 lb/in² (Table 5-2).

Solution:

$$\sigma = \frac{F}{A} = \frac{600\,00 \text{ lb}}{50 \text{ in}^2} = 12\,000 \text{ lb/in}^2$$

Therefore

$$\text{factor of safety} = \frac{70\,000 \text{ lb/in}^2}{12\,000 \text{ lb/in}^2} = 5.8$$

Usually, in buildings, the factor of safety of structural steel must not be less than 4, but this value will differ for different materials.

When a material is subjected to repeated applications of varying stresses over a long period of time, it gradually loses its strength. This is known as *fatigue*, and it occurs sooner if the material has a flaw. Fatigue is a common cause of failure in machinery.

If a metal is subjected to a constant stress for a long period of time while it is maintained at a relatively high temperature, it may undergo a plastic deformation called *high temperature creep*. This type of deformation increases with time and may eventually cause failure of the metal. High temperature creep must be considered in devices, such as boilers and blades of turbines, where metals are stressed at relatively high temperatures. In some materials, such as lead and glass, creep deformations even occur at normal room temperatures.

When a constant stress is applied to a "hot" metal, the creep deformation initially increases rapidly, but it is soon followed by a long period during which the deformation (strain) increases at a constant rate. Just before the metal fails, the rate of increase in strain becomes very rapid.

The *hardness* of a material is specified in terms of its ability to resist scratching or penetration by other materials. In the *Moh scale*, ten standard materials are assigned values from one to ten in order of their hardness. The materials with the greater hardness are given larger values; they can scratch materials with smaller values. For example, diamond (the hardest material) is assigned the value 10; other materials are compared with the ten standards. The Moh hardness scale has some applications in nondestructive testing of materials.

Most hardness tests consist of applying a standard load to force a standard object into the material. In the *Brinell hardness test* standard loads are used to press a hardened steel ball into the material; the area of the resulting indentation is related to the hardness of that material. The *Rockwell hardness test* involves measuring the depth of penetration when a standard indenter is forced into the material by a standard load.

Problems

8. If an elastic wire elongates by 1.02 in when it is subjected to a tensile force of 20.0 lb, what is its force constant? By how much would it elongate if it were subjected to a tensile force of 50.0 lb?

9. A coil elastic spring is stretched 2.00 cm when a 5.00 kg load is suspended from it. What is its force constant? By how much would it elongate if it were subjected to a 12.0 kg load?

10. A 1.00 cm diameter steel cable is used to lift a 3 000 kg load. What is the factor of safety?

11. A wrought iron girder is designed to support tensile loads of 1.00×10^5 lb. If the factor of safety is 5.00, what is the cross-sectional area of the girder?

12. Determine the shearing force that would be necessary to shear a steel rivet that has a diameter of 8.0 mm.

13. If a steel member of a bridge truss is expected to support tensile loads of 2.5×10^6 kg with a factor of safety of 6.0, what should its cross-sectional area be?

14. What maximum load can be lifted by a crane that has a 1.5 cm diameter steel cable if the factor of safety is 5.0?

15. What is the total cross-sectional area of the steel cables supporting a 5 000 kg elevator that has a maximum acceleration of 4.0 m/s² if the factor of safety is 6.0?

5-3 YOUNG'S MODULUS

Consider a uniform bar or a wire whose original length L_0 is much greater than its cross-sectional area A. The application of small tensile or compressive forces F to the ends of the bar will leave the area A approximately constant* but will cause a change in length ΔL in accordance with Hooke's law. Thus the stress $\sigma = F/A = k\Delta L/A$; multiplying the numerator and denominator by L_0, since the strain $\epsilon = \Delta L/L_0$, we obtain

$$\sigma = \frac{kL_0}{A}\frac{\Delta L}{L_0} = E\epsilon \quad \text{or} \quad \frac{F}{A} = \frac{E\,\Delta L}{L_0} \quad (5\text{-}12)$$

where $E = kL_0/A$ is a constant for that object.

Therefore, Hooke's law may be restated as: the stress is directly proportional to the strain for elastic materials. The proportionality constant E is called *Young's modulus* (or the *modulus of elasticity*) for the material and is independent of any dimensions. A knowledge of its value allows us to determine the elastic deformation of any body constructed from that material.

Since strain is a dimensionless quantity, the units of Young's modulus are the same as those of stress, i.e. lb/in², lb/ft², and Pa (N/m²) in SI.

Example 5-6

An aluminum wire with an initial length $L_0 = 25.000$ m and a diameter $d = 2.0$ mm is used to lift a 10 kg load. Determine (a) the final length L_1 of the wire, (b) the factor of safety, and (c) the maximum load that the wire can support.

Solution:

The cross-sectional area of the wire $A = \pi(d/2)^2$. The force exerted is due to the weight of the load $F = w = mg = (10 \text{ kg})(9.8 \text{ m/s}^2) = 98$ N.

---

*In reality, the bar will become thinner if it is being stretched and thicker if it is being compressed; however, the change in the cross-sectional area is very small compared with the changes in length and to a first approximation may be ignored.

(a) From Eq. 5-12 and Table 5-2, the elongation of the wire

$$\Delta L = \frac{FL_0}{AE} = \frac{FL_0}{[\pi(d/2)^2]E}$$
$$= \frac{(98 \text{ N})(25 \text{ m})}{\pi(10^{-3} \text{ m})^2(7.0 \times 10^{10} \text{ Pa})} = 1.1 \times 10^{-2}\,\text{m}$$

Therefore

$$L_1 = L_0 + \Delta L = 25.000 \text{ m} + 0.011 \text{ m}$$
$$= 25.011 \text{ m}$$

(b)
$$\sigma = \frac{F}{A} = \frac{mg}{\pi(d/2)^2} = \frac{98 \text{ N}}{\pi(10^{-3} \text{ m})^2}$$
$$= 3.1 \times 10^7 \text{ Pa}$$

The ultimate strength $= 1.4 \times 10^8$ Pa. Thus

$$\text{The factor of safety} = \frac{\text{ultimate strength}}{\text{allowable stress}}$$
$$= \frac{1.4 \times 10^8 \text{ Pa}}{3.1 \times 10^7 \text{ Pa}} = 4.5$$

(c) The wire will break when the applied stress is equal to the ultimate strength, i.e. when

$$F/A = 1.4 \times 10^8 \text{ Pa}$$

Consequently, it will break when the applied force

$$F = A(1.4 \times 10^8 \text{ Pa})$$
$$= (\pi \times 10^{-6} \text{ m}^2)(1.4 \times 10^8 \text{ Pa})$$
$$= 4.4 \times 10^2 \text{ N}$$

or when the mass of the load

$$m = \frac{F}{g} = \frac{(4.4 \times 10^2 \text{ N})}{(9.8 \text{ m/s}^2)} = 45 \text{ kg}$$

Note that this is equal to the product of the factor of safety and the actual load: 4.5(10 kg) = 45 kg.

Problems

16. A 100 ft long wire with a cross-sectional area of 0.10 in² stretches by 0.21 in when a tensile force of 500 lb is applied at its ends. Determine Young's modulus for the wire.

17. A brass rod, with a length of 100 m and a cross-sectional area of 1.0 cm², is subjected to a tensile force of 1 000 N. Find

(a) the force constant of the rod, (b) the stress, (c) the strain, (d) the final length of the rod.

18. A steel cable with a length of 100 m and a cross-sectional area of 1.0 cm² is subjected to a tensile force of 500 N. Find (a) the force constant, (b) the stress, (c) the strain, (d) the final length of the wire. (e) Is the elastic limit exceeded?

19. A 10 ft steel wire and a 20 ft aluminum wire with equal cross-sectional areas of 0.20 in² are fastened end to end and subjected to a tensile force of 50 lb. What is the final length of the combined wires?

Table 5-3
Bulk Moduli of Liquids (20°C)

Substance	Bulk Modulus K	
	10^{10} Pa	10^6 lb/in²
Alcohol ethyl	0.09	0.13
Benzene	0.104	0.15
Glycerin	0.386	0.56
Mercury	2.62	3.8
Oil	0.18	0.26
Petroleum	0.13	0.19
Water	0.23	0.33

5-4 BULK MODULUS

It is possible to exert the same magnitude of compressive stress (or pressure) p on a volume of a fluid (liquid or gas), or over the entire surface area of a solid by immersing it in a fluid. This will result in a reduction ΔV of the original volume V_0, and, from the definition, the *volume strain*

$$\epsilon = -\Delta V/V_0 \qquad (5\text{-}13)$$

(The negative sign indicates a reduction in volume.)
Thus

$$\frac{\text{stress}}{\text{strain}} = \frac{p}{\left(\dfrac{-\Delta V}{V_0}\right)} = \frac{-pV_0}{\Delta V} = K \quad (5\text{-}14)$$

The constant K is called the *bulk modulus* (or the *modulus of volume elasticity*) of the material (Tables 5-2 and 5-3); it is independent of the dimensions of that material and has units of Pa, lb/ft², or lb/in². The bulk moduli of liquids and solids are numerically larger than those of gases because gases are more easily compressed.

Example 5-7

If 1.0 L of pure water is subjected to a pressure of 5.0 kPa, what is the change in its volume?

Solution:

From Eq. 5-14 and Table 5-3, we obtain:

$$-\Delta V = \frac{pV_0}{K} = \frac{(5.0 \times 10^3 \text{ Pa})(1.0 \times 10^3 \text{ cm}^3)}{(2.3 \times 10^9 \text{ Pa})}$$
$$= 2.2 \times 10^{-3} \text{ cm}^3$$

Problems

20. A 100 in³ volume of material is reduced by 3.30×10^{-3} in³ when subjected to a 225 lb/in² pressure. What is the bulk modulus of the material?

21. 1 000 cm³ of benzene is subjected to a pressure of 5.3×10^8 Pa. What is the change in the volume of the benzene?

22. 300 in³ of ethyl alcohol is subjected to a pressure of 20 atm (1 atm = 14.7 lb/in²). What is the change in volume of the alcohol?

5-5 SHEAR MODULUS

When a shearing force F is applied tangentially to a surface of area A, it produces a strain ϵ in the body such that:

$$\frac{\text{stress}}{\text{strain}} = \frac{F/A}{\epsilon} = G = \frac{F/A}{x/y} \qquad (5\text{-}15)$$

The constant G is called the *shear modulus* (or the *modulus of rigidity*) of the material. It is constant for a particular material and has units of Pa, lb/ft^2 or lb/in^2.

Example 5-8

What is the shearing force required to distort a block of steel by 0.010 in, if its thickness is 12 in and its surface area is 9.0 in^2?

Solution:

From Eq. 5-15 and Table 5-2, we obtain

$$F = \frac{GAx}{y}$$
$$= \frac{(12 \times 10^6 \text{ lb/in}^2)(9.0 \text{ in}^2)(1.0 \times 10^{-2} \text{ in})}{(12 \text{ in})}$$
$$= 9.0 \times 10^4 \text{ lb}$$

Problems

23. Determine the shearing force required to distort a block of aluminum by 1.5 cm if its thickness is 30 cm and its surface area is 12 cm^2.

24. Determine the minimum shearing force that would break an aluminum block of thickness 6.0 in and surface area 15.0 in^2.

REVIEW PROBLEMS

1. What is the density in SI base units of an object whose volume is 7.4 cm^3 if its mass is 63 g?

2. What is the volume of a 20.0 lb block of steel?

3. Determine the stress when a 25 lb load rests on a horizontal surface area of 2.0 in^2.

4. What minimum force is required to shear an aluminum ingot that has a cross-sectional area of 4.0 cm^2?

5. An irregularly shaped block of metal has a mass of 15 kg. When it is lowered into a container, which is completely full of water, 1.8 L of water overflows. Determine the density of the metal.

6. An elastic wire elongates by 2.6 cm when it supports a 1.2 kg load. (a) What is the force constant? (b) By how much would the wire elongate if it supported a 3.0 kg load?

7. Determine the tensile stress in a steel cable that has a diameter of 8.0 mm and is used to lift a 90.0 kg load.

8. What minimum force is required to punch a 2.0 in diameter hole in an aluminum plate that is $\frac{1}{2}$ in thick?

9. Determine the minimum cross-sectional area of a steel member of a structure if it is used to support tensile loads of 1.5×10^7 kg with a factor of safety of 5.0.

10. (a) What maximum shearing force can a steel girder withstand if its cross-sectional area is 0.15 m^2? (b) If the factor of safety is 5.0, what maximum force may be applied?

11. What maximum tensile load can an aluminum rod withstand if the factor of safety is 5.8 and its cross-sectional area is 2.0 in^2?

12. A 3.0 m long wire with a diameter of 1.5 mm is elongated by 0.60 mm when a 6.0 kg load is hung from one end. Determine Young's modulus for the wire.

13. A copper sphere with an initial volume

of 500 cm³ is subjected to a pressure of 8.0×10^6 Pa. What is the change in its volume?

14. (a) Determine the minimum shearing force required to distort a block of steel by 5.0 mm if its thickness is 15 cm and its surface area is 25 cm². (b) What minimum force would shear the steel block?

15. A 10 ft steel girder with a cross-sectional area of 1.1 ft² is used to support a load of 2.0×10^5 lb. (a) What is the final length of the column? (b) What is the factor of safety if the ultimate compressive strength of the girder is 4.7×10^4 lb/in²? (c) What is the maximum load that the girder can withstand? Ignore the weight of the girder.

We have seen that forces produce accelerations and therefore changes in the motion of objects, but in some cases there is no acceleration even though very large forces are present. This state of no acceleration is called *equilibrium*; it arises because the forces balance each other. For example, in structures, such as buildings and bridges, very large forces are present, yet the structure must remain in equilibrium. Engineers design these structures so that they remain rigid even when they support loads.*

6-1 CONCURRENT COPLANAR FORCES IN EQUILIBRIUM

Forces whose lines of action pass through a common point are called *concurrent*; if they also lie in the same geometric plane, they are said to be *coplanar* (Fig. 6-1). These forces tend to produce accelerations in a straight line.

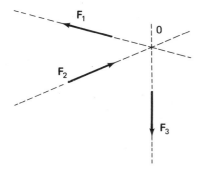

Figure 6-1
Three coplanar forces which are concurrent at 0.

6

Equilibrium

In accordance with Newton's laws of motion, a rigid body will remain at rest or in a state of uniform motion in a straight line

---

*In fact, structures actually deform slightly when they support loads (Chapter 5).

only if no unbalanced external force acts upon it. Therefore we say that two or more concurrent forces acting simultaneously are in *equilibrium* only if their resultant is identically zero since this corresponds to a state of no acceleration. Concurrent forces are said to be in a state of *static equilibrium* if their resultant is identically zero and the system is at rest with respect to the observer. This type of equilibrium is present in most structures, such as buildings and bridges, where any significant motion with respect to the earth would be disastrous.

Even moving objects can be in equilibrium as long as they do not accelerate, i.e. they must move in a straight line at a constant speed. For example, an aircraft in straight and level flight at a constant velocity is in equilibrium. In this case, the upward force on the wings (*lift*) balances the weight, and the forward force (*thrust*) due to the engines balances the retarding force due to the surrounding air (*drag*) (Fig. 6-2). This is called *dynamic equilibrium*.

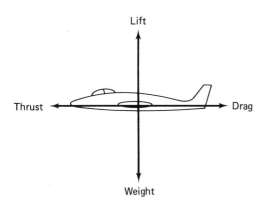

Lift

Thrust

Drag

Weight

Figure 6-2

An aircraft in straight and level flight at a constant velocity.

To obtain a resultant force of zero, the *vector sum* of all component forces must also be zero. Therefore the condition for a system

of concurrent forces to be in equilibrium may be written as:

$$\underset{\text{vector sum}}{\Sigma^*}\ \mathbf{F} = 0 \qquad (6\text{-}1)$$

Since any force $\mathbf{F}$ in a coplanar system may be resolved into two independent perpendicular components (F_x and F_y), an equivalent description of a system of concurrent coplanar forces in equilibrium is that the algebraic sum of their components in each direction must equal zero.

$$\Sigma F_x = 0\dagger \quad \text{and} \quad \Sigma F_y = 0 \qquad (6\text{-}2)$$

Note that the conditions described in Eq. 6-2 apply to any perpendicular set of axes that we choose since the equilibrium condition implies that the algebraic sum of the components of the forces *in any direction* equals zero (see Problem 3).

Equilibrant

If a system of concurrent forces is not in equilibrium, there is always some other force called the *equilibrant* that would give the system equilibrium if it were applied to the concurrence point. In order to attain equilibrium (zero resultant force), the equilibrant must have equal magnitude but opposite direction to the resultant of the original force system. For example, if a force of 70.0 N inclined at an angle of 30° with the horizontal and a 60.0 N horizontal force are applied at a common point, their resultant, which represents their combined effect, is a force of 126 N inclined at an angle of 16° with the horizontal. Therefore, to obtain equilibrium for the two original forces, we would have to apply the equilibrant of 126 N in the opposite direction

---

*The symbol Σ will be used to indicate a sum. Thus $\Sigma \mathbf{F} = \mathbf{F}_1 + \mathbf{F}_2 + \mathbf{F}_3 + \cdots$ etc.

†If the system of forces is not coplanar, a third condition is required to describe the equilibrium of the third perpendicular axis, i.e., $\Sigma F_z = 0$.

to their resultant. This would give the final system a resultant of zero, which is represented by a closed polygon of forces (Fig. 6-3).

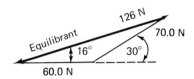

Figure 6-3

Since the sum of a system of concurrent coplanar forces may be represented by a closed polygon, which has sides of lengths representing the magnitude of those forces to scale, we may use scale diagrams to solve

equilibrium problems. The procedure is as follows:

1. Draw the free-body diagram of the system.

2. Add the forces graphically by constructing the polygon of forces to scale.

3. Measure the unknown values directly from the polygon.

Example 6-1 _____

An 80 N traffic light is suspended in equilibrium from two wires, which make angles of 20° and 30° with the horizontal (Fig. 6-4a). Determine the tensions T_1 and T_2 in the wires.

Solution:

Draw the free-body diagram (Fig. 6-4b). Choose an appropriate scale, 1 cm ≡ 20 N for example. Construct the vertical vector representing the 80 N weight to scale, and

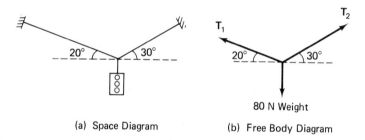

(a) Space Diagram (b) Free Body Diagram

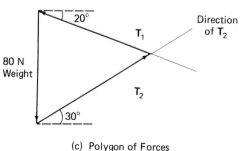

(c) Polygon of Forces
Scale 1 cm ≡ 20 N

Figure 6-4

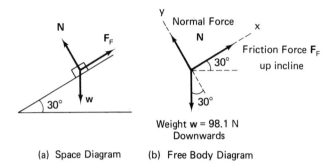

(a) Space Diagram   (b) Free Body Diagram

Figure 6-5

from the terminal point draw a line parallel to the tension T_2. Since the polygon (triangle) of forces must be closed, the terminal point of the remaining tension vector T_1 must coincide with the origin of the load vector. Construct a line parallel to the tension vector T_1 from the origin of the load vector to intersect the line drawn parallel to T_2 (Fig. 6-4c). The intersection of these lines completes the polygon of forces. The magnitudes of T_1 and T_2 may be measured directly from the diagram (to the same scale as the load vector). In this case, the magnitudes of T_1 and T_2 are 90 N and 98 N, respectively. Note that the 80 N weight is the equilibrant of T_1 and T_2.

The equilibrium conditions in Eq. 6-2 are used in the analytical approach. The procedure is as follows:

1. Draw a free-body diagram of the system.

2. Resolve all forces into a *convenient* set of perpendicular components. Since the system is in equilibrium, we may choose *any* set of perpendicular axes. Often, horizontal and vertical axes are not the most

convenient because a different choice will enable us to solve the problem with less work.

3. Apply the equilibrium conditions Eq. 6-2 in order to obtain equations involving the unknown items.

4. Solve the equations obtained in Step 3.

*Example 6-2*_____

A 10.0 kg block is held at rest on a 30° incline by a force (the friction force F_f) that is directed up the incline (Fig. 6-5a). Determine the magnitudes of the normal force N (perpendicular to the incline) and the friction force.

Solution:

Sketch the free-body diagram. Since the normal force N and the friction force F_f are perpendicular to each other, for convenience choose the axes of the coordinate system in the directions parallel and perpendicular to the incline (Fig. 6-5b). This choice will reduce the algebraic manipulations, but *any* coordinate system will give the same results (see Problem 9). Note that the weight of the 10.0 kg block $w = mg = (10.0 \text{ kg})(9.81 \text{ m/s}^2) = 98.1 \text{ N}$.

Force	x-component	y-component
weight $w = mg$	$-(98.1 \text{ N}) \sin 30 = -49.1 \text{ N}$	$-(98.1 \text{ N}) \cos 30 = -85.0 \text{ N}$
friction force F_f	F_f	0
normal force N	0	N

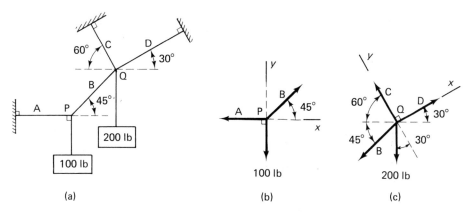

Figure 6-6

Applying the equilibrium conditions, we obtain

$\Sigma F_x = -49.1\,\text{N} + F_f = 0$ and $F_f = 49.1\,\text{N}$

and

$\Sigma F_y = -85.0\,\text{N} + N = 0$ and $N = 85.0\,\text{N}$

To solve more complex systems in equilibrium, repeat the procedure *at each concurrence point* since each point is in equilibrium.

Example 6-3

Determine the magnitudes A, B, C and D of the tensions $\mathbf{T}_A$, $\mathbf{T}_B$, $\mathbf{T}_C$ and $\mathbf{T}_D$ in the strings A, B, C and D, respectively, of the system in Fig. 6-6a. Assume three significant figures.

Solution:

Draw a free-body diagram for each point of concurrence (see Figs. 6-6b and 6-6c). Consider the concurrence point P; choose the

horizontal and vertical directions as the perpendicular axes.

Force	x-Component	y-Component
$\mathbf{T}_A$	$-A$	0
$\mathbf{T}_B$	$B \cos 45°$	$B \sin 45°$
100 lb load	0	-100 lb

Thus

$\Sigma \mathbf{F}_y = B \sin 45 - 100\,\text{lb} = 0$ or $B = 141\,\text{lb}$

and

$$\Sigma \mathbf{F}_x = B \cos 45 - A = 0$$

or

$$A = (141\,\text{lb}) \cos 45 = 100\,\text{lb}$$

Now consider the concurrence point Q; for simplification, choose the direction of the strings C and D as the perpendicular axes. Using the calculated value for B, we obtain

Force	x-Component	y-Component
$\mathbf{T}_C$	0	$+C$
$\mathbf{T}_D$	D	0
200 lb load	$-(200\,\text{lb}) \sin 30$ $= -100\,\text{lb}$	$-(200\,\text{lb}) \cos 30$ $= -173\,\text{lb}$
$\mathbf{T}_B$	$-B \cos 15 = -136\,\text{lb}$	$-B \sin 15 = -36\,\text{lb}$

Then

$$\Sigma F_x = D - 100 \text{ lb} - 136 \text{ lb} = 0$$

or

$$D = 236 \text{ lb}$$

and

$$\Sigma F_y = C - 173 \text{ lb} - 36 \text{ lb} = 0$$

or

$$C = 209 \text{ lb}$$

Problems

Assume three significant figures.

1. A horizontal force of 20 N is applied to a body whose weight is 50 N. What is the force required to keep the system in equilibrium?

2. A 10 lb load is suspended in equilibrium from two ropes, one inclined at 20° with the horizontal and the other at 35° with the horizontal. Determine the tensions in the ropes (a) by scale diagram and (b) analytically.

3. A 500 N load is suspended in equilibrium from two ropes, one horizontal and the other inclined at an angle of 45° with the horizontal. Determine analytically the tensions in the ropes by resolving all forces into (a) horizontal and vertical components, (b) components inclined at 20° with the horizontal and vertical.

4. To pull a car from a ditch, the driver ties one end of a rope to the car and the other to a tree 50 ft away. He then pulls sideways at the midpoint of the rope with a 100 lb force. How much force is exerted on the car when the man has pulled the rope 3.0 ft to one side?

5. If a 25 lb coat is hung from the center of a 50 ft clothesline, causing the clothesline to sag 1.0 ft from the horizontal at that point. What is the tension in the clothesline?

6. Determine the magnitude of **A** and the direction of the 80 lb force if the system in Fig. 6-7 is in equilibrium.

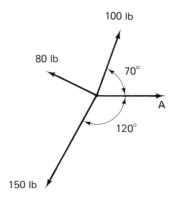

Figure 6-7

7. A 10 kg mass is kept at rest on a plane, inclined at an angle of 30° with the horizontal, by the application of a force parallel to the plane. What is the magnitude of that force?

8. A 10 lb body is kept at rest on a perfectly smooth plane by the application of a 2.0 lb force parallel to the plane. What is the inclination of the plane with respect to the horizontal?

9. Do Example 6-2 by resolving forces into their horizontal and vertical components.

10. A load of 500 N is applied at the junction of two members of a structure. If the members are perpendicular to each other and the load is inclined at 30° from one member and 60° from the other (Fig. 6-8), determine the forces exerted in the members.

11. A toggle is a pair of jointed bars, which are used to exert large forces (Fig. 6-9). If the bars are of equal length and the angle α between them is 160°, what are the magnitudes of the horizontal forces

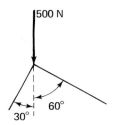

Figure 6-8

F_1 and F_2 required to keep the system in equilibrium when a load **w** of magnitude 50 N is applied at the joint? *Hint:* There are 3 concurrence points.

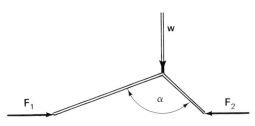

Figure 6-9

12. Repeat Problem 11 for a toggle having one bar half the length of the other.

13. A 25 N steel ball is resting in a groove whose sides are inclined at an angle of 45° and 60° with the horizontal (Fig. 6-10). Determine the reaction forces at

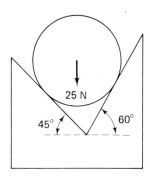

Figure 6-10

the points of contact between the ball and the groove.

14. Determine the tensions in each of the strings of the system in Fig. 6-11.

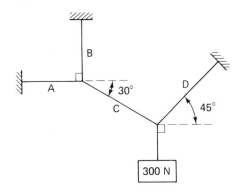

Figure 6-11

6-2 NONCONCURRENT COPLANAR FORCES

Forces whose lines of action never cross are called *nonconcurrent*. These forces have parallel or antiparallel lines of action, and they tend to cause rotations, such as those in wheels, drive shafts and gears.

Moment of Force and Torque

In order to describe the "turning effect" of a force, we define two terms called *moment of force* and *torque*. Even though their definitions are identical, we normally use the term torque when there is an actual rotation; moments of force are used in calculations when the system is in equilibrium.

From experience we know that less effort is required to open or close a door when we apply the force to the handle that is on the side away from the axis of rotation (the hinges). It is much more difficult to turn the door by pushing it near the hinges! Thus the turning effect of a force depends on both its magnitude and the distance of its line of action from the axis of rotation.

Consider a force **F**, which acts on a rigid body that has an axis of rotation (perpendicular to the line of action) at a point 0 (Fig. 6-12). The perpendicular distance s from the line of action to the axis of rotation is called the *moment arm*. We define the *moment of force* (and *torque*) M_0 (τ_0) as the product of

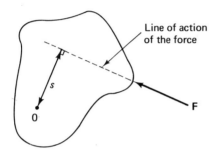

Figure 6-12

the moment arm s and the magnitude F of the force.

$$M_0 = sF \quad (\text{or } \tau_0 = sF) \quad (6\text{-}3)$$

The subscript 0 indicates the axis of rotation.

Torques and moments of force have the same dimensions as work and energy, but in SI they are written in units of newton meters (N·m) to indicate the mechanical nature. The British Engineering System unit is the pound foot (lb·ft).

Applied torques may cause a system to rotate in a clockwise or an anticlockwise direction; since these oppose each other, they are given opposite signs. We shall use the convention that a torque that tends to produce an anticlockwise rotation is given a positive value and that a torque that tends to produce a clockwise rotation is given a negative value.

Example 6-4

Determine the torques on the systems in Fig. 6-13 about the axes of rotation 0.

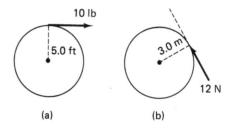

(a) (b)

Figure 6-13

Solution:

(a) $\tau_0 = sF = -(5.0 \text{ ft})(10 \text{ lb}) = -50 \text{ lb·ft}$
(b) $\tau_0 = sF = (3.0 \text{ m})(12 \text{ N}) = 36 \text{ N·m}$

Example 6-5

Determine the magnitude of the torque exerted on a wheel of diameter $d = 0.75$ m by a tangential force $F = 480$ N on its rim.

Solution:

In this case the moment arm $s =$ radius of the wheel because it rotates about its center. Thus

$$M = sF = \left(\frac{0.75 \text{ m}}{2}\right)(480 \text{ N}) = 180 \text{ N·m}$$

Resultant

As in the case of concurrent forces, we may replace a system of nonconcurrent forces by a single force (the *resultant*) that produces the same effect. However, we must now specify the point of application of this force.

Example 6-6

Determine the resultant of the system in Fig. 6-14.

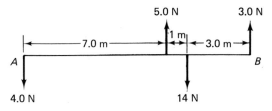

Figure 6-14

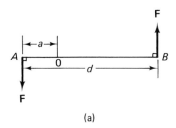

(a)

Solution:

To obtain the magnitude R of the resultant, sum the forces in the vertical direction:

$$\Sigma F_y = -4.0 \text{ N} + 5.0 \text{ N} - 14.0 \text{ N} + 3.0 \text{ N}$$
$$= -10.0 \text{ N} = R$$

If R is applied at a distance of d from point A, then the resultant torque is given by the algebraic sum of the other torques:

$$Rd = \Sigma \tau_A$$

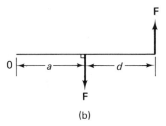

(b)

Figure 6-16

Therefore,

$$d = \frac{\Sigma \tau_A}{R} = \frac{[(7.0 \text{ m})(5.0 \text{ N})] - [(8.0 \text{ m})(14.0 \text{ N})] + [(11.0 \text{ m})(3.0 \text{ N})]}{-10.0 \text{ N}} = 4.4 \text{ m}$$

Hence the resultant is a force 10 N downwards applied at a point 4.4 m to the right of A.

Suppose a force **F** is applied at a point whose displacement is **r** from the axis of rotation 0, such that the angle between **r** and **F** is θ (Fig. 6-15). The component of **r**, which is perpendicular to **F**, has a magnitude of $r \sin \theta$; this corresponds to the moment arm s. From the definition of torque:

$$\tau_0 = sF = (r \sin \theta)F \text{ *}$$

A *couple* is defined as two, equal-magnitude, nonconcurrent forces applied in opposite directions (Fig. 6-17). The magnitude of

the torque produced by a couple is the product of the magnitude F of *one* of the forces and the perpendicular distance d between the lines of action of the forces.

$$\Sigma \tau = Fd \qquad (6\text{-}4)$$

Suppose the distance of the nearest force from the axis of rotation 0 is a.

In Fig. 6-16a,

$$\Sigma \tau_0 = aF + (d - a)F = Fd.$$

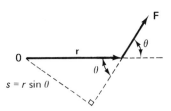

Figure 6-15

---

*Note that this can also be written in terms of a vector product $\mathbf{M_0} = \mathbf{r} \times \mathbf{F}$ or $\boldsymbol{\tau}_0 = \mathbf{r} \times \mathbf{F}$.

In Fig. 6-16(b),

$$\Sigma\tau_0 = -aF + (a + d)F = Fd.$$

Therefore, since a never appears in the answer, Eq. 6-4 must be valid for any axis of rotation.

Problems_____

Assume three significant figures.

15. A string is wrapped around a cylinder with a radius of 6.0 in. If a force of 80 lb is applied to the string, what is the magnitude of the torque that is developed about the axis of the cylinder?

16. Determine the magnitude of the torque, in SI units, that is exerted on a spark plug when a 250 N force is applied to the end of a 30.0 cm torque wrench.

17. Determine the force that must be applied to the end of a 15 cm long torque wrench in order to produce a torque of 25 N·m.

18. Calculate the magnitude of the torque produced by two 25 N forces that are 3.0 m apart and in opposite directions.

19. A 3 200 lb car with 14 in radius wheels is accelerated from rest to 60 mi/h in 10 s. Find (a) the force required and (b) the torque supplied to each of the two rear wheels.

20. An 1 800 kg car with 48 cm diameter wheels is accelerated from rest to 80 km/h in 15 s. Find (a) the net force and (b) the torque supplied to each of the two rear wheels.

21. What tangential force would produce a torque of 700 N·m on a solid cylindrical drive shaft that has a radius of 2.5 cm?

6-3 CENTER OF GRAVITY

Every particle of matter on the surface of the earth is attracted towards the center of the earth by the force of gravity. Most objects

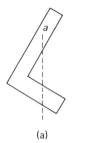
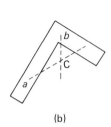

(a)   (b)

Figure 6-17
Determining the center of gravity.

are composed of large numbers of particles; thus if the object is small compared with the earth, the forces on all of its constituent particles are nearly parallel. The *weight* of the object is mainly due to these forces of gravity and is usually represented by a single force vector **w**, which always passes through a certain point (not necessarily in the object) called the *center of gravity C*. In fact the object can be supported by a single upward force (the *equilibrant*), which is equal in magnitude to the weight; the force's line of action passes through the center of gravity. The location of the center of gravity in structures is extremely important.

To determine the center of gravity of a body experimentally, suspend the body freely from one point a, and let it attain stable equilibrium. Then, use a plumb bob to draw a straight line on the body from the point of suspension towards the center of the earth. Repeat the process, suspending the body at some other point b. The intersection of the two lines drawn will correspond to the center of gravity C of the body* (Fig. 6-18).

Moment of Area & Centroids_____

If a body has a uniform density and a uniform thickness, then the area of the body is directly

---

*A third point suspension is required for three dimensional objects.

proportional to its weight. In this case, the center of the area, which is called the *centroid*, coincides with the center of gravity of the body; the area may be used instead of forces in the determination of the position of the center of gravity. Therefore, in analogy to the moment of force, we define the moment of area M_0^A as the product of the magnitude of the total area A of the body and the distance s from the axis of rotation 0 to the centroid of the area.

$$M_0^A = sA \qquad (6-5)$$

The centroids of some simple regular areas are listed in Table 6-1. To determine the position of the centroid of more complex regular areas, proceed as follows:

1. Separate the body into simple areas in which the centroids are either known or may be easily determined.

2. Construct a convenient perpendicular coordinate system on the diagram of the body.

3. Determine the coordinates of the centroids and calculate the magnitudes of the simple areas.

4. Determine the net area A of the body by subtracting the sum of the areas of the holes from the sum of the solid areas. Since the weight is proportional to the area, solid areas are directed downwards whereas holes are directed upwards.

5. The centroid of the body is where the net area A appears to act. If the coordinates of the centroid are $(\bar{x}, \bar{y})$, take moments about the y-axis; the resultant moment of area is the algebraic sum of the moments of the simple areas.

$$\bar{x}A = \Sigma M_y^A$$
$$= \Sigma \text{ (simple area)}(x\text{-coordinate})$$

Therefore the x-coordinate of the centroid

$$\bar{x} = \frac{\Sigma M_y^A}{A} \qquad (6-6)$$

Similarly the y-coordinate of the centroid

$$\bar{y} = \frac{\Sigma M_x^A}{A} \qquad (6-7)$$

Example 6-7

Determine the position of the centroid of the regular area in Fig. 6-18a. Assume three significant figures.

Solution:

Choose the areas and the coordinate system as in Fig. 6-18(b) and use Table 6-1.

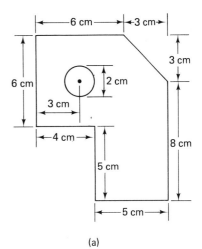

(a)

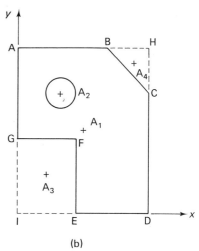

(b)

Figure 6-18

Table 6-1
Properties of Simple Areas

Description of the area	Diagram	Area	Centroid c $\overline{x}$ $\quad$ $\overline{y}$	Moment of area about neutral axis
Rectangle with base b, height h		bh	$b/2$ $\quad$ $h/2$	$I_x = \dfrac{bh^3}{12}$
Isosceles triangle height h, base b		$\dfrac{bh}{2}$	$b/2$ $\quad$ $h/3$	$I_x = \dfrac{bh^3}{36}$
Circle with radius R and diameter D		$\pi R^2 = \dfrac{\pi D^2}{4}$	R $\quad$ R	$I_x = \dfrac{\pi R^4}{4} = \dfrac{\pi D^4}{64}$
Semicircle of radius R and diameter D		$\dfrac{\pi R^2}{2} = \dfrac{\pi D^2}{8}$	R $\quad$ $\dfrac{4R}{3\pi}$	$I_x = 0.11R^4$ $I_y = \dfrac{\pi R^4}{8}$
Quadrant of radius R		$\dfrac{\pi R^2}{4}$	$\dfrac{4R}{3\pi}$ $\quad$ $\dfrac{4R}{3\pi}$	$I_x = 0.055R^4$

Area	Magnitude of Area	Coordinates of Centroid/cm
A_1(A H D I)	(9 cm)(11 cm) $= 99.0$ cm²	$(4\frac{1}{2}, 5\frac{1}{2})$
A_2(circle)	π(1 cm)² $= 3.14$ cm²	(3, 8)
A_3(G F E I)	(4 cm)(5 cm) $= 20.0$ cm²	$(2, 2\frac{1}{2})$
A_4(B C H)	$(\frac{1}{2})$(3 cm)(3 cm) $= 4.50$ cm²	(8, 10)

$$A = A_1 - A_2 - A_3 - A_4 = 99.0 \text{ cm}^2 - 3.14 \text{ cm}^2 - 20.0 \text{ cm}^2 - 4.5 \text{ cm}^2 = 71.4 \text{ cm}^2$$

Since the holes act in the opposite direction to the solid area, from Eq. 6-6, the x-coordinate of the centroid

$$\bar{x} = \frac{\Sigma M_y^A}{A} = \frac{(4\frac{1}{2} \text{ cm})(99.0 \text{ cm}^2) - (3 \text{ cm})(3.14 \text{ cm}^2) - (2 \text{ cm})(20.0 \text{ cm}^2) - (8 \text{ cm})(4.50 \text{ cm}^2)}{71.4 \text{ cm}^2}$$

$$\bar{x} = 5.04 \text{ cm}$$

From Eq. 6-7, the y-coordinate of the centroid

$$\bar{y} = \frac{\Sigma M_x^A}{A} = \frac{(5\frac{1}{2} \text{ cm})(99.0 \text{ cm}^2) - (8 \text{ cm})(3.14 \text{ cm}^2) - (2\frac{1}{2} \text{ cm})(20.0 \text{ cm}^2) - (10 \text{ cm})(4.50 \text{ cm}^2)}{71.4 \text{ cm}^2}$$

$$\bar{y} = 5.94 \text{ cm}$$

Center of Gravity of Regular Solids

If a solid has a uniform density but a non-uniform thickness, its center of gravity may be determined by the procedure outlined in Example 6-7, except that the areas are now replaced by volumes and that moments must be taken about three axes instead of two. However, many of these problems may be simplified if the solid is symmetrical since the center of gravity will lie along the axis of symmetry.

Example 6-8

A 6.00 ft long concrete cylinder of radius 6.00 in has a 4.00 ft long rectangular hole along its axis (Fig. 6-19). If the rectangle has 3.00 in and 6.00 in sides, determine the center of gravity of the cylinder.

Solution:

Since the system is symmetrical, the center of gravity must lie along the axis of the cylinder. Therefore it is necessary only to determine the distance to the center of gravity along the axis from one of the faces of the cylinder. Measuring the distances from the face with the hole, we obtain

Volume	Magnitude	Distance to Center of Gravity
total cylinder	$\pi(\frac{1}{2} \text{ ft})^2(6.00 \text{ ft}) = 4.71$ ft³	3.00 ft
rectangular hole	$(\frac{1}{4} \text{ ft})(\frac{1}{2} \text{ ft})(4.00 \text{ ft}) = 0.500$ ft³	2.00 ft

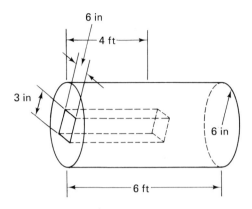

Figure 6-19

Therefore the net volume of the solid

$$V = 4.71 \text{ ft}^3 - 0.50 \text{ ft}^3 = 4.21 \text{ ft}^3$$

and the distance to the center of gravity

$$\bar{z} = \frac{\Sigma M}{V}$$

$$= \frac{(3.00 \text{ ft})(4.71 \text{ ft}^3) - (2.00 \text{ ft})(0.500 \text{ ft}^3)}{4.21 \text{ ft}^3}$$

$$= 3.12 \text{ ft}$$

If the density of the solid varies, the weights of the component parts must be considered instead of the areas or volumes.

Example 6-9

Determine the center of gravity of the cylinder in Example 6-8 if the hole is filled with steel. The weight densities of steel and concrete are 488 lb/ft³ and 150 lb/ft³, respectively (Table 5-1).

Solution:

The center of gravity must still lie along the axis of symmetry. Determine the net volume of the concrete and its center of gravity as in Example 6-8. Then:

Therefore the total weight $w = 632 \text{ lb} + 244 \text{ lb} = 876 \text{ lb}$; the distance to the center of gravity

$$\bar{z} = \frac{\Sigma M}{w}$$

$$= \frac{(3.12 \text{ ft})(632 \text{ lb}) + (2.00 \text{ ft})(244 \text{ lb})}{876 \text{ lb}}$$

$$= 2.81 \text{ ft}$$

Problems

Assume three significant figures.

22. Determine the centroid (with respect to point A) of (a) the L-section in Fig. 6-20 without the circular hole and (b) the same L-section with the circular hole.

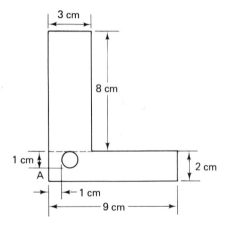

Figure 6-20

23. Determine the centroid (with respect to point A) of (a) the T-section in Fig. 6-21 without the circular hole and (b) the same T-section with the circular hole.

24. Determine the center of gravity of a 4

Component	Volume	Weight	Distance to Center of Gravity
net concrete	4.21 ft³	$(4.21 \text{ ft}^3)(150 \text{ lb/ft}^3) = 632 \text{ lb}$	3.12 ft
steel	0.500 ft³	$(0.500 \text{ ft}^3)(488 \text{ lb/ft}^3) = 244 \text{ lb}$	2.00 ft

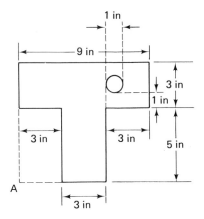

Figure 6-21

in × 6 in × 18 in block of wood (oak) if a 2.0 in diameter circular hole is drilled 5.0 in deep through its long axis.

25. Determine the center of gravity of a 12 in × 8 in × 6 in steel block if a 2.0 in diameter circular hole is drilled 8.0 in through its longest axis.

26. Determine the center of gravity of a 18 cm × 12 cm × 3 cm concrete block if a 10 cm diameter circular hole is drilled 2 cm through its shortest axis.

27. Determine the center of gravity of the block of wood in Problem 24 if the hole is filled with copper. The densities are given in Table 5-1.

28. If the wood block in Problem 24 has a 2 in diameter hole drilled completely through its long axis, determine the center of gravity (a) if a 3 in lead plug is inserted in one end and a 7 in copper plug is inserted in the other, (b) if a 5 in aluminum plug is inserted in one end and an 11 in iron plug is inserted in the other.

6-4 OBJECTS IN EQUILIBRIUM

When the sum of the clockwise moments is not balanced by the sum of the anticlockwise moments, the system will rotate with a resultant torque equal to the difference between the clockwise and anticlockwise torques. Therefore the conditions given in Eq. 6-1 or Eq. 6-2 for a system of concurrent forces to be in equilibrium are not sufficient for nonconcurrent forces since a rotational effect may be present. In Fig. 6-22, for example, the system is not in equilibrium because it will rotate, even though the vector sum of the forces is zero. An additional equilibrium condition is required for nonconcurrent forces.

Figure 6-22

A system of two or more forces acting simultaneously is said to be in equilibrium if:

1. The vector sum of the forces is zero.

$$\Sigma F = 0 \qquad (6\text{-}8)$$

or $\qquad \Sigma F_x = 0 \quad$ and $\quad \Sigma F_y = 0 \qquad (6\text{-}9)$

2. The algebraic sum of the clockwise (negative) moments and the anticlockwise (positive) moments is zero about any point.

$$\Sigma M = 0 \qquad (6\text{-}10)$$

Using these conditions, we may solve problems involving systems in equilibrium.

Example 6-10

The system in Fig. 6-23 is in equilibrium. Determine the magnitude of the force **F** and the distance *d*.

Solution:

$$\Sigma F_y = 200 \text{ N} - F + 50 \text{ N} = 0$$

or

$$F = 250 \text{ N}.$$

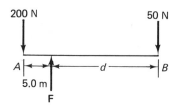

Figure 6-23

(a) Stable equilibrium   (b) Unstable equilibrium

Taking moments about point 0, condition 2 implies that

$$\Sigma M_0 = (5.0 \text{ m})(200 \text{ N}) - d(50 \text{ N}) = 0$$

Therefore

$$d = \frac{(5.0 \text{ m})(200 \text{ N})}{50 \text{ N}} = 20 \text{ m}$$

Since the system is in equilibrium, we may check the solution by taking moments about any other point; if the answers are correct, we should obtain a total moment of zero.

$$\Sigma M_A = (5.0 \text{ m})F - (5.0 \text{ m} + d)50 \text{ N}$$
$$= (5.0 \text{ m})(250 \text{ N}) - (5.0 \text{ m} + 20 \text{ m})(50 \text{ N})$$
$$= 0$$

Whenever a body is in a state of equilibrium, the equilibrium conditions must apply, but the nature of the equilibrium is not always the same. There are three kinds of equilibrium:

1. *Stable:* The center of gravity C is directly below the point of support 0 (Fig. 6-24a). In this case the application of a force will usually disturb the body, but the body will eventually return to its original position after the force has terminated.

2. *Unstable:* The center of gravity C is directly above the point of support 0 (Fig. 6-24b). An applied force will usually destroy this form of equilibrium, and the body will change its position to attain stable equilibrium after the force has ceased.

(c) Neutral equilibrium

Figure 6-24
Kinds of equilibrium.

3. *Neutral:* The center of gravity C coincides with the point of support 0 (Fig. 6-24c), and the body will remain in any position.

In Example 6-6, we saw that several non-concurrent forces may be replaced by a single force (*resultant*) acting at some specific location without changing the net effect. As before, the *equilibrant* is a force of equal magnitude and opposite direction that acts at the same point. In any object the resultant of all the forces of gravity on its constituent particles is the weight vector that passes through the center of gravity. It should be noted that this is always true, regardless of the point about which we take moments. As a result, we may assume that *the total weight of any object always acts at the center of gravity, regardless of where the object is supported.*

Example 6-11

One end of a horizontal uniform bar is embedded in a wall so that 5.0 m projects (Fig. 6-25). If the bar weighs 60 N per linear meter, what is the moment of force tending to break the bar at the wall?

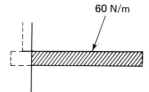

60 N/m

Figure 6-25

Solution:

Since the bar is uniform, we may assume that all of its weight acts at the center of gravity 2.5 m from the wall. The total weight $w = (5.0 \text{ m})(60 \text{ N/m}) = 300$ N. Thus

$$M_{\text{wall}} = sF = (2.5 \text{ m})(300 \text{ N}) = 750 \text{ N} \cdot \text{m}$$

*Example 6-12*_____

A 20.0 ft uniform beam AB rests on two supports, one at A and the other 4.00 ft from B. If the beam weighs 200 lb, determine the reaction forces R_1 and R_2 of the supports (Fig. 6-26).

Solution:

Since the beam is uniform, its weight may be considered to act at the center of gravity,

which is 10 ft from A. Taking moments about point A, we obtain:

$$\Sigma M_A = (16.0 \text{ ft})R_2 - (10.0 \text{ ft})(200 \text{ lb}) = 0$$

Therefore $R_2 = 125$ lb.
Summing the forces in the vertical direction, we obtain:

$$\Sigma F_y = R_1 + R_2 - 200 \text{ lb}$$
$$= R_1 + 125 \text{ lb} - 200 \text{ lb} = 0$$

Therefore $R_1 = 75$ lb.

*Example 6-13*_____

An 8.00 m long uniform beam AB weighing 1 200 N is hinged at one end A to a vertical wall. It is held in a horizontal position by a tie, which is attached to end B and makes an angle of 37° with the beam (Fig. 6-27a). If a 500 N load is suspended from the beam at point B, determine the tension **T** in the tie and the reaction force **R** of the wall on the beam at point A.

Solution:

Assume **R** makes an angle ϕ with the beam and draw the free body diagram (see Fig. 6-27b). The weight of the beam is considered to act at the center of gravity C.

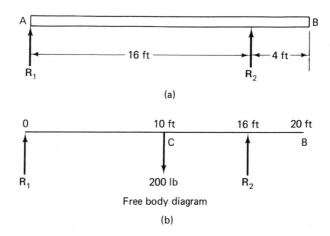

(a)

Free body diagram

(b)

Figure 6-26

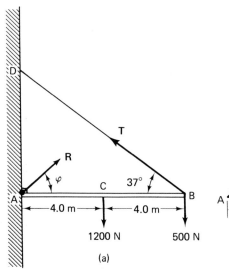

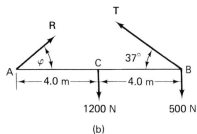

Figure 6-27

Force	Horizontal x-Component	Vertical y-Component
R	R_x	R_y
T	$-T\cos 37°$	$T\sin 37°$
w beam	0	$-1\,200$ N
w load	0	-500 N

Take moments about A.

$$\Sigma M_A = -(4.00 \text{ m})(1\,200 \text{ N}) - (8.00 \text{ m})(500 \text{ N})$$
$$+ (8.00 \text{ m})(T\sin 37°) = 0$$

Therefore $T = 1\,830$ N.

$$\Sigma F_x = R_x - T\cos 37°$$
$$= R_x - (1\,830 \text{ N})(\cos 37°) = 0$$

Therefore $R_x = 1\,460$ N.

$$\Sigma M_B = (4.00 \text{ m})(1\,200 \text{ N}) - (8.00 \text{ m})R_y = 0$$

Therefore $R_y = 600$ N.
Thus

$$R = \sqrt{R_x^2 + R_y^2}$$
$$= \sqrt{(1\,460 \text{ N})^2 + (600 \text{ N})^2} = 1\,580 \text{ N}$$

and

$$\tan \phi = \frac{R_y}{R_x} = \frac{600 \text{ N}}{1\,460 \text{ N}} = 0.411 \quad \text{or} \quad \phi = 22.3°$$

Problems

29. One end of a horizontal uniform bar is embedded in a wall so that 9.0 ft of the bar projects from the wall. If the bar weighs 15 lb per linear foot, what is the moment of force tending to break the bar at the wall?

30. If the bar in Problem 29 is inclined at 30° with the horizontal, determine the moment of the force tending to break the bar.

31. One end of a horizontal uniform beam is embedded in a wall so that 12 m of the beam project. If the beam has a weight per unit length of 50 N/m and if a 300 kg load is hung at the projecting end of the beam, determine the resultant moment of force tending to break the beam at the wall.

32. Repeat Problem 31 for a beam that is inclined at an angle of 45° with the horizontal.

33. A horizontal uniform beam 20 ft long and weighing 100 lb is supported at each end. If loads of 700 lb are hung 3.0 ft and 8.0 ft from one end, determine the reaction forces at the supports.

34. Two men are carrying a 12 m uniform beam on their shoulders. Man A is 1.0 m from one end, and man B is 3.0 m from the other end. If the beam has a mass of 200 kg, determine the load supported by each man.

35. If a 300 kg load is hung 3.0 m from man A in Problem 34, where would man B have to be in order to carry the same load as man A?

36. Each front wheel of a car with a 4.2 m wheelbase (the distance between the front and rear axles) presses on the ground with a 3 500 N force and each rear wheel with a 4 000 N force. Determine the center of gravity.

37. A crate, which is uniformly packed, is 2.5 m high, 1.5 m wide and 1.0 m deep. If the total mass of the crate is 500 kg and it does not slide, determine the minimum horizontal force that must be applied to the top of the crate in order to tip it.

38. A horizontal 80 ft long bridge weighing 120 000 lb is supported at each end. If the center of gravity of the bridge is at its center and a 45 000 lb truck is on the bridge 30 ft from one end, calculate the forces on the supports.

39. A 15 ft long uniform beam, which weighs 50 lb/ft, is supported at end A and at 2.0 ft from the other end B. If loads of 800 lb, 1 100 lb and 300 lb are placed at points 1.0 ft, 3.0 ft and 14 ft from A

respectively, determine the reaction forces at the supports.

40. A 20 m uniform beam of linear mass density 10 kg/m is supported at points 3.0 m from each end. Concentrated loads of 5.0 kg and 8.0 kg are located at points a distance of 2.0 m and 12 m from end A, and a uniform load of 5.0 kg/m is placed on the 3.0 m span at end B (Fig. 6-28). Determine the reaction forces R_1 and R_2 at the supports.

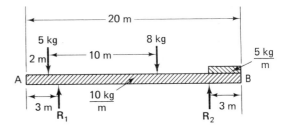

Figure 6-28

41. A 10 m uniform ladder AB rests with end A on rough horizontal ground and end B against a perfectly smooth (frictionless) vertical wall 7.0 m above the ground. If the ladder has a mass of 200 kg, determine the reaction forces of the ground and the wall on the ladder.

42. Repeat Problem 42 if an 80 kg man is one-third of the way up the ladder from point A.

43. A 12 ft long uniform boom AB, which weighs 120 lb, is hinged at A to a vertical mast and is held in position by a tie that is attached to the mast 6.0 ft above the hinge (Fig. 6-29). If the boom is horizontal, find the tension in the tie and the reaction of the mast on the boom.

44. Repeat Problem 43 if a load of 300 lb is suspended at point B.

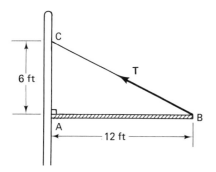

Figure 6-29

45. Repeat Problem 44 if the boom is suspended so that the tie is horizontal (Fig. 6-30).

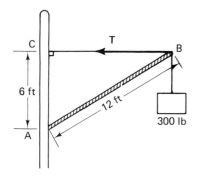

Figure 6-30

6-5 TRUSSES

Trusses are frame structures, such as roofs, bridges and booms of cranes, that are com-

posed of straight members arranged to form triangles. All of the forces in a truss act along these members, resulting only in tensions and compressions: no shearing forces exist. Therefore it is necessary to calculate the force in each member in order to select members that will withstand the load without buckling. Since the truss is in equilibrium, the forces at each end of a member have equal magnitude but opposite directions. The joints of the truss are points at which the forces in the members are concurrent. Therefore the equilibrium condition $\Sigma F = 0$ may be used at each joint to determine the unknown forces in the truss. We may also take moments when it is only necessary to determine the forces in a few of the members.

Example 6-14

Determine the forces in members BD, BC, AC and AB of the bridge truss in Fig. 6-31a.

Solution:

The height of the truss $h = \sqrt{(10 \text{ ft})^2 - (8 \text{ ft})^2}$ $= 6$ ft; therefore, $\sin \theta = 6/10 = 3/5$, and $\cos \theta = 8/10 = 4/5$.

$$\Sigma M_G = (8 \text{ ft})(2\,000 \text{ lb}) + (24 \text{ ft})(5\,000 \text{ lb}) + (40 \text{ ft})(5\,000 \text{ lb}) - (48 \text{ ft})R_1 = 0$$

Thus $R_1 = 7\,000$ lb. It is not necessary to solve for R_2. Draw the free-body diagram of point A (Fig. 6-31b). Note that AB pushes down on A and that AC pulls to the right.

Force	x-Component	y-Component
R_1	0	7 000 lb
AB	$-AB \cos \theta = -\frac{4}{5}AB$	$-AB \sin \theta = -\frac{3}{5}AB$
AC	AC	0

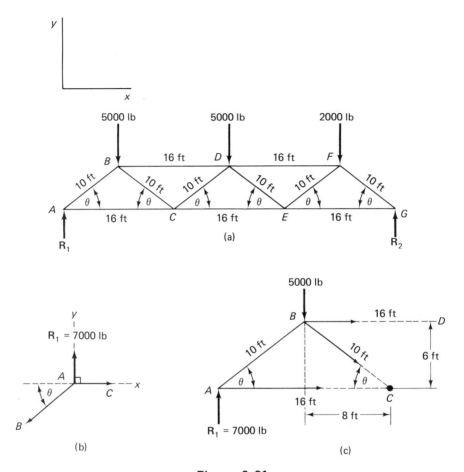

Figure 6-31

$$\Sigma F_y = 7\,000 \text{ lb} - \tfrac{3}{5}AB = 0$$

or $\qquad AB = \tfrac{35\,000}{3}$ lb

$$\Sigma F_x = -\tfrac{4}{5}AB + AC = 0$$

or $\qquad AC = \tfrac{4}{5}AB = \tfrac{28\,000}{3}$ lb

We could now draw the free-body diagrams of points B (with AC pushing up) and C (with AC pulling to the left) to determine the forces in members BC, BD and AC. However, we shall use the *method of sections* by moments.

Take a section through the members BD, BC and AC, and assume a force in either direction along each of these members. The directions are not important because, if the

forces are reversed, negative answers will be obtained.

Draw the free-body diagram of the left section of the truss (Fig. 6-31c). The forces in the members BD, BC and AC must be included in this free-body diagram. Note that two of the unknown forces (AC and BC) pass through the point C; therefore they have no moment about that point. Thus $\Sigma M_c =$ (8 ft)(5 000 lb) — (16 ft)(7 000 lb) — (6 ft)BD = 0; $BD = -72\,000$ lb/6 = $-12\,000$ lb, i.e. 12 000 lb in the opposite direction to that indicated; BD is under a compressive stress. Two of the unknown forces (BD and BC)

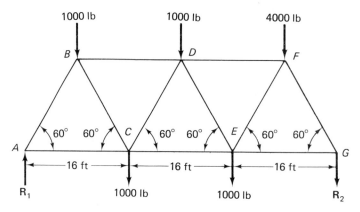

Figure 6-32

pass through the point B. Thus $\Sigma M_B = (6\text{ ft})AC - (8\text{ ft})(7\,000\text{ lb}) = 0$; $AC = 56\,000\text{ lb}/6$ (tension). Then

$$\Sigma F_y = 7\,000\text{ lb} - 5\,000\text{ lb} - BC\sin\theta = 0$$

$$BC = \frac{2\,000\text{ lb}}{3/5} = \frac{10\,000\text{ lb}}{3}\text{ (tension)}$$

Problems

46. Determine the forces in the members EF, FD, FG and ED of the bridge truss in Fig. 6-32.

47. Determine the forces in all of the members of the Fink roof truss in Fig. 6-33.

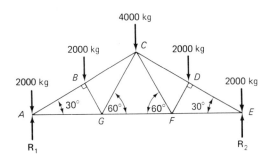

Figure 6-33
The Fink roof truss.

REVIEW PROBLEMS

Assume three significant figures.

1. A rope is tied to two fixed points 15.0 m apart. A vertical load of 5.1 kg is suspended from the rope at a point 5.0 m from one end; this depresses the rope 2.0 m from the horizontal at that point. Determine the tensions in the rope.

2. Determine the tension in the string and the magnitude of the horizontal force in the system illustrated in Fig. 6-34.

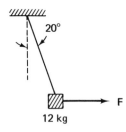

Figure 6-34

3. Determine the tension in each string in the system illustrated in Fig. 6-35.

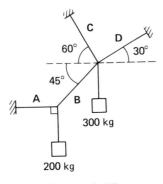

Figure 6-35

4. Determine the equilibrant of two perpendicular forces of 85 lb and 60 lb.

5. A boy sits on a swing that is pulled to one side by a horizontal force until the ropes make an angle of 30° with the vertical. If the tension in each rope is 300 N (a total of 600 N), find (a) the mass and weight of the boy and (b) the horizontal force.

6. Find the magnitude of **R** and the angle θ if the system illustrated in Fig. 6-36 is in equilibrium.

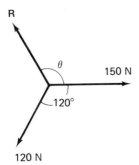

Figure 6-36

7. A 15 ft long uniform beam, which weighs 750 lb, is supported at end A and at 2.0 ft from end B. If loads of 800 lb, 1 100 lb and 300 lb are placed at distances of 1.0 ft, 3.0 ft and 14 ft from A, respectively, find the reaction forces at the supports.

8. Determine the centroid with respect to point A of the area in Fig. 6-37.

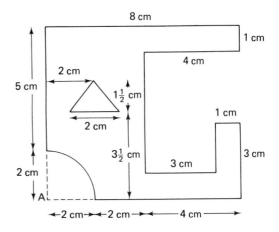

Figure 6-37

9. A 12 in × 8 in × 6 in steel block has a 2 in diameter circular hole drilled completely through its longest axis. Determine the center of gravity (a) if a 7.0 in brass plug is inserted in one end of the hole and (b) if a 7.0 in brass plug is inserted in one end and a 5.0 in copper plug is inserted in the other end.

10. A uniform beam 20 ft long balances at a point 5.0 ft from one end when a 75 lb load is hung 2 ft from that end. What is the weight of the beam?

11. A 2 500 kg truck with a wheel base of 4.0 m has its center of gravity 2.5 m in front of the rear axle. If a 1 500 kg load is located at the center of the truck at a distance 1.5 m from the rear axle, find the force on each tire of the truck.

12. A horizontal uniform beam AB is 4 ft long weighs 10 lb. If a tie is attached 3 ft from the hinged end A and a 20 lb load is hung

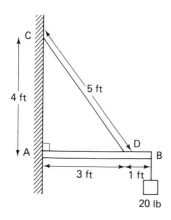

Figure 6-38

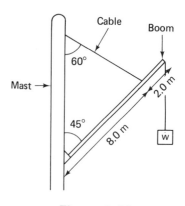

Figure 6-39
Derrick.

from end B (Fig. 6-38), calculate (a) the tension in the tie rod CD, (b) the compressive force on the boom, (c) the reaction of the hinge on the boom.

13. In the derrick illustrated in Fig. 6-39, the boom weighs 1 250 N and the load 1 400 N. If the center of mass of the boom is 7.5 m from the axle, the cable is tied 8.0

m from the axle, and the load is tied 10 m from the axle, determine (a) the tension **T** in the cable and (b) the reaction force **R** of the mast on the axle.

14. Find the forces in members AB, AC, FG, FH and EG of the bridge truss in Fig. 6-40.

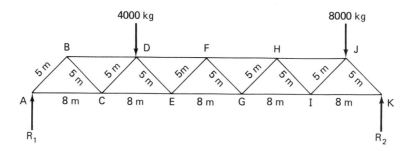

Figure 6-40
Bridge truss.

There is no such thing as a perfectly smooth material; all known materials have some irregularities in their surfaces. Consequently, when two objects are in contact, these irregularities interlock, and the surfaces adhere to each other (Fig. 7-1). If a force is applied in such a way that these objects slide over each other, the adhesion between the surfaces results in a resistance to the relative motion of the bodies. This resistance to the relative motion is called *friction*.

Figure 7-1
Surface irregularities interlock to produce friction.

Friction is used to start fires, allows us to move and to stop, holds nails in wood, and can even be used to weld two surfaces together. However friction is often undesirable since it produces unwanted heat, causes wear, and reduces the efficiency of machines. Its effects can usually be minimized by lubrication, by smoothing the surfaces, and by the use of bearings.

7
Friction

7-1 THE ANALYSIS OF FRICTION

Consider a block of one material on a flat horizontal surface of some other material. The weight **w** of the block is exactly balanced by the reaction force **R** of the surface so that the net force on the block is zero and it remains at rest. The block will usually remain at rest even when a small horizontal force $\mathbf{F}_A$ is applied. Consequently $\mathbf{F}_A$ must be exactly balanced by a force $\mathbf{F}_F$ in the opposite direction; this opposing force is due to the friction between the surfaces. The free-body diagram (Fig. 7-2) must include this friction force.

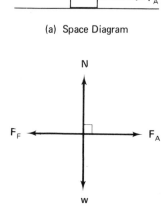

(a) Space Diagram

(b) Free Body Diagram

Figure 7-2
Block on a horizontal surface.

Friction forces act tangentially to the surface of contact. Each increase in the applied force $\mathbf{F}_A$ is exactly balanced by an increase in the friction force $\mathbf{F}_F$ between the surfaces, until the applied force reaches some critical value, called the *limiting friction force* $\mathbf{F}_l$. There is relative motion between the surfaces when the applied force $\mathbf{F}_A$ exceeds the limiting friction force. Note that the friction force $\mathbf{F}_F$ can never exceed the magnitude of the applied force $\mathbf{F}_A$. If it could, there would be a net unbalanced force in the opposite direction to $\mathbf{F}_A$, and, according to Newton's second law, we would have perpetual motion because the block would move in the direction of the unbalanced friction force.

Friction forces are approximately independent of the relative speed between the block and the surface. Once a body has started to slide, it takes less force to keep it sliding at a constant speed than it did to start its motion. In order to make the block move with a constant speed, $\mathbf{F}_A$ must now be reduced. The force required to maintain the relative motion at a constant speed is equal to the *kinetic friction force* $\mathbf{F}_k$ between the two surfaces.

It is also found that both the *limiting and kinetic friction forces are approximately independent of the areas of contact.*

Since friction is due to the adhesion of irregularities in the surfaces in contact, we should expect that the greater the force pushing the surfaces together (the normal force **N**), the greater the friction force $\mathbf{F}_F$. For two particular nonlubricated surfaces in contact, the magnitude of the limiting friction force F_l is directly proportional to the magnitude of the normal force N.

$$F_l \propto N \quad \text{or} \quad F_l = \mu_s N \qquad (7\text{-}1)$$

The dimensionless proportionality constant μ_s is called the *coefficient of static friction.* Similarly, the magnitude of the kinetic friction force F_k is directly proportional to the magnitude of the normal force N. The dimensionless proportionality constant, in this case μ_k, is called the *coefficient of kinetic (sliding) friction.*

$$\mu_k = \frac{F_k}{N} \qquad (7\text{-}2)$$

The coefficients of static and kinetic friction depend only on the nature of the two materials in contact. Some typical values for the coefficients of friction are listed in Table 7-1. Note that, since the limiting friction force $\mathbf{F}_l$ is greater than the kinetic friction force $\mathbf{F}_k$, the coefficients of static friction μ_s are greater than corresponding coefficients of kinetic friction μ_k.

Example 7-1

Determine the force required to start a 500 lb steel block sliding over a horizontal concrete floor.

Solution:

From Table 7-1, $\mu_s = 0.50$. Since the floor is horizontal, the weight **w** of the block is numerically equal to the normal force **N**. Thus,

$$F = \mu_s N = (0.50)(500 \text{ lb}) = 250 \text{ lb}$$

Table 7-1
Typical Values of Friction Coefficients Between Two Surfaces

Surface	μ_s	μ_k
Glass on glass	0.95	0.80
Steel on steel	0.58	0.25
Wood on wood	0.35	0.30
Wood on metal	0.40	0.20
Wood on brick	0.60	0.25
Steel on wood	0.55	0.40
Steel on concrete	0.50	0.33
Rubber tire on dry concrete	0.95	0.71
Rubber tire on wet concrete	0.72	0.52
Wood on concrete	0.55	0.35

Suppose that a block of weight **w** is placed on an inclined plane and that the angle of inclination of the plane with the horizontal is gradually increased. When the angle of inclination is increased beyond a certain value α, called the *angle of friction*, the block will begin to slide down the plane. Consequently, when the angle of inclination is equal to α, the component of the weight down the plane exactly balances the limiting friction force: the block is in a state of static equilibrium.

Example 7-2

Determine the coefficient of static friction between a block and an incline if the angle of friction $\alpha = 30°$.

Solution:

Construct the free-body diagram of the block at the angle of friction, and for convenience resolve the forces into components parallel and perpendicular to the incline (Fig. 7-3).

Force	x-Component	y-Component
w	$-w \sin \alpha$	$-w \cos \alpha$
Normal force N	0	N
limiting friction force $\mathbf{F}_l$	F_l	0

Since the block remains at rest at the angle of friction, its acceleration is zero in both the x and y directions, i.e. it is in equilibrium. Thus

$$\Sigma F_y = -w \cos \alpha + N = ma_y = 0,$$

or

$$N = w \cos \alpha$$

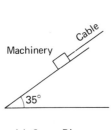

Machinery

35°

(a) Space Diagram

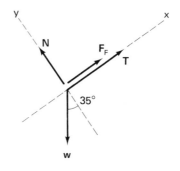

35°

(b) Free Body Diagram

Figure 7-3

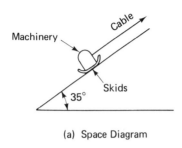

(a) Space Diagram

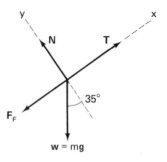

(b) Free Body Diagram

Figure 7-4

and

$$\Sigma F_x = -w \sin \alpha + F_l = ma_x = 0,$$

or

$$F_l = w \sin \alpha$$

Therefore

$$\mu_s = \frac{F_l}{N} = \frac{w \sin \alpha}{w \cos \alpha} = \tan \alpha = \tan 30° = 0.577$$

since $\sin \alpha / \cos \alpha = \tan \alpha$.

Example 7-3

A 1 500 kg piece of machinery on wooden skids is lowered at a constant velocity down a 35° concrete ramp by means of a cable that is attached to the machinery and pulls up the incline (Fig. 7-4). What is the tension in the cable?

Solution:

The weight of the machinery $w = mg = $ (1 500 kg)(9.81 m/s²) = 14 700 N. Sketch the free-body diagram, and resolve the forces into

Force	x-Component	y-Component
Normal force N	0	N
Friction force F_k	$F_k = \mu_k N$	0
Weight **w**	$-w \sin 35°$	$-w \cos 35°$
Tension in cable **T**	T	0

components parallel and perpendicular to the ramp (Fig. 7-4b). Note that the friction force F_k acts in the opposite direction to the motion. Since the velocity is constant, the acceleration of the machinery is zero, i.e. it is in equilibrium even though it is moving. Thus

$$\Sigma F_y = N - w \cos 35° = ma_y = 0,$$

or

$$N = w \cos 35° \qquad (1)$$

Also

$$\Sigma F_x = \mu_k N - w \sin 35° + T = ma_x = 0,$$

or

$$T = w \sin 35° - \mu_k N \qquad (2)$$

Substituting from (1) into (2), we obtain,

$$T = w \sin 35° - \mu_k(w \cos 35°)$$
$$= w(\sin 35° - \mu_k \cos 35°)$$

But

$$\mu_k = 0.35 \qquad \text{(Table 7-1)}$$

Hence

$$T = (14\ 700\ \text{N})(\sin 35° - 0.35 \cos 35°)$$
$$= 4\ 220\ \text{N}$$

Problems

1. What is the force required to start a 200 lb block of steel moving over a horizontal concrete surface?

2. What is the minimum force required to keep a 100 lb wooden crate moving over a horizontal wooden floor?

3. What is the minimum force required to start a 10.0 kg block of wood sliding over a horizontal metal surface?

4. What is the minimum force required to keep a 20.0 kg steel box sliding over a concrete floor?

5. Determine the angle of friction between the following surfaces: (a) steel on steel, (b) wood on wood, (c) wood on metal, (d) wood on brick, (e) steel on concrete.

6. A 5.00 kg steel block begins to slide down an adjustable incline when the incline exceeds 25°. Determine (a) the limiting friction force, (b) the normal force between the block and the plane at the angle of friction, (c) the coefficient of static friction between the surfaces.

7. A 50.0 lb crate begins to slide down an adjustable ramp when the incline exceeds 28°. What is the limiting friction force?

8. What is the minimum force required to accelerate a 30.0 kg steel block at 3.00 m/s² up a 37° concrete slope?

9. What is the acceleration of a 20.0 kg wooden crate down a concrete slope if the slope is inclined at 43° with the horizontal?

10. A 3 000 lb car is travelling at 60 mi/h on a dry horizontal concrete surface when the driver applies the brakes. (a) How far does the car travel before it stops? (b) How long does it take the car to stop?

11. Repeat Problem 10 for a wet horizontal concrete surface.

12. A 10.0 kg steel block is placed at the top of a concrete ramp that is 2.40 m long and inclined at 37° with the horizontal. If the block is allowed to slide, (a) what is its acceleration down the ramp? (b) How long does it take the block to reach the bottom? (c) What is its speed at the bottom of the ramp?

13. A man pushes a 10 kg steel block up a 30° incline by exerting a force of 75 N parallel to the incline. The coefficient of kinetic friction between the block and the incline is 0.30, and the initial speed of the block is 1.0 m/s. What is the speed of the block up the incline after 50 s?

14. A runner is travelling at a constant speed of 25 ft/s on a flat horizontal road when a car travelling at 30 mi/h passes him. If the driver of the car slams on the brakes at the instant that he passes the runner and the coefficient of kinetic friction between the tires and the road is 0.60, (a) how long does it take for the runner to overtake the car? (b) How far does the car go before the runner passes it?

15. What is the acceleration of the system in Fig. 7-5 if the coefficient of kinetic friction between the 5.0 kg block and the horizontal surface is 0.40 and the pulley is frictionless? What is the tension in the string? *Hint:* Draw a free-body diagram of each block.

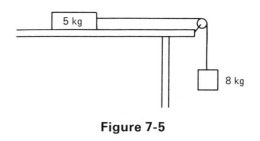

Figure 7-5

7-2 THE REDUCTION OF FRICTION

Friction reduces the efficiency of machines, and it can cause considerable damage to moving parts; lubricants and bearings are usually employed to reduce these effects. Lubricants, such as mineral oils, greases, graphite and gases, reduce friction because they are easily deformed or sheared and consequently allow surfaces to slide easily over them. In many modern machines, enormous stresses are produced in very small regions; lubricants and additives make compounds that shear easily and reduce the stresses on the machine itself.

In spite of these lubricants, about 20% of the power developed by an automobile engine is actually lost to friction for example. The control of dust and dirt and the use of new long-lasting, hermetically sealed greases are also used to control friction. Gases, such as air, are excellent lubricants. For example, hovercraft float on a cushion of air, which reduces the friction between the craft and the surface.

Rolling Friction

The invention of the wheel was one of man's greatest achievements because it enabled him to move heavy objects with much less effort. Friction effects between a rolling object (a wheel or a sphere) and a flat surface are much less than the friction between two flat surfaces. Therefore wheels and bearings allow us to achieve greater efficiency and are used to great extent in machinery.

When a round object rolls over a flat surface, the flat surface is deformed by the weight of the round object; at the same time, the round surface is deformed by the normal force of the flat surface. These deformations have the effect that the round object must continually run uphill (Fig. 7-6). The normal force of the flat surface on the round object is greater on the front surface; this results in a backward component. These effects may be reduced by using very hard surfaces, such as in bearings.

A *bearing* is a part of a machine that acts as a support for some other part, such as a rotating *shaft* or *journal*. The journal may turn or slide inside the bearing, or the bearing may rotate about a fixed journal. There are

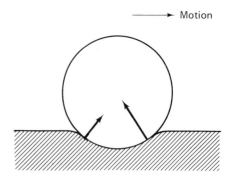

Motion

Figure 7-6
The surface is deformed by a round object, and a greater normal force is present on the front surface of a rolling object.

several kinds of bearings in common use (Fig. 7-7).

Sleeve bearings consist of a shell made from a low friction metal, such as cast iron, bronze or babbitt (88% tin, 4% copper and 8% antimony), which wears to form a smooth surface. A steel journal rotates by sliding over the inside of the sleeve on a film of a lubricant, such as graphite or oil. These bearings are used in lathes, milling machines, electric motors, crankpins and axles of trains and cylinder blocks of internal combustion engines.

Ball and roller bearings consist of hardened balls or rollers in *separators* located between two hardened steel sleeves, called *races*. Since these bearings operate in a rolling action, they are more efficient than sleeve bearings; however they are more expensive. They can also be made to operate with axial or radial loads. They are commonly used in wheels and as thrust bearings.

Needle bearings are similar to roller bearings, but the rollers are longer and have smaller diameters. They are often made without a separator or cage.

Since the effects of friction become excessive in a vacuum, the space age has presented many problems concerning the control of

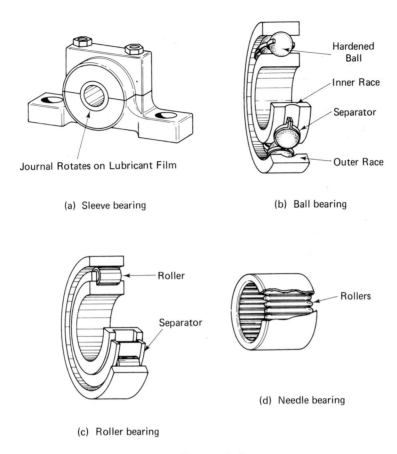

(a) Sleeve bearing

Journal Rotates on Lubricant Film

(b) Ball bearing

Hardened Ball

Inner Race

Separator

Outer Race

(c) Roller bearing

Roller

Separator

(d) Needle bearing

Rollers

Figure 7-7
Bearings

friction. In the near vacuum conditions and excessive cold of outer space, many materials even fuse together when they touch each other. Rockets themselves operate under extreme conditions of heat and therefore require special oils with low vapor pressures or even solid lubricants, such as certain metal crystals or plastics (polytetra fluorethaline), to reduce friction. One of the most promising developments is the magnetic bearing in which the shaft is supported only by a magnetic field and never actually touches the bearing itself. The magnetic bearing is extremely efficient; the shaft can be turned with only the slightest touch.

Recently, new tires for vehicles have been developed from high hysteresis rubber. These tires are able to produce a high friction on wet surfaces and consequently are safer.

REVIEW PROBLEMS

1. What minimum force is required to start a 60 kg steel block moving over a horizontal concrete surface?

2. Determine the minimum force required to keep a 75 lb wooden crate sliding over a horizontal concrete floor.

3. Determine the angle of friction between wood and concrete.

4. What is the acceleration of the blocks in the system in Fig. 7-8 if the coefficient of kinetic friction between the blocks and the inclines is 0.45 and the pulley is light and frictionless? What is the tension in the strings?

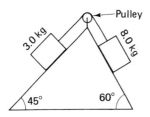

Figure 7-8

5. A man pushes a 200 lb crate up a 30° incline by exerting a 240 lb force parallel to the incline. If the coefficient of kinetic friction is 0.30, what is the acceleration of the crate?

6. An object slides from rest down a 28° inclined plane in 4.8 s. If the coefficient of kinetic friction is 0.41, what is the length of the plane?

7. A 12 kg block is pushed up a 20° incline that is 15 m long. If the coefficient of kinetic friction is 0.30 and the block was initially at rest at the bottom of the incline, determine the average force that will give the block a final speed of 6.0 m/s at the top of the incline.

The concepts of work, power and energy are frequently used in technology. Machinery is used to increase man's physical abilities and to convert natural fuels into useful energy forms, such as electricity, heat, light and mechanical energy. The widespread use of machinery in industry and the home has produced an enormous requirement for energy sources; as a result, forms of fuel, such as gasoline, coal and natural gas, are being rapidly depleted. Machinery is constantly being developed to produce greater output power and efficiency and improved to utilize other energy forms, such as solar and nuclear energy. The reduction of emissions, which pollute our environment, is also an area of considerable research.

8-1 WORK

When a force is applied, it may accelerate an object, maintain it at a constant velocity, or even have no visible effect if it does not overcome other forces such as friction. In our everyday lives, we generally consider the term work to be synonymous with a physical exertion. In physics and engineering sciences, work is a scalar quantity that is used to describe the *effects* of an applied force. Even though a force may be the result of some effort or exertion, the application of that force does not always result in the accomplishment of work on some objects.*

Work W is defined as the product of the magnitudes of the applied force $\mathbf{F}_A$, the displacement $\mathbf{s}$ of the object and the cosine of the angle θ between them, i.e. the scalar product between $\mathbf{F}_A$ and $\mathbf{s}$ (Fig. 8-1).

$$W = \mathbf{F}_A \cdot \mathbf{s} = F_A s \cos \theta \qquad (8\text{-}1)$$

Thus, *work is only done on an object if it is displaced from its original position.* Note that the term $(F_A \cos \theta)$ represents the component

8

Work, Power and Energy

---
*Of course, work may be done on the muscles, bones, etc., of our bodies.

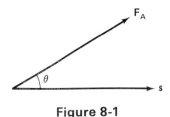

Figure 8-1

of the applied force $\mathbf{F}_A$ in the direction of the displacement s of the object.

The SI unit of work is called a *joule* (J), which is the work done when a force of one newton is applied through a displacement of one meter (1 J = 1 N·m). In the British Engineering System, the unit of work is the *foot pound* (ft·lb), which is the work done when a one pound force is applied through a one foot displacement.

Example 8-1

A man moves a trolley of machinery 12.0 ft over a flat horizontal surface by pulling on the handle with a 50.0 lb force. If the handle is inclined at 30° with the horizontal, how much work does the man do?

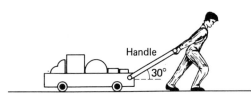

Handle

30°

Figure 8-2

Solution:

$$W = F_A s \cos \theta = (50.0 \text{ lb})(12.0 \text{ ft}) \cos 30$$
$$= 520 \text{ ft·lb}$$

Note that the force involved in the calculation of work done is the *applied force*, and this is not necessarily equal to the *net force*. Retarding forces, such as friction, are only important in the sense that they must be overcome in order to displace the object. If the

object moves, opposing forces are not considered in the calculation of the work done, but they are important with respect to energy losses.

Example 8-2

A man pushes horizontally on a large crate with a force $F_A = 150$ N (Fig. 8-3). Determine the work done (a) if the friction force $F_f = 140$ N and the man moves the crate a horizontal distance of 6.0 m, (b) if the friction force $F_f = 150$ N and the crate is initially at rest, (c) if the friction force $F_f = 150$ N and the crate moves 6.0 m in the direction of the applied force.

Solution:

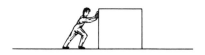

(a) Space Diagram

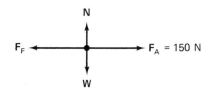

(b) Free Body Diagram

Figure 8-3

(a)

$$W = F_A s \cos \theta = (150 \text{ N})(6.0 \text{ m}) \cos 0 = 900 \text{ J}$$

Even though the friction force was present, it is not considered in the calculation of the work done; it only affects the acceleration according to Newton's second law.

(b) No work is done because the net force is zero; the crate is initially at rest and therefore will remain at rest. There is no displacement even though the man exerts himself.

(c)

$$W = F_A s \cos \theta = (150 \text{ N})(6.0 \text{ m}) \cos 0 = 900 \text{ J}$$

In this case, the net force is zero, but the crate is displaced with no acceleration; therefore work is done.

Example 8-3

A motor is used to drag a 1 200 kg container 15 m up a 20° ramp at a constant velocity (Fig. 8-4a). If the coefficient of kinetic friction $\mu_k = 0.30$, (a) what is the tension in the hoisting cable? (b) How much work is done by the motor?

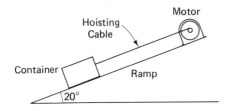

(a) Motor Pulling Container up a Ramp

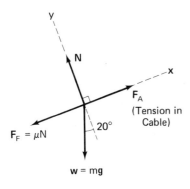

(b) Free Body Diagram

Figure 8-4

Solution:

Draw the free-body diagram of the container, and resolve the forces into components paral-

lel and perpendicular to the incline (Fig. 8-4b).

Force	x-Component	y-Component
Applied force $\mathbf{F}_A$	F_A	0
Normal force $\mathbf{N}$	0	N
Friction force $\mathbf{F}_f$	$-F_k = -\mu_k N$	0
Weight $\mathbf{w} = m\mathbf{g}$	$-mg \sin 20°$	$-mg \cos 20°$

(a) Since the acceleration is zero the container is in equilibrium. Thus

$$\Sigma F_y = N - mg \cos 20° = 0$$

or

$$N = mg \cos 20° \qquad (1)$$

$$\Sigma F_x = F_A - \mu_k N - mg \sin 20° = 0$$

Solving for F_A and substituting for N from (1), we obtain,

$$F_A = \mu_k mg \cos 20° + mg \sin 20°$$
$$= mg(\mu_k \cos 20° + \sin 20°)$$

Thus

$$F_A$$
$$= (1\ 200 \text{ kg})(9.8 \text{ m/s}^2)\{(0.30)(\cos 20°) + \sin 20°\}$$
$$= 7.3 \times 10^3 \text{ N}$$

(b) Since the applied force is parallel to the ramp, the work done

$$W = F_A s \cos \theta = (7.3 \times 10^3 \text{ N})(15 \text{ m}) \cos 0$$
$$= 1.1 \times 10^5 \text{ J}$$

When the smallest angle between the applied force and the displacement is greater than 90°, the cosine of that angle has a negative value, resulting in a negative value for work. Either this corresponds to the force opposing the motion of the object, or, if the object has already been displaced and is now at rest, the new applied force tends to return the object to its original position. For example, the brakes of a car do *negative work* because they slow the car down; the gradual lowering of an object to the floor constitutes *negative work* since the object would fall naturally if it were not slowed by the applied force.

Problems_____

1. How much work is done by a construction worker if he raises 20 lb of bricks at a constant velocity through a vertical distance of 60 ft?

2. A boy pulls a sled along a horizontal surface by exerting a 5.0 lb force at an angle of 45°. How much work does he do in pulling the sled 20 ft (a) if there is no friction? (b) If there is a 1.0 lb friction force?

3. How much work is done by a pump if it raises 4.0 kg of water vertically through a distance of 5.0 m?

4. A 1 500 lb piece of machinery is mounted on skids and pulled 20 ft up a 15° incline at a constant velocity. If the coefficient of kinetic friction between the skids and the incline is 0.25, determine: (a) the tension in the hoisting cable and (b) the work done.

5. How much work is done in 11 s by a 3 200 lb car if it accelerates uniformly from rest to 60 mi/h on a flat horizontal surface and 60% of the total force is used to overcome friction and other retarding forces?

6. How much work is done by an electric field in 150 ns if it accelerates an electron from rest to a final speed of 6.0×10^6 m/s during that time and there are no retarding forces?

8-2 POWER

A particular amount of work may be performed by many different agents, but some of these will do that work at a faster rate than others.

Power is defined as the time rate at which work is performed. Since the power output of an agent may vary, it is convenient to define the average power $\bar{P}$ as the total work W performed divided by the total time t taken.

$$\bar{P} = \frac{\text{total work performed}}{\text{total time taken}} = \frac{W}{t} = \frac{F_A s \cos \theta}{t}$$
(8-2)

The SI unit of power is called a **watt** (W); it is defined as the power delivered when work is done at a rate of one joule per second (1 W = 1 J/s). In British Engineering System units, power is expressed in foot pounds per second (ft·lb/s) or **horsepower** (hp) (1 hp = 550 ft·lb/s = 746 W).

Example 8-4_____

What average power must be produced by a crane in order to raise a 2 500 lb load through a vertical height of 120 ft in 30 s?

Solution:

The minimum applied force upwards must equal the weight of the load. Thus

$$\bar{P} = \frac{W}{t} = \frac{F_A s \cos \theta}{t} = \frac{(2\ 500\ \text{lb})(120\ \text{ft}) \cos 0}{30\ \text{s}}$$

$$= 10\ 000\ \frac{\text{ft·lb}}{\text{s}} = \frac{10\ 000}{550}\ \text{hp} = 18\ \text{hp}$$

Example 8-5_____

What minimum horsepower must a pump deliver in order to raise 38.5 kL of water per minute from the bottom of a 35 m well? Neglect losses of power.

Solution:

The mass of 38.5 kL of water is 38.5 kg, and the pump must exert a minimum upward force equal to the weight of this water. Thus

$$P = \frac{F_A s \cos \theta}{t} = \frac{(38.5\ \text{kg})(9.81\ \text{m/s}^2)(35\ \text{m})}{60\ \text{s}}$$

$$= 220\ \text{W}$$

or

$$P = (220\ \text{W})(1.34 \times 10^{-3}\ \text{hp/W}) = 0.30\ \text{hp}$$

The power of an agent may vary, but in many cases the applied force F_A remains constant with respect to time. In some time interval t, if the applied force $\mathbf{F}_A$ is constant and inclined at some angle θ to the corresponding displacement $\mathbf{s}$, the average power developed

$$\bar{P} = \frac{F_A s \cos \theta}{t} = F_A \cos \theta \left(\frac{s}{t}\right) = F_A v \cos \theta$$

or

$$P = \mathbf{F}_A \cdot \mathbf{v} \qquad (8\text{-}3)$$

where $\mathbf{v}$ is the velocity.

*Example 8-6*_____

A car travels in a straight line at a constant speed of 80.0 km/h. If the sum of all retarding forces is 720 N, what is the average power developed by the engine?

Solution:

The engine must develop a force that balances the retarding forces $F_A = 720$ N in the direction of motion. Therefore

$$P = F_A v \cos \theta = (720 \text{ N}) \left\{\frac{80.0 \times 10^3 \text{ m}}{3\,600 \text{ s}}\right\} \cos 0$$
$$= 1.60 \times 10^4 \text{ W}$$

or

$$P = (1.60 \times 10^4 \text{ W})(1.34 \times 10^{-3} \text{ hp/W}) = 21.4 \text{ hp}$$

Electric power is usually measured with a wattmeter that utilizes a dynamometer principle. Dynamometers are also used to measure the mechanical output power of a motor, but first all the mechanical power is transformed into electric power in a heavy-duty generator.

Mechanical output power of a motor is often used to produce a rotation in a drive shaft; this type of output power may be measured with a *prony brake*. A drum of radius r is rigidly attached so that it rotates with the drive shaft, and a belt is suspended from two balances so that it makes firm contact with the drum (Fig. 8-5). As the drum rotates at a fixed rate, the friction force between the belt and the drum produces a difference ΔF

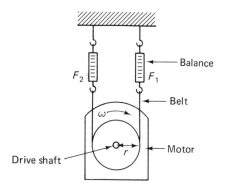

Figure 8-5
The prony brake.

in the readings F_1 and F_2 indicated on the balances. This difference in force

$$\Delta F = (F_1 - F_2)$$

is equivalent to the force exerted on the circumference of the drum by the motor. When the drum has completed one rotation, the work done by the motor $W = \Delta F \times$ (circumference of the drum) $= \Delta F(2\pi r)$. If the drive shaft (drum) makes ω complete revolutions per minute, the average power developed in each minute

$$\bar{P} = (\Delta F)(2\pi r)\omega \qquad (8\text{-}4)$$

In the British Engineering System

$$1 \text{ hp} = 550 \text{ ft} \cdot \text{lb/s} = 33\,000 \text{ ft} \cdot \text{lb/min}$$

Therefore the rotational horsepower developed (which is called the *brake horsepower*) is given by:

$$P = \frac{\Delta F 2\pi r \omega}{33\,000} \text{ bhp} \qquad (8\text{-}5)$$

where ΔF is in pounds, r is in feet and ω is expressed in revolutions per minute.

*Example 8-7*_____

The drum of a prony brake has a radius of 1.50 ft. When the drum rotates at a rate $\omega = 3\,000$ rev/min, the readings on the balances are 100 lb and 420 lb. Determine the brake horsepower of the motor.

Solution:

$$\bar{P} = \frac{\Delta F(2\pi r)\omega}{33\ 000}$$

$$= \frac{(420 - 100)(2\pi \times 1.50)(3\ 000)}{33\ 000} \text{bhp}$$

$$= 274 \text{ bhp}$$

Problems

7. If 12.0 kg of water are raised to a height of 30.0 m in 1.00 min by a pump, (a) how much work is done? (b) What is the average power output of the pump?

8. A conveyor belt can pull a 100 lb box at a constant velocity 30.0 ft up a 30° incline in 5.00 s. (a) What is the work done? (b) What is the average power developed? Ignore friction.

9. A 3 200 lb car climbs a hill at a constant speed of 30 mi/h. If the grade of the hill is 5% and 60% of the power developed by the car is used to overcome friction, determine the total output horsepower of the motor.

10. A train travels along a straight track with a constant speed of 45 mi/h. If the total retarding force is 12 000 lb, what is the average power developed?

11. Determine the power required to lift a 100 kg load vertically through 25 m in 20 s.

12. What minimum horsepower must a pump deliver in order to raise 12 U.S. gallons (1 gal of water weighs 8.33 lb) of water per minute from the bottom of a 90 ft well? Neglect losses of power.

13. How many liters of water per minute can a 0.50 hp pump raise from the bottom of a 28 m well? Neglect power losses.

14. An electric motor is used to raise an 850 kg loaded elevator vertically through 75 m in 45 s. What minimum power is produced by the motor if 20% is lost?

15. What average power is developed by the engines of a 28 t aircraft as it accelerates from rest to a takeoff speed of 175 km/h in 25 s if 40% of the power is used to overcome retarding forces?

16. Determine the power required to pull a 125 kg load along a horizontal surface at a constant speed of 12 m/s if the coefficient of kinetic friction is 0.25.

17. The drum of a prony brake has a diameter of 2.5 ft. What is the output brake horsepower of a motor if the spring balances read 300 lb and 150 lb when the drum rotates at 2 000 rev/min?

8-3 ENERGY

A flash of light, a nuclear explosion, a compressed spring, boiling water and sound are just a few examples of the many different forms of an entity called energy, which is common to all matter. *Energy* is defined as the capacity to do work; it is therefore closely related to the concept of work and is expressed in the same units, i.e. ft·lb and joules. However in electricity special units of work and energy are frequently used. A *kilowatt hour* (kW·h) is the work done (or energy delivered) in one hour by an agent which delivers an average power of one kilowatt.

$$1 \text{ kW·h} = (1 \times 10^3 \text{ W})(3\ 600 \text{ s}) = 3.6 \times 10^6 \text{ J}$$

Similarly a *horsepower hour* (hp·h) is the work done in one hour by an agent delivering an average of one horsepower.

In general, power is the time rate of doing work or the time rate of change of energy. Thus

$$P = \frac{W}{t} \quad \text{or} \quad P = \frac{E}{t} \qquad (8\text{-}6)$$

Example 8-8

Determine the cost of running a 1 200 W electric heater for 3.0 h if electricity costs 5.3 ¢/(kW·h).

Solution:

The energy consumed

$$E = Pt = (1.2 \text{ kW})(3.0 \text{ h}) = 3.6 \text{ kW} \cdot \text{h}$$

Thus the cost $= (3.6 \text{ kW} \cdot \text{h})\{5.3 \text{ ¢}/(\text{kW} \cdot \text{h})\} = 19 \text{ ¢}$

In mechanics, objects may do work by virtue of their motion, state or position. Falling water possesses energy by virtue of its motion. This energy may be used to do work by turning turbines that generate hydroelectric power. A raised pile driver has the ability to do work because of its position; if it is allowed to fall, it can drive a pile into the ground. Finally a spring in a compressed state can do work when it is released.

There are two forms of mechanical energy, kinetic and potential.

Kinetic Energy

Kinetic energy is the energy that an object possesses as a result of its motion. When work is performed on an object, some of that work may accelerate it, changing its speed. If a body of mass m is displaced a distance s in the direction of an applied force $\mathbf{F}_A$, the work done on the body

$$W = F_A s \cos 0° = F_A s \qquad (8\text{-}7)$$

According to Newton's second law, if there are no frictional losses, the applied force results in an acceleration $\mathbf{a}$ of that body

$$\text{net } \mathbf{F} = \mathbf{F}_A = m\mathbf{a} \qquad (8\text{-}8)$$

Combining Eqs. 8-8 and 8-9, we obtain:

$$W = mas \qquad (8\text{-}9)$$

The work done by the applied force accelerates the body from an initial speed v_0 to a final speed v_1, which may be determined from Eq. 3-13

$$v_1^2 = v_0^2 + 2as$$

Thus

$$as = \frac{v_1^2 - v_0^2}{2} \qquad (8\text{-}10)$$

Substituting Eq. 8-10 in Eq. 8-9, we obtain the work done

$$W = m\left(\frac{v_1^2 - v_0^2}{2}\right) = \frac{mv_1^2}{2} - \frac{mv_0^2}{2} \qquad (8\text{-}11)$$

The product of one-half of the mass m and the square of the speed v of the body is a scalar quantity, which is called its *kinetic energy* E_k:

$$E_k = \tfrac{1}{2}mv^2 \qquad (8\text{-}12)$$

Therefore the body acquires a change in kinetic energy as a consequence of the work done by the applied force. Work must be done on an object in order to increase its kinetic energy. An object may reduce its kinetic energy by doing work on its surroundings.

Example 8-9

An electron in an oscilloscope is uniformly accelerated from rest to a final speed $v_1 = 6.00 \times 10^6$ m/s in a distance $s = 1.50$ cm. Determine (a) the final kinetic energy E_{k1} of the electron, (b) the average force F_A on the electron, (c) the power P required.

Solution:

(a) $E_{k1} = \tfrac{1}{2}mv_1^2$
$= \tfrac{1}{2}(9.11 \times 10^{-31} \text{ kg})(6.00 \times 10^6 \text{ m/s})^2$
$= 1.64 \times 10^{-17}$ J

(b) The work done W is equal to the change in kinetic energy.

$$W = F_A s = \tfrac{1}{2}mv_1^2 - 0$$

Thus

$$F_A = \frac{E_{k1}}{s} = \frac{1.64 \times 10^{-17} \text{ J}}{1.50 \times 10^{-2} \text{ m}} = 1.09 \times 10^{-15} \text{ N}$$

(c) $$s = \left(\frac{v_1 + v_0}{2}\right)t$$

Thus

$$t = \frac{2s}{v_1 + v_0} = \frac{2(1.50 \times 10^{-2} \text{ m})}{(6.00 \times 10^6 \text{ m/s} + 0 \text{ m/s})}$$
$$= 5.00 \times 10^{-9} \text{ s}$$

Therefore

$$P = \frac{W}{t} = \frac{E_{k1}}{t} = \frac{1.64 \times 10^{-17} \text{ J}}{5.00 \times 10^{-9} \text{ s}}$$
$$= 3.28 \times 10^{-9} \text{ W}$$

Potential Energy

Potential energy is the energy possessed by virtue of position or state.

An object can do work if it is allowed to fall in a gravitational field, which is said to exist at any location where a mass would experience a gravitational force (Chapter 10). If an object is raised vertically above the earth's surface, its potential energy increases by virtue of its new position because it can do work by falling back to its original location. Consider the work W done in raising an object of mass m at a constant velocity through a relatively small height h above the earth. Since there is no acceleration, the mass is in equilibrium; if there is no friction or air resistance, the applied force F_A is equal and opposite to the weight $\mathbf{w} = m\mathbf{g}$ of the object. Thus the work done

$$W = F_A s \cos \theta = wh \cos 0 = mgh \quad (8\text{-}13)$$

The *change* in the *gravitational potential energy* ΔE_p of the object is defined as

$$\Delta E_p = mgh = mg(h_f - h_i) = wh \quad (8\text{-}14)$$

where h_i and h_f are the initial and final heights above some reference surface, respectively. Therefore the work done, which is equal to the change in potential energy, is the product of the magnitudes of the weight and vertical displacement of the object. In the absence of air resistance, if the object is allowed to fall freely back to its original level, it will increase its speed and kinetic energy as it loses potential energy.

The choice of a reference level is purely arbitrary because only changes in potential energy are considered. Levels below the reference level are the result of negative vertical displacements; therefore the lowest position involved is often chosen as a reference because this choice gives positive values for the potential energy changes (see Problem 8-21). Horizontal changes of position with respect to the earth's surface result in no gravitational potential energy changes.

Example 8-10

Determine the change in the gravitational potential energy of a 2.00 kg mass (a) when it is raised to an altitude of 30.0 m from the surface of the earth, (b) when it is lowered to 20.0 m below the surface of the earth.

Solution:

For convenience, choose the reference level as the surface of the earth.

(a) $\Delta E_p = mg(h_f - h_i)$
$= (2.00 \text{ kg})(9.81 \text{ m/s}^2)(30.0 - 0) \text{ m}$
$= 589 \text{ J}$

(b) $\Delta E_p = mg(h_f - h_i)$
$= (2.00 \text{ kg})(9.81 \text{ m/s}^2)(-20.0 - 0) \text{ m}$
$= -392 \text{ J}$

The negative sign indicates a loss of potential energy as the object loses altitude.

In the previous discussion of gravitational potential energy, since the change in height h was small, the acceleration due to gravity $\mathbf{g}$ was approximately constant in magnitude. If the change in height is relatively large, the acceleration due to gravity changes significantly in accordance with the universal law of gravitation. This will be discussed in Chapter 10.

Compressed or extended springs and wires may also do work (by firing projectiles for example). They possess elastic potential energy due to their state of distortion. In accordance with Hooke's law, the magnitude of the external force F_A required to distort the spring or wire is directly proportional to the amount of the distortion s:

$$F_A = ks$$

where k is the spring constant. The force required to distort the spring or wire by an amount s varies from zero (no distortion) to a maximum of $F_{max} = ks$ in the direction of the distortion. Thus the average applied force

$$\bar{F}_A = \tfrac{1}{2}(0 + F_{max}) = \tfrac{1}{2}ks$$

The work done

$$W = F_A s \cos \theta = (\bar{F}_A s) \cos 0 = \tfrac{1}{2} F_{max} s$$
$$= \tfrac{1}{2}(ks)s = \tfrac{1}{2} ks^2 = E_p \qquad (8\text{-}15)$$

which is equivalent to the *elastic potential energy* stored.

Problems

18. What is the kinetic energy of a 200 lb man if he is running with a speed of 10 mi/h?

19. How much work must be done in order to throw a 0.50 kg object with a speed of 30 m/s?

20. What is the potential energy of a 5.0 kg mass with respect to the earth's surface if it is at a height of 30 m?

21. Determine the change in potential energy when a 50 lb object is (a) raised 100 ft above the earth's surface, (b) lowered 50 ft below the earth's surface. Choose the lowest position as the reference level.

22. A 3 000 lb car is accelerated uniformly from rest to 45 mi/h in a distance of 1 000 feet. (a) What was the average net accelerating force? (b) If it took 14 s, what was the average power developed?

23. How much does it cost to run a 100 W light bulb for 24 h if electricity costs 5.0¢/(kW·h)?

24. If electricity costs 6.0¢/(kW·h), how much does it cost to run a $\tfrac{1}{2}$ hp pump for 12 h?

25. How much work is done when a 5.0 kg load stretches a wire by 2.3 mm?

26. Determine the change in potential energy when a 5.0 lb object is raised from an altitude of 70 ft to an altitude of 150 ft above the earth's surface. Take the earth as the reference level.

27. Determine how much work is done when a 300 lb load compresses a metal rod by 0.25 in.

8-4 CONSERVATION OF ENERGY

There are many forms of energy; some examples are mechanical, electrical, heat, light, sound and nuclear. Energy may be used to heat water; a nuclear explosion converts mass, which is itself a form of energy, into heat, light, sound and radiation. The investigation of energy transformations led to the discovery of one of the most useful physical laws, the *law of conservation of energy*. The law may be stated as:

> *The total energy in an isolated system is a constant. In other words, in a system that is isolated from external influences, energy is neither created not destroyed but may change from one form to another.*

There are many direct applications of conservation of energy in mechanics since systems often interchange their kinetic and potential energies. In these systems, mechanical energy is conserved, and the energy conservation law may be expressed as:

$$\Delta E_k + \Delta E_p + E_f = 0 \qquad (8\text{-}16)$$

where ΔE_k and ΔE_p represent changes in kinetic and potential energy, respectively, and E_f is the energy lost to friction, which usually results in heating the system.

The conservation of energy may be used to solve some relatively complex motions.

Example 8-11

A simple pendulum is constructed by suspending a bob from the end of a string. If the string is attached to a fixed point and the bob is displaced through an arc so that it is 3.0 in above its lowest position (Fig. 8-6) and then released, what is the speed of the bob at the lowest point of its swing? Assume no losses to friction.

Solution:

If the lowest position of the bob is chosen as the reference level for the gravitational potential energy, from the conservation of energy

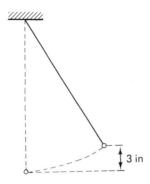

3 in

Figure 8-6

$\Delta E_k + \Delta E_p = \frac{1}{2}m(v_1^2 - v_0^2) + mg(h_f - h_i) = 0$

At the top of the arc, the initial speed $v_0 = 0$; at the bottom of the swing, $h_f = 0$. Therefore

$$\frac{1}{2}m(v_1^2 - 0) + mg(0 - h_i) = 0$$

and

$$v_1 = \sqrt{2gh_i} = \sqrt{2(32 \text{ ft/s}^2)(\tfrac{1}{4} \text{ ft})} = 4.0 \text{ ft/s}.$$

Example 8-12

Each minute, 5.25×10^6 L of water fall 35.0 m through a penstock of a dam onto a water wheel of an electric generator. If 20% of the kinetic energy of the water is converted into electric energy, what power does the generator develop?

Solution:

1.0 L of water has a mass of 1.0 kg. If mechanical energy is conserved, the water would acquire a kinetic energy equal to its loss of potential energy. Thus the power developed in the generator

$$P = (0.20)\frac{E_k}{t} = (0.20)\frac{mgh}{t}$$

$$= \frac{(0.20)(5.25 \times 10^6 \text{ kg})(9.81 \text{ m/s}^2)(35.0 \text{ m})}{60.0 \text{ s}}$$

$$= 6.01 \times 10^6 \text{ W} \quad \text{or} \quad 6.01 \text{ MW}.$$

Problems

28. What is the change in potential energy of 50 U.S. gallons of water when it falls down a 100 ft waterfall? One gallon of water weighs 8.3 lb.

29. Determine the speed at which water hits the bottom of a 120 m waterfall. Assume no friction.

30. A stone is dropped from a height of 250 ft and falls to the earth. What is its speed just before it lands if there are no friction losses?

31. A large simple pendulum is constructed by suspending a cannon ball from the end of a strong wire. The cannon ball is displaced through an arc so that it is 2.5 m above its original lowest position and then is released. If there are no friction losses, determine (a) the maximum speed of the cannon ball and (b) the speed of the cannon ball when it is 1.2 m above its lowest position.

32. A 20.0 kg block is released from rest at the top of a 35° incline; after sliding 15.0 m down the incline, it has a speed of 10.5 m/s. Determine (a) the average kinetic friction force and (b) the coefficient of kinetic friction between the block and the incline.

33. An 8.0 lb block is released from rest at the top of a 30° incline; after sliding 20 ft down the incline, it has a speed of 20 ft/s. Determine (a) the average kinetic friction force F_k and (b) the coefficient of kinetic friction μ_k between the block and the incline.

34. A 25 lb block is pushed up a 20° incline that is 15 ft long. If the coefficient of kinetic friction between the block and the incline is 0.32 and the block was initially at rest at the bottom of the incline, determine the minimum average force that will give the block a final speed of 8.0 ft/s at the top of the incline.

35. A 1.0 ton pile driver falls 15.0 ft to a pile that is driven 0.65 ft into the ground by the impact. Determine the average force exerted by the driver on the pile.

REVIEW PROBLEMS

1. A pump raises 40 L of water per minute from the bottom of a 35 m well. (a) How much work is done in one minute? (b) What horsepower is developed by the pump?

2. A motor pulls a 1 500 kg load on skids 25 m up a 15° incline in 12 min at a constant velocity. If the coefficient of kinetic friction is 0.25, (a) how much work is done? (b) What minimum horsepower does the motor deliver?

3. If the density of water is 62.4 lb/ft³, what is the minimum power of a pump if it raises 5.2 ft³ of water through 25 ft in 15 s?

4. Calculate the power required to pull a 3 200 lb crate along a horizontal road at 30 mi/h if the coefficient of kinetic friction is 0.20.

5. What average power is developed by the engines of a 3.5 t aircraft as it accelerates from rest to a take-off speed of 180 km/h in 30 s if 40% of the power is used to overcome retarding forces?

6. Determine the cost of running a 1.5 kW electric heater for 5.0 h if electricity costs 6.2 ¢/(kW·h).

7. What is the kinetic energy of an electron, which has a mass of 9.1×10^{-31} kg, if its speed is 6.5×10^6 m/s?

8. How much work must be done in order to throw a 2.0 oz ball with a speed of 15 ft/s?

9. What is the change in the potential energy of 500 kg of water when it falls down a 100 m waterfall?

10. (a) How much work is done when a 12 lb load stretches a wire by 0.48 in? (b) What is the spring constant of the wire?

11. A 400 lb block is pushed with a constant velocity up a 15° incline through a vertical height of 6.0 ft in 1.0 min. If the coefficient of kinetic friction between the block and the incline is 0.25, find (a) the minimum applied force parallel to the incline, (b) the work done, (c) the average power expended, (d) the change in gravitational potential energy of the block, (e) the energy lost due to the friction.

Work must be done by some applied force in order to change the motion of any object. All moving objects possess kinetic energy by virtue of their motion. Momentum is another property that a moving object has by virtue of its mass and velocity.

Momentum is different from kinetic energy because it is an indication of the effect that it will have on some other object during an interaction. For example, though interactions between objects often occur during very brief time intervals, they may involve extremely large forces. The conservation of energy enables us to solve many types of motion, but in many interactions, while the total energy is conserved, many of the energy forms are changed, and mechanical energy is not always conserved. In a collision between two objects, some mechanical energy may be converted into heat, sound, light, etc. However, collisions, recoils and even rocket propulsion may be described in terms of momentum changes and of an important conservation law called the *conservation of linear momentum*. This conservation law is actually a form of Newton's third law of motion, and it involves the principles of action and reaction.

9

Momentum

9-1 IMPULSE AND MOMENTUM

The *linear momentum* **p** of an object is a vector quantity that is defined as the product of its mass m and velocity **v**:

$$\mathbf{p} = m\mathbf{v} \qquad (9\text{-}1)$$

This term is an indication of the impetus that the moving object possesses; it can be used to determine the effect that the object has during a collision or interaction with some other object. For example, a heavy truck, even moving at a slow speed, has a relatively large linear momentum, and it can produce large forces during a collision.

In Chapter 4, we introduced Newton's second law of motion for the special case where the mass m was constant. Newton actually expressed the second law of motion as: the net force **F** acting on an object is equal to its time rate of change of linear momentum.*

$$F = \frac{\Delta p}{t} = \frac{\Delta(mv)}{t} \qquad (9-2)$$

where $\Delta p = \Delta(mv)$ is the change in linear momentum during the elapsed time t. Therefore

$$Ft = \Delta p = \Delta(mv) \qquad (9-3)$$

The product of the net force **F** and the time t for which it acts is called the *impulse*. Hence

Impulse = change in linear momentum

Therefore the units of impulse and linear momentum both have dimensions of (mass)(length)/time or (force)(time), i.e. $kg \cdot m/s = N \cdot s$ in SI and $slug \cdot ft/s = lb \cdot s$ in the British Engineering System. If the mass m is constant and an object changes its velocity from v_0 to v_1 in an elapsed time t, Eq. 9-3 reduces to:

$$Ft = m(v_1 - v_0) \qquad (9-4)$$

Example 9-1

A 0.250 kg ball is thrown with an initial speed of 20.0 m/s to a batter who hits it back along its original path with a speed of 30.0 m/s. If the duration of the impact was 10.0 ms, what was the magnitude of the average force on the ball?

Solution:

The ball changes direction after impact; therefore, if the final direction of the ball's motion is taken as positive, the final velocity $v = 30.0$ m/s, and the initial velocity $u = -20.0$ m/s. From Eq. 9-4, we obtain

---

*Newton called linear momentum "the quantity of motion." In calculus, this expression is written as:

$$F = \frac{dp}{dt} = \frac{d(mv)}{dt}$$

$$F = \frac{m(v-u)}{t} = \frac{(0.250\,kg)[(30.0\,m/s) - (-20.0\,m/s)]}{0.0100\ s}$$
$$= 1.25 \times 10^3\ N$$

Problems

1. Determine the magnitude of the linear momentum of an electron moving at 6.0×10^6 m/s.

2. Calculate the linear momentum of a 3 200 lb car that travels east at 60 mi/h.

3. A 250 lb pile driver is released from a height of 20 ft above a pile. (a) Determine the magnitude of the linear momentum of the pile driver just before it strikes the pile. (b) If the impact lasts 0.030 s, what is the average force exerted on the pile?

4. A 6.0 oz ball is thrown with an initial speed of 60 ft/s to a batter who hits it back along its original path with a speed of 90 ft/s. If the duration of the impact was 12 ms, what was the average force on the ball?

9-2 CONSERVATION OF LINEAR MOMENTUM

Consider an interaction between two objects with masses m_1 and m_2, which have velocities v_1 and v_2, respectively. Newton's third law (action and reaction) implies that the average force $\mathbf{F}_{21}$ exerted by m_2 on m_1 is equal and opposite to the average force $\mathbf{F}_{12}$ exerted by m_1 on m_2.

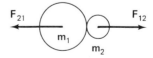

Figure 9-1

Thus

$$\mathbf{F}_{12} + \mathbf{F}_{21} = 0 \quad \text{or} \quad \mathbf{F}_{12} = -\mathbf{F}_{21}$$

Since the duration t of the interaction must be the same for each object

$$F_{12}t + F_{21}t = 0 \quad \text{or} \quad F_{12}t = -F_{21}t$$

That is, the impulse given to m_1 is equal and opposite to that on m_2. Therefore, using Eq. 9-3, we obtain:

$$F_{12}t + F_{21}t = \Delta p_1 + \Delta p_2$$
$$= \Delta(m_1 v_1) + \Delta(m_2 v_2) = 0 \quad (9-5)$$

where $\Delta p_1 = \Delta(m_1 v_1)$ and $\Delta p_2 = \Delta(m_2 v_2)$ are the changes in linear momentum. Consequently, if m_1 and m_2 change momentum values from p_1^0 and p_2^0 to p_1^1 and p_2^1 during the interaction, respectively,

$$(p_1^1 - p_1^0) + (p_2^1 - p_2^0) = 0$$

and

$$p_1^1 + p_2^1 = p_1^0 + p_2^0 \quad (9-6)$$

Equation 9-6 represents an extremely important physical law called the law of *conservation of linear momentum*. In general, this law may be stated as: *in a closed system, the total linear momentum is constant*. In other words, if there is no net external force, the total linear momentum before an action is equal to the total linear momentum after the action.

$$(\Sigma m_i v_i)_{\text{before}} = (\Sigma m_i v_i)_{\text{after}} \quad (9-7)$$

This law may be used to describe interactions, such as collisions and recoils, between any number of objects in a system. It must be remembered that linear momentum is a *vector quantity*; therefore the directions of motion must be considered.

In the special case where the masses m_1 and m_2 are constant and the velocities change from u_1 and u_2 to v_1 and v_2 during the interaction for two masses, these equations reduce to:

$$m_1 v_1 + m_2 v_2 = m_1 u_1 + m_2 u_2$$

or

$$m_1(v_1 - u_1) + m_2(v_2 - u_2) = 0 \quad (9-8)$$

The weights of the objects are $w_1 = m_1 g$ and $w_2 = m_2 g$; therefore Eq. 9-8 can also be written as:

$$w_1(v_1 - u_1) + w_2(v_2 - u_2) = 0 \quad (9-9)$$

since g cancels.

Example 9-2

An 8.0 lb rifle fires a 0.10 oz bullet with a muzzle velocity of 2 560 ft/s towards a target. If the rifle was initially at rest, what is its recoil velocity?

Solution:

The initial velocities of the bullet and gun u_B and u_g, respectively, were zero. If the direction of motion of the bullet is taken as positive, the final velocity of the bullet $v_B = 2\ 560$ ft/s. Thus in Eq. 9-9,

$$w_B(v_B - 0) + w_g(v_g - 0) = 0$$

or

$$v_g = \frac{-w_B v_B}{w_g} = -\frac{\left(\frac{0.10}{16}\text{ lb}\right)(2\ 560\text{ ft/s})}{8.0\text{ lb}} = -2.0\text{ ft/s}$$

That is, the gun recoils at 2.0 ft/s in the opposite direction to the bullet.

Rocket and jet propulsion involves the ejection of hot gases at high speeds from the rear of the craft, causing it to recoil in the forward direction. Air resistance, variations of the acceleration due to gravity and changes in the total mass of the craft as it ejects fuel must be considered. In most cases, calculus must be used to solve for the motion. However, if a mass Δm of gas is ejected at a constant speed v *relative to the rocket* in an elapsed time t, the average thrust that is developed by the engines

$$F = v\frac{\Delta m}{t} \quad (9-10)$$

Example 9-3

A rocket with a fuel has a total mass of 2.80 t is launched vertically upwards. Determine the average thrust developed by its engines as they exhaust gases at a rate of 1.60×10^4 kg/s and at a constant speed of 2.13×10^3

m/s relative to the rocket. (See Fig. 9-6 at the end of the chapter.)

Solution:

Since v is constant:

$$F = \frac{\Delta(mv)}{t}$$

$$= v\frac{\Delta m}{t} = (2.13 \times 10^3 \text{ m/s})(1.60 \times 10^4 \text{ kg/s})$$

$$= 3.41 \times 10^7 \text{ N}$$

Problems

5. A 4.50 kg rifle fires a 15.0 g bullet with a muzzle velocity of 800 m/s towards a target. If the rifle was initially at rest, what is its recoil velocity?

6. While completing a space walk, an astronaut, whose total mass including equipment is 140 kg, throws a 5.0 kg object at a speed of 5.0 m/s. If the astronaut was initially at rest, determine his final speed.

7. A 90 lb boy and a 70 lb girl are at rest on frictionless ice skates. If the boy pushes the girl away at a speed of 4.0 ft/s, what happens to the boy?

8. A rocket with fuel has a total mass of 2.2 kt when it fires its engines. (a) Determine the average thrust developed by its engines as they exhaust gases at a rate of 13 000 kg/s and at a constant speed of 1.5 km/s relative to the rocket. (b) What is the instantaneous acceleration of the rocket after 1.5 min? Assume that there are no gravitational forces present.

9-3 COLLISIONS

While momentum and energy are always conserved in a closed system, some mechanical kinetic energy is often converted into other energy forms during an interaction. In some cases, the total kinetic energy of the system actually increases as particles are projected at high speeds from their original positions as a result of explosions or radioactive decay. However, in collisions the total kinetic energy usually is either constant or reduced.

$$(E_k)_{\text{before}} \geq (E_k)_{\text{after}} \qquad (9\text{-}11)$$

A collision in which kinetic energy is conserved is called an *elastic collision*; this corresponds to the equality in Eq. 9-11. If kinetic energy is not conserved, the collision is termed *inelastic*, and the final total kinetic energy of the system is usually less than the initial total value. Atomic, nuclear or fundamental particles occasionally collide elastically, and collisions between very hard spheres are approximately elastic; but all other objects collide inelastically to some extent.

Consider a one dimensional collision between two nonrotating masses m_1 and m_2, which move with nonrelativistic velocities $\mathbf{u}_1$ and $\mathbf{u}_2$ before and $\mathbf{v}_1$ and $\mathbf{v}_2$ after the collision, respectively. Since the collision is one dimensional, the motion of the masses is always parallel or antiparallel along an extended line joining their centers (Fig. 9-2).

In Case 1, the masses always move in the same direction. If this direction is chosen as the positive reference, the speeds of the masses may be considered instead of their velocities. Thus, since linear momentum is conserved, from Eq. 9-8

$$m_2(u_2 - v_2) = m_1(v_1 - u_1)$$

and from Eq. 9-11,

$$\tfrac{1}{2}m_1u_1^2 + \tfrac{1}{2}m_2u_2^2 \geq \tfrac{1}{2}m_1v_1^2 + \tfrac{1}{2}m_2v_2^2$$

The negative ratio of the relative speed of approach $(v_1 - v_2)$ after the interaction to the relative speed of approach $(u_1 - u_2)$ before the interaction is defined as the *coefficient of restitution e* for the collision:

$$e = -\frac{v_1 - v_2}{u_1 - u_2} \qquad (9\text{-}12)$$

The coefficient of restitution has a value between zero and one; $e = 1$ for a perfectly

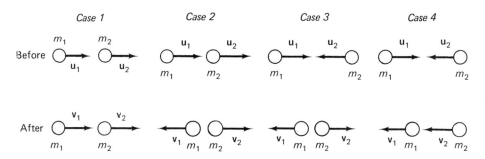

Figure 9-2

elastic collision, and $e = 0$ for a perfectly inelastic collision in which the objects adhere to each other after the collision has taken place.

For the other three cases in Fig. 9-2, if the direction of motion is reversed, the appropriate speed is given a negative value in Eq. 9-12.

Example 9-4

A 0.50 kg object is moving in a straight line with an initial speed of 2.0 m/s when it collides head-on with a a 1.0 kg object moving in the opposite direction with an initial speed of 5.0 m/s (Fig. 9-3). If the coefficient of restitution $e = 0.80$, what are the final velocities of the two objects?

Solution:

If the initial direction of motion of the 0.50 kg load is taken as positive, the velocity of the 1.0 kg load is given a negative value. Thus, from Eq. 9-12,

$$e(u_1 - u_2) = v_2 - v_1$$

or

$$0.80[2.0 - (-5.0)]\,\text{m/s} = 5.6\,\text{m/s} = v_2 - v_1 \quad (1)$$

From the conservation of linear momentum Eq. 9-8,

$$m_1(v_1 - u_1) + m_2(v_2 - u_2)$$
$$= 0.50\,\text{kg}(v_1 - 2.0\,\text{m/s}) + 1.00\,\text{kg}[v_2$$
$$- (-5.0\,\text{m/s})] = 0$$

or

$$0.50v_1 + v_2 + 4.0\,\text{m/s} = 0 \quad (2)$$

Substituting for v_2 from (1) into (2), we obtain

$$0.50v_1 + (5.6\,\text{m/s} + v_1) + 4.0\,\text{m/s} = 0$$

Hence

$$v_1 = \frac{-9.6\,\text{m/s}}{1.5} = -6.4\,\text{m/s}$$

Using this value in (1), we obtain

$$v_2 = 5.6\,\text{m/s} + v_1 = 5.6\,\text{m/s} - 6.4\,\text{m/s}$$
$$= -0.8\,\text{m/s}$$

Example 9-5

A ballistic pendulum consists of a 1.0 kg block that is suspended at rest from two light strings. When a 1.5 g bullet embeds itself in the block, the block swings through a vertical

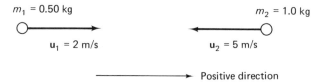

$m_1 = 0.50\,\text{kg}$ $u_1 = 2\,\text{m/s}$ $m_2 = 1.0\,\text{kg}$ $u_2 = 5\,\text{m/s}$

Positive direction

Figure 9-3

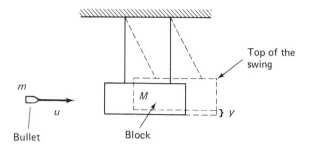

Figure 9-4
The ballistic pendulum.

displacement $y = 7.3 \text{ cm}$ (Fig. 9-4). Determine the speed u of the bullet just before it strikes the block.

Solution:

This is a perfectly inelastic collision; therefore after the collision the block and the bullet have the same speed v. If the initial level of the block is taken as the reference for gravitational potential energy, m is the mass of the bullet and M is the mass of the block, *once the bullet is embedded in the block*, mechanical energy is conserved as the block swings. Thus

$$(E_k)_{\text{before}} + (E_p)_{\text{before}} = (E_k)_{\text{after}} + (E_p)_{\text{after}}$$

or

$$\tfrac{1}{2}(m + M)v^2 + 0 = 0 + (m + M)gy$$

Therefore $v = \sqrt{2gy}$. Also from the conservation of momentum before and at the instant after the bullet embeds itself in the block

$$mu + 0 = (m + M)v$$

or

$$u = \left(\frac{m + M}{m}\right)v = \left(\frac{m + M}{m}\right)\sqrt{2gy}$$

$$u = \frac{(1.0 \text{ kg} + 1.5 \times 10^{-3} \text{ kg})}{(1.5 \times 10^{-3} \text{ kg})}$$
$$\cdot \sqrt{2(9.8 \text{ m/s}^2)(7.3 \times 10^{-2} \text{ m})}$$

$$= 800 \text{ m/s}$$

*Problems*_____

9. A $6.68 \times 10^{-27} \text{ kg}$ helium atom moves with a speed of 2.50×10^5 m/s and makes a head-on elastic collision with a station-ary atom whose mass is 1.67×10^{-27} kg. Determine the final speeds of both atoms.

10. A 0.25 lb steel ball moves with a speed of 20 ft/s and makes a head-on elastic collision with a 0.20 lb brass ball moving in the opposite direction with a speed of 15 ft/s. Determine the speeds of both balls after the collision.

11. A 6.68×10^{-27} kg helium atom moves with a speed of 6.20×10^6 m/s and makes a head-on elastic collision with a lithium atom whose mass is 11.7×10^{-27} kg. Determine the final speeds of both atoms (a) if the lithium was initially at rest, (b) if the lithium atom initially moves at a speed of 5.00×10^5 m/s in the same direction as the helium, (c) if the lithium initially moves at a speed of 4.80×10^6 m/s in the opposite direction to the helium.

12. A 0.20 kg ball is moving in a straight line with a speed of 5.0 m/s when it collides head-on with a 0.35 kg ball that is moving at a speed of 30 m/s. If the coefficient of restitution is 0.60, determine the final speed of each ball (a) if they initially travel in the same direction, (b) if they initially travel in opposite directions.

13. A rubber ball is allowed to fall freely in the earth's gravitational field from a height of 1.8 m above a flat horizontal surface. If the coefficient of restitution

between the ball and the surface is 0.40, determine how high the ball will bounce.

14. A 0.100 oz bullet is moving at 2 000 ft/s when it strikes the 6.25 lb block of a ballistic pendulum. (a) Determine the maximum vertical displacement of the block after the bullet is embedded. (b) Compare the initial and final mechanical energies of the system.

15. A 3 200 lb car is moving due east at a speed of 30 mi/h when it collides head-on with a 4 000 lb car; the two cars come to a dead stop. Determine the initial *velocity* of the 4 000 lb car.

9-4 TWO DIMENSIONAL ELASTIC COLLISIONS

The vast majority of collisions between two or more objects are not "head-on" but are glancing collisions where the particles move along different directions after the interaction. These collisions are of special importance in nuclear physics because the tracks of elementary particles may be photographed, and the analysis of these tracks enables a description of the collision and the particles

involved. The total change in the *components* of momentum must be zero *in any direction*.

Consider a particle of mass m_1 and initial velocity u_1 that impinges on a *stationary* mass m_2. After an *elastic* collision with m_2, m_1 has a velocity v_1, and m_2 a velocity v_2 (Fig. 9-5). Choose a convenient set of perpendicular axes parallel and perpendicular to the initial motion of m_1. Resolving the linear momentum into the perpendicular components, we obtain

Momentum	x-Component	y-Component
$m_1 u_1$ before	$m_1 u_1$	0
$m_1 v_1$ after	$m_1 v_1 \cos \theta_1$	$-m_1 v_1 \sin \theta_1$
$m_2 u_2$ before	0	0
$m_2 v_2$ after	$m_2 v_2 \cos \theta_2$	$m_2 v_2 \sin \theta_2$

Linear momentum is conserved in all directions. Thus equating the components of momentum before and after the interaction, we obtain

$$m_1 u_1 + 0 = m_1 v_1 \cos \theta_1 + m_2 v_2 \cos \theta_2 \tag{9-13}$$

$$0 + 0 = -m_1 v_1 \sin \theta_1 + m_2 v_2 \sin \theta_2 \tag{9-14}$$

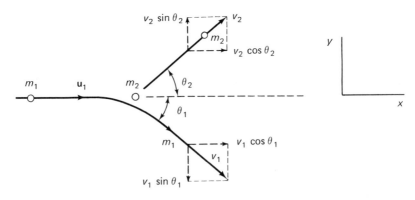

Figure 9-5
A two dimensional elastic collision.

The conservation of kinetic energy implies that:

$$m_1 u_1^2 + 0 = m_1 v_1^2 + m_2 v_2^2 \qquad (9\text{-}15)$$

Three unknown quantities may be determined from these three simultaneous equations.

REVIEW PROBLEMS

1. Determine the momentum of an 8.0 oz hockey puck travelling at 100 mi/h.

2. Calculate the impulse given a 30 kg projectile if it is accelerated from 20 m/s to 30 m/s.

3. A 6.0 lb rifle fires a 0.10 oz bullet with a muzzle velocity of 1 500 ft/s towards a target. Calculate (a) the impulse given the bullet and (b) the recoil velocity of the rifle.

4. A 0.036 kg golf ball is hit off a tee with an initial speed of 75 m/s. What magnitude impulse was it given?

5. A 6.4 oz hockey puck travels at 160 ft/s after it has been hit with a hockey stick. If the puck was initially at rest and the duration of the impact was 10.0 ms, (a) what was the impulse given the puck? (b) What was the average force during the impact?

6. A 2.50 g bullet is fired into a 2.00 kg block of a ballistic pendulum. If the bullet embeds itself in the block and the pendulum swings so that its center of gravity is raised vertically through 5.40 cm, determine the speed of the bullet just before it strikes the block.

7. A neutron of mass m and speed 1.0×10^6 m/s makes a head-on elastic collision with a stationary helium atom of mass $4m$. What are the final speeds of the neutron and the helium atom?

8. A 3 200 lb car moving due east at 60 mi/h collides with a 4 000 lb car moving due

Figure 9-6
Saturn 1B launch vehicle.
(NASA)

north at 45 mi/h. Determine the final velocity of the wreckage if it sticks together.

9. A rocket with a gross weight of 5.0×10^6 lb is launched vertically from the surface of the earth. If the engines exhaust gases at a rate of 1.2×10^3 slug/s at a speed of 1.0 mi/s relative to the rocket, what is the average thrust developed?

The special case of uniformly accelerated motion corresponds to motion with constant acceleration; the derived equations enable us to solve for straight line and projectile motions. However, many objects, such as planets orbiting the sun, electrons orbiting an atomic nucleus and aircraft, frequently accelerate in order to change the direction of their motion. If an object moves in a circle about some fixed point, it is said to be in *circular motion*, and an object performs *uniform circular motion* if it moves in a circle at a constant speed. Even though its speed is constant, an object in uniform circular motion continually changes its direction of motion, and therefore its velocity changes.

10-1 CENTRIPETAL FORCE AND ACCELERATION

Consider an object in a circular motion at distance r about some fixed point 0. If the object moves from some point A at time t_0 and arrives at some other point B at time t_1, its displacement **s** is the straight line from A to B (Fig. 10-1) and its average velocity:

$$\bar{\mathbf{v}} = \frac{\mathbf{s}}{t_1 - t_0} \qquad (10\text{-}1)$$

The direction of the average velocity is the same as that of the displacement **s**; the arc length AB through which the object moves will depend on the time interval $(t_1 - t_0)$. As we decrease the time interval, the displacement decreases and its direction approaches that of the tangent* to the circle at point A. By taking the limit as the time interval $(t_1 - t_0)$ approaches zero, we find that the *instantaneous velocity* of the object at instant t_0 at point A is

$$\mathbf{v} = \lim_{t_1 \to t_0} \frac{\mathbf{s}}{t_1 - t_0} \qquad (10\text{-}2)$$

---

*That is, the perpendicular to line OA (the radius).

10

Circular Motion

and **v** is in a direction that is tangential to the circle at A.

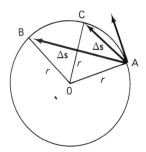

Figure 10-1
Instantaneous velocity is tangential to the circle.

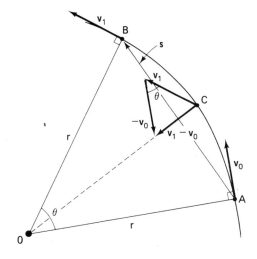

Figure 10-2

If an object moves in a circle at a constant speed, it is said to be in *uniform circular motion*. This type of motion, which is repeated at regular time intervals, is called a *periodic motion*. Other types of periodic motions are oscillations and vibrations (Chapter 17). In each case the time required to complete one *cycle* of the motion, e.g. a circle, or a complete to and fro vibration or oscillation, is called the *period T* of the motion. For uniform circular motion of radius r at a constant speed v, the period

$$T = \frac{\text{distance travelled}}{\text{speed}} = \frac{\text{circumference}}{\text{speed}} = \frac{2\pi r}{v}$$
$$(10\text{-}3)$$

Objects in uniform circular motion are continually changing their direction of motion as they orbit some fixed point; therefore their velocity changes, and they must be continually accelerated.

Consider an object that moves in uniform circular motion with a constant speed v at a distance r about a fixed point 0 (Fig. 10-2). If the object moves in a circular arc from point A at time t_0 where its instantaneous velocity was $\mathbf{v}_0$, to point B at time t_1 where

its instantaneous velocity was $\mathbf{v}_1$, its average acceleration

$$\bar{\mathbf{a}}_c = \frac{(\mathbf{v}_1 - \mathbf{v}_0)}{t_1 - t_0} = \frac{\Delta \mathbf{v}}{\Delta t} \qquad (10\text{-}4)$$

In order to find the *average acceleration*, we shall consider the vector difference $\Delta \mathbf{v} = (\mathbf{v}_1 - \mathbf{v}_0)$ at the middle of the time interval when the object is at point C. Since the instantaneous velocities $\mathbf{v}_0$ and $\mathbf{v}_1$ are always tangential to the circle, the angle θ between them is equal to the angle subtended at 0 by the arc AB. Also, in uniform circular motion, the magnitude of $\mathbf{v}_0$ and $\mathbf{v}_1$ is constant and equal to the speed v; therefore the triangle representing the vector subtraction of $\mathbf{v}_0$ from $\mathbf{v}_1$ is similar to the triangle OAB.* Thus $\Delta v/v = s/r$ and $\Delta v = vs/r$, where Δv is the magnitude of the vector difference $\Delta \mathbf{v} = \mathbf{v}_1 - \mathbf{v}_0$ and s is the magnitude of the displacement **s**. Thus the average acceleration is directed along the vector $(\mathbf{v}_1 - \mathbf{v}_0)$, and it has magnitude

---

*Equal-angled isosceles triangles.

$$\bar{a}_c = \frac{|\mathbf{v}_1 - \mathbf{v}_0|}{t_1 - t_0} = \frac{\Delta v}{(t_1 - t_0)} = \frac{v}{r} \frac{s}{(t_1 - t_0)}$$
$$(10\text{-}5)$$

Note that the direction of the change in velocity $\mathbf{v}_1 - \mathbf{v}_0$ and therefore the direction of the average acceleration $\mathbf{a}_c$ is towards the center 0 of the circle. Again, the instantaneous acceleration $\mathbf{a}_c$ at the instant t_0 is directed towards the center of the circle, and its magnitude is found by taking the limit as t_1 approaches t_0.

$$a_c = \underset{t_1 \to t_0}{limit} \frac{v}{r} \frac{s}{(t_1 - t_0)} = \frac{v}{r} \underset{t_1 \to t_0}{limit} \frac{s}{(t_1 - t_0)} = \frac{v^2}{r}$$
$$(10\text{-}6)$$

Since v and r are constant and the instantaneous *speed* at t_0 is

$$\underset{t_1 \to t_0}{limit} \frac{s}{(t_1 - t_0)} = v$$

This acceleration is called *centripetal acceleration* because it is directed towards the center of the circular orbit. Because the object is continually accelerating towards the center 0 of the circle, according to Newton's first law of motion, a corresponding force, called the *centripetal force* $\mathbf{F}_c$, must also exist. From Newton's second law, if the object performing circular motion has a mass m and a centripetal acceleration $\mathbf{a}_c$, it experiences a centripetal force

$$\mathbf{F}_c = m\mathbf{a}_c \qquad (10\text{-}7)$$

that is also directed towards the center 0 of the orbit; it has a magnitude

$$F_c = ma_c = \frac{mv^2}{r} \qquad (10\text{-}8)$$

where v is the tangential speed of the mass m and r is the radius of the orbit. For example, in order to whirl an object in a circle at the end of a string, it is necessary to keep pulling inwards on the string. If the string is released, the object flies off at a tangent to the circle. The faster the object is whirled, the greater the inward force required to maintain the circular motion. Note that, when an object of mass m performs uniform circular motion, its kinetic energy $E_k = \frac{1}{2} mv^2$ is also

constant since its speed v is constant. The centripetal force always acts perpendicular to the direction of motion; therefore, if friction is neglected, centripetal forces do no work.

$$W = F_A s \cos \theta = F_c s \cos 90° = 0$$

Centripetal forces must be exerted on any object if it is to perform a circular motion. These centripetal forces may be due to friction, gravitation, electric fields, magnetic fields and many other sources. They all have one thing in common however, they are directed towards the center of the orbit.

Cars are able to turn corners on flat roads only because of the friction force between the wheels and the road surface. If the magnitude of the friction force is not sufficient, the car will skid and not make the turn.

Example 10-1

Determine the minimum coefficient of friction μ between the tires and a flat horizontal road that would enable a 3 200 lb car to make a turn of radius $r = 500$ ft at a constant speed $v = 30$ mi/h.

Solution:

$v = 30$ mi/h $= 44$ ft/s. Draw the free-body diagram (Fig. 10-3). Since the car is in equilibrium in the vertical direction

$$\Sigma F_y = N - w = 0$$

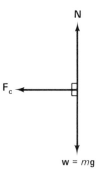

Figure 10-3

Free body diagram of a car on a horizontal road.

and

$$N = w = mg$$

The friction force $F_f = \mu N$ supplies the necessary centripetel force.

$$F_f = \mu N = \mu mg = \frac{mv^2}{r}$$

Thus

$$\mu = \frac{F_f}{N} = \frac{mv^2}{rmg} = \frac{v^2}{rg} = \frac{(44 \text{ ft/s})^2}{(500 \text{ ft})(32 \text{ ft/s}^2)} = 0.12$$

Note that μ is independent of the weight or mass of the car!

In order to reduce the possibility of cars skidding off a road because of a slippery surface or excessive speeds, roads are usually banked at some angle θ with respect to the horizontal (Fig. 10-4). In this case, the unbalanced *horizontal* component of the normal force $\mathbf{N}$ of the road on the car supplies at least a part of the centripetal force $\mathbf{F}_c$.

*Example 10-2*_____

At what optimum angle must a curve of radius 100 m be banked so that cars can turn the curve at a speed of 60.0 km/h without skidding (Fig. 10-4a)?

Solution:

Draw the free-body diagram of the car (Fig. 10-4b). The speed

$$60.0 \text{ km/h} = \frac{60.0 \times 10^3 \text{ m}}{3\ 600 \text{ s}} = 16.7 \text{ m/s}$$

Resolving the forces into their *horizontal* and *vertical* components, we obtain

Force	Horizontal x-Component	Vertical y-Component
Weight of car $\mathbf{w} = m\mathbf{g}$	0	$-mg$
Normal force $\mathbf{N}$	$N \sin \theta$	$N \cos \theta$

Since the car is required to turn the curve without changing its vertical position, it must be in equilibrium *vertically*. Thus

$$\Sigma F_y = -mg + N \cos \theta = 0$$

or

$$N \cos \theta = mg \qquad (1)$$

The net *horizontal* force must supply the centripetal force; therefore

$$\Sigma F_x = N \sin \theta = (mv^2)/r \qquad (2)$$

Dividing (2) by (1), we obtain,

$$\frac{N \sin \theta}{N \cos \theta} = \frac{(mv^2)/r}{mg}$$

Thus

$$\tan \theta = \frac{v^2}{rg} \qquad (10\text{-}9)$$

since $\sin \theta / \cos \theta = \tan \theta$.

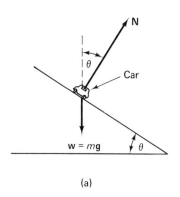

w = mg

(a)

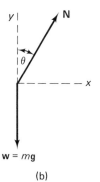

w = mg

(b)

Figure 10-4
Banking of a curve.

Note that the angle in Eq. 10-9 is independent of the mass or weight of the car. Thus

$$\tan \theta = \frac{v^2}{rg} = \frac{(16.7 \text{ m/s})^2}{(100 \text{ m})(9.81 \text{ m/s}^2)} = 0.284$$

and $\theta = 15.9°$.

If the driver moves either too fast or too slow around the curve, an additional source of centripetal force is required to prevent skidding. This force is usually supplied by the friction between the tires and the road.

Similar principles also apply to aircraft in flight. The wing of the aircraft is shaped so that, when it is propelled through the air, an upward force, called *lift*, is exerted on its wings (Chapter 13). When the aircraft flies in a straight horizontal path over the earth's surface, the lift exactly balances the weight of the aircraft.

In order to complete a turn without changing its altitude, the aircraft banks, dipping its wing in the direction of the turn (Fig. 10-5). Since the lift is always perpendicular to the wing, this results in an unbalanced horizontal force on the wings, which provides the centripetal force necessary for the turn. The vertical component of the lift balances the weight of the aircraft.

According to Newton's third law, a reaction force called a *centrifugal force*, which acts away from the center of the circle, is often introduced to balance the centripetal force. The centrifugal force does not act on the object in circular motion since these objects are not in equilibrium but are continually accelerating, changing their direction of motion. Newton's laws are not valid for accelerating systems, but they may be used to describe the motion of these systems.

The centrifugal force is due to the tendency of the orbiting object to continue in a straight line. For example, when an object is whirled on a string, the string exerts a centripetal force on that object, but the object exerts an equal and opposite centrifugal force on the string.

Problems_____

1. A model of a hydrogen atom pictures an electron in a circular orbit of radius 5.3×10^{-11} m about the atomic nucleus. If the speed of the electron is 2.2×10^6 m/s, determine the period of its orbit.

2. Determine the average speed of the earth about the sun if the average radius of its orbit is 1.50×10^{11} m and each orbit takes 365 days.

3. A model of the hydrogen atom pictures an electron of mass 9.1×10^{-31} kg in a circular orbit of radius 5.3×10^{-11} m about the nucleus. If the centripetal force due to the electrostatic attraction between the nucleus and the electron is 8.2×10^{-8} N, determine (a) the centripetal acceleration of the electron and (b) the speed of the electron.

4. A bicycle wheel of diameter 28 in rotates 10 times in 6.0 s. Determine (a) the speed of a point on its rim and (b) the centripetal acceleration of its rim.

5. An 8.0 oz ball is whirled in a horizontal circle at the end of a 3.0 ft long string. If the string has a tensile strength of 12 lb, determine the maximum speed that the

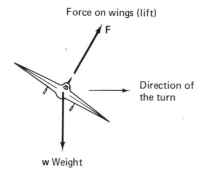

Force on wings (lift)

F

Direction of the turn

w Weight

Figure 10-5
Turning aircraft.

ball can attain without breaking the string.

6. Determine the maximum possible speed of the ball in the previous problem if it is whirled in a vertical circle.

7. A 50 g ball is attached to a 75 cm long string. What speed must the ball maintain if it is whirled in a horizontal circle and the string is inclined at an angle of 30° with respect to the horizontal?

8. (a) Determine the minimum coefficient of friction between the tires and a flat horizontal road that enables a 1 600 kg car to make a turn of radius 300 m at a constant speed of 25 km/h. (b) Calculate the acceleration of the car.

9. Calculate the optimum angle at which a curve of radius 700 ft should be banked so that cars can move around the curve at 60 mi/h without skidding.

10. A curve of radius 100 m is banked at an angle of 10° with respect to the horizontal. If the coefficient of friction between the road and the tires is 0.40, determine the maximum speed at which the car can travel and still complete the turn without skidding.

11. Determine the maximum speed of the car in the Problem 10 if the curve is banked the wrong way.

10-2 THE SOLAR SYSTEM AND KEPLER'S LAWS

The study of stars, planets and other celestial bodies is called *astronomy*. To an observer on the earth the planets and stars appear to move from east to west across the sky. There are five planets that are visible to the naked eye, and they appear to move in rather erratic paths. As a result of this apparent motion, early astronomers believed that the heavens rotated about the earth, which was the center of the universe. They were able to develop a calendar which synchronized these apparent celestial motions with the seasons on the earth.* In spite of some successes, the motions of the heavens could only be described by very complicated orbits.

After studying the theories of the ancient Greeks, Nicholas Copernicus (1475–1545) suggested that a simpler description of celestial motions could be obtained by assuming that the earth and other planets were in circular orbits around the sun and that the earth rotated about its axis.

The first person to make and record accurately the positions of the planets was the Danish astronomer Tycho Brahe (1546–1601). Using homemade sextants (telescopes were not invented until after his death), Brahe was able to achieve an accuracy of about 1/6 of a minute of arc. This information proved to be invaluable.

Brahe's observations of the planet Mars were analyzed after his death by his assistant Johannes Kepler (1571–1630). Originally Kepler adopted the Copernican system, but he found that the calculated positions of Mars differed from Brahe's observations by 8 minutes of arc. Since Kepler firmly believed the accuracy of Brahe's observations, he rejected Copernicus' theory and eventually determined that the planetary orbits were elliptical rather than circular. He developed three laws to describe planetary motions:

1. *The orbital law:* All planets move about the sun in elliptical orbits with the sun at one focus.** The closest orbital distance between the planet and the sun is called the *perihelion* and the farthest distance is the *aphelion.*

---

*The Chinese emperor Yao officially adopted 365¼ days as one year as early as 2317 BC!

**Even though the oribits are elliptical, they are nearly circular.

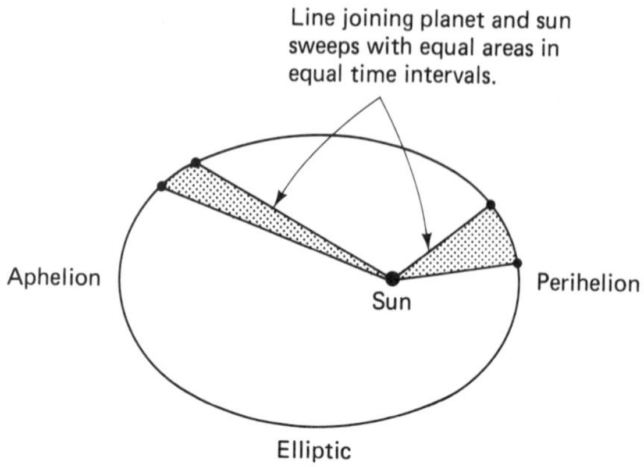

Figure 10-6
Planetary orbit about the sun which is at one focus.

2. *The area law:* A straight line joining the sun and any planet sweeps out equal areas in equal time intervals. This implies that the speed of the planet is greater when it is closer to the sun.

3. *The harmonic law:* The cube of the average distance d of any planet from the sun varies as the square of the period T of its orbit.

$$d^3 \propto T^2, \quad \text{or} \quad \frac{d^3}{T^2} = \text{constant} \quad (10\text{-}10)$$

Example 10-3

If the orbital period of the earth $T_e = 365.26$ d and its average distance from the sun $d_s = 1.495 \times 10^{11}$ m, what is the average distance d_m of Mars from the sun if its orbital period $T_m = 687.93$ d?

Solution:

$$\frac{d_m^3}{T_m^2} = \frac{d_e^3}{T_e^2}$$

Thus

$$d_m = \sqrt[3]{d_e^3 \left(\frac{T_m}{T_e}\right)^2}$$

$$= \sqrt[3]{(1.495 \times 10^{11} \text{ m})^3 \left(\frac{687.93 \text{ d}}{365.26 \text{ d}}\right)^2}$$

$$= 2.280 \times 10^{11} \text{ m}$$

Problems

12. If the orbital period of the earth is 365.26 d and its average distance from the sun is 1.495×10^{11} m, (a) what is the average distance of Venus from the sun if its orbital period is 255 d? (b) What is the orbital period of Mercury if its average distance from the sun is 5.79×10^{10} m? (c) What are the orbital speeds of Mercury and Venus about the sun?

10-3 NEWTON'S UNIVERSAL LAW OF GRAVITATION

While Kepler's laws describe planetary motions, they do not contain theoretical reasons for the orbits. Even after Kepler formulated these laws, scientists still believed that the motions of the heavens were quite different from the motions of objects at the earth's surface. It was not until several years later that Isaac Newton (1642–1727) discovered the basic law that governs these motions.

Newton began with the theories developed by Galileo Galilei (1564–1642) to analyze the flight of a projectile. However, he considered the projectile as if it were launched

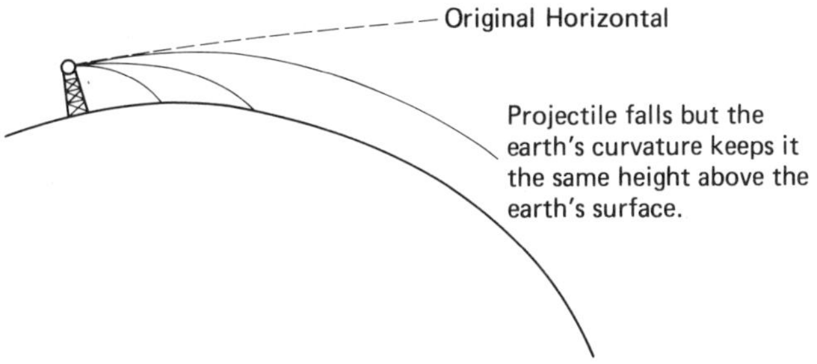

Figure 10-7
Trajectories of projectiles.

horizontally from a sphere (the earth) and returned to the sphere. He found that the trajectory would be elliptical rather than parabolic. If the projectile could be launched with a sufficient speed, it would orbit the earth because the earth's curvature would compensate for the amount that it falls (Fig. 10-7). Newton had finally discovered the connection between planetary motions and the motions of objects over the earth! He then used Kepler's laws to establish a formula.

Consider a planet of mass m in a circular orbit of redius r and period T at a constant speed v about the sun. From Kepler's third law:

$$r^3 = kT^2$$

where k is a constant. But the period

$$T = \frac{2\pi r}{v}$$

Therefore

$$r^3 = k\left(\frac{2\pi r}{v}\right)^2 = \frac{4\pi^2 r^2 k}{v^2}, \quad \text{or} \quad v^2 = \frac{4\pi^2 k}{r}$$

Consequently the centripetal force,

$$F_c = \frac{mv^2}{r} = \frac{4\pi^2 km}{r^2}$$

That is, the force of gravity varies directly as the mass m and inversely as the square of the distance r from the sun.

Newton then checked his theory for the orbit of the moon about the earth and decided that the distances should be measured from the centers of mass of the objects (rather than

the surfaces), but to prove this he had to invent calculus! Finally he applied his third law of motion (action and reaction) and reasoned that the gravitational force must depend on both masses involved.

Newton's Universal Law of Gravitation states that:

Every particle in the universe attracts every other particle in the universe with a force **F** *that is directly proportional to the product of the masses* m_1 *and* m_2 *of the particles and inversely proportional to the square of the distance d between the centers of these masses.*

$$\mathbf{F} \propto \frac{m_1 m_2}{d^2} \quad \text{or} \quad \mathbf{F} = \frac{Gm_1 m_2}{d^2} \text{ (attractive)} \quad (10\text{-}11)$$

The proportionality constant G is called the *gravitation constant*. This law is valid for all masses. The forces involved represent an action-reaction pair since each mass attracts the other; it is therefore another example of Newton's third law.

It was not until 1798 that Henry Cavendish (1731–1810) used a torsion balance to accurately measure the gravitation constant. He attached a small horizontal rod that had small spherical masses at each end to the bottom of a vertical thin quartz fiber. (Fig. 10-8). Large spherical masses were then placed in the proximity of these smaller masses, attracting them and causing the fiber to twist. By measuring the twisting forces, Cavendish determined that the gravitation constant G

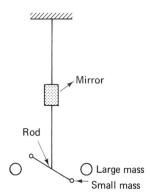

Figure 10-8
A torsion balance.

had a value of 6.67×10^{-11} N·m²/kg² or 3.44×10^{-8} lb · ft²/slug².

The mass of the earth could now be determined from the Universal Law of Gravitation. Consider a small mass m at the surface of the earth. Since the mass m is small, the radius R of the earth, which is 2.09×10^7 ft, is approximately the distance between the center of the mass and the center of the earth. Also, the weight w of the mass m is due to the attraction of that mass by the earth.

$$w = mg = \frac{GM_e m}{R^2}$$

where M_e is the mass of the earth. Thus

$$g = \frac{GM_e}{R^2}$$

and

$$M_e = \frac{gR^2}{G} = \frac{(32.2 \text{ ft/s}^2)(2.09 \times 10^7 \text{ ft})^2}{3.44 \times 10^{-8} \text{ lb·ft}^2/\text{slug}^2}$$
$$= 4.08 \times 10^{23} \text{ slug}$$

Example 10-4

Determine the attractive force between a 30.0 kg mass and a 50.0 kg mass if their centers are 4.00 m apart.

Solution:

$$F = \frac{Gm_1 m_2}{d^2}$$
$$= \frac{(6.67 \times 10^{-11} \text{ N} \cdot \text{m}^2/\text{kg}^2)(30.0 \text{ kg})(50.0 \text{ kg})}{(4.00 \text{ m})^2}$$
$$= 6.25 \times 10^{-9} \text{ N}$$

It should be noted that Newton's law of gravitation was used to explain the origin of tides (the attraction of the waters towards the moon) and the paths of comets; it even led to the discovery of the planet Neptune in 1846.

Artificial Satellites

Orbiting satellites continually fall towards the earth, but, due to its curvature, the surface of the earth falls away at the same rate so that the height of the satellite above the earth remains approximately constant (Fig. 10-9).

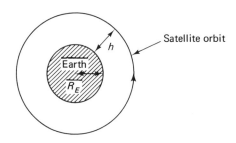

Figure 10-9

The centripetal force of all orbiting satellites is due to the gravitational attraction between the satellite and the celestial body that it orbits.

Example 10-5

A satellite is in a circular orbit at a height $h = 3.20 \times 10^5$ m above the earth's surface. If the earth has a mass $M_E = 5.98 \times 10^{24}$ kg and a mean radius $R_E = 6.35 \times 10^6$ m, (a)

what is the orbital speed v of the satellite?
(b) What is the period T of its orbit?

Solution:

(a) If m is the mass of the satellite and the radius of its orbit $r = R_E + h$, the centripetal force

$$F_c = \frac{GM_Em}{r^2} = \frac{GM_Em}{(R_E + h)^2} = \frac{mv^2}{(R_E + h)}$$

Thus

$$v = \sqrt{\frac{GM_e}{(R_E + h)}}$$

$$= \sqrt{\frac{(6.67 \times 10^{-11}\,\text{N} \cdot \text{m}^2/\text{kg}^2)(5.98 \times 10^{24}\,\text{kg})}{(6.35 \times 10^6\,\text{m}) + (3.20 \times 10^5\,\text{m})}}$$

$$= 7.73 \times 10^3\,\text{m/s}$$

(b) $T = \dfrac{2\pi r}{v} = \dfrac{2\pi(R_E + h)}{v}$

$$= \frac{2\pi[(6.35 \times 10^6\,\text{m}) + (3.20 \times 10^5\,\text{m})]}{7.73 \times 10^3\,\text{m/s}}$$

$$= 5.42 \times 10^3\,\text{s}$$

Artificial satellites have been placed in orbits about the earth by many countries. (See Fig. 10-10.) These satellites have many uses, some of which are:

1. Weather satellites are placed in orbits so that they continually circle the earth and send us meteorological information, such as cloud formations.

2. Research satellites, such as Skylab, are put into orbits to investigate radiation, solar winds, the effects of zero gravity, etc. With special equipment and telescopes, they can take pictures of comets, eclipses, other planets and stars without the interference of the earth's atmosphere. In 1983, a new telescope is to be put into orbit. This telescope will be able to detect very faint objects and will probe deeper into space than ever before. Finally, scientists are investigating the possibility of manufacturing items, such as perfect spheres and optical fibers, while in orbit.

3. Surveillance satellites are used to spy on military installations of other countries.

Others can monitor agricultural and gather geological information.

4. Seasat is a special ocean survey satellite, which is used to monitor the oceans. It orbits the earth 14 times a day, sending us information about winds, temperatures, storms, ice, waves and pollution.

5. Communications satellites such as Intelsat,* Anik and Westar are used to relay microwave signals around the world. These microwaves are used as the carrier waves for radio, telephone and television signals. Communications satellites are placed in precise orbits above the earth's equator at altitudes of approximately 3.59×10^7 m. At this altitude, their position above the earth's surface is fixed because their orbital period is the same as the period of the earth's rotation. This type of orbit is called *geosynchronous*. At present one or two such satellites enable us to communicate with any other part of the North American continent. (See Fig. 10-11.)

Example 10-6

If the mean radius of the earth $R = 6.37 \times 10^6$ m and its mass $M_e = 5.975 \times 10^{24}$ kg, show that the stable orbital period of a communications satellite at a height $h = 35\,900$ km above the earth's surface is 24 h. Assume a circular orbit.

Solution:

The radius of the orbit $r = R + h = 6.37 \times 10^6$ m $+ 3.59 \times 10^7$ m $= 4.22 \times 10^7$ m. Since the force of gravity is the centripetal force,

$$F_c = \frac{mv^2}{r} = \frac{GM_em}{r^2} \quad \text{or} \quad v = \sqrt{\frac{GM_e}{r}}$$

---

*Early Bird (Intelsat I) was launched on April 6, 1965.

(a) Canadian domestic communications satellite Intelsat-C. (NASA)

(c) Oceanographic-geodetic satellite, the 350 kg GEOS-C. (NASA)

(b) Communications satellite Intelsat IV-A. (NASA)

Figure 10-10

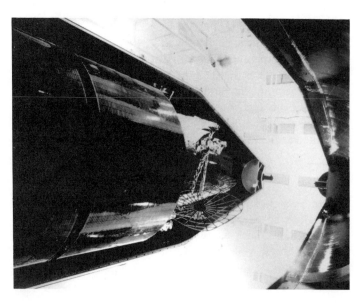

Thus the period

$$T = \frac{2\pi r}{v} = 2\pi r \sqrt{\frac{r}{GM_e}} = 2\pi(4.22 \times 10^7 \text{ m}) \sqrt{\frac{4.22 \times 10^7 \text{ m}}{(6.67 \times 10^{-11} \text{ N} \cdot \text{m}^2/\text{kg}^2)(5.975 \times 10^{24} \text{ kg})}}$$
$$= 8.63 \times 10^4 \text{ s, or } 24 \text{ h.}$$

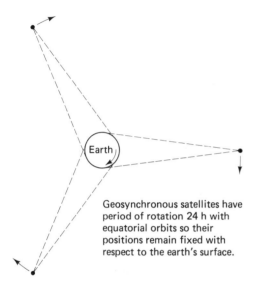

Geosynchronous satellites have period of rotation 24 h with equatorial orbits so their positions remain fixed with respect to the earth's surface.

Figure 10-11

At present it is a very complicated process to put astronauts into orbit. Multistage rockets are used, and each stage is discarded after it has burned its fuel. The National Aeronautics and Space Administration (NASA) is now developing a space shuttle that will be able to take astronauts to and from their orbiting satellites. This craft must be able to withstand temperatures of about 1 260°C during reentry into the earth's atmosphere!

In the near future satellites will probably be used to collect solar energy and then to beam it back to the earth (probably as microwaves). However this source of energy will require relatively large satellites, which will have to be assembled while in orbit. The development of new receivers will enable us to receive satellite signals directly without the complicated earth receiving stations.

Problems_____

13. Determine the gravitational force between a 200 lb man and a 130 lb woman when they are 2.0 ft apart.

14. The mass of the earth is 5.98×10^{24} kg, the mass of the sun is 1.99×10^{30} kg, and the average distance between their centers is 1.50×10^{11} m. Assuming the earth's orbit of the sun to be uniform circular motion, determine (a) the gravitational force of the sun on the earth, (b) the speed of the earth in its orbit, (c) the time it takes the earth to complete one orbit of the sun.

15. Determine the acceleration due to gravity on the moon if its mass is 0.012 3 times the mass of the earth and its radius is 0.273 times that of the earth.

16. If the diameter of the earth is 1.27×10^7 m and its mass is 5.98×10^{24} kg, determine the average orbital speed of a satellite that is in a stable orbit 300 000 m above the earth's surface.

17. If the mean radius of the moon is 1.73×10^6 m and its mass is 7.35×10^{22} kg, determine (a) the average speed and (b) the period of orbit of a satellite at a height of 5.00×10^6 m above the moon's surface. Assume a circular orbit.

Gravitational Fields and Energy_____

It is often convenient to describe the effects of gravity in terms of a *gravitational field*, which is said to exist at any location where a mass would experience a gravitational force. Two or more masses interact even if they do

Figure 10-12
Delta launch rocket.
(NASA)

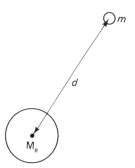

Figure 10-13

that its displacement is always perpendicular to the gravitational force, i.e. approximately parallel to the earth's surface, no work is done against gravitational forces, and the gravitational potential energy is constant.

Gravitational fields are called *conservative fields* because the work required to displace an object in a gravitational field is independent of the path taken. For example, the amount of work required to raise an object through a height h in a straight line is equal to the work required to raise the object to the same position through a variable path. Only the vertical component of the displacement is considered in the calculation of the work done.

When an object is pushed over a surface, friction continually produces heat losses all along its path; therefore less work is required to move an object between two points if the object is pushed in a straight line. In this case work is dependent on the path taken, and the system is *nonconservative*.

When an object is a distance r from the *center* of the earth, equating its weight $m\mathbf{g}$ to the attractive gravitational force of the earth, we obtain

$$\mathbf{F} = m\mathbf{g} = \frac{GM_e m}{r^2} \quad \text{(attractive)}$$

where M_e is the mass of the earth and G is the gravitation constant. Thus the magnitude of the acceleration due to gravity at this position:

not make contact because gravitational forces act over distances. Since gravitational forces exist between two or more masses, gravitational fields are set up by masses, and other masses are affected by that field.

Consider the motion of some mass m in the gravitational field set up by an object such as the earth. The force of gravity varies inversely with the square of the distance between the center of the earth and the center of the mass m (Fig. 10-13.)

Work must be done by some external agent against gravitational forces in order to raise slowly an object vertically above the earth, but that object then has a greater gravitational potential energy because it can do an equal amount of work as it returns to its original position. If the object is moved so

$$g = \frac{GM_e}{r^2} \qquad (10\text{-}12)$$

As the distance r increases, the acceleration due to gravity g and the weight (or attractive force to the earth) decreases in magnitude.

At very large distances the force of gravity is small, and very little work is required to move the mass m in the earth's gravitational field. Therefore we define the mass to have zero gravitational potential energy when it is at an infinite distance from the earth where the force of gravity would be zero.

The *gravitational potential energy* E_p of

earth is zero; positive work must be done on the mass to remove it to infinity.

Example 10-7

If the earth's mass M_e is 6.00×10^{24} kg and its radius R is 6.40×10^6 m, how much work must be done in order to remove a 1.00 kg mass completely from the earth's gravitational field? Neglect air resistance.

Solution:

The work done $W_{r \to \infty}$ is equal to the change in gravitational potential energy. Thus

$$W_{r \to \infty} = E_{P\infty} - E_{PR} = 0 - \left(\frac{-GM_e m}{R} \right) = \frac{GM_e m}{R}$$

$$= \frac{(6.67 \times 10^{-11} \text{ N} \cdot \text{m}^2/\text{kg}^2)(6.00 \times 10^{24} \text{ kg})(1.00 \text{ kg})}{(6.40 \times 10^6 \text{ m})} = 6.25 \times 10^7 \text{ J}$$

a mass m at a distance r from the center of the earth is defined as the work that must be done in order to bring the mass m from infinity (∞) to the distance r. It may be shown by calculus that this work

$$W_{\infty \to r} = \frac{-GM_e m}{r} = E_{pr} \qquad (10\text{-}13)$$

The gravitational potential energy has a negative value because the force between the earth and the mass is attractive, and in the absence of an external force the mass will do work in falling towards the earth.

If air resistance is neglected, the work that must be done on a mass m in order to raise it vertically from a distance r_1 to a distance r_2 from the earth's center is equal to the change in its gravitational potential energy.

$$W_{r_1 \to r_2} = E_{pr_2} - E_{pr_1} = \frac{-GM_e m}{r_2} - \left(\frac{-GM_e m}{r_1} \right)$$

$$= GM_e m \left(\frac{1}{r_1} - \frac{1}{r_2} \right) \qquad (10\text{-}14)$$

It is positive valued if r_2 is greater than r_1. Note that the gravitational potential energy of any mass at an infinite distance from the

Weightlessness

Astronauts have experienced the phenomenon of weightlessness even though they were not an infinite distance from the earth or other celestial bodies. In orbits about the earth the condition of weightlessness arises because the force of gravity supplies the necessary centripetal force for the orbit and the astronaut and satellite are in a continual state of free fall towards the earth!

A state of zero gravity can also be obtained without going into orbit. This can be accomplished by doing an outside loop in an aircraft so that the weight supplies the centripetal force required for the loop.

Example 10-8

If the moon's mass $M = 7.38 \times 10^{22}$ kg and its radius $r_m = 1.73 \times 10^6$ m, with what minimum initial velocity (*the escape velocity*) must a rocket be launched so that it can escape the moon's gravitational field?

Solution:

From the conservation of energy, we obtain

$$\Delta E_k + \Delta E_p = \tfrac{1}{2}m(v_1^2 - v_0^2) + (E_{p\infty} - E_{pr}) = 0$$

At an "infinite distance" from the moon, $E_{p\infty} = 0$; for minimum escape velocity, the final speed $v_1 = 0$ m/s. Thus

$$-\tfrac{1}{2}mv_0^2 - E_{pr} = -\tfrac{1}{2}mv_0^2 - \left(\frac{-GM_m m}{r_m}\right) = 0,$$

$$v_0 = \sqrt{\frac{2GM_m}{r_m}} = \sqrt{\frac{2(6.67 \times 10^{-11} \text{ N} \cdot \text{m}^2/\text{kg}^2)(7.38 \times 10^{22} \text{ kg})}{1.73 \times 10^6 \text{ m}}} = 2.38 \times 10^3 \text{ m/s}$$

Problems

18. What is the potential energy of a 100 kg mass at the surface of the earth?

19. A 30 000 lb rocket is shot 2.0 mi vertically above the earth's surface. If the mass of the rocket is a constant, how much work is done?

20. If the mass of the moon is 0.012 3 times that of the earth and its center is 3.844 × 10^8 m from the center of the earth, what is its gravitational potential energy?

21. Determine the minimum escape velocity of a rocket ship from the surface of the moon. The mass of the moon is 7.35 × 10^{22} kg and its diameter is 3.47 × 10^6 m.

REVIEW PROBLEMS

1. What is the gravitational force between an electron and a proton if they are 5.0 × 10^{-9} m apart?

2. The force between two masses is 4.5 × 10^{-12} N when they are 2.0 m apart. What is the force between the masses when they are 8.0 m apart?

3. A proton moves at 7.2 × 10^6 m/s in a circular orbit of radius 0.75 m in a cyclo-

tron. Determine (a) the period of the orbit, (b) the centripetal acceleration, (c) the centripetal force on the proton.

4. An aircraft performs a loop at a constant acceleration of 5.0 g, i.e. 5 times the acceleration due to gravity. If the aircraft has a speed of 650 km/h, what is the radius of the loop?

5. If the radius of the earth at the equator is 6.4 × 10^6 m and it rotates about its axis in 24 h, what is the centripetal acceleration of an object on the earth's surface at the equator? What produces this acceleration?

6. A 50 kg boy is on a swing with ropes 3.5 m long. If the swing goes to a maximum height of 2.0 m above its lowest position, calculate (a) the speed of the boy at the bottom of the swing and (b) the tension in each of the two ropes at the bottom of the swing.

7. A satellite is in circular orbit at a height of 150 mi above the earth's surface. If the radius of the earth is 4 000 mi and its mass is 4.09 × 10^{23} slug, (a) what is the orbital speed of the satellite? (b) What is the period of its orbit?

8. What is the period of revolution of a satellite in a circular orbit of radius 8 000 km about the earth if the mass of the earth is 5.98 × 10^{24} kg?

9. A 2 400 lb car rounds a level curve of radius 400 ft on an unbanked road at 40 mi/h. (a) What is the minimum coefficient of friction between the tires and the road

if the car does not skid? (b) At what optimum angle should the road be banked?

10. If the orbital period of the earth is 365.26 d and its average distance from the sun is 1.495×10^{11} m, (a) what is the orbital period of Jupiter if its average distance from the sun is 7.77×10^{11} m? (b) What is the orbital speed of Jupiter about the sun?

11. What is the gravitational potential energy of a 70 kg man at the earth's surface?

12. If the mass of Venus is 4.86×10^{24} kg and its mean diameter is 1.232×10^7 m, (a) how much work must be done in order to remove a 1.2 t rocket from its gravitational field? (b) What is the escape velocity? Neglect all resistances.

Many rigid objects, such as drive shafts, wheels, cams and gears, are designed to spin or rotate about some axis. A rigid object is said to move in pure *rotary motion* if all of its constituent particles* move in circles, the centers of which are located on the same fixed straight line, called the *axis of rotation.* In general, if the axis of rotation is not fixed, the motion of the rigid object is a mixture of translational and rotational motion. The translational motion is represented by the motion of the center of mass of the system, and the rotational motion about some axis is superimposed on this. For example, the wheels of a car rotate about the center of mass (the axle), but the axle also moves linearly with the car. Also, the earth spins on its axis as it orbits the sun.

11-1 RIGID BODY ROTATION

Rotations are usually measured in terms of angles, but the various definitions are quite similar to those for translational motion. In SI, angles are measured in radians (rad). A *radian* is defined as the angle subtended at the center of a circle by an arc length equal to the radius r (Fig. 11-1). Since the circum-

11

Rotary Motion

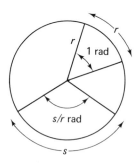

Figure 11-1

*Except those particles actually on the axis of rotation.

ference of a circle has a length $2\pi r$, the total circumference subtends an angle of $2\pi r/r = 2\pi$ rad at the center of the circle. Thus

$$1 \text{ rad} = \frac{360°}{2\pi} = \frac{180°}{\pi} \qquad (11\text{-}1)$$

Similary, any other arc length s must subtend an angle:

$$\theta = \frac{s}{r} \text{ rad} \qquad (11\text{-}2)$$

at the center of the circle. One *revolution* of an object is equivalent to a complete rotation of 2π rad.

Since a rigid object is composed of a large number of small particles, its rotation about an axis 0 may be described in terms of the motion of its constituent particles about that axis. The *angular displacement* θ of a rigid object is the angle through which it rotates; angular displacement is normally expressed in radians or revolutions. The time required for one complete rotation of a rigid object is called its *period* of rotation.

By analogy with linear velocity, the *angular velocity* ω of a rotating object is defined as its time rate of change of angular displacement. Thus, if an object rotates through some angular displacement θ in an elapsed time t, its average angular velocity is:

$$\bar{\omega} = \frac{\theta}{t} \qquad (11\text{-}3)$$

Note that, since θ is expressed in radians or revolutions in Eq. 11-3, angular velocity must be expressed in rad/s or rev/min (rpm). The conversion from rpm to rad/s is:

$$1 \text{ rpm} = \frac{2\pi}{60} \text{ rad/s} \qquad (11\text{-}4)$$

In Fig. 11-2 as the particle at A moves in an arc to B, the particle at C moves in an arc to D, but the different arc lengths AB and CD both subtend the same angle θ at 0. Consider a rigid object which rotates through some angle θ in an elapsed time t. If some particle in the rigid object is located at a distance r from the axis of rotation and moves

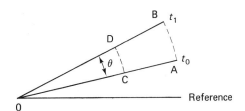

Figure 11-2

in an arc through a distance s, its average speed is:

$$\bar{v} = \frac{s}{t} = \frac{r\theta}{t} = r\bar{\omega} \qquad (11\text{-}5)$$

since $s = r\theta$ and the average angular velocity of the rigid object $\bar{\omega} = \theta/t$.

Example 11-1

The earth has a diameter of 1.276×10^7 m, and it completes one revolution about its axis in 24 h. (a) What is the average angular velocity of the earth? (b) What is the average speed of a point on the earth's equator?

Solution:

(a) The earth completes one revolution (2π radians) in 24 h; thus

$$\bar{\omega} = \frac{\theta}{t} = \frac{2\pi \text{ rad}}{(24 \text{ h})(3\,600 \text{ s/h})} = 7.27 \times 10^{-5} \text{ rad/s}$$

(b) The equatorial radius of the earth is $(1.276 \times 10^7 \text{ m})/2$. Thus, the average speed of a point at the equator is:

$$v = r\bar{\omega} = \left(\frac{1.276 \times 10^7 \text{ m}}{2}\right)(7.27 \times 10^{-5} \text{ rad/s})$$
$$= 464 \text{ m/s}$$

All particles in a rotating rigid object experience a centripetal acceleration $\mathbf{a}_c$ directed towards the axis of rotation; at the same time, they may also be subjected to an acceleration $\mathbf{a}_T$ tangential to their circular path (Fig. 11-3). Their total acceleration is the *vector sum* of $\mathbf{a}_c$ and $\mathbf{a}_T$; since the centripetal and tangential accelerations are perpendicular, they are independent of each other. Tan-

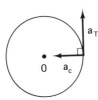

Figure 11-3

gential accelerations a_T change the angular velocity of the rotating object, and centripetal accelerations a_c change the direction of motion.

The *angular acceleration* α of an object is defined as the time rate of change of its angular velocity. If an object changes from an angular velocity ω_0 at time t_0 to an angular velocity ω_1 at time t_1, its average angular acceleration is:

$$\bar{\alpha} = \frac{\omega_1 - \omega_0}{t_1 - t_0} = \frac{\Delta\omega}{\Delta t} \qquad (11\text{-}6)$$

Its instantaneous angular acceleration at the instant t_0 is:

$$\alpha = \underset{t_1 \to t_0}{limit} \frac{\omega_1 - \omega_9}{t_1 - t_0} \qquad (11\text{-}7)$$

Angular acceleration is usually expressed in units of rad/s² or rev/min².

At any instant t_0, the magnitude of the tangential acceleration of a particle that is located at a distance r from the axis of rotation of the rigid object is given by:

$$a_T = \underset{t_1 \to t_0}{limit} \frac{v_1 - v_0}{t_1 - t_0}$$

But $v = r\omega$, and r is a constant. Thus

$$a_T = \underset{t_1 \to t_0}{limit} \left(\frac{r\omega_1 - r\omega_0}{t_1 - t_0}\right)$$

$$= r \underset{t_1 \to t_0}{limit} \left(\frac{\omega_1 - \omega_0}{t_1 - t_0}\right) = r\alpha \qquad (11\text{-}8)$$

*Example 11-2*_____

A car starts from rest and accelerates uniformly to a final speed of 50.0 km/h in an elapsed time $t = 18.0$ s. If the diameter of the wheels $d = 70.0$ cm, calculate (a) the acceleration of the car, (b) the final angular veloc-

ity of the wheels, (c) the angular acceleration of the wheels.

Solution:

$$v_0 = 0 \text{ m/s}, v_1 = 50.0 \text{ km/h} = \frac{50.0 \times 10^3 \text{ m}}{3\ 600 \text{ s}}$$

$$= 13.9 \text{ m/s}, t = 18.0 \text{ s}$$

and $\qquad r = d/2 = 0.350$ m.

(a) The acceleration of the car is equal to the acceleration of the rims of the wheels.

$$a_t = \frac{v_1 - v_0}{t} = \frac{(13.9 \text{ m/s}) - (0 \text{ m/s})}{18.0 \text{ s}}$$

$$= 0.772 \text{ m/s}^2$$

(b) $\qquad \omega_1 = \frac{v_1}{r} = \frac{13.9 \text{ m/s}}{0.350 \text{ m}} = 39.7 \frac{\text{rad}}{\text{s}}$

$$= (39.7)\frac{(60 \text{ rpm})}{(2\pi)} = 379 \text{ rpm}$$

(c) $\qquad \alpha = \frac{a_t}{r} = \frac{0.772 \text{ m/s}^2}{0.350 \text{ m}} = 2.20 \text{ rad/s}^2$

Four equations of rotational motion with a constant angular acceleration, which are completely analogous to the equations of uniformly accelerated translational motion, may be obtained (Table 11-1).

Table 11-1

Equations of Motion with Constant Translational or Rotational Acceleration

Translational Motion (Straight Line)	Rotational Motion (Fixed Axis of Rotation)	
$s = \left(\frac{v_1 + v_0}{2}\right)t$	$\theta = \left(\frac{\omega_1 + \omega_0}{2}\right)t$	(11-9)
$v_1 = v_0 + at$	$\omega_1 = \omega_0 + \alpha t$	(11-10)
$s = v_0 t + \frac{1}{2}at^2$	$\theta = \omega_0 t + \frac{1}{2}\alpha t^2$	(11-11)
$v_1^2 = v_0^2 + 2as$	$\omega_1^2 = \omega_0^2 + 2\alpha\theta$	(11-12)

Note the similarity between the sets of equations. The rotational equations may be obtained by substituting angular displacement θ for linear displacement s, angular velocity

ω for linear velocity v, and angular acceleration α for linear acceleration a in the translational equations.

Example 11-3

A flywheel, which is initially rotating with an angular velocity $\omega_0 = 50.0$ rad/s, is subjected to an angular acceleration $\alpha = 5.00$ rad/s² for an elapsed time $t = 10.0$ s. (a) What is its final angular velocity $\dot\omega_1$? (b) How many revolutions does it perform while it is being accelerated?

Solution:

(a) $\omega_0 = 50.0$ rad/s, $\alpha = 5.00$ rad/s², $t = 10.0$ s
and $\omega_1 = ?$
$\omega_1 = \omega_0 + \alpha t$
$= (50$ rad/s$) + (5.00$ rad/s²$)(10.0$ s$)$
$= 100$ rad/s

(b) $\theta = \omega_0 t + \frac{1}{2}\alpha t^2$
$= (50.0$ rad/s$)(10.0$ s$)$
$+ \frac{1}{2}(5.00$ rad/s²$)(10.0$ s$)^2$
$= 500$ rad $+ 250$ rad $= 750$ rad

or

$\theta = (750/2\pi)$ rev $= 119$ rev.

Example 11-4

An aircraft engine idling at 1 200 rpm accelerates uniformly at 60.0 rad/s² to a final angular velocity of 7 800 rpm. Determine (a) the time taken and (b) the number of revolutions made while accelerating.

Solution:

$\omega_0 = 1\,200$ rpm $= 1\,200\left(\dfrac{2\pi \text{ rad}}{60 \text{ s}}\right) = 126$ rad/s,

$\omega_1 = 7\,800$ rpm $= 7\,800\left(\dfrac{2\pi \text{ rad}}{60 \text{ s}}\right) = 817$ rad/s

and $\alpha = 60.0$ rad/s²

(a) $t = ?$, $\qquad \alpha = \dfrac{\omega_1 - \omega_0}{t}$

Thus

$t = \dfrac{\omega_1 - \omega_0}{\alpha} = \dfrac{817 \text{ rad/s} - 126 \text{ rad/s}}{60.0 \text{ rad/s}^2}$

$= 11.5$ s

(b) $\theta = ?$, $\qquad \omega_1^2 = \omega_0^2 + 2\alpha\theta$

Therefore

$\theta = \dfrac{\omega_1^2 - \omega_0^2}{2\alpha}$

$= \dfrac{(817 \text{ rad/s})^2 - (126 \text{ rad/s})^2}{2(60.0 \text{ rad/s}^2)}$

$= 5\,430$ rad $= \dfrac{5\,430}{2\pi}$ rev $= 864$ rev

Problems

1. (a) Convert 70° to radians. (b) Convert 10 rad to degrees. (c) Convert 75.4 rad to revolutions. (d) How many radians are subtended by an arc length of 12.5 ft at the center of a circle of radius 1.50 ft?

2. Show that 1 rpm $= (2\pi/60)$ rad/s.

3. A flywheel has a diameter of 50 cm, and it makes 1 400 rev in 1.2 min. Determine (a) its average angular velocity in rad/s and (b) the average speed of a point on its rim.

4. A car starts from rest and accelerates uniformly to a speed of 50.0 km/h in 18.0 s. If the wheels of the car are 70.0 cm in diameter, determine (a) the acceleration of the car, (b) the final angular velocity of the wheels, (c) the angular acceleration of the wheels, (d) the number of revolutions of the wheels while the car is accelerating.

5. The driveshaft of a motor starts from rest and accelerates uniformly to an angular velocity of 1 200 rpm in 20 s. Determine (a) the angular acceleration and (b) the number of revolutions completed by the driveshaft while it accelerates.

6. A wheel is rotating with an angular velocity of 25 rad/s when it is subjected to an angular acceleration of 6.0 rad/s². (a) How long does it take the wheel to reach an angular velocity of 40 rad/s? (b) How many revolutions does it complete in the first 5.0 s that it is accelerating?

7. An aircraft engine accelerates uniformly from 1 200 rpm to 5 000 rpm in 8.0 s. Find

(a) the angular acceleration and (b) the number of revolutions completed while it accelerates.

8. The shaft of a generator is turning at 1 500 rpm and decelerates at 5.0 rad/s² for 30 s. Determine (a) its final angular velocity and (b) the number of revolutions that it makes while decelerating.

11-2 TORQUE AND MOMENT OF INERTIA

Since a torque (moment of force) tends to cause an object to rotate, it is reasonable to expect that a torque will produce a rotational acceleration. This is the rotational equivalent of Newton's first law of motion.

Consider a rigid body that is constrained to rotate about some axis 0 (Fig. 11-4). A particle of mass m that is a distance r from 0 can only change its speed if there is a tangential component of acceleration a_T; centripetal accelerations only change the direction of its motion.

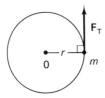

Figure 11-4

According to Newton's second law, a tangential acceleration is produced by a tangential component of force F_T such that $F_T = ma_T$. But, from definition, the application of a force F_T at a perpendicular distance r from the axis of rotation 0 constitutes a torque τ_0. Thus

$$\tau_0 = rF_T = rma_T \qquad (11\text{-}13)$$

Rigid objects are composed of large numbers of particles of masses m_i that are located at different distances r_i from the axis of rotation. The net torque on a rigid object is equal to the sum of the torques on the individual particles. Thus

$$\text{net } \tau_0 = \Sigma_i m_i r_i a_{T_i} = \Sigma_i m_i r_i^2 \alpha = (\Sigma_i m_i r_i^2)\alpha \qquad (11\text{-}14)$$

since the angular acceleration of the rigid object $\alpha = a_{T_i}/r_i$ is the same for all particles in the object.

The term $\Sigma_i m_i r_i^2$ is called the *moment of inertia* I_0 of the rigid object; it is analogous to the mass (inertia) in linear motion. Thus

$$I_0 = \Sigma_i \, m_i r_i^2 = m_1 r_1^2 + m_2 r_2^2 + m_3 r_3^2 + \cdots \qquad (11\text{-}15)$$

where m_i is the mass of the ith particle, which is located at a distance r_i from the axis of rotation 0. Note that the moment of inertia of any object depends on its mass distribution and the position of the axis of rotation. Moments of inertia are usually expressed in units of slug-ft² in the British Engineering System and in kg·m² in SI.

Thus, from Eq. 11-14, the rotational equivalent of Newton's law of motion may be written as:

$$\text{net } \tau_0 = I_0 \alpha \qquad (11\text{-}16)$$

where net τ_0 is the net torque, I_0 is the moment of inertia and α is the angular acceleration.

The calculation of the moment of inertia is often quite complex and involves integral calculus. Some formulae for the moments of inertia of common objects are listed in Table 11-2.

Example 11-5

A solid 2.50 kg cylindrical drive shaft with a radius of 1.00 cm and a length of 1.00 m accelerates from rest to an angular velocity of 600 rpm in 10.0 s. Calculate the average torque applied to the shaft.

Table 11-2
Moments of Inertia of Common Objects

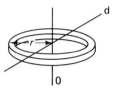

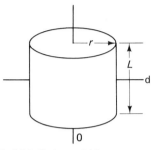

Ring or Hoop

(a) About Central Axis
$$I_0 = Mr^2$$

(b) About Diameter d
$$I_d = \frac{1}{2} Mr^2$$

Solid Cylinder or Disk

(a) About Central Axis 0
$$I_0 = \frac{1}{2} Mr^2$$

(b) About Central Perpendicular Axis d
$$I_d = \frac{1}{4} Mr^2 + \frac{1}{12} ML^2$$

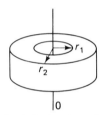

Annular Cylinder About Central Axis
$$I_0 = \frac{1}{2} M(r_2^2 + r_1^2)$$

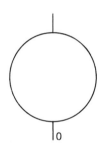

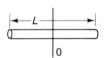

Thin Uniform Rod About Central Perpendicular Axis
$$I_0 = \frac{1}{12} ML^2$$

(a) Solid Sphere About any Diameter
$$I_0 = \frac{2}{5} Mr^2$$

(b) Thin Hollow Sphere About any diameter
$$I_0 = \frac{2}{3} Mr^2$$

Solution:

$$\omega_1 = 600 \text{ rpm} = (600)\left(\frac{2\pi}{60} \frac{\text{rad}}{\text{s}}\right) = 20\pi \text{ rad/s}$$

and, from Table 11-2,

$$I_0 = \frac{Mr^2}{2}$$

Thus

$$\text{Net } \tau_0 = I_0\alpha = \frac{Mr^2}{2}\left(\frac{\omega_1 - \omega_0}{t}\right)$$

$$= \frac{(2.50 \text{ kg})(1.00 \times 10^{-2} \text{ m})^2}{2}$$

$$\times \left(\frac{20\pi \text{ rad/s} - 0}{10.0 \text{ s}}\right)$$

$$= 7.85 \times 10^{-4} \text{ N} \cdot \text{m}$$

Example 11-6

The axle of a solid cylinder of mass $M = 3.00$ kg and radius $r = 10.0$ cm is mounted on

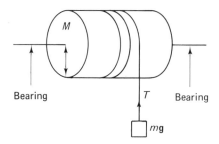

(a) Solid cylinder mounted on bearings

w = mg

(b) Free body diagram of the mass m

Figure 11-5

frictionless bearings. One end of a light string is tied and wrapped around the cylinder, and a mass $m = 0.500$ kg is suspended from the free end (Fig. 11-5a). Determine (a) the translational acceleration a_T of the 0.500 kg mass, (b) the angular acceleration α of the cylinder, (c) the tension T in the string.

Solution:

(a) Draw the free-body diagram of the 0.500 kg mass (Fig. 11-5b). Applying Newton's second law, we obtain:

$$\text{net } F = mg - T = ma_T \qquad (1)$$

If I is the moment of inertia of the cylinder, since the string tension T acts at a distance r from the axis of rotation, the rotational form of Newton's second law for the cylinder is:

$$\text{net } \tau_0 = rT = I\alpha$$

Thus, from Table 11-2, $rT = (\tfrac{1}{2}Mr^2)\alpha = \tfrac{1}{2}Mra^T$ since $r\alpha = a_T$. Hence

$$T = Ma_T/2 \qquad (2)$$

Substituting (2) into (1), we obtain:

$$mg - \frac{Ma_T}{2} = ma_T \quad \text{or} \quad a_T = \frac{2mg}{M + 2m} \qquad (3)$$

Thus

$$a_T = \frac{2(0.500 \text{ kg})(9.81 \text{ m/s}^2)}{(3.00 \text{ kg}) + 2(0.500 \text{ kg})} = 2.45 \text{ m/s}^2$$

(b) $\qquad \alpha = \dfrac{a_T}{r} = \dfrac{2.45 \text{ m/s}^2}{0.100 \text{ m}} = 24.5 \text{ rad/s}^2$

(c) From (1) and (2), we obtain:

$$T = m(g - a_T) = m\left[g - \frac{2T}{M}\right]$$

Thus

$$T = \frac{mMg}{M + 2m} = \frac{(0.500 \text{ kg})(3.00 \text{ kg})(9.81 \text{ m/s}^2)}{(3.00 \text{ kg}) + 2(0.500 \text{ kg})}$$
$$= 3.68 \text{ N}$$

Obviously, Eq. 11-15 for the moment of inertia is often very difficult to apply. However, if the total mass M of a rigid object could be located at a distance k from its axis of rotation, then its moment of inertia

$$I = Mk^2$$

For convenience it is often simplified by defining a distance k, called the *radius of gyration*, as the distance from the axis of rotation 0 at which the total mass M of the object could be located without changing its moment of inertia. The radius of gyration k may be written as:

$$k = \sqrt{\frac{I_0}{M}} \qquad (11\text{-}17)$$

where I_0 is the total moment of inertia about the axis 0.

If the moment of inertia I_c of an object about its center of mass is known, its moment of inertia about a parallel axis may be determined from the *parallel axis theorem*, which states:

The moment of inertia I of an object about any axis is equal to the sum of the moment of inertia I_c of the object about a parallel axis through its center of mass and the product of the mass M of the object and the square of the perpendicular distance h between the axes.

$$I = I_c + Mh^2 \qquad (11\text{-}18)$$

Example 11-7

Determine the radius of gyration k of a thin uniform rod of mass M and length L about an axis passing through one end perpendicular to its length.

Solution:

From the parallel axis theorem and Table 11-2, we obtain:

$$I = I_c + Mh^2 = \frac{ML^2}{12} + M\left(\frac{L}{2}\right)^2 = \frac{ML^2}{3}$$

Therefore

$$k = \sqrt{\frac{I}{M}} = \sqrt{\frac{ML^2/3}{M}} = \frac{L}{\sqrt{3}}$$

Problems

9. Determine the moment of inertia of a 150 g solid sphere about a diameter if its radius is 0.60 cm.

10. Determine the moment of inertia of a 200 lb solid cylinder about its central axis if its diameter is 8.0 in.

11. A solid 6.0 lb cylindrical shaft with a diameter of 1.8 in and a length of 1.0 ft accelerates from rest to an average angular velocity of 1 000 rpm in 30 s. Determine the average torque applied to the shaft.

12. Determine the angular acceleration of a 0.50 kg, thin uniform rod about a perpendicular axis through its center if it is 2.0 m long and is subjected to an average torque of 0.25 N·m.

13. Calculate the torque required to accelerate a flywheel from 80 rpm to 1 500 rpm in 15 s. The flywheel is a 1.0 kg solid disc with a diameter of 30 cm.

14. A 200 g solid cylinder has a radius of 15 cm and a length of 20 cm. Determine its radius of gyration (a) about its central axis along its length and (b) about a central axis perpendicular to its length.

15. Determine the torque required to accelerate a 15 lb object from 50 rpm to 1 800 rpm in 30 s if the radius of gyration about its axis is 8.0 in.

16. One end of a light string is wound on a 15.0 kg solid cylinder of radius 10.0 cm, and the other end of the string is attached to a 3.50 kg load (Fig. 11-5). If the cylinder is mounted with its axis horizontal on frictionless bearings and the 3.50 kg load falls to the earth, determine the angular acceleration of the cylinder.

17. The center of a light string is wrapped around the circumference of a 9.00 lb solid cylinder of radius 6.00 in; one free end of the string is attached to a 12.0 lb load, and the other free end is attached to a 23.0 lb load (Fig. 11-6). The axle of the cylinder is mounted horizontally on frictionless bearings, and the 23.0 lb load falls towards the earth. Determine (a) the angular acceleration of the cylinder and (b) the acceleration of the loads. *Hint:* Draw the free-body diagrams.

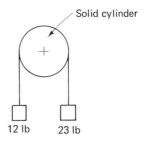

Solid cylinder

12 lb   23 lb

Figure 11-6

18. Determine the moment of inertia of a solid sphere of mass M and radius r about an axis tangential to its surface.

19. Determine the moment of inertia of a thin 4.0 oz rod of length 5.0 ft about a perpendicular axis one-third of the way from one end.

11-3 ROTATIONAL KINETIC ENERGY, WORK AND POWER

Rigid objects are composed of a large number of individual particles of masses m_i that are located at distances r_i from the axis of rotation. The rotational kinetic energy of a rigid object is the sum of the kinetic energies of all of its constituent particles.

$$E_{kr} = \tfrac{1}{2}\{m_1 v_1^2 + m_2 v_2^2 + m_3 v_3^2 + \cdots\}$$
$$= \tfrac{1}{2} \Sigma_i \, m_i v_i^2$$

If ω is the angular velocity, then $v_i = r_i \omega$. Therefore the total rotational kinetic energy is:

$$E_{kr} = \tfrac{1}{2} \Sigma_i \, m_i v_i^2 = \tfrac{1}{2} (\Sigma_i \, m_i r_i^2) \omega^2 = \tfrac{1}{2} I_0 \omega^2$$

$$(11\text{-}19)$$

Example 11-8

Determine the rotational kinetic energy of an 80.0 kg solid cylindrical flywheel with a radius $r = 15.0$ cm if it rotates at 3 250 rpm.

Solution:

$$E_{kr} = \tfrac{1}{2} I_0 \omega^2 = \tfrac{1}{2}(\tfrac{1}{2} M r^2)\omega^2$$

$$= \tfrac{1}{4}(80.0 \text{ kg})(0.150 \text{ m})^2 \left(3\,250 \times \frac{2\pi}{60} \text{ rad/s}^2\right)^2$$

$$= 5.21 \times 10^4 \text{ J}$$

In general, the total kinetic energy of an object is the sum of its translational and rotational kinetic energies:

$$E_{kTOT} = \tfrac{1}{2} m v^2 + \tfrac{1}{2} I_c \omega^2 \qquad (11\text{-}20)$$

where v is the speed of the center of mass and I_c is the moment of inertia about the *center of mass*.

The law of conservation of mechanical energy must be modified to include a rotational kinetic energy term. If a mass m initially had a speed v_0, angular velocity ω_0 and a height h_0 above a reference and finally has a speed v_1, angular velocity ω_1, and height h_1, then

$$\tfrac{1}{2} m v_0^2 + \tfrac{1}{2} I_0 \omega_0^2 + mgh_0$$
$$= \tfrac{1}{2} m v_1^2 + \tfrac{1}{2} I_0 \omega_1^2 + mgh_1 + E_f \quad (11\text{-}21)$$

where E_f is the energy lost to friction, and g is the acceleration due to gravity.

Example 11-9

A solid cylinder and a thin hoop with equal masses m and equal radii of $r = 6.00$ in are allowed to roll freely from rest down an incline through a vertical height of 6.00 ft. If there are no frictional losses, compare their speeds at the bottom of the incline.

Solution:

In both cases, $E_f = 0$ and potential energy is converted into kinetic energy. Initially v_0 and ω_0 are zero, and, if the bottom of the incline is taken as the reference level, $h_1 = 0$. Thus from Eq. 11-21:

$$mgh_0 + 0 + 0 = \tfrac{1}{2} m v_1^2 + \tfrac{1}{2} I_0 \omega_1^2 + 0$$

or

$$mgh_0 = \tfrac{1}{2} m v_1^2 + \tfrac{1}{2} I_0 (v_1/r)^2$$

since $\omega_1 = v_1/r$. From Table 11-2, the moment of inertia of the solid cylinder $I_0 = \tfrac{1}{2} m r^2$. Therefore

$$mgh_0 = \tfrac{1}{2} m v_1^2 + \tfrac{1}{2}(\tfrac{1}{2} m r^2)(v_1/r)^2 = \tfrac{3}{4} m v_1^2$$

or

$$v_1 = \sqrt{\tfrac{4}{3} g h_0} = \sqrt{\tfrac{4}{3}(32.2 \text{ ft/s}^2)(6.00 \text{ ft})} = 16.0 \text{ ft/s}$$

For the hoop

$$mgh_0 = \tfrac{1}{2} m v_1^2 + \tfrac{1}{2}(m r^2)(v_1/r)^2 = m v_1^2$$

or

$$v_1 = \sqrt{g h_0} = \sqrt{(32.2 \text{ ft/s}^2)(6.00 \text{ ft})} = 13.9 \text{ ft/s}$$

Note that the speed of the solid cylinder is greater because its mass distribution gives it a smaller moment of inertia.

Consider point A that is located at a distance r from the axis of rotation 0 of a rigid object. If the rigid object performs pure rotary motion, then point A moves in a circular motion about 0. Since centripetal forces act perpendicular to the motion of A, centripetal forces do no rotational work; all rotational work must be accomplished by an applied tangential force F_T. If the tangential force acts at A and moves through an infinitesimally small arc Δs along the circular path as the rigid object rotates through an infinitesimally small angle $\Delta\theta$, then $\Delta s = r\,\Delta\theta$. Since F_T is parallel to Δs, the work done by the tangential force is:

$$\Delta W = F_T \Delta s = F_T(r\Delta\theta) = \tau_0\Delta\theta$$

since $\tau_0 = rF_T$ is the torque developed by F_T about the axis of rotation 0. Consequently, if the magnitude of the tangential force is constant, τ_0 is also constant, and the total work done as the rigid object rotates through an angle θ is given by:

$$W = \tau_0\theta \qquad (11\text{-}22)$$

The rotational power P developed by the applied torque τ_0 in causing the rigid object to rotate through an infinitesimally small angle $\Delta\theta$ in an equally small time interval Δt is:

$$P = \frac{\Delta W}{\Delta t} = \frac{\tau_0\Delta\theta}{\Delta t}$$

Since τ_0 is constant

$$P = \tau_0\frac{\Delta\theta}{\Delta t} = \tau_0\omega = F_T r\omega \qquad (11\text{-}23)$$

where ω is the instantaneous angular velocity.

Example 11-10

The differential gear ratio in a car is 3.0 to 1.0 (the drive shaft rotates 3.0 times for each rotation of the rear wheels). If the rear wheels have a diameter of 28 in and the efficiency (i.e. the ratio of output power to input power) of the system is 80%, calculate the force exerted by the rear wheels when the engine develops 140 hp and causes the drive shaft to rotate at 3 000 rpm.

Solution:

$$\omega = \frac{3\,000\text{ rpm}}{\text{differential ratio}} = \frac{3\,000\text{ rpm}}{3}$$

$$= \frac{3\,000}{3}\left(\frac{2\pi}{60}\frac{\text{rad}}{\text{s}}\right) = 105\text{ rad/s}$$

Since the efficiency $\eta = 80\%$

$$P_{\text{out}} = \eta P_{\text{in}} = (0.80)(140\text{ hp}) = 112\text{ hp}$$

or

$$P_{\text{out}} = (112)(550\text{ ft}\cdot\text{lb/s}) = 6.16\times10^4\text{ ft}\cdot\text{lb/s}.$$

Thus from Eq. 11-23,

$$F = \frac{P_{\text{out}}}{r\omega} = \frac{(6.16\times10^4\text{ ft}\cdot\text{lb/s})}{\left(\frac{14}{12}\text{ ft}\right)(105\text{ rad/s})} = 503\text{ lb}$$

Rotational energy is often transmitted from one point to another by rotating shafts. A torque is applied to one end of the shaft, and a load rotates at the other end; this produces a shearing strain called *torsion* on the shaft. As a result, one end of the shaft is displaced through some angle θ with respect to the other end. The magnitude of this distortion θ is directly proportional to the applied torque:

$$\tau = k\theta$$

where k is a constant for the shaft (Fig. 11-7).

It can be shown that for a cylindrical rod or drive shaft of radius r and length L, the

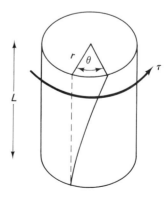

Figure 11-7
Torsion in a cylinder.

torque τ_0 on one end about the axis 0 of the rod is given by:

$$\tau_0 = \frac{(G\pi r^4)\theta}{(2L)} \qquad (11\text{-}24)$$

where G is the shear modulus of the rod and θ is the angular distortion.

*Example 11-11*_____

A solid steel cylindrical drive shaft is 3.0 m long and has a radius of 2.0 cm. Determine the torque delivered and the angular distortion of the drive shaft when it delivers 240 hp and rotates at 1 000 rpm.

Solution:

(a) $\quad \tau = \dfrac{P}{\omega} = \dfrac{(240 \text{ hp})(746 \text{ W/hp})}{1\,000 \left(\dfrac{2\pi}{60} \dfrac{\text{rad}}{\text{s}}\right)}$

$\quad = 1\,710 \text{ N} \cdot \text{m}$

(b) From Table 5-2 and Eq. 11-24:

$\theta = \dfrac{2L\tau}{G\pi r^4} = \dfrac{2(3.0 \text{ m})(1\,710 \text{ N} \cdot \text{m})}{(8.4 \times 10^{10} \text{ Pa})\pi(2.0 \times 10^{-2} \text{ m})^4}$

$\quad = 0.24 \text{ rad} = 14°$

*Problems*_____

20. Determine the rotational kinetic energy of a 20.0 lb cylindrical drive shaft if its diameter is 4.00 in and it rotates at 1 200 rpm.

21. Determine the rotational kinetic energy of a 2.0 kg object if its radius of gyration is 30 cm and its angular velocity is 450 rpm.

22. A solid sphere and a thin hollow sphere with equal masses and equal radii of 3.0 cm are allowed to roll freely from rest down a 20° incline through a vertical height of 1.0 m. If there are no frictional losses, compare their speeds at the bottom of the incline.

23. A drive shaft is coupled directly to a disc of diameter 30.0 cm. If an electric motor delivers 85.0 kW and causes the drive shaft

to rotate at 1 000 rpm, determine the force exerted at the rim of the disc.

24. The differential gear ratio of a truck is 5.0 to 1.0; the rear wheels have a diameter 70 cm; and the efficiency of the system is 60%. Determine the power delivered if the drive shaft rotates at 1 000 rpm and the rear wheels exert a total force of 200 N.

25. A solid cylindrical drive shaft is 3.0 m long and has a radius of 2.0 cm. If the shear modulus is 2.4×10^{10} Pa, determine the torque delivered and the angular distortion of the drive shaft when it delivers 150 kW and rotates with an angular velocity of 2 300 rpm.

11-4 ANGULAR MOMENTUM

The rotational analog of linear momentum is called *angular momentum b*. It is used in atomic physics and in interactions between rotating objects.

Consider an object of mass m and instantaneous velocity $\mathbf{v}$ moving at a distance r from a reference axis 0 (Fig. 11-8); the object has an instantaneous linear momentum $\mathbf{p} = m\mathbf{v}$. The angular momentum of the object with respect to the axis 0 is defined as the product

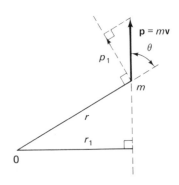

Figure 11-8

of p and the perpendicular distance r_1 to the direction of **p**.

$$b = r_1 p$$

or

$$b = rp \sin \theta = rmv \sin \theta \qquad (11\text{-}25)$$

where θ is the angle between the straight line joining m with 0 and the direction of the linear momentum **p**. Angular momentum is expressed in kg·m²/s in SI and in slug·ft²/s in the British Engineering System.

When a rigid object performs pure rotary motion about some axis 0, all the particles in that object (except those along 0) are in pure circular motion, and their velocity is always tangent to the circle. The total angular momentum b_{TOT} of the rigid object is the sum of the angular momenta b_i, possessed by its constituent particles.

$$b_{TOT} = \Sigma_i b_i = \Sigma_i r_i m_i v_i = \Sigma_i m_i r_i^2 \omega$$
$$= (\Sigma_i m_i r_i^2)\omega = I\omega \qquad (11\text{-}26)$$

where I is the moment of inertia of the object and ω is the angular velocity of the rigid object, since $v = r\omega$.

Example 11-12

Determine the angular momentum of a 96 lb solid cylindrical drive shaft with a 4.0 in diameter if it rotates with an angular velocity of 300 rpm.

Solution:

$$b = I\omega = \left(\frac{mr^2}{2}\right)\omega$$

$$= \frac{(96/32 \text{ slug})(1/6 \text{ ft})^2}{2} \cdot (300 \text{ rpm})\left(\frac{2\pi \text{ rad/s}}{60 \text{ rpm}}\right)$$

$$= 1.3 \text{ slug-ft}^2/\text{s}$$

Consider a particle of mass m that is located in a rotating rigid object at a distance r from the axis of rotation. When a tangential force $\mathbf{F}_T$ is applied for a very short time interval Δt, it produces a change $\Delta \mathbf{p}$ in the particle's momentum such that $\mathbf{F}_T \Delta t = \Delta \mathbf{p}_i$. The resulting torque is:

$$\tau_0 = rF_T = r\frac{\Delta p_i}{\Delta t} = \frac{\Delta b}{\Delta t}$$

where Δb is the change in the angular momentum of the mass m. Thus, for the rotating rigid object, the total torque is equal to the sum of the torques on each of its constituent particles.

$$\tau_0 = \Sigma\frac{\Delta b}{\Delta t} = \frac{\Delta b_{TOT}}{\Delta t} \qquad (11\text{-}27)$$

where Δb_{TOT} is the change in the total angular momentum of the rigid object. If there is no net external torque, the total angular momentum of the rotating rigid object is constant. This is a direct result of Newton's third law as it pertains to rotational motion.

It should be noted that since

$$\tau_0 = I_0\alpha = I_0\frac{\Delta \omega}{\Delta t}$$

if there is no net external torque, any change in angular momentum must be compensated for by a change in angular acceleration. For example, when an ice skater executes a spin, he brings his arms closer to his body, reducing his moment of inertia and consequently increasing his angular velocity.

Problems

26. Determine the angular momentum of a 100 kg solid cylindrical drive shaft of diameter 8.0 cm if it rotates with an angular velocity of 800 rpm.

27. Determine the angular momentum of a 9.1×10^{-31} kg electron in a hydrogen atom if it is in a circular orbit of radius 5.3×10^{-11} m with a tangential speed of 2.2×10^6 m/s.

28. Determine the angular momentum of a 5.0 lb cylindrical flywheel of radius 6.0 in when it rotates with an angular velocity of 1 600 rpm.

11-5 COMPARISON BETWEEN LINEAR AND ANGULAR MOTION

While many of the equations of rotational motion may appear to be new, they are easily obtained from the equations governing translational motion. In order to obtain the rotational equation from the translational equation, start with the translational equation, and replace displacement s by angular displacement θ, velocity v by angular velocity ω, acceleration a by angular acceleration α, mass m by moment of inertia I, force F by torque τ and momentum p by angular momentum b (Table 11-3).

Table 11-3
Comparison of Translational and Rotational Equations

	Translational	Rotational
Newton's second law	net $\mathbf{F} = m\mathbf{a}$	net $\tau = I\alpha$
Kinetic energy	$E_k = \frac{1}{2}mv^2$	$E_k = \frac{1}{2}I\omega^2$
Work	$W = F_T s$	$W = \tau\theta$
Power	$P = F_T v$	$P = \tau\omega$
Impulse	$F\Delta t = \Delta p$	$\tau\Delta t = \Delta b$

REVIEW PROBLEMS

1. A flywheel requires 3.0 s to rotate through 40 rev. If its angular velocity after the 3.0 s is 108 rad/s, find (a) the initial angular velocity and (b) the average angular acceleration.

2. A flywheel rotating at 1 500 rpm decelerates uniformly to rest in 25 s. Calculate (a) the angular acceleration and (b) the number of revolutions that it makes while decelerating to rest.

3. A drive shaft of a car rotating at 500 rpm accelerates uniformly at 25 rad/s² for 3.0 s. What is the final angular velocity?

4. Determine the moment of inertia of a 3.0 lb solid cylinder about its axis if its radius is 0.50 in.

5. Determine the torque required to accelerate a 3.0 kg flywheel from 120 rpm to 1 500 rpm in 22 s if the radius of gyration about its axis is 2.0 cm.

6. A uniform solid disc with a mass of 60 kg and a diameter of 50 cm is used as a flywheel in an engine. What torque must be supplied to accelerate the flywheel from rest to 2 500 rpm in 12.0 s?

7. The flywheel of an engine has a moment of inertia of 75 kg·m². If the flywheel produces an angular acceleration of 0.65 rad/s² in the drive shaft, what is the torque on the drive shaft?

8. A car engine delivers 60 hp at 3 200 rpm; if there is no slippage between the tires and the road; the differential gear ratio is 6:1, and the wheels are 24 in diameter, calculate the thrust at each wheel.

9. Calculate the torque that must be applied to a solid disc-shaped flywheel with a mass of 80 kg and radius of 20 cm in order to increase its angular velocity from 500 rpm to 1 200 rpm in 14 s.

10. What is the power generated by a drive shaft turning at 3 200 rpm if it delivers a torque of 2 100 N·m?

11. Determine the power delivered by an electric motor that exerts a torque of 1 500 N·m on a drum, turning it at 15 rpm.

12. What is the tangential speed of the tip of a propeller that rotates at 4 000 rpm in a circular arc with a diameter of 8.0 ft?

13. Determine (a) the moment of inertia and (b) the applied torque required to accelerate a steel roller that has a diameter of

50 cm and a length of 3.0 m from rest to 6.0 rpm in 50 s. *Hint:* Compute the mass from the density and volume.

14. A heavy 120 kg flywheel is in the form of an annular cylinder with inner and outer radii of 45 cm and 60 cm, respectively. (a) What torque is required to accelerate it from rest to 1 200 rpm in 5.0 min? (b) If the flywheel is then loaded and comes to rest in an additional 12.0 min, what torque does it deliver?

15. If a steel drive shaft of a ship is 15 m long and has a diameter of 15 cm, find the angle of twist when it delivers 20 000 kW at 250 rpm to the propellers.

16. The engine of a car develops 120 hp at 3 200 rpm while travelling along a flat horizontal road. If the wheel diameter is 28 in and the differential gear ratio is 4.5 to 1, calculate the thrust developed at each rear wheel. Assume no losses.

17. A car wheel in the shape of an annular cylinder with inner and outer radii of 8.0 in and 14 in, respectively, has a total weight of 20 lb. Neglecting the spokes, calculate the torque required to accelerate the wheel from rest to 76 rad/s in 15 s.

18. Determine the moment of inertia of a solid disc-shaped flywheel that has a total mass of 5.0 kg and a diameter of 30 cm.

19. Calculate the angular momentum of a solid cylindrical drive shaft that has a mass of 12.0 kg, a diameter of 10.0 cm and rotates at 1 200 rpm.

20. A light string hangs over a frictionless, solid cylindrical "pulley" that has a mass of 2.0 kg and a radius of 30 cm. One end of the string is attached to a 3.0 kg load that is 2.0 m above the floor, and the other end is attached to a 2.0 kg load that is initially at rest on the floor (similar to Fig. 11-6). If the 3.0 kg load turns the pulley as it falls to the floor, determine (a) the acceleration of the 3.0 kg load, (b) the tensions in the string, (c) the change in potential energy of the system, (d) the angular velocity of the pulley just before the 3.0 kg load strikes the floor.

21. Determine the rotational kinetic energy of a 1.2 kg cylindrical drive shaft if its diameter is 2.0 cm and it rotates at 3 000 rpm.

22. A solid circular cylinder rolls from rest down an incline through a vertical height of 12.0 ft. What is its speed at the bottom of the incline?

23. A solid cylindrical drive shaft is 3.0 ft long and has a radius of 1.5 in. If the shear modulus is 9.0×10^6 lb/in², determine the torque delivered and the angular distortion of the drive shaft when it delivers 200 hp and rotates at 3 000 rpm.

24. Calculate the angular momentum of a 3.0 lb flywheel that has a radius of gyration of 4.0 in and rotates at 2 000 rpm.

Man by himself is a rather puny being, but his ability to construct and use tools and machinery has enabled him to dominate many of the physically more powerful forces in his environment. The development of machinery has contributed considerably to the rapid advances of many of his technologies.

A *machine* is a device that may be used to transmit or modify applied forces and energies advantageously in a predetermined manner. With machinery, man can use his own power or the energy from some fuel to lift heavy objects, bend or break the strongest of materials, travel at great speeds and perform many other tasks that he would otherwise find impossible.

Even though many machines in use today appear to be quite complex in construction, essentially they are all composed of one or more of three basic simple machines: the lever, the inclined plane and the hydraulic press (Chapter 13).

12-1 MECHANICAL ADVANTAGE AND EFFICIENCY

Man has found through experience that he can perform certain tasks more easily in some ways than in others. It is easier to pull down than to pull up on a rope; it is easier to push rather than pull objects; in order to do a particular amount of work, it is easier to exert a small force through a large displacement than a large force through a small displacement. Machinery is used to change the magnitude, direction and point of application of required forces in order to make tasks easier.

A measure of a machine's ability to assist the user is called the *mechanical advantage* of that machine; it is defined as the ratio of output force to the input force. Thus, if a total *effort* (input force) $\mathbf{F}$ must be applied to the machine in order to overcome a particular *load* (output force) $\mathbf{w}$ on the machine, the *actual mechanical advantage* (A.M.A.) of the machine is the ratio of the magnitudes of output force to input force.

12

Simple Machines

$$\text{A.M.A.} = \frac{\text{output force}}{\text{input force}} = \frac{\text{load}}{\text{effort}} = \frac{w}{F} \quad (12\text{-}1)$$

The output of useful work from any machine can never exceed the total input of work and energy. Even machines are not capable of perpetual motion! Thus the actual mechanical advantage gained due to the difference between input and output forces is always lost elsewhere, mainly in the distances through which these forces must act.

Friction between the moving parts of a machine always results in a loss of mechanical energy or power. In complex machines, which may have many moving parts, these losses may be very substantial. The effects of friction are accounted for in a term called the *efficiency* (η) of the machine, which is defined as the ratio of useful work output W_{out} to total work W_{in} or energy supplied to that machine.

$$\eta = \frac{\text{useful work output}}{\text{total work input}} = \frac{W_{out}}{W_{in}} \quad (12\text{-}2)$$

Thus, if s is the distance moved by the effort (input force) $\mathbf{F}$, h is the corresponding distance moved by the load (output force) $\mathbf{w}$, and the forces act in the same directions as their displacements, then from the definition of work:

$$\eta = \frac{W_{out}}{W_{in}} = \frac{w \cdot h}{F \cdot s} = \frac{w/F}{s/h} \quad (12\text{-}3)$$

For an *ideal machine* in which there are no frictional losses, the efficiency $\eta = 1$, and therefore its mechanical advantage is:

$$\text{M.A.} = \frac{w}{F} = \frac{s}{h}$$

Since the linkages and moving parts of a machine are usually rigid, the ratio of the distance moved by the effort to the useful distance moved by the load (s/h) represents the maximum possible mechanical advantage for that machine; it is called its *ideal mechanical advantage* (I.M.A.) or *velocity ratio*. Frictional losses are accounted for in the A.M.A.

Thus the efficiency of any machine is:

$$\eta = \frac{W_{out}}{W_{in}} = \frac{w/F}{s/h} = \frac{\text{A.M.A.}}{\text{I.M.A.}} \quad (12\text{-}4)$$

or, expressed as a percentage,

$$\eta = \frac{W_{out}}{W_{in}} \times 100\% = \left(\frac{w/F}{s/h}\right) \times 100\%$$
$$= \left(\frac{\text{A.M.A.}}{\text{I.M.A.}}\right) \times 100\% \quad (12\text{-}5)$$

Since the elapsed time is the same for input and output of work by the machine, the efficiency may also be expressed as the ratio of its output power P_{out} to the input power P_{in}:

$$\eta = \frac{P_{out}}{P_{in}} \quad \text{or} \quad \eta = \left(\frac{P_{out}}{P_{in}}\right) \times 100\% \quad (12\text{-}6)$$

Example 12-1

A pump is used to raise water at a rate of 150 L/min from the bottom of an 80.0 m well. If the efficiency is 80%, what power is supplied to the pump?

Solution:

Since 1.0 L of water has a mass of 1.0 kg, the rate at which the water is raised = 150 kg/min = 150 kg/60 s = 2.50 kg/s. Thus

$$P_{out} = \frac{W_{out}}{t} = \frac{mgh}{t}$$
$$= \left(2.50 \frac{\text{kg}}{\text{s}}\right)(9.81 \text{ m/s}^2)(80.0 \text{ m}) = 1\,960 \text{ W}$$

Therefore

$$P_{in} = \frac{P_{out}}{\eta} = \frac{1\,960 \text{ W}}{0.80} = 2\,450 \text{ W}$$
$$= \frac{2\,450}{746} \text{ hp} = 3.28 \text{ hp}$$

Problems

1. Determine the efficiency of a machine if its A.M.A. is 3.5 and its I.M.A. is 8.2.

2. How much input power is required to operate a laser with an output of 2.5 mW if its efficiency is 4%?

3. A mechanical pump lifts 10 kg of water through a vertical height of 10 m in 15 s. If 100 W input power is supplied to the pump, what is its efficiency?

4. A 3 200 lb car climbs a 15° hill at a constant speed of 20 mi/h. If only 55% of the total horsepower developed by the engine drives the wheels, determine the total horsepower developed by the engine.

5. A mechanical hoist raises 2 000 lb through a vertical height of 2.0 ft in 15 s. If the efficiency is 60%, determine the minimum horsepower that must be supplied to the hoist.

12-2 THE INCLINED PLANE

It usually requires less force to push a load up a shallow incline than to lift the load vertically; therefore the incline itself is a form of a machine. Some common examples in which inclined planes are utilized as machines are ramps, staircases, wedges, screws, cams and the hills of roads.

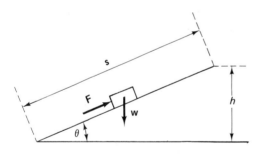

Figure 12-1

If a load w is pushed a distance s by a force F to a vertical height h (Fig. 12-1), then from the definitions:

$$\text{A.M.A.} = \frac{\text{load}}{\text{input force}} = \frac{w}{F}$$

and

$$\text{I.M.A.} = \frac{\text{distance moved by the effort}}{\text{useful distance moved by load}}$$

$$= \frac{s}{h} = \frac{1}{\sin \theta}$$

where θ is the incline of the plane. Thus, the efficiency is:

$$\eta = \frac{\text{A.M.A.}}{\text{I.M.A.}} = \frac{w/F}{s/h} = \frac{w \sin \theta}{F} \quad (12\text{-}7)$$

Friction forces are always present to some degree between the incline and the load; they act parallel to the incline, resulting in efficiencies of less than 100%.

Example 12-2

A man must exert a minimum force of 75 lb in order to push a 300 lb load up a 10° inclined plane. What is (a) the A.M.A., (b) the I.M.A., (c) the efficiency of this inclined plane as a machine?

Solution:

(a) $\quad \text{A.M.A.} = \frac{w}{F} = \frac{300 \text{ lb}}{75 \text{ lb}} = 4.0$

(b) $\quad \text{I.M.A.} = \frac{s}{h} = \frac{1}{\sin \theta} = \frac{1}{\sin 10°} = 5.76$

(c) $\quad \eta = \frac{\text{A.M.A.}}{\text{I.M.A.}} = \frac{4.0}{5.76} = 0.69 \quad \text{or} \quad 69\%$

Wedges, as machines, are commonly used as cutting instruments, such as knives and axes, or as inserts capable of raising large loads through small vertical displacements (Fig. 12-2). The I.M.A. $= s/h$, but large friction losses between wedges and their loads usually give the wedge a relatively low efficiency as a machine; even so, it is still very useful. Rotary wedges find frequent uses (as *cams*) in complex machines. If r and R are the minimum and maximum distances, respectively, from the rim of the cam to the pivot and the circumference of the cam is $2L$ (Fig. 12-3), then its

$$\text{I.M.A.} = \frac{\text{distance moved by effort}}{\text{useful distance moved by load}}$$

$$= \frac{L}{R-r} \quad (12\text{-}8)$$

Screws and screw jacks consist of an inclined plane that spirals around a cylinder. The distance between successive crests of the thread is called the pitch p (Fig. 12-4). One

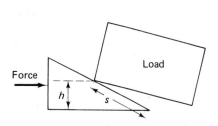

(a) Wedge used as a
cutting instrument

(b) Wedge used as an insert
to raise a large load

Figure 12-2
Wedges.

complete rotation of the screw or screw jack moves the load through a distance equal to the pitch. Thus for a screw,

$$\text{I.M.A.} = \frac{\text{circumference of circle moved by screwdriver handle}}{\text{pitch of the thread}}$$

and for a jack,

$$\text{I.M.A.} = \frac{2\pi L}{p} \qquad (12\text{-}9)$$

where L is the length of the handle.

*Example 12-3*_____

A screw jack with a thread pitch of 0.200 cm and a handle of length 50.0 cm is used to lift a 2 000 kg load. If the efficiency of the jack is 10%, what is the minimum force exerted at the end of the jack handle?

Solution:

The minimum force must be applied perpendicularly to the handle in the direction of the rotation. From Eqs. 12-4 and 12-9, we obtain:

$$\frac{w}{F} = \text{A.M.A.} = \eta(\text{I.M.A}).$$

and

$$\text{I.M.A.} = \frac{2\pi L}{p}$$

Thus

$$F = \frac{w}{\eta(\text{I.M.A.})} = \frac{mg}{\eta}\left(\frac{p}{2\pi L}\right)$$
$$= \frac{(2\,000\text{ kg})(9.81\text{ m/s}^2)}{0.100}\left(\frac{0.200\text{ cm}}{2\pi \times 50.0\text{ cm}}\right)$$
$$= 1.25 \times 10^2\text{ N}$$

In order to push a load up an inclined plane, the applied force must not only overcome the component of the load down the

Figure 12-3
The cam.

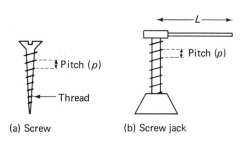

(a) Screw

(b) Screw jack

Figure 12-4

plane but also the friction force between the load and that plane. Therefore, the efficiency of an inclined plane as a machine is dependent on the incline of the plane and the nature of its surface.

Example 12-4

A 120 kg load is raised vertically through 2.0 m by being pushed at a constant speed up a ramp which is inclined at 15° with the horizontal (Fig. 12-5a). If the coefficient of kinetic friction between the ramp and the load is 0.200, (a) what force must be exerted parallel to the ramp? (b) What is the efficiency of the ramp?

Solution:

Draw the free-body diagram (Fig. 12-5b) and resolve the forces into components parallel and perpendicular to the incline:

Force	x-Component	y-Component
Aplied force F_A	F_A	0
Weight of load $w = mg$	$-mg \sin \theta$	$-mg \cos \theta$
Normal force N	0	N
Friction force $F_F = \mu N$	$-\mu N$	0

Since the load is not accelerating, it is in equilibrium. Thus

$$\Sigma F_y = -mg \cos \theta + N = 0$$

or

$$N = mg \cos \theta \quad (1)$$

$$\Sigma F_x = F_A - mg \sin \theta - \mu N = 0$$

or

$$F_A = mg \sin \theta + \mu N \quad (2)$$

Substituting for N from (1) into (2), we obtain:

(a)
$$F_A = mg(\sin \theta + \mu \cos \theta)$$
$$= (120 \text{ kg})(9.81 \text{ m/s}^2)$$
$$\times [\sin 15° + (0.200) \cos 15°]$$
$$= 532 \text{ N}$$

(b)
$$\eta = \frac{\text{A.M.A.}}{\text{I.M.A.}} = \frac{w/F_A}{s/h} = \left(\frac{mg}{F_A}\right)\left(\frac{h}{s}\right)$$

$$= \left[\frac{mg}{mg(\sin \theta + \mu \cos \theta)}\right] \sin \theta$$

Thus

$$\eta = \frac{\sin \theta}{\sin \theta + \mu \cos \theta}$$

$$= \frac{\sin 15°}{\sin 15° + (0.200) \cos 15°}$$

$$= 0.57 \quad \text{or} \quad 57\%$$

Note that only the incline of the plane and the coefficient of kinetic friction appear in this expression of efficiency.

Problems

6. A man pushes a 100 kg frictionless trolley with a constant speed up a 15° incline. (a) What minimum force must he exert?

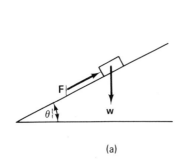

(a)

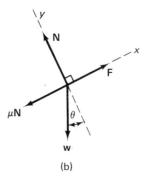

(b)

Figure 12-5

(b) Determine the A.M.A. and I.M.A.
(c) What is the efficiency of the incline?

7. A minimum force of 100 lb is required to push a 200 lb load up a 20° incline. Determine (a) the A.M.A., (b) the I.M.A., (c) the efficiency of the incline. If the load is raised vertically through 3.0 ft, determine (d) the work done on the load, (e) the energy lost to friction.

8. A screw jack with a thread pitch of 0.120 in and a handle of length 24.0 in is used to lift a 500 lb load. If the efficiency of the jack is 18%, determine the minimum force exerted at the end of the jack handle.

9. In order to raise a 150 kg load vertically through 3.0 m, a man pushes it with a constant speed up a 15° incline. If the coefficient of kinetic friction between the incline and the load is 0.18, determine (a) the efficiency of the incline, (b) the minimum force exerted by the man, (c) the increase in potential energy of the load, (d) the total energy lost to friction.

12-3 THE LEVER

The *lever* is a very efficient, yet simple machine. It consists of a rigid bar that is free to rotate about a fixed point, called the *fulcrum*. Levers are categorized into three classes according to the position (P) of the fulcrum with respect to the effort (input force) $\mathbf{F}$ and the load (output force) $\mathbf{w}$ (Fig. 12-6).

Friction losses are usually quite small in lever actions because of the small area of contact at the fulcrum. Therefore, when the lever is in equilibrium, the equilibrium conditions for nonconcurrent forces may be used to solve for unknown quantities. Thus, if l_{in} and l_{out} are the lengths of the lever arms from the fulcrum to the effort and fulcrum to load, respectively, taking moments about the fulcrum for all three classes in Fig. 12-6, we obtain:

$$\Sigma M_p = l_{in}F - l_{out}w = 0$$

Therefore

$$\frac{w}{F} = \frac{l_{in}}{l_{out}} = \text{A.M.A.} \qquad (12\text{-}10)$$

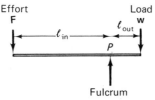

(a) *Class (1)*—Fulcrum between effort and load.
Examples: scissors, pliers, valve rocker arms.

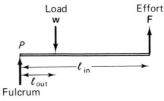

(b) *Class (2)*—Load between fulcrum and effort.
Examples: wheelbarrow, nutcrackers.

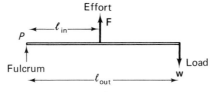

(c) *Class (3)*—Effort between fulcrum and load.
Example: human forearm, tweezers.

Figure 12-6
The lever.

If the effort moves through an arc length s and the load through a corresponding arc length h, they subtend the same angle θ at the fulcrum. Thus

$$\theta = \frac{s}{l_{in}} = \frac{h}{l_{out}} \quad \text{and} \quad \text{I.M.A.} = \frac{s}{h} = \frac{l_{in}}{l_{out}}$$

Example 12-5

A man inserts one end of a 7.0 ft long steel bar beneath a 720 lb rock and places a smaller rock, which acts as a fulcrum, one foot from the load (Fig. 12-7). (a) What is the minimum force that the man must apply to the other end of the steel bar in order to pry up the rock? (b) What is the I.M.A. of this machine?

Solution:

(a) Taking moments about the fulcrum, we obtain:

$$\Sigma M_p = (6.0 \text{ ft})F - (1.0 \text{ ft})(720 \text{ lb}) = 0$$

Therefore

$$F = \frac{(1.0 \text{ ft})(720 \text{ lb})}{6.0 \text{ ft}} = 120 \text{ lb}$$

(b) $\quad \text{I.M.A.} = \dfrac{s}{h} = \dfrac{l_{in}}{l_{out}} = \dfrac{6.0 \text{ ft}}{1.0 \text{ ft}} = 6.0$

Problems

10. A 2.0 m long crowbar is used as a first-class lever in order to raise a 120 kg load. If the fulcrum is located 30 cm from the load, (a) what minimum force must be applied to the other end of the crowbar? (b) What is the I.M.A.?

11. The center of gravity of a 100 lb load is located 1.5 ft from the axle (which acts as the fulcrum) of a wheelbarrow. (a) What minimum force must be exerted on the handles, 5.0 ft from the axle, in order to raise the load? (b) What is the I.M.A.?

12-4 PULLEYS

A *pulley* is a wheel that is pivoted so that it is free to rotate about its center. The circumference of a pulley is often grooved or notched in order to give it a greater hold on belts or other pulleys. Pulleys use the principle of the lever, but they usually are not as efficient because of the frictional losses in their bearings and the mass of the pulley itself. However, systems of one or more pulleys can be used to change the direction and magnitude of required forces and also to transmit torques. They have many applications as hoists, belt drives, drive shafts and gears.

A simple pulley is often used to change the direction of a required force so that the user can pull down rather than up on a belt (chain or rope). In this case, since equal distances are moved by the effort and the load, the I.M.A. $= 1$. If friction effects and the mass of the pulley are negligible, the tension in the belt must be the same on each side of the pulley (Fig. 12-8).

The *wheel and axle* consists of a large diameter pulley (the *wheel*) and a small diameter pulley (the *axle*), which are joined and rotate together about the same axis. The effort F is supplied to a belt, which is wrapped

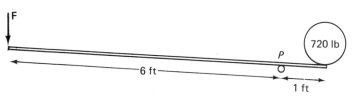

Figure 12-7

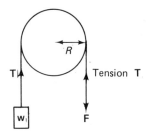

Figure 12-8
A simple pulley.

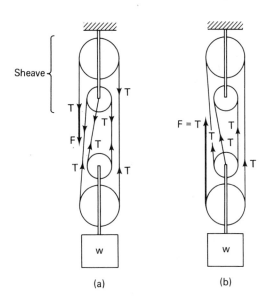

Figure 12-10
The block and tackle.

around the wheel, and the load *w* is raised by a belt wrapped around the axle (Fig. 12-9). The distances moved by the effort and the load correspond to the amount of the belt that winds off or on the wheel and axle, respectively. Since both pulleys rotate through the same angle θ:

$$\text{I.M.A.} = \frac{\text{distance moved by effort}}{\text{distance moved by load}}$$
$$= \frac{R\theta}{r\theta} = \frac{R}{r} \qquad (12\text{-}11)$$

where *R* and *r* are the radii of the wheel and the axle respectively.

The *block and tackle* is a more complex system of groups (*sheaves*) of pulleys, which are connected by the same belt. The load is suspended from the lower sheave (Fig. 12-10).

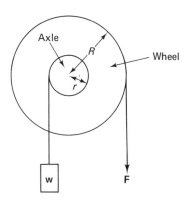

Figure 12-9
The wheel and axle.

Usually, it is desirable that the effort pull down. Therefore, if there are equal numbers of pulleys in each sheave, the belt is attached to the upper sheave; however, if the top sheave has more pulleys, the belt is attached to the lower sheave. It is also possible to have the effort pulling upwards (Fig. 12-10b).

In order to calculate the I.M.A. of a block and tackle, it is assumed that all of the strings between the pulleys are approximately parallel. Therefore, if friction and the weights of the strings and lower pulleys are neglected (since each pulley merely changes the direction of the force), the tension *T* is the same in each string and is numerically equal to the effort *F*. The I.M.A. is determined by considering the tensions in the strings, which act on the lower sheave, or by the principle of work. For example, in the system in Fig. 12-10a, each string that acts on the lower sheave pulls up on the load with a force equal in magnitude to the effort *F*. Thus, in equilibrium, since four strings act on the lower sheave:

$$w = 4F$$

Since friction is ignored, the efficiency is 100%; thus

$$\text{A.M.A.} = w/F = \text{I.M.A.} = 4$$

Similarly in Fig. 12-10b, since five strings pull up on the load, the I.M.A. = 5. In general

I.M.A. = number of strings that are
pulling upwards on the lower   (12-12)
sheave

*Example 12-6*_____

If the block-and-tackle system illustrated in Fig. 12-11 is used to raise a 180 kg steel girder through a vertical height of 12 m, (a) what is the I.M.A.? (b) How much of the belt must be pulled through the system? (c) What minimum effort is required if the efficiency is 95%?

Solution:

(a)   I.M.A. = number of strings pulling up
              on the load
       = 6

(b) $$\text{I.M.A.} = \frac{s}{h}$$

Thus

$$s = h(\text{I.M.A.}) = (12.0 \text{ m})(6) = 72 \text{ m}$$

(c) $$\eta = \frac{\text{A.M.A.}}{\text{I.M.A.}} = \frac{w/F}{\text{I.M.A.}}$$

Thus

$$F = \frac{w}{\eta(\text{I.M.A.})} = \frac{(180 \text{ kg})(9.8 \text{ m/s}^2)}{(0.95)(6)}$$
$$= 310 \text{ N}$$

A *chain hoist* (*Weston differential pulley*) consists of an upper block of two pulleys (one with a larger diameter than the other) and a single lower pulley to which the load w is attached (Fig. 12-12). An "endless belt" passes over the larger (A) of the upper block pulleys, around the lower pulley (B) and then around the smaller upper block pulley (C)

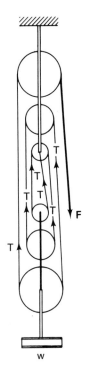

Figure 12-11

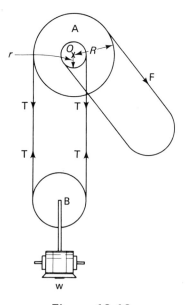

Figure 12-12
The chain hoist.

leaving some slack. The effort F is applied as shown. If friction and the weight of the belt and the lower pulley are neglected and the system is in equilibrium, the tensions **T** in the belt support the load **w**. Thus

$$2T = w \qquad (1)$$

If R and r are the radii of the larger and smaller pulleys of the upper block, respectively, taking moments about the axis of rotation (0) of the upper block, we obtain:

$$\Sigma M_0 = RF + rT - RT = 0$$

or from (1),

$$RF = (R - r)T = (R - r)\left(\frac{w}{2}\right)$$

Thus, if the system is frictionless and the mass of the chain hoist is neglected:

$$\text{I.M.A.} = \frac{w}{F} = \frac{2R}{R - r} \qquad (12\text{-}13)$$

*Example 12-7*_____

A chain hoist that has upper pulleys of diameters of 24.0 cm and 20.0 cm and an efficiency of 92% is used to raise a 200 kg machine through 2.00 m. (a) How much of the belt must be pulled through the system by the applied force? (b) What minimum applied force is required?

Solution:

(a) $\text{I.M.A.} = \dfrac{s}{h} = \dfrac{2R}{R - r}$

$$= \frac{2(24.0 \text{ cm}/2)}{(24.0 \text{ cm}/2) - (20.0 \text{ cm}/2)} = 12.0$$

Thus

$$s = h(\text{I.M.A.}) = (2.00 \text{ m})(12.0) = 24.0 \text{ m}$$

(b) $\qquad \eta = \dfrac{\text{A.M.A.}}{\text{I.M.A.}} = \dfrac{w/F}{\text{I.M.A.}}$

Therefore

$$F = \frac{w}{\eta(\text{I.M.A.})} = \frac{(200 \text{ kg})(9.81 \text{ m/s}^2)}{(0.92)(12.0)}$$
$$= 178 \text{ N}$$

_____

Gears are used extensively in machinery. They are essentially pulleys that have notches

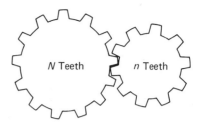

(a) Gear Ratio $= \dfrac{N}{n}$

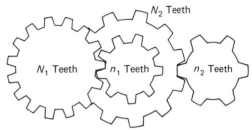

(b) Gear Ratio of Gear Train $= \dfrac{N_1}{n_1} \times \dfrac{N_2}{n_2}$

Figure 12-13
Gears.

around their circumference in order to give them a greater hold on other gears (Fig. 12-13). The same principles of mechanical advantage and efficiency are used as for pulleys. Gears are capable of delivering large torques because the teeth of the gears intermesh; there is no slippage of a belt or chain involved. The rate of rotation may be controlled by fixing the relative numbers of teeth in interacting gears. The *gear ratio* is the ratio of the numbers of teeth in two interacting gears:

$$\text{gear ratio} = \frac{\text{number of teeth in one gear}}{\text{number of teeth in the other gear}}$$

If a gear with a large number of teeth is used to drive a gear with a smaller number of teeth, the rotation rate is increased. *Gear trains*, which are composed of a number of interacting gears, are used to change rotation rates by even greater amounts. The total gear ratio of a gear train is the product of the individual gear ratios of the interacting pairs of gears in that train.

Gears, chains and belt drives are used to transmit mechanical power from one rotating shaft to another. When gears are used, these shafts may be parallel or inclined at angles to each other. During the transmission of power, a gear may be used to increase or decrease the torque or angular velocity of the rotation. However, increases in torque result in decreases in the angular velocity and vice versa.

Spur gears are cylindrically shaped and may have teeth on either the inner or outer surface (Fig. 12-14a and b). They are used to connect shafts that have parallel axes of rotation; their main disadvantage is that the load is transferred suddenly from one tooth to another. This type of gear is used extensively.

Helical gears have teeth that are not parallel to the shaft (Fig. 12-14c). As a result, there is no sudden transfer of the load. Consequently these gears are quieter and stronger than spur gears, and they can operate at higher angular velocities.

Mitre and *bevel gears* are used to connect shafts that are not parallel. They are basically cone shaped and have either straight or spiral teeth (Fig. 12-14d and e). The alignment of these cone-shaped gears specifies the angle of intersection of the shafts. The gear is called a *mitre gear* when there are an equal number of teeth and the shafts are perpendicular.

Worm gears are mainly used to increase the torque (decrease the angular velocity) between perpendicular shafts (Fig. 12-14f and g). These gears may have single or multi-thread worms. Their gear ratio is:

$$\text{gear ratio} = \frac{\text{number of teeth on gear}}{\text{number of teeth on worm}}$$

and may be very high. However, their efficiency is relatively low because sliding rather than rolling actions are involved. Worm gears are frequently used in hoists and winches.

Problems

12. Determine the I.M.A. of a wheel and axle if the wheel has a diameter of 2.0 ft and the axle has a diameter of 2.0 in.

13. Determine the I.M.A.'s of the pulley systems in Fig. 12-15 and find the input force required to raise a 50 N load if their efficiency is 100%. *Hint:* Consider free-body diagrams.

14. A large gear with 50 teeth interacts with a smaller gear with 15 teeth. If the larger gear rotates at 100 rev/min, determine (a) the gear ratio and (b) the angular velocity of the smaller gear.

15. Consider the gear train in Fig. 12-13b. (a) If $N_1 = 50$, $n_1 = 10$, $N_2 = 60$ and $n_2 = 10$, determine the gear ratio of the train. (b) If N_1 rotates with an angular velocity of 80 rev/min, what is the angular velocity of N_2?

12-5 COMPOUND MACHINES

Compound machines are constructed by combining two or more simple machines. The total efficiency of the compound machine is equal to the product of the efficiencies of its individual simple machine components. For example, if a compound machine is constructed from a block and tackle, which has an efficiency $\eta_1 = 0.90$, and an inclined plane, which has an efficiency $\eta_2 = 0.40$, the total efficiency $\eta = \eta_1 \times \eta_2 = (0.90)(0.40) = 0.36$.

Problem

16. A compound machine is constructed from a pulley system with an efficiency of 90% and an inclined plane with an efficiency of 65%. What is the total efficiency of the compound machine?

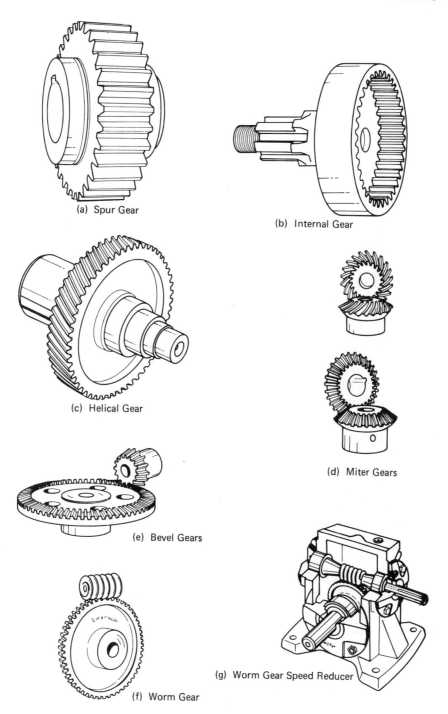

(a) Spur Gear

(b) Internal Gear

(c) Helical Gear

(d) Miter Gears

(e) Bevel Gears

(f) Worm Gear

(g) Worm Gear Speed Reducer

Figure 12-14
Gears.

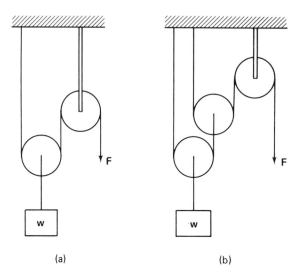

(a)   (b)

Figure 12-15

REVIEW PROBLEMS

1. Determine the efficiency of a machine if its A.M.A. is 4.8 and its I.M.A. is 6.0.

2. What is the output power from an electric motor if its efficiency is 85% and the input power is 2.0 kW?

3. A pump raises 25 lb of water through a vertical height of 12 ft in 1.6s. If the input power to the pump is 1.0 hp, what is its efficiency?

4. A pump raises 120 L of water in 1.0 min through a vertical height of 50 m. If the input power is 1.5 hp, what is the efficiency of the pump?

5. A 1 500 kg car climbs a 10° incline at a constant speed of 40 km/h. If 60% of the total power developed by the engine drives the wheels, what is the total power developed by the engine?

6. A 50 kg block is pushed with a constant speed up a 30° incline through a vertical height of 5.0 m in 20.0 s. If the coefficient of kinetic friction between the block

and the incline is 0.30, find (a) the magnitude of the applied force parallel to the incline, (b) the work done by the applied force, (c) the average power developed, (d) the change in the gravitational potential energy of the block, (e) the A.M.A. of the incline, (f) the I.M.A. of the incline, (g) the efficiency of the incline as a machine.

7. Calculate the I.M.A. of a cam if its circumference is 22 cm, its larger radius is 4.0 cm and its smaller radius is 1.5 cm.

8. A screw jack has a thread pitch of 0.40 cm and a handle of length 80 cm. If the efficiency of the jack is 15%, what minimum force must be applied to the end of the handle to lift a 1 500 kg car?

9. A 10 ft steel rod is used as a first-class lever to raise a 450 lb load. If the load is located at one end of the rod and the fulcrum is 1.0 ft from the load, (a) what minimum force must be applied to the other end of the rod to lift the load? (b) What is the I.M.A.?

10. The center of gravity of a 75 kg load is located 0.60 m from the axle of a wheel-

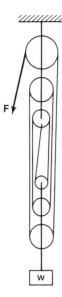

Figure 12-16

barrow. (a) What minimum force must be exerted on the handles 1.6 m from the axle to raise the load? (b) What is the I.M.A.?

11. Determine the I.M.A. of a wheel and axle if the wheel has a diameter of 85 cm and the axle diameter is 5.0 cm.

12. If the efficiency of the block and tackle in Fig. 12-16 is 96%, (a) what is the I.M.A.? (b) What is the A.M.A.? (c) How much work input is required to raise a 1 500 kg car through a vertical height of 3.0 m at a constant speed?

13. A large gear with 40 teeth interacts with a smaller gear with 8 teeth. If the smaller gear rotates at 3 000 rpm, determine (a) the gear ratio and (b) the angular velocity of the larger gear.

14. A compound machine is constructed from a lever system with a total efficiency of 90% and a series of inclined planes with a total efficiency of 40%. What is the total efficiency of the compound machine?

13

Fluids

All matter is composed of small particles called atoms, which may exist alone or in combination with one or more other atoms in the form of molecules. The state of these particles, which depends on the conditions of temperature and pressure, determines whether the matter exists as a solid, liquid or gas.

All molecules attract each other. The attraction between like molecules is called *cohesion*; the attraction between unlike molecules is called *adhesion*. Solids tend to maintain a definite shape and volume because relatively strong cohesive forces fix the relative positions of the atoms. Since the molecules of a liquid are able to flow, a liquid will take the shape of any container to a height that depends on the volume. In general a liquid will tend to have a level surface. It is not easily compressed because the molecules are quite close together.

On the other hand, the molecules of a gas are relatively far apart and flow readily. Consequently gases take the shape of any container and are easily compressed.

The difference among solids, liquids and gases are usually quite obvious, but in some cases they are not easily distinguished. For example, glass appears to be a solid; however, during the manufacturing process, there is no latent heat of fusion (Chapter 14). In fact, the molecules of glass actually flow, but only very slowly.

A *fluid* is defined as any matter that flows. Therefore both liquids and gases are fluids. Many liquids and gases approximate perfect fluids because there is almost no friction between the molecules.

Since we are surrounded by fluids, such as air and water, their uses and properties are of considerable importance. Aircraft, cars and boats are streamlined so that they are able to move through fluids with the least possible resistance. Pipelines and pumps are used to move fluids, such as gas, oil and water, over large distances. In machinery, fluids, such as oil and air, reduce friction and wear between moving parts. Hovercraft even "float" on a cushion of air. Fluids are used as the agent to transfer heat in cooling and heating systems. Because of their incompressibility, liquids are used in machinery, such as the hydraulic press, in order to transfer forces and to gain a mechanical advantage.

13-1 HYDROSTATIC PRESSURE

Solids and liquids are only slightly compressible. Therefore their volumes and hence their densities are approximately constant under normal conditions. However gases are easily compressed, and their densities vary.

Fluids cannot withstand either shearing or tensile stresses because their molecules simply flow. Only compressive stresses, where the applied force is perpendicular to the surface and avoids shear, are possible. The compressive stress in a fluid is called *pressure p*; since the direction of the compressive force depends entirely on the orientation of the surface, pressure is a scalar quantity. The *average pressure* $\bar{p}$ is defined as the total perpendicular force F per unit surface area A:

$$\bar{p} = \frac{F}{A} \qquad (13\text{-}1)$$

Thus the units of pressure may be expressed in terms of a force divided by an area; some common examples are N/m^2, which is called the *pascal* (Pa) in SI, lb/ft^2 and lb/in^2 (psi) in the British Engineering System. Special units are also frequently used: a *millibar* (mb) is equal to 100 Pa.

Significant pressure is also produced by the weight of the air above the earth's surface. The air pressure exerted at sea level on the earth's surface under standard conditions of temperature is called a *standard atmosphere* (atm); it is equal to 101.325 kPa. Occasionally pressures are expressed in terms of the pressure exerted by a column of mercury or water. A standard atmosphere of pressure can support a 760 mm high column of mercury.

Pressures are caused by the weight of fluid above a surface area, by the motion of the fluid molecules, and by the application of a force to a particular area. The molecules of a gas are in continual and rapid motion, often colliding with each other and the walls of any container. It is the constant bombardment of the walls of a container that gives rise to pressure because the molecules change their momentum as a result of these collisions. Thus there are action and reaction forces (according to Newton's third law) on the walls of the container. Since gases are easily compressed, their density varies considerably with depth; the calculation of pressure is relatively complex.

A fluid always exerts forces perpendicular to the walls of any container. The pressure exerted by a column of a fluid is due to the weight that the fluid exerts per unit area. Liquids are essentially incompressible, and their density remains approximately constant with depth.

Consider a horizontal area A at a depth h in a *liquid* of density ρ (Fig. 13-1). The

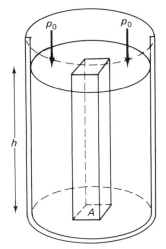

Figure 13-1
The total pressure at a depth of h is the sum of the surface pressure p_0 and the weight **w** per unit surface area A of the liquid column.

force **F** exerted on A is equal to the weight **w** of the column of liquid directly above A. Since the volume of the liquid column $V = Ah$ and density ρ is the ratio of mass m to volume V, the mass of the liquid column $m = \rho Ah$. This column exerts a force of magnitude:

$$F = w = mg = (\rho Ah)g$$

on A, hence the pressure due to the liquid column is:

$$p = \frac{F}{A} = \rho gh = Dh \qquad (13\text{-}2)$$

where g is the acceleration due to gravity and D is the weight density $= \rho g$. If there is also an external pressure p_0 at the surface, it must be added in order to obtain the total pressure p at any depth h in the liquid.

$$p = p_0 + \rho gh = p_0 + Dh \qquad (13\text{-}3)$$

Example 13-1

If the surface of water in a reservoir is 85.0 m above a tap and the atmospheric pressure at the reservoir is 101 kPa, (a) calculate the water pressure at the tap. (b) If the tap nozzle has a diameter $d = 1.25$ cm, what is the force exerted on the tap?

Solution:

The density of water $\rho = 10^3$ kg/m³

(a) $p = p_0 + \rho gh$
$$= 1.01 \times 10^5 \text{ Pa}$$
$$+ (10^3 \text{ kg/m}^3)(9.81 \text{ m/s}^2)(85.0 \text{ m})$$
$$= 9.35 \times 10^5 \text{ Pa} \quad \text{or} \quad 9.26 \text{ atm}$$

(b) $p = \dfrac{F}{A}$

Therefore
$$F = pA = p\left(\frac{\pi d^2}{4}\right)$$
$$F = \frac{(9.35 \times 10^5 \text{ Pa})\pi(1.25 \times 10^{-2} \text{ m})^2}{4}$$
$$= 115 \text{ N}$$

Note that, if the tap is turned on, atmospheric pressure would oppose the water pressure at the tap. The height of the surface of water in the reservoir above the outlet is called the *pressure head*.

Pressure exists at all points in a *liquid* and has the same value at all points located at the same depth, regardless of the size or shape of the container. If this were not so, then an unbalanced force would exist, and the liquid would simply flow until the pressure equalized. *Stationary liquids tend to find a common surface level throughout any container* (Fig. 13-2).

Problems

1. Determine the force exerted on an area of 3.0 ft² at a point in a liquid where the pressure is 30 lb/in².

2. Show that an atmospheric pressure of 14.7 lb/in² is approximately equivalent to 101 kPa.

3. Determine the pressure at the bottom of a 5.0 ft deep tank due to the liquid if it is filled with (a) water, (b) mercury, (c) a liquid with a relative density of 2.4.

4. Determine the pressure on the hull of a diving bell at a depth of 1 000 ft below the surface of sea water, which has an average relative density of 1.35. Assume that atmospheric pressure is 14.7 lb/in².

5. The surface of water in a reservoir is 80.0 m above a tap, and the atmospheric pressure at the reservoir is 101 kPa. Calculate (a) the water pressure at the tap and (b) the force that the water exerts on the tap if the nozzle has a diameter of 1.20 cm.

6. A rectangular water tank with an open top has a depth of 2.00 m and a bottom with dimensions of 3.00 m and 2.50 m. If the atmospheric pressure is 101 kPa, determine (a) the pressure at the bottom of the tank, (b) the pressure on the sides of the tank 1.50 m below the water surface, (c)

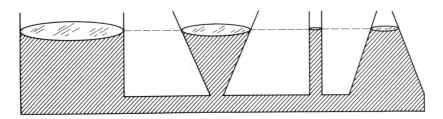

Figure 13-2
Stationary liquids tend to find a common
surface level.

the total force on the bottom of the tank,
(d) the total force on each vertical side of
the tank.

7. A 1.0 ft deep layer of gasoline floats on a
2.0 ft deep body of oil in an open tank.
If the atmospheric pressure is 15 lb/in²,
determine (a) the pressure at the bottom
of the tank and (b) the pressure at the
interface between the two liquids.

13-2 MEASUREMENT OF PRESSURES IN GASES

Gas pressures may be measured with a variety
of different devices. A *mercury barometer*
measures pressures in terms of the height of
a liquid column above an open surface. (Fig.
13-3). The liquid used is usually mercury
because of its high density and low vaporization at room temperatures. A simple mercury
barometer may be constructed from a glass
tube that is closed at one end. The tube is
filled with mercury, the open end is sealed,
the tube is inverted in an open mercury bath,
and the seal is removed beneath the surface
of the bath. The mercury column then drops
until it exerts a pressure that is equal to the
magnitude of the external pressure at the
surface of the bath. Except for a small amount
of mercury vapor, this leaves a vacuum between the closed end of the glass tube and the

top of the mercury column. Thus the external
pressure $p = \rho g h$, where ρ is the density of
mercury, g is the acceleration due to gravity
and h is the height of the mercury column.
The pressure may also be quoted in terms of
the height h of the mercury column (1 atm is
equivalent to 760 mm of mercury).

An *aneroid barometer* consists of one or
more partially evacuated metallic capsules
connected by levers to a cable that winds
around a spindle to a spring. (Fig. 13-4). A
pointer is attached to the spindle, and it
moves over a calibrated scale. The capsules
expand and contract as the air pressure

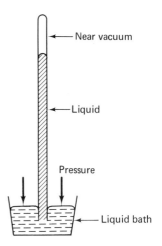

Figure 13-3
The barometer.

varies; these changes are amplified by the system of levers so that there is a change in the tension of the cable, which produces a movement of the pointer over a scale.

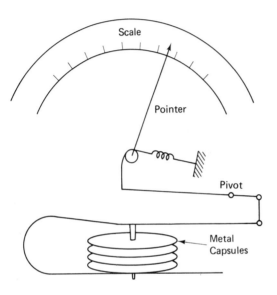

Figure 13-4
Aneroid barometer.

other side is open to atmospheric pressure p_0. When the system is in equilibrium, the difference in the height of mercury in both arms of the tube is related to the difference in pressure between p and p_0 (Fig. 13-5). Thus, $p = p_0 + \rho g h$ if the mercury is higher on the side open to the atmospheric pressure. The value of atmospheric pressure may be determined from a barometer, and therefore the unknown pressure p may be calculated.

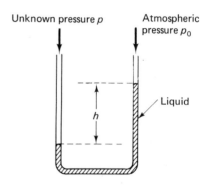

Figure 13-5
The manometer.

Since atmospheric pressure decreases with height above the earth's surface, a barometer may also be used to measure altitudes. A variation of an aneroid barometer, called a *pressure altimeter*, is calibrated in terms of altitudes rather than pressures, and it can be easily adjusted to the prevailing atmospheric pressure. These useful devices are found in all aircraft. However, they do not respond to rapid changes in altitude or the terrain below the aircraft.

A *manometer* is a device that measures the difference between an unknown pressure and atmospheric pressure. This device consists of a U-shaped glass tube that is partially filled with mercury. One side of the tube is open to the unknown pressure p, and the

A *bourdon gauge* is frequently used to measure gas or steam pressures. It consists of a hollow tube that is bent into an arc; the end of the tube is connected via a linkage to a pointer (Fig. 13-6). When gas enters, the tube tends to straighten out under the pressure, and the motion of the tube is transmitted to the pointer.

Gauge pressures indicate the magnitude of the pressure above atmospheric pressure. The total or *absolute pressure* is found by adding the atmospheric pressure to the gauge pressure:

absolute pressure = gauge pressure
+ atmospheric pressure

$$p = p_{\text{gauge}} + p_{\text{atmospheric}} \qquad (13\text{-}4)$$

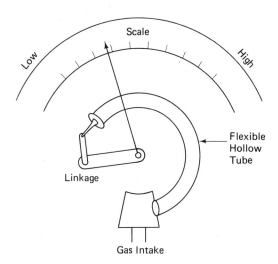

Figure 13-6
Bourdon gauge.

*Problems*_____

8. A manometer is used to measure an unknown pressure. If the atmospheric pressure is 1.01×10^5 Pa, determine the unknown pressure if it supports a 120 cm column of water.

9. If the atmospheric pressure is 15 lb/in², what is the magnitude of a pressure that supports a 24 in column of water in a manometer?

10. Determine the unknown pressure in a manometer if the water level is 24 in *below* the level on the side open to an atmospheric pressure of 15 lb/in².

13-3 PASCAL'S PRINCIPLE

Pressures in liquids may be produced by the weight of the liquid itself or by means of an external mechanical force. The basic principle regarding the transmission of pressures in liquids was discovered by Blaise Pascal (1623–1662). The principle is based on the fact that liquids are practically incompressible.

> *Pascal's principle: When a pressure is applied to a confined liquid, that pressure is transmitted undiminished throughout the extent of the liquid.*

If this were not true, a pressure difference would exist, and the liquid would simply flow until the pressure equalized.

Liquids have the ability to transmit pressures, and they are used in a basic machine called a *hydraulic press* (Fig. 13-7). If a force F_1 is applied perpendicular to the area A_1 of the smaller piston 1, it produces a pressure:

$$p = \frac{F_1}{A_1}$$

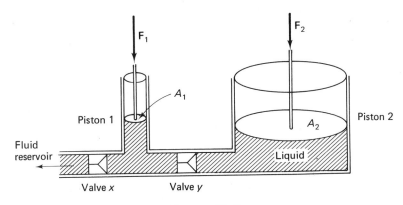

Figure 13-7
The hydraulic press.

which is transmitted undiminished to all parts of the confined fluid. This pressure produces a perpendicular force $F_2 = pA_2$ on the area A_2 of the larger piston 2. Thus

$$p = \frac{F_1}{A_1} = \frac{F_2}{A_2} \quad \text{or} \quad \frac{F_2}{F_1} = \frac{A_2}{A_1} \quad (13\text{-}5)$$

If a load F_2 is placed on the larger piston 2, it may be held in equilibrium by a smaller effort F_1 applied at the smaller piston 1, as long as the system can withstand the pressure. Therefore, the hydraulic press may be used to gain a mechanical advantage. The actual mechanical advantage is:

$$\text{A.M.A.} = \frac{\text{load}}{\text{effort}} = \frac{F_2}{F_1} = \frac{A_2}{A_1} \quad (13\text{-}6)$$

The efficiency of a hydraulic press is usually relatively high; it depends on the compressibility of the liquid and the losses in pressure due to leakage in the valves and along the sides of the pistons. If the effort F_1 at the small piston 1 moves through a distance s_1 and raises the load F_2 at the larger piston 2 a distance s_2, the efficiency of the hydraulic press is:

$$\eta = \frac{\text{useful work output}}{\text{work input}} = \frac{F_2 s_2}{F_1 s_1} = \frac{A_2 s_2}{A_1 s_1} \quad (13\text{-}7)$$

since $F_2/F_1 = A_2/A_1$. Therefore, for an ideal machine in which the efficiency is 100%,

$$F_2 s_2 = F_1 s_1 \quad \text{or} \quad \frac{F_2}{F_1} = \frac{s_1}{s_2} = \frac{A_2}{A_1} \quad (13\text{-}8)$$

and the smaller effort must move through a larger distance than the load.

The load may be moved through greater distances by a series of "pumps" by the effort on piston 1 using the valves to introduce more liquid to the confined area. In order to raise the load, valve x is closed and valve y is opened. The effort moves piston 1 down forcing fluid through valve y as the load is raised by a relatively small amount. Valve y is then closed to prevent fluid from returning to the smaller cylinder and valve x is opened. The smaller piston 1 is then raised again drawing fluid from the reservoir. Valve x is then closed, valve y is opened and the process is repeated.

Example 13-2

The smaller piston of a hydraulic press has a diameter $d_1 = 5.00$ cm, and the larger piston has a diameter $d_2 = 30.0$ cm. (a) If the hydraulic press has an efficiency $\eta = 100\%$, what minimum force is required to lift a 1 600 kg car? (b) If the smaller piston can be pumped through 15.0 cm, how many times must it be pumped to raise the car 3.00 m? (c) If the pump develops an average power of 10.0 hp, how long does it take to raise the load?

Solution:

(a) The force on the larger piston is the weight $w = mg$ of the car. Thus, from Eq. 13-8 we obtain:

$$F_1 = \frac{F_2 A_1}{A_2} = (mg)\frac{(\pi d_1^2/4)}{\pi d_2^2/4} = mg\left(\frac{d_1}{d_2}\right)^2$$

$$= (1\ 600 \text{ kg})(9.81 \text{ m/s}^2)\left(\frac{5.00 \text{ cm}}{30.0 \text{ cm}}\right)^2$$

$$= 436 \text{ N}$$

(b)
$$s_1 = s_2\left(\frac{A_2}{A_1}\right) = s_2\frac{(\pi d_2^2/4)}{(\pi d_1^2/4)} = s_2\left(\frac{d_2}{d_1}\right)^2$$

$$s_1 = (3.00 \text{ m})\left(\frac{30.0 \text{ cm}}{5.00 \text{ cm}}\right)^2 = 108 \text{ m}$$

The length of each stroke of the smaller piston is 15.0 cm; therefore it must be pumped $\dfrac{108 \text{ m}}{0.150 \text{ m}} = 720$ times.

(c) Since $\eta = 100\%$,

$$P_{\text{out}} = W_{\text{out}}/t = P_{\text{in}} = W_{\text{in}}/t$$

Therefore

$$t = \frac{W_{\text{out}}}{P} = \frac{F_2 s_2}{P}$$

$$= \frac{(1\ 600 \text{ kg})(9.81 \text{ m/s}^2)(3.00 \text{ m})}{(10.0 \text{ hp})(746 \text{ W/hp})} = 6.31 \text{ s}$$

A *hydraulic jack* (Fig. 13-8) combines a hydraulic press with a second-class lever. For the lever

$$\text{A.M.A.} = \frac{F_1}{F} = \frac{l_{\text{in}}}{l_{\text{out}}}$$

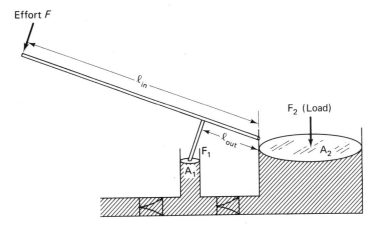

Figure 13-8
Hydraulic jack.

Therefore

total A.M.A. $= (\text{A.M.A.})_{\text{lever}} \times (\text{A.M.A.})_{\text{press}}$

$$\text{A.M.A.}_{\text{TOT}} = \left(\frac{F_1}{F}\right)\left(\frac{F_2}{F_1}\right) = \frac{F_2}{F}$$

*Problems*_____

11. The smaller piston of a hydraulic press has a diameter of 4.00 cm, and the larger piston has a diameter of 80.0 cm. (a) If the hydraulic press has an efficiency of 100%, determine the minimum force required to lift a 1 500 kg load. (b) If the smaller piston can be pumped through 20.0 cm, how many times must it be pumped to raise the load 2.50 m? (c) If the pump develops an average power of 4.50 kW, how long does it take to raise the load?

12. The smaller piston of a hydraulic press has a diameter of 3.0 in, and the larger system has a diameter of 2.5 ft. If the efficiency of the press is 95%, determine the minimum force required to balance a 4 000 lb car.

13. A hydraulic press has a smaller piston of diameter 2.5 in and a larger piston of diameter 1.5 ft. (a) If the efficiency of the press is 100%, determine the minimum force required to lift a 3.0 ton (6 000 lb) load. (b) If the smaller piston can be pumped through 8.0 ft, how many times must it be pumped to raise the load 6.0 ft? (c) What minimum power should the pump develop if the load must be raised 6.0 ft in 12 s?

14. The smaller piston of a hydraulic press has a 2.00 cm diameter, and the larger piston has a 20.0 cm diameter. (a) If the press has a 100% efficiency, what minimum force is required to lift a 2 100 kg load? (b) If the smaller piston can be pumped through 12.5 cm, how many times must it be pumped to lift the load 2.50 m? (c) How long does it take to raise the load if the pump delivers an average power of 5.00 kW?

13-4 BUOYANCY AND ARCHIMEDES' PRINCIPLE

Objects weigh less in water than in air. Both air and water exert an upward *buoyant* force on the object, but the object has a greater buoyancy in the denser water.

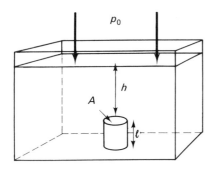

Figure 13-9
Analysis of buoyant force.

Consider a regular-shaped block that is submerged to a depth h in a fluid of density ρ so that the cross-sectional area A of the block is parallel to the fluid surface (Fig. 13-9). If the block has a length l and the pressure at the fluid surface is p_0, then, from Eq. 13-3, the pressure on the upper face of the block is:

$$p_u = p_0 + \rho g h \quad \text{at a depth } h$$

and the pressure on the lower surface (at a depth $h + l$) is:

$$p_L = p_0 + \rho g(h + l)$$

Therefore the difference in pressure between the two faces is:

$$\Delta p = p_L - p_u = \rho g h$$

This gives rise to a net upward or buoyant force:

$$F_{up} = \Delta p A = \rho g h A \qquad (13-9)$$

Where hA equals the volume V of the block and ρg equals the weight density $D = w/V$ of the fluid. Therefore the net upward force is equal to the weight **w** of the fluid that is displaced by the block. The horizontal pressures on the sides of the block balance. This result was first stated by the Greek mathematician Archimedes (287–212 B.C.); it is known as *Archimedes' principle*:

When an object is wholly or partially immersed in a fluid, it experiences a buoyant force or loss in weight that is equal to the weight of the displaced fluid.

Consequently when an object displaces a quantity of fluid equal to its own weight, it floats in that fluid. The position of the center of gravity C *of the displaced fluid* is called the *center of buoyancy* since the buoyant force always appears to act through this point. When an object floats, its center of gravity C and the center of buoyancy C^1 lie in the same vertical line with C^1 below C (Fig. 13-10).

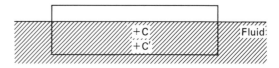

Figure 13-10
An object floats with its center of gravity C directly above the center of buoyancy C^1.

*Example 13-3*_____

A right circular cylinder with a mass $m = 75$ kg, diameter $d = 40$ cm and length $l = 100$ cm floats with its axis vertical in water. Determine the length h of the cylinder below the water level.

Solution:

Since it floats, the cylinder displaces its own weight $w = mg$ in water. The volume of water displaced $= Ah$, and the density of water $\rho = 10^3$ kg/m^3; from Eq. 13-9, the buoyant force is

$$F_{up} = w = mg = \rho g A h = \rho g\left(\frac{\pi d^2}{4}\right)h$$

Therefore

$$h = \frac{4m}{\rho \pi d^2} = \frac{4(75 \text{ kg})}{(10^3 \text{ kg/m}^3)\pi(0.40 \text{ m})^2} = 0.60 \text{ m}$$

The *relative density (specific gravity)* ρ_r of a solid (or liquid) is taken as the ratio of its density to that of pure water at a reference temperature (which should be stated) (Chapter 5).

$$p_r = \frac{\text{mass of substance}}{\text{mass of an equal volume of water}} \quad (13\text{-}10)$$

$$= \frac{\text{weight of substance}}{\text{weight of an equal volume of water}} \quad (13\text{-}11)$$

If the object does not float on water since its loss in weight is equal to the weight of water displaced, its relative density (specific gravity) is:

$$p_r =$$
$$\frac{\text{weight of substance}}{\text{loss of weight when immersed completely in water}}$$

Example 13-4

If a sample of brass weighs 10.800 lb in air and 9.552 lb when totally immersed in water, determine (a) the relative density and (b) the weight density of brass.

Solution:

From Table 5-1, $D_{\text{water}} = 62.4 \text{ lb/ft}^3$.

(a) The loss in weight $= 10.800 \text{ lb} - 9.552 \text{ lb}$
$$= 1.248 \text{ lb}$$

   Therefore

$$p_r = \frac{10.800 \text{ lb}}{1.248 \text{ lb}} = 8.654$$

(b) $D = p_r D_{\text{water}} = (8.654)(62.4 \text{ lb/ft}^3)$
$$= 540 \text{ lb/ft}^3$$

A *hydrometer* is a device that utilizes Archimedes' principle of flotation to measure relative densities of liquids. It consists of a graduated glass stem with two bulbs, the lower of which is weighted to ensure that the hydrometer floats vertically (Fig. 13-11). Since the weight of the hydrometer is fixed, the volume of liquid that it displaces depends on the relative density of that liquid. The hydrometer floats with its stem higher above the surface in denser liquids.

Problems

15. A rectangular block of wood is 40 cm wide, 60 cm long and 30 cm deep. If its relative density is 0.70 and it floats with its 30 cm side vertical, determine (a) the

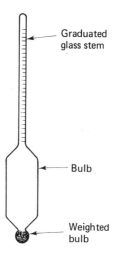

Figure 13-11
The hydrometer.

length of the block above the surface of the water and (b) the minimum force required to completely submerge the block.

16. A uniform 120 lb right circular cylinder with a diameter of 14 in and a length of 30 in floats with its axis vertical in salt water of relative density 1.3. Determine the length of the cylinder above the water level.

13-5 SURFACE TENSION AND CAPILLARITY

Both cohesive and adhesive forces exist at a boundary between two different materials, and these forces give rise to several phenomena. For example, some liquids bead whereas others spread out over a solid surface; it is possible to float a steel needle on the surface of water even though steel is much denser than water; some insects are even able to walk on the surface of water.

Even though liquid molecules are free to move, they are relatively close together;

therefore the cohesive forces between individual liquid molecules must be quite large. No *net* cohesive force exists on a molecule in the interior of the liquid because other liquid molecules are positioned symmetrically all around it and the cohesive force must have the same magnitude in all directions. However, at a boundary, with air for example, the surface liquid molecules experience a net force due to the cohesive forces from other liquid molecules below that is not balanced by the adhesive forces between the liquid and air molecules. This net force acts perpendicularly to the surface and towards the interior of the liquid, tending to compress it. Since liquids are virtually incompressible, the net effect of the unbalanced cohesive forces is to reduce the surface area of the liquid (Fig. 13-12). Consequently, liquid drops, such as rain, are approximately spherical because a sphere represents a minimum surface area. Also, the pressure is greater inside than outside a liquid drop.

Figure 13-12
Forces on a liquid drop tend to reduce the surface area.

By virtue of the cohesive forces between surface and interior molecules, the surface molecules of liquids and solids possess potential energy. This is more apparent in liquids because liquid molecules are able to move. The potential energy per unit surface area is called the *surface energy*. Thus, when a bubble bursts, the surface energy is converted into kinetic energy as parts of the liquid scatter in all directions.

The cohesive forces between surface

molecules act tangentially to the surface. If the surface is uniform, flat, and in equilibrium, these forces are balanced at every point; but if the surface is distorted, they tend to oppose the distortion and to cause the surface to behave like a film. When a heavy object floats on water, it distorts the surface; as a result, the cohesive forces between surface molecules act to oppose the cause of that distortion, namely the weight of the object (Fig. 13-13).

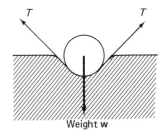

Figure 13-13
Vertical components of the surface tension **T** oppose the weight **w**.

The tangential molecular force F per unit surface length L is defined as the surface tension T of the liquid.

$$T = F/L \qquad (13\text{-}12)$$

Its value depends on temperature and the types of matter in contact (see Table 13-1).

Table 13-1
Typical Surface Tensions of Liquids at a Surface with Air (at 20°C)

Liquid	Surface Tension $T/(mN/m)$
Benzene	28.9
Ethyl alcohol	22.3
Oil	40
Mercury	4.65
Water	72

The units of surface tension are mN/m in SI and lb/ft in the British Engineering System.

Consider a liquid film formed in a region ABCD of a rectangular wire whose segment AB is free to move (Fig. 13-14). In order to

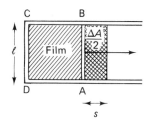

Figure 13-14
Surface tension of a liquid film.

maintain equilibrium, an external force F must be applied perpendicularly to AB to balance the cohesive forces between the surface molecules of the film. Therefore, if l is the length of AB, the surface tension in the film is

$$T = \frac{F}{2l} \qquad (13\text{-}13)$$

The factor of two is required because the film has two sides. If the wire AB is now pulled slowly through a small distance s by the constant force of magnitude F perpendicular to the wire, the work done $W = Fs = (2lT)s$. If both sides of the film are included, $2ls$ is the total change ΔA in the area. Thus $W = T\Delta A$ or

$$T = \frac{W}{\Delta A} = \frac{F}{2l} \qquad (13\text{-}14)$$

The surface tension T is therefore the work required per unit area increase ΔA against the molecular forces. Also, since work causes changes in energy (if the temperature remains constant), the surface tension T is equivalent to the change in the surface potential energy per unit area increase ΔA. Surface tension may be measured by determining the force required to pull a platinum ring from the liquid surface.

Example 13-5

At 23°C, a vertical force $F = 6.3 \times 10^{-3}$ N is required to pull a horizontal thin platinum ring of radius $r = 2.0$ cm free of a liquid surface. Determine the surface tension of the liquid.

Solution:

Since both sides of the wire break the liquid surface, the total length of surface broken is $L = 2 \times (2\pi r) = 4\pi r$. Thus, from Eq. 13-12, we obtain:

$$T = \frac{F}{L} = \frac{F}{4\pi r} = \frac{6.3 \times 10^{-3} \text{ N}}{4\pi \times (2.0 \times 10^{-2} \text{ m})}$$
$$= 2.5 \times 10^{-2} \text{ N/m}$$

When a sheet of glass is inserted vertically in water, the water is "drawn up" the glass. This happens because the adhesive force between glass and water molecules is stronger than the cohesive forces between the water molecules at the surface; as a result, the water is said to "wet" the glass. When a glass sheet is inserted vertically in mercury, the surface of the mercury is depressed at the surface with glass. In this case the cohesive forces between mercury molecules is greater than the adhesive forces between mercury and glass; therefore the mercury does not "wet" the glass. These phenomena occur between other surfaces in contact. The angle between the solid surface and the tangent to the liquid surface at the boundary is called the *angle of contact* ϕ. If ϕ is less than 90°, the adhesive forces are greater than the cohesive forces, and the liquid "wets" the solid surface; if ϕ is greater than 90°, the cohesive forces are larger, and the liquid does not "wet" the solid (Fig. 13-15).

The effects of cohesive and adhesive forces at the boundary with a solid are prominent in small bore tubes called *capillaries*. Some liquids rise up capillary tubes so that their surface is higher inside than outside; others are depressed in capillary tubes so that

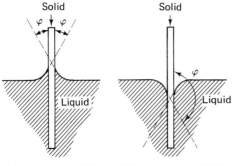

(a) φ is less than 90°. Adhesive forces are strongest, and the liquid rises up the solid.

(b) φ greater than 90°. Cohesive forces are strongest, and liquid is depressed at the solid.

Figure 13-15
Angles of contact.

their surface is lower inside than outside the tube. This rise or fall of liquid in the capillary tube is called *capillarity*.

Consider a liquid of density ρ and surface tension T, which rises to an elevation h inside a vertical capillary tube of radius r (Fig. 13-16a). If ϕ is the angle of contact between

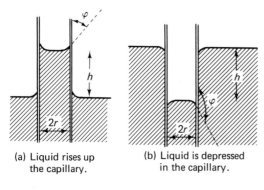

(a) Liquid rises up the capillary.

(b) Liquid is depressed in the capillary.

Figure 13-16
Capillarity.

the liquid and the tube, the net upward force F on the liquid column is due to the total vertical component of the surface tension. Thus

$$F = (T \cos \phi)L = (T \cos \phi)(2\pi r) \quad (13\text{-}15)$$

Since the system is in equilibrium, this force must balance the weight w of the liquid column. If m is the mass and V is the volume of the liquid column, $w = mg = (\rho V)g = \rho(\pi r^2 h)g$; equating with Eq. 13-15, we obtain

$$h = \frac{2T \cos \phi}{\rho r g} \quad (13\text{-}16)$$

If ϕ is greater than 90°, $\cos \phi$ is negative valued, and h becomes negative. This is equivalent to a depression of the liquid inside the capillary (Fig. 13-16b). For glass and water $\phi \approx 0°$, and for glass and mercury $\phi = 128°$.

Waxes and other waterproofing agents are often used to prevent water from wetting a surface. These agents change the nature of the surface so that the cohesive forces between water molecules exceed the adhesive forces between the solid and water; thus the angle of contact becomes greater than 90°, and the water beads on the surface. Detergents alter the surface of the solid so that adhesive forces dominate, the angle of contact is less than 90°, and the water then wets the surface. The threat of vast oil spills in our oceans, lakes and rivers has instigated a very active research program to develop substances that can contain and also break up the oil, chemically changing it so that it wets an absorbing material.

Problems

17. Determine the vertical force at 20°C that is required to pull a thin horizontal platinum ring of radius 1.2 cm free of (a) liquid benzene, (b) liquid water, (c) liquid mercury.

18. At 30°C a vertical force of 5.0×10^{-3} N is required to pull a thin horizontal platinum ring of radius 2.5 cm free of a liquid surface. Determine the surface tension of the liquid.

19. A clean glass capillary with a diameter of 1.2 mm is inserted vertically into mer-

cury at 20°C. Determine the depression of the mercury in the capillary.

13-6 BERNOULLI'S THEOREM

The principles of fluid flow are encountered in many applications of modern technology. The flow of fluids may be controlled by pipes and channels, such as sewers, drains, household plumbing, oil, gas and water pipelines. Weather variations are caused by the flow of hot air masses relative to cold air masses. The stresses caused by the flow of fluids, such as the current of a river or the wind, must be considered in the design of structures, such as bridges, dams and buildings.

Many aspects of technology do not involve the actual motion of a fluid but are concerned with the motion of solid objects through a fluid, which produces a similar effect. To reduce retarding forces ("drag") due to friction, cars, boats and aircraft are "streamlined" so that they are able to proceed through the fluid without causing turbulence.

Fluid molecules have mass, and, according to Newton's second law, some net force is required to set them in motion; this force is usually produced by hydrostatic pressure or by pumps. Once they are in motion, the fluid molecules possess momentum, and a force is required to arrest them; therefore the molecules are able to exert dynamic pressures by virtue of their motion. Note that energies of the *individual molecules give rise to interactions that produce hydrostatic pressures;* this is an internal property of a fluid. For example, hydrostatic pressure exists in a coordinate system that moves with the center of mass of the fluid where the dynamic pressure is zero.

A fluid is said to be in *steady flow* if its velocity past any fixed position is a constant in time. The velocity may vary at different positions in the fluid, but every molecule has the same velocity when it arrives at some particular position. The fluid molecules tend to follow each other in paths called *streamlines*. Streamlines cannot cross each other, and they are constant in time if the fluid is in steady flow.

Consider a fluid in steady flow through an irregularly shaped tube that has cross-sectional areas of A_1 at X and A_2 at Y. Suppose that, in a small time interval Δt, a volume V_1 of the fluid enters the tube at X with a velocity v_1 and density ρ_1 while a volume V_2 leaves the tube at Y with a velocity v_2 and a density ρ_2 (Fig. 13-17). Since the

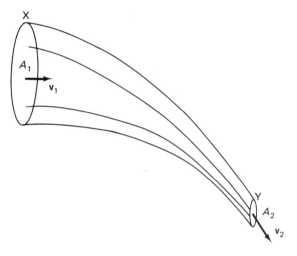

Figure 13-17
Streamlines are farther apart at X than at Y, indicating that v_2 is greater than v_1.

flow is steady and mass is conserved, the mass m_1 of fluid entering at X must be equal to the mass m_2 leaving at Y. Thus

$$m_1 = \rho_1 V_1 = m_2 = \rho_2 V_2$$
$$\frac{m_1}{\Delta t} = \frac{\rho_1 V_1}{\Delta t} = \frac{m_2}{\Delta t} = \frac{\rho_2 V_2}{\Delta t}$$

In the limit as Δt approaches zero, since ρ_1, ρ_2, A_1 and A_2 are constants, and

$$limit_{\Delta t \to 0} \frac{V}{\Delta t} = limit_{\Delta t \to 0} A \frac{\Delta s}{\Delta t} = Av$$

then

or

$$p_1 A_1 v_1 = p_2 A_2 v_2$$
$$pAv = \text{constant} \qquad (13\text{-}17)$$

This is called the *equation of continuity*. If the density p of the fluid is constant, then

$Av = R$ (the rate of volume flow) is constant

$$(13\text{-}18)$$

Thus the speed is slower when the streamlines are farther apart.

This equation is used for sizing ducts in air conditioning systems. The rate of volume flow required depends on the size of the room and the number of occupants. Normally a conditioned space should be completely replenished with air approximately every 10 min, but the air speed should be slow enough to avoid excessive noise and drafts.

Example 13-6

What is the minimum duct size required to deliver 810 ft³ of air per min (cfm) at 15 ft/s to an air conditioned room? Assume that the air has a constant density.

Solution:

From Eq. 13-18:

$$A = \frac{R}{v} = \frac{810 \text{ ft}^3/\text{min}}{15 \text{ ft/s}} = \frac{(810 \text{ ft}^3/60 \text{ s})}{15 \text{ ft/s}}$$
$$= 0.90 \text{ ft}^2$$

In 1738, David Bernoulli (1700–1782) applied the conservation of mechanical energy to incompressible liquids in steady flow. Consider an incompressible liquid in steady flow through the section of pipe in Fig. 13-18. Suppose that the liquid enters the section with a velocity v_1 at point X where the pressure is p_1 and the pipe has a cross-sectional area A_1, the center of which is a vertical height h_1 above a horizontal reference. The liquid leaves the section with a velocity v_2 at point Y where the pressure is p_2 and the pipe has a cross-sectional area A_2, the center of which is a vertical height h_2 above the same reference level. If the pressure p_1 applied at X results in a displacement Δs_1, it also results in a displacement Δs_2 against the pressure p_2 at Y such that the total volume V displaced is the same at both X and Y. Thus the positive work done on the system at X:

$$W_1 = p_1 A_1 \Delta s_1 = p_1 V$$

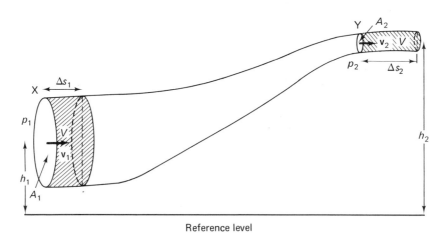

Reference level

Figure 13-18
Fluid flow in a variable tube.

is transmitted by the system to point Y where negative work W_2 is done *by* the system against the external pressure p_2:

$$W_2 = -p_2 A_2 \, \Delta s_2 = -p_2 V$$

Thus, the total positive work done on the system is:

$$W = (p_1 - p_2)V = (p_1 - p_2)\frac{m}{\rho}$$

where m is the mass displaced and ρ is the density of the liquid. From the work-energy principle, this work done is equal to the change in mechanical energy (kinetic and potential) of the liquid as it is displaced.

$$W = \Delta E_k + \Delta E_p$$
$$= (\tfrac{1}{2}mv_2^2 - \tfrac{1}{2}mv_1^2) + (mgh_2 - mgh_1)$$

Therefore

$$(p_1 - p_2)\frac{m}{\rho} = \frac{1}{2}m(v_2^2 - v_1^2) + mg(h_2 - h_1)$$

Rearranging the terms, we obtain:

$$p_1 + \tfrac{1}{2}\rho v_1^2 + \rho g h_1 = p_2 + \tfrac{1}{2}\rho v_2^2 + \rho g h_2$$

or

$$p + \frac{\rho v^2}{2} + \rho g h = \text{constant} \qquad (13\text{-}19)$$

This is known as *Bernoulli's equation*. Note that Bernoulli's equation is valid for incompressible fluids in steady flow and may be used in conjunction with Eq. 13-18 to solve many types of technical problems. Bernoulli's equation may also be written as:

$$\frac{p}{\rho g} + h + \frac{v^2}{2g} = \text{constant} \qquad (13\text{-}20)$$

The vertical distance h below the liquid surface is called the *pressure head*, and the term $h_v = v^2/2g$ is called the *velocity head*.

According to Pascal's law, pressure is transmitted undiminished to all points in a liquid at rest. Therefore, in any system of liquid-carrying pipes, the pressure is the same at all points as long as the liquid is stationary. However, if a valve or faucet is opened the liquid flows and friction (viscosity) reduces the pressure at points along the pipe. This loss in pressure due to friction is called the *friction head*. For this reason, pumping sta-

tions must be installed at regular intervals along long pipelines. The *total head* is the sum of the pressure head h_p, velocity head h_v and friction head h_f,

$$h_{\text{total}} = h_p + h_v + h_f \qquad (13\text{-}21)$$

Example 13-7

Water is pumped at an initial pressure $p_1 = 40.0 \text{ lb/in}^2$ and speed $v_1 = 2.00 \text{ ft/s}$ through a vertical height of 75.0 ft from the basement to the eighth floor of an apartment building through a pipe that tapers from an inner diameter $d_1 = \tfrac{3}{4}$ in at the basement to a diameter $d_2 = \tfrac{1}{2}$ in at the eighth floor. If the weight density of water is 62.4 lb/ft³, what is the speed v_2 and pressure p_2 of the water as it flows in the pipe at the eighth floor?

Solution:

From Eq. 13-18,

$$A_1 v_1 = A_2 v_2 \quad \text{or} \quad \pi (d_1/2)^2 v_1 = \pi (d_2/2)^2 v_2$$

Thus

$$v_2 = \frac{d_1^2 v_1}{d_2^2} = \frac{(\tfrac{3}{4}\text{ in})^2 (2.00 \text{ ft/s})}{(\tfrac{1}{2}\text{ in})^2} = 18.0 \text{ ft/s}$$

From Bernoulli's Equation (Eq. 13-19), we obtain:

$$p_2 = p_1 + \rho g (h_1 - h_2) + \frac{\rho}{2}(v_1^2 - v_2^2)$$
$$= (40.0 \times 144 \text{ lb/ft}^2) + (62.4 \text{ lb/ft}^3)(0 - 75.0 \text{ ft})$$
$$+ \left(\frac{62.4}{32.2} \text{ slug/ft}^3\right)\left(\frac{2.00^2 - 18.0^2}{2}\right) \text{ft}^2/\text{s}^2$$

Therefore

$$p_2 = 770 \text{ lb/ft}^2 \quad \text{or} \quad 5.35 \text{ lb/in}^2$$

Problems

20. Determine the number of cubic meters of air that can pass through a 400 cm² duct at 6.0 m/s in 10 s. Assume the air has a constant density.

21. Calculate the minimum size of a duct required to replenish completely the air in a 15 000 ft³ room every 10 min if the air moves at 3.0 ft/s. Assume the density of air is constant.

22. How long does it take to fill a $1.5\,\text{ft}^3$ tank with water from a faucet that has a diameter of $\frac{1}{2}$ in if the water is emitted with a speed of $5.0\,\text{ft/s}$?

23. Oil flows with a speed of $3.0\,\text{m/s}$ through a pipe with a diameter of 40 cm. Determine the speed of the oil through a constriction in the pipe where the diameter tapers to 10 cm.

24. Oil (density $900\,\text{kg/m}^3$) flows at $1.50\,\text{m/s}$ at a pressure of 200 kPa through a horizontal pipe with a diameter of 12.0 cm. Determine the velocity and pressure at a constriction in the pipe where the diameter tapers to 4.00 cm.

25. Water enters a pipe with an inner diameter of 1.5 in at a speed of $2.0\,\text{ft/s}$ and a pressure of $70\,\text{lb/in}^2$. It is delivered to a point that is 80 ft higher through a pipe with an inner diameter of 0.50 in. Determine (a) the speed and pressure of the water in the higher pipe and (b) the velocity head at each level.

26. Water from a storage tank enters a pipe with an inner diameter of 3.0 cm at a speed of 80 cm/s and a pressure of 500 kPa. It is delivered at a point that is 100 m lower through a pipe with an inner diameter of 1.0 cm. Determine (a) the speed and pressure of the water in the lower pipe and (b) the final velocity head.

13-7 APPLICATIONS OF BERNOULLI'S THEOREM

Even though Bernoulli's equation was derived for incompressible liquids in steady flow, it may be used as a close approximation for the flow of many fluids. There are many devices that use the principles of Bernoulli's theorem in order to measure the flow speed of a fluid or the speed of an object as it passes through a fluid.

Flow Through an Orifice—Torricelli's Theorem

When a small hole or orifice is located a distance h below the surface of liquid in a large open container, the liquid flows out of the orifice (Fig. 3-19). If the container has a much

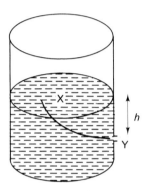

Figure 13-19
Streamline from the liquid surface to the orifice.

larger cross-sectional area than the orifice, the level of the liquid does not vary rapidly with time and may be approximated to a constant. The net speed of the molecules at the liquid surface is approximately zero. Since the container is open, the pressures at the liquid surface and the orifice are both approximately equal to the atmospheric pressure. Bernoulli's equation for a streamline from the liquid surface X to the orifice Y reduces to

$$gh = v^2/2 \quad \text{or} \quad v = \sqrt{2gh} \quad (13\text{-}22)$$

where v is the speed of the liquid as it leaves the orifice. This is known as *Torricelli's theorem*.

Example 13-8

Determine the rate at which liquid flows through an orifice of diameter $d = 1.20$ cm located in a large tank at a depth $h = 5.00$ m below the liquid surface.

Solution:

From Eqs. 13-18 and 13-22,

$$R = Av = \frac{\pi d^2}{4}\sqrt{2gh}$$

$$= \frac{\pi(1.20 \times 10^{-2} \text{ m})^2}{4}\sqrt{2(9.81 \text{ m/s}^2)(5.00 \text{ m})}$$

$$= 1.12 \times 10^{-3} \text{ m}^3/\text{s} = 1.12 \text{ L/s}$$

The Venturi Meter

The *Venturi meter* is a simple device that is commonly used to determine the flow speed of a liquid or gas. It consists of a horizontal tube with a constriction at its center. The pressure difference between the wide and the constricted portion of the tube is measured with a manometer (Fig. 13-20). Note that *the*

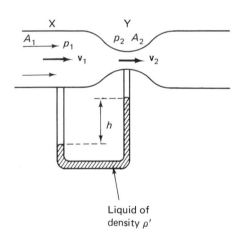

Figure 13-20
The Venturi meter.

pressure is largest at points where the speed is least. If fluid of constant density ρ enters the Venturi tube with a speed v_1 at X where the pressure is p_1 and the cross-sectional area is A_1 and passes with a speed v_2 through the constriction where the pressure is p_2 and the cross-sectional area is A_2, then from the continuity equation (Eq. 13-17), $v_2 = (v_1 A_1)/A_2$.

Since both points are located at the same level, Bernoulli's equation becomes

$$p_1 - p_2 = \frac{\rho}{2}(v_2^2 - v_1^2) = \frac{\rho v_1^2}{2}\left(\frac{A_1^2}{A_2^2} - 1\right)$$

and

$$v_1 = A_2\sqrt{\frac{2(p_1 - p_2)}{\rho(A_1^2 - A_2^2)}} \qquad (13\text{-}23)$$

If the liquid in the manometer tube has a density ρ^1 and the difference in the levels of the manometer tube is h, then

$$p_1 - p_2 = \rho^1 gh - \rho gh = (\rho^1 - \rho)gh \quad (13\text{-}24)$$

Note that the term ρgh must be included because of the weight of the liquid, which moved from the Venturi tube into the left arm of the manometer.

Example 13-9

A Venturi meter tapers from a diameter of $d_1 = 5.00$ cm to $d_2 = 1.50$ cm at the constriction, and a mercury manometer is attached between the two regions. When water flows through the Venturi meter, the difference in height of the mercury in the manometer $h = 7.00$ cm. Determine (a) the difference in pressure between the wide section and the constriction and (b) the rate at which the water flows.

Solution:

(a) The density of mercury $\rho^1 = 1.36 \times 10^4$ kg/m³, and the density of water $\rho = 10^3$ kg/m³; thus $\rho^1 - \rho = 12.6 \times 10^3$ kg/m³ and the difference in pressure

$$p_1 - p_2 = (\rho^1 - \rho)gh$$
$$= (12.6 \times 10^3 \text{ kg/m}^3)(9.81 \text{ m/s}^2)$$
$$\times (7.00 \times 10^{-2} \text{ m})$$
$$= 8.65 \times 10^3 \text{ Pa}$$

(b) From Eqs. 13-18 and 13-23, the rate of volume flow is:

$$R = A_1 v_1 = A_1 A_2 \sqrt{\frac{2(p_1 - p_2)}{\rho(A_1^2 - A_2^2)}}$$

where

$$A_1 = \pi d_1^2/4 \quad \text{and} \quad A_2 = \pi d_2^2/4.$$

Thus

$$R = \frac{\pi d_1^2 d_2^2}{4} \sqrt{\frac{2(p_1 - p_2)}{\rho(d_1^4 - d_2^4)}}$$

$$= \frac{\pi (5.00 \times 10^{-2} \text{ m})^2 (1.50 \times 10^{-2} \text{ m})^2}{4} \sqrt{\frac{2(8.65 \times 10^3 \text{ Pa})}{(10^3 \text{ kg/m}^3)[(5.00 \times 10^{-2} \text{ m})^4 - (1.50 \times 10^{-2} \text{ m})^4]}}$$

$$= 7.38 \times 10^{-4} \text{ m}^3/\text{s} = 0.738 \text{ L/s}$$

The Pitot Tube

Pitot tubes are usually used to measure the speed with which an object moves through a fluid, e.g. a plane through air or a boat through water. This simple device compares static pressure from point X, where the fluid flow is parallel to an opening, with the pressure from point Y, where the opening is perpendicular to the fluid flow (Fig. 13-21). This difference in pressure Δp is measured either with a manometer or in terms of the distortion of a diaphragm. Since points X and Y are approximately at the same level, Bernoulli's equation reduces to:

$$\Delta p = \frac{\rho v^2}{2} \quad \text{or} \quad v = \sqrt{\frac{2\Delta p}{\rho}} \quad (13\text{-}25)$$

where ρ is the density of the fluid and v is the speed of the fluid relative to the pitot head.

Airfoils

An *airfoil* is usually shaped so that the upper surface is curved more than the lower surface. As a result, the speed of fluid flow over the upper surface is slightly greater than that over the lower surface; therefore the pressure is greater on the lower surface, and it produces some contribution to a total net upward force called *lift* on the airfoil.

Lift is also produced because the airfoil deflects the air molecules and changes their momentum as it passes. The airfoil exerts a force on the air molecules, and, according to Newton's third law, the air molecules exert an equal and opposite force on the airfoil.

Lift is due to the flow of fluid over the airfoil being faster on the top and also the reaction forces as the airfoil deflects air molecules downwards as it passes.

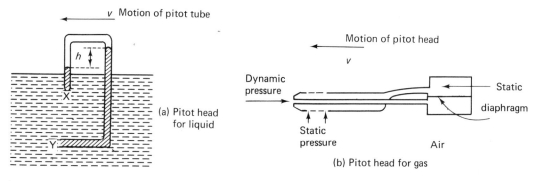

Figure 13-21
Pitot tubes.

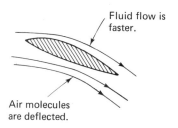

Fluid flow is faster.

Air molecules are deflected.

Figure 13-22
The airfoil.

Problems_____

27. Determine the rate of liquid flow through an orifice of diameter 3.0 cm if it is located in a large tank at a depth of 12 m below the liquid surface.

28. If 12 ft³/min of liquid escapes through a 1.5 in diameter orifice in a large tank, what is the depth of the orifice below the liquid surface?

29. The diameter of a Venturi meter tapers from 8.0 in to 1.5 in at the constriction, and a mercury manometer is attached between the two regions. When oil with a relative density of 0.78 flows through the Venturi meter, the difference in the height of mercury in the manometer is 14 in. Determine the rate of oil flow through the system.

30. If the weight density of water is 62.4 lb/ft³, determine the speed of a boat relative to the water if it produces a pressure difference of 8.4×10^{-2} lb/in² in a pitot tube.

31. If the density of air is 1.3 kg/m³, determine the speed of an aircraft relative to air if it produces a pressure difference of 3.15×10^4 Pa in a pitot tube.

13-8 VISCOSITY

The speed of a fluid is greater at the center of a pipe or channel than at the edges because friction between the vessel and the adjacent fluid layers slows the flow at the edges. The slower moving fluid layers in turn retard other adjacent layers so that there is a gradual transition in the speed of the consecutive layers of fluid. This resistance to fluid flow due to the internal friction between adjacent fluid layers is called *viscosity*; its effects in liquids may be reduced by increasing the temperature. Viscosity has many important consequences in technology. Pumps are required to transport fluids over large distances through pipelines; low viscosity lubricants allow machinery to run efficiently; and some fluids are used to transmit torques (in automatic transmissions of cars for example).

Consider two imaginary flat horizontal plates of fluid with surface area A, which are vertical distance y apart with more fluid between them. Fluids cannot withstand a shearing force. Therefore, if a horizontal shearing force F is applied to the top plate, it flows, and the viscosity between the successive fluid layers causes the bottom plate to flow, but at a reduced speed (Fig. 13-23). It has been determined experimentally that the speed v of the top plate relative to the bottom plate depends on the type of fluid, but it is directly proportional to the shearing stress $\sigma = F/A$ and the distance y between the plates. Thus

$$v \propto \frac{F}{A}y \quad \text{or} \quad \frac{F}{A} = \eta \frac{v}{y} \qquad (13\text{-}26)$$

The proportionality constant η for the particular fluid is called the *coefficient of viscosity*, and the term v/y is called the *velocity gradient*. In SI, the coefficient of viscosity is expressed in units of Pa·s.*

When the speed of flow of a fluid or the speed of a solid object through a fluid becomes large, the flow of the fluid becomes turbulent, and eddies are produced in the

_____

*The *poise* unit has been abandoned in SI.

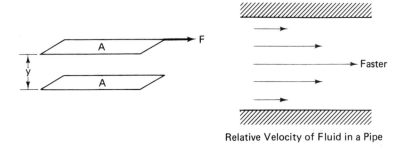

Relative Velocity of Fluid in a Pipe

Figure 13-23

fluid and behind moving objects. These eddies produce a pressure difference, which results in a force that resists the flow. Therefore, objects are streamlined to reduce the eddies and to allow faster motion through the fluid.

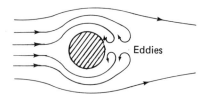

Figure 13-24

Eddies produced when the fluid flow is turbulent.

REVIEW PROBLEMS

1. Determine the force on a 25 cm² plate at a point in a liquid where the pressure is 1.2 MPa.

2. Determine the total force on the shell of a spherical diving bell of radius 4.0 m at a depth where the pressure is 2.0 MPa.

3. At what depth in water is the pressure equal to 2.0 atm?

4. The surface of water in a reservoir is 250 ft above a tap, and the atmospheric pressure at the reservoir is 14.5 lb/in². (a)

Calculate the water pressure at the tap. (b) What force does the water exert on the tap nozzle if it has a diameter of 0.50 in?

5. (a) An unknown pressure supports a 30 in column of water in a manometer when the atmospheric pressure is 15 lb/in². (a) What is the unknown pressure? (b) What is the unknown pressure if the water level is 30 in below the level on the side open to atmospheric pressure?

6. Water at a pressure of 80 lb/in² is used to operate a hydraulic lift. If the piston area is 1.80 ft² and the efficiency is 95%, what is the maximum load that can be raised?

7. The smaller piston of a hydraulic press has a diameter of 5.0 cm, and the larger piston diameter is 75 cm. (a) If the efficiency is 100%, determine the minimum force required to lift a 2 500 kg load. (b) If the smaller piston can be pumped through 15 cm, how many times must it be pumped to raise the load 3.0 m? (c) What minimum power should the pump develop if the load is raised 3.0 m in 15 s?

8. A rectangular raft is 5.0 m wide, 8.0 m long and 20 cm deep. If the weight of the raft is 13 500 N, what is the depth of the raft below the surface of the water?

9. Determine the relative density of a block of material if it weighs 28.4 lb in air and

21.6 lb when it is totally immersed in water.

10. When a clean glass capillary is inserted vertically into mercury at 20°C, the mercury level falls 4.0 mm inside the tube. What is the radius of the capillary?

11. What is the minimum duct size required to deliver 30.0 m³ of air per minute at 5.00 m/s into an air conditioned room?

12. Water flows at a rate of 500 ft³/min through an inclined pipe. If the pressure at end A of the pipe is 20.0 lb/in², where the diameter is 10.0 in, what is the pressure at the end B, where the diameter is 5.00 in and its center is 50.0 ft lower than at A?

13. The diameter of a Venturi meter tapers from 5.0 cm to 1.2 cm at the constriction, and a mercury manometer is attached between the two regions. When gasoline with relative density of 0.68 flows through the Venturi meter, the difference in the height of mercury in the manometer is 8.0 cm. Determine the rate at which the oil flows.

14

Heat
and
Temperature

Heat is a form of energy that is transmitted to individual atoms and molecules, causing them to vibrate more rapidly and with greater amplitudes as the object becomes "hotter." The *internal energy* of a system is the total kinetic and potential energy possessed by the atoms and molecules. When an object is heated, its internal energy increases.

Temperature is a measurement of the level of internal energy possessed by the individual molecules; it is *not* a measure of the total heat content or internal energy of the whole system. In an isolated system heat flows naturally from higher to lower temperature areas and tends to equalize internal energies within the system; thus temperatures also indicate the direction of natural heat flow. There is a continual exchange of thermal energy between two objects that are in contact even if their temperatures are the same; however, if their temperatures are the same, there is no net heat transfer, and the objects are said to be in *thermal equilibrium*. The term *heat transfer* applies to the transmission of nonmechanical molecular energy from one point to another.

The properties of heat and heat transfer are of great importance to us. Engineers must consider these properties when they design air conditioning systems for buildings, engines and even electronic devices.

The *state* of a system is specified in terms of its pressure, volume, temperature and mass. The system is said to change its state when it changes the value of one or more of these quantities. However, when a material is heated, it does not necessarily change its temperature. Heat energy may also produce a *phase* change between solid, liquid and gaseous states *at constant temperature*.

14-1 TEMPERATURE SCALES

In order to establish a temperature scale, a minimum of two reproducible standard points are required so that a series of graduations, called *degrees*, may be made between them. The two most common temperature scales, *Fahrenheit* (F) the *Celsius* (C), use the properties of pure water at a pressure of one standard atmosphere to define the two reproducible points.

The *freezing* or *ice point* of water is defined as the temperature at which an ice and liquid water mixture is in thermal equilibrium. If the system is isolated from external influences, both the mass of ice and the mass of liquid remain constant. At a pressure of one standard atmosphere, the ice point of water is taken as 0° on the Celsius scale and 32° on the Fahrenheit scale. Similarly, the *boiling* or *steam point* is the temperature at which a liquid water and steam mixture is in thermal equilibrium. When the external pressure is one standard atmosphere, the steam point is taken as 100° on the Celsius scale and 212° on the Fahrenheit scale. These points are then divided into even graduations, which are extended both above the steam point and below the ice point.

Since 0°C is equivalent to 32°F and 100°C is equivalent to 212°F, a temperature *change* of
$$100°C = 180°F \quad \text{or} \quad 5°C = 9°F \quad (14\text{-}1)$$

The actual conversion between the two temperature scales is given by:
$$°F = (9/5)°C + 32 \quad \text{and} \quad °C = (5/9)(°F - 32) \quad (14\text{-}2)$$

*Example 14-1*_____

Convert 86°F to the equivalent Celsius temperature.

Solution:

$$°C = \frac{5}{9}(°F - 32) = \frac{5}{9}(86 - 32)°C = 30°C$$

The SI unit for temperature, the *kelvin* (K), is actually related to the Celsius scale. The relationship between kelvin and Celsius temperatures is approximately:

$$K = °C + 273 \quad (14\text{-}3)$$

Thus 20°C is equivalent to $(20 + 273)$ K $= 293$ K.*

*Problems*_____

1. Convert the following Fahrenheit temperatures to the equivalent Celsius temperatures: (a) 80°F, (b) 10°F, (c) −20°F, (d) −135°F, (e) 360°F, (f) 3 000°F.

2. Convert the following Celsius temperatures to the equivalent Fahrenheit temperatures: (a) 5°C, (b) 83°C, (c) 240°C, (d) −72°C, (e) −48°C, (f) 2 100°C.

3. Determine the temperature at which the Celsius and the Fahrenheit temperatures have the same magnitude.

4. Determine the temperatures at which the Celsius and Fahrenheit temperatures have the same magnitude but opposite signs.

5. Convert the following to kelvin temperatures: (a) 32°C, (b) −38°C, (c) 362°C, (d) 48°F, (e) −72°F.

6. Convert the following to Celsius temperatures: (a) 298 K, (b) 1 096 K, (c) 24 K.

14-2 HEAT

Temperature is only a measure of the *level* of internal energy that an object possesses by virtue of the kinetic and potential energy of its individual atoms and molecules. The heat content of the object is not only a function of the temperature, it also depends on the mass and the nature of the material present since the energy of each atom and molecule must be included as part of the total heat energy. *Heat* is that energy which is transfered from higher to lower temperature areas via molecular interactions solely because of the

_____

*Note that no degree sign is written with the kelvin symbol.

temperature difference. It is not equal to the internal energy.*

Before it was known to be a form of energy, heat was measured in terms of its effect on pure water, and special units were defined to describe it.

A *calorie* (cal) is the amount of heat required to raise the temperature of one gram of water by 1°C (from 14.5°C to 15.5°C)†, and in the British Engineering System, a *British thermal unit* (BTU) is the amount of heat required to raise the temperature of a one pound mass of water by 1°F (from 58.5°F to 59.5°F).† Note that common engineering practice defines thermal units in terms of the *absolute unit*, 1 lb *mass*, which is equivalent to (1/32.2) slug or 2.2 kg. The SI unit for heat is the same as for all other energies, i.e. the *joule* (J); the calorie should not be used.

James Joule (1818–1889) originally determined the relationship between heat units and energy units by measuring the amount of heat generated when fixed amounts of work were done. He found that in an isolated system (such as the contents of a well insulated container called a *calorimeter*) the heat Q generated was directly proportional to the mechanical work W done. These results are completely independent of the method used to do work. Thus,

$$W \propto Q \quad \text{or} \quad W = JQ \qquad (14\text{-}4)$$

The proportionality constant J is called the *mechanical equivalent* of heat; it has a value of 4.186 J/cal or 778 ft·lb/BTU.

It is a relatively simple task to convert mechanical energy into heat by means of friction; however special devices called heat engines are required to convert heat into mechanical energy (Chapter 16). No known machine can convert a fixed quantity of heat

*See Chapter 16.
†The precise temperature range is often ignored.

into an equivalent amount of mechanical energy because some heat is always lost.

Example 14-2

A 3 200 lb car travelling at 60.0 mi/h brakes uniformly to rest. If all the mechanical energy is converted into heat, how much heat is produced?

Solution:

The work done is the loss in kinetic energy. Thus

$$Q = \frac{W}{J} = \frac{\frac{1}{2}mv^2}{J} = \frac{1}{2}\left(\frac{3\ 200\ \text{lb}}{32.2\ \text{ft/s}^2}\right)\left(\frac{(88.0\ \text{ft/s})^2}{778\ \text{ft·lb/BTU}}\right)$$
$$= 495\ \text{BTU}$$

Problems

7. Convert 3 600 kcal to joules.

8. Convert 1 250 BTU to ft·lb.

9. A 1 500 kg car travelling at 50 km/h brakes uniformly to rest. How much heat is produced if all of the mechanical energy is converted into heat?

10. A heat engine develops an average power of 1.5 kW for 30 min, and 30% of this is converted into heat. How much heat is produced?

11. A heat engine develops an average power of 500 hp for 10 min. If 40% of the energy is converted into heat, what amount of heat is produced?

14-3 SPECIFIC HEAT CAPACITY

Even when different materials of the same mass and temperature are subjected to the same quantity of heat, their temperature changes may be quite different. If there are no phase changes, the change ΔT in temperature of a mass is directly proportional to the quantity of heat Q supplied, but it also depends on the nature and dimensions of the material. Thus

$$Q \propto \Delta T \quad \text{or} \quad Q = C\Delta T \qquad (14\text{-}5)$$

The proportionality constant C is called the *heat capacity* of the object. Therefore the *heat capacity of an object* is defined as the amount of heat required to produce a temperature change of one degree in the total mass. It has units of J/°C, J/K in SI, or BTU/°F in the British Engineering System.

If there is no phase change, when a homogeneous material is subjected to a fixed quantity of heat Q, its increment ΔT in temperature is inversely proportional to the mass m of material. If the mass is doubled, the heat Q must be evenly redistributed so that the temperature change is halved. Since $\Delta T \propto 1/m$ and $Q \propto \Delta T$:

$$\frac{Q}{m} \propto \Delta T \quad \text{or} \quad Q = mc\Delta T = mc(T_1 - T_0)$$
$$(14\text{-}6)$$

where T_1 is the final temperature and T_0 is the initial temperature of the mass m. The proportionality constant c is commonly called the *specific heat capacity* of the material; it is defined as the heat required to produce a unit temperature change in a unit mass of material. Consequently, specific heat capacity c has a characteristic value which is independent of the dimensions of the material and is expressed in units of J/(kg·°C) or J/(kg·K) in SI and BTU/(lb·°F) in the British Engineering System. The specific heat capacities of some materials are listed in Table 14-1.

The heat capacity of *any object* of mass m and specific heat capacity c is given by:

$$C = mc \qquad (14\text{-}7)$$

If the mass m and the specific heat capacity c of a material is known, then Eq. 14-6 may be used to determine the temperature change produced by the input of any quantity of heat Q, *provided the material does not change its phase* between solid, liquid and gas (or vapor)

states. A heat input is called *sensible heating* if it produces a change in temperature without a phase change.

Compared with most other substances, water has a relatively high specific heat capacity; it is also inexpensive and abundant. Therefore, many cooling and heating systems utilize water as a heat transfer agent to "carry" heat from one place to another.

Example 14-3

How much heat is required to raise the temperature of 5.00 kg of ice from $-20°C$ to $0°C$?

Solution:

From Table 14-1 and Eq. 14-6

$$
\begin{aligned}
Q &= mc(T_1 - T_0) \\
&= (5.00\,\text{kg})[2.13 \times 10^3\,\text{J/(kg} \cdot °\text{C)}][0°\text{C} - (-20°\text{C})] \\
&= 4.26 \times 10^4\,\text{J}
\end{aligned}
$$

Heat is a form of energy, and as such it may be conserved in an isolated system, such as the contents of a well insulated calorimeter. Thus the heat energy gained by some components of an isolated system must equal the heat energy lost by the other components as long as heat is not converted into some other energy form.

If heat is not transformed into some other energy form, the algebraic sum of all heat changes in an isolated system is equal to zero. Thus, if there are no phase changes or external heat influences,

$$\sum_{\text{algebraic}} Q_i = \sum_i m_i c_i (T_1 - T_{oi}) = 0 \quad (14\text{-}8)$$

where Q are the individual heat changes, m_i are the masses of different materials with specific heat capacities c_i and initial temperatures T_{oi}, and T_1 is the final temperature of the whole system. Note that the final temperature is the same for all components when thermal equilibrium is reached. If a component receives heat, its final temperature T_1 is greater than its initial temperature T_{oi}, and

Table 14-1
Typical Heat Constants

Material	Melting Point		Boiling Point		Specific Heat Capacity c		Specific Latent Heat of Fusion l_f		Specific Latent Heat of Vaporization l_v	
	°C	°F	°C	°F	$\times \dfrac{10^3 J}{kg \cdot °C}$	$\dfrac{BTU}{lb \cdot °F}$	$\times \dfrac{10^5 J}{kg}$	$\dfrac{BTU}{lb}$	$\times \dfrac{10^6 J}{kg}$	$\dfrac{BTU}{lb}$
Air	—	—	—	—	1.0	0.24	1.04	46	0.21	92
Alcohol—ethyl	-130	-202	78	172	2.4	0.58	3.2	139	0.858	369
Aluminum	660	1 220	1 800	3 270	0.921	0.22	1.8	76	8.3	3 591
Brass	900	1 652	—	—	0.376	0.090	—	—	—	—
Copper	1 083	1 980	2 296	4 164	0.390	0.093	—	—	7.3	3 158
Glass	1 100	2 012	—	—	0.880	0.21	—	—	—	—
Hydrogen	-259	-434	-253	-432	14.3	3.4	—	—	0.45	195
Ice	0	32	—	—	2.13	0.51	—	—	—	—
Iron	1 535	2 790	3 000	5 430	0.50	0.12	0.23	9.9	—	—
Lead	328	622	1 644	2 990	0.135	0.032	0.23	9.9	—	—
Mercury	-39	-38	357	675	0.140	0.033	0.12	5.0	0.297	128
Oxygen	-218	-360	-183	-297	—	—	0.13	5.8	0.21	90
Steam	—	—	—	—	2.0	0.48	—	—	—	—
Water—liquid	0	32	100	212	4.186	1.00	3.35	144	2.26	970
Zinc	419	786	907	1 664	0.377	0.090	1.2	51	—	—

the heat change Q is positive; if the final temperature T_1 of a component is less than its initial temperature, it loses heat, and its heat change is negative. The signs must be included in Eq. 14-8.

The specific heat capacity c of an unknown component or the final temperature T_1 of a system may be determined from calorimetric experiments and the conservation of heat energy equation (Eq. 14-8).

Example 14-4

If 0.430 kg of copper at 200°C are added to a 100 g glass calorimeter that contains 500 g of water at 20°C and there are no external heat influences, what is the final temperature in equilibrium?

Solution:

In thermal equilibrium the final temperature T_1 is the same for all components. Thus from Table 14-1 and Eq. 14-8:

$$\sum_{algebraic} Q = Q_{copper} + Q_{glass} + Q_{water}$$
$$= \sum_i m_i c_i (T_1 - T_{oi}) = 0$$

or

(0.430 kg)[390 J/(kg·°C)](T_1 − 200°C)
+ (0.100 kg)[880 J/(kg·°C)](T_1 − 20°C)
+ (0.500 kg)[4186 J/(kg·°C)](T_1 − 20°C) = 0

Solving for the final temperature, we obtain:

$$T_1 = 33°C.$$

Problems

12. Determine the quantity of heat required to raise the temperature of 20 kg of copper by 30°C from 20°C to 50°C.

13. Determine the heat required to raise the temperature of 30 lb of air by 10°F.

14. A 100 g block of aluminum at 260°C is added to a 200 g glass calorimeter that contains 400 g of water at 35°C. What is the final temperature of the system?

15. A 5.0 kg block of lead at 200°C and a 3.0 kg block of iron at 150°C are added to a large 3.0 kg copper container that holds 15 kg of water at 5°C. If there are no heat losses, determine the final temperature of the system.

16. 100 g of a metal at 150°C are added to a 75.0 g aluminum calorimeter that contains 200 g of water at 10°C. If the final temperature of the system is 15.5°C, determine the specific heat capacity of the metal.

17. How much heat is required to increase the temperature of a 450 kg iron rail from 20°C to its melting point?

18. An 1 800 kg automobile travelling at 80 km/h brakes uniformly to rest. If 40% of the energy produces an increase in the temperature of the two 5.0 kg iron brake drums, what is their increase in temperature?

19. If all the potential energy is converted into heat, what is the change in the temperature of water when it falls 55 m over a waterfall?

20. A 500 g piece of iron is placed inside a furnace until it reaches thermal equilibrium. It is then dropped in a 250 g aluminum calorimeter containing 1.5 kg of water at 5°C. What was the temperature of the furnace if the final temperature of the water was 12°C?

14-4 PHASE CHANGES

When a solid is heated, its internal energy, and hence its temperature increases [according to the sensible heat equation (Eq. 14-6)] until the molecular vibrations in the solid become excessive. At this point, the application of more heat energy gradually breaks down the structure of the solid, and it "melts" *at a constant temperature.*

The heat input per unit mass required to

change a mass from solid to liquid at a constant temperature and pressure is called the *specific latent heat* of fusion ℓ_f of the material:

$$\ell_f = \frac{Q_f}{m} \qquad (14\text{-}9)$$

where Q_f is the heat required to melt a mass m of the material. The specific latent heat of fusion ℓ_f is usually expressed in units of J/kg or BTU/lb.

If the material freezes from liquid to solid form, it reduces its energy by losing heat. Consequently, the specific latent heat ℓ_f has a negative value. Equation 14-9 may be used to calculate the amount of heat required to melt a mass of material or to determine the heat evolved when a mass reduces its energy by solidifying at constant temperature and pressure.

Example 14-5

How much heat is liberated when 12.0 kg of water solidifies into ice at 0°C at constant pressure?

Solution:

Using Table 14-1 and Eq. 14-9, we obtain:

$$Q_f = \ell_f m = (-3.35 \times 10^5 \text{ J/kg})(12.0 \text{ kg})$$
$$= -4.02 \times 10^6 \text{ J}$$

The minus sign implies that 4.02×10^6 J are *lost* by the liquid.

If, after all of a solid has been liquified by the application of heat, the liquid is heated, it gains internal energy, and its temperature increases according to the sensible heat equation until the *boiling* or *vaporization* point is reached. At this point some liquid molecules have sufficient kinetic energy to escape through the liquid surface tension, and continued application of heat energy causes the liquid to boil and vaporize *at a constant temperature*.

The amount of heat per unit mass required to change a mass of material from liquid to vapor (or gas) at a constant tem-

perature and pressure is called the *specific latent heat of vaporization* ℓ_v of that material.

$$\ell_v = \frac{Q_v}{m} \qquad (14\text{-}10)$$

where Q_v is the heat required and m is the mass of material vaporized. The specific latent heat of vaporization is also expressed in units of J/kg or BTU/lb, and it is negative valued if the material condenses from the vapor to liquid. Also, Eq. 14-10 may be used to calculate the heat required to vaporize a mass of material or to determine the heat evolved when a mass of vapor is condensed at a constant temperature and pressure. It requires far more heat energy to vaporize than to melt a fixed mass of the same material (Table 14-1).

Specific latent heats may be determined from calorimetric experiments, but both sensible and latent heats must be accounted for in the conservation of heat energy. As in the case of sensible heating, heat gains (when masses convert from solid to liquid or liquid to vapor) are positive valued, and heat losses (when masses condense from vapor to liquid or solidify from liquid to solid) are negative valued. In general, the conservation of heat energy equation may be written as

$$\sum_{\text{algebraic}} Q = \sum Q_s + \sum Q_L = 0 \quad (14\text{-}11)$$

where Q_s and Q_L are the sensible and latent heats, respectively.

In order to solve problems involving mixtures and phase changes of different materials, it is important to proceed in stages:

1. Check *each* material involved to see if it passes through either its boiling or freezing point in its possible temperature range.

2. Calculate all the possible latent heats Q_L involved for each material.

3. Determine the sensible heats Q_s involved for *each phase* of each substance.

4. Use the conservation of heat energy equation to solve for the unknown quantity.

5. Check to see if the answer falls within the expected range; if not, ensure that all phase changes actually occur.

Example 14-6

A 0.010 0 kg of ice at $-10°C$ is added to a 0.200 kg calorimeter [$c = 600$ J/(kg·°C)] that contains 0.500 kg of water at 20°C. If 0.010 0 kg of steam at 100°C is then condensed in the system, calculate the final equilibrium temperature if there are no external heat influences.

Solution:

(a) As it increases its temperature from $-10°C$ to $0°C$, its melting point, the ice receives a sensible heat:

$Q_1 = mc(T_1 - T_0)$
$= (0.010\ 0\ \text{kg})[2\ 130 \text{J}/(\text{kg·°C})][0°C - (-10°C)]$
$= 213$ J

It then melts at a constant temperature of $0°C$ using latent heat:

$Q_2 = m\ell_f = (0.010\ 0\ \text{kg})(3.35 \times 10^5\ \text{J/kg})$
$= 3\ 350$ J

It then receives a sensible heat Q_3 as its temperature changes from $0°C$ to the final temperature T_1:

$Q_3 = mc(T_1 - T_0)$
$= (0.010\ 0\ \text{kg})[4\ 186\ \text{J}/(\text{kg·°C})](T_1 - 0°C)$
$= (41.86\ T_1)\ \text{J/°C}$

(b) The calorimeter is heated from 20°C to the final temperature T_1 by a sensible heat:

$Q_4 = mc(T_1 - T_0)$
$= (0.200\ \text{kg})[600\ \text{J}/(\text{kg·°C})](T_1 - 20°C)$
$= (120\ T_1)\ \text{J/°C} - 2\ 400$ J

(c) The water in the calorimeter is also heated from 20°C to the final temperature T_1 by a sensible heat:

$Q_5 = mc(T_1 - T_0)$
$= (0.500\ \text{kg})[4\ 186\ \text{J}/(\text{kg·°C})](T_1 - 20°C)$
$= (2093T_1)\ \text{J/°C} - 41\ 860$ J

(d) The steam condenses at 100°C into water at 100°C, losing latent heat; thus, its specific latent heat ℓ_v is negative valued.

$Q_6 = m\ell_v$
$= (0.010\ 0\ \text{kg})(-2.26 \times 10^6\ \text{J/kg})$
$= -22\ 600$ J

The condensed water then cools from 100°C to the final temperature T_1, releasing sensible heat:

$Q_7 = mc(T_1 - T_0)$
$= (0.010\ 0\ \text{kg})[4\ 186\ \text{J}/(\text{kg·°C})](T_1 - 100°C)$
$= (41.86T_1)\ \text{J/°C} - 4\ 186$ J

Therefore, in the conservation of energy equation

$$\sum_{\text{algebraic}} Q = Q_1 + Q_2 + Q_3 + Q_4 + Q_5 + Q_6 + Q_7$$
$= 213\ \text{J} + 3\ 350\ \text{J} + (41.86T_1)\ \text{J/°C}$
$+ [(120T_1)\ \text{J/°C} - 2\ 400\ \text{J}]$
$+ [(2\ 093T_1)\ \text{J/°C} - 41\ 860\ \text{J}]$
$- 22\ 600\ \text{J} + [(41.86T_1)\ \text{J/°C}$
$- 4\ 186\ \text{J}] = 0$

Solving, we obtain:

$$T_1 = 29.4°C.$$

In some problems, the phase change of one component may not be completed before the final equilibrium temperature is reached. The final temperature is then equal to the temperature at which that phase change occurs.

Under certain conditions, many substances, such as solid carbon dioxide (dry ice), change directly from solid to the vapor state without entering the liquid phase; this is called *sublimation*. Also, liquids need not boil in order to vaporize; molecules at the liquid surface often receive extra energy from molecules below the surface. If their energy is sufficient, the surface molecules may escape the surface tension; this process is called *evaporation*. Evaporation results in a loss of heat energy from the liquid because the more energetic molecules escape; thus evaporation cools a liquid. The rate of evaporation depends on the surface area, the temperature and pressure of the liquid.

Liquid molecules are continually evaporating from the surface, but some vapor molecules return to the liquid. When the rate of evaporation is equal to the rate of return of vapor molecules, the vapor is said to be *saturated*. If vapor molecules are continually removed from the proximity of the liquid surface (by ventilation), the net evaporation rate increases.

Evaporation may occur even at relatively low temperatures, but the evaporation rate increases as the temperature (and hence the internal energy) of the liquid increases. When a liquid boils, bubbles of saturated vapor form within the liquid body and rise to the surface. The internal pressure of these vapor bubbles must be equal to the total pressure at their depth in the liquid; otherwise they would collapse. Near the surface of the boiling liquid, the internal vapor pressure of the bubbles is approximately equal to the surface pressure. Thus a liquid boils when its saturated vapor pressure is equal to the surface pressure. A reduction in the surface pressure causes the liquid to boil at a lower temperature. If the surface pressure is increased, the liquid will boil at a higher temperature because more evaporation is required to increase the vapor pressure. Evaporation of water is responsible for much of our weather phenomena and is an important factor in air conditioning systems.

Pressure also affects the freezing or fusion point of a material, although not to such a degree. When pressure is applied to a liquid, it tends to reduce the spacing between the molecules. Therefore most materials contract when they solidify, and an increase in pressure favors a change to their solid state. Water expands when it freezes; thus ice floats on water because its density is less than water. Therefore an increase in pressure causes ice to melt.

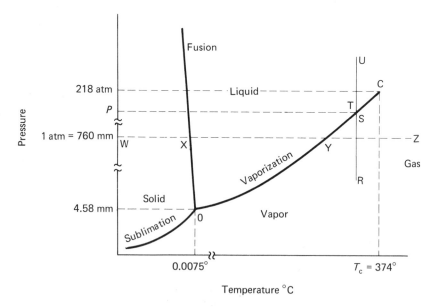

Figure 14-1
Phase diagram for water.

Phase changes are dependent on the external conditions of both temperature and pressure. A graphical representation of the corresponding values of temperatures and pressures at which phase changes occur is called a *phase diagram* for a material (Fig. 14-1). The lines represent corresponding temperatures and pressures where two phases are in equilibrium. The intersection of these lines is a precisely defined reproducible point called the *triple point O*. At the triple point the solid, liquid and saturated vapor exist simultaneously *in equilibrium*. If the system is isolated, the masses of the solid, liquid and vapor states all remain constant. The triple point o of pure water occurs at a pressure of 4.58 mm of mercury and a temperature of 0.007 5°C.

Point C, the terminal point of the vaporization curve, is called the *critical point*. At temperatures in excess of the corresponding *critical temperature* T_c, the substance cannot be liquified no matter how much pressure is applied. The gaseous phase above the critical temperature is called a *gas*; below the critical temperature it is called a *vapor*.

Consider a process in which a fixed mass of ice is heated at constant pressure (see Figs. 14-1 and 14-2). As the solid ice is heated from W to X, its temperature increases according to Eq. 14-6, which describes sensible heating. The change in temperature is directly proportional to the heat supplied. When the temperature of the ice reaches 0°C (corresponding to point X), the ice begins to melt at a constant temperature; the heat energy is utilized as latent heating in order to break down the structure of the solid. Once all the ice has melted, further heat input produces sensible heating from 0°C to 100°C from X to Y; the change in temperature of the liquid is directly proportional to the heat supplied. At point Y (100°C), the water starts to boil at a constant temperature as the latent heating increases the kinetic energy of the molecules, enabling them to escape from the liquid surface. Once all the liquid has vaporized, the vapor is superheated by additional heat input, and its change in temperature is directly proportional to the heat supplied.

Problems

21. How much heat (in joules) is required to melt 30 kg of ice at 0°C?

22. How much heat (in joules) is evolved when 10 kg of steam are condensed at 100°C?

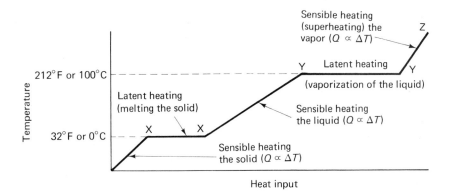

Figure 14-2
Heat input versus temperature of a fixed mass at a constant pressure of 1 atmosphere.

23. How much heat is required to change 1.0 kg of ice at $-20°C$ to steam at $100°C$?

24. How much heat is evolved when 2.0 lb (mass) of steam at $212°F$ are converted into ice at $32°F$?

25. If there are no external heat influences, determine the final temperature when 20 g of steam at $100°C$ are mixed with 5.0 kg of water at $10°C$.

26. When 15.0 g of steam at $100°C$ are condensed in a 100 g aluminum calorimeter containing 150 g of water at $30°C$, the final temperature of the system is $79.5°C$. Calculate the specific latent heat of vaporization of steam. Assume no external heat influences.

27. How much heat is evolved when 80.0 kg of iron solidify in a mold at a constant temperature?

28. How much heat is required to melt 10 lb mass of aluminum if its initial temperature is $70°F$?

29. If 100 g of steam at $100°C$ are passed into 200 g of water at $30°C$, determine the final temperature of the system. Assume that there are no external heat influences.

30. If a heating system delivers 12 000 BTU/h, how much air can it heat from $30°F$ to $70°F$ in 10 min?

31. If the weight density of air is 0.073 lb/ft³, how much heat is required to raise the temperature of the air in a $15 \times 20 \times 8$ ft³ room by 20 F°? Assume that there are no external heat influences.

32. At what rate must heat be removed from an air conditioning system if it cools 310 m³ of air per minute from $30°C$ to $22°C$ and also condenses and removes water from the air at a rate of 1.0 kg/h? The density of air is 1.2 kg/m³.

33. A 20 kg ice block at $0°C$ is pushed at a constant speed across a floor. If the coefficient of kinetic friction between the ice and the floor is 0.10 and all the heat attributed to the friction is used to melt some of the ice, how much ice melts when the block has been pushed 50 m in a straight line?

14-5 HEAT OF COMBUSTION

Since heat is a form of energy, other energy forms, such as light, sound, mechanical energy and chemical energy, may be transformed into heat. Mass is also an energy form; the conversion of mass directly into heat and other energies occurs in nuclear reactions.

One of the most common and useful sources of heat energy is the chemical energy liberated by the direct combustion of a fuel, such as coal, oil and gasoline. The *heat of combustion H* of a fuel is the heat evolved when a unit mass (or volume) of fuel is completely burned.

Table 14-2
Typical Heats of Combustion
of Common Fuels

Material	Heat of Combustion	
	J/kg	BTU/lb
Coal	3.26×10^7	14 000
Wood	1.4×10^7	6 000
Gasoline	4.7×10^7	20 200
Diesel oil	4.5×10^7	19 400
Domestic fuel oil	4.6×10^7	19 800
Kerosene	4.6×10^7	19 800
Gases	J/m³	BTU/ft³
Natural gas	4.35×10^7	1 200
Propane	8.8×10^7	2 400

$$H = Q/m \quad \text{or} \quad H = Q/V \quad (14\text{-}12)$$

where Q is the heat liberated, m is the mass of fuel burned and V is the volume. This heating capacity of a fuel is usually determined experimentally by its heating effects on water (Table 14-2).

Example 14-7

How many cubic meters of natural gas are required to heat 12 L of water from 20°C to 90°C if 40% of the heat is lost to the surroundings?

Solution:

Since 1.0 L of water has a mass of 1.0 kg, and 40% of the heat is lost, using Eqs. 14-12 and 14-6

$$0.6V = \frac{Q}{H} = \frac{mc\Delta T}{H}$$
$$= \frac{(12 \text{ kg})[4\ 186 \text{ J/(kg} \cdot °\text{C})](90°\text{C} - 20°\text{C})}{4.35 \times 10^7 \text{ J/m}^3}$$
$$= 8.1 \times 10^{-2} \text{ m}^3$$

Therefore

$$V = \frac{8.1 \times 10^{-2} \text{ m}^3}{0.6} = 0.14 \text{ m}^3$$

Problems

34. How much heat in joules is evolved when 20 kg of coal are burned?

35. If 2.5 kg of wood are completely burned and 25% of the heat evolved is used to heat a 20 kg iron boiler containing 300 kg of water at 10°C, what is the final temperature of the water?

36. A motor burns gasoline at a rate of 8.0 lb/h. If 20% of the heat energy evolved is converted into useful power, what power is developed by the motor?

37. A camper uses a propane stove to boil 500 g of water in a 100 g aluminum pot. If the efficiency of the stove is 15% and the initial temperature of the water in the

pot was 20°C, how much propane is required?

38. A heating system delivers 20 m³ of air per minute at 25°C to a room. If the air was initially at a temperature of 10°C and the conditioned air has a density of 1.2 kg/m³, (a) what power is delivered by the heating system? (b) If the heating system utilizes natural gas and is 60% efficient, at what rate does it consume the fuel?

14-6 PSYCHROMETRICS AND AIR CONDITIONING

Water vapor is always present in air, but it is the most variable component. It enters the air mainly by evaporation from bodies of water, such as oceans, lakes and rivers. Air is said to be *saturated* at a certain temperature when it cannot hold any more water vapor. If the temperature is increased, however, the air can hold more water vapor. Conversely, a decrease in the temperature of the saturated air causes condensation of some water vapor from the air.

The water content of air is extremely important to comfort because all animals control their metabolism to some extent by perspiration, i.e. evaporation and hence cooling from the skin, or by absorbing water vapor. There are many methods of defining the water content in a body of air. The *absolute humidity* h_A is the mass of water vapor present per unit volume of air. The actual quantity (weight or mass) of water vapor present per unit quantity of dry air is called the *specific humidity* h_s.

$$h_s = \frac{\text{quantity of water vapor present}}{\text{quantity of dry air involved}} \quad (14\text{-}13)$$

Since the water vapor content in the air is usually quite small, it is measured in grains or grams. One *grain* is equals to 1/7 000 lb.

The cooling process of evaporation from the skin is related to the degree of saturation of the air but not to its actual water content. If the air is saturated, it cannot hold any more moisture at that temperature, and moisture cannot evaporate from the skin. When the degree of saturation is low, air can hold more moisture, and the evaporation rate from the skin increases. The degree of saturation is expressed in terms of *relative humidity* h_r, which is defined as the ratio of the actual quantity of water vapor present to the quantity of water vapor required to saturate that mass of air at that temperature.

$$h_r = \frac{\text{quantity of water vapor present}}{\text{quantity of water vapor required for saturation}} \quad (14\text{-}14)$$

or

$$h_r = \frac{\text{actual vapor pressure}}{\text{saturated vapor pressure}} \quad (14\text{-}15)$$

since vapor pressure is directly proportional to the mass of the vapor present. Relative humidity is usually expressed as a percentage.

Either a *wet and dry bulb thermometer* or a *sling psychrometer* (Fig. 14-3) may be used to measure relative humidity; both instruments use the same principle. The instruments consist of two identical thermometers; one, the *wet bulb*, has its bulb covered with a water-soaked wick. As water evaporates, it removes heat from the wet bulb at a rate that depends on the quantity of water vapor

Table 14-3
Water Content of Saturated Air

Temperature (°C)	Water Content g/m^3
−10	2.2
−5	3.3
0	4.8
5	6.8
10	9.3
15	12.7
20	17.2
25	22.8
30	30.1

present in the air. Therefore, the wet bulb thermometer reads lower than the dry bulb thermometer; the temperature difference is directly related to the relative humidity h_r. Vapor pressure tables (Table 14-3) or a psychrometric chart (Fig. 14-4) may be used to determine the relative humidity.

The *dew point* is the temperature at which condensation begins to occur when an air mass is cooled. At the dew point, the air is saturated, and consequently the relative humidity is 100%.

Example 14-8

If the temperature of an air mass is 25°C and its dew point is 15°C, determine (a) its relative humidity at 25°C and (b) the mass of

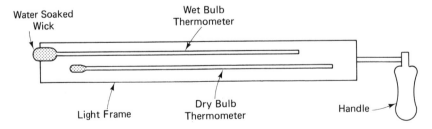

Figure 14-3
The sling psychrometer.

water that condenses when the air is cooled to 10°C.

Solution:

(a) From Table 14-3, the water vapor content of saturated air at 25°C, $m_{25} = 22.8 \text{ g/m}^3$, and at 15°C, $m_{15} = 12.7 \text{ g/m}^3$. Thus

$$h_r = \frac{m_{15}}{m_{25}} = \frac{12.7 \text{ g/m}^3}{22.8 \text{ g/m}^3} = 0.56 \quad \text{or} \quad 56\%$$

(b) The water content of saturated air at 10°C, $m_{10} = 9.3 \text{ g/m}^3$; since the dew point is 15°C, the air originally held a water vapor content of $m_{15} = 12.7 \text{ g/m}^3$. Thus, the condensed water:

$$m_{15} - m_{10} = 12.7 \text{ g/m}^3 - 9.3 \text{ g/m}^3$$
$$= 3.4 \text{ g/m}^3$$

The psychrometric chart is essentially a graphical analysis of heat and humidity. It is frequently used in air conditioning calculations. Dry bulb temperatures are plotted horizontally, and specific humidity is plotted vertically. The quantity of water vapor present in *saturated air* is determined experimentally for each dry bulb temperature, and the values are plotted as the 100% relative humidity or dew-point line. Other relative humidity lines are obtained from Eq. 14-15. For example, saturated air contains 132 grains of water vapor per pound at 75°F; therefore at 10% relative humidity, air contains 13.2 grains/lb. The wet bulb lines (left to right and falling) are obtained by adding moisture to an air mass until it becomes saturated. If the air is originally at 85°F dry bulb and the specific humidity is 60 grains/lb, the addition of moisture with no external heating causes the air to saturate at a temperature of 65°F, and it has a specific humidity of 93 grains/lb. These two points are joined by the wet bulb temperature line.

Example 14-9

A sling psychrometer gives a dry bulb temperature of 72°F and a wet bulb temperature of 58°F. Determine (a) the relative humidity, (b) the specific humidity, (c) the dew point.

Solution:

Determine the point of intersection B of the 72°F dry bulb temperature line and the 58°F wet bulb temperature line.

(a) The relative humidity is approximately 42%.

(b) From the vertical axis, this point corresponds to a specific humidity of 50 grains/lb.

(c) The water vapor content and thus the specific humidity is constant; therefore condensation occurs if the air is cooled to below a dry bulb temperature of 48°F.

Air conditioning is concerned mainly with the control of heat, humidity and air circulation, but it also involves the control of noise and dust. The load of air conditioning equipment depends on the building materials, the size and use of the conditioned space and other sources of heat losses and gains.

The type of air conditioning depends on the external weather conditions. Winter air conditioning is mainly associated with heating and humidifying air in the conditioned space whereas summer air conditioning is mainly concerned with cooling and dehumidifying the air.

Changes in dry bulb temperature are caused by sensible heating at a constant moisture content or specific humidity. Therefore sensible heat changes correspond to horizontal changes on the psychrometric chart. In order to change the moisture content or specific humidity of the air, more water must be evaporated or condensed, and latent heat is involved. These latent heat changes correspond to vertical changes on the psychrometric chart. In general, heat changes are involved in both the sensible and latent heat aspects of air conditioning.

The heat content (or enthalpy) scale is

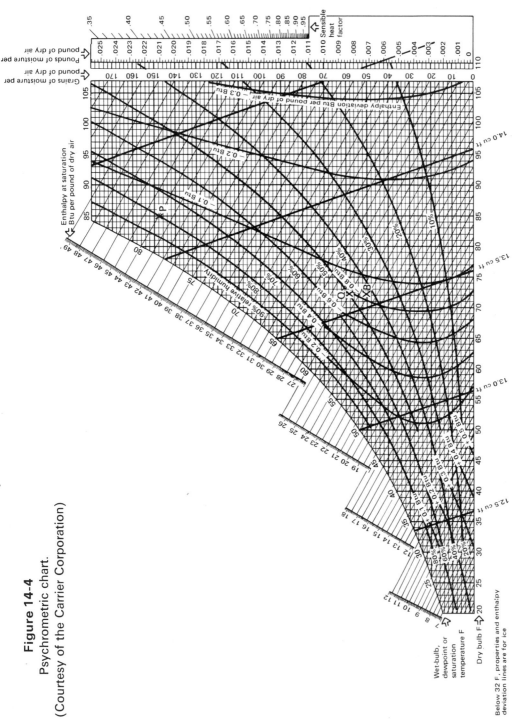

Figure 14-4
Psychrometric chart.
(Courtesy of the Carrier Corporation)

(Copyright © 1947, 1956, 1960, 1961, Carrier Corporation. Reproduced by permission of Carrier Corporation)

located at the projection of the wet bulb temperature lines since these lines correspond to changes in specific humidity and dry bulb temperature with no external heat influences. Finally, the density of the air may be obtained from the specific volume lines (left to right and falling steeply). If the volume of the conditioned space and the rate of air replacement are known, the required heating or cooling capacity of the air conditioning equipment may be calculated.

Example 14-10

Determine the operating capacity of an air conditioner if it converts 1 000 ft³ of air per minute from 85°F dry bulb and 80% relative humidity to 72°F dry bulb and 50% relative humidity.

Dry Bulb Temperature	Wet Bulb Temperature	Relative Humidity	Specific Humidity	Dew Point
72°F	53°F	—	—	—
80°F	—	40%	—	—
—	—	60%	90 grains/lb	—
85°F	—	—	—	60°F
—	50°F	30%	—	—
70°F	—	—	80 grains/lb	—

Solution:

On the psychrometric chart, the air conditioning process corresponds to a change from point P (85°F and 80% relative humidity) to point Q (72°F and 50% relative humidity). The heat content at P is 43.7 BTU/lb, and the heat content at Q is 26.5 BTU/lb. Thus

$$43.7 \text{ BTU/lb} - 26.5 \text{ BTU/lb} = 17.2 \text{ BTU/lb}$$

must be removed, i.e. 17.2 BTU per pound of dry air.

Since the specific volume of the initial air

at P is 14.2 ft³/lb, the heat removed per unit volume:

$$\frac{Q}{V} = \frac{17.2 \text{ BTU/lb}}{14.2 \text{ ft}^3/\text{lb}} = 1.21 \text{ BTU/ft}^3$$

Since 10^3 ft³ of air flows each minute, the air conditioner must be able to remove:

$$(1.21 \text{ BTU/ft}^3)(10^3 \text{ ft}^3/\text{min}) = 1\,210 \text{ BTU/min}$$

Problems

39. The temperature of an air mass is 30°C, and its dew point is 10°C. Determine (a) its relative humidity at 30°C and (b) the mass of water that condenses when the air is cooled to 0°C.

40. Determine the values of the blanks in the following table, using the psychrometric chart.

41. Determine the operating capacity of an air conditioner that converts 500 ft³ of air per min from 90°F dry bulb and 75% relative humidity to 70°F dry bulb and 30% relative humidity.

42. How much heat input must a heating system deliver in order to convert 1 000 ft³ of air per minute from 35°F dry bulb and 20% relative humidity to 70°F dry bulb and 50% relative humidity? If the heating system utilizes domestic fuel oil and is 80% efficient, determine the rate of fuel oil consumption.

REVIEW PROBLEMS

1. An engine develops an average power of 3.2 kW for 5.0 min, and 40% of this is converted into heat. How much heat is produced?

2. How much heat is required to raise the temperature of a 120 kg aluminum girder from 23°C to its melting point?

3. A 1.8 lb piece of iron is heated in a furnace to 2 000°F. It is then dropped in an 8.0 oz aluminum calorimeter containing 8.0 lb of water at 40°F. What is the final temperature of the water?

4. How much heat in joules must be added to a 3.5 kg copper rod in order to raise its temperature from 20°C to 100°C?

5. How much water at 10°C must be mixed with 350 g of water at 80°C to bring the mixture to a final temperature of 48°C? Assume that no heat is lost or gained from the surroundings.

6. What is the final temperature when 1.4 kg of water at 30°C is mixed with 2.1 kg of ethyl alcohol at 5°C? Assume that no heat is lost or gained from the surroundings.

7. If 0.25 lb of a substance at 250°F is added to a 1.0 lb aluminum calorimeter that contains 3.0 lb of water at 50°F, what is the specific heat capacity of the substance if there are no external heat influences and the final temperature in thermal equilibrium of the whole system is 56°F?

8. If 10 lb of steam at 212°F are passed into 500 lb of water at 40°F, what is the final temperature of the mixture? Assume that no heat is lost or gained from the surroundings.

9. If 1.0 lb of ice at 12°F is added to a 2.0 lb calorimeter with a specific heat capacity of 0.25 BTU/(lb·°F) that contains 3.0 lb of water at 42°F, how much ice melts? Assume that there are no external heat influences.

10. If 500 g of water and 100 g of ice are in equilibrium at 0°C, what is the final temperature if 200 g of steam at 100°C are added? Assume that there are no external heat influences.

11. If 10 g of ice at −10°C are added to a 200 g metal calorimeter containing 500 g of water at 20°C and 10 g of steam are then passed into the system, determine the final temperature of the system. Assume that there are no external heat influences and that the specific heat capacity of the calorimeter is 1 047 J/(kg·°C).

12. How much heat in joules is evolved when 2.5 m³ of propane is completely burned?

13. How much alcohol with a heat of combustion of 4.10×10^7 J/kg is needed to change 1.5 kg of ice at −10°C to liquid water at 70°C if only 60% of the generated heat is applied to the ice?

14. If the temperature of an air mass is 30°C and its dew point is 15°C, determine (a) its relative humidity and (b) the mass of water that condenses when 30 m³ of the air is cooled to 10°C.

15. Use the psychrometric chart to fill in the blanks in the table following.

16. An air conditioner handles 1 000 ft³/min of outside air at 90°F dry bulb and 60% relative humidity. If the apparatus dew point is 65°F, (a) how much water is condensed each hour by the cooling coils? (b) How much cooling is done? (c) What is the relative humidity if the dry bulb temperature in the room is 75°F? Assume no latent heat losses.

Dry Bulb (°F)	Wet Bulb (°F)	Percent Relative Humidity	Dew Point (°F)	Specific Humidity (grains/lb)	Specific Volume (ft³)	Enthalpy (BTU/lb)
78	65					
72		40				
68				80		
	65		45			
90						34.1

15

Thermal Expansion and Heat Transfer

Our senses enable us to estimate whether an object is "hot" or "cold," but reliable temperature scales and measuring devices, called *thermometers*, must be used in order to obtain more precise measurements. Many physical items, such as volume, pressure, electrical resistance and color, change when an object is heated; these changes may be used to determine the temperature variation that caused them.

Thermal expansion has many important consequences in structures, engines, electrical components and many other aspects of technology. The properties of thermal expansion may be utilized to advantage in some devices, such as circuit breakers, thermometers and thermal switches; it must be allowed for in other areas such as structures, engines, molds and cables, which must operate within prescribed tolerances for all expected temperature ranges. While small variations in the temperature of liquids and solids only cause small changes in their dimensions, these expansions or contractions may result in enormous stresses. If these thermal stresses are not allowed for, they may cause mechanical failure of the system.

Heat energy flows from areas of higher temperature to areas of lower temperature at a rate that depends on the nature of the material (if any) through which it travels. Some materials, called *thermal insulators*, offer a large resistance to heat flow; they are used in clothing, refrigerators, buildings (to improve the air conditioning) and in many other areas where hotter regions must be isolated from cooler regions. Other materials, called *thermal conductors*, readily allow heat to flow through them; they are used when heat must be transported from a hotter area to a cooler area.

15-1 LINEAR EXPANSION

It is a well known fact that most substances expand when they are heated and that this expansion always occurs in three dimensions. However, in solids, it is often necessary only to consider the consequences of expansion in one dimension. When a solid is heated, its temperature and hence the internal vibrational energy possessed by its atoms and molecules increases; as the amplitude (distance through which the atoms vibrate) increases, the material expands.

Consider a solid rod that has a length L_0 at a temperature T_0 and expands to a length L when it is heated to a temperature T. The *change* in length $L - L_0 = \Delta L$ of the rod is approximately proportional both to the temperature *change* $T - T_0 = \Delta T$ and to its initial length L_0.

The proportionality constant α is called the *coefficient of linear expansion* and is defined as the fractional change in length per degree increase in temperature:

$$\alpha = \frac{\Delta L}{L_0 \Delta T} = \frac{L - L_0}{L_0(T - T_0)} \qquad (15\text{-}1)$$

It is approximately constant for a particular material within a small temperature range (see Problem 15-8). The fractional elongation term $\Delta L/L_0$ is known as the *thermal strain* in the rod (Chapter 5). The coefficient of linear expansion is usually expressed in units of per kelvin (K^{-1}) (or per degree Celsius,/°C) in SI and per degree Fahrenheit in the British Engineering System. Some typical values are listed in Table 15-1.

Solving Eq. 15-1 for the final length L of the rod, we obtain:

$$L = L_0 + L_0\alpha(T - T_0) = L_0[1 + \alpha(T - T_0)] \qquad (15\text{-}2)$$

Even though thermal changes in length are usually relatively small compared with the original length, the thermal strain may be quite substantial, and very large thermal stresses may arise. These thermal stresses are quite capable of buckling even steel girders. Consequently thermal elongation must be allowed for by special expansion joints.

Table 15-1

Typical Coefficients of Linear Expansion of Solids at Room Temperature

	$\alpha/(10^{-5}/K)$	$\alpha/(10^{-5}/°F)$
Aluminum	2.3	1.3
Brass	1.9	1.1
Copper	1.7	1.1
Glass	0.91	0.51
Pyrex Glass	0.32	0.18
Iron (soft)	1.2	0.67
Lead	2.9	1.6
Platinum	0.90	0.50
Quartz	0.055	0.031
Silver	1.9	1.1
Steel	1.2	0.65
Tungsten	0.45	0.25

Linear expansion also affects the fitting of engine parts, the length of a surveyor's measuring tape, the suspension of cables, the wire connections between circuit elements, etc.

*Example 15-1*_____

A steel I-beam in a bridge truss is exactly 12 m long at 5°C. (a) What is its length when the temperature rises to 35°C? (b) If its cross-sectional area is 35 cm², what force is produced by the linear expansion if the beam is constrained?

Solution:

Using Eq. 15-2 and Tables 15-1 and 5-2, we obtain:

(a) $L = L_0[1 + \alpha(T - T_0)]$
$= 12 \text{ m}[1 + (1.2 \times 10^{-5}/°C)(35°C - 5°C)]$
$= 12.004\ 32 \text{ m}$

(b) $E = \dfrac{FL_0}{A\Delta L}$ and $\alpha = \dfrac{\Delta L}{L_0\Delta T}$,

or $\dfrac{\Delta L}{L_0} = \alpha T$

Therefore

$$F = EA\left(\frac{\Delta L}{L_0}\right) = EA\,(\alpha\Delta T)$$

$$F = (20 \times 10^{10}\ \text{Pa})(35 \times 10^{-4}\ \text{m}^2)$$
$$\times\ (1.2 \times 10^{-5}/°\text{C})(35°\text{C} - 5°\text{C})$$
$$= 252\,000\ \text{N}$$

Thermal expansion also allows shrink fitting of one material over another.

Example 15-2

A wagon wheel has a diameter of 36.04 in, and a steel rim has an inside diameter of 36.00 in at 65°F. To what minimum temperature must the rim be heated so that it will fit over the wheel?

Solution:

The inner circumference of the rim must expand sufficiently so that it will be at least equal to the outer circumference of the wheel. The initial circumference of the rim $L_0 = \pi \times 36.00$ in, and the final circumference $L = \pi \times 36.04$ in. Therefore the elongation $= L - L_0 = 0.04\,\pi$ in.

From Table 15-1 and Eq. 15-2, the minimum final temperature is:

$$T = T_0 + \frac{L - L_0}{L_0\alpha}$$

$$= 65°\text{F} + \frac{(0.04\pi\ \text{in})}{(36.00\pi\ \text{in})(6.5 \times 10^{-6}/°\text{F})}$$

$$= 65°\text{F} + 171°\text{F} = 236°\text{F}$$

A bimetallic strip is composed of 2 different metal rods that are welded or riveted together lengthwise. Since each metal has a different coefficient of linear expansion, one metal rod expands more than the other when the strip is heated. If the composite rod is initially straight, it bends when it is heated. Thus a bimetallic strip may be used to measure temperature changes, e.g. as the temperature-sensitive element of a thermostat, or to compensate for the effects of thermal expansion (Fig. 15-1).

Problems

1. Determine the coefficient of linear expansion of a metal rod if it increases its length from 15.000 m to 15.016 m when the temperature increases from 0°C to 35°C.

2. An aluminum girder is 20.000 m long at 30°C. What is its length when the temperature drops to −10°C?

3. What minimum gap should be left between steel girders that are 15 ft long at

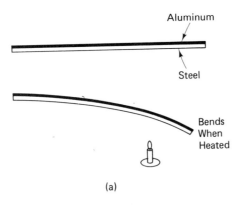

Aluminum

Steel

Bends When Heated

(a)

Contact

In a thermostat a bimetallic strip will tend to straighten out when heated and make contact with a circuit element

(b)

Figure 15-1
A bimetallic strip.

32°F if they will be subjected to a maximum temperature of 110°F?

4. A surveyor has a steel tape that reads correctly at 23°C. If he measures a length of 23.00 m at −3°C, what is the error?

5. In a circuit, a copper wire is heated from 20°C to 40°C. Determine the percentage increase in its length.

6. A steel pushrod is 25 cm long at 20°C. What clearance should be allowed for if it is expected to operate at 100°C?

7. A wheel has a diameter of 1.002 m, and a copper rim has an inside diameter of 1.000 m at 20°C. To what minimum temperature must the rim be heated in order to fit it on the wheel?

8. Determine the percentage difference between the coefficients of linear expansion at 0°C and 100°C for the same material. *Hint:* Write expressions for the coefficients based on each temperature in terms of initial and final lengths.

9. A steel I-beam has a cross-sectional area of 40 cm² at 20°C. If the temperature increases to 40°C, (a) what is the thermal strain? (b) What force would the beam exert along its length if it were constrained?

10. Two vertical steel girders are rigidly connected by a steel I-beam that has a cross-sectional area of 20 in². If the beam was installed when the temperature was 85°F, what force does it exert on the girders when the temperature drops to −25°F?

11. The end of a 2.0 ft long steel rod is rigidly joined to the end of 3.0 ft long aluminum rod with an equal cross-sectional area $A = 3.0$ in². Determine the total elongation of the system when the temperature is increased by 50°F.

15-2 VOLUME EXPANSION

The maximum density of water occurs at 4°C under standard conditions. Water actually increases its volume, and consequently reduces its density when it is cooled from 4°C to 0°C; this is why ice can float on water. All other materials (and water in other temperature ranges) increase their volume when they are heated.

By analogy with the definition of the coefficient of linear expansion α, the *thermal coefficient of volume expansion* β is defined as the fractional change in volume per degree increase in temperature.

$$\beta = \frac{\Delta V}{(V_0 \Delta T)} = \frac{(V - V_0)}{[V_0(T - T_0)]} \quad (15\text{-}3)$$

A linear relationship exists between the volume V and the temperature T:

$$V = V_0[1 + \beta(T - T_0)] \quad (15\text{-}4)$$

where V_0 is the initial volume at the temperature T_0, V is the final volume at the temperature T and ΔV is the change in volume when the temperature increases by ΔT. The coefficient of volume expansion β is also expressed in units of per kelvin (K^{-1}) or per degree Celsius (1/°C) in SI and per degree Fahrenheit (1/°F) in the British Engineering System. Some typical values are listed in Table 15-2.

Consider an isotropic material (which has the same value for the thermal coefficient of linear expansion α in all directions). Suppose that the object is subjected to a temperature change ΔT and that it changes its dimensions a, b and c by Δa, Δb and Δc, respectively. From Eq. 15-3, $\Delta a = a\alpha \Delta T$, $\Delta b = b\alpha \Delta T$ and $\Delta c = c\alpha \Delta T$. Therefore the final volume $V = (a + \Delta a)(b + \Delta b)(c + \Delta c)$,

or $\qquad V = abc(1 + \alpha \Delta T)^3$.

But abc is the initial volume V_0. Since α is usually much less than one, terms involving α^2 and α^3 may be neglected in the expansion.

Table 15-2
Typical Volume Expansion Coefficients

	$\beta/(10^{-3}\ K^{-1})$	$\beta/(10^{-4}/°F)$
Liquids		
Benzene	1.18	6.6
Ethyl alcohol	1.04	5.78
Gasoline	1.1	6.0
Mercury	0.182	1.01
Water	0.37	2.0
(above 10°C)		
Gases		
Air	3.67	20.4
Carbon dioxide	3.7	20.6
Hydrogen	3.66	20.3
Nitrogen	3.67	20.3
Oxygen	4.86	27
Solids		
Glass (Pyrex)	0.026	0.144

Thus

$$V \approx V_0(1 + 3\alpha\ \Delta T) \qquad (15\text{-}5)$$

Comparing with Eq. 15-4, we obtain:

$$3\alpha \approx \beta \quad \text{for isotropic materials} \quad (15\text{-}6)$$

Example 15-3

A steel fuel tank is filled to the brim with exactly 70 L of gasoline at 20°C. If the temperature rises to 35°C, how much gasoline overflows?

Solution:

The amount of gasoline that overflows equals the final difference ΔV between the volume V_G of the gasoline and the volume V_T of the tank. The final volume V_T of the tank may be determined by assuming that the cavity of the tank is also filled with steel. Thus, as the steel expands, the cavity expands. Since the initial volumes were the same:

$$\Delta V = V_G - V_T$$
$$= V_0(1 + \beta\ \Delta T) - V_0(1 + 3\alpha\ \Delta T)$$
$$= V_0\ \Delta T(\beta - 3\alpha).$$

Therefore, referring to Tables 15-1 and 15-2, we obtain:

$$\Delta V = (70\ \text{L})(35°C - 20°C)$$
$$\times (1.1 \times 10^{-3}/°C - 3 \times 1.2 \times 10^{-5}/°C)$$
$$= 1.12\ \text{L}$$

Problems

12. A Pyrex glass container is filled with 1.0 L of benzene at 0°C. How much benzene overflows when the temperature is raised to 25°C?

13. If the density of mercury at 0°C is 13 595.5 kg/m³, what is its density at 30°C?

14. If the density of benzene at 0°C is 899 kg/m³, what is its density at −20°C?

15. If 10.00 cm³ of gasoline and 10.00 cm³ of benzene fill a 20.00 cm³ Pyrex container at 0°C, what happens if the temperature is raised to 30°C?

15-3 THE MOLE

All elements are composed of very small particles called *atoms*, which cannot be chemically decomposed. Atoms of different elements and even some atoms, called *isotopes* (Chapters 31 and 34), of the same element have different masses. The relative atomic masses were originally related to the mass of the "lightest" element, hydrogen, but by international agreement they are now specified relative to the most abundant isotope, carbon 12.

A *unified atomic mass unit* (u) is permitted for use with SI; it is defined as $\frac{1}{12}th$ of the mass of a neutral atom of the nuclide carbon 12: or $1\ u = 1.660\ 44 \times 10^{-27}$ kg. The *mole* (mol) is an SI base unit used to describe quantity of matter. It is defined as the amount of a substance that contains the same number of entities as there are atoms in 12 g of carbon 12. These elementary entities may be atoms, molecules, ions, electrons or other particles or groups of particles. Since $1 \text{ u} = 1.660\ 44 \times 10^{-24}$ g, $1 \text{ g} = 6.022\ 50 \times 10^{23}$ u. The number $N_0 = 6.022\ 50 \times 10^{23}$ is called *Avogadro's number*, and it has a special significance.

Note that 1 atom of carbon 12 has a mass of 12 u; therefore 12 g of that isotope contains Avogadro's number of atoms. A *mole* of any substance must contain Avogadro's number of entities of that substance. Thus a mole of hydrogen atoms is Avogadro's number of atoms, which has a mass of $1.008\ 145\ 6$ g; a mole of hydrogen gas contains Avogadro's number of molecules or $2.016\ 291\ 2$ g (two atoms per molecule).

Gas molecules are very small, therefore even though large numbers of gas molecules may be present, under normal conditions the average distance between them is very large compared with their size. Consequently, in many respects, molecules of different gases behave physically in the same way.

Under standard conditions of temperature and pressure (S.T.P., which corresponds to 0°C and 1 atm pressure), one *mole* of any gas contains Avogadro's number of molecules and occupies a volume of 22.4 L.

*Example 15-4*_____

If one mole of carbon dioxide gas has a mass of 44.0 g, how many carbon dioxide molecules are there in a volume of 5.60 L at S.T.P.?

Solution:

Since 1.0 mol is 22.4 L and it is occupied by Avogadro's number of molecules, 5.60 L are occupied by $\dfrac{5.60}{22.4}$ mole, which contains:

$$\left(\frac{5.60 \text{ mol}}{22.4}\right)\left(6.023 \times 10^{23}\ \frac{\text{molecules}}{\text{mol}}\right)$$
$$= 1.51 \times 10^{23} \text{ molecules}$$

*Problems*_____

16. A mole of silicon has a mass of 28.06 g. How many silicon atoms are there in 83.05 g?

17. A mole of nitrogen gas has a mass of 28.02 g. How many nitrogen molecules are there in a volume of 12.2 L at S.T.P.?

15-4 THE GAS LAWS AND ABSOLUTE TEMPERATURE

Gases are fundamentally different from liquids and solids. A fixed mass of gas has an indefinite volume since it occupies all of the available space. The spacing between gas molecules is much greater than the spacing between the molecules of liquids and solids. Therefore gases are more easily compressed and are lighter.

Gas molecules are in continual rapid motion, colliding frequently with each other and the walls of any container. The pressure exerted by a confined gas is mainly due to this molecular bombardment of the container walls. If the distance between the walls of a container of gas is reduced, the volume of the gas is also reduced; however, if the temperature is constant, the molecules still have the same average speed, and they collide more frequently with the walls, producing an increased pressure.

The experimental investigation of the relationship between the pressure and volume of a confined gas was first performed by Robert Boyle (1627–1691) in 1662. His results, now called *Boyle's law*, are:

If a fixed mass of gas is maintained at a constant temperature, then its absolute pressure p is inversely proportional to its volume V.

Thus

$$p \propto 1/V \quad \text{or} \quad pV = \text{constant} \qquad (15\text{-}7)$$

or

$$p_1 V_1 = p_2 V_2 \text{ (at constant temperature)}$$

where p_1 is the absolute pressure when the volume is V_1 and p_2 is the absolute pressure when the volume is V_2.

Gauge pressures must be converted into absolute pressures by the addition of atmospheric pressure before they can be used in Boyle's law.

*Example 15-5*_____

What volume of air at atmospheric pressure 101 kPa is required to fill a 15.0 L cylinder to a gauge pressure of 505 kPa at a constant temperature?

Solution:

The final absolute pressure $p_2 = 505$ kPa + 101 kPa = 606 kPa. Thus the volume required is:

$$V_1 = \frac{p_2 V_2}{V_1} = \frac{(606 \text{ kPa})(15.0 \text{ L})}{101 \text{ kPa}} = 90.0 \text{ L}$$

Boyle's law describes the relationship between pressure and volume of a fixed mass of gas at constant temperature. However, if the gas liquifies, Boyle's law is invalid because liquids are virtually incompressible. For example, when a cylinder of gas is subjected to changes in pressure and volume at a constant temperature in excess of the critical temperature, the kinetic energy of the gas molecules is too great to allow the gas to liquify. Thus at this temperature Boyle's law is valid, and the gas pressure is inversely proportional to its volume. The constant temperature curve, called an *isotherm*, of the pressure-volume relationship in this region is represented by AB in Fig. 15-2.

However at a constant temperature t,[*] which is less than the critical temperature t_c, Boyle's law is only valid in the low pressure region R to S. At some pressure P, the vapor begins to liquify, and it decreases its volume

_____

[*]Celsius temperatures are represented by t while T represents absolute temperatures.

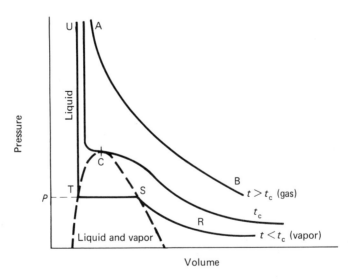

Figure 15-2
Isotherms for a gas and vapor.

at that constant pressure. As the vapor condenses, the volume of the liquid increases, and the volume of the gas decreases (between S and T) until only the liquid phase remains. Since liquids are nearly incompressible, any further increase in pressure results in negligible volume changes (between T and U). At the critical temperature t_c, the material cannot form a meniscus because the densities of the liquid and vapor are equal.

Like liquids and solids, gases expand when they are heated; however, gases are far more sensitive to pressure and volume changes, and their coefficients of thermal expansion are much larger.

Consider a container that has a volume V_0 and contains a fixed number n of gas molecules at an initial temperature t_0 and pressure p_0. The pressure on the walls of the container is due to their constant bombardment by the gas molecules. If the gas is heated, its temperature increases and the kinetic energy of the molecules increases. Therefore these molecules move more rapidly, and they collide more frequently with the container walls. Thus, if the volume V_0 of the container is constant, the pressure of the gas increases as its temperature increases.

In order to maintain a constant pressure p_0 as the temperature is increased, the walls of the container must be moved farther apart, increasing the volume so that the faster moving molecules must travel a greater average distance between two successive collisions with the walls. Thus, if the pressure is constant, the volume of a fixed mass of the gas increases as its temperature increases.

If its pressure is constant, the change in the volume of a fixed mass of gas is directly proportional to its change in temperature according to Eq. 15-3:

$$V - V_0 = \Delta V = V_0 \beta (t - t_0)$$

where V is its final volume at the temperature t, V_0 is its volume at the temperature t_0 and

β is the coefficient of volume expansion. Therefore, if V_0 is the volume of the gas at $t_0 = 0°C$, a linear relationship exists between its volume V and the corresponding temperature t:

$$V = V_0[1 + \beta(t - t_0)] = V_0 + V_0\beta t \quad (15\text{-}8)$$

Since gas molecules are very small and relatively far apart and molecules of different gases all behave in approximately the same way, the thermal coefficient of volume expansion (β_0) at $0°C$ or $32°F$ also has approximately the same value of $\frac{1}{273}°C$ for all low pressure gases. Therefore, Eq. 15-8 reduces to:

$$V = V_0 + \frac{V_0 t}{273°C} = V_0\left(1 + \frac{t}{273°C}\right)$$

Thus a graph of the volume V versus the temperature t of a fixed mass of gas under a constant low pressure yields a straight line (Fig. 15-3a). The extension of the line through the temperature axis coincides with a temperature of $-273.15°C$ or $-460.27°F$.

If the volume of the fixed mass of gas is kept constant as its temperature increases, the kinetic energy and speed of the molecules increase, and the gas molecules collide more frequently with any constraining walls. Therefore the change in pressure is also directly proportional to the change in temperature (Fig. 15-3b). By analogy with the coefficient of volume expansion, the coefficient of pressure change γ is the fractional pressure change per unit increase in temperature.

$$\gamma = \frac{p - p_0}{p_0(t - t_0)} \quad \text{or} \quad p - p_0 = \Delta p = p_0\gamma(t - t_0)$$
$$(15\text{-}9)$$

where p is the absolute pressure at a temperature t and p_0 is the absolute pressure at temperature t_0. Again, low pressure gases behave similarly, and the coefficient of pressure change at $0°C$ is $\gamma \approx \frac{1}{273}°C$.* The cor-

---

*This may also be proved from Boyle's law.

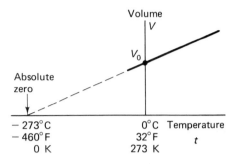

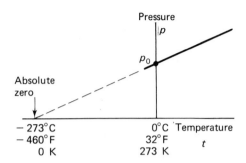

(a) Temperature versus volume of a fixed mass of gas at constant pressure

(b) Temperature versus absolute pressure of a fixed mass of gas at constant volume

Figure 15-3

Temperature versus volume and temperature versus pressure graphs.

responding graph of absolute pressure versus the temperature t of a fixed mass of gas at constant volume is a straight line, which extends through the temperature axis at $-273.15°C$ or $-460.27°F$ (Fig. 15-3b).

This temperature $-273.15°C$ is called *absolute zero*. Of course, in practice the gas would liquify or even solidify before absolute zero is reached. However, Lord Kelvin proposed a new temperature scale called the *absolute* or *thermodynamic temperature*.† The corresponding temperature unit is called a *kelvin* (K) and it is taken as a base unit in SI. This temperature scale is based on *absolute zero*, which is taken as 0 K, and the *triple point of water*, which is assigned the value 273.16 K.* Since absolute zero corresponds to $-273.15°C$ and the triple point of water is $0.01°C$, a temperature *change* of 1 K is equivalent to a temperature change of $1°C$. Thus the relationship between the kelvin and Celsius temperatures is approximately:

$$K = °C + 273 \qquad (15\text{-}10)$$

---

†See the thermodynamic definition in Chapter 16.

*The word and symbol for degree is *not* included for kelvin temperatures.

Another absolute temperature scale called the *Rankine scale* has *some* common usage. The Rankine temperature (R) is also based on absolute zero (0 R) and the triple point of water (492 R). Since absolute zero is $-460°F$, a temperature change of 1 R is approximately equal to a temperature change of $1°F$.

$$R = °F + 460 \qquad (15\text{-}11)$$

Note that we can redraw the graphs in Fig. 15-3 with absolute temperature T versus volume V and absolute pressure p. In each case we obtain a straight line that passes through the origin; therefore:

a. *If a fixed mass of gas is maintained at a constant pressure, its volume V is directly proportional to its absolute temperature T:*

$$V \propto T \quad \text{or} \quad \frac{V_2}{T_2} = \frac{V_1}{T_1} = \text{constant} \quad (15\text{-}12)$$

This relationship is known as *Charles' law*.

b. *If the volume of a fixed mass of gas is constant, its absolute pressure p is directly proportional to its absolute temperature T:*

$$p \propto T \quad \text{or} \quad \frac{p_2}{T_2} = \frac{p_1}{T_1} = \text{constant} \quad (15\text{-}13)$$

Both relationships are valid for low pressure gases.

Combining Charles' law with Boyle's law, we obtain the more useful *ideal gas law*:

$$\frac{p_1 V_1}{T_1} = \frac{p_2 V_2}{T_2} = \text{constant} \qquad (15\text{-}14)$$

where p is the absolute pressure, V is the volume and T is the *absolute temperature* (kelvin or Rankine). Since the density of a fixed mass m of gas with a volume V is $\rho = m/V$, $V = m/\rho$, and we may rewrite the ideal gas law as:

$$\frac{pm}{\rho T} = \text{constant} \quad \text{or} \quad \frac{p}{\rho T} = \text{constant} \quad (15\text{-}15)$$

since m is constant.

Example 15-6

A car tire is inflated to a gauge pressure of 28.0 lb/in² when the temperature is 32°F; if the volume of the tire does not change and atmospheric pressure is 14.7 lb/in², what is the gauge pressure when the temperature is 99°F?

Solution:

The absolute conditions are: $p_1 = 28.0$ lb/in² $+ 14.7$ lb/in² $= 42.7$ lb/in², $T_1 = 32°F = (32 + 460)$ R $= 492$ R and $T_2 = 99°F = (99 + 460)$ R $= 559$ R. Thus

$$\frac{p_1}{T_1} = \frac{p_2}{T_2} \quad \text{or} \quad p_2 = \frac{p_1 T_2}{T_1} = \frac{(42.7 \text{ lb/in}^2)(559 \text{ R})}{492 \text{ R}}$$
$$= 48.5 \text{ lb/in}^2 \text{ (absolute)}$$

Therefore, the final gauge pressure is:

$$p = 48.5 \text{ lb/in}^2 - 14.7 \text{ lb/in}^2 = 33.8 \text{ lb/in}^2$$

Example 15-7

A weather balloon is filled to a volume of $V_1 = 2\,000$ cm³ with helium gas at a temperature $t_1 = -3°C$ and a density of $\rho_1 = 0.20$ kg/m³. If the density of helium at a temperature $t_2 = 0°C$ and an absolute pressure $p_2 = 1.0$ atm is $\rho_2 = 0.18$ kg/m³, determine (a) the initial absolute pressure p_1 of the helium in the balloon and (b) the volume of the balloon at an altitude where the temperature $t_3 = -33°C$ and the absolute pressure $p_3 = 0.70$ atm.

Solution:

(a) $T_1 = (273 - 3)$ K $= 270$ K, $T_2 = 273$ K and $T_3 = (273 - 33)$ K $= 240$ K.

$$\frac{p_1}{\rho_1 T_1} = \frac{p_2}{\rho_2 T_2} = \text{constant}$$

Therefore

$$p_1 = \frac{p_2 \rho_1 T_1}{\rho_2 T_2}$$
$$= \frac{(1.0 \text{ atm})(0.20 \text{kg/m}^3)(270 \text{ K})}{(0.18 \text{ kg/m}^3)(273 \text{ K})} = 1.1 \text{ atm}$$

(b) $$\frac{p_1 V_1}{T_1} = \frac{p_3 V_3}{T_3} = \text{constant}$$

Therefore

$$V_3 = \frac{p_1 V_1 T_3}{T_1 p_3} = \frac{(1.1 \text{ atm})(2\,000 \text{ cm}^3)(240 \text{ K})}{(270 \text{ K})(0.70 \text{ atm})}$$
$$= 2\,800 \text{ cm}^3$$

If the volume and temperature of a gas are kept constant, the pressure that it exerts depends on the number of gas molecules present. The frequency of bombardment of the container walls increases as the number of gas molecules is increased. Therefore the pressure p of a gas is also directly proportional to the number of moles n present if the volume and temperature are constant. Thus $p \propto n$, and combining this with the ideal gas law, Eq. 15-14, we obtain:

$$\frac{pV}{nT} = R \quad \text{(a constant)} \qquad (15\text{-}16)$$

where p is the absolute pressure of the gas, V is the volume, T is the absolute temperature and n is the number of moles of gas present. The constant R is called the *general gas constant*; it has a value of 8.314 J/(mol·K) or 0.082 05 L·atm/(mol·K).

Example 15-8

How many moles of helium gas produce a gauge pressure of 800 kPa in a 5.60 L cylinder at a temperature of 20°C?

Solution:

$$p = 800 \text{ kPa} + 101 \text{ kPa} = 901 \text{ kPa}$$
$$T = (273 + 20) \text{ K} = 293 \text{ K}$$

and
$$V = 5.60 \text{ L} = 5.60 \times 10^{-3} \text{ m}^3$$

Thus

$$n = \frac{pV}{RT} = \frac{(9.01 \times 10^5 \text{ Pa})(5.60 \times 10^{-3} \text{ m}^3)}{[8.314 \text{ J/(mol} \cdot \text{K)}](293 \text{ K})}$$

$$= 2.07 \text{ mol}$$

Problems

18. A fixed mass of gas exerts a gauge pressure of 15 lb/in² when its volume is 10 ft³. What volume does it occupy when its gauge pressure is 32 lb/in² if the temperature remains constant?

19. Determine the volume of oxygen at an atmospheric pressure of 1.01×10^5 Pa that is required to fill a cylinder of volume 2.5 L to a gauge pressure of 8.0×10^5 Pa.

20. If 200 L of water are pumped into a pressure tank that contains 600 L of air at standard atmospheric pressure, what is the resultant pressure in the tank?

21. A cylinder of nitrogen gas has an initial gauge pressure of 800 kPa. If the gauge pressure drops to 150 kPa after some nitrogen has been used, what fraction of the gas remains in the tank?

22. Outside air at $-15°$C is drawn into a furnace and is heated to 23°C at constant pressure. What is the percentage change in the volume?

23. A tank holds 8.5 L of nitrogen gas at a gauge pressure of 700 kPa at 20°C. Additional nitrogen is pumped in, raising the pressure to 1 200 kPa and the temperature to 60°C. How much gas was added?

24. A diver at the bottom of a lake where the temperature is 10°C releases an air bubble that rises to the surface, increasing its volume by a factor of 1.5. If atmospheric pressure is 100 kPa and the temperature

at the surface is 25°C, how deep was the diver?

25. Convert the following to kelvin temperatures: (a) 20°C, (b) $-30°$C, (c) 2 010°C, (d) 32°F, (e) 72°F, (e) $-220°$F.

26. A cylinder is filled with gas at a gauge pressure of 600 lb/in² at 20°C. If the cylinder is then used in a location where the temperature is $-15°$C, under what absolute pressure is the gas maintained? Assume that the atmospheric pressure = 14.7 lb/in².

27. A car tire is inflated to a gauge pressure of 32.0 lb/in² when the temperature is 75°F. If the volume of the tire does not change and atmospheric pressure is 14.7 lb/in², what is the gauge pressure when the temperature is $-10°$F?

28. A sample of gas has a density of 1.25 kg/m³ at 20°C and an absolute pressure of 750 mm of mercury. If 0.80 kg of the gas is forced into a 30 L cylinder at 30°C, what is the absolute pressure of the gas in atmospheres?

29. A tank contains 3.0 ft³ of oxygen at a temperature of 60°F and an absolute pressure of 16 lb/in². Determine the absolute pressure of the oxygen if its volume is increased to 9.0 ft³ and the temperature is raised to 420°F.

30. If 3 600 cm³ of neon gas at an absolute pressure of 700 mm of mercury and a temperature of 0°C are heated until the temperature is 277°C and the absolute pressure is 1 500 mm of mercury, determine the final volume of the gas.

31. Determine the percentage decrease in the pressure of a fixed mass of gas when its volume is doubled and its temperature is reduced by one-third.

32. How many atoms of the monatomic gas helium occupy a volume of 8.6 L at 20°C and a pressure of 0.56 atm?

15-5 THERMOMETERS

There are many methods of measuring temperature with some precision; most of these methods depend on the variation of some other physical property as the temperature changes.

The most common thermometer is the liquid (usually mercury) in glass thermometer. It consists of an evacuated glass capillary tube that is attached to a bulb full of the liquid. As its temperature rises, the volume of liquid in the bulb expands by an amount that is proportional to the change in temperature, and the liquid moves up the capillary tube. Since the cross-sectional area of the capillary is approximately constant, the volume change of the liquid (and hence the temperature change) is directly proportional to the distance that the liquid moves up or down the capillary. To calibrate the thermometer, the liquid level in the capillary is determined at the boiling and freezing points of water under standard pressure, and the interval between these points is divided into 100°C or 180°F.

The precision of liquid in glass thermometers is limited by slight imperfections in their construction and by their orientation since the pressure of the liquid in the capillary gives different readings between vertical and horizontal positions. Generally liquid in glass thermometers are accurate to approximately 0.01 K in the temperature range from 238 K to 628 K. Outside this temperature range, the liquid usually freezes or boils.

Gas thermometers, while bulky and somewhat awkward, are more accurate (0.005 K), and they may be used over a larger temperature range (approximately from 25 K to 2 000 K). A gas thermometer consists of a bulb of gas that is connected by a flexible U-tube to an open glass column of mercury (Fig. 15-4). It operates on the principle that, if the volume of a fixed mass of gas is con-

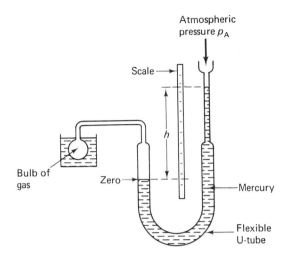

Figure 15-4
The gas thermometer.

stant, the absolute pressure p is directly proportional to the absolute temperature T. The mercury level is adjusted to coincide with the zero (by raising or lowering one side of the U-tube) so that the volume of gas is constant before each reading. The gas pressure p at any temperature T is equal to the sum of the atmospheric pressure p_0 and the pressure due to the height h of the mercury column: $p = p_0 + h$ (in mm of mercury). Note that h is negative if the mercury column is below the zero marker.

The gas thermometer may be calibrated according to the absolute pressure p_T exerted by the constant volume of gas at the temperature of the triple point of water, 273.16 K. Then, if p is the absolute pressure of the fixed volume of gas at the absolute temperature T, from Eq. 15-13:

$$\frac{p_T}{273.16 \text{ K}} = \frac{p}{T}$$

thus

$$T = (273.16 \text{ K})\left(\frac{p}{p_T}\right) \qquad (15\text{-}17)$$

Example 15-9

The bulb of a gas thermometer is placed in bath A, which is maintained at the temperature of the triple point of water. The arms of the U-tube are then adjusted until the mercury levels are even, and this level is taken as the zero for the constant volume of gas. When the bulb is placed in some other liquid bath B and the level of the mercury is adjusted to the zero, a 20.0 cm mercury column is above the zero marker. If the atmospheric pressure is 750 mm of mercury, determine the temperature of the liquid in the bath B.

Solution:

In bath A, since the mercury levels are the same in both arms of the U-tube, the gas pressure is equal to the atmospheric pressure, $p_0 = p_A = 750$ mm. The pressure of the gas when it is immersed in bath B is:

$$p_B = p_A + h = 750 \text{ mm} + 200 \text{ mm} = 950 \text{ mm}$$

Thus

$$T = (273.16 \text{ K})\left(\frac{p_0}{p_A}\right) = (273.16 \text{ K})\left(\frac{950 \text{ mm}}{750 \text{ mm}}\right)$$
$$= 346 \text{ K}$$

Other devices, such as resistance thermometers, thermocouples and thermistors, are also used to measure temperatures (Chapters 21 and 32).

Problems

33. If a gas exerts an absolute pressure of 360 mm of mercury at the temperature of the triple point of water, what pressure does it exert at 100 K?

34. The bulb of a gas thermometer is placed in a bath A, which is maintained at the temperature of the triple point of water. The arms of the U-tube are then adjusted until the mercury levels are even, and this level is taken as the zero for the constant volume of gas. When the bulb is placed in some other bath B and the level of

mercury is adjusted to the zero, the top of the other mercury column is 28.0 cm *below* the zero marker. If the atmospheric pressure is 755 mm of mercury, determine the temperature of the liquid in bath B.

15-6 THE KINETIC THEORY OF GASES

While individual molecules are not visible even under a light microscope, it is possible to see the effects of their continual bombardment on other small particles. Robert Brown (1773–1858) observed the erratic motions of pollen grains in water as they recoiled when they were bombarded by molecules. This motion, which is called Brownian motion, was the first experimental evidence of the existence of molecules.

It is not practical to describe pressure in a gas in terms of motion of individual molecules because of the vast number of molecules present in even small volumes of gas. The speeds and directions of motion of the individual molecules vary significantly; therefore statistical averages must be used.

In a sample of an *ideal monatomic gas*, which is composed of a large number of negligibly small gas molecules in continual rapid motion, the average translational kinetic energy of an individual gas molecule of mass m is

$$\bar{E}_k = \frac{m\overline{v^2}}{2} = \frac{3}{2}\frac{RT}{N_0} = \frac{3}{2}kT \quad (15\text{-}18)$$

where $\overline{v^2}$ is the average value of the square of the speed v of the molecules, R is the universal gas constant, T is the absolute temperature, N_0 is Avogadro's number and k is the universal constant. Note that the kinetic energy is proportional to the absolute temperature.

The universal constant

$$k = \frac{R}{N_0} = \frac{8.314 \text{ J/(mol·K)}}{6.023 \times 10^{23}/\text{mol}}$$
$$= 1.381 \times 10^{-23} \text{ J/K}$$

is called *Boltzmann's constant*. Equation 15-18 shows the relationship between the absolute temperature and kinetic energy of gas molecules: Lighter molecules must move faster in order to have the same absolute temperature. The *root-mean-square* (rms) speed of the gas molecules is defined as:

$$v_{rms} = \sqrt{\overline{v^2}} \qquad (15\text{-}19)$$

Example 15-10

If 1.00 mol of helium is 4.00 g, calculate the rms speed of helium molecules at 27°C.

Solution:

The mass of Avogadro's number of helium molecules is 4.00 g; therefore each molecule has a mass $m = 4.00 \text{ u} = (4.00 \text{ u})(1.660 \times 10^{-27} \text{ kg/u}) = 6.64 \times 10^{-27} \text{ kg}$. Thus

$$v_{rms} = \sqrt{\overline{v^2}} = \sqrt{\frac{3kT}{m}}$$
$$= \sqrt{\frac{3(1.381 \times 10^{-23} \text{ J/K})(300 \text{ K})}{(6.64 \times 10^{-27} \text{ kg})}}$$
$$= 1\,370 \text{ m/s}$$

Problems

35. Determine the rms speed of an argon molecule at 27°C if its mass is 40 u.

36. Determine the kinetic energy of an ideal gas molecule at 25°C.

15-7 HEAT TRANSFER

The control of heat flow has always been an important but often complex problem. In some aspects of technology, such as the air conditioning of buildings, large quantities of insulating materials are used in order to reduce the heat flow whereas heat flow is promoted in other areas, such as heating or cooling systems. There are three basic mechanisms of heat transfer: conduction, convection and radiation. Two or more of these mechanisms often occur simultaneously.

Conduction

The temperature of a material is related to the internal energy possessed by its molecules: the higher the temperature, the greater the internal energy. This molecular energy may be transferred from *one molecule to its neighbors* by a direct interaction process called *conduction*. The ability of a material to conduct heat depends largely on its structure. Since conduction requires the transfer of internal energy from one molecule to another, the proximity of these molecules is important. Gases are good insulators because the average spacing between gas molecules is relatively large. Metals are the best conductors of heat because their electrons are free to move and can be used to transfer heat from molecule to molecule.

Consider a bar of material that has a length L and a cross-sectional area A. If the two ends of the bar are maintained at different temperatures by heating one end at temperature T_1 and cooling the other at temperature T_2, then heat flows continually by a conductive process from the "hotter" to the "cooler" areas (Fig. 15-5). If the temperatures T_1 and T_2 are constant, then the rate of heat flow $\phi = Q/t$ varies directly as the product of the temperature difference $(T_1 - T_2)$ and the cross-sectional area A and inversely as the length (or thickness) L. Q is the heat conducted in an elapsed time t.

$$\phi = \frac{Q}{t} \propto \frac{A(T_1 - T_2)}{L}$$

or

$$\phi = \frac{Q}{t} = \lambda\frac{A(T_1 - T_2)}{L} \qquad (15\text{-}20)$$

The average temperature difference per unit length or thickness $(T_1 - T_2)/L$ is called the

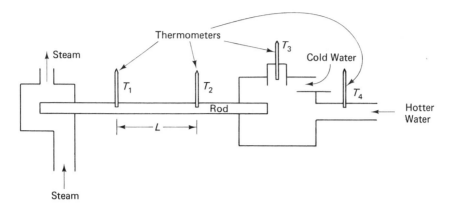

Figure 15-5

temperature gradient; the proportionality constant λ is called the *thermal conductivity*, and it is approximately constant for a particular material (Table 15-3). Thus thermal conductivity λ is defined as the rate of heat flow per unit temperature gradient per unit area:

$$\lambda = \frac{\phi}{A(T_1 - T_2)/L} = \frac{\phi L}{A(T_1 - T_2)}$$
$$= \frac{QL}{tA(T_1 - T_2)} \qquad (15\text{-}21)$$

It is expressed in units of W/(m·K) or BTU/(ft·s·°F). However, in many calculations mixed units, such as BTU·in/(ft²·h·°F), are common.

Example 15-11

A 8.0 mm thick glass window of a house is 2.0 m wide and 1.5 m high. If the outside temperature is −20°C and the inside temperature is 23°C, how much heat does it conduct in one hour? Express the answer in kilowatt hours.

Table 15-3
Typical Thermal Conductivities

Substance	Unit	
	$W/(m \cdot K)$	$BTU \cdot in/(h \cdot ft^2 \cdot °F)$
Aluminum	209	1 460
Copper	385	2 670
Iron	46	320
Brick	0.65	4.5
Concrete	1.08	7.5
Corkboard	0.043	0.30
Fiberglass	0.038	0.29
Glass	0.65	4.96
Wood— pine (across grain)	0.113	0.78
Air	0.025	0.17
Hydrogen	0.18	1.3
Water	0.599	4.15

Solution:

$$Q = \lambda \frac{A(T_1 - T_2)}{L} t = \frac{[0.65 \text{ W}/(m \cdot K)](2.0 \text{ m} \times 1.5 \text{ m})[23°C - (-20°C)](3\ 600 \text{ s})}{8.0 \times 10^{-3} \text{ m}}$$
$$= 3.8 \times 10^7 \text{ J} \quad \text{or} \quad 11 \text{ kW·h}$$

Example 15-12

An aluminum heat sink is 3.0 cm thick and has a cross-sectional area of 48 cm². How much heat does it conduct in 10 min if one side is attached to an amplifier at 70°C and the other is maintained at 20°C?

Solution:

From Table 15-3 and Eq. 15-20:

$$Q = \lambda \frac{A(T_1 - T_2)}{L} t = \frac{[209 \text{ W/(m·K)}](48 \times 10^{-4} \text{ m}^2)(70°C - 20°C)(600 \text{ s})}{3.0 \times 10^{-2} \text{ m}}$$

$$= 1.0 \times 10^6 \text{ J}$$

surface is kept at 72°F and the outer surface is at −20°F.

Solution:

$$Q = AU(T_1 - T_2)t = \frac{A(T_1 - T_2)t}{R}$$

Thus

$$Q = \frac{(25.0 \text{ ft} \times 8.00 \text{ ft})[72°F - (-20°F)](3.00 \text{ h})}{4.50 \text{ h·ft}^2 \cdot °F/BTU}$$

$$= 12\,300 \text{ BTU}$$

In heat load and air conditioning calculations, manufacturers specify thermal properties of many standard insulating materials (which may or may not have uniform structure) in terms of a *thermal transmission coefficient* or *coefficient of heat transfer U*. The coefficient of heat transfer U is defined as the rate of heat flow per unit area per temperature difference between the two surfaces. It is expressed in W/(m²·K) or BTU/(h·ft²·°F). The rate of heat flow is:

$$\phi = \frac{Q}{t} = AU(T_1 - T_2) \qquad (15\text{-}22)$$

where A is the surface area and $(T_1 - T_2)$ is the temperature difference between the material surfaces. The reciprocal of the coefficient of heat transfer is called the *thermal resistance R*:

$$R = \frac{1}{U} \qquad (15\text{-}23)$$

In order to calculate the thermal resistance (*R*-value) of such structures as walls, the R values of all the components are added (including air films, etc.).

Example 15-13

Determine the heat lost in 3.00 h through a 25.0 ft × 8.00 ft wall that has a total thermal resistance $R = 4.50$ h·ft²·°F/BTU if the inner

Conductive heat losses and gains must be computed for each surface used in a building. Storm windows and insulation in the walls and roof reduce conductive heat losses substantially.

Most electronic devices convert some electrical energy into heat. If the device is not able to dissipate the heat that it generates, its temperature increases, and, as a result of melting of components and expansion, the device fails. Therefore most devices, especially semiconductors, have operating limits of power dissipation. As it converts electrical energy into heat, if the operating limit of the device is not exceeded, the temperature of the device rises until the internal heat generated is equal to the rate of heat dissipation.

Heat transfer in semiconductors and many other devices is mainly a conductive process, but in electronics the rate of heat flow equation is modified to:

$$P = \frac{T_1 - T_2}{\theta_{12}} \quad \text{or} \quad T_1 = T_2 + \theta_{12}P \quad (15\text{-}24)$$

where P is the power dissipation in watts, $(T_1 - T_2)$ is the temperature difference between the ambient (surroundings) and the interior of the device and θ_{12} is the thermal resistance. In electronics thermal resistance is expressed in units of K/W or °C/W, and its

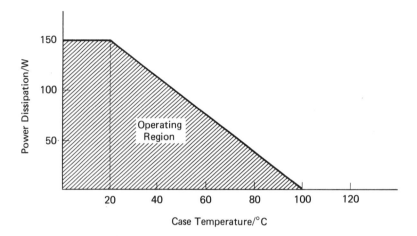

Figure 15-6
Power derating curve.

magnitude usually depends on the size of the device.

Manufacturers may express the thermal characteristics of a device graphically in terms of a temperature-derating curve (Fig. 15-6), which shows the operating limits in terms of power dissipation and the case temperature. The slope of the line is called the *derating factor*, and it is equal to the inverse of the thermal resistance. The maximum internal operating temperature T_1 (max) occurs when the case temperature T_2 is a maximum value. "Heat sinks" (which conduct heat away from the device) are often used to increase the operating limits of many devices, such as transistors. Normally T_1 (max) is approximately 200°C for silicon and 100°C for germanium transistors.

Example 15-14

Given the power derating curve (Fig. 15-6), determine (a) the derating factor, (b) the maximum junction temperature (c) the thermal resistance, (d) the maximum power dissipation at 60°C.

Solution:

(a)   derating factor $=$ slope $= \dfrac{150 \text{ W}}{80°C}$

$= 1.88\text{W}/°C$

(b) $T_{\max} = 100°C$

$=$ intercept on temperature axis

(c) $\theta_{12} = \dfrac{1}{\text{derating factor}} = \dfrac{1}{1.88 \text{ W}/°C}$

$= 0.53°C/W$

(d) At 60°C, the maximum power dissipation is 75 W.

Convection

Molecules of fluids are able to flow; therefore heat may be transmitted by the motion of the molecules that possess the most kinetic energy. This heat flow, caused by the motion of higher energy molecules, is called *convection*. Since most fluids expand when they are heated, the density of a warmer section of fluid is less than that of the cooler areas, and the warmer fluid rises while the cooler fluid sinks. This circulation of fluid, which is due to density variations, is called *natural convection*. Many heating systems utilize natural convection in order to circulate hot air or water, and heat transfer inside most buildings is mainly the result of convection currents of air. Some heating systems use fans and pumps in order to force the hotter molecules to circulate; this is called *forced convection*.

Radiation

Heat transfer by conduction and convection requires a material medium, but heat is also transmitted through areas that are devoid of any medium. For example, heat reaches the earth directly from the sun through the near vacuum of space. In this heat flow process, called *radiation*, energy is transmitted at the speed of light as an electromagnetic wave. While radiative heat transfer does not require a material medium, it can travel through media. Radiative heat passes through air and glass, for example, and it must be considered in air conditioning load calculations. When a material absorbs radiation, it is heated.

All objects radiate heat at a rate that depends on their temperature and the nature of their surface. As they radiate heat, their molecules lose energy, and the objects cool. When an object absorbs radiation, the energy is transmitted to its molecules, increasing the internal energy and temperature.

A *blackbody radiator* is defined as an object that absorbs all radiation that is incident on it. An approximation of a blackbody is a small hole in a black box or a dull lampblack surface.

Good absorbers of radiation must also be good emitters;* if they were not, they

---

*A possible exception to this is a *black hole* (the remnants of a very large collapsed star), if they exist. It has been proposed that a black hole absorbs matter and all incident radiation and that its gravitational field is too strong to allow even radiation, such as light, to escape.

would absorb more energy than they emit, and their energy and temperature would increase uncontrolled until the object eventually would melt or blow up. When the temperature and hence the internal energy of an object are constant, it must emit and absorb radiation at the same rate since energy is conserved.

When an object cools in air, it loses heat by a mixture of conductive, convective and radiative processes. The relationship that governs the radiative losses of a blackbody was originally stated by Josef Stefan (1835–1893) after examination of experimental data; later the same results were derived mathematically by Ludwig Boltzmann (1844–1906). They found that the rate of heat energy radiated by a blackbody Q/t is directly proportional to its surface area A and its absolute temperature T raised to the fourth power.

$$\phi = \frac{Q}{t} \propto AT^4 \quad \text{or} \quad \phi = \frac{Q}{t} = \sigma AT^4 \quad (15\text{-}25)$$

This is known as the *Stefan-Boltzmann law*; the constant $\sigma = 5.67 \times 10^{-8}$ W/(m²·K⁴) is called *Stefan's constant*. While the blackbody loses radiative heat energy at a rate that is proportional to the fourth power of its absolute temperature, it gains radiative heat from its surroundings. Therefore, if the surroundings have an absolute temperature T_0, the rate of radiative heat gain is σAT_0^4, and the *net* rate of radiative heat loss (or cooling) of a blackbody is:

$$\phi = \text{net } \frac{Q}{t} = \sigma A(T^4 - T_0^4) \quad (15\text{-}26)$$

All objects radiate and absorb heat energy, but not to the same extent as a blackbody. The color and finish of their surface affects their radiative properties. Bright shiny surfaces are poor emitters and absorbers but good reflectors of radiation whereas dull dark surfaces are good emitters and absorbers but poor reflectors. Therefore, white clothing and bright shiny surfaces are common in tropical climates because they reflect heat. The *emissivity* (or *emissive power*) e of a surface is defined as the ratio of its power ϕ_1 radiated per unit surface area A_1 to the power ϕ_2 radiated per unit surface area A_2 of a blackbody at the same temperature.

$$e = \frac{\phi_1/A_1}{\phi_2/A_2} \quad \text{or} \quad \frac{\phi_1}{A_1} = e\left(\frac{\phi_2}{A_2}\right)$$

If the areas A_1 and A_2 are equal, from Eq. 15-25 the rate of radiative heat loss ϕ from *any* surface of area A, emissivity e and absolute temperature T is given by:

$$\phi = e\sigma A T^4 \qquad (15\text{-}27)$$

where σ is Stefan's constant.

The *absorptivity* (*absorptive power*) is defined as the fraction of the total incident radiation absorbed by the surface. Therefore, if the temperature of the surface is constant and energy is conserved, the emissivity is equal to the absorptivity.

Example 15-15

The filament of an evacuated light bulb has a length $L = 10.0$ cm, a diameter $d = 0.200$ mm and an emissivity $e = 0.200$. What power does it radiate at 2 000 K? Neglect conductive losses.

Solution:

The *surface* area of the filament $A = \pi dL$. Therefore, since room temperature is much less than the filament temperature, the power radiated:

$$\phi \approx e\sigma A T^4 = e\sigma \pi d L T^4$$
$$= (0.200)[5.67 \times 10^{-8} \text{ W/(m}^2 \cdot \text{K}^4)]$$
$$\times \pi(2.00 \times 10^{-4} \text{ m})(10.0 \times 10^{-2} \text{ m})(2\,000 \text{ K})^4$$
$$= 11.4 \text{ W}$$

When an object cools, it loses heat to its surroundings by a mixture of conductive, convective and radiative processes. The rate of heat loss ϕ is directly proportional to the temperature difference $(T - T_0)$ between the object and its surroundings. Thus

$$\phi \propto T - T_0 \quad \text{or} \quad \phi = K(T - T_0) \quad (15\text{-}28)$$

where K is a constant. This is called *Newton's law* of cooling. It is only valid if the temperature difference is small and other conditions remain constant, but it is sufficiently accurate for use in heat load and air conditioning calculations.

Problems

37. How much heat in BTU is conducted in 12 h through a concrete wall that is 6.0 in thick and has an area of 160 ft² if the temperature difference is 50°F?

38. A glass window is 3.0 mm thick and has dimensions of 3.0 m by 2.0 m. If the outside temperature is $-15°$C and the inside temperature is 22°C, how much heat in joules does it conduct in each hour?

39. Determine the rate of heat flow in watts through a 1.0 cm thick corkboard that has an area of 35 cm² if the temperature difference between its surfaces is 30°C.

40. An aluminum pot that is 0.12 in thick holds boiling water. The base of the pot has an area of 65 in², and the stove maintains an average temperature of 300°F. If only 6.0% of the heat is transferred to the water, how much water evaporates in 5.0 min?

41. A wall is constructed from a layer of brick 4.0 in thick and a layer of fiberglass 6.0 in thick. If the total surface area of the wall is 100 ft² and the outside and the inside temperatures are $-10°$F and 72°F, respectively, determine (a) the temperature at the interface between the layers and (b) the rate of heat flow through the wall.

42. Determine the relationship between conductivities in W/(m·K) and BTU·in/ (h·ft²·°F).

43. The inside dimensions of a refrigerator are 1.0 m by 0.90 m by 2.3 m. If all the

walls are constructed of a material 1.0 cm thick that has a thermal conductivity of 2.2×10^{-2} W/(m·K) and the temperatures inside and outside are $-5°C$ and $23°C$, respectively, determine the rate of heat flow into the refrigerator.

44. (a) How many BTUs are conducted in one hour through a 1.0 ft thick concrete wall (transmission coefficient U = 0.35 BTU/(h·ft²·°F)) of surface area 75 ft² if the outside temperature is $-28°F$ and the inside temperature is 72°F. (b) What is the thermal resistance?

45. (a) A 120 ft² masonry wall of a house is 6.0 in thick and has a heat transfer coefficient of 0.49 BTU/(h·ft²·°F). Determine the rate of heat flow through the wall when the outside temperature is 20°F and the inside temperature is 70°F. (b) What is the thermal resistance?

46. Determine the heat lost in 12.0 h through a 3.0 m × 8.0 m wall that has a total thermal resistance of 0.36 m²·K/W if the inner surface is kept at 22°C and the outer surface is at $-15°C$.

47. From the power temperature derating curve in Fig. 15-7, determine (a) the derating factor, (b) the maximum junction temperature, (c) the thermal resistance, (d) the maximum power dissipation at 30°C.

48. Determine the power radiated from the surface of a spherical blackbody if it has a diameter of 30.0 cm and is maintained at a temperature of 1 500°C.

49. The tungsten filament of a light bulb has a length of 15 cm, a diameter of 0.15 mm and an emissivity of 0.31. What power does it radiate at 3 000 K? Neglect radiative losses.

50. The filament of a lamp maintains a temperature of 1 500 K when it is supplied with 150 W. What temperature would it maintain if it were supplied with 200 W?

REVIEW PROBLEMS

1. Determine the coefficient of linear expansion of a metal rod if its length increases from 9.800 0 ft to 9.813 0 ft when the temperature increases from 40°F to 160°F.

2. A steel cable is 300.000 m long at 23°C. What is its length when the temperature drops to $-40°C$?

3. An iron steam line is 350.000 ft long at 40°F. What is its length at 250°F?

4. The steel cable of a suspension bridge is exactly 2 200 ft long at 32°C. What is its length when the temperature drops to $-40°C$?

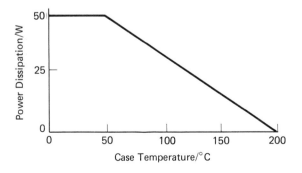

Figure 15-7

5. At 20°C a bare wheel has a diameter of 71.00 cm, and the inside diameter of a steel rim is 70.90 cm. To what minimum temperature must the rim be heated before it can fit over the wheel?

6. A metal rod (Young's modulus = 3.0×10^7 lb/in² and coefficient of linear expansion $6.0 \times 10^{-6}/°F$) has a cross-sectional area of 10.0 in². If the temperature of the rod drops from 80°F to 30°F, (a) what is the thermal strain? (b) What tensile force has to be applied to the ends of the rod to return it to its original length?

7. A steel tape measures the length of a copper rod as 1.000 000 m at 15°C. What will it indicate the length of the copper rod to be at 40°C?

8. A Pyrex container is filled with 1.5 L of gasoline at 10°C. What volume of gasoline overflows when the temperature is increased to 30°C?

9. If 1.0 mole of copper has a mass of 63.57 g, how many copper atoms are there in 120.00 g?

10. If 50 m³ of air are heated from 0°C to 75°C at constant pressure, what is its new volume?

11. A cylinder of helium gas has a gauge pressure of 4.0×10^5 Pa at 15°C. What is its gauge pressure at 80°C?

12. The absolute pressure in a cylinder of a car engine rises from 4.00×10^5 Pa to 1.25×10^6 Pa during combustion of the gas and air mixture. If the initial temperature of the gas air mixture was 200°C, what is the temperature after combustion?

13. If 4.0 L of helium are contained in a cylinder at 20°C, to what temperature must the helium be heated to triple the pressure?

14. Determine the rms speed of a neon molecule at 23°C if its mass is 20.2 u.

15. Determine the kinetic energy of an ideal gas molecule at 60°C?

16. A plate glass window measuring 1.20 m by 2.40 m is 5.00 mm thick. If the inside and outside surfaces of the window are kept at 22°C and −10°C, respectively, and the thermal conductivity is 2.00 W/(m·K), how much heat is lost through the window by conduction in one hour?

17. How many BTU are lost in 10.0 min through a $\frac{1}{4}$ in glass window with dimensions 2.0 ft by 4.0 ft if the outside temperature is 30°F and the inside temperature is 72°F?

18. An iron frying pan is 4.0 mm thick and its cross-sectional area is 120 cm². If the pan contains boiling water and it is on a stove that maintains a temperature of 150°C, (a) what is the rate of heat transfer through the pan? (b) If only 30% of this heat is transferred to the water, how much water is vaporized in 10 s?

19. Determine the heat lost in 24 h through a 3.0 m × 12 m wall that has a thermal resistance of 0.42 m²·K/W if there is a 35°C temperature difference between its surfaces.

20. Given the power temperature derating current for a germanium power transistor (Fig. 15-8), find (a) the derating factor, (b) the maximum junction temperature, (c) the thermal resistance, (d) the maxi-

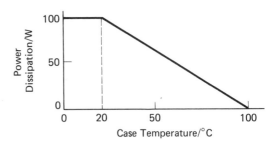

Figure 15-8

mum power dissipation at 40°C. (e) Shade the operating region on the graph.

21. How many BTUs are conducted in 1.0 h through a concrete wall (heat transfer coefficient U = 0.35 BTU/(h·ft²·°F)) 1.0 ft thick of surface area 50 ft² if the outside temperature is −3°F and the inside temperature is 72°F, respectively?

22. The tungsten filament of a light bulb has a length of 8.0 cm, a diameter of 0.40 mm and an emissivity of 0.30. What power does it radiate at 4 000 K? Neglect radiative losses.

23. The filament of a lamp maintains a temperature of 1 600 K when it is supplied with 150 W. What temperature would it maintain if it were supplied with 200 W?

24. A shed is constructed from materials of the following dimensions and transmission coefficients:

Material	Dimensions	Transfer Coefficient U W/(m²·K)
Floor concrete	4.0 m × 3.0 m	1.09
Sloped roof— wood and insulation	6.0 m × 3.0 m	2.8
Door— wood	1.0 m × 3.0 m	4.0
Walls— wood and insulation	13.0 m × 3.0 m	6.0

At what rate must an electric heater deliver heat to the room in order to maintain an inside temperature of 23°C when the outside air temperature is −2°C and the temperature of the earth (below the floor) is −10°C? Neglect all but conductive heat losses.

Heat is a common form of energy, and many aspects of technology are concerned with its control and application as a source for doing mechanical work. *Thermodynamics* is the study of the conversion of heat energy into other energy forms and mechanical work.

Natural energy sources, such as the fuels oil, gasoline, natural gas and coal, are used to do mechanical work through an intermediate stage in devices called *heat engines*. The fuel is burned, and the heat evolved is applied to a gas, which performs mechanical work as it expands. Most heat engines operate in a sequence of thermodynamic processes called a *cycle* that eventually returns the system to its original state.

Heat engines, such as steam engines, gasoline engines and rockets, have produced enormous changes in our way of life. Food-processing techniques and human comfort have been improved because of the development of refrigerators, which are essentially heat engines that operate in reverse. These devices do work in order to remove heat from a material.

16

Thermo-dynamics

16-1 THE FIRST LAW OF THERMODYNAMICS

The first law of thermodynamics is a special case of the conservation of energy principle; it is merely a restatement of work-energy relationships with regard to heat, work and internal energy. The law states:

The quantity of heat energy ΔQ that is supplied to a system is equal to the sum of the work W done by that system and its increase in internal energy ΔE.

$$\Delta Q = W + \Delta E \qquad (16\text{-}1)$$

Note that the heat ΔQ that is applied *to* the system, the work W done *by* the system and an *increase* in the internal energy of the system are all *positive valued*. Heat losses *by* the system, work done *on* the system and a *reduction* in the internal energy of the system are all *negative valued*.

Most thermodynamic processes involve changes in all variables of state simultaneously. However, in a few cases, thermodynamic processes may be reversed so that the system is restored to its initial state by the imposition of infinitesimally small changes in conditions without the input of additional external work or energy. These processes are called *reversible*. For example, work must be done in order to raise slowly an object vertically from the earth's surface, but the object can then do an approximately equal amount of work if it is allowed to return slowly to the earth. This process is approximately reversible; there are minute losses. Most processes are irreversible because they require extra external influences to return the system to its original state. Any process involving energy losses because of friction, viscosity or electrical resistance is irreversible.

Sometimes one or more of the variables of state remains approximately constant during an entire thermodynamic process, and the analysis of the process is simplified.

Constant Volume Process

In order to do mechanical work, the volume of a substance must change so that it produces a displacement. No work is done in a constant volume process. Therefore, from the first law of thermodynamics

$$\Delta Q = \Delta E \text{ (constant volume)} \qquad (16\text{-}2)$$

But according to kinetic theory, the internal energy of an ideal monatomic gas that contains N molecules at an absolute temperature T is given by:

$$E = \frac{3}{2} NkT = \frac{3}{2} nRT \qquad (16\text{-}3)$$

where k is Boltzmann's constant, R is the universal gas constant and n is the number of moles of the gas. Since the change in the internal energy ΔE is directly proportional to the change in the temperature ΔT of the gas,

if no phase change occurs, the relationship between heat and the corresponding temperature change may be expressed in terms of a sensible heat equation:

$$\Delta Q = mc_v \Delta T = \Delta E = \frac{3}{2} nR \Delta T$$

where m is the mass of the substance and c_v is its specific heat capacity at constant volume. Comparing the expressions, we see that the specific heat capacity of an ideal monatomic gas at constant volume is:

$$c_v = \frac{3}{2} \frac{nR}{m} = \frac{3}{2} \frac{R}{M}$$

where M is the mass of one mole of the gas.*

In thermodynamics it is often convenient to consider gases in terms of their *molar specific heat capacity C*. If a quantity of heat Q produces a temperature change ΔT in n moles of gas, the molar specific heat capacity is:

$$C = \frac{\Delta Q}{n \Delta T} \qquad (16\text{-}4)$$

Therefore

$$\Delta Q_v = nC_v \Delta T = \Delta E \qquad (16\text{-}5)$$

where C_v is the molar specific heat at constant volume and ΔQ_v is the heat input. Comparing Eqs. 16-3 and 16-5, we obtain:

$$C_v = \frac{3}{2} R$$

Isobaric Process

A thermodynamic process is called *isobaric* if it is performed at a constant pressure.

Consider a substance that is confined in the sealed cavity between the closed end of a rigid cylinder and a movable piston (Fig. 16-1). In an isobaric process as the substance is heated it expands at some constant pressure p and does work W to displace the piston through some distance Δs. The substance

---

*The molar mass.

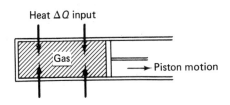

Heat ΔQ input

Gas

Piston motion

Figure 16-1

exerts a constant perpendicular force of magnitude F on the piston such that $F = pA$, where A is the area of the piston. Therefore the work done is:

$$W = F \Delta s = pA \, \Delta s = p \, \Delta V$$

where $\Delta V = A \, \Delta s$ is the volume change of the substance.

Significant volume changes usually occur when substances change their phase between the liquid and vapor states. If a liquid is heated, its temperature increases until the boiling point is reached. Additional heat input then vaporizes the liquid at a constant temperature. When a mass m of liquid is vaporized at a constant pressure p and a constant temperature T, the heat absorbed is:

$$\Delta Q = ml_v$$

where l_v is the specific latent heat of vaporization. If the volume of the mass in the liquid and vapor states is V_{Liq} and V_{vap}, respectively, the work done as the substance vaporizes is:

$$W = p \, \Delta V = p(V_{\text{vap}} - V_{\text{Liq}})$$

Therefore from the first law of thermodynamics, the change in the internal energy of the mass m is:

$$\Delta E = \Delta Q - W = ml_v - p(V_{\text{vap}} - V_{\text{Liq}}) \quad (16\text{-}6)$$

Example 16-1 _____

Determine the change in the internal energy of 1.00 kg of water when it vaporizes at a constant temperature of 100°C at a constant pressure of 101 kPa. Assume that the liquid density $\rho_{\text{Liq}} = 1\,000$ kg/m³ and that the steam density $\rho_s = 0.598$ kg/m³.

Solution:

$$V_{\text{Liq}} = \frac{m}{\rho_{\text{Liq}}}, \quad V_s = \frac{m}{\rho_s}$$

and

$$l_v = 2.26 \times 10^6 \text{ J/kg} \quad (\text{Table 14-1})$$

Thus

$$\begin{aligned}
\Delta E &= ml_v - p(V_s - V_{\text{Liq}}) \\
&= ml_v - pm\left(\frac{1}{\rho_s} - \frac{1}{\rho_{\text{Liq}}}\right) \\
&= (1.00 \text{ kg})(2.26 \times 10^6 \text{ J/kg}) - (1.01 \times 10^5 \text{ Pa}) \\
&\quad \times (1.00 \text{ kg})\left(\frac{1}{0.598 \text{ kg/m}^3} - \frac{1}{1\,000 \text{ kg/m}^3}\right) \\
&= 2.09 \times 10^6 \text{ J}
\end{aligned}$$

_____

During an isobaric expansion, a gas does work $W = p \, \Delta V$ as its volume increases by an amount ΔV at a constant pressure p; the temperature of the gas also changes because a part of the total heat input is converted into internal energy. If the molar specific heat capacity at constant pressure is C_p, then, according to the first law of thermodynamics, the total heat input to n moles of a gas is:

$$\Delta Q_p = nC_p \, \Delta T = W + \Delta E \quad (16\text{-}7)$$

where ΔT is the corresponding temperature increase. However, if there is no phase change, the change ΔE in the internal energy of the gas depends only on the temperature change ΔT. Consequently from Eq. 16-5:

$$\Delta E = nC_v \, \Delta T = \Delta Q_v$$

Thus

$$\Delta Q_p = nC_p \, \Delta T = p \, \Delta V + nC_v \, \Delta T \quad (16\text{-}8)$$

The equation of state for n moles of an ideal gas that has a volume V, an absolute pressure p and an absolute temperature T is given by $pV = nRT$, where R is the universal gas constant. If the mass of gas remains constant in an isobaric process, then $p \, \Delta V = nR \, \Delta T$, and Eq. 16-8 becomes:

$$nC_p \, \Delta T = nR \, \Delta T + nC_v \, \Delta T$$

or

$$C_p = R + C_v \quad \text{and} \quad C_p > C_v \quad (16\text{-}9)$$

For an ideal monatomic gas, $C_v = \frac{3}{2}R$; therefore, $C_p = \frac{5}{2}R$.

The ratio of molar specific heat capacities $\gamma = C_p/C_v$ varies for different gases. For monatomic gases (one atom per molecule) such as helium and argon, $\gamma \simeq 5/3$; for diatomic gases (two atoms per molecule) such as nitrogen, hydrogen and oxygen, $\gamma \approx 7/5$ (Table 16-1). When each of the gas molecules contains two or more atoms, vibrational and rotational energy constitutes a part of the total internal energy; therefore their specific heats are different.

*Example 16-2*_____

When 2.00 mol of helium gas expand at a constant pressure $p = 1.00 \times 10^5$ Pa, the temperature increases from 2.0°C to 112°C. If the initial volume of the gas was 45.0 L, determine (a) the work W done by the gas as it expands, (b) the total heat ΔQ_p applied to the gas, (c) the change in its internal energy. Assume that helium is an ideal gas.

Solution:

(a) Since the pressure is constant:

$$\frac{V_1}{T_1} = \frac{V_2}{T_2}$$

or

$$V_2 = V_1 \frac{T_2}{T_1} = \frac{(45.0 \text{ L})(273 + 112) \text{ K}}{(273 + 2) \text{ K}}$$
$$= 63.0 \text{ L}$$

Thus

$$W = p(V_2 - V_1)$$
$$= (1.00 \times 10^5 \text{ Pa})(63.0 \text{ L} - 45.0 \text{ L})$$
$$\times (10^{-3} \text{ m}^3/\text{L})$$
$$= 1.80 \times 10^3 \text{ J}$$

From Table 16-1, $C_p = 20.8$ J/(mol·K) and $C_v = 12.6$ J/(mol·K).

(b) $\Delta Q_p = nC_p \Delta T$
$$= (2.00 \text{ mol})[20.8 \text{ J/(mol·K)}](112 - 2) \text{ K}$$
$$= 4.58 \times 10^3 \text{ J}$$

(c) $\Delta E = \Delta Q_v = nC_v \Delta T$
$$= (2.00 \text{ mol})[12.6 \text{ J/(mol·K)}](112 - 2) \text{ K}$$
$$= 2.78 \times 10^3 \text{ J}$$

Note that

$$\Delta Q_p = W + \Delta E.$$

Isothermal Process_____

Theremodynamic processes that occur at a constant temperature are called *isothermal processes*. Phase changes at constant pressure are also isothermal processes.

Table 16-1
Typical Molar Specific Heat Capacities of Some Gases

Gas	Symbol	$C_p/J/(mol·K)$	$C_v/J/(mol·K)$	$\gamma = C_p/C_v$
Monatomic				
Helium	He	12.6	20.8	1.65
Argon	Ar	12.6	20.9	1.66
Diatomic				
Oxygen	O_2	20.8	29.1	1.40
Nitrogen	N_2	20.8	29.1	1.40
Triatomic				
Carbon dioxide	CO_2	28.2	36.6	1.30
Steam	H_2O	27.0	35.4	1.31

When a fixed mass of gas expands isothermally, its pressure also changes. If the gas is ideal, then $pV = nRT = $ constant. It can be shown by integral calculus that when n moles of an ideal gas expand isothermally at an absolute temperature T from an initial volume V_1 to a final volume V_2, the work done by the gas is

$$W = nRT \ln \left(\frac{V_2}{V_1}\right)$$

$$= 2.30 nRT \log_{10} \left(\frac{V_2}{V_1}\right) \qquad (16\text{-}10)$$

or

$$W = p_1 V_1 \ln \left(\frac{V_2}{V_1}\right)$$

since $p_1 V_1 = nRT = $ constant, where R is the universal gas constant. If a graph of the absolute pressure versus the volume of the gas is plotted, the work done by the gas is equivalent to the area between the curve and the volume axis (Fig. 16-2).

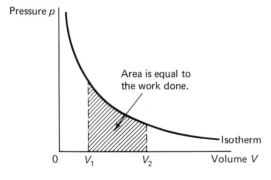

Figure 16-2
Isothermal expansion of a gas.

Example 16-3

A volume $V_1 = 89.6$ L of an ideal gas at a temperature $t = 0°C$ and an initial absolute pressure $p_1 = 101$ kPa is compressed isothermally to a volume $V_2 = 33.6$ L. Determine (a) the final pressure p_2 and (b) the work done by the gas.

Solution:

(a) Since the temperature is constant, according to Boyle's law:

$$p_2 = \frac{p_1 V_1}{V_2} = \frac{(1.01 \times 10^5 \text{ Pa})(89.6 \text{ L})}{33.6 \text{ L}}$$

$$= 2.69 \times 10^5 \text{ Pa}$$

(b) At $0°C = 273$ K and an absolute pressure $p_1 = 101$ kPa, 1 mol of an ideal gas occupies a volume of 22.4 L. Thus in 89.6 L there are $\frac{89.6}{22.4}$ mol $= 4.00$ mol of gas. As it is compressed, the work done is:

$$W = nRT \ln \left(\frac{V_2}{V_1}\right)$$

$$= (4.00 \text{ mol})\left(\frac{8.314 \text{ J}}{\text{mol} \cdot \text{K}}\right)(273 \text{ K}) \ln \left(\frac{33.6 \text{ L}}{89.6 \text{ L}}\right)$$

$$= -8.90 \times 10^3 \text{ J}$$

The negative sign merely implies that work was done on the gas rather than by the gas.

Adiabatic Process

In an *adiabatic process* the system is isolated, or the changes occur very rapidly so that there is no net heat transfer between the system and its surroundings. The net heat input ΔQ to the system is equal to zero. Therefore from the first law of thermodynamics, the increase in the internal energy of the system is equal to the work done on the system, i.e. negative work:

$$\Delta E = -W = nC_v \Delta T = mc_v \Delta T \quad (16\text{-}11)$$

Note that, if a gas does work by expanding adiabatically, its temperature decreases.

Since the temperature of the gas is not constant, Boyle's law is not valid for an adiabatic process. If the ratio of molar specific heat capacities $\gamma = C_p/C_v$ is independent of temperature, the absolute pressure p and the volume V of an ideal gas follow the relationship:

$$pV^\gamma = \text{constant} \quad \text{or} \quad p_1 V_1^\gamma = p_2 V_2^\gamma \quad (16\text{-}12)$$

Using the general gas equation $pV = nRT$, we obtain:

$$TV^{(\gamma-1)} = \text{constant}$$

$$\text{or} \qquad T_1 V_1^{(\gamma-1)} = T_2 V_2^{(\gamma-1)} \qquad (16\text{-}13)$$

where T is the absolute temperature.

Example 16-4

5.00 g of hydrogen gas at an initial pressure $p_1 = 3.0$ atm and an initial temperature $t_1 = 27°C$ are compressed adiabatically to one half the initial volume V_1. If the specific heat capacities for hydrogen gas are $c_v = 1.03 \times 10^4$ J/(kg·K) and $c_p = 1.44 \times 10^4$ J/(kg·K), determine (a) the final pressure p_2, (b) the final temperature T_2, (c) the work W done by the gas.

Solution:

(a) Since the specific heat capacities are directly proportional to the molar specific heats and the mass of gas is constant:

$$\gamma = \frac{c_p}{c_v} = \frac{1.44 \times 10^4 \text{ J/(kg·K)}}{1.03 \times 10^4 \text{ J/(kg·K)}} = 1.40$$

and

$$V_2 = V_1/2$$

Thus from Eq. 16-12:

$$p_2 = p_1\left(\frac{V_1}{V_2}\right)^\gamma = (3.0 \text{ atm})\left(\frac{V_1}{V_1/2}\right)^{1.40}$$

$$= (3.0 \text{ atm})(2)^{1.40} = 7.9 \text{ atm}$$

(b) $\quad T_2 = T_1\left(\dfrac{V_1}{V_2}\right)^{\gamma-1} = [(273 + 27) \text{ K}](2)^{0.40}$

$$= 396 \text{ K} \quad \text{or} \quad 123°C$$

(c) $\quad W = -mc_v(T_2 - T_1) = -(5.00 \times 10^{-3} \text{ kg})$

$$\times [1.03 \times 10^4 \text{ J/(kg·K)}](396 - 300) \text{ K}$$

$$= -4.9 \times 10^3 \text{ J}$$

That is 4.9×10^3 J of work is done *on* the gas.

Throttling or Joule-Thomson Process

In a *throttling process* a fluid is forced from a region of some constant high pressure p_1 through a well insulated throttling valve, such as a porous plug or a long narrow valve, into a region of some lower constant pressure p_2. The throttling valve must be effectively insulated from its surroundings so that the process occurs adiabatically.

Consider a mass M of fluid that has a volume V_1 and internal energy E_1 on the higher pressure side, and a volume V_2 and internal energy E_2 after it passes through the throttling valve to the lower pressure side. On the higher pressure side of the throttling valve, the fluid does isobaric work $W_1 = p_1(0 - V_1)$ as its volume changes from V_1 to zero at a constant pressure p_1. Similarly, as its volume increases from zero to V_2 at a constant pressure p_2 on the lower pressure

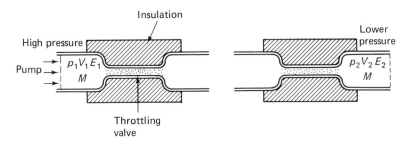

Figure 16-3
Throttling or Joule-Thomson process.

side, the fluid does isobaric work $W_2 = p_2(V_2 - 0)$. Therefore the total isobaric work done by the fluid as it passes through the valve is:

$$W = W_1 + W_2 = p_2 V_2 - p_1 V_1$$

Since the throttling valve is well insulated from its surroundings, the process is essentially adiabatic, and the net heat flow into the system $\Delta Q = 0$. Thus, from the first law of thermodynamics, for a throttling process:

$$\Delta Q = \Delta E + W$$
$$= (E_2 - E_1) + (W_2 + W_1) = 0$$

or

$$E_2 + p_2 V_2 = E_1 + p_1 V_1$$
$$= \text{constant} = H \quad (16\text{-}14)$$

The constant H is called the *enthalpy* of the fluid.

A "nearly boiling" liquid partially vaporizes and its temperature decreases during a throttling process; this property is utilized to cool the liquid refrigerant in a refrigeration cycle (Section 16-5).

For an ideal monatomic gas, $pV = nRT$ and $E = \frac{3}{2}nRT$; therefore the enthalpy $H = E + pV$ is only a function of the absolute temperature T, and an *ideal gas* does not change its temperature during a throttling process because enthalpy is constant. However, real gases may either increase or decrease their temperature in a throttling process; their temperature change depends on the nature of the gas and the initial temperature and pressure. Throttling processes are often used to cool or liquify gases.

Problems

1. Determine the amount of heat that is required to raise the temperature of 5.0 mol of helium by 50°C at constant volume.

2. Calculate the work done by a gas when its volume increases from 30 ft³ to 80 ft³ under a constant pressure of 15 lb/in².

3. The temperature of 2.4 mol of steam is increased from 100°C to 160°C at constant volume. Calculate (a) the work done by the steam, (b) the increase in its internal energy, (c) the total heat applied to the steam.

4. Calculate the work done when the volume of a gas is compressed from 75.0 L to 15.0 L at a constant pressure of 3.0×10^5 Pa.

5. Liquid water has a weight density of 62.4 lb/ft³, and steam has a weight density of 3.73×10^{-2} lb/ft³ at 212°F under a pressure of 14.7 lb/in². If the specific latent heat of vaporization of water is 970 BTU/lb, compute (a) the work done and (b) the change in internal energy when 5.0 lb of water are vaporized at a constant temperature of 212°F and a constant pressure of 14.7 lb/in².

6. The temperature of 3.0 mol of carbon dioxide is increased from 20°C to 85°C at constant pressure. Compute (a) the total heat supplied, (b) the change in internal energy, (c) the work done by the gas.

7. Determine the heat liberated when the temperature of 20 mol of nitrogen is decreased from 100°C to 30°C (a) at constant volume, (b) at constant pressure.

8. How much work is done by an ideal gas when it expands isothermally from an initial volume of 7.2 L at a pressure of 3.0 atm to a final volume of 26 L?

9. Determine the work done on an ideal gas when it is compressed from an initial volume of 80 m³ and pressure of 6.0×10^4 Pa to a final volume of 15 m³ if the temperature remains constant at 0°C.

10. Calculate the work done by 3.0 mol of an ideal gas when its volume is tripled at a constant temperature of 0°C.

11. If 12.0 L of carbon dioxide gas at 0°C and a pressure of 101 kPa are adiabatically compressed to 3.00 L, calculate (a) the final pressure, (b) the final temperature, (c) the work done by the gas.

12. If 25.0 L of helium gas at 0°C and 1.00 atm expand adiabatically to a final volume of 120 L, compute (a) the final pressure, (b) the final temperature, (c) the work done by the gas.

16-2 THE SECOND LAW OF THERMODYNAMICS

It is relatively easy to convert mechanical energy and work into an equivalent amount of heat, but it is not possible to reverse the process. The first law of thermodynamics is concerned with the conservation of energy, but it does not indicate how heat energy is utilized in a heat engine. Practical heat engines usually operate in a continuous cycle, taking heat from a hot source, converting some of the heat into useful work, but always ejecting the rest of the heat to some cooler reservoir (Fig. 16-4). A perfect heat engine would be able to convert all of the heat acquired from a hot source into an equivalent amount of work.

There are several equivalent statements of the *second law of thermodynamics*. One form that is attributed to Kelvin states:

It is impossible to construct a perfect heat engine because some heat is always rejected to a cooler reservoir.

Heat flows naturally from high to low temperature areas, and work must be done on a system to reverse the direction of heat flow. A heat engine can only perform work if a heat source with a temperature in excess of the surroundings or a heat sink at a lower temperature than the surroundings is avail-

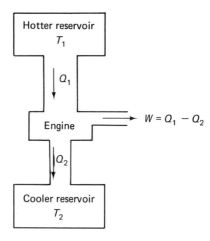

Figure 16-4
Heat flow diagram for a heat engine.

able. The oceans and the atmosphere both contain considerable quantities of heat, but they cannot be utilized adequately as heat sources because a practical, lower temperature reservoir is not available. The earth itself may have some future application as a heat source, because its interior is quite hot. If this heat could be tapped, it would rise naturally to the lower temperature area at the earth's surface.

16-3 THE CARNOT CYCLE AND EFFICIENCY

The efficiency of heat engines is usually quite poor because only a relatively small amount of the total input heat energy is converted into mechanical energy. Some of the heat energy is used to drive the moving parts of the engine, and more is lost because of friction, but the majority of the heat that is supplied to the engine is either conducted and lost to the surroundings or is rejected with the exhaust.

In 1824 a French engineer Sadi Carnot

(1796–1832) analyzed the theoretical efficiency of an ideal heat engine in which there were no losses due to friction or the conduction of heat to the surroundings. Carnot's discoveries proved to be extremely important because they defined the operating limits of heat engines in terms of the proportion of the total heat input that could be converted into mechanical work. A Carnot heat engine operates in a continuous *Carnot cycle*, which consists of a series of reversible adiabatic and isothermal processes.

Consider a Carnot heat engine that consists of a nonconducting piston in a cylinder that has nonconducting walls and a perfectly conducting head. The motion of the piston in the cylinder is frictionless, and n moles of an ideal gas (the working substance) are located in the sealed cavity between the piston and the cylinder (Fig. 16-5). After each complete Carnot cycle, the ideal gas is returned to its original thermodynamic state so that its net change in internal energy is equal to zero and the cycle can be repeated.

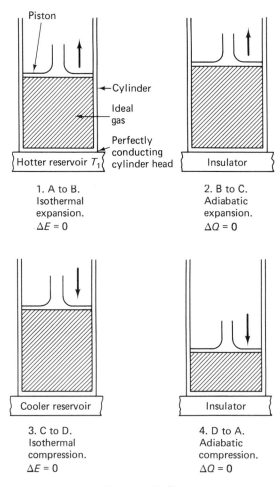

Figure 16-5
The Carnot cycle of an ideal heat engine.

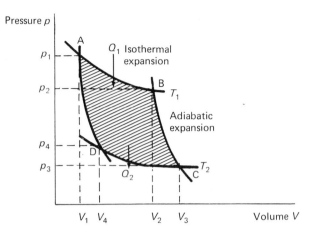

Figure 16-6
The Carnot cycle.

There are four steps in a complete Carnot cycle (Figs. 16-5 and 16-6):

1. A to B. When the cylinder head is placed in thermal contact with a heat source at some constant temperature T_1, the ideal gas absorbs heat energy Q_1 and does work W_1 as it expands *isothermally* from thermodynamic state A (represented by pressure p_1, volume V_1 and absolute temperature T_1) to state B (represented by p_2, V_2 and T_1). During this process, the internal energy is constant. Consequently the work done by the gas is equal to the total heat input: $Q_1 = W_1 = (nRT_1) \ln (V_2/V_1)$ and according to Boyle's law $p_2 V_2 = p_1 V_1$.

2. B to C. The cylinder head is insulated from the surroundings, and the ideal gas does work W_2 as it undergoes a slow adiabatic expansion, decreasing its temperature and pressure until it reaches state C (represented by p_3, V_3 and T_2). Thus, $p_2 V_2^\gamma = p_3 V_3^\gamma$ where γ is the ratio of molar specific heats.

3. C to D. When the cylinder head is placed in thermal contact with the cooler heat reservoir at a constant absolute tempera-

ture T_2, the heat Q_2 that is rejected to the cooler reservoir is equal to the negative work $-W_3$ done by the gas as its volume decreases isothermally until it reaches state D (represented by p_4, V_4 and T_2). Thus

$$Q_2 = -W_3 = -nRT_2 \ln \left(\frac{V_4}{V_3} \right)$$
$$= nRT \ln \left(\frac{V_3}{V_4} \right)$$

and

$$p_4 V_4 = p_3 V_3$$

4. D to A. To complete the Carnot cycle, the cylinder head is insulated, and the gas does negative work W_4 as it is compressed adiabatically to its original state A. Then $p_4 V_4^\gamma = p_1 V_1^\gamma$.

In the pressure-versus-volume graph (Fig. 16-6), the total work done by the ideal gas is equal to the area ABCD that is bounded by the four lines representing the steps of the Carnot cycle. Also, since the gas returns to its original thermodynamic state, its net change in internal energy is zero. From the first law of thermodynamics, the total work done is:

$$W = \Delta Q = Q_1 - Q_2$$

Thus the efficiency of a Carnot heat engine is:

$$\eta = \frac{\text{useful work output}}{\text{total input energy}} = \frac{W}{Q_1}$$
$$= \frac{Q_1 - Q_2}{Q_1} = 1 - \frac{Q_2}{Q_1} \qquad (16\text{-}15)$$

Carnot also proved that the efficiency of any engine operating in a Carnot cycle is independent of its construction and the nature of its working substance.

No heat engine that operates within the same given temperature limits can be more efficient than a Carnot engine.

Carnot Efficiency and Absolute Temperature

Lord Kelvin (1824–1907) used the efficiency of a Carnot heat engine to establish an absolute temperature scale that does not depend on the nature of any substance. It can be shown that for a Carnot cycle:

$$\frac{Q_2}{Q_1} = \frac{T_2}{T_1}$$

Thus the efficiency of a Carnot heat engine is:

$$\eta = 1 - \frac{Q_2}{Q_1} = 1 - \frac{T_2}{T_1} = \frac{T_1 - T_2}{T_1} \quad (16\text{-}16)$$

where T_1 and T_2 are the absolute temperatures of the heat source and cooler reservoir, respectively.

Absolute zero (0 K) is defined as the temperature T_2 to which a Carnot heat engine will reject no heat ($Q_2 = 0$). If the temperature of the cooler reservoir is absolute zero, a Carnot heat engine has an efficiency of 100% because all of the input heat Q_1 is converted into useful work. This definition of absolute zero is completely independent of the working substance, and it does not require the existence of the gaseous state at low temperatures.

The triple point of water is assigned the temperature $T_T = 273.16$ K, and any other absolute (*thermodynamic*) temperature T in *kelvin* is defined by the relationship

$$T = (273.16 \text{ K})\left(\frac{Q}{Q_T}\right) \qquad (16\text{-}17)$$

where Q and Q_T are the heats that flow when a Carnot heat engine operates between reservoirs at temperatures T and 273.16 K, respectively.

Example 16-5

Determine the efficiency of a Carnot heat engine that operates between a heat source at 227°C and a cooler reservoir at −73°C.

Solution:

$$\eta = \frac{T_1 - T_2}{T_1} = \frac{(227 + 273) \text{ K} - (273 - 73) \text{ K}}{(227 + 273) \text{ K}}$$
$$= 0.60 \quad \text{or} \quad 60\%$$

Note that theoretically it is not possible to attain a temperature of absolute zero.

Problems

13. Determine the efficiency of a Carnot heat engine that absorbs 3 200 J of heat from a high temperature source and rejects 800 J of heat to a lower temperature reservoir.

14. Determine the work done by a Carnot heat engine that has an efficiency of 60% when it receives 1.5×10^4 J of heat from a high temperature source.

15. Calculate the efficiency of a Carnot heat engine that operates between temperatures of 800°C and −150°C.

16. Compute the efficiency of a Carnot heat engine that operates between temperatures of 700°F and 80°F.

17. Determine the temperature of the heat source if a Carnot heat engine has a 75% efficiency and it rejects heat to a cooler reservoir at 60°C.

16-4 PRACTICAL HEAT ENGINES

Many different types of heat engines have been developed to convert heat into mechanical energy. An assembly of simple machines, such as levers, cams and gears, are used to

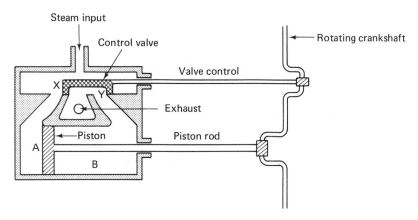

Figure 16-7
The reciprocating steam engine.

transmit and apply the mechanical energy to perform work. While heat engines are normally designed to meet a particular requirement, other factors, such as efficiency, size, power, fuel to be used, pollution, cost, reliability and noise, must also be considered.

The Reciprocating Steam Engine

The operation of a reciprocating steam engine depends on the actions of a piston and a sliding control valve (Fig. 16-7). Steam enters the engine and flows through X into cavity A where it cools as it does work to displace the piston. As the piston moves, the volume of cavity B decreases, and any spent steam in B is forced through Y to the exhaust. At the same time the reciprocating motion of the piston (as it moves back and forth in a straight line) is converted into a rotary motion by means of a crankshaft or a flywheel. The rotating crankshaft is also used to perform external work, and the control valve is linked to the crankshaft so that, when the piston stroke is complete, the control valve shifts to seal intake X. Steam then enters the system through Y and forces the piston to move in the opposite direction at the same time that the spent gas in A passes to the exhaust.

At the end of the piston stroke, the control valve shifts again to seal intake Y, and the process is repeated.

Reciprocating steam engines are not very efficient, and they have been almost completely replaced by the more reliable turbines and internal combustion engines.

Steam Turbines

A steam turbine consists of a set of vanes called *rotors*, which are attached to a movable rotor shaft, and an alternate series of oppositely curved vanes called *stators*, which are rigidly attached to the casing of the heat engine (Fig. 16-8). High pressure steam from a boiler is forced past the vanes. As the steam strikes the rotor vanes, it deflects them and turns the rotor shaft. The deflected steam then strikes the stators, which are orientated so that they redirect the steam at the correct angle onto the next set of rotors. Finally, the spent steam passes to a condenser, where it liquifies, reducing the pressure so that the steam flows naturally from the high pressure intake to the low pressure exhaust.

Many modern power plants utilize steam turbines in the production of electricity. The rotor shaft of a gas turbine is often connected

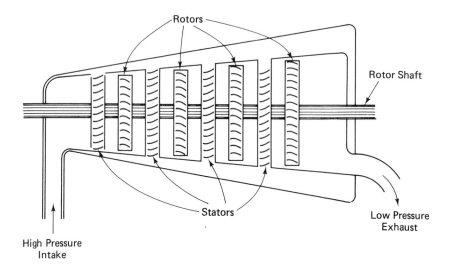

Figure 16-8
The steam turbine.

directly to the coils of a generator. As the rotor shaft turns, the generator coils rotate about their axis in a magnetic field. Fossil fuel, such as coal, oil or natural gas, or nuclear energy is used to produce steam by heating the boiler, and the condenser is normally cooled by water taken from a nearby lake or ocean.

In reciprocating steam engines and steam turbines the fuel is burned and steam is produced in a boiler external to the actual heat engine. Consequently they are sometimes called *external combustion engines*.

Internal Combustion Engines

Heat engines that burn fuel in the interior of the engine mechanism are called *internal combustion engines*.

Reciprocating Internal Combustion Engine

In a simple reciprocating internal combustion engine the combustion chamber is the cavity between the top of a movable piston and the sealed end of a metal cylinder. The piston slides loosely up and down in the cylinder; it is given room for thermal expansion. Expandable aluminum rings are attached around the diameter of the piston to provide a seal between the piston and the cylinder wall.

Most automobiles are powered by reciprocating internal combustion engines that contain four or more cylinders. In operation each piston performs a four stroke cycle called an *Otto cycle*. Two spring-loaded valves control the intake and exhaust of gases, and an electric discharge in a spark plug serves to ignite the fuel in the combustion chamber. The sequence of actions in each cylinder (Fig. 16-9) is as follows:

1. *Intake stroke.* During the intake stroke, the intake valve is open, and the exhaust valve is closed. As the piston moves down, it draws a mixture of gasoline and air into the combustion chamber from a carburetor.

2. *Compression stroke.* After the piston passes through the lower limit of its motion, the intake valve closes, and the air and gaso-

line mixture is compressed as this piston moves upwards. When the piston reaches the upper limit of its motion, the fuel mixture has been compressed to a fraction of its initial volume. This is called the compression ratio; it is normally $\approx 7:1$.

3. *Power stroke.* A spark from the spark plug ignites the fuel mixture, which burns gradually, producing a substantial quantity of heat. As the pressure in the combustion

chamber increases (to $\approx 600 \text{ lb/in}^2$), the piston is forced downwards, and it can do work.

4. *Exhaust stroke.* At the end of the power stroke, the exhaust valve opens, and the piston moves upwards, forcing the spent gas out of the exhaust port. After completion of the exhaust stroke, the exhaust valve closes, the intake valve opens, and the cycle is repeated.

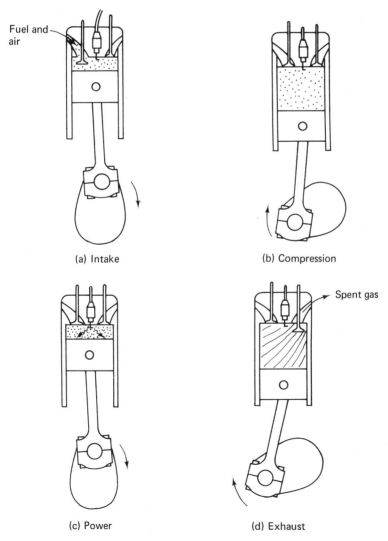

(a) Intake

(b) Compression

(c) Power

(d) Exhaust

Figure 16-9
The four stroke cycle.

Most reciprocating internal combustion engines have four or more cylinders to produce smoother operation and to reduce stresses on the moving parts. In most automobiles the cylinders are arranged in a line, but in aircraft they are normally arranged radially about the propeller crankshaft.

In multicylinder reciprocating engines the pistons are connected to a common crankshaft by means of piston rods and cranks (Fig. 16-10). Each piston only delivers power on one stroke, and energy is required to move it on the other three strokes. Therefore piston actions are arranged to be in sequence so that their reciprocating (to and fro) motion continually delivers power to the rotating crankshaft.

As the crankshaft rotates, it performs work, and its rotary motion is transmitted to a camshaft that contains two cams for each cylinder. The rotating cams control the action of the push rods. When a push rod moves up,

it tips the rocker arm, and the valve moves down, opening the vent into the combustion chamber. As the cam rotates farther, a spring returns the valve to its original position, sealing the vent, and the push rod moves down. Actions of the cams are synchronized with the piston strokes so that the valves open and close the vents at the correct times.

Some reciprocating internal combustion engines operate on a two-stroke cycle that delivers power to the crankshaft on alternate strokes of the piston (Fig. 16-11). As the piston moves up, it compresses the fuel and air mixture in the combustion chamber; at the same time the pressure is reduced in the crankcase, and the leaf valve opens. New fuel and air from the carburetor enter the crankcase under pressure through the leaf valve. At the top of the stroke, the spark plug is fired, and the fuel in the combustion chamber is ignited; the burning fuel expands and forces the piston down, delivering power to the

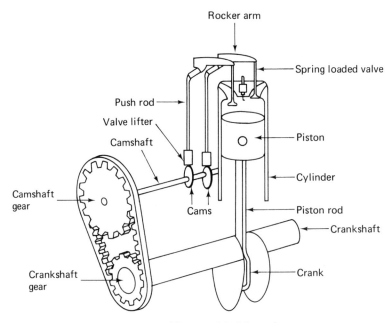

Figure 16-10
The multicylinder reciprocating engine.

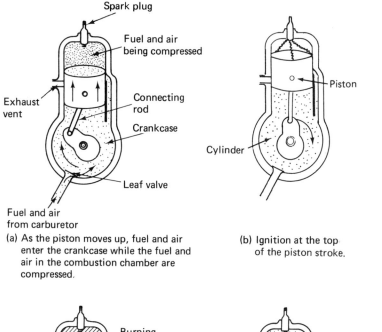

Spark plug

Fuel and air being compressed

Exhaust vent

Connecting rod

Crankcase

Leaf valve

Fuel and air from carburetor

(a) As the piston moves up, fuel and air enter the crankcase while the fuel and air in the combustion chamber are compressed.

Piston

Cylinder

(b) Ignition at the top of the piston stroke.

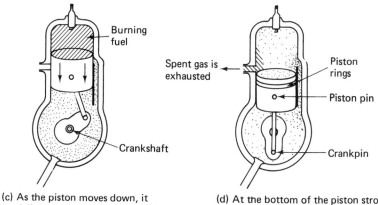

Burning fuel

Crankshaft

(c) As the piston moves down, it delivers power to the crankshaft and compresses the new fuel in the crankcase.

Spent gas is exhausted

Piston rings

Piston pin

Crankpin

(d) At the bottom of the piston stroke, new fuel enters the combustion chamber and the spent gas is expelled through the exhaust vent.

Figure 16-11
The two stroke cycle.

crankshaft. As it moves down, the piston compresses the new fuel and air mixture in the crankcase, and the leaf valve closes. When the top of the piston passes the exhaust vent, new fuel flows under pressure from the crankshaft into the combustion chamber, forcing the spent gas out through the exhaust.

The *stroke* of a piston is the distance that it travels in one stroke, and the *bore* is the diameter of an engine cylinder. The *compression ratio* of an engine is the ratio of the maximum to the minimum volumes of the combustion chamber. The more efficient the engine, the higher its compression ratio.

Unfortunately the temperature of the system also increases as the compression ratio increases. If the temperature or compression ratio is too high for the fuel, *preignition* occurs, and the fuel burns before the spark. New fuels, which have high preignition temperatures, are now being developed in order to increase the efficiency of the internal combustion engine.

Diesel Engines

Diesel engines are similar to gasoline internal combustion engines, but they do not require spark plugs. On the intake stroke of a diesel engine, only air is drawn into the cylinder. During the compression stroke, the air is highly compressed (compression ratios are often higher than $15:1$), and this produces high temperatures ($\approx 900°C$). Fuel oil is then sprayed into the cylinder, and it immediately ignites in the high temperature, delivering power to the engine.

Diesels use lower-priced fuels, and they are designed to operate on four-stroke or two-stroke cycles. They are used extensively to power trucks, buses, trains, ships and generating plants.

Gas Turbine

The operation of a gas turbine is similar to that of a steam turbine. There are four main components: an air intake, compressor, combustion chamber and a turbine rotor. An auxiliary power supply is required to start the cycle. As air enters through the intake, it is compressed by a series of rotors and stators and then is sprayed with fuel and ignited in the combustion chamber. The hot gases move under pressure through the turbine, where they deflect the rotors and turn the rotor shaft, which in turn drives a propeller, generator, etc. (See Fig. 16-12.)

These devices are more efficient than steam turbines, and they run on low-grade

Figure 16-12
The GE CJ610 turbojet, with one shaft, eight-stage axial flow compressor, annular combustor, and two-stage turbine.
(Courtesy of Smithsonian Institution)

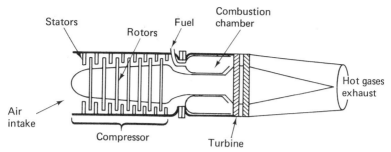

Figure 16-13
The turbojet.

fuels or natural gas. They are often used in small generating and industrial plants.

The Turbojet

A turbojet engine utilizes a gas turbine to drive an air compressor. As the air enters the front of the engine, it is compressed, mixed with fuel (usually kerosene) and burned. The hot gas produced passes through a turbine (which drives the compressor) and is ejected at high speed from the rear of the engine (Fig. 16-13). The thrust is developed from the reaction force, which drives the engine in the opposite direction to the escaping gas. Turbojet engines have a higher power-to-weight ratio than reciprocating internal combustion engines, and they are more efficient at high speeds and in reduced atmospheric pressure.

A *turboprop* engine consists of a turbojet with a propellor connected to the shaft of the turbine. The propellor provides most of the forward thrust; therefore turboprop engines are more efficient than turbojets at lower speeds and altitudes (Fig. 16-14).

Figure 16-14
Garrett-type 331 turboprop engine.
(Courtesy of Smithsonian Institution)

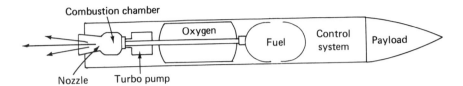

Figure 16-15
Rocket engine.

Ram-Jet Engines

Air enters a ram-jet engine through a wide intake, and it is forced down a tapered channel, which increases the pressure. Fuel is then injected, the mixture is ignited, and the hot gases are ejected from the rear of the engine, supplying a forward thrust from the reaction force. An auxiliary engine is required to start the device.

Rocket

Rockets are usually relatively large because they must carry large quantities of both fuel and oxygen (in solid or liquid form) in order to operate in space where there is no atmosphere. They are also constructed in stages so that, as fuel and oxygen tanks empty, they are discarded to reduce the total weight. The fuel and oxygen are pumped separately into a combustion chamber where they are mixed and burned. The hot gases that are produced are ejected at enormous speeds from the rear of the rocket, propelling it forward (Fig. 16-15).

Electron Bombardment Thruster

The electron bombardment thruster (Fig. 16-16) is a type of ion engine that is usable in space only. Mercury or Cesium is vaporized by electron bombardment in the discharge chamber. The electric field between the accel-

erator and the screen attracts the ions and accelerates them to speeds up to 50 km/s before they pass through the screen. The electrons are restored to keep the exhaust neutral. This type of engine is much more efficient than chemical propulsion.

16-5 REFRIGERATORS

A refrigerator operates as a heat engine in reverse; it takes heat Q_2 from a low temperature reservoir and does work W to reject heat equivalent to $W_1 + Q_2$ into a higher temperature reservoir (Fig. 16-17).

The main components of a typical refrigerator are:

1. The *refrigerant* is a vapor at normal room temperature and pressure, but it has a low boiling point so that it is easily liquified by relatively small pressure increases or temperature decreases. There are many refrigerants; some of the most common are ammonia, sulphur dioxide, methyl chloride and freon-12.*

2. A *mechanical compressor* performs the work on the refrigerant.

3. A *condenser* may be air or water cooled.

---

*The trade name given to a chemical compound that was developed by Dupont De Nemours Inc.

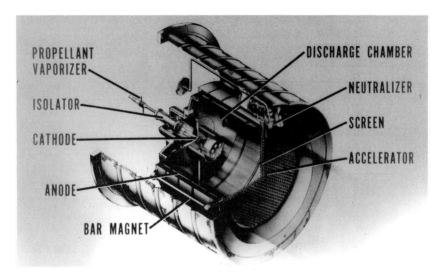

Figure 16-16
Electron bombardment thruster.
(Courtesy of Smithsonian Institution)

4. An *expansion valve*.

5. The *refrigerator box (evaporator)* constitutes the cooler reservoir.

Consider the operation of a typical refri-

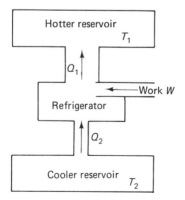

Figure 16-17
Flow diagram for a refrigerator.

gerator (Fig. 16-18). The cool liquid refrigerant absorbs heat Q_2 from the surroundings and vaporizes as it passes through the cooling coils in a refrigerator box (the low temperature reservoir). Under low pressure the vaporized refrigerant is drawn through valve A into the compressor, which does work W to compress and superheat the vapor. Valve B then opens, and the superheated vapor is pumped at relatively high pressure into the condenser (the hotter reservoir), where it liquifies by transferring heat Q_1 to the circulating air or water. The hot liquid then flows at high pressure to the expansion valve, and, as it moves through the valve to the lower pressure, it expands rapidly and cools. Finally the cool liquid refrigerant proceeds under low pressure to the refrigeration box, and the process is repeated. The temperature of the cooler reservoir (the refrigeration box) is controlled by means of a heat sensor that

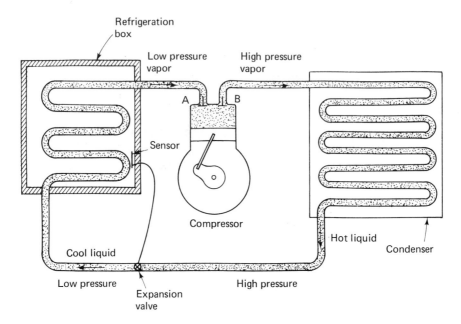

Figure 16-18
A typical refrigeration cycle.

governs the expansion valve. When the temperature of the cooler reservoir fluctuates, an impulse is transmitted from the sensor to the expansion valve, which automatically adjusts the flow of the refrigerant to counteract the fluctuations.

In a complete cycle the net change in the internal energy of the refrigerant is zero. Consequently from the first law of thermodynamics:

$$W = Q_1 - Q_2 \quad \text{or} \quad Q_2 + W = Q_1$$

where W is the work done by the compressor, Q_1 is the heat rejected to the hotter reservoir and Q_2 is the heat removed from the cooler reservoir.

Refrigerators are normally rated in terms of their *coefficient of performance* μ, which is defined as the ratio of the heat Q_2 removed from the cooler reservoir to the work W that is done by the compressor.

$$\mu = \frac{Q_2}{W} = \frac{Q_2}{Q_1 - Q_2} \qquad (16\text{-}18)$$

For an ideal (Carnot) refrigerator that removes heat Q_2 from a reservoir at an absolute temperature T_2 and rejects heat Q_1 to a hotter reservoir at an absolute temperature T_1:

$$\frac{Q_2}{Q_1} = \frac{T_2}{T_1}$$

The coefficient of performance is:

$$\mu = \frac{Q_2}{Q_1 - Q_2} = \frac{T_2}{T_1 - T_2} \qquad (16\text{-}19)$$

Example 16-6

An ideal refrigerator removes heat from an area where the temperature $T_2 = -13°C$ and rejects heat to an area at 27°C. Determine (a) the coefficient of performance and (b) the work required to remove 13.0 J of heat from the cooler area.

Solution:

(a) $\quad \mu = \dfrac{T_2}{T_1 - T_2} = \dfrac{260 \text{ K}}{300 \text{ K} - 260 \text{ K}} = 6.5$

(b) $\quad W = \dfrac{Q_2}{\mu} = \dfrac{13.0 \text{ J}}{6.5} = 2.0 \text{ J}$

The rate at which a refrigeration or air conditioning system removes heat is called its *cooling capacity*. It is usually expressed in terms of *tons of refrigeration*. One ton of refrigeration is equivalent to 12 000 BTU/h.

Example 16-7

Determine the input power to a refrigerator if it has an operating capacity of 10 tons and a coefficient of performance equal to 5.0.

Solution:

Suppose that the compressor does work W to remove heat Q_2 in an elapsed time t. The coefficient of performance

$$\mu = \frac{Q_2}{W} \quad \text{or} \quad W = \frac{Q_2}{\mu}$$

and the average rate of heat removal = $Q_2/t = 10$ t. Therefore the power supplied is:

$$P = \frac{W}{t} = \frac{Q_2}{\mu t} = \frac{(10 \text{ ton})[12000 \text{ BTU}/(h \cdot ton)]}{5.0}$$

$$= 24\,000 \, \frac{\text{BTU}}{\text{h}}$$

But 1 BTU = 778 ft·lb and 1 hp = 550 ft·lb/s. Therefore

$$P = 24\,000 \, \frac{\text{BTU}}{\text{h}}$$

$$= \left(24\,000 \, \frac{\text{BTU}}{\text{h}}\right) \frac{(778 \text{ ft} \cdot \text{lb}/\text{BTU})}{(3\,600 \text{ s}/\text{h})} \left(\frac{1 \text{ hp}}{550 \text{ ft} \cdot \text{lb}/\text{s}}\right)$$

$$= 9.4 \text{ hp}$$

Problems

18. Determine (a) the coefficient of performance and (b) the work done by the compressor when a refrigerator removes 30.0 J of heat from a cooler reservoir and deposits 35.5 J to a hotter reservoir.

19. An ideal refrigerator removes heat from an evaporator at 5.0°C and exhausts heat at 98°C. Calculate (a) the coefficient of performance and (b) the work required to remove 3.35 kJ of heat from the evaporator.

20. Compute the input power in horsepower

to a refrigerator that has an operating capacity of 4.5 t and a coefficient of performance equal to 5.4.

21. Determine the cooling capacity of an air conditioning system that requires an input power of 16.0 hp if the coefficient of performance is 6.0.

REVIEW PROBLEMS

1. A gas does 1 200 ft·lb of work when it receives 4.00 BTU of heat. What is its change in internal energy?

2. If the internal energy of a gas decreases by 1 500 J when it does 3 000 J of work, what is the change in its heat content?

3. What is the amount of heat required to raise the temperature of 42.0 L of helium from 0°C to 30°C at constant volume?

4. How much work is done by a gas when its volume decreases from 20.0 L to 4.0 L under constant pressure of 100 kPa?

5. The temperature of 8.0 mol of nitrogen is increased from 20°C to 100°C at constant volume. Calculate (a) the work done by the nitrogen, (b) the increase in its internal energy, (c) the heat applied.

6. The temperature of 5.0 mol of steam is increased from 120°C to 300°C at constant pressure. Determine: (a) the heat supplied, (b) the change in internal energy, (c) the work done by the steam.

7. How much work is done by an ideal gas when it expands isothermally from an initial volume of 210 L at a pressure of 200 kPa to a final volume of 640 L?

8. If 12 L of helium gas at an initial pressure of 300 kPa and an initial temperature of −120°C are compressed adiabatically to a volume of 5.0 L, determine (a) the final pressure, (b) the final temperature, (c) the work done by the gas.

9. Determine the efficiency of a Carnot heat engine that absorbs 2 800 J from a high temperature source and rejects 1 200 J to a lower temperature reservoir.

10. How much work is done by a Carnot heat engine when it receives 3.4×10^5 J from a high temperature source if its efficiency is 72%?

11. Calculate the efficiency of a Carnot heat engine that operates between temperatures of 1 200°C and −80°C.

12. Determine (a) the coefficient of performance and (b) the work done by a compressor when a refrigerator removes 85 J of heat from a cooler reservoir and deposits 120 J in a hotter reservoir.

13. An ideal refrigerator removes heat from an evaporator at 3.0°C and exhausts heat at 78°C. Calculate (a) the coefficient of performance and (b) the work required to remove 4.2×10^5 J from the evaporator.

All materials are deformed to some extent by the application of an external force. Even if the force is not large enough to destroy the material's structure, it still changes the spacing between the various atoms. When the external force is removed from the material, the internal forces tend to make the atoms vibrate about their equilibrium positions in much the same way as an object vibrates at the end of a spring.

Vibrations may have disastrous effects on smooth-running machinery and relatively rigid structures. Many machines have been severely damaged or destroyed because internal vibrations produced unwanted friction between moving parts and even caused some machines to break free of their mounts. In Tacoma, Washington, a large bridge was completely destroyed by excessive vibrations which literally tore it apart.

The energy produced by one of the many available sources must often be transmitted to some other position. This transmission of energy from one place to another is usually accomplished either by the direct transfer of matter, which has kinetic energy due to its motion, or by means of a vibratory disturbance called a *wave*. The motion of a wave involves no net transfer of matter, yet waves possess both kinetic energy and momentum. Many familiar entities, such as sound, light and radio and television signals, are transmitted as waves. Some kinds of waves, such as sound, require a material medium (solid, liquid or gas) in order to propagate, but *electromagnetic waves*, such as light, radio and television signals, do not require any material medium and are able to propagate through a vacuum (Chapter 26). Even though many different kinds of waves exist, they all behave alike in many respects. Consequently, the analysis of one type of wave, such as a water wave, enables us to describe and make predictions about the behavior of all other waves.

17

Vibrations and Waves

17-1 SIMPLE HARMONIC MOTION

We have previously considered the special types of motion in which the acceleration was constant in magnitude or in both magnitude and direction. In many naturally occurring motions however, the acceleration varies both in magnitude and direction; the results are often quite complex.

There is one form of accelerated motion that is not too complex but is of great importance to technology because of its frequent occurrence. This motion is a periodic motion called *simple harmonic motion (SHM)*; it occurs when the acceleration **a** is directly proportional to the displacement **s** but acts in the opposite direction. Thus

$$\mathbf{a} \propto -\mathbf{s} \text{ (for SHM)} \qquad (17\text{-}1)$$

Simple harmonic motion is involved to some extent in many vibrating systems, such as alternating electric current, mechanical vibrations and waves. Most vibrating systems do not perform pure simple harmonic motion: They lose energy because of friction and other external influences by a process called *damping*. However in some vibrating systems the effects of damping are relatively small, and the motion is approximately simple harmonic motion.

Consider a perfectly elastic spring of negligible mass. One end of the spring is attached to a fixed support, and a mass m is suspended from the free end so that in equilibrium it is located at some position O (Fig. 17-1). If an external force **F** is applied to the free end parallel to the axis of the spring, the spring distorts by an amount **s** in the direction of the force, and according to Hooke's law:

$$\mathbf{F} = k\mathbf{s} \qquad (17\text{-}2)$$

where k is the *elastic* or *spring constant*.

If the external force **F** is maintained on the system, the spring is in equilibrium in its distorted state; therefore some other force, called the restoring force $\mathbf{F}_R$ (which is equal and opposite to the applied force **F**), must also act on the spring. This restoring force is due to the changes in interatomic forces between the molecules when the material is distorted. Therefore in the distorted state $\Sigma\mathbf{F} = \mathbf{F} - \mathbf{F}_R = 0$ and

$$\mathbf{F}_R = -\mathbf{F} = -k\mathbf{s} \qquad (17\text{-}3)$$

The restoring force $\mathbf{F}_R$ is directly proportional to the displacement **s**, but it acts in the opposite direction.

If the external force **F** is removed, the unbalanced restoring force $\mathbf{F}_R$ tends to return the mass at the free end of the spring to its original position O. According to Newton's second law, $\mathbf{F}_R = m\mathbf{a} = -k\mathbf{s}$ and

$$\frac{\mathbf{a}}{\mathbf{s}} = \frac{-k}{m} = \text{constant} \qquad (17\text{-}4)$$

since k and m are constants. Consequently from the definition (Eq. 17-1), the mass m performs simple harmonic motion about the position O. As the mass m accelerates towards its equilibrium position O, the restoring force decreases, but the speed and therefore the momentum of the mass increases. Consequently the mass overshoots and continues past O. As soon as it passes O however, the mass begins to distort the spring in the opposite direction. The restoring force acts to oppose the distortion, slowing the mass and bringing it momentarily to rest at Q before accelerating it back towards the equilibrium position O. The mass again overshoots the equilibrium position and continues past O, decelerating until it comes momentarily to rest at P before moving once again in the original direction. This process is repeated continually as the mass oscillates to and fro in a periodic motion about its equilibrium position.

The maximum displacement from the equilibrium position O is called the *amplitude*

A of the vibration. In the example the amplitude is equal to the distance OP and also OQ.

One complete to and fro oscillation of the mass is called a *cycle*. In a complete cycle, the mass must pass through its equilibrium position twice and each maximum displacement once.

The *period T* of the vibration is the time taken to complete one cycle. The *frequency f* of the vibration is the number of complete cycles performed per unit time. Therefore

$$f = \frac{1}{T} \qquad (17\text{-}5)$$

In SI the unit of frequency is called a *hertz* (Hz); it is equivalent to the reciprocal of time in seconds.

Consider an object that moves in uniform circular motion with a tangential speed v^1

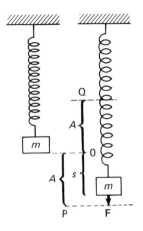

Figure 17-1

at a radius *r* about some fixed point C (Fig. 17-2). If parallel rays of light are incident to

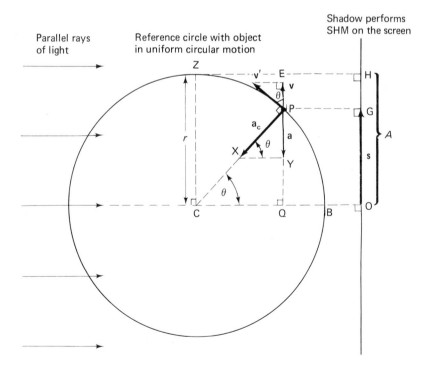

Figure 17-2
Reference circle and SHM.

the system such that the shadow of the object is projected onto a perpendicular screen, then the *shadow* performs simple harmonic motion. Suppose that the object begins its motion at point B while its shadow starts at point O such that the line CBO is parallel to the light rays. After elapsed time t the object has moved through arc BP, which subtends an angle θ at C while the shadow moved up the screen a displacement **s** to position G, $s =$ OG $=$ QP. In the right angled triangle CPQ, the displacement of the shadow:

$$s = CP \sin \theta = r \sin \theta$$

But when the object is located at position Z, its shadow is at H, the angle $\theta = 90°$, and the displacement of the shadow is a maximum value equal to the amplitude A; it is also equal to the radius r of the reference circle.

$$s = r = A \quad \text{when } \theta = 90°$$

Therefore, if $\omega = \theta/t$ is the angular velocity, the displacement of the shadow is:

$$s = r \sin \theta = A \sin \theta = A \sin \omega t \quad (17\text{-}6)$$

The component of the tangential velocity v^1 parallel to the screen corresponds to the velocity v of the shadow. Therefore, when the object is at P, the shadow is at G, and, in the right-angled triangle PED, the shadow is moving up the screen with a speed $v = v^1 \cos \theta = v^1 \cos \omega t$. Since $v^1 = r\omega = A\omega$,

$$v = v^1 \cos \omega t = A\omega \cos \omega t = A\omega \cos \theta \quad (17\text{-}7)$$

At the same time the object undergoes a centripetal acceleration $\mathbf{a}_c$ towards the center C of the circle. The acceleration **a** of the shadow corresponds to the component of $\mathbf{a}_c$ that is parallel to the plane of the screen. From right-angled triangle PXY, $a = -a_c \sin \theta = -a_c \sin \omega t$; the negative sign is required because **a** is in the opposite direction to **s** and **v**.

Since $a_c = r\omega^2 = A\omega^2$,

$$a = -A\omega^2 \sin \omega t = -A\omega^2 \sin \theta \quad (17\text{-}8)$$

Comparing Eqs. 17-8 and 17-6, we obtain:

$$a = -\omega^2 s \quad (17\text{-}9)$$

Since the angular velocity ω is constant from the definition (Eq. 17-1), $a \propto -s$, and the shadow performs simple harmonic motion.

Even though Eqs. 17-6, 17-7 and 17-8 were derived for a particular example, they are valid in general for all pure simple harmonic vibrations. The angle $\theta = \omega t$ is called the *phase* of the vibration.

Since the object completes one revolution (2π rad) in an elapsed time equal to the period T, the angular velocity equals:

$$\omega = \frac{2\pi}{T} = 2\pi f = \frac{\theta}{t} \quad (17\text{-}10)$$

where f is the frequency of the vibration or rotation. Thus the object in SHM has a displacement of magnitude:

$$s = A \sin \theta = A \sin (\omega t) = A \sin (2\pi f t)$$
$$= A \sin \left(\frac{2\pi t}{T} \right) \quad (17\text{-}11)$$

a speed:

$$v = A\omega \cos \theta = \left(\frac{2\pi A}{T} \right) \cos \left(\frac{2\pi t}{T} \right) \quad (17\text{-}12)$$
$$- 2\pi f A \cos (2\pi f t)$$

and an acceleration:

$$a = -A\omega^2 \sin \theta = \left(\frac{-4\pi^2 A}{T^2} \right) \sin \left(\frac{2\pi t}{T} \right) \quad (17\text{-}13)$$
$$= (-4\pi^2 f^2 A) \sin (2\pi f t)$$

These relationships are plotted in Fig. 17-3.

Note that **s** and **a** are both zero, and **v** has a maximum magnitude:

$$v_{max} = \pm \frac{2\pi A}{T} = \pm 2\pi f A \quad (17\text{-}14)$$

when $\theta = n\pi$ (or $t = nT/2$), where n is zero or an integer, since $\sin n\pi = 0$ and $\cos n\pi = \pm 1$. Similarly **v** is zero while **s** has a maximum magnitude:

$$s_{max} = \pm A \quad (17\text{-}15)$$

and **a** has a maximum value:

$$a_{max} = \mp \frac{4\pi^2 A}{T^2} = \mp 4\pi^2 f^2 A \quad (17\text{-}16)$$

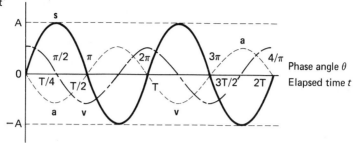

Figure 17-3
Displacement, velocity and acceleration of particle in SHM versus phase angle or elapsed time.

when $\theta = (2n - 1)\dfrac{\pi}{2}$ or $t = (2n - 1)\dfrac{T}{4}$ where n is an integer, since $\sin (2n - 1)\pi/2 = \pm 1$ and $\cos (2n - 1)\pi/2 = 0$.

*Example 17-1*_____

Determine (a) the displacement, (b) the velocity, (c) the acceleration of a particle that vibrates in SHM with an amplitude $A = 5.0$ cm and a frequency $f = 20$ Hz at an instant 0.040 s after it passed its equilibrium position moving in the positive direction.

Solution:

(a) $s = A \sin (2\pi ft)$
$= (5.0 \text{ cm}) \sin [2\pi(20 \text{ Hz})(0.040 \text{ s})]$
$= -4.8 \text{ cm}$

(b) $v = 2\pi fA \cos (2\pi ft)$
$= 2\pi(20 \text{ Hz})(5.0 \text{ cm})$
$\times \cos [2\pi(20 \text{ Hz})(0.040 \text{ s})] = 194 \text{ cm/s}$

(c) $a = -\omega^2 s = -4\pi^2 f^2 s$
$= -4\pi^2(20 \text{ Hz})^2(-4.8 \text{ cm})$
$= 7.6 \times 10^4 \text{ cm/s}$

Note that the angles are expressed in radians.

_____

Period and Frequency_____

If some mass m vibrates in SHM at a frequency f and period T by distorting some elastic material, such as a spring, that has an elastic constant k, then from Eq. 17-4

$$-\frac{s}{a} = \frac{m}{k} \qquad (17\text{-}4)$$

When the displacement is s, the acceleration $a = -\omega^2 s$, where $\omega = 2\pi/T$. Therefore $T = 2\pi/\omega$, $\omega = \sqrt{-a/s}$ and the period of a simple harmonic oscillator is:

$$T = 2\pi \sqrt{\frac{m}{k}} = 2\pi \sqrt{\frac{w}{gk}} = \frac{1}{f} \qquad (17\text{-}17)$$

where w is the weight and g is the acceleration due to gravity. Note that the period of vibration depends only on the mass m vibrating and the elastic constant k: it is completely independent of the amplitude. If the amplitude A of the displacement is increased, the acceleration and the speed of the mass increase to maintain the same period.

*Example 17-2*_____

One end of a very light spring is attached to a fixed support, and the other end is allowed to hang freely. When a mass $m = 0.200$ kg is attached to its free end, the spring elongates by $s_1 = 5.00$ cm. If the mass is then displaced an additional 10.0 cm and released, determine (a) the elastic constant k of the spring, (b) the period T of the resulting vibration, (c) the acceleration a of the 0.200 kg mass when it is 4.00 cm from its equilibrium position, (d) the maximum speed v_{max}, (e) the maximum acceleration a_{max} of the mass. Ignore the mass of the spring.

Solution:

(a) The external force F is due to the weight w of the 0.200 kg load; therefore from Eq. 17-2:

$$k = \frac{w}{s_1} = \frac{mg}{s_1} = \frac{(2.00 \times 10^{-1} \text{ kg})(9.81 \text{ m/s}^2)}{(5.00 \times 10^{-2} \text{ m})}$$
$$= 39.2 \text{ N/m}$$

(b) $T = 2\pi \sqrt{\dfrac{m}{k}} = 2\pi \sqrt{\dfrac{0.200 \text{ kg}}{39.2 \text{ N/m}}} = 0.449 \text{ s}$

(c) Using Eqs. 17-9 and 17-10, we obtain:

$$a = -\omega^2 s = \frac{-4\pi^2 s}{T^2} = \frac{-4\pi^2(4.00 \times 10^{-2} \text{ m})}{(0.449 \text{ s})^2}$$
$$= -7.85 \text{ m/s}^2$$

(d) $v_{max} = \dfrac{2\pi A}{T} = \dfrac{2\pi(0.100 \text{ m})}{(0.449 \text{ s})} = 1.40 \text{ m/s}$

(e) $a_{max} = -\dfrac{4\pi^2}{T^2} A = \dfrac{-4\pi^2}{(0.449 \text{ s})^2}(0.100 \text{ m})$
$$= -19.6 \text{ m/s}^2$$

Problems

1. A rod vibrates with a frequency of 20 Hz. What is its period?

2. Determine the force constant of a spring if a 3.0 kg load causes it to elongate elastically by 30 cm.

3. Determine the force required to elongate an elastic spring by 3.0 in if its force constant is 8.0 lb/ft.

4. A simple harmonic oscillator has a period of 1.50 ms and vibrates with an amplitude of 10.0 μm. What are the maximum speed and acceleration?

5. Determine the maximum speed and acceleration of a crystal oscillator that performs SHM with a period of 1.5 μs and an amplitude of 5.0 nm.

6. A tuning fork vibrates in SHM at a frequency of 256 Hz and an amplitude of 1.2 mm at the tips. Determine the maximum speed and acceleration of the tips of the fork.

7. If the period of SHM vibration of a spring with a 50.0 g load is 0.50 s, (a) what is the spring constant? (b) By how much would the spring elongate if a 100 g load were suspended from it? Ignore the mass of the spring.

8. A 2.0 lb weight is attached to a vertical spring, causing it to elongate by 5.0 in. If the system is then made to vibrate with SHM, (a) what is the frequency of the vibration? (b) If the amplitude of the vibration is 5.0 in, determine the maximum speed and acceleration of the weight. Ignore the mass of the spring.

17-2 THE SIMPLE PENDULUM

A *simple pendulum* consists of a relatively heavy mass, called a *bob*, which is suspended vertically from a fixed rigid support by means of a long light string. If the bob is displaced in a small arc and then released, its oscillations are approximately simple harmonic.

Consider a simple pendulum consisting of a bob (mass m) at the end of a light string of length l. When the bob is displaced through some arc length s (which subtends an angle θ at the support) and then released, the restoring force $\mathbf{F}_R$ is the vector sum of the tension $\mathbf{T}$ in the string and the weight $\mathbf{w}$ of the bob. This restoring force $\mathbf{F}_R$ acts perpendicularly to the string and tangentially to the arc (Fig. 17-4) because the bob moves in this direction when it is released. Therefore $F_R = -mg \sin \theta$, where g is the acceleration due to gravity. If the angle θ is small (less than 5°, for example), $\sin \theta \simeq \tan \theta \simeq \theta$ radians and thus $F_R \simeq -mg\theta$. Also the arc length $s = l\theta$; therefore the restoring force has a magnitude:

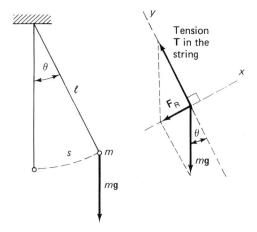

(a) A simple pendulum   (b) Free body diagram
                                of the bob

Figure 17-4

$$F_R = -mg\frac{s}{l} = ma$$

where a is the magnitude of the tangential acceleration of the bob. Thus since g and l are constants, from Eq. 17-4:

$$\frac{s}{a} = \frac{-l}{g} = \frac{-m}{k} = \text{constant}$$

and the bob performs SHM.

Substituting the period of the simple pendulum in Eq. 17-17, we obtain:

$$T = 2\pi\sqrt{\frac{l}{g}} \qquad (17\text{-}18)$$

Note that the period is completely independent of the mass of the bob; it depends only on the length l of the string and the acceleration due to gravity $\mathbf{g}$. By measuring the period of a simple pendulum when the bob is displaced through a small arc, we are able to determine the acceleration due to gravity with reasonable accuracy.

Example 17-3

Determine the length l of a simple pendulum that has a period $T = 0.500$ s at the earth's surface, where the acceleration due to gravity $g = 9.81$ m/s².

Solution:

$$T = 2\pi\sqrt{\frac{l}{g}}$$

Therefore

$$l = \frac{T^2 g}{4\pi^2} = \frac{(0.500 \text{ s})^2 (9.81 \text{ m/s}^2)}{4\pi^2}$$
$$= 6.21 \times 10^{-2} \text{ m}$$

Problems

9. Determine the period of a simple pendulum at the surface of the earth if the string is 1.5 m long.

10. Determine the acceleration due to gravity on a celestial body if a pendulum with 3.0 ft string has a period of 2.3 s when it is located on the surface of that celestial body.

11. If a 2.0 m long pendulum has a period of 1.5 s, what is the period of an 8.0 m long pendulum in the same location?

17-3 ENERGY AND SIMPLE HARMONIC MOTION

If external work is done against elastic restoring forces in order to displace a mass from its initial equilibrium position, the potential energy of the mass increases. A force $\mathbf{F} = k\mathbf{s}$ is required to produce a displacement $\mathbf{s}$ against elastic forces in a system that has an elastic constant k. As it increases from zero to a maximum value ks, the average applied force has a magnitude $F = ks/2$; therefore, when an average force $\bar{\mathbf{F}}$ produces a total displacement $\mathbf{s}$ of a mass m, the elastic potential energy:

$$E_P = W = \bar{\mathbf{F}} \cdot \mathbf{s} = \tfrac{1}{2}ks^2 \qquad (17\text{-}19)$$

Just before it is released, the mass has no vibrational speed, and its displacement is equal to the amplitude A of the subsequent vibration. Therefore the vibrational potential energy is a maximum value, which is equal to the total vibrational energy:

$$E_T = E_{P\,max} = \tfrac{1}{2}kA^2 \qquad (17\text{-}20)$$

When the mass is released, it begins a simple harmonic motion, accelerating towards its initial equilibrium position, gaining kinetic energy at the expense of its potential energy. If there are no external influences, the total vibrational energy E_T of the mass is a constant, which is equal to the sum of the vibrational potential energy E_P and the vibrational kinetic energy E_K. Thus if at some instant the mass m has a vibrational speed v when its displacement is s, its total vibrational energy:

$$E_T = E_P + E_K = \tfrac{1}{2}kA^2 = \tfrac{1}{2}ks^2 + \tfrac{1}{2}mv^2 \quad (17\text{-}21)$$

Solving for the speed of vibration, we obtain:

$$v = \sqrt{\frac{k}{m}(A^2 - s^2)} = \frac{2\pi}{T}\sqrt{(A^2 - s^2)} \qquad (17\text{-}22)$$
$$= 2\pi f\sqrt{A^2 - s^2}$$

since

$$\sqrt{k/m} = \omega = 2\pi/T = 2\pi f.$$

Example 17-4

A 2.00×10^{-27} kg atom in an elastic crystal vibrates in SHM with a period $T = 4.00\ \mu s$ and an amplitude $A = 20.0$ nm. Determine (a) the total vibratory energy E_T of the atom, (b) its maximum vibratory speed v_{max}, (c) the acceleration a and speed v when it is 16.0 nm from its equilibrium position.

Solution:

(a) From Eq. 17-17:

$$k = \frac{4\pi^2 m}{T^2}$$

Thus

$$E_T = \frac{1}{2}kA^2 = \frac{1}{2}\left(\frac{4\pi^2 m}{T^2}\right)A^2$$
$$= \frac{(4\pi^2)(2.00 \times 10^{-27}\ kg)(2.00 \times 10^{-8}\ m)^2}{2(4.00 \times 10^{-6}\ s)^2}$$
$$= 9.87 \times 10^{-31}\ J$$

(b)
$$v_{max} = \frac{2\pi A}{T} = \frac{2\pi(2.00 \times 10^{-8}\ m)}{4.00 \times 10^{-6}\ s}$$
$$= 3.14 \times 10^{-2}\ m/s$$

(c) $a = -\omega^2 s = -\left(\frac{2\pi}{T}\right)^2 s$

$$= -\left(\frac{2\pi}{4.00 \times 10^{-6}\ s}\right)^2 (1.60 \times 10^{-8}\ m)$$
$$= -3.95 \times 10^4\ m/s^2$$
$$v = \frac{2\pi\sqrt{A^2 - s^2}}{T}$$
$$= \frac{2\pi\sqrt{(2.00 \times 10^{-8}\ m)^2 - (1.6 \times 10^{-8}\ m)^2}}{4.00 \times 10^{-6}\ s}$$
$$= 1.88 \times 10^{-2}\ m/s$$

Problems

12. An object vibrates in SHM with a frequency of 70 Hz and an amplitude of 3.0×10^{-2} cm. Determine (a) the maximum acceleration and speed of the object and (b) its acceleration and speed when it is displaced 1.5×10^{-2} cm from its equilibrium position.

13. A 500 g load causes an elastic spring to elongate by 20 cm. If the system with the load is then made to vibrate with an amplitude of 3.0 cm, determine (a) the force constant of the spring, (b) the frequency of the vibration, (c) the maximum acceleration of the load, (d) the maximum speed of the load, (e) the speed and acceleration of the load when it passes through a point 1.0 cm above its distorted equilibrium position. Ignore the mass of the spring.

14. An object vibrates in SHM with a frequency of 2 000 Hz and an amplitude of 1.5×10^{-3} in. Determine (a) the maximum acceleration and speed of the object and (b) its acceleration and speed when it is 1.0×10^{-3} in from its equilibrium position.

17-4 RESONANCE

Objects often vibrate with some natural frequency that depends on their mass and the elastic constant. If energy is lost by the vibrating object, it is said to be *damped*, and the amplitude of its vibration diminishes. Additional vibrational energy may be imposed on the object if it is subjected to a fluctuating force; if the frequency of the fluctuating force is not the same as the natural frequency of the object, the additional energy is usually dissipated as heat, and, even though the object vibrates with the induced frequency, the amplitude of the vibration is small. When the frequency of the imposed force is the same as the natural frequency, the energy and amplitude of the vibration may increase substantially; this is called *resonance*. This may be illustrated by considering a child on a swing. In order to increase the amplitude of the swing and to cause the child to go higher, the swing must be pushed with the same frequency as the natural frequency of oscillation. Note that, if the natural frequency is an integer multiple of the applied frequency, the amplitude of the vibration may still increase; for example, the input energy may be applied to every second vibration.

Resonance is an important phenomenon that is used in electrical timing, oscillation, and memory circuits, musical instruments and a variety of mechanical vibrations. Damping is used to reduce unwanted mechanical vibrations and electrical "noise."

17-5 WAVES

Energy may be transmitted from one place to another either by the motion of matter (as kinetic energy) or by means of a vibratory disturbance called a *wave*. Waves are used to transmit information in terms of sound, radio and television signals, etc., from a source to a receiver without the net transfer of matter.

Water Waves

Possibly the most familiar form of a wave disturbance is the ordinary water wave that occurs when an object, such as a stone, is dropped into a body of water. As the object strikes, it depresses the surface of the water until it finally breaks the surface tension. During this process, the water surface is distorted, and the cohesive forces between the surface molecules act to restore the surface to its original position. Since liquids are almost incompressible, as the object depresses the surface, some molecules beneath that surface move to one side. This raises the level of the surface at points around the object (Fig. 17-5). Once the object has passed through the liquid surface, these cohesive forces spring the liquid back towards its normal position,

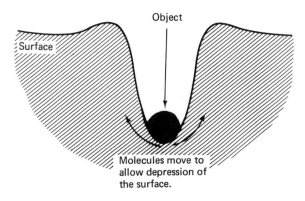

Figure 17-5
Production of a water wave.

but the liquid surface continues past this position and vibrates approximately in simple harmonic motion. The vibration of the water molecules near the entry point of the stone is transmitted to the adjacent points as molecules move beneath the surface, and in turn they transmit the vibration to their neighbors. Thus the vibratory disturbance is transmitted from molecule to successive molecule throughout the surface of the body of water. Molecules at and near the surface vibrate about their normal equilibrium positions, but the wave is transmitted horizontally over the water surface; therefore there is no net transfer of water molecules or matter. Vibrating water molecules behave in much the same way as a cork floating on the surface: The cork vibrates in a nearly elliptical path as a wave passes, but it does not move with the wave.

Transverse and Longitudinal Waves

As a wave passes through a perfectly elastic medium, the particles of that medium are displaced and vibrate in SHM about their equi-

librium positions. A wave can only propagate if the particles of the medium are close together so that the vibrational energy is transmitted from particle to successive particle. Therefore the medium must be relatively dense and elastic.

Waves may be classified in terms of the direction in which the medium particles vibrate as the wave propagates.

In a *transverse wave* the particles of the transmitting medium vibrate *perpendicularly* to the direction in which the energy (or wave) propagates (Fig. 17-6). Transverse waves may be illustrated by the vibrations produced in a stretched string when it is laterally distorted and then released. The string segments vibrate laterally as the wave travels along the length of the string. Electromagnetic radiation is a very important example of a transverse wave. Note that the transmitting medium must be able to withstand a shearing stress in order to transmit a transverse wave.

In a *longitudinal* or *compression wave* the particles of the transmitting medium vibrate parallel to the direction of the energy flow and wave motion. Consider successive parti-

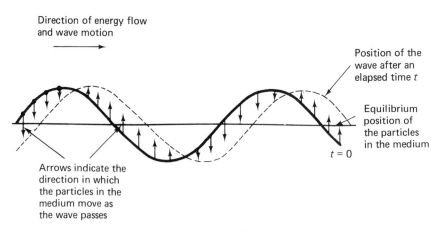

Figure 17-6
A transverse wave. The particles of the transmitting medium vibrate perpendicular to the wave and energy flow.

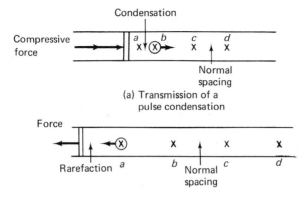

Condensation

(a) Transmission of a
pulse condensation

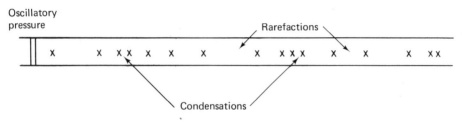

(b) Transmission of a pulse rarefaction

(c) Continuous transmission of rarefactions and
condensations produced by oscillatory pressure

Figure 17-7

cles *a*, *b*, *c* and *d* in a dense, elastic transmitting medium. An external compressive force causes a crowding of particles, called *condensation* (Fig. 17-7a). The crowded particles repel each other as they attempt to regain their equilibrium spacing; as a result, since *a* cannot move in the direction of the compressive force, particle *b* is pushed closer to particle *c*, producing a condensation at *b* and *c*. Particle *b* then "pushes" particle *c* closer to particle *d*, repeating the process; as a result a pulse condensation is transmitted through the medium. When the external force is moved in the opposite direction, it creates a region of reduced density called a *rarefaction* (Fig. 17-7b). As particle *a* moves into the region of reduced density, it produces a rarefaction between particles *a* and *b*. Particle *b* then moves to regain its equilibrium spacing

with *a*, creating a rarefaction between *b* and *c*. This process is repeated between successive particles of the medium; as a result a pulse rarefaction is transmitted. When an oscillatory pressure is applied to a dense elastic medium, a series of condensations and rarefactions is transmitted (Fig. 17-7c).

Longitudinal waves may be illustrated by the vibrations produced in a stretched spring when the coils are distorted longitudinally (compressed, for example) and then released. Other important examples of longitudinal waves are sound waves, ultrasonic waves and shock waves.

There are many kinds of waves: some, such as the transverse electromagnetic waves, do not require a material medium; others, such as water waves, are essentially a complex mixture of longitudinal and transverse waves.

17-6 WAVE CHARACTERISTICS

When a transverse or a longitudinal wave is transmitted through a dense elastic medium, the particles of that medium vibrate in simple harmonic motion. If a particle at a reference point has zero displacement when the elapsed time is zero, and if it moves towards the negative direction as the wave moves from left to right, its displacement from Eq. 17-6 after an elapsed time t is given by:

$$s = -A \sin (2\pi ft) \qquad (17\text{-}23)$$

where A is the amplitude and f is the frequency of the vibration. The wave will have the same frequency f and period T as the SHM. At any instant the displacement of different particles in the medium may have different values (Fig. 17-8).

Particles in the transmitting medium are said to have the same *phase of vibration* if they have the same displacement (magnitude and direction) and are moving with the same velocity (speed and direction). Thus in Fig. 17-8, particles a and e are in phase, b and f are in phase, and c and g are in phase. Even though a and b have the same displacement, they are not in phase because their velocities are different (Fig. 17-6, for example).

The *wave speed* v is the time rate of propagation of a particular wave characteristic, such as a peak or a trough, through the medium.

The *wavelength* λ is the distance between any two successive particles of the medium that are vibrating in phase (Fig. 17-8). It is also equivalent to the distance that the wave propagates in an elapsed time equal to the period T. Thus the wave speed

$$v = \frac{\lambda}{T} = f \qquad (17\text{-}24)$$

where v is the wave speed and f is the frequency. Many common waves, such as light, X-rays and γ-rays have very small wavelengths, which are usually expressed in nanometers.

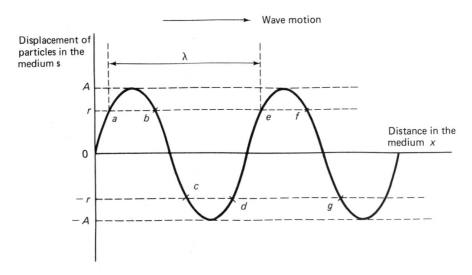

Figure 17-8
Displacement of medium particles as a function of their position.

*Example 17-5*_____

The speed of light is 3.00×10^8 m/s. What is the frequency of light emitted from a helium-neon gas laser if its wavelength is 633 nm?

Solution:

$$v = f\lambda.$$

Thus

$$f = \frac{v}{\lambda} = \frac{3.00 \times 10^8 \text{ m/s}}{633 \times 10^{-9} \text{ m}} = 4.74 \times 10^{14} \text{ Hz}.$$

Note that when a wave passes from one medium into another of different density, the wave speed and wavelength change, but the *frequency of the wave remains the same.* Some waves, such as sound waves, speed up, and their wavelength increases when they enter a denser medium. However, electromagnetic waves slow down, and their wavelength is reduced when they pass into a denser medium.

*Problems*_____

15. If the speed of sound in air is 330 m/s, determine the wavelength of a sound that has a frequency of 500 Hz.

16. Light is an electromagnetic wave that has a speed of 3.0×10^8 m/s in a vacuum. (a) Determine the period of light that has a wavelength of 500 nm. (b) Calculate its frequency.

17. Microwaves are a form of electromagnetic radiation that travel at a speed of 3.0×10^8 m/s in a vacuum. What is the frequency of microwaves that have a wavelength of 3.0 cm?

18. A source emits light with a frequency of 5.0×10^{14} Hz. If the speed of light is 1.86×10^5 mi/s, how many vibrations does the light source make while the light travels from the sun to the earth, a distance of 9.3×10^7 mi?

17-7 THE WAVE EQUATION

Consider a wave of wavelength λ, frequency f, period T and speed v, that moves from left to right through a dense elastic transmitting medium (Fig. 17-9). The displacement of the particle at $x = 0$ reaches a particle located at some distance x after an elapsed time $t = x/v$.

In general, the displacement at $x = 0$ after an elapsed time t is equivalent to the displacement at x after an elapsed time $[(x/v) - t]$ since x is ahead of $x = 0$. Therefore, if the wave moves from *left to right*, the

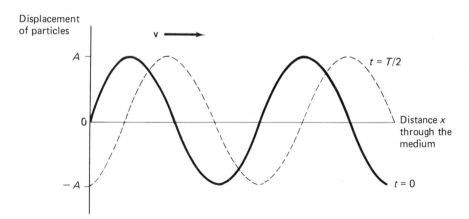

Figure 17-9

general equation describing the displacement at x after an elapsed time t is:

$$s = A \sin\left[2\pi f\left(\frac{x}{v} - t\right)\right]$$
$$= A \sin\left[2\pi\left(\frac{x}{\lambda} - \frac{t}{T}\right)\right]$$

(17-25)

since $f = 1/T$ and $v = f\lambda$. If the wave moves from *right to left*, the displacement at x is:

$$s = A \sin\left[2\pi f\left(\frac{x}{v} + t\right)\right]$$
$$= A \sin\left[2\pi\left(\frac{x}{\lambda} + \frac{t}{T}\right)\right]$$

(17-26)

since the displacement at the origin leads the displacement at x by an elapsed time of (x/v). The *phase difference* between particles located at x_1 and x_2 is given by the ratio:

$$\phi = 2\pi\frac{(x_1 - x_2)}{\lambda}$$

(17-27)

Example 17-6

A wave passes between two points that are 0.60 cm apart. If the wavelength $\lambda = 3.0$ cm, determine the phase difference between the points.

Solution:

$$\phi = 2\pi\frac{(x_1 - x_2)}{\lambda} = 2\pi\frac{(0.60\text{ cm})}{(3.0\text{ cm})}$$
$$= 0.40\pi \text{ radians} = 72°$$

Example 17-7

(a) Write the equation of a wave that has a wavelength $\lambda = 65$ cm, a speed $v = 330$ m/s and an amplitude $A = 2.00$ mm if it moves from right to left. Assume that the displacement at the origin $x = 0$ is zero when the elapsed time $t = 0$ s. (b) Determine the displacement at $x = 175$ m after an elapsed time $t = 40$ ms.

Solution:

(a)
$$s = A \sin\left[2\pi\left(\frac{x}{\lambda} + ft\right)\right]$$
$$= A \sin\left[\frac{2\pi}{\lambda}(x + vt)\right]$$

since $v = f\lambda$. Thus

$$s = (2.00\text{ mm}) \sin\left\{\frac{2\pi}{(0.65\text{ m})}\left[x + \left(330\,\frac{\text{m}}{\text{s}}\right)t\right]\right\}$$

(b) $x = 175$ m and $t = 40 \times 10^{-3}$ s. Thus

$$s = (2.00\text{ mm}) \sin\left\{\frac{2\pi}{(0.65)\text{ m}}\right.$$
$$\times [175\text{ m} + (330\text{ m/s})(40 \times 10^{-3}\text{ s})]\}$$
$$= -0.48\text{ mm.}$$

Problems

19. A particle at $x = 0$ has zero displacement and moves in the negative direction when the elapsed time $t = 0$. If the wave moves from left to right and its amplitude is 2.0 mm, determine the displacement at $x = 5\lambda/4$ after an elapsed time $t = T/5$ where λ is the wavelength and T is the period of the wave.

20. A particle at $x = 0$ has zero displacement and moves in the negative direction when the elapsed time $t = 0$. If the wave has an amplitude of 2.0 mm, a wavelength of 6.0 m and a period of 2.0×10^{-10} s, determine the displacement at a point $x = 34.5$ cm after an elapsed time of 3.5×10^{-10} s (a) if the wave moves from left to right, (b) if the wave moves from right to left.

21. A wave passes between two points that are 5.0 m apart. If the wavelength is 30 m, determine the phase difference between the points.

22. Determine the phase difference between two points that are 0.50 cm apart in a wave of wavelength 0.75 m.

23. (a) Write the equation of a wave if its wavelength is 3.0 cm, its speed is 50 m/s, its amplitude is 3.0×10^{-2} m, and it moves from left to right. (b) Determine the displacement at a distance $x = 20$ m after an elapsed time of 0.20 s. Assume

that the displacement at the origin $x = 0$ is zero when the elapsed time $t = 0$.

24. A wave is represented by the equation:

$$s = 2.5 \sin \left[\frac{2\pi}{200} (x - 7\,500t) \right]$$

where distances are in centimeters and elapsed times are in seconds. Determine (a) the wavelength, (b) the wave speed, (c) the frequency, (d) the period, (e) the displacement at a point $x = 2\,500$ cm after an elapsed time of 0.25 s.

17-8 ENERGY TRANSMISSION

As a wave propagates through a medium, vibrational energy is transmitted from particle to successive particle. The particles in a dense elastic medium vibrate in simple harmonic motion with a constant total energy; it is this vibrational energy that is transmitted by the wave disturbance.

Consider a single particle of mass m that performs SHM with a frequency f and an amplitude A. At the instant when the vibrating particle has a maximum displacement equal to its amplitude A, its speed and therefore its kinetic energy are zero. Consequently all of its vibrational energy is potential energy.

$$E_{P\,\text{max}} = \text{maximum potential energy} = \tfrac{1}{2}kA^2$$

where k is the elastic constant of the medium. Since the particle vibrates with SHM, from Eq. 17-17:

$$k = 4\pi^2 mf^2$$

and its total vibrational energy:

$$E_{\text{TOT}} = E_{P\,\text{max}} = \tfrac{1}{2}kA^2 = 2\pi^2 mf^2 A^2 \quad (17\text{-}28)$$

If there are N such vibrating particles per unit volume V in the medium, the total vibrational energy per unit volume (*energy density*):

$$\frac{E_{\text{TOT}}}{V} = (2\pi^2 mf^2 A^2)N = 2\pi^2 \rho f^2 A^2 \quad (17\text{-}29)$$

since $Nm = \rho$, the density of the medium.

The *intensity* I of a wave is defined as the time rate of transfer of energy (i.e. power $P = E/t$) per unit area A_1 perpendicular to the direction of the energy flow.

$$I = \frac{E}{tA_1} = \frac{P}{A_1} \quad (17\text{-}30)$$

If v is the wave speed, the wave will travel a distance $d = vt$ in an elapsed time t. Thus, from Eq. (17-29), the total energy transmitted by the wave through a perpendicular area A_1 (Fig. 17-10) is:

$$E_{\text{TOT}} = (2\pi^2 \rho f^2 A^2)(V) = (2\pi^2 \rho f^2 A^2)(vtA_1)$$

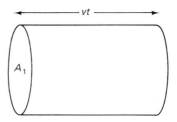

Figure 17-10

Therefore, the intensity of the wave is:

$$I = \frac{E_{\text{TOT}}}{tA_1} = \frac{P}{A_1} = 2\pi^2 \rho f^2 A^2 v \quad (17\text{-}31)$$

Note that the wave intensity is proportional to the density ρ of the medium, the wave speed v and the squares of the frequency f and amplitude A of the vibration.

The direction of energy flow is often represented by a straight line called a *ray*; rays are always perpendicular to the wavefront.

If the medium is isotropic and the source radiates uniformly in all directions, the wave speed is the same in all directions, and the wave front spreads out spherically from the source. Consequently, if the wave speed is v, after an elapsed time t the spherical wavefront has a radius vt and a surface area $A_1 = 4\pi r^2 = 4\pi v^2 t^2$. Therefore, since the energy flow is perpendicular to the wavefront, the intensity of the wave after an elapsed time t is given by:

$$I = \frac{E}{A_1 t} = \frac{E}{4\pi r^2 t} = \frac{P}{4\pi r^2} = \frac{E}{4\pi v^2 t^3} = \frac{P}{4\pi v^2 t^2}$$

$$(17\text{-}32)$$

where $P = E/t$ is the power transmitted by the source. If the power of the source is constant, the intensity is inversely proportional to the square of the distance from the source; this is another example of an *inverse square law*.

Consider a small source S of wave energy that transmits its energy at a rate P uniformly in all directions. If the medium is isotropic and there is no damping, the wave front spreads out spherically from the source. Thus the rate of energy transmission through any two imaginary concentric spheres, both centered at S, must be the same (Fig. 17-11).

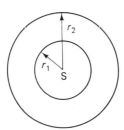

Figure 17-11
Wavefronts spread out spherically from the source. Therefore, if there are no energy losses, the rate of energy flow is the same through any two imaginary spherical surfaces of radii r_1 and r_2.

From Eq. 17-32, the intensity I_1 of the waves when they have travelled a distance r_1 from S is given by $I_1 = P/4\pi r_1^2$; when they have travelled a distance r_2 from S, their intensity $I_2 = P/4\pi r_2^2$. Since $P/4\pi$ is constant:

$$\frac{P}{4\pi} = I_1 r_1^2 = I_2 r_2^2$$

or

$$\frac{I_1}{I_2} = \frac{r_2^2}{r_1^2} = \left(\frac{r_2}{r_1}\right)^2 \qquad (17\text{-}33)$$

Example 17-8

A small source radiates sound with a frequency of 1.00 kHz uniformly in all directions. The speed of sound is 330 m/s, and the density of the medium (air) is 1.29 kg/m³. If there is no damping and the amplitude of the wave is 5.49×10^{-7} m at a distance 10.0 m from the source, determine (a) the intensity I_1 of the sound vibration at a distance $r_1 = 10.0$ m from the source, (b) the power P of the source, (c) the intensity I_2 of the wave at a distance $r_2 = 20.0$ m from the source.

Solution:

(a) $I_1 = 2\pi^2 \rho f^2 A^2 v$
$= 2\pi^2(1.29 \text{ kg/m}^3)(1.00 \times 10^3 \text{ Hz})^2$
$\times (5.49 \times 10^{-7} \text{ m})^2(330 \text{ m/s})$
$= 2.53 \times 10^{-3} \text{ W/m}^2$

(b) $P = 4\pi r_1^2 I_1$
$= 4\pi(10.0 \text{ m})^2(2.53 \times 10^{-3} \text{ W/m}^2)$
$= 3.18 \text{ W}$

(c) $I_2 = \frac{I_1 r_1^2}{r_2^2} = \frac{(2.53 \times 10^{-3} \text{ W/m}^2)(10.0 \text{ m})^2}{(20.0 \text{ m})^2}$
$= 6.33 \times 10^{-4} \text{ W/m}^2$

Some waves, such as radio and television signals, emanate from cylindrical-shaped surfaces, such as long, straight wire antennae. These waves tend to diverge from the source with distorted cylindrical-shaped wavefronts. If the ends of the cylinder are ignored, the wave energy is transmitted through the curved surface of imaginary cylinders of length l and radius r. Thus at distance r from the source, the intensity of the wave is inversely proportional to the distance r from the source:

$$I = \frac{P}{2\pi r l} \qquad (17\text{-}34)$$

*Problems*_____

25. Determine the intensity of a wave at a distance of 50 m from a small source that radiates wave energy at a rate of 10 W uniformly in all directions. Assume no losses.

26. A small 50 kW source emits waves with a frequency of 1 500 Hz uniformly in all directions. Determine the intensity of the waves at a distance 20 m from the source. Assume no losses.

27. An antenna emits waves that diverge with a cylindrical wavefront of length 100 m. If there are no losses and the power of the source is 50 kW, determine the intensity of the wave at 5.0 km from the source.

17-9 INTERFERENCE

Two or more waves may pass simultaneously through the same points in a medium. Each wave continues unaffected by the other waves that are present, but the medium is affected since each wave produces a displacement in the medium. The effect on the medium when two or more similar waves occur simultaneously at the same point is called *interference*.

The total displacement of the particle in the medium is the *vector sum* of the displacements that the individual waves would produce alone at that instant. Thus, if a wave A alone would produce a displacement s_A at some instant and a wave B alone would produce a displacement s_B at the same position and time, when they occur simultaneously at that position, their combined effect is to produce a displacement:

$$s = s_A + s_B \qquad (17\text{-}35)$$

This is called the *superposition theorem*.

If the waves arrive simultaneously in phase at some position in the medium, they are said to *interfere constructively* at that point since the resultant displacement of the medium is greater than the individual displacements due to each wave. When two waves arrive simultaneously 180° out of phase at some position, they are said to *interfere destructively* at that point because the resultant displacement of the medium is less than at least one of the individual displacements due to each wave (Fig. 17-12).

In contrast to pure constructive or destructive interference, if the two waves have slightly different frequencies, a mixture of constructive and destructive interference patterns is produced (Fig. 17-13). When the two waves are in phase, they support each other; when they are 180° out of phase, they suppress each other, producing a phenomenon called *beats*. If sound waves of the same amplitude are used, these beats are usually audible since the intensity of the sound increases and a "beat" is heard when the waves interfere constructively. The intensity is reduced and the beat dies when destructive interference occurs. The frequency of the beats corresponds to the *difference* in frequency between the two waves.

In many electronic devices, such as radio receivers, incoming high frequency signals are mixed with some other sinusoidal signal, producing a lower frequency signal (the *beat frequency*) that is easier to amplify. This process is called *heterodyning*.

Interference of waves constitutes the basis for radio and television communication systems. A constant high frequency electromagnetic wave which is called the *carrier* (radio, television, microwaves, etc.) is used to carry the signal from the source to receiver. The sound or signal to be transmitted has a lower variable frequency; therefore, if this signal is mixed with the carrier wave, the amplitude of the carrier is modified (Fig. 17-14a). This process is called *amplitude modulation* (AM). The modulated carrier signal is

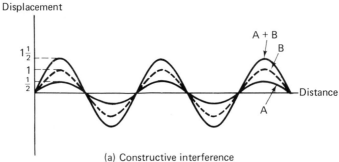

(a) Constructive interference
between waves A and B
which have the same speed,
frequency, and phase, and
emanate from the same point.

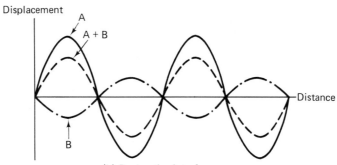

(b) Destructive interference
between waves A and B
which have the same
frequency and speed, and
emanate from the same point
but differ in phase by 180°.

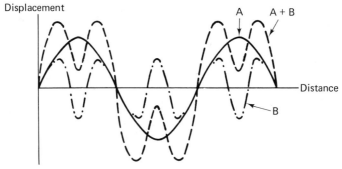

(c) Interference between
two waves of different
frequency.

Figure 17-12
Interference.

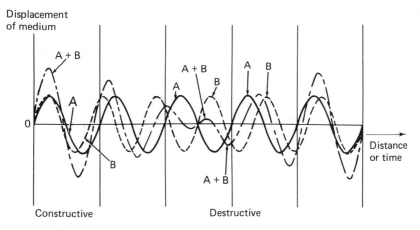

Figure 17-13
A mixture of constructive and destructive
interference patterns.

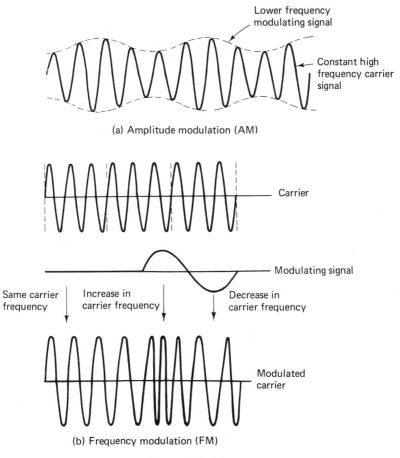

(a) Amplitude modulation (AM)

(b) Frequency modulation (FM)

Figure 17-14

then transmitted to the receiver, where it is demodulated by a similar heterodyne process.

Consider a carrier wave of frequency 500 kHz that is amplitude modulated by a sound signal normally varying in frequency from 0 to 5 kHz. The transmitter must be capable of sending all frequencies within the range from 495 kHz to 505 kHz, i.e. *upper and lower sidebands* of 5 kHz on each side of the 500 kHz carrier frequency. Therefore the *bandwidth* of the transmitter is 10 kHz, and different stations must have carrier frequencies that differ by at least 10 kHz in order to avoid interference.

Sounds varying in frequency from 20 Hz to 20 kHz are audible to the human ear: Because of their relatively small bandwidth, AM stations cannot transmit the higher frequencies required for high fidelity.

Frequency modulation (FM) involves the variation of the carrier wave frequency while the amplitude of the carrier wave remains constant. The *magnitude* of the frequency variation is related to the amplitude of the modulating signal, and the *rate* of carrier frequency variation corresponds to the *frequency* of the modulating signal (Fig. 17-14b). Frequency-modulated carrier waves are less susceptible to unwanted static, which produces fluctuations in the amplitude of the carrier, and the bandwidth of FM stations is usually large enough to transmit all frequencies required for high fidelity.

Very high frequency carrier waves, such as laser beams and microwaves, are capable of carrying large numbers of signals simultaneously. Each signal to be transmitted first modulates a different high frequency oscillation, and then each modulated high frequency oscillation is used to modulate the very high frequency carrier wave. This process is called *multiplexing*. The receiver is tuned to receive only the frequency of the oscillation that was originally modulated by the appropriate signal.

If the phase difference between waves is constant in time, the waves are said to be *coherent*. Coherence is an extremely important property of *laser* light (Chapter 31).

Problems

28. Two waves A and B pass simultaneously through the same medium. Wave A has an amplitude A and a frequency f, and wave B has an amplitude $A/2$ and a frequency $3f$. Sketch the resultant disturbance produced by the two waves in the medium if they both have zero displacement at the origin.

29. Discuss why the frequency of a carrier wave must be very high if it is to transmit several signals simultaneously.

17-10 HUYGENS' PRINCIPLE AND DIFFRACTION

Waves do not necessarily flow in straight lines from a source; they also have the ability to flow around barriers. This "bending" of the waves, delivering energy to regions behind barriers, is called *diffraction*.

Christian Huygens (1629–1695) was able to predict the position and shape of a wave front after a wave had travelled for some time in a medium, even if the wave passed a series of obstacles! Huygens' principle states that:

> *Each point on a wave front acts as a new source of a secondary wavelet that moves forward with the same speed as the original wave. The new wave front at some instant is drawn as the tangent to the surfaces of all secondary wavelets.*

If the medium is isotropic, the wavelets will have a hemispherical shape. For example, if a wave moves with a speed v through an isotropic medium, in order to predict the

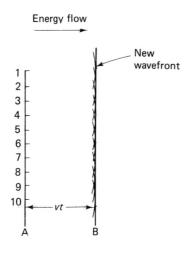

(a) Plane wave.

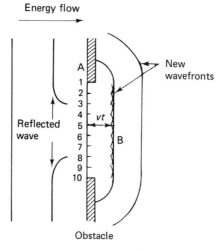

(b) Plane wave after it passes through an obstacle. Choose the sources of secondary wavelets in the region of the gap; the tangent to these wavelets must include the circular portions of the hemispheres drawn from the extremities (points 1 and 10) of the gap.

Figure 17-15

new wavefront after an elapsed time t proceed as follows (Fig. 17-15).

Choose a large number of points (represented by points 1 to 10 for example) on the old wavefront A as sources of new secondary wavelets. From each of these points, draw hemispheres (semicircles in two dimensions) of radii vt in the forward direction. The new wavefront B is the tangent to the hemispheres.

It should be noted that, even though waves can diffract around corners, the intensity of the wave is greatly reduced in regions directly behind obstacles. Therefore, while the center of the wave in Figure 17-15b continues with relatively high intensity, the intensity of the curved portions is reduced.

*Problems*_____

30. Sketch the shape of the wave front after a plane wave passes a small circular barrier.

31. Sketch the shape of the transmitted wave front after a plane wave impinges on a barrier that has a very small hole in it.

17-11 REFLECTION OF WAVES

Waves may be completely reflected or partially reflected and partially transmitted at a boundary between two different media. The nature of the reflected wave is determined by the type of boundary that exists.

If a plane wave AB is incident at some angle on a smooth, flat reflecting surface, different points on the wave front are reflected at different instants. For example, at the instant when point A of the wave front is incident on the reflecting surface, point B must still travel some extra distance s before it too can be reflected (Fig. 17-16). Since the speeds v of the incident and reflected waves are

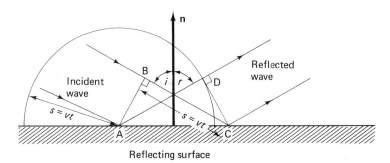

Figure 17-16
Reflection of waves.

equal, an extra elapsed time $t = s/v$ is required before point B reaches the reflecting surface at C; during this elapsed time, the reflected wave from point A travels a distance $s = vt$. According to Huygens' principle, after the elapsed time t the new wavefront CD is the straight line drawn through C and tangential to a circle of radius $s = vt$ that is centered at point A.

Waves travel in directions that are perpendicular to their wavefronts, and these "directions of energy flow" are represented by *rays*. There is a symmetrical relationship between incident and reflected waves. The *angle of incidence i* is the angle between the normal **n** (perpendicular) to the reflecting surface and the incident ray. Similarly the *angle of reflection r* is the angle between the normal **n** and the reflected ray. From the symmetry in Fig. 17-16, it can be seen that:

1. The angle of incidence i is equal to the angle of reflection r.

2. The normal **n** to the surface and the directions in which the incident and reflected waves travel all lie in the same plane.

These statements are called *laws of reflection*, and they apply to both longitudinal and transverse waves.

Auditoriums must be designed so that the sound waves that are reflected from the walls, ceiling and floor do not interfere with the direct waves from the source. The interference between the direct and reflected waves would produce unwanted distortions of the sound.

Special curved reflectors are frequently used to control the concentration and direction of waves. For example, parabolic reflectors are used to direct microwaves for transmission and to concentrate any incident microwaves.

Consider a transverse pulse moving along a string that is rigidly attached to a solid wall (Fig. 17-17a). As the pulse reaches the wall, it exerts a force that tries to displace the wall in the same direction, but the wall cannot move. The wall exerts an equal and opposite reaction force on the string, displacing it in the opposite direction and producing the 180° phase change.

In general, any wave that travels towards a denser material receives a 180° phase change as it is reflected at the material boundary.

If a transverse pulse moves along a string towards an end that is free to move (Fig. 17-17b), the reflected pulse has the same phase as the incident pulse. The end of the string moves when it is affected by the pulse, and no reaction forces are produced.

In general, waves that are reflected from a less dense material boundary do not undergo a phase change.

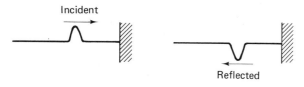

Incident

Reflected

(a) Reflection of a wave in a string
    from a fixed point.

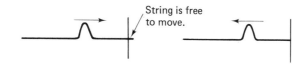

String is free
to move.

(b) Reflection of a wave in a string
    from a free end.

Figure 17-17
Reflection of waves.

17-12 STANDING WAVES

When two similar waves with the same frequency and amplitude move in opposite directions through the same medium, they interfere to produce *standing* (or *stationary*) *waves*. Standing waves may be produced by the interference of waves from two separate sources, but more often they are produced by a single wave and its own reflection. In musical instruments, standing waves are produced in vibrating strings and air columns. Standing waves are also important in electronic devices, such as wave guides and coaxial cables.

Consider two similar waves X and Y that have the same amplitude A, frequency f and speed v and travel simultaneously in opposite directions through the same medium (Fig. 17-18). The superposition of these two waves produces the standing wave Z. Certain positions called *nodes* N are never displaced by the standing wave whereas other positions called *antinodes* A fluctuate with maximum displacements equal to twice the amplitude of each constituent wave. The distance between any two successive nodes is equal to one-half

the wavelength of the standing wave. The antinodes are positioned at the midpoint between successive nodes; therefore the distance between successive node and antinode is one-quarter the wavelength of the standing wave. The amplitude of the standing wave = $2A$.

Example 17-9

In order to determine the wavelength of microwaves, a reflector is placed 1.0 m in front of the source, and a diode is moved in a straight line between them. If the diode starts at a node and detects 28 additional nodes (null points) in a distance of 42 cm, (a) what is the wavelength of the microwaves? (b) What is the frequency if the speed of microwaves is 3.0×10^8 m/s?

Solution:

(a) The distance between successive nodes is one-half the wavelength; therefore there are 14 wavelengths in 42 cm or 3.0 cm per wavelength.

(b) $f = \dfrac{v}{\lambda} = \dfrac{3.0 \times 10^8 \text{ m/s}}{3.0 \times 10^{-2} \text{ m}} = 1.0 \times 10^{10} \text{ Hz}$

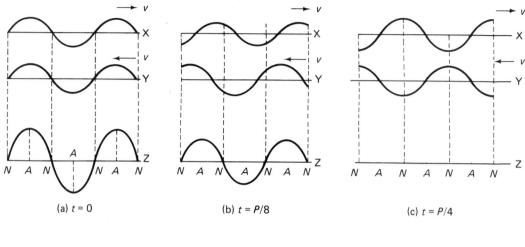

(a) $t = 0$ (b) $t = P/8$ (c) $t = P/4$

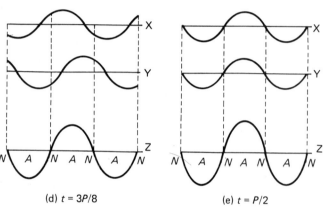

(d) $t = 3P/8$ (e) $t = P/2$

Figure 17-18
Superposition of two waves X and Y to form a standing wave Z. The nodes N are never displaced; the antinodes A fluctuate with maximum displacements.

Problems

32. Determine the distance between two successive nodes of an AM radio wave with a frequency of 1.05 kHz.

33. Determine the wavelength and frequency of an FM radio wave if a reflector produces a standing wave with nodes 1.45 m apart.

REVIEW PROBLEMS

1. A spring vibrates with a period of 0.42 s. What is its frequency?

2. A 300 g load is attached to a light vertical spring, causing it to elongate by 2.4 cm. The load is then displaced another 2.0 cm

and released. If it vibrates in SHM, find (a) the frequency of the vibration and (b) the maximum acceleration and speed.

3. Determine the period of a pendulum 2 ft 8 in long on the moon where the acceleration due to gravity is approximately one-sixth that on earth.

4. If the period of a pendulum is 4.0 s, by what factor should the length of the string be changed in order to increase the period to 7.0 s?

5. An atom in a crystal has a mass of 5.0 × 10⁻²⁶ kg and vibrates in SHM with a period of 2.5 × 10⁻⁶ s and an amplitude of 75 nm. Find (a) the total vibratory energy of the atom, (b) the maximum speed of the atom, (c) the acceleration of the atom when it is 2.5 nm from its equilibrium position.

   An atom in a crystal has a mass of 5.0×10^{-26} kg and vibrates in SHM with a period of 2.5×10^{-6} s and an amplitude of 75 nm. Find (a) the total vibratory energy of the atom, (b) the maximum speed of the atom, (c) the acceleration of the atom when it is 2.5 nm from its equilibrium position.

6. An object vibrates in SHM with a frequency of 15 kHz and an amplitude of 2.5×10^{-3} cm. Determine (a) the maximum acceleration of the object, (b) the maximum speed of the object, (c) the acceleration and speed when the object is 1.5×10^{-3} cm from its equilibrium position.

7. A 4.00×10^{-27} kg atom in an elastic crystal vibrates in SHM with a period of 5.00 μs and an amplitude of 30.0 nm. Compute (a) the total vibratory energy of the atom, (b) the elastic constant, (c) the maximum vibrational speed of the atom, (d) the vibrational acceleration when it is 20.0 nm from its equilibrium position, (e) the speed of the atom when it is 20.0 nm from its equilibrium position.

8. What is the wavelength of a wave if its period is 50 ms and its speed is 500 m/s?

9. If a speaker produces a sound wave with a wavelength of 0.155 m and the speed of sound in air is 330 m/s, what is (a) the frequency and (b) the period of the wave?

10. What is the wavelength of light with a frequency of 6.0×10^{14} Hz?

11. A wave with an amplitude of 2.0×10^{-3} m, a wavelength of 5.0×10^{-2} m and a speed of 300 m/s travels from left to right. If the displacement is zero at the origin $x = 0$ when the elapsed time $t = 0$, (a) write the wave equation. (b) What is the displacement of a particle at a distance $x = 4.0$ cm after an elapsed time of 200 μs?

12. A wave is represented by the equation

$$s = 4.0 \sin \left[\frac{2\pi}{500} (x - 3\,000t) \right]$$

where distances are in centimeters and times are in seconds. Determine (a) the wavelength, (b) the wave speed, (c) the frequency, (d) the displacement at a point $x = 1\,500$ cm after an elapsed time $t = 1.30$ s.

13. Determine the intensity of a wave at a distance of 30 m from a small source that radiates 50 W uniformly in all directions. Assume no losses.

14. A small source radiates sound with a frequency of 2 500 Hz uniformly in all directions. The speed of sound in air is 335 m/s, and the density of air is 1.29 kg/m³. If there is no damping and the wave amplitude is 7.4×10^{-7} m at a distance 15.0 m from the source, determine (a) the intensity of the wave 15 m from the source, (b) the power of the source, (c) the wave intensity at 25 m from the source.

15. A 150 kW antenna emits waves with a cylindrical wave front of length 120 m. Determine the intensity of the wave 8.0 km from the source if there are no losses.

16. Two waves pass simultaneously through the same medium. Wave A has an amplitude A and a frequency f while wave B has an amplitude $A/2$ and a frequency $5f$. Sketch the resultant disturbance if they both have zero displacement at the origin.

17. Determine the wavelength and frequency of a radio wave if a reflector produces nodes (null points) that are 80.0 m apart.

The term *sound* is used to describe the psychological sensation produced when a vibratory disturbance reaches the ear and also to denote the physical cause of that disturbance.

In order to experience a sound, a source of vibratory energy (such as a speaker or vibrating string), a material medium (usually air), and a receiver (the ear) are required. The energy produced by a vibrating source is transmitted as a longitudinal mechanical wave through the material medium to the ear; sound waves will not propagate through a vacuum. The ear is sensitive to the small pressure changes caused by the longitudinal sound wave, and it is able to convert these small pressure variations into small electrical impulses, which are transmitted by auditory nerves to the brain where they produce the sensation of sound.

Sound waves with frequencies higher than those audible by the human ear are called *ultrasonic waves*. Ultrasonic waves with frequencies above 10^8 Hz have been produced by the resonant elastic vibrations of a quartz crystal. Ultrasonic waves are utilized in ultrasonic cleaners, which vibrate dirt from materials; they also have applications in medicine, such as to soften scar tissue. These longitudinal waves possess very good directional properties; therefore high energy ultrasonic waves are used to drill holes of any shape (since no rotation is necessary). Soldering of materials, such as aluminum, is normally an extremely difficult process because the metal surface oxidizes, and the solder cannot wet the metal; however when exposed to ultrasonic waves, the solder immediately wets the metal.

Sound sources are present almost everywhere in our environment, and the control of sound is a major problem of technology. Human beings find some sounds quite pleasing and others quite irritating. Unwanted sounds are called *noise*. The control of noise is becoming an important consideration in our everyday lives. Excessive noise is known to produce drastic changes in our personalities, it is fatiguing and may even cause a reduction in our life span.

18

Sound and Acoustics

In recent years the production and recording of sound has developed into an important industry, which utilizes vast quantities of electronic apparatus. *Acoustics* is the term that is used to denote the science of the production, transmission, reception and effects of sound. In order to utilize sound to the fullest extent, the acoustic properties of buildings must be considered.

18-1 SOURCES OF SOUND

All sources of sound produce vibratory energy with frequencies within the audible range of the ear. Some of the most common sound sources are vibrating strings (stringed musical instruments), vibrating air columns (wind musical instruments), vibrating membranes (speakers, drums) and vibrating rods. Many sound sources may even vibrate with different frequencies. Each different vibratory state is called a *mode* of vibration.

Vibrating Strings

Consider a string of mass m that is stretched under a tension T between two fixed points a distance L apart. When the string is displaced laterally and then released, it vibrates with some natural frequency, which depends on its tension T and length L. Transverse standing waves are produced in the string since the wave is continually reflected from the fixed ends. Note that, if the string does not vibrate at a natural frequency, the standing wave is rapidly damped.

Nodes represent points that are not displaced from their equilibrium positions by the standing wave; therefore nodes must always occur at fixed points in vibrating strings (fixed ends).

Points at which a maximum displacement occurs are called *antinodes*, and the distance between a successive node and antinode is equal to a quarter wavelength. Some of the possible natural modes of vibration are illustrated in Fig. 18-1. In each case the length L

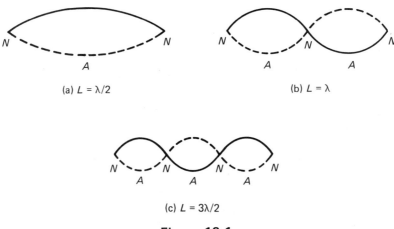

(a) $L = \lambda/2$

(b) $L = \lambda$

(c) $L = 3\lambda/2$

Figure 18-1
The three simplest natural modes of vibration of a string of length L which is fixed at each end, showing the positions of the nodes N and the antinodes A.

is kept constant, and a node occurs at the fixed ends. The tension T in the string is varied in order to produce the different natural frequencies. Since the distance between node and antinode represents one-quarter of the wavelength λ of the standing wave, in general the natural resonant frequencies of vibration occur when:

$$L = n\lambda/2 \quad \text{or} \quad \lambda = 2L/n$$

where n is a positive integer.

However, the speed of the transverse wave in a string of mass m and mass per unit length $\mu = m/L$ under a tensile force T is:

$$v = \sqrt{\frac{T}{\mu}} = \sqrt{\frac{TL}{m}} \qquad (18\text{-}1)$$

Therefore the natural frequencies of vibration are:

$$f_n = \frac{v}{\lambda} = \frac{nv}{2L} = \frac{n}{2L}\sqrt{\frac{T}{\mu}} \qquad (18\text{-}2)$$

where n represents positive integers (1, 2, 3, 4, ...). These natural frequencies of vibration are actually resonant frequencies. If energy is applied to the vibrating string at a natural frequency, the amplitude of the vibration increases.

The lowest possible natural frequency of vibration of any source is called its *fundamental frequency*. Thus the fundamental frequency of the vibrating string (corresponding to $n = 1$ in Eq. 18-2 is:

$$f_1 = \frac{v}{2L} = \frac{1}{2L}\sqrt{\frac{T}{\mu}}$$

Harmonic frequencies are integer multiples of the fundamental frequency. In the vibrating string the frequency of the nth harmonic is obtained merely by substituting the value for n in Eq. 18-2. For example, the frequency of the third harmonic corresponds to $n = 3$. That is,

$$f_3 = \frac{3v}{2L} = \frac{3}{2L}\sqrt{\frac{T}{\mu}}$$

Note that the fundamental frequency is the first harmonic.

Overtones are the higher harmonics with frequencies greater than the fundamental; they are "counted" in the order of the *actual* appearance of harmonics greater than the fundamental. Thus for the vibrating string, the second harmonic is also the first overtone, the third harmonic the second overtone, etc.

Example 18-1

A string that is 2.00 m long and has a total mass of 1.50 g is fixed at each end and is subjected to a tension of 30.0 N. Determine (a) the speed of a transverse wave in the string, (b) the fundamental frequency, (c) the frequency of the third harmonic, (d) the frequency of the 5th overtone of the vibrating string.

Solution:

(a) $v = \sqrt{\dfrac{TL}{m}} = \sqrt{\dfrac{(30.0\ \text{N})(2.00\ \text{m})}{1.50 \times 10^{-3}\ \text{kg}}} = 200\ \text{m/s}$

(b) $\qquad f_1 = \dfrac{v}{2L} = \dfrac{200\ \text{m/s}}{2(2.00\ \text{m})} = 50\ \text{Hz}$

(c) $\qquad f_3 = 3f_1 = 3(50\ \text{Hz}) = 150\ \text{Hz}$

(d) The 5th overtone is the same as the 6th harmonic; therefore

$$f_6 = 6f_1 = 6(50\ \text{Hz}) = 300\ \text{Hz}$$

Vibrating Air Columns in a Pipe That Is Closed at One End

In a longitudinal (compression) wave the particles of the transmitting medium vibrate parallel to the direction of energy flow; therefore the medium is not subjected to a shearing stress. The speed of the longitudinal wave in a fluid medium that has a bulk modulus G and a density ρ is given by:

$$v = \sqrt{\frac{G}{\rho}} \qquad (18\text{-}3)$$

Similarly the speed of a longitudinal impulse through a solid bar with Young's modulus E is

$$v = \sqrt{\frac{E}{\rho}} \qquad (18\text{-}4)$$

*Example 18-2*_____

How long does it take to transmit a sound a distance $d = 750$ m in water?

Solution:

From Tables 5-1 and 5-3 for water:

$\rho = 1.0 \times 10^3$ kg/m³ and $G = 2.3 \times 10^9$ Pa

Thus

$$t = \frac{d}{v} = \frac{d}{\sqrt{G/\rho}} = d\sqrt{\frac{\rho}{G}}$$

$$= (750 \text{ m})\sqrt{\frac{1.0 \times 10^3 \text{ kg/m}^3}{2.3 \times 10^9 \text{ Pa}}} = 0.49 \text{ s}$$

Consider an air column in a tube of length L that is closed at one end only. The air in the tube can be made to vibrate with a natural frequency that depends on the length L and the temperature. Longitudinal standing waves are formed as the wave reflects from the closed end of the pipe. A node exists at the closed end of the pipe because the air

molecules there cannot vibrate longitudinally. Also, when the air column vibrates with a natural frequency, an *antinode* exists approximately* at the open end of the pipe where the air molecules receive a maximum displacement. A few of the possible natural modes of vibration are illustrated in Fig. 18-2. Note that the displacements of the air molecules are plotted vertically for convenience. Since longitudinal standing waves are produced, the molecules are really displaced parallel to the length of the pipe. Thus, if λ is the wave-length of the standing wave, in general the natural frequencies of vibration occur when

$$L = \frac{n\lambda}{4} \quad \text{or} \quad \lambda = \frac{4L}{n}$$

_____

*The antinode is slightly beyond the open end of the pipe by an amount that depends on the pipe's dimensions.

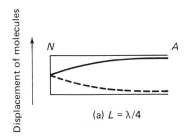

(a) $L = \lambda/4$

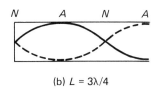

(b) $L = 3\lambda/4$

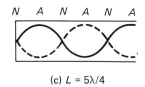

(c) $L = 5\lambda/4$

Figure 18-2

The three simplest natural modes of vibration of an air column in a pipe of length L that is closed at one end. *N.B.*: Air molecules are really displaced parallel to the length of the pipe.

where n is a positive *odd* integer (i.e. 1, 3, 5, 7, ...). Therefore the natural frequencies of vibration of the air column are:

$$f_n = \frac{v}{\lambda} = \frac{nv}{4L} = \frac{n}{4L}\sqrt{\frac{G}{\rho}} \qquad (18\text{-}5)$$

The fundamental frequency (or first harmonic) of the vibrating air column (corresponding to $n = 1$) is

$$f_1 = \frac{v}{4L} = \frac{1}{4L}\sqrt{\frac{G}{\rho}}$$

The second harmonic (corresponding to $n = 2$) does not exist because n can only take odd integer values. The third harmonic ($n = 3$) is the first higher frequency that actually appears above the fundamental; therefore, the *third harmonic* is the *first overtone*.

$$f_3 = \frac{3v}{4L} = \frac{3}{4L}\sqrt{\frac{G}{\rho}}$$

Similarly the fifth harmonic is the second overtone, etc.

*Example 18-3*_____

(a) If the speed of sound in air is 332 m/s, what is the length of a vibrating air column in a pipe that is open at one end only if its fundamental frequency is 1 024 Hz? (b) What is the frequency of the second overtone?

Solution:

(a) For the fundamental, $n = 1$; thus from Eq. 18-5:

$$L = \frac{nv}{4f_n} = \frac{v}{4f_1} = \frac{(332 \text{ m/s})}{4(1\,024 \text{ Hz})} = 8.11 \times 10^{-2} \text{ m}$$

(b) Since n is an odd integer, the second overtone corresponds to the 5th harmonic. Thus

$$f_5 = 5f_1 = 5(1\,024 \text{ Hz}) = 5\,120 \text{ Hz}$$

A simple experiment that shows the effects of resonance is illustrated in Fig. 18-3. As the hollow pipe and vibrating tuning fork are moved vertically, the sound increases considerably when resonance occurs.

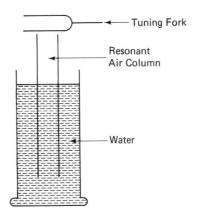

Figure 18-3

Simple experiment to show resonance in air columns.

It should be noted that, even though we have considered sound, the definitions of fundamental frequency, harmonics and overtones are of considerable importance in electronics. These terms are used extensively in wave shaping, frequency response, etc., and the operation of antenna systems are analogous to vibrating air columns.

Vibrating Rods_____

Rods may vibrate longitudinally or transversely. Consider a vibrating rod of length L that is clamped at some position along its length. Many modes of vibration are possible with *nodes at the clamped position(s)* and *antinodes at the free end(s)*.

There are many other sources of sound, each of which is capable of transferring vibrational energy to a material medium, where it is transmitted as a longitudinal wave. Energy must be continually applied in order to sustain the vibration of a sound source. In many cases, one vibrating object is able to transmit its vibrational energy to a second object, causing it to vibrate with the same frequency; this is called a *forced vibration*. For example, the air inside a speaker enclosure is forced to vibrate at the same frequency as the speaker.

Forced vibrations may be produced in an object at one of its natural frequencies. When this occurs, the object vibrates at the same frequency as the applied periodic impulses, and the amplitude of its vibration may be greatly enhanced. The object is then said to be in a state of *resonance* with the applied periodic impulses.

While the effects of resonance are desirable in many musical instruments because of the greatly enhanced amplitudes of vibration, in some cases they are quite irritating because they produce large distortions of normal sound. Resonance effects can also cause parts of a mechanical system to vibrate, and this can eventually lead to serious damage.

Problems

1. Determine the wavelength of a sound if its frequency is 540 Hz and its speed is 330 m/s.

2. If the speed of sound in air is 1 100 ft/s, what is the wavelength of a sound if its frequency is 1 024 Hz?

3. Determine the speed of a transverse pulse along a 3.0 g string of length 5.0 m if it is maintained under a tension of 15 N.

4. Compute the speed of a transverse wave in a metal wire of mass 2.5 g and length 2.0 m if it is maintained under a tension of 80 N.

5. One end of a string is attached to a rigid support, and the other end is passed over a fixed pulley to a 3.0 kg load. (a) Determine the speed of a transverse wave in the string if its linear mass density is 0.75 g/m. (b) What happens if the attached load is increased to 12.0 kg?

6. What is the speed of sound in oil, which has a bulk modulus of 1.7×10^9 Pa and a density of 910 kg/m³?

7. Ultrasonic waves (longitudinal) with a frequency of 40 kHz are transmitted through water, which has a bulk modulus of 2.2×10^9 Pa. Determine the speed and wavelength of the waves.

8. Determine the speed and wavelength of a 500 Hz sound wave in a metal rod that has a density of 2.5×10^3 kg/m³ and a Young's modulus of 6.5×10^{10} Pa.

9. A string, which is 1.50 m long and has a mass of 0.70 g, is fixed at each end and is subjected to a tension of 50.0 N. Determine (a) the speed of a transverse wave in the string, (b) its fundamental frequency, (c) the second harmonic frequency, (d) the third overtone frequency of the vibrating string.

10. A rope, which is 10.0 ft long and weighs 1.5 lb, is fixed at each end and is subjected to a tension of 100 lb. Determine (a) the speed of a transverse wave in the string, (b) its fundamental frequency, (c) the fourth harmonic frequency, (d) the seventh overtone frequency of the vibrating string.

11. A string has a length 50.0 cm and a mass 0.15 g. One end of the string is attached to a mechanical vibrator that makes 60 vibrations per second. What tension will cause the string to vibrate at its fundamental frequency?

12. If the speed of sound in air is 335 m/s, determine the length of a vibrating air column in a pipe that is open at one end if its fundamental frequency is 540 Hz.

13. Determine all possible natural modes of vibration of an air column in a pipe that is open at both ends.

14. The speed of sound in a vibrating metal rod is 5 400 m/s. If the rod is 2.40 m long, determine the fundamental frequency, the frequency of the third harmonic, and the frequency of the third overtone (a)

if the rod is clamped at its center, (b) if the rod is clamped at a point 60.0 cm from one end.

18-2 THE MATERIAL MEDIUM

Sound propagates as a longitudinal wave and requires a material medium; sound waves cannot travel through a vacuum.

In a gaseous medium, such as air, where γ is the ratio of the molar specific heat capacities, p is the absolute pressure, ρ is the density, T is the absolute temperature, M is the mass of one mole of the gas and R is the universal gas constant, the speed of sound is:

$$v = \sqrt{\frac{\gamma p}{\rho}} = \sqrt{\frac{\gamma RT}{M}} \qquad (18\text{-}6)$$

Since γ, R and M are constant, the speed of sound is:

$$v \propto \sqrt{T} \quad \text{or} \quad \frac{v_2}{v_1} = \sqrt{\frac{T_2}{T_1}} \qquad (18\text{-}7)$$

*Example 18-4*_____

Determine the speed of sound in oxygen at 30°C.

Table 18-1
Typical Speeds of Sound at 0°C (32°F)

Medium	Speed	
	ft/s	*m/s*
Fluids		
Air (dry)	1 086	331
Carbon dioxide	853	260
Helium	3 170	965
Hydrogen	4 220	1 285
Oxygen	1 033	315
Solid Rods		
Aluminum	16 400	5 000
Iron	16 900	5 150
Steel	16 600	5 050

Solution:

Using data from Table 18-1, we obtain:

$$v_2 = v_1 \sqrt{\frac{T_2}{T_1}} = (315 \text{ m/s}) \sqrt{\frac{(273 + 30) \text{ K}}{273 \text{ K}}}$$
$$= 332 \text{ m/s}$$

Objects that travel at speeds greater than the speed of sound are called *supersonic*. At subsonic speeds a longitudinal (compression) wave preceeds the object. At supersonic speeds there is a sudden formation of a compression wave as the object passes; this compression wave is called a *shock wave*. When the shock wave from a supersonic aircraft reaches the ground, it is heard as a *sonic boom*; because of its high energy, it can cause damage by breaking windows, etc.

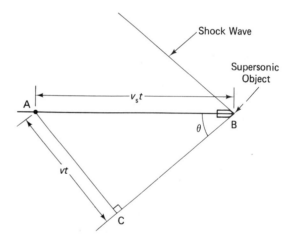

Figure 18-4
An object moving at a supersonic speed v_s travels a distance $v_s t$ in an elapsed time t while sound (speed v) travels a distance vt. This produces a shock wave.

The speed of supersonic objects is often quoted in terms of a *mach number*, which is the ratio of the object's speed to the speed of sound.

$$\text{Mach number} = \frac{\text{speed of object}}{\text{speed of sound}} \qquad (18\text{-}8)$$

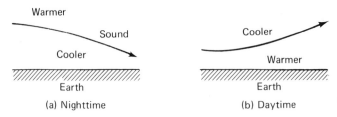

Figure 18-5
Refraction due to a temperature gradient.

Refraction and Interference

Refraction. Sound waves are refracted as their speed changes when they pass into a different medium or even when they enter a region with different properties in the same medium. Note that, even though the speed and wavelength of sound changes when it passes into the different medium, the *frequency remains constant.*

The refraction of sound in air may be caused by a temperature gradient above the earth's surface or by the effects of wind. Sounds usually carry farther over the earth's surface at night because the sound is refracted towards the cooler air near the earth. The reverse effect occurs in the daytime (Fig. 18-5). At a fixed temperature the speed of sound is constant relative to the medium (air); therefore wind velocities change the speed of sound relative to the ground (Fig. 18-6).

Interference. Two or more sound waves produce interference when they pass simultaneously through the same medium. The resultant disturbance is the algebraic sum of the individual disturbances caused by each wave.

If the frequency of two sound waves is slightly different, audible beats in sound may be heard as a result of the interference pattern that they produce. The number of beats corresponds to the frequency difference between the sound waves. For example, if sounds with frequencies of 500 Hz and 504 Hz are produced simultaneously, there are 4 beats per second.

Problems

15. Determine the speed of sound in air at (a) 60°F and (b) −20°C.

16. Determine the speed of sound in helium at 10°C.

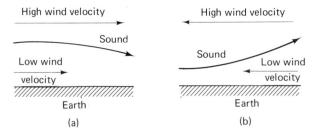

Figure 18-6
Refraction due to wind effects.

17. Determine the mach number of a jet that travels at 1 200 km/h at sea level where the temperature is 30°C.

18. If two sounds with frequencies of 1 026 Hz and 1 031 Hz are produced simultaneously, determine the frequency of the beats.

19. Two sources produce sounds simultaneously, and they produce beats with a frequency of 2 Hz. If one of the sources has a frequency of 700 Hz, discuss the possible frequency (or frequencies) of the other source.

18-3 THE DOPPLER EFFECT

Sources and receivers of sound are not always stationary; as a result, the sound frequency detected by the receiver is not always the same as that transmitted by the source. The frequency difference depends on the relative motion between the source and the observer. For example, there is a noticeable decrease in the *pitch* (frequency) of sound when a train moves towards and then away from a stationary observer on the platform of a station.

Frequency changes produced because of the relative motion of the source with respect to a receiver were predicted by Johann Doppler (1803–1853); the phenomenon is called the *Doppler effect*. It can be understood in terms of a swimmer in the ocean: when the swimmer swims into the waves, they pass at a higher frequency than when the swimmer is stationary; similarly the frequency at which the waves pass decreases when the swimmer swims away from them.

Consider a source S that emits sound with a frequency f and a velocity c (relative to the surrounding air) towards an observer 0. The source S moves with a velocity u towards the observer, and the observer 0 moves with a velocity v away from the source (Fig. 18-7). The source moves in the same

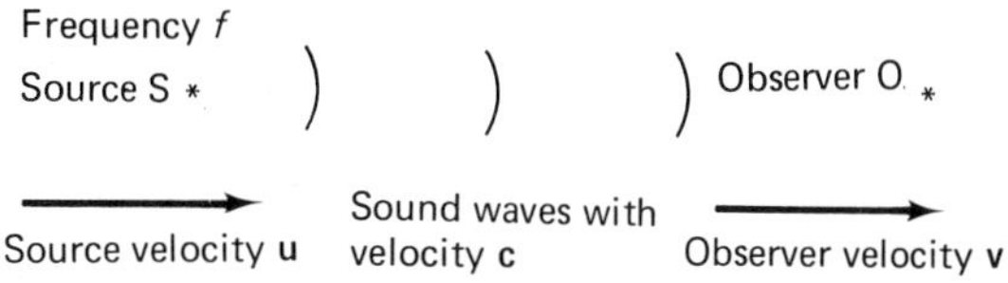

Frequency f

Source S *   )   )   ) Observer 0 *

Source velocity u ————→   Sound waves with velocity c   Observer velocity v ————→

Figure 18-7
The Doppler effect.

direction as the sound, and as a result the wavelength is contracted. The distorted wavelength is:

$$\lambda = \frac{\text{speed of separation of the wave from the source}}{\text{frequency of the source}} = \frac{c - u}{f}$$

The observer moves away from the source; therefore the waves pass him less frequently than they would if he were stationary. The frequency received by the observer is:

$$F = \frac{\text{speed of approach of wave and observer}}{\text{distorted wavelength}} = \frac{c - v}{\lambda}$$

Combining the equations, we obtain:

$$F = f\left(\frac{c - v}{c - u}\right) \qquad (18\text{-}4)$$

Note that the velocities are vector quantities; therefore, if the source moves away from the observer, u is replaced by $-u$, and, if the observer moves towards the source, v is replaced by $-v$. Equation 18-9 should always be related to Fig. 18-7!

*Example 18-5*_____

Two trains A and B head towards each other on the same straight track. The driver of train A sees train B approaching, and in panic he blows the whistle, which has a frequency of 1 000 Hz. If train A is moving at 10 m/s, train B is moving at 25 m/s, and the speed of sound in air is 330 m/s, what frequency does the driver of train B hear?

Solution:

The source heads towards the observer, but the observer also heads towards the source. Therefore Eq. 18-9 becomes:

$$F = f\frac{(c + v)}{(c - u)} = (10^3 \text{ Hz})\frac{(330 \text{ m/s} + 25 \text{ m/s})}{(330 \text{ m/s} - 10 \text{ m/s})}$$
$$= 1\,110 \text{ Hz}$$

The Doppler effect also has applications in detection and navigation devices, such as radar* and sonar. The transmitter sends a wave to the target, where it is reflected; the reflected wave returns to a receiver, which is located at the transmitting station. Directional waves and the elapsed time between transmission and reception of the reflected wave give the location of the target. The change in frequency between the transmitted and the reflected wave is used to determine the velocity of the target.

Problems_____

20. A train is travelling at 80 mi/h when the engineer blows a whistle that has a frequency of 1 500 Hz. If the speed of sound in air is 1 100 ft/s, determine the frequency that a stationary observer on the platform hears (a) before the train enters the station and (b) after the train leaves the station at the same speed.

21. A boat is moving at 8.0 m/s towards a second boat B that is moving in the same direction with a speed of 6.0 m/s. If the foghorns of the two boats have identical frequencies of 250 Hz and the speed of sound in air is 335 m/s, compute (a) the frequency of the foghorn that an observer on A hears coming from B and (b) the

---

*Electromagnetic waves, such as light, radar and microwaves, move at the speed of light, and the effects of relativity must be considered. The relativistic Doppler effect is discussed in Chapter 26.

frequency that an observer on B hears coming from A.

22. A car is moving at 60 mi/h directly towards a large wall. If the driver sounds the horn that has a frequency of 1 200 Hz, what is the frequency of the reflected sound that he hears? The speed of sound in air is 1 100 ft/s.

23. Two ships A and B head directly towards each other on a collision course at the same speed. The captain of ship A blows his foghorn, which has a frequency of 300 Hz, and the captain of ship B hears the sound with a frequency of 308 Hz. Determine the speeds of the two ships if the speed of sound in air is 330 m/s.

18-4 INTENSITY AND ATTENUATION

The time rate of emission of sound energy from a source is measured in watts. Different sources emit sound energy at different rates. For example, most hi-fi's emit energy at a rate of approximately 1 μW.

The *intensity I* of a sound at some point is the time rate of transmission of sound energy P per unit area A_1 that is perpendicular to the direction of energy flow. Thus

$$I = \frac{\text{sound power}}{\text{perpendicular area}} = \frac{P}{A_1} \quad (18\text{-}10)$$

As the sound wave spreads out from the source its intensity decreases, and the wave is said to be *attenuated*.

Consider a small sound source S that emits sound energy uniformly in all directions. If there is no absorption of sound energy and there are no reflecting surfaces, the sound wave has a spherical wave front. Thus, if P is the power of the source (the time rate at which it emits sound energy), the intensity of the sound at a distance r_1 from the source is:

$$I_1 = \frac{P}{4\pi r_1^2} \qquad (18\text{-}11)$$

Similarly, the intensity of the sound wave at a distance r_2 from the source $I_2 = P/4\pi r_2^2$. Thus

$$P = (4\pi r_1^2)I_1 = (4\pi r_2^2)I_2$$
$$\frac{I_1}{I_2} = \frac{r_2^2}{r_1^2} = \left(\frac{r_2}{r_1}\right)^2 \qquad (18\text{-}12)$$

This is called the *intensity ratio*. The intensity ratio is often used in acoustics to express amplification and attenuation of sound.

*Example 18-6*_____

A small source emits sound energy at a rate of 2.0 μW uniformly in all directions. Determine (a) the intensity at 10 m from the source and (b) intensity ratio between points at distances 20 m and 40 m from the source.

Solution:

(a) $\quad I = \dfrac{P}{4\pi r^2} = \dfrac{2.0 \times 10^{-6} \text{ W}}{4\pi(10 \text{ m})^2}$

$\qquad\quad = 1.6 \times 10^{-9} \text{ W/m}^2$

(b) $\quad \dfrac{I_1}{I_2} = \left(\dfrac{r_2}{r_1}\right)^2 = \left(\dfrac{40 \text{ m}}{20 \text{ m}}\right)^2 = \dfrac{4}{1}$

---

*Problems*_____

24. A small sound source emits sound energy at a rate of 8.2×10^{-7} W uniformly in all directions. Determine (a) the intensities of the sound at distances of 3.0 m and 5.0 m from the source and (b) the intensity ratio between these points.

25. A small sound source emits sound energy at a constant rate uniformly in all directions. Determine the intensity ratio between two points if one point is three times farther from the source than the other.

18-5 DECIBEL SCALE

The human ear is sensitive to variations in the intensity I and pressure p of a sound,

but the audio level or loudness is not in direct proportion. Ernst Weber (1795–1878) and Gustav Fechner (1801–1887) determined that the human ear is approximately attuned to logarithmic intensity and pressure changes; therefore we define the *sound intensity level* in *decibels* (dB) (which should not be confused with intensity) of a sound as:

$$L_I = 10 \log_{10}\left(\frac{I}{I_0}\right) \qquad (18\text{-}13)$$

where I is the intensity of the sound and
$$I_0 = 10^{-12} \text{ W/m}^2$$

is a constant. The human ear is able to distinguish between two sounds with intensity levels differing by approximately 1 dB although this varies somewhat with the frequency of the sound.

Since the intensity I of a sound varies as the square of the pressure p, Eq. 18-13 may also be written as a *sound pressure level* in decibels:

$$L_p = 10 \log_{10}\left(\frac{p}{p_0}\right)^2 = 20 \log_{10}\left(\frac{p}{p_0}\right) \quad (18\text{-}14)$$

where p is the sound pressure and $p_0 = 20$ μPa is a reference pressure. Sound level meters are usually calibrated to read sound pressure levels in decibels. Similarly, the *sound power level* of a sound in decibels is defined as:

$$L_p = 10 \log_{10}\left(\frac{P}{P_s}\right) \qquad (18\text{-}15)$$

where P is the given sound power and P_s is a reference power, *which must be stated*, i.e. kW, mW, μW or pW.

*Example 18-7*_____

If the intensity of a sound is
$$I = 1.6 \times 10^{-4} \text{ W/m}^2,$$
what is its intensity level in decibels?

Solution:

$$L_I = 10 \log_{10}\left(\frac{I}{I_0}\right)$$
$$= 10 \log_{10}\left(\frac{1.6 \times 10^{-4} \text{ W/m}^2}{10^{-12} \text{ W/m}^2}\right) = 82 \text{ dB}$$

Attenuations and amplifications are also stated and measured in decibel units. For example, if the intensity level is L_{I1} (dB) when the intensity is I_1 at a distance r_1 from a small source and the intensity level is L_{I2} (dB) when the intensity is I_2 at a distance r_2 from the same source, the attenuation or amplification in decibels is:

$$L_{I2} - L_{I1} = 10 \log_{10}\left(\frac{I_2}{I_0}\right) - 10 \log_{10}\left(\frac{I_1}{I_0}\right)$$

$$= 10 \log_{10}\left(\frac{I_2/I_0}{I_1/I_0}\right)$$

Thus

$$L_{I2} - L_{I1} = 10 \log_{10}\left(\frac{I_2}{I_1}\right) \qquad (18\text{-}16)$$

If the source radiates uniformly in all directions and there are no losses, $I_2/I_1 = (r_1/r_2)^2$; thus

$$L_{I2} - L_{I1} = 10 \log_{10}\left(\frac{r_1}{r_2}\right)^2 = 20 \log_{10}\left(\frac{r_1}{r_2}\right)$$

$$(18\text{-}17)$$

Example 18-8_____

If the sound intensity level is $L_{I1} = 75$ dB at a distance $r_1 = 5.0$ m from a small source radiating uniformly in all directions, determine the intensity level L_{I2} at a distance $r_2 = 25.0$ m from the same source if there are no losses.

Solution:

From Eq. 18-17:

$$L_{I2} = L_{I1} + 20 \log_{10}\left(\frac{r_1}{r_2}\right)$$

$$= 75 \text{ dB} + 20 \log_{10}\left(\frac{5.0 \text{ m}}{25.0 \text{ m}}\right) = 61 \text{ dB}$$

In electronics, the *terminal rating* of communications and audio equipment is often expressed in decibels as:

$$L_p = 10 \log_{10}\left(\frac{P_{out}}{P_s}\right) \qquad (18\text{-}18)$$

where P_{out} is the output power and P_s is a reference power. The reference power is normally taken as 1.0 mW to a 600 Ω load to give a *terminal rating in dBm*.

$$L_p \text{ (dBm)} = 10 \log_{10}\left(\frac{P_{out}}{1.0 \text{ mW}}\right)\bigg|_{600\,\Omega} \qquad (18\text{-}19)$$

Since the power $P = V^2/R$, if the resistance of the load remains constant and the voltage changes from V_1 to V_2, the difference in sound power levels.*

$$L_{p2} - L_{p1} = 20 \log_{10}\left(\frac{V_2}{V_1}\right) \qquad (18\text{-}20)$$

The decibel scale is frequently used to indicate *gain* or *loss* in devices.

$$L_{p2} - L_{p1} = 10 \log_{10}\left(\frac{P_2}{P_1}\right) \qquad (18\text{-}21)$$

Example 18-9_____

Determine the gain in decibels in an audio amplifier that delivers 8.0 W output when the input signal power is 1.0 mW.

Solution:

$$L_{p2} - L_{p1} = 10 \log_{10}\left(\frac{P_2}{P_1}\right)$$

$$= 10 \log_{10}\left(\frac{8.0 \text{ W}}{1.0 \times 10^{-3} \text{ W}}\right) = 39 \text{ dB}$$

If a system amplifies (or attenuates) a sound in several stages, the total amplification (or attenuation) is equal to the *algebraic sum* of the individual amplifications (or attenuations) produced by each stage.

Problems_____

26. The intensity of a sound is 1.2×10^{-6} W/m². Determine the sound intensity level.

27. Determine the intensity of a sound if the intensity level is (a) 70 dB and (b) 84 dB.

28. Determine the terminal rating of an audio amplifier if its output power is 30 W and $P_{standard}$ is taken as 1 mW.

29. Determine the attenuation of a system in decibels if it decreases the intensity of a sound from 12.2×10^{-5} W/m² to 8.4×10^{-6} W/m².

_____

*See Chapter 20.

30. If the sound pressure level is 84 dB at 10 m from a small source, determine the pressure level at 20 m if there is no absorption.

31. Determine the intensity ratio that corresponds to a 5 dB change in intensity level.

32. Determine the gain in decibels in an amplifier that delivers an output power of 30 W when the input signal power is 1.2 mW.

33. If an amplifier has a gain of 42 dB, determine its output power when the input signal power is 1.0 mW.

18-6 REFLECTION, ABSORPTION AND REVERBERATION

When a sound wave impinges on a boundary between two media of different density, some of the sound energy may be reflected while the rest is transmitted or absorbed. From the conservation of energy, the total incident sound energy E_i must equal the sum of the reflected sound energy E_r and the energy absorbed or transmitted E_A.

$$E_i = E_r + E_A$$

Reflection

Rigid smooth surfaces are better reflectors of sound energy than soft irregular surfaces, but sound waves are reflected to some extent from most surfaces. As in the case of other waves, the maximum intensity of a reflected sound from a regular surface occurs at an angle of reflection that is equal to the angle of incidence.

The reflective properties of sound are used to measure distances and to focus and transmit sounds in a preferred direction.

Explorations for oil and minerals are aided by "soundings" of the earth. Explosions at the earth's surface produce pulses of sound, parts of which are reflected from the areas of different density below the surface of the earth, and the reflected sounds return to a series of receivers. By analysis of the elapsed times between the initial sounds and the reception of the reflections, scientists can estimate the earth's structure. The terrain and depths of ocean beds are measured by a similar process.

Sonar devices transmit a sound pulse through a body of water. This pulse is reflected from submarines, ships and the ocean beds, and part of the reflected sound returns to the transmitter. Since the speed of sound in water is known, the range of the reflecting object may be determined by measuring the elapsed time between the transmission and the reception of the reflected signal.

In auditoriums, the reflection of sound may be used to advantage by locating a curved surface of good reflective properties behind the stage or behind the speaker. The curved surface directs the reflected sound to all parts of the audience.

Reverberation and Absorption

Reflection of sound may be quite critical in closed areas, such as auditoriums and rooms. Usually some reflection is required, but if the surfaces continually reflect a large percentage of incident sound energy, the original sound will persist and interfere with any subsequent sounds. This continual reflection of audible sound is called *reverberation*, and the time required for a sound to reduce in intensity level by 60 dB is called the *reverberation time*. In effect, the reverberation time is approximately the time required for a single sound to become inaudible; it usually varies from 0.5 s to 2.0 s in rooms and auditoriums.

The reverberation time of an enclosed space depends on the nature of the reflecting surfaces; different materials have different reflective properties. Each time a sound is incident on a particular surface, some sound energy is absorbed, and some is transmitted. The *absorption coefficient* β of a surface is defined as the fractional amount of the total incident sound energy that is *not* reflected by that surface. Therefore, if E_i is the incident sound energy and E_r is the sound energy reflected, the absorption coefficient of the surface is:

$$\beta = \frac{E_i - E_r}{E_r} \qquad (18\text{-}22)$$

The absorption coefficient is a dimensionless quantity that depends on the nature of the material surface and to some extent on the frequency of the incident sound. Some typical absorption coefficients (for 500 Hz) are listed in Table 18-2. An open window has an absorption coefficient of 1.00 because all incident sound energy is transmitted through the window.

The *absorbing power* a_p of a surface is the product of the surface area A and the absorption coefficient.

$$a_p = A\beta \qquad (18\text{-}23)$$

In an auditorium or a room, the total absorbing power or *Sabine absorption S* is the sum of the absorbing powers of all the individual surfaces.

$$S = \Sigma\, a_p = \sum_i A_i \beta_i \qquad (18\text{-}24)$$

W. Sabine determined experimentally that in the British Engineering System the reverberation time (in seconds) is given by:

$$T = \frac{V}{10S} = \frac{V}{20\Sigma a_p} \qquad (18\text{-}25)$$

where all dimensions are measured in feet. Similarly, in SI, the reverberation time (in seconds) is:

$$T = \frac{V}{6.25S} = \frac{V}{6.25\Sigma a_p} \qquad (18\text{-}26)$$

and all dimensions are measured in meters.

*Example 18-10*_____

Determine the approximate reverberation time of an auditorium of dimensions 10 m $\times$ 30 m $\times$ 50 m if it has the following surfaces: 1 500 m^2 of plaster ceiling, 3 000 m^2 of plywood panels, 500 m^2 of heavy drapes, 200 m^2

Table 18-2
Typical Absorption Coefficients

Brick	0.03	Marble	0.01
Carpet (heavy)	0.38	Open window	1.00
Concrete block	0.30	Plaster	0.02
Concrete floor	0.02	Plywood panels	0.18
Drapes (heavy)	0.50	Linoleum	0.02
Glass	0.04	Wood (floor)	0.09
Gypsum	0.05		

Absorbing Power/(sabines)

Seated person	1.7
Occupied chairs	1.5
Unoccupied chairs	
wood	0.23
upholstered	1.6

of glass, 500 m² of brick, 800 m² of heavy carpet and 300 m² of linoleum. There are also 800 occupied seats and 200 unoccupied upholstered seats.

Solution:

From Table 18-2, the absorbing power of each surface is as follows:

Surface	Area (A)	Absorption Coefficient β	Absorbing Power (βA)
plaster	1 500 m²	0.02	30
plywood panels	3 000 m²	0.18	540
drapes	500 m²	0.5	250
glass	200 m²	0.04	8
brick	500 m²	0.03	15
carpet	800 m²	0.38	304
linoleum	300 m²	0.02	6
800 occupied seats		800 × 1.5	1 200
200 unoccupied seats		200 × 1.6	320
Total Sabine absorption $S =$			2 673

Thus from Eq. 18-26, the reverberation time is:

$$T \simeq \frac{V}{6.25S} = \frac{10 \text{ m} \times 30 \text{ m} \times 50 \text{ m}}{(6.25)(2\,673 \text{ sabines})} = 0.90 \text{ s}$$

Note that the reverberation time depends on the volume of the room and the nature of all reflecting surfaces. It may be reduced by covering the walls, floors and ceiling with absorbing materials, such as drapes, cork, acoustic tiles and carpets. Some auditoriums are even constructed with movable baffles, which control the volume of the space.

Problems_____

34. Determine the absorption coefficient of a surface if it reflects 30% of incident sound energy.

35. Determine the reverberation time of an auditorium of dimensions 30 ft × 80 ft ×

150 ft if its surfaces are composed of the following materials:

80 ft × 150 ft plaster
300 ft × 30 ft plywood panels
80 ft × 20 ft brick
80 ft × 10 ft marble
300 ft × 30 ft heavy carpet
80 ft × 10 ft drapes
80 ft × 20 ft glass

and it contains 800 occupied chairs and 200 unoccupied upholstered chairs.

18-7 THE HUMAN EAR

The human ear is sensitive to the minute pressure changes produced by the mechanical vibrations of a sound wave, and it amplifies these pressure changes internally.

In the *outer ear* region, the external ear channels the mechanical sound vibrations through the *auditory canal* to the *ear drum*. The ear drum transmits the vibration to the *middle ear* region, which consists of three small bones, called the *hammer, anvil* and *stirrup,* and a smaller membrane S. In the middle ear region, the pressure changes are amplified by a factor of approximately 50

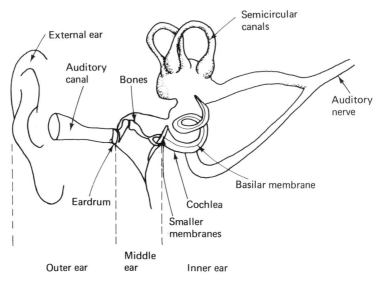

Figure 18-8
The human ear.

due to the lever action of the three bones and the relative sizes of the ear drum and the smaller membrane S. Mechanical vibrations are converted into small electrical impulses in the *cochlea* of the *inner ear*; these impulses are then transmitted through *auditory nerves* to the brain. The spiral-shaped cochlea is filled with a liquid and also contains the *basilar membrane*, which is connected to the ends of the auditory nerves. The smaller membrane S causes the liquid in the cochlea to vibrate; this vibration is channelled through the cochlea to the sensitive nerve endings in the basilar membrane. (See Fig. 18-8.)

The *semicircular canals* of the inner ear region are not used in the auditory process, but they are important organs that enable us to keep our balance.

The response of a human ear to a particular sound varies with the individual. In most cases the human ear is only sensitive to a relatively small frequency range of approximately 20 Hz to 20 kHz. Outside this range, a sound is usually inaudible, regardless of its intensity.

The sensation of *loudness* depends on the intensity and the frequency of a sound. Most human beings are more sensitive to sounds in the frequency range from 2 000 Hz to 4 000 Hz. In this most sensitive frequency range, the human ear responds to a relatively large range of pressure variations approximately from 10^{-5} Pa to 10 Pa. Low intensity sounds in the most sensitive frequency range are usually judged to be as loud as sounds of much higher intensity outside this frequency range (Fig. 18-9). Therefore, while sound pressure level in decibels is a relatively accurate indication of loudness at frequencies of 1.0 kHz, it cannot be used in all frequency ranges.

In order to describe the sensation of loudness in other frequency ranges, the loudness level of a sound is compared with the sound pressure level in decibels of an equally loud sound at a frequency of 1.0 kHz. The unit of loudness level is a *phon*. Thus the loudness level in *phons* is equal to the sound pressure level in decibels of an equally loud sound at 1.0 kHz. For example, a sound with a fre-

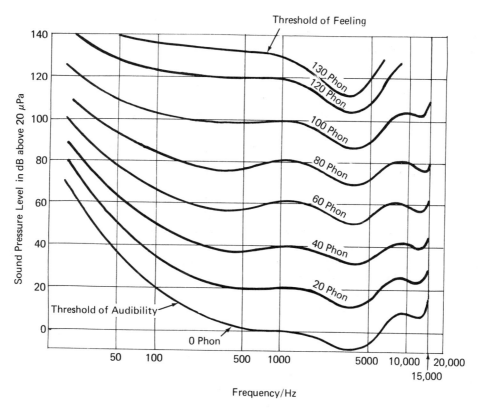

Figure 18-9
The frequency response of a human ear showing loudness in phons.

quency of 100 Hz and a pressure level of 63 dB is as loud as 1.0 kHz sound with a pressure level of 20 dB; therefore the loudness level of the 100 Hz sound is 20 phon. Lines of equal loudness levels for different frequencies are illustrated in Fig. 18-9.

Only sounds with a loudness greater than the *threshold of audibility* (0 phon) are audible. Above the *threshold of feeling* (approximately 125 phon), a sound will start to hurt the ear of a human being. Normally, most people can hear sounds comfortably within the range between the threshold of audibility and the threshold of feeling.

Within the audible range the human ear

is relatively sensitive to frequency variations. The term *pitch* is mainly associated with the frequency of a sound, but it also varies to a lesser degree with the sound intensity level. A sound with a high pitch has a greater frequency than a sound with a low pitch.

A single sound source may vibrate simultaneously in many different ways, producing a very complex wave form that contains the fundamental and many harmonics. The *tone* of a sound is related to the complexity of the total wave form. The human ear can often distinguish between two sounds that produce the same basic note because their tone is different. For example, a trumpet and a piano may both play the same note, but the two sounds are quite different because the waves produced have different complexities.

Problem

36. A sound at 10.0 kHz has a pressure level of 50 dB. What is its loudness level in phons?

18-8 THE REPRODUCTION OF SOUND

Modern technology enables us to reproduce accurately all sounds within the audible range of the human ear; stereo and multichannel systems even reproduce the relative locations of the individual sound sources.

A *transducer* is any device that is able to convert signals from one form of energy into another. In sound recording or reproduction processes, transducers, such as microphones and speakers, utilize piezoelectric crystals or electromagnets (Chapters 21 and 23) to convert mechanical vibrations into electrical impulses and vice versa.

Records

In a recording process, a microphone converts sound vibrations into electrical impulses that are amplified and then used to drive an electromagnetic cutting tool. The vibrating cutting tool makes a continuous spiral groove of variable dimensions in a rotating disk, and a master die is made from the finished disk. Plastic sheets are heated and then pressed to shape in the die to form records that are identical to the original disk.

Sound is reproduced from the record by the reverse process. The record is rotated on a turntable and a needle or stylus is located so that it vibrates in the continuous groove. This vibration is transmitted to a piezoelectric crystal or an electromagnet in the pickup, where it is converted into small electric impulses. These impulses are amplified and used to drive an electromagnetic speaker cone that is attached to a diaphragm; the vibrating diaphragm produces longitudinal sound waves.

Tapes and Film Sound Tracks

Tapes are composed of a long plastic strip that contains small magnetic particles of ferric oxide. During a recording process, the tape is passed through the recording head of the tape recorder. Electric impulses from the microphone via an amplifier produce varying magnetic fields in the gap of the core of the electromagnetic recording head. These fields reorganize the magnetic particles into various patterns in the tape.

Small electric impulses are reproduced when the magnetized tape is passed over the electromagnetic playback head; these impulses are amplified and used to drive a speaker.

The principles of magnetic tape recording are also utilized in film sound tracks although some older films reproduce sounds optically by variations in the intensity of the light that is allowed to pass through the edge of the

REVIEW PROBLEMS

1. A string that is 80 cm long and has a mass of 3.0 g is stretched under a tensile force of 16.0 N. Determine: (a) the speed of a transverse wave in the string, (b) the fundamental frequency, (c) the third harmonic frequency, (d) the frequency of the third overtone.

2. The speed of sound in an air column in a 1.0 m long pipe that is closed at one end is 330 m/s. Find the frequency of (a) the 3rd overtone and (b) the 5th harmonic.

3. A metal bar 10.0 m long that is clamped at its center vibrates longitudinally in such a way that it gives its second over-

tone when it vibrates in unison with a tuning fork of frequency 1 300 Hz. Compute the speed of sound in the metal.

4. The speed of sound in a vibrating metal rod is 5.0 km/s. If the rod has a length of 1.00 m and it is clamped at its center, determine (a) the fundamental frequency and (b) the frequency of the third overtone.

5. A 1.00 m metal rod is clamped at its center and vibrates at its fundamental frequency. A small diaphragm at the end of the rod sends out longitudinal waves that produce a series of nodes in cork dust. If the distance between successive nodes in the cork dust is 8.00 cm and the speed of sound in air is 330 m/s, what is the speed of sound in the metal rod?

6. Two sounds with frequencies of 2 000 Hz and 1 989 Hz are produced simultaneously. What is the frequency of the beats?

7. The foghorns of two ships have identical frequencies of 200 Hz. The speed of sound in air is 332 m/s. Ship B is stationary, and ship A moves along the line joining the two ships; the captain of B hears the frequency of 196 Hz coming from ship A. (a) Calculate the speed of A relative to B. (b) Determine if A moves towards or away from B.

8. A target moves with a velocity v along a line towards or away from a stationary sound source that transmits a wave with a frequency f and wave speed c. If the source receives the reflected signal from the target at a frequency F, derive an expression for the target velocity. *Hint:* Assume that the target is a source when it reflects the waves.

9. A small sound source emits sound energy at 4.0 μW uniformly in all directions. Determine (a) the intensity of sound 8.0

m and 15.0 m from the source and (b) the intensity ratio between these points.

10. If the intensity of a sound is 4.2×10^{-5} W/m^2, what is its pressure level in dB?

11. If a sound has an pressure level of 84 dB, what is its intensity?

12. The sound pressure level of a source is 72 dB at a distance of 10 m from the source. What is the sound pressure level at 25 m from the source?

13. The sound pressure level of a small source is 50 dB at a distance of 100 m. What is the pressure level at 10 m from the same source?

14. The sound pressure level of a source is 145 dB at a distance of 10 m from a source. Determine the pressure level 1 000 m from the source.

15. Determine the amplification of a system in dB if it increases the intensity of a sound from 5.0 μW/m^2 to 150 μW/m^2.

16. Determine the attenuation of a system if it decreases the intensity of a sound from 122 μW/m^2 to 8.4 μW/m^2.

17. Determine the gain in decibels in an audio amplifier if its output power is 3.6 W when the input signal power is 1.0 mW.

18. A classroom is 30 ft wide, 50 ft long and 10 ft high. The floor (1 500 ft^2) is wood (absorption coefficient $\beta = 0.08$). The walls are 1 000 ft^2 smooth plaster ($\beta = 0.03$), 200 ft^2 wood ($\beta = 0.04$), 200 ft^2 cotton draperies ($\beta = 0.5$), 100 ft^2 glass ($\beta = 0.03$) and 100 ft^2 chalkboard ($\beta = 0.06$). The ceiling is 1 500 ft^2 of acousticelotex ($\beta = 0.80$). (a) What is the absorbing power of the room? (b) What is the reverberation time?

19. What is the loudness level in phons of a sound at 10 kHz that has a pressure level of (a) 78 dB? and (b) 40 dB?

19

Electrostatics

All fundamental properties of electricity and magnetism may be traced to the state or motion of an entity called *electric charge*. The exact nature of an electric charge is not known, but it is responsible either directly or indirectly for many phenomena.

Magnetism applies only to the phenomena produced between moving charges as a result of their motion. The term *electricity* is used to denote other characteristic properties of electric charges and the electric currents produced by the flow of these charges.

Modern households and industries require vast quantities of energy, which must often be transported with minimal losses over large distances from the source to the consumer. This energy transmission is usually accomplished most conveniently with electricity. At the source, energy (such as the energy of waterfall) is converted into electrical energy; it is then transmitted to the consumer where it is transformed into the useful energy required (light, heat and mechanical energy).

Many of the rapid developments in technology may be directly or indirectly attributed to advances in the controlled use of electricity.

19-1 ELECTRIC CHARGE AND ATOMIC STRUCTURE

The ancient Greeks (before 600 B.C.) noticed that, after it had been rubbed with fur, amber attracted light objects, such as small pieces of straw and cork.* This property may be induced in other objects if they are rubbed with a suitable material; these objects are then said to be *electrified*, and they possess a net electric charge.

There are two different kinds of electric charge. A glass rod is electrified when it is

---

*The term *electricity* is derived from "elektron," the Greek word for amber.

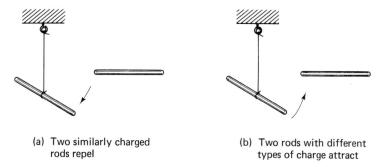

(a) Two similarly charged rods repel

(b) Two rods with different types of charge attract

Figure 19-1

rubbed with silk, and a lucite rod is electrified when it is rubbed with fur, but the rods possess different kinds of electric charges. The existence of two different types of electric charge may be demonstrated by a simple experiment. If the electrified glass rod is suspended freely from a long light thread, it is repelled by a second similarly charged glass rod: As the second rod moves closer, the suspended rod moves away. However, when the electrified lucite rod is brought close to the suspended glass rod, they attract each other, and the glass rod moves towards the lucite rod (Fig. 19-1). Two similarly electrified lucite rods repel each other.

In the eighteenth century, Benjamin Franklin (1706–1790) named the two types of charge. A glass rod is said to possess a positive charge when it has been rubbed with silk; after a hard rubber rod (or a lucite rod) has been rubbed with fur, it has a negative charge. Using this convention, we may determine a net charge on any object simply by finding the algebraic sum of positive and negative charges. If the net charge is zero, the object is said to be *uncharged* or *electrically neutral*. Thus *objects with the same type of net charge repel each other, and objects with unlike net charges attract each other.*

As far as we know, an electric charge is not a form of matter or energy although it is associated with particles, such as protons and electrons, and we may use charges to store

or transfer energy. For example, work must be done to accumulate a net charge such as on the plates of a capacitor, and energy can be transferred from one place to another by means of a flow of electric charge, which is called an *electric current*. In SI electric charge is represented by a derived unit called a *coulomb* (C).

In order to explain many of the basic phenomena of electricity and magnetism, we must first investigate the structure of matter. All matter is composed of one or more fundamental substances called *elements*, which cannot be chemically decomposed. An atom is the smallest complete subdivision of an element.

For many years atoms were thought to be indivisible, but early in the twentieth century scientists determined that individual atoms consist of a small, very dense, positively charged *nucleus* and one or more light, negatively charged particles called *electrons*, which orbit the nucleus at very high speeds (Chapter 31). Individual electrons have a mass of 9.108×10^{-31} kg and a charge of -1.602×10^{-19} C.

The actual structure of an atomic nucleus is still an active area of research. Atoms of different chemical elements have different atomic nuclei, but all atomic nuclei contain one or more positively charged particles called *protons*. The number of protons in the atomic nucleus is called its *atomic number Z*. Different chemical elements have different atomic numbers, but atoms of the same element always have the same atomic number. For example, nitrogen atoms always have seven nuclear protons and an atomic number of seven whereas oxygen has eight protons and an atomic number of eight. A single proton has a mass of 1.673×10^{-27} kg and a charge of $+1.602 \times 10^{-19}$ C. Therefore electrons and protons have charges of equal magnitude but oppostie signs and in an electrically neutral atom the number of nuclear protons is equal to the number of orbital electrons.

Other particles, called *neutrons*, also exist in most atomic nuclei. Individual neutrons have no electric charge, and their mass is 1.675×10^{-27} kg (almost the same as the mass of a proton).

Electrons of the same atom may perform their orbits at different distances from the nucleus in much the same way that the planets have different orbits about the sun. Like the solar system, most of an atom is actually empty space. However, while the orbits of the planets are relatively precise, electron orbits can only be described in terms of statistics because we can only give probabilities that electrons are in a given location.

Electrons with orbits that are farthest from the nucleus are not as tightly bound to the atom as the electrons with closer orbits. Unlike charges attract each other; therefore negatively charged electrons are attracted towards the positively charged protons of all atomic nuclei. However, some nuclei exert greater attractive forces than others. When two materials are rubbed together, electrons with the farthest orbits from the nuclei in one material may be more strongly attracted to atoms in the other material. As a result, some negatively charged electrons may flow from the first material (leaving it with a deficiency of electrons and thus a positive charge) to the second material (giving it an excess of electrons and hence a net negative charge). Thus electric charges are not created or destroyed; the negatively charged electrons merely flow from one material to the other. As the negative charge builds up on one material, a positive charge forms on the other so that the total net charge of the whole system remains the same. This fact is expressed in terms of the *law of conservation of electric charge*, which states:

The algebraic sum of all electric charges in any isolated system is a constant.

For example, when one object has a net charge of $+5.0$ C and another has a charge of -3.0 C, the total net charge is $+2.0$ C. If the objects are isolated and the charge on one changes to $+8.0$ C, the other must become -6.0 C so that the total net charge remains $+2.0$ C.

Electrons and protons possess the smallest known quantity of charge, an *elementary charge* $e = 1.602 \times 10^{-19}$ C. An object acquires a net charge by either gaining or losing charged particles, normally electrons; therefore *a net electric charge always exists in discrete and definite amounts, and it is said to be quantized*. That is, net charges exist in integer (whole number) multiples of an elementary charge e. For example, net charges of $501\ e$ and $-208\ e$ exist but not of $2.5\ e$ or $-5.2\ e$ because these are not integer multiples.

Problems

1. In a dry climate hair becomes unmanageable when it is combed or brushed with a nylon comb or brush. Explain.

2. Mild electric shocks are common when a person walks across a nylon carpet and touches a metal, such as a TV knob. Explain.

3. What is the net charge in a system of the following charges $2.600\ \mu C$, -925 nC and $-0.003\ 700$ mC?

4. How many electrons are required to produce a net charge of $-5.0\ \mu C$?

19-2 CHARGE AND ELECTRIC CURRENT

Solids are composed of large numbers of atoms ($\sim 10^{21}$ atoms/cm³). In some materials, called *conductors*, it is very easy to remove some of the electrons from individual atoms; these electrons are relatively free to move through the material. Other materials are poor conductors because their electrons are tightly bound to the individual atoms; these materials are called *insulators*.

An *electric current* is a flow of electric charge. By international agreement, the direction of *conventional current* is taken as the direction of positive charge flow; if electrons flow one way in a material, the conventional electric current is in the opposite direction.

In SI, electric current in *amperes* (A) is taken as a base unit because it is easier to measure than charge. The ampere is defined in terms of the forces between two parallel electric currents. Consider two, long, parallel conducting wires that are exactly 1 m apart in a vacuum and carry identical electric currents. The magnitude of the electric current in each wire is defined as one *ampere* (*A*) when the attractive force* per unit length of wire is exactly 2×10^{-7} N/m.

The SI unit of electric charge is a derived unit called a *coulomb*.† A *coulomb* (C) is defined as the quantity of charge that flows past a fixed point in a conductor in each second if there is a constant current of one ampere in that conductor. Thus

$$Q = It \qquad (19\text{-}1)$$

where Q is the charge in coulombs, I is the current in amperes and t is the elapsed time in seconds.

---

*The origin of this force is discussed in Chapter 23.

†A coulomb is actually a very large charge. Normally we would find micro, nano or even picocoulombs.

Example 19-1

How many electrons flow past a given point in a copper wire in 1.0 min if that wire carries a steady current of 1.5 mA?

Solution:

$$Q = It = (1.5 \times 10^{-3}\ \text{A})(60\ \text{s}) = 9.0 \times 10^{-2}\text{C}$$

Therefore, the number of electrons is equal to:

$$\frac{9.0 \times 10^{-2}\ \text{C}}{e} = \frac{9.0 \times 10^{-2}\ \text{C}}{1.6 \times 10^{-19}\ \text{C}} = 5.6 \times 10^{17}$$

Problems

5. If the current in a wire is 1.8 A, how many electrons flow past a fixed point in 3.0 s?

6. Determine the current in a wire if 4.0×10^{-6} C of charge flows past a fixed point in 0.15 s.

19-3 CHARGING PROCESSES AND GROUND

A leaf electroscope is a simple device that is used to compare and study electric charges. It consists of a metal conducting rod with a plate or a knob at the top and a pair of metal foil leaves (aluminum or gold) that hang freely at the other end. The rod is suspended vertically through an insulating ring at the top of a container that has glass sides (Fig. 19-2).

If the electroscope is uncharged and there are no net charges nearby, the metal foil leaves hang limply; however, if a charged object is brought close to the plate, the leaves diverge.

When a positive charge is brought close to an uncharged electroscope, electrons are attracted to the unlike positive charge, and they flow from the leaves and conducting rod onto the metal plate. A negative charge is then said to have been *induced* on to the metal plate. However this results in a net

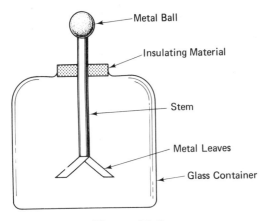

Figure 19-2
Leaf electroscope.

positive charge on each of the metal leaves because they have a deficiency of electrons. Since like charges repel, the leaves diverge. When the positive charge is removed, electrons on the metal plate repel each other, and the excess electrons redistribute themselves, neutralizing the charge on the leaves, and the leaves collapse.

Similarly when a negative charge is brought close to an uncharged electroscope, electrons are repelled from the metal plate and flow to the leaves, causing them to diverge. The metal plate of the electroscope now contains an induced net positive charge. When the negative charge is removed, the electrons are redistributed, and the leaves collapse.

An object may be charged by two different methods, conduction and induction.

Charge by Conduction. Objects may be charged merely by touching them with some other charged object. This process, called *charging by conduction*, results in the same polarity (or sign) of net charge on each object. For example, to give a conductor a negative charge by a conduction process, we could simply touch it with a negatively charged rod. As the rod approaches the conductor, it repels

some electrons to the farthest surface, leaving the closer surface with a net positive charge, which actually attracts the rod. When the rod touches the conductor, some electrons flow from the rod, and they are distributed over the surface of the conductor. If the rod is now removed, the conductor is left with an excess of electrons and hence a negative charge. During this charging process, the rod actually loses some electrons, and the magnitude of its net charge decreases (Fig. 19-3).

The electroscope may also be positively charged by conduction if it is touched with a positively charged object. In this case electrons flow from the electroscope to the object, leaving a positive charge on the electroscope.

The magnitude of the divergence of the metal leaves is related to the net charge on the object that approaches or makes contact with the metal plate of the electroscope. The larger the net charge, the greater the divergence of the leaves.

Charge by Induction. The earth is a relatively massive object, and it contains an extremely large number of atoms. When electrons flow between an object and the earth, the change in the charge of the object may be quite noticeable, but, because of its size, the change in the earth's net charge is insignificant. Consequently the earth may be considered as a source or drain of electrons.

An *electrical ground* utilizes the earth as a reservoir of charge and it is represented schematically by the symbols:

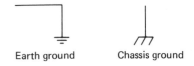

Earth ground   Chassis ground

When a charged object is connected to the earth by a conductor, electrons flow between the earth and that object in a direction which tends to neutralize the charge.

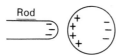

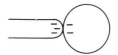

(a) As rod approaches, the electrons in the object are repelled away leaving the closer surface with a net positive charge

(b) When the rod touches, electrons are transferred to the object

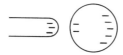

(c) Object has a net negative charge when the rod is removed

Figure 19-3

We may also use our bodies to ground a charge (as long as it is not very large) merely by touching it.

By utilizing an electrical ground, it is possible to charge objects by a process called *induction*. When a negatively charged object is brought close to an uncharged conductor, it repels some electrons to the farthest surface of that conductor, leaving the front surface with an induced net positive charge (Fig. 19-4). If the rear surface is then grounded, some of the excess electrons flow from the conductor to the earth. When the ground is removed, the conductor is left with a net

positive charge. The negatively charged object can then be removed so that the remaining electrons redistribute over the surface of the conductor, but it is still left with a net positive charge.

An object may be given a net negative charge by a similar process. Note that in this case the net charge on the rod remains unchanged during the charging process.

Problems

7. Discuss the formation of a positive charge by conduction on a conductor.

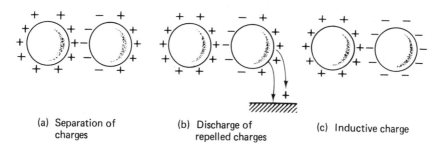

(a) Separation of charges

(b) Discharge of repelled charges

(c) Inductive charge

Figure 19-4
Charging by induction.

8. Explain how you would give a conductor a positive charge by induction.

19-4 COULOMB'S LAW

Electric forces always act along a line joining the centers of two charges. According to Newton's third law, in an isolated system the force experienced by one charge is equal and opposite to the force experienced by the other. The magnitude of the electric forces depends on the nature of the medium (it is greatest if the particles are in a vacuum), the magnitude of the two charges and their distance apart.

Charles Augustin de Coulomb (1736–1806) used a torsion balance to investigate the magnitudes of the electric forces between pairs of charged objects. A simple torsion balance is illustrated in Fig. 19-5. The forces between the charged pith balls are determined from the torsion of the quartz fiber.

Coulomb determined that the magnitude of an electric force F between two charges Q_1 and Q_2 is directly proportional to the product of the charges and inversely proportional to the square of their distance d apart.

$$F \propto \frac{Q_1 Q_2}{d^2} \quad \text{or} \quad F = \frac{k Q_1 Q_2}{d^2} \quad (19\text{-}2)$$

This statement is called *Coulomb's law*; it is valid only if the dimensions of the charged objects are much less than the distance d between their centers. The proportionality constant k is called the *Coulomb constant*; its value depends on the nature of the medium surrounding the particles. In a vacuum* (or approximately in air), $k = 9.00 \times 10^9$ N·m²/C².

If two charges have opposite polarity, they attract each other, but in Coulomb's law

---

*Free space.

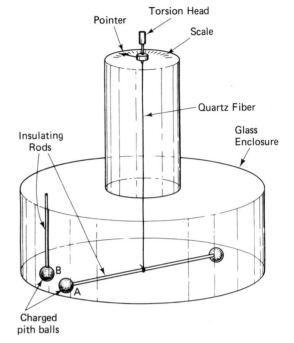

Figure 19-5
Torsion balance.

this gives rise to a negative value for the electric force. Similarly like charges repel each other, and the electric force between them is positive.

The Coulomb constant is often expressed as $k = 1/4\pi\epsilon$, where ϵ is called the *permittivity* of the surrounding medium; its value varies for different media, and its significance is discussed in Chapter 22. In a vacuum (or air), the permittivity $\epsilon_0 = 8.85 \times 10^{-12}$ C²/N·m², and Coulomb's law becomes:

$$F = \frac{k Q_1 Q_2}{d^2} = \left(\frac{1}{4\pi\epsilon_0} \right) \frac{Q_1 Q_2}{d^2} \quad (19\text{-}3)$$

Example 19-2

An electron and a proton in a hydrogen atom are 5.30×10^{-11} m apart. Determine (a) the electric force between them, (b) the force of gravity between them, (c) the ratio of these forces.

Solution:

(a)
$$F_{elec} = \frac{kQ_1Q_2}{d^2} = \frac{(9.00 \times 10^9 \text{ N} \cdot \text{m}^2/\text{C}^2)(-1.60 \times 10^{-19} \text{ C})(1.60 \times 10^{-19} \text{ C})}{(5.30 \times 10^{-11} \text{ m})^2}$$
$$= 8.20 \times 10^{-8} \text{ N}$$

(b)
$$F_{grav} = \frac{Gm_1m_2}{d^2} = \frac{(6.67 \times 10^{-11} \text{ N} \cdot \text{m}^2/\text{kg}^2)(9.11 \times 10^{-31} \text{ kg})(1.67 \times 10^{-27} \text{ kg})}{(5.30 \times 10^{-11} \text{ m})^2}$$
$$= 3.61 \times 10^{-47} \text{ N}$$

(c)
$$\frac{F_{elec}}{F_{grav}} = \frac{8.20 \times 10^{-8} \text{ N}}{3.61 \times 10^{-47} \text{ N}} = 2.27 \times 10^{39}$$

The electric force is considerably greater!

Distribution of Charge

Charge cannot flow very easily in insulators; therefore insulated areas of net charge tend to remain isolated. However, any net electric charge is always located on the outer surface of a conductor. This can be seen in terms of the electric forces between like charges (Fig. 19-6). The vector sum of the repulsive forces is towards the outer surface of the conductor.

By investigating conductors which have different shapes, we find that excess charges tend to accumulate in regions where the surface curvature is the greatest. Any net charge tends to concentrate on the surface in regions where the conductor comes to a point (Fig.

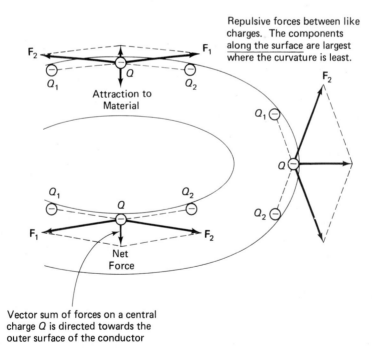

Repulsive forces between like charges. The components along the surface are largest where the curvature is least.

Vector sum of forces on a central charge Q is directed towards the outer surface of the conductor

Figure 19-6
Forces on charges in a conductor.

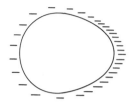

Figure 19-7

Distribution of net charge over the surface of a conductor. Charge tends to accumulate more in regions of greatest curvature.

19-7). This can be seen in terms of the electric forces between charges. The components of the repulsive forces tangent to the surface are smallest (and charges can move closest together) where the curvature is the sharpest. Note that the components of the forces *away* from the surface are strongest where the curvature is the sharpest. This force between the like charges may result in a leakage of charge from the sharply curved surface of a conductor to the surroundings. This phenomenon is utilized in some electrical devices in order to transfer a charge from one conductor to another.

Problems

9. Determine the electric force between two very small particles in free space if they have charges of 2.7×10^{-8} C and -1.6×10^{-5} C and they are 45 cm apart.

10. Determine the change in electric force between two charged particles if the distance between their centers is tripled.

11. What is the change in the electric force between two charges if each charge is doubled and the distance between the centers of charge is reduced by one third?

12. Figure 19-8 illustrates a system of three very small charged particles in a vacuum. Determine the net electric force on the 2.0 μC charge.

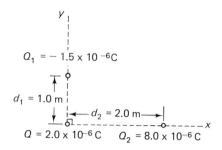

The system of three charges

Figure 19-8

13. Two identical pith balls with masses of 0.80 g are suspended from the same point by light threads 40 cm long. When equal charges are placed on the pith balls, they repel each other so that their strings make an angle of 16° at the point of support (Fig. 19-9). Determine the charge on the pith balls. *Hint:* Draw the free-body diagrams.

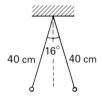

Figure 19-9

The suspended pith balls.

14. Determine (a) the electric force, (b) the gravitational force, (c) the ratio of these forces between two protons that are 12 cm apart in a vacuum.

19-5 ELECTRIC FIELDS

Two or more charges interact even if they do not make contact. Objects that possess net charges are affected by the presence of any other charge because electric forces act over distances.

An *electric field* is said to exist in any region where a charge experiences an electric force. Since electric forces only occur between charges, electric fields are directly associated with charge. Electric fields are produced by charges, and charges in turn are affected by the presence of an electric field.

The electric field at any location is described by a vector quantity called the *electric field strength* (or *intensity*) **E**, which is defined as the electric force per unit positive test charge at that location. However, because the presence of a positive test charge would produce a distortion of the electric field that is being measured, the magnitude of the test charge is assumed to be extremely small. Thus at any location where some small positive test charge $+Q_0$ experiences an electric force **F**, the electric field strength is:

$$E = \frac{F}{+Q_0} \qquad (19\text{-}4)$$

In SI, electric field strength is expressed in units of newtons per coulomb (N/C).

*Example 19-3*_____

Determine the electric force on the electrons in an oscilloscope if they are accelerated in a uniform electric field strength of magnitude 5.0×10^3 N/C.

Solution:

Since an electron has a charge $Q = -1.6 \times 10^{-19}$ C, from Eq. 19-4:

$F = QE = (-1.6 \times 10^{-19} \text{ C})(5.0 \times 10^3 \text{ N/C})$
$\quad = -8.0 \times 10^{-16} \text{ N}$

The minus sign implies that the force is antiparallel to **E**.

_____

The electric field strength at some distance d from an *isolated point charge Q* in a vacuum may be determined by considering its effects on some positive test charge Q_0 at that location. According to Coulomb's law, the test charge Q_0 experiences an electric force of magnitude:

$$F = \frac{kQQ_0}{d^2} = \left(\frac{1}{4\pi\epsilon_0}\right)\frac{QQ_0}{d^2}$$

Thus from Eq. 19-4 the electric field strength at that location has a magnitude:

$$E = \frac{F}{Q_0} = \frac{kQ}{d^2} = \left(\frac{1}{4\pi\epsilon_0}\right)\frac{Q}{d^2} \qquad (19\text{-}5)$$

Since like charges repel and unlike charges attract, the direction is *away* from the positive point charges or *towards* negative point charges.

In systems of two or more point charges, the total electric field strength is the *vector sum* of the field strengths due to the individual point charges Q_i.

$$\mathbf{E}_{TOT} = \mathbf{E}_1 + \mathbf{E}_2 + \cdots$$
$$= \frac{1}{4\pi\epsilon_0} \sum_i \frac{Q_i}{d_i^2} \text{(vector sum)} \quad (19\text{-}6)$$

It should be noted that Eqs. 19-5 and 19-6 apply only to point charges. Other equations must be used if the net charge is distributed over some area.

Lines of Force_____

It is often convenient to represent electric fields around areas of charge in terms of *lines of force*, which originate at areas of net positive charge and terminate at areas of net negative charge. Lines of force have the following characteristics:

1. *The direction of the electric field strength* **E** *is tangential to the line of force at any point, and a small positive test charge always tends to move tangentially to a single line of force.* The lines of force about isolated charges and a system of two charges are illustrated in Fig. 19-10. These lines of force are plotted by considering the direc-

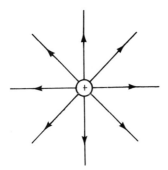

(a) An isolated positive charge

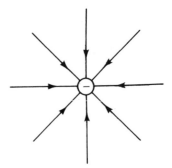

(b) An isolated negative charge

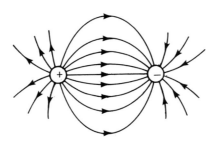

(c) Two unlike charges

Figure 19-10
Electric lines of force.

tion in which a small positive test charge would tend to move if it were placed at the various points. For example, an isolated positive charge would repel a positive test charge; thus the lines of force radiate away from isolated positive charges.

2. *Lines of force never intersect.* If they did, it would imply that the net electric field strength could have two different directions at the same point in space.

3. *An electric field is produced at any area of net charge.* Lines of force originate from areas of net positive charge and are continuous until they terminate at an area that has a net negative charge of equal magnitude.

4. The *spacing* between the various lines of force *is an indication of the magnitude of the electric field strength.* The electric field strength is large in regions where there is a high concentration of the lines of force.

Problems

15. Determine the electric force on an electron if it is located at a point where the electric field strength has a magnitude of 3.0×10^4 N/C.

16. Electrons are accelerated in a uniform electric field strength of magnitude 2.0×10^4 N/C for a distance of 2.0 cm. If the electrons were initially at rest, determine (a) their final speed and (b) the time that it takes for them to travel the 2.0 cm.

17. Determine the electric field strength in air at a point 5.0 cm from an isolated positive charge of 3.0 μC.

18. At what distance from an isolated 4.0×10^{-8} C charge is the electric field strength equal to 2.0×10^5 N/C?

19. Determine the change in the magnitude of the electric field strength if the distance from a small charge is tripled.

20. (a) Determine the total electric field strength at point A due to the system of charges in Fig. 19-11 if they are in a vacuum.
    (b) What electric force would an electron experience if it were placed at point A?

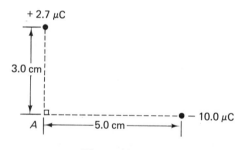

Figure 19-11

21. Sketch the electric lines of force about a very large flat circular plate that has a net negative charge.

22. Sketch the electric lines of force about an infinitely long straight wire that has a net positive charge. *Hint:* Consider a point that is symmetrically located with respect to two equal wire segments, and determine the electric field strength at that point due to those segments.

19-6 ELECTRIC POTENTIAL AND POTENTIAL DIFFERENCE

The vector quantity called electric field strength **E** completely describes an electric field, and it may be used to determine the electric force **F** on a charge Q_0 at any location ($\mathbf{F} = Q_0\mathbf{E}$). However, in many cases, it is more convenient to describe the effects of an electric field in terms of a scalar quantity called *electric potential*, which is directly related to electric potential energy. The concepts of electric potential energy are analogous to gravitational potential energy. Both electric and gravitational fields are *conservative fields* in which any work done is completely independent of the path taken.

Consider the motion of a very small positive test charge Q_0 from A to B to C through an electric field due to some other negative point charge $-Q$ (Fig. 19-12). If

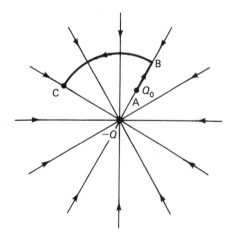

Figure 19-12
Motion of a test charge Q_0 in a field due to a charge $-Q$.

they are not distorted by the small test charge Q_0, the lines of electric force are straight lines that are directed towards the charge $-Q$.

The magnitude of the coulomb attractive force between the two unlike charges is inversely proportional to the square of the distance between their centers. When the test charge Q_0 is moved from point A to point B along a line of force, work must be done by some external agent in order to overcome the coulomb attraction. This work may be retrieved if the charge Q_0 is allowed to return freely to A by virtue of the coulomb attraction. Consequently the potential energy of the test charge Q_0 is greater at point B than at point A. Work done by some external agent in moving a positive charge against the field is usually regarded as positive work, and work done by an electric field (because of the coulomb forces) is taken as negative. Positive work produces an increase in electric potential energy.

As in the case of gravitational potential energy, usually only changes in electric potential energy with respect to some reference are considered. Note that the coulomb force and therefore the electric potential energy depend on the magnitude and polarity (sign) of the test charge Q_0. However, we wish to describe the electric field in a way that is independent of the test charge. Therefore, we define the *electric potential difference* U between two points A and B as the work $W_{A \to B}$ that must be done by some external agent per unit positive test charge Q_0 in order to move that charge from A to B.

$$U = \frac{W_{A \to B}}{Q_0} \qquad (19\text{-}7)$$

In SI, potential differences are expressed in terms of a special derived unit called a *volt* (V): $1\text{ V} = 1\text{ J/C}$, and a joule of work is required to move a charge of 1 C through a potential difference of 1 V. Note that potential difference is a *scalar* quantity that is related to a change in energy (or work done).

In electric circuits and in many problems, the electric ground (earth) is usually chosen as the zero reference electric potential, and other electric potentials are determined relative to the ground.

Example 19-4

An electron in an oscilloscope accelerates from rest through a potential difference $U = 250$ V. Determine (a) the work W done by the field and (b) the final speed v_1 of the electron.

Solution:

(a)
$$U = \frac{W}{Q}$$

Thus
$$W = QU = (1.6 \times 10^{-19}\text{ C})(250\text{ V})$$
$$= 4.0 \times 10^{-17}\text{ J}$$

(b) Since the initial kinetic energy was zero, the work done is equal to the final kinetic energy of the electron.

$$W = QU = \tfrac{1}{2}mv_1^2$$

or
$$v_1 = \sqrt{\frac{2QU}{m}} = \sqrt{\frac{2(1.6 \times 10^{-19}\text{ C})(250\text{ V})}{9.1 \times 10^{-31}\text{ kg}}}$$
$$= 9.4 \times 10^6\text{ m/s*}$$

If the test charge Q_0 is moved in a circular arc from B to C about the point charge $-Q$ (Fig. 19-12), the coulomb force is always perpendicular to the path taken; therefore no work is done against the electric forces. Consequently, the electric potential energy of Q_0 is the same at all points along the arc from B to C, and there is no potential difference between any of these points. Similarly, if we consider three dimensions, there is no potential difference between any points at the same distance from an isolated point change; these points are said to form an *equipotential surface*.

The same amount of work is required to move the test charge Q_0 from A to C *by any path* because electric fields are conservative.

$$^* \sqrt{\frac{C \cdot V}{kg}} = \sqrt{\frac{J}{kg}} = \sqrt{\frac{kg \cdot m^2/s^2}{kg}} = m/s$$

Electric forces vary inversely as the square of the distance between two charges; therefore there are virtually no coulomb forces between two charges that are very far apart. As a result, no work would have to be done against the corresponding electric forces in order to move the test charge Q_0 if it is an infinite distance from the charge $-Q$. Thus at an infinite distance from $-Q$, the potential energy of Q_0 would be zero, and the electric potential is said to be zero. The *electric potential* V_B at any other point B is defined as the work $W_{\infty \to B}$ that must be done by an external agent per unit positive test charge $+Q_0$ in order to move it from an infinite distance to point B.

$$V_B = \frac{W_{\infty \to B}}{+Q_0} \qquad (19\text{-}8)$$

Similarly, the electric potential of any charged surface is defined as the work done per unit positive test charge as the test charge is brought from infinity to that surface.

Since unlike charges attract each other, negative work is required to move a unit positive test charge from infinity to a surface that has a net negative charge. Therefore *lines of force are directed towards net negative charges, and these charges have negative electric potentials.* Similarly, since positive work must be done by some external agent in order to move a unit positive test charge from infinity to some surface that has a net positive charge, *objects with net positive charges have positive electric potentials, and lines of force are directed away from them.*

Using calculus, we can see that the elec-

tric potential at a distance r from an isolated point charge Q is given by:

$$V = \frac{1}{4\pi\epsilon} \frac{Q}{r} \qquad (19\text{-}9)$$

where ϵ is the permittivity of the surrounding medium. Similarly, the electric potential of a spherical conductor of radius r that has a charge Q uniformly distributed over its surface is given by:

$$V = \frac{1}{4\pi\epsilon} \frac{Q}{r} \qquad (19\text{-}10)$$

Electric potential is a scalar quantity that is related to potential energy; therefore at a single point the total electric potential due to a system of charges is equal to the *algebraic sum* of the potentials due to each charge in the system.

$$V_{\text{TOT}} = \frac{1}{4\pi\epsilon} \left\{ \frac{Q_1}{r} + \frac{Q_2}{r_2} + \frac{Q_3}{r_3} + \cdots \right\} \qquad (19\text{-}11)$$

$$= \frac{1}{4\pi\epsilon} \sum_i \frac{Q_i}{r_i}$$

where r_i is the distance of the charge Q_i from the point. Finally, the potential *difference* between two points A and B where the potentials are V_A and V_B respectively is given by:

$$U = V_B - V_A \qquad (19\text{-}12)$$

Example 19-5

Point charges $Q_1 = +2.0 \times 10^{-8}$ C and $Q_2 = -1.5 \times 10^{-8}$ C are 50.0 cm apart in a vacuum (Fig. 19-13). Determine the potential difference between a point A, which is midway between the charges, and a point B, which is 20.0 cm from Q_1 and 30.0 cm from Q_2.

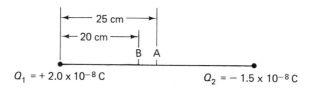

Figure 19-13

Solution:

$$V_A = k\left(\frac{Q_1}{r_{1A}} + \frac{Q_2}{r_{2A}}\right) = (9.0 \times 10^9 \text{ N·m}^2/\text{C}^2)\left(\frac{2.0 \times 10^{-8} \text{ C}}{0.25 \text{ m}} + \frac{-1.5 \times 10^{-8} \text{ C}}{0.25 \text{ m}}\right) = 180 \text{ V}$$

$$V_B = k\left(\frac{Q_1}{r_{1B}} + \frac{Q_2}{r_{2B}}\right) = (9.0 \times 10^9 \text{ N·m}^2/\text{C}^2)\left(\frac{2.0 \times 10^{-8} \text{ C}}{0.20 \text{ m}} + \frac{-1.5 \times 10^{-8} \text{ C}}{0.30 \text{ m}}\right) = 450 \text{ V}$$

Thus,

$$U = V_B - V_A = 450 \text{ V} - 180 \text{ V} = 270 \text{ V}$$

Problems

23. Determine the work that must be done in order to move an electron between two points that differ in electric potential by 300 V.

24. An electron is accelerated through a potential difference of 30 kV in an X-ray tube. Determine its gain in kinetic energy.

25. Determine the potential difference between two points that are 20 cm and 30 cm from an isolated 3.0×10^{-8} C point charge in a vacuum.

26. Calculate the electric potential at a point that is 15 cm from an isolated -3.0×10^{-7} C point charge in a vacuum.

27. Point charges $Q_1 = +5.0 \times 10^{-7}$ C and $Q_2 = -1.2 \times 10^{-6}$ C are 30 cm apart in a vacuum. (a) Determine the potential difference between a point A, which is midway between the charges, and a point B, which is 10 cm from Q_1 and 20 cm from Q_2. (b) At what point(s) is the electric potential equal to zero along the extended line through Q_1 and Q_2?

28. Four equal charges of $+2.0 \mu$C are located at the corners of a square that has sides of length 2.0 m. Determine the electric potential (a) at the center of the square and (b) at the center of one of the sides of the square. (c) What is the potential difference between these points?

29. Charges of $Q_1 = +1.0 \mu$C, $Q_2 = -2.0 \mu$C and $Q_3 = +3.0 \mu$C are located in order 2.0 m apart in a straight line. (a) Determine the potential difference between point A, midway between Q_1 and Q_2, and point B, midway between Q_2 and Q_3. (b) At what point(s) is the electric potential equal to zero along the same extended straight line?

30. Two point charges of $+3.0 \times 10^{-8}$ C and -5.0×10^{-8} C are 80 cm apart in a vacuum. Determine (a) the electric field strength and (b) the electric potential at a point midway between the charges.

31. A charge of $Q_1 = 1.5 \times 10^{-8}$ C is located 50 cm from another charge $Q_2 = -2.5 \times 10^{-8}$ C in a vacuum. How much work is required to move Q_1 so that it is 80 cm from Q_2?

19-7 POTENTIAL GRADIENT AND ELECTRIC FIELD

It is usually more convenient to determine electric potentials rather than electric field strengths. These electric potentials at various positions in an electric field are often represented graphically in terms of a series of surfaces (lines in two dimensions) called *equipotential surfaces*. All points on a single equipotential surface have identical electric potentials. Therefore the potential difference between any two points on the same equipotential surface is equal to zero, and no work is required to overcome electric forces when a charge is displaced over an equipotential surface. This implies that the *electric field strength* **E** *must be perpendicular to the equipotential surface at all points* (Fig. 19-14). If

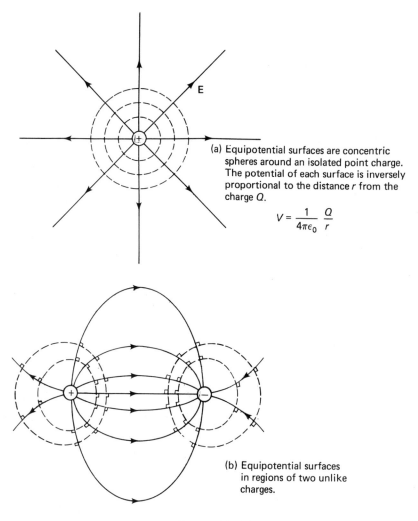

(a) Equipotential surfaces are concentric spheres around an isolated point charge. The potential of each surface is inversely proportional to the distance r from the charge Q.

$$V = \frac{1}{4\pi\epsilon_0} \frac{Q}{r}$$

(b) Equipotential surfaces in regions of two unlike charges.

Figure 19-14
Equipotential surfaces (dotted lines).

this were not true, work would have to be done in order to overcome any component of the electric field parallel to the surface, and the potential difference between points on that surface could not be equal to zero.

The surface of any conductor is an equipotential surface. If a potential difference did exist, charges would flow over the surface.

Consider the motion of a small positive test charge Q_0 in a uniform electric field

where the field strength is E. If there are no external constraints, e.g. magnetic forces, on the charge, its displacement is parallel to E, and the work done by the field (negative valued) when the displacement of the charge is s is given by:

$$W = -Fs \cos 0 = -Q_0 Es \quad \text{since} \quad E = \frac{F}{Q_0}$$

But the potential difference between the origin

and the terminal point of the displacement is $U = W/Q_0$; thus

$$W = Q_0U = -Q_0Es$$

or

$$E = \frac{-U}{s} \qquad (19\text{-}13)$$

Thus the electric field strength has a magnitude of U/s, and it points in the direction of decreasing electric potential. The term U/s is called the *potential gradient* of the electric field; it corresponds to the rate of change of electric potential with distance along the line of force. In SI the potential gradient is expressed in units of

$$\frac{V}{m} = \frac{J/C}{m} = \frac{Nm/C}{m} = N/C$$

Example 19-6

Determine the electric field strength in an insulator with a thickness $s = 0.80$ mm if it is used to isolate areas of charge producing a potential difference $U = 320$ V.

Solution:

$$E = \frac{U}{s} = \frac{320 \text{ V}}{8.0 \times 10^{-4}\text{m}} = 4.0 \times 10^5 \text{ V/m}$$

Example 19-7

An electron is accelerated from a grounded cathode to a positive potential anode. If the potential difference between the anode and cathode is 240 V and they are 2.0 cm apart, determine (a) the average electric field strength and (b) the average electric force on the electron in the region between the anode and cathode. (c) If the electron was initially at rest, with what speed v_1 does it strike the anode? (d) Compare the electric and gravitational forces on the electron.

Solution:

(a) $E = \dfrac{U}{s} = \dfrac{240 \text{ V}}{2.0 \times 10^{-2} \text{ m}} = 1.2 \times 10^4 \text{ N/C}$

That is, 1.2×10^4 N/C directed from the anode to the cathode.

(b) The charge of an electron $= -1.6 \times 10^{-19}$ C; thus from cathode to anode:

$$\begin{aligned}F &= Q_0E \\ &= (-1.6 \times 10^{-19} \text{ C})(-1.2 \times 10^4 \text{ N/C}) \\ &= 1.9 \times 10^{-15} \text{ N}\end{aligned}$$

(c) The work done is equal to the gain in kinetic energy $W = QU = \frac{1}{2}mv^2$. Thus

$$v = \sqrt{\frac{2QU}{m}} = \sqrt{\frac{2(1.6 \times 10^{-19} \text{ C})(240 \text{ V})}{9.1 \times 10^{-31} \text{ kg}}}$$

$$= 9.2 \times 10^6 \text{ m/s}$$

(d) The weight of an electron is:

$$\begin{aligned}w = mg &= (9.1 \times 10^{-31} \text{ kg})(9.8 \text{ m/s}^2) \\ &= 8.9 \times 10^{-30} \text{ N}\end{aligned}$$

Consequently the electric force is much larger.

Problems

32. Sketch the equipotential surfaces about two like charges that are separated by some distance.

33. In what distance will an electron increase its kinetic energy by 5.0×10^{-17} J as it moves through a uniform electric field of 3 000 N/C?

34. Determine the electric field strength between the anode and cathode of a vacuum tube if they have a potential difference of 320 V and they are 3.6 cm apart.

35. An electron is accelerated between two electrodes that are 1.5 cm apart and have a potential difference of 120 V. Determine (a) the average electric force on the electron and (b) the acceleration of the electron. (c) If the electron is initially at rest, how long would it take to move through 1.0 cm and what maximum speed would it reach in the 1.0 cm?

REVIEW PROBLEMS

1. How many electrons flow past a fixed point in a copper wire in 15 min if the wire carries a steady current of 250 mA?

2. Determine the electric force between two very small particles in a vacuum if they have charges of 8.2×10^{-8} C and -1.6×10^{-5} C and they are 35 cm apart.

3. Determine the electric force on a proton at a point where the electric field strength has a magnitude of 2 500 N/C.

4. (a) Determine the magnitude of the force on an electron when it is accelerated from rest through 4.5 cm in a uniform electric field strength of 3.5×10^3 N/C. (b) What is the final speed of the electron? (c) How long does it take the electron to travel the 4.5 cm?

5. (a) Determine the total electric field strength at A in Fig. 19-15. (b) What would be the electric force on an electron if it is placed at A?

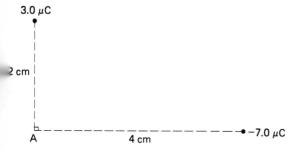

3.0 μC

2 cm

A

4 cm

−7.0 μC

Figure 19-15

6. Charges $Q_1 = 8.0\ \mu C$, $Q_2 = 15.0\ \mu C$ and $Q_3 = 4.0\ \mu C$ are located in a straight line 4.0 m apart. Determine the potential difference between point A, midway between Q_1 and Q_2, and point B, midway between Q_2 and Q_3.

7. Charges $Q_1 = 4.0\ \mu C$, $Q_2 = -6.0\ \mu C$ and $Q_3 = -5.0\ \mu C$ are located in order 50 cm apart in a straight line. Point A is midway between Q_1 and Q_2, and point B is 20 cm from Q_2 and 30 cm from Q_3. (a) Deter-

mine the potential difference between A and B due to the charges. (b) What is the total electric field strength at point A? (c) What electric force would an electron experience if it were placed at point A?

8. How much work is required to move a 15 μC charge from a point where the potential is −12 V to a point where the potential is 20 V?

9. (a) Find the kinetic energy of an electron that accelerates through a 12 V potential difference. (b) What is the final speed of the electron if it started from rest?

10. How much work is required to move a 7.0 μC charge from a point where the potential is −5.0 V to a point where the potential of +150 V?

11. Find the final kinetic energy and the final speed of an electron that accelerates from rest through a 5.0 V potential difference.

12. (a) Determine the electric field strength between the anode and cathode of a vacuum tube if their potential difference is 320 V and they are 3.6 cm apart. (b) What force would be exerted on an electron if it were placed between the plates? (c) How much work must be done in order to move the electron from anode to cathode?

13. An electron is between two parallel plates 1 cm apart. One plate is at a potential of 1 000 V above the other. What is the potential gradient between the plates?

14. An electron is accelerated from a cathode to a positive potential anode. If the potential difference between the cathode and anode is 300 V and they are 2.0 cm apart, determine (a) the average electric field strength between the cathode and anode and (b) the average electric force that would be exerted on the electron. (c) If the electron was at rest at the cathode, with what speed does it strike the anode?

15. The plates of a cathode ray tube are parallel and 0.60 cm apart, and there is a 15 kV potential difference between them. Find the force on an electron between the plates.

16. Two large parallel metal plates are 2.0 cm apart and at a potential difference of 800 V. (a) What is the force exerted on an electron when it is placed midway between the plates? (b) What speed would be acquired by the electron if it moves from rest at the negative plate to the positive plate?

17. (a) What is the potential difference between two points 20 cm and 40 cm from a $-20\ \mu C$ charge? (b) How much work must be done on a $+2.0\ \mu C$ charge to move it from the point of lower potential to the point of higher potential?

Most practical applications of electricity utilize the properties and effects of moving charges. A flow of electric charge constitutes an *electric current*.

Conductors of electricity contain some charges that are free to move when they experience an electric force; in order to maintain a flow of charge, some source of electrical energy (such as a battery, power supply or generator) is required in the circuit. Electrons are the free charged particles in metallic conductors; they tend to drift through the metal in the direction opposite to the applied electric field. The term *electronics* is used to denote the study of the emission, properties and effects of electrons in circuits and in devices that operate directly or indirectly because of electron flow.

20-1 ELECTRIC CURRENT

Materials are often classified in terms of their properties as a medium for electric current.

Conductors of electricity are materials that contain some charges that are free to move when they experience an electric force. In *metallic conductors* the free charges are the atomic electrons that have the farthest orbits from the nuclei. In metals most electrons are tightly bound in individual atoms where they orbit the atomic nucleus at high speeds. However, some electrons, which are called *valence* or *conduction electrons*, are not tightly bound to any individual atom. They are relatively free to move through a conductor when they experience a force. The magnitude of an electric current depends on the number of available free charges, and their ability to move through the material.

As it accelerates in the electric field, a free electron acquires a velocity, called its *drift velocity*, in excess of its random thermal motion. Drift velocities are normally quite small, but the actual speeds of the free electrons in their random motions may be quite high. Valence electrons tend to drift in the

20

Electric Current and Resistance

direction opposite to the applied electric field and to flow towards the region of higher electric potential.

Ions are atoms or groups of atoms that have a net positive or a net negative charge because of an excess or a deficiency of electrons. Some conductors, called *electrolytes*, contain "free" ions, which move when they experience an electric force. Other conductors contain both free ions and free electrons.

An electrical *insulator* is a very poor conductor of electricity that has no free charges under normal conditions. Very large electric fields are required to produce a charge flow through an insulator.

Semiconductors have few free charges at low temperatures, but only a relatively small energy input is required to produce more free charges, which can then move through the material. In many semiconductors, as an electron is "freed" from an atom, it leaves that atom with a positive charge; the space vacated by the electron is called a *hole*. The hole may then be occupied by an electron from some neighboring atom, creating a positive ion and a hole at that atom. As the electron moves one way to fill the hole and to neutralize the positive ion, the hole is effectively transferred in the opposite direction, and a positive ion is created at the neighbor. Note that only the negatively charged electrons are actually moving through the semiconductor, but the holes appear to move and carry a positive charge in the opposite direction (Chapters 32 and 33). The apparent positive charge of a hole is equal in magnitude to the charge of an electron ($+1.602 \times 10^{-19}$ C), and it represents a deficiency of an electron in the atom. Some special semiconductor materials are even manufactured with more holes than free electrons, and others have more free electrons than holes. Therefore in semiconductors the "free" charges may be electrons, holes, or both electrons and holes.

When a negative charge moves in one direction, it is equivalent to the motion of a positive charge in the opposite direction; both positive and negative charges can contribute to the total electric current. In a material that has both free positive and free negative charges, the positive charges move parallel to any electric field, and the negative charges move in the opposite direction. By convention, the direction of motion of the positive charge is taken as the direction of *conventional electric current*; negative charges move in the direction opposite to conventional current. The magnitude of the total electric current is equal to the sum of the currents produced by the motion of each type of charge.

Electric current is defined as the time rate of flow of electric charge. If a positive charge Q_1 and a negative charge Q_2 flow in opposite directions past a fixed point in an elapsed time t, the average electric current has a magnitude:

$$I = \frac{|Q_1| + |Q_2|}{t} *\qquad (20\text{-}1)$$

In SI electric current is a fundamental physical quantity, and it is expressed in terms of a base unit called an *ampere* (A). A current of one ampere is equivalent to a charge flow of one coulomb per second past a fixed point.

*Example 20-1*_____

Charges of $+2.0~\mu$C and $-3.0~\mu$C flow in opposite directions past a fixed point in an elapsed time of 2.5 ms. Determine the total average electric current.

Solution:

$$I = \frac{|Q_1| + |Q_2|}{t}$$
$$= \frac{2.0 \times 10^{-6}~C + 3.0 \times 10^{-6}~C}{2.5 \times 10^{-3}~s}$$
$$= 2.0 \times 10^{-3}~A = 2.0~mA$$

_____

*That is absolute values: the values are always taken as positive regardless of the sign.

Conventional electric current is always directed from a region of higher (more positive) electric potential towards a region of lower (more negative) electric potential. Negative charges flow in the opposite direction.

*Problems*_____

1. Determine the average electric current when 5.0 μC of positive charge and 3.0 μC of negative charge flow in opposite directions past a fixed point in a semiconductor in 2.0 ms.

2. Determine the electric current in a conductor if 5.0×10^{18} electrons flow past a fixed point in 20 s.

3. (a) Determine the electric current in a semiconductor if 8.0×10^{12} holes flow past a fixed point in 64 μs. (b) What is the total electric current if an equal number of electrons flow simultaneously in the opposite direction?

20-2 RESISTANCE

Any area that possesses a net charge has an electric potential, and it exerts an electric force on other neighboring charges, tending to make them move.

Virtually no charge can flow through an insulator; therefore, if an insulator is used to isolate an area of net positive charge from an area of net positive (or even less positive) charge, the electric field strength and potential difference between the charged regions remain constant (Fig. 20-1). This principle is used in devices, called capacitors, that are used to store electric energy.

However, if the two regions are joined by a conductor, the coulomb attraction between unlike charges and repulsion between like charges produces an electric current in the conductor. This current is directed from the higher potential (the region of greater positive or less negative charge) to the lower potential. As positive charge flows to the negatively charged region and/or negative charge flows to the positively charged region, the net charges and hence the potential difference between the regions decrease. Usually, after a brief elapsed time the net charges are neutralized, the regions acquire equal electric potentials, and the current ceases. This very brief current is called a *transient current*.

To maintain a current in a conductor, some source of energy, such as a battery, power supply or generator, must do work to maintain a potential difference and to displace charges continually (Fig. 20-2). These

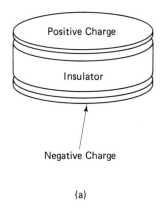

(a)

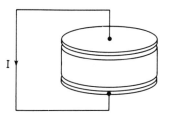

(b) Areas of charge joined by a conductor. The current tends to neutralize the charges and potential difference

Figure 20-1

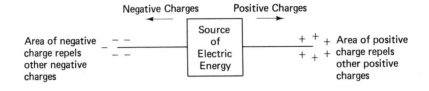

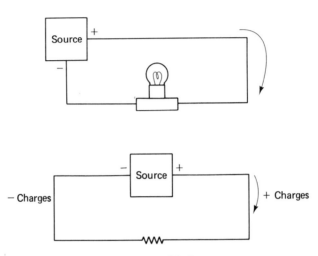

Figure 20-2

charges must also be able to flow around a *closed path* called a *circuit*. If the path is not closed, the displaced charges tend to accumulate at the ends of a conductor. This charge accumulation gives rise to coulomb forces and a potential difference, which opposes the displacement of additional charges. Consequently the (transient) current ceases after a short period of time. By completing a circuit, we avoid this accumulation of charge, but a source of energy is still required to displace the charges because energy is dissipated in the circuit.

With the possible exception of superconductors, all objects offer some opposition to the flow of electric charges. This opposition is described by a term called *electrical resistance*. *R* the electrical resistance of an object is defined as the ratio of the applied potential difference *U* to the corresponding electric current *I* in the object.

$$R = \frac{U}{I} \qquad (20\text{-}2)$$

The SI unit for resistance is the *ohm* (Ω). An applied potential difference of one volt produces a one ampere current in an object that has a resistance of one ohm ($1\,\Omega = 1\,\text{V/A}$).

The reciprocal of the resistance is called the *electric conductance G*.

$$G = \frac{1}{R} \qquad (20\text{-}3)$$

The corresponding SI unit is the *siemen* (S).

*Example 20-2*_____

Determine the resistance of a conductor if it carries 2.5 mA current when subjected to a potential difference of 5.0 V.

Solution:

$$R = \frac{U}{I} = \frac{5.0 \text{ V}}{2.5 \times 10^{-3} \text{ A}} = 2\,000\ \Omega$$

or $\qquad\qquad 2.0\ k\Omega$

The electric resistance and conductance of any object depends on its dimensions and the nature of its material as well as external conditions, such as temperature. Many objects have resistances that also vary with the potential difference. That is, under the same external conditions, the ratio of their potential difference to the electric current varies. The resistance of some circuit elements, such as diodes, even depends on the direction of the electric current!

In semiconductors and insulators the electrical resistance is mainly due to the absence of free charges; their electrical resistance decreases as their temperature increases. The electrical resistance of a metallic conductor is due to deformations and impurities in its structure and to the thermal vibrations of its atoms and molecules. As the temperature increases, the resistance of a metallic conductor increases.

Resistors (Fig. 20-4) are devices that conduct some electricity but also dissipate some electric energy as heat. They are represented schematically by the symbols:

Fixed

or

Variable

The most common inexpensive resistors are *composition resistors*; they consist of small, cylindrical-shaped rods of hot, pressed carbon granules. *Film resistors* are made by depositing a thin film of carbon granules or a metal on an insulating surface (substrate), such as ceramic or glass. *Wire-wound resistors* are

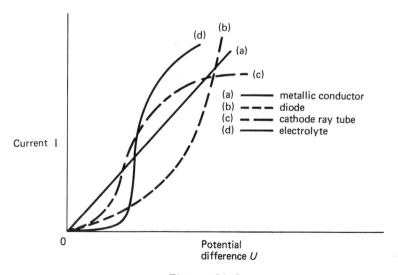

(a) ——— metallic conductor
(b) – – – diode
(c) – · – cathode ray tube
(d) ——— electrolyte

Current I

Potential difference U

Figure 20-3
Current-voltage characteristics.

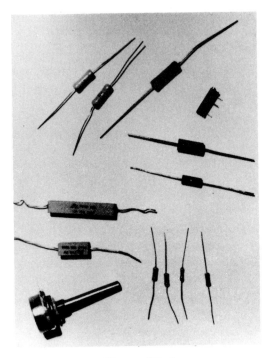

Figure 20-4
Resistors.

composed of a coil of high resistance wire, which is wound on an insulator. *Cermet resistors* consist of a mixture of metal or metal oxide and glass on a ceramic substrate. Variable resistors may be wire-wound or carbon composition resistors that have a central terminal attached to a slider, which can be moved in order to make contact with a different part of the resistor.

Heater elements and light bulb filaments are resistors that dissipate electric energy as heat and light. Resistors are also used to control electric currents and potential differences in circuits. When a potential difference U exists between the ends of a conductor that has a resistance R, the conductor must carry electric current $I = U/R$. The potential difference can only exist when the conductor carries the current. In circuits the potential difference between the ends of a conductor is often referred to as the *voltage, potential* or *IR drop.*

Example 20-3

Determine the potential drop acress a 220 Ω resistor when it carries 320 mA.

Solution:

$$U = IR = (0.320 \text{ A})(220 \text{ Ω}) = 70.4 \text{ V}$$

While the resistance of many conductors and devices varies with the applied potential difference, Georg Simon Ohm (1787–1854) observed that the electrical resistance of *metallic conductors* remains constant over wide ranges of potential difference. Thus in metallic conductors the electric current I is directly proportional to the applied potential difference U.

$$I \propto U \quad \text{or} \quad \frac{U}{I} = R = \text{constant} \quad (20\text{-}4)$$

where R is the resistance of the metallic conductor. This statement is known as *Ohm's law*; it should not be confused with the definition of electrical resistance. The resistance of all objects is defined by Eq. 20-3; *the ratio of potential difference to the electric current is only a constant for metallic conductors and some special resistors.* Note that Ohm's law is not valid if strong electric fields exist in the metallic conductor or if these fields fluctuate rapidly.

Example 20-4

A 110 V potential difference maintains a 2.00 A current in the metallic filament of a light bulb. Determine (a) the resistance of the filament and (b) the potential difference that would produce a 2.20 A current if the temperature is constant.

Solution:

(a) $$R = \frac{U}{I} = \frac{110 \text{ V}}{2.00 \text{ A}} = 55.0 \text{ Ω}$$

(b) Since the filament is metallic and the temperature is constant, its resistance is constant. Thus

$$U = IR = (2.20 \text{ A})(55.0 \ \Omega) = 121 \text{ V}$$

Problems

4. Determine the resistance of an electric heater when it draws 12.0 A from a 110 V source.

5. What is the resistance of the field coil of an electric motor that draws 6.5 A from a 115 V source?

6. Calculate the potential drop across a 220 Ω resistor when it carries 520 mA.

7. Determine the current in a 5.60 kΩ resistor when there is a 16.8 V potential drop across it.

8. What is the potential drop across a 150 μΩ bus bar in a circuit box when it carries 90.0 A?

9. What resistance is required to limit the current to 500 mA when the applied potential difference is 25 V?

10. A 28.2 V potential difference maintains a 65.0 mA current in a metallic resistor. (a) What is the resistance of the resistor? (b) What potential difference would be required to maintain a 120 mA current at the same temperature?

11. (a) What is the resistance of a wire coil if it draws 300 mA from a source that delivers 12 V? (b) What is the current in the coil when it is connected to a source that delivers 16 V if the temperature remains constant?

20-3 RESISTIVITY

The resistance of an object depends on its physical dimensions and the material from which it is constructed.

Ohm also investigated variations of the electric resistance in metallic conductors of different dimensions. He found that, for conductors constructed of the same material, if the temperature is constant, their electric resistance R is directly proportional to their length l and inversely proportional to their cross-sectional area A

$$R \propto \frac{l}{A} \quad \text{or} \quad R = \frac{\rho l}{A} \qquad (20\text{-}5)$$

The proportionality constant ρ is called the *electric resistivity* (or *specific resistance*). Its value depends on the temperature and the nature of the material present and is independent of the physical dimensions of the material. In SI, resistivities are expressed in ohm meters. Some typical values for the resistivity of various materials are listed in Table 20-1. The inverse of the resistivity is called the *conductivity* of the material:

$$\sigma = 1/\rho$$

The conductivity σ of any material depends on the number of free charges that it contains and the ease with which these charges are able to flow.

Example 20-5

The tungsten filament of a light bulb is 15 cm long and has a diameter $d = 0.20$ mm. What is its resistance at 20°C?

Solution:

From Table 20-1: $\rho = 5.5 \times 10^{-8} \ \Omega \cdot \text{m}$. Thus,

$$\begin{aligned} R &= \frac{\rho l}{A} = \frac{\rho l}{\pi d^2/4} \\ &= \frac{4(5.5 \times 10^{-8} \ \Omega \cdot \text{m})(0.15 \text{ m})}{\pi(2.0 \times 10^{-4} \text{ m})^2} = 0.26 \ \Omega \end{aligned}$$

Most wires have small, but circular, cross-sectional areas. Consequently for convenience a new unit for the cross-sectional area of a wire is defined in the British Engineering System: a *circular mil* (CM) is the

Table 20-1
Typical Electronic Properties of Materials (20°C)

Material	Resistivity ρ, $\Omega \cdot m \times 10^{-8}$	Resistivity ρ, $\Omega \cdot CM/ft$	Temperature Coefficient of Resistivity α, $(1/°C) \times 10^{-3}$
Aluminum	2.6	15.6	3.9
Carbon	3 500	21 000	−0.50*
Copper	1.7	10.2	3.9
Constantan	50	300	0.001 9
Nichrome	115	690	0.16
Platinum	11	66	3.9
Silver	1.6	9.6	3.8
Tin	11	66	4.7
Tungsten	5.5	33	4.7

*Note that the value is negative because carbon is a semiconductor and its resistance decreases as the temperature increases (see Chapter 32).

cross-sectional area of a wire that has a diameter of 1 *mil* = 10^{-3} in. The cross-sectional area of any other wire that has a diameter of d mils is d^2 CM. This eliminates the $\pi/4$ factor that is incorporated into the resistivity term. Using this system of units, we usually express the resistivity in $\Omega \cdot CM/ft$.

Example 20-6

Determine the resistivity of an aluminum wire that has a length $l = 2.5$ ft, a diameter $d = 0.020$ in and a resistance $R = 1.0 \, \Omega$.

Solution:

$d = 0.020$ in $= 20 \times 10^{-3}$ in $= 20$ mils. Consequently the cross-sectional area of the wire $A = d^2$ CM $= 400$ CM. Therefore

$$\rho = \frac{RA}{l} = \frac{(1.0 \, \Omega)(400 \, CM)}{25 \, ft} = 16 \, \Omega \cdot CM/ft$$

Variations of Resistivity with Temperature

The electric resistivity of most materials varies as its temperature changes. In metallic con-

ductors the resistivity usually increases as the temperature increases, but the resistivity of semiconductors and insulators decreases as the temperature increases (Chapter 32). Some special alloys, such as constantan, have approximately constant resistivities over relatively wide ranges of temperature. In this section we shall only consider the resistance and resistivity of metallic conductors.

As free electrons flow through a metallic conductor, their motion is impeded when they interact with the atomic ion cores; the electric resistivity of the metal is mainly due to these interactions. The electric resistivity of a metal is due to impurities and irregularities in its structure and to the thermal vibrations of its atoms.

When a metal is heated, its atoms increase their energy, and they vibrate with larger amplitudes. These vibrating atoms present greater obstacles in the path of any flowing free electrons, and effective interactions occur more frequently. Consequently, as the temperature of a metal increases, its resistivity also increases. For pure metals at relatively high temperatures, the change in resistivity

$\Delta\rho$ varies directly as the change in its temperature ΔT and the initial resistivity ρ_0. Thus

$$\Delta\rho \propto \rho_0 \, \Delta T \quad \text{or} \quad \Delta\rho = \alpha\rho_0 \, \Delta T \quad (20\text{-}6)$$

The proportionality constant α is called the *temperature coefficient of resistivity*. It is defined as the fractional change in resistivity per degree change in temperature and is expressed in units of per degree. Therefore, if ρ_0 is the resistivity at temperature T_0 and ρ is the resistivity at temperature T,

$$\rho - \rho_0 = \alpha\rho_0(T - T_0)$$
or
$$\rho = \rho_0[1 + \alpha(T - T_0)] \quad (20\text{-}7)$$

$$\alpha = \frac{\rho - \rho_0}{\rho_0(T - T_0)} \quad (20\text{-}8)$$

Note that the value of α depends on the choice of the temperature T_0; it is not the same for all temperatures. Some typical values for α are listed in Table 20-1.

If a metallic conductor has a length l, a cross-sectional area A, and it is composed of a metal that has a resistivity ρ at some temperature, its electric resistance at the same temperature is:

$$R = \frac{\rho l}{A}$$

Consequently, if the length l and the cross-sectional area A do not change significantly with temperature, the resistance is directly proportional to the resistivity and equations similar to Eqs. 20-7 and 20-8 are valid for resistance of a metallic conductor. The resistance at a temperature T:

$$R_T = R_0[1 + \alpha(T - T_0)] \quad (20\text{-}9)$$

where R_0 is the resistance of the conductor at the temperature T_0 and α is the temperature coefficient of resistivity (or resistance). Also

$$\alpha = \frac{R_T - R_0}{R_0(T - T_0)} \quad (20\text{-}10)$$

These equations are approximately correct, provided that the temperature variation is not large. The relationship between resistivity and temperature is similar for all pure metals at high temperatures (Fig. 20-5); the resistivity is approximately proportional to the absolute temperature. At 0°C for example, all pure metals have temperature coefficients of resistivity $\alpha \approx 1/273°\text{C}$.

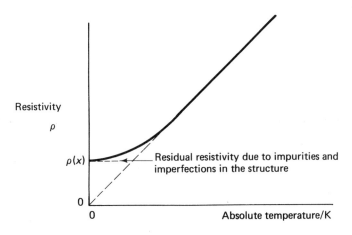

Figure 20-5
Variation of resistivity with temperature for a typical metal.

Example 20-7

A metallic wire has a resistance $R_0 = 1.80\ \Omega$ at a temperature $T_0 = 20°C$ where its temperature coefficient of resistance $\alpha = 3.75 \times 10^{-3}/°C$. Determine its resistance at (a) 50°C and (b) 0°C.

Solution:

(a) $\quad R_T = R_0[1 + \alpha(T - T_0)]$
$\quad\quad = 1.80\ \Omega[1 + (3.75 \times 10^{-3}/°C)$
$\quad\quad\quad \times (50 - 20)°C] = 2.00\ \Omega$

(b) $\quad R_T = R_0[1 + \alpha(T - T_0)]$
$\quad\quad = 1.80\ \Omega[1 + (3.75 \times 10^{-3}/°C)$
$\quad\quad\quad \times (0 - 20)°C] = 1.67\ \Omega$

Variations in the resistance of a metallic conductor are also used to measure corresponding temperature changes. *Resistance thermometers* are composed of a coil of metallic wire, such as platinum, which has been calibrated so that its resistance is accurately known for a wide range of temperatures. These devices can measure temperatures very accurately.

Example 20-8

A resistance thermometer has a resistance $R_0 = 25\ \Omega$ and a temperature coefficient of resistance $\alpha = 3.6 \times 10^{-3}/°C$ at 20°C. When it is immersed in a liquid, the resistance drops to 7.0 Ω. Calculate the temperature of the liquid.

Solution:

From Equation 20-10:

$$T = T_0 + \frac{R_T - R_0}{R_0\alpha}$$

$$= 20°C + \frac{(7.0\ \Omega - 25\ \Omega)}{(25\ \Omega)(3.6 \times 10^{-3}/°C)}$$

$$= 20°C - 200°C = -180°C.$$

Thermistors

A *thermistor* is a temperature-sensitive resistance element. Most thermistors have a nega-

tive temperature coefficient of resistance, but some are available with positive temperature coefficients. These devices are usually made of oxides of nickel, strontium, cobalt, manganese and magnesium, and they are available in various shapes (rods, beads, discs, washers, etc.).

Problems

12. Determine the resistivity of a wire that is 50.0 m long if it has a diameter of 0.800 mm and its resistance is 2.50 Ω.

13. Calculate the resistance of 250 m of copper wire at 20°C if its diameter is 0.80 mm.

14. A 4.5 m long nichrome wire with a diameter of 0.50 mm is made into a resistance coil for a small heater. (a) What is its resistance at 20°C? (b) What is its resistance at 80°C?

15. What length of 0.80 mm diameter copper wire has a resistance of 5.0 Ω at 20°C?

16. A 250 m length of a conductor with a diameter of 1.2 mm has a resistance of 102 Ω. Determine the diameter of a 65 m long wire of the same material that will have a resistance of 130 Ω.

17. Determine the resistivity of copper at (a) 400°C and (b) −50°C.

18. At what temperature is the resistivity of aluminum equal to $5.2 \times 10^{-8}\ \Omega \cdot m$?

19. A carbon resistor has a resistance of 560 Ω at 20°C. What is its resistance at (a) 150°C and (b) −40°C?

20. If the tungsten filament of a light bulb has a resistance of 68 Ω at 3 000°C, what is its resistance at 20°C?

21. Determine the resistance of a 25 m long copper wire at 110°C if its diameter is 1.5 mm. Assume that the dimensions are constant.

22. What length of nichrome wire with a dia-

meter of 0.75 mm has a resistance of 64 Ω at 500°C?

23. A resistance thermometer has a 6.0 Ω resistance and a temperature coefficient of resistance of $4.0 \times 10^{-3}/°C$ at 20°C. When the thermometer is placed in a furnace, the resistance increases to 30 Ω. What is the temperature of the furnace?

20-4 SUPERCONDUCTIVITY

It was not until 1908 when H. Kamerlingh Onnes liquified helium (at a temperature of 0.7 K) that it became possible to investigate the effects of very low temperatures on the electrical properties of materials. Using his liquid helium to maintain low temperatures, Onnes measured the electric resistance of conductors as a function of their temperature. In 1911, he made a remarkable discovery: As he cooled a sample of mercury its resistance decreased relatively uniformly as expected; however, when the temperature was reduced to below 4.15 K, the electric resistance dropped suddenly to zero or so close to zero that it could not be measured (Fig. 20-6). Below 4.15 K, which is called its *critical temperature* T_c, the mercury is said to be in a *superconducting state* because its dc resistivity is essentially zero. If an electric current is produced in a ring-shaped superconductor, that current will continue indefinitely without any measurable losses.

Subsequent investigations have shown that many metals, alloys, compounds and semiconductors become superconductors at sufficiently low temperatures. Superconducting states have not been found for all materials; for example, lithium, sodium, copper, silver and gold do not become superconductors even when they are cooled below 0.1 K. The superconducting state is destroyed when the temperature of the material is increased to some value in excess of the critical tem-

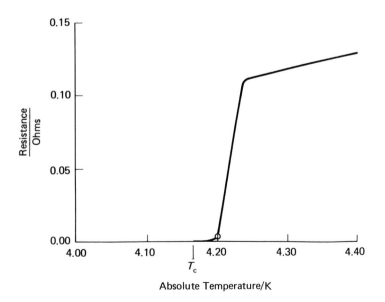

Absolute Temperature/K

Figure 20-6

Electric resistance versus absolute temperature of a mercury sample.

perature or when the material is subjected to a magnetic field of sufficient intensity.

There are many possible uses of super-conductivity. Superconductors are used in low temperature switches called *cryotrons* and in special superconductor magnets. They are also used in the memory circuits of some computers; the superconducting state corresponds to the "on" state because it has no

electrical resistance wereas the normal resistive state is the "off" state.

Superconducting transmission lines consisting of a three-phase cable are now being used in some locations to transfer electric energy from one place to another. These lines can transmit a bulk power of more than 2.5 GW with virtually no transmission losses; however, the superconductor must contin-

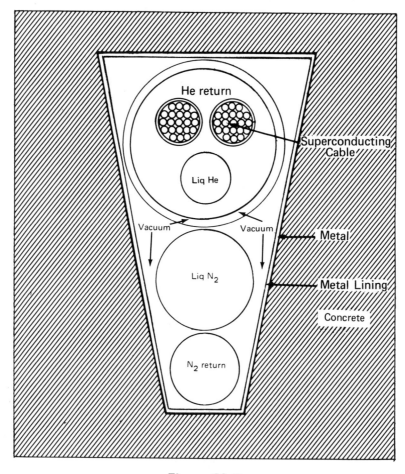

Figure 20-7
Cross-section of a superconducting transmission line.

ually be cooled with liquid helium to $-269°C$ (Fig. 20-7).

20-5 ENERGY TRANSFORMATIONS AND ELECTRIC POWER

Most circuit elements tend to dissipate electric energy when they carry an electric current. For example, electric motors are used to convert electric energy into mechanical work, whereas resistors and heating elements transform electric energy into heat.

When a potential difference is maintained across a conductor, free electrons tend to drift from points of lower to points of higher (more positive) electric potential. Between interactions with atomic ion cores, the drift velocity of each free electron increases, and it gains kinetic energy, but during an interaction these electrons transfer some of their energy to the atomic ion cores. As a result the vibratory energy of the atomic ion core increases, and the kinetic energy that was lost by the electron during the interactions is effectively transformed into heat. At first the heat increases the temperature of the conductor; if it does not melt, after a very brief time interval the conductor reaches an equilibrium condition where it is able to dissipate the heat to its surroundings at the same rate that the heat is generated. In an *electric welding process*, the conductor melts and unites with some other material before it reaches an equilibrium state. *Fuses* are devices containing a piece of wire that is able to carry some maximum current; if the current exceeds this value the heat generated melts the wire and this opens the circuit (Fig. 20-8).

Consider some circuit element or group of elements XY that carries an electric current I when there is a potential difference U between its ends (Figure 20-9). When a charge Q flows from X to Y through the element, the work W done by the electric field is equal to the energy lost by the charge as it decreases its electric potential.

$$W = QU$$

This energy is transformed into heat, mechanical work, chemical energy, light, etc. in the circuit element. But from the definition of charge, in some elapsed time t an electric charge $Q = It$ flows past a fixed point in the conductor. Consequently the work done by the electric field is:

$$W = QU = ItU$$

Therefore the rate of energy dissipation or the

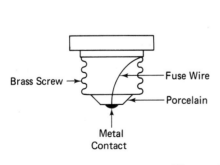

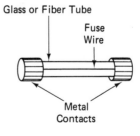

Figure 20-8
Fuses.

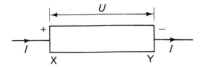

Figure 20-9

Energy transformation in a circuit element. The power dissipated $P = IU$.

power dissipated (in units of watts) by *any* circuit element is:

$$P = W/t = IU \qquad (20\text{-}11)$$

Example 20-9

Determine the electric power rating of a lamp that draws 1.25 A from a 120 V source.

Solution:

$$P = IU = (1.25\ \text{A})(120\ \text{V}) = 150\ \text{W}$$

The cost of electricity is usually determined in terms of the number of kilowatt hours of electric energy that are consumed. A *kilowatt hour* (kW·h) is the electric energy used when a kilowatt of electric power is dissipated for one hour. If a resistor dissipates electric energy at a rate of one kilowatt, in one hour it dissipates a total energy of:

$$1\ \text{kW·h} = (1\ 000\ \text{W})(3\ 600\ \text{s}) = 3.6 \times 10^6\ \text{J}$$

In the special case when the circuit element obeys Ohm's law, $U = IR$ where R is the electric resistance, the power dissipated is:

$$P = IU = I^2R = \frac{U^2}{R} \qquad (20\text{-}12)$$

Example 20-10

What minimum power rating must a 2.2 kΩ resistor have if it is located in a circuit where the current is 15 mA?

Solution:

$$P = I^2R = (15 \times 10^{-3}\ \text{A})^2(2.2 \times 10^3\ \Omega)$$
$$= 0.50\ \text{W}$$

If Q is the total heat energy produced in a resistor during an elapsed time t, the average power that the resistor dissipates is:

$$\bar{P} = \frac{Q}{t} = I^2R \qquad (20\text{-}13)$$

Example 20-11

A 220 Ω electric heater draws an average current of 2.0 A for 3.0 h. (a) Determine the total heat produced. (b) If electric energy costs 5.3 ¢/kW·h, calculate the cost.

Solution:

(a) $Q = I^2Rt = (2.0\ \text{A})^2(220\ \Omega)(3.0 \times 3\ 600\ \text{s})$
$$= 9.5 \times 10^6\ \text{J}$$

(b) $\qquad 9.5 \times 10^6\ \text{J} = \dfrac{9.5 \times 10^6}{3.6 \times 10^6}\ \text{kW·h}$
$$= 2.64\ \text{kWh}$$

Thus

$$\text{cost} = (2.64\ \text{kW·h})(5.3\ ¢/\text{kW·h}) = 14\ ¢$$

Most electric circuits consist of a source of electric energy and a load composed of one or more devices that dissipate the electric energy. Normally many components of an electric circuit dissipate electric energy. Consequently they contribute to the total load. The load of a circuit is often represented schematically by a single resistor that dissipates the equivalent amount of electric power.

Problems

24. Determine the electric power dissipated in a small motor that draws 0.50 A from a 110 V power supply.

25. What minimum power rating must a 560 Ω resistor have if it is located in a circuit (a) where the current is 120 mA? (b) Where it must withstand a 35 V potential difference?

26. What maximum voltage can be safely applied to a 470 Ω, 0.50 W resistor?

27. Determine the power consumption of a portable radio that draws 3.0 mA from a 9.0 V source?

28. Determine the current in a 60 W light bulb when it is connected to a 120 V source.

29. What current will a 12 kW load draw from a 120 V source?

30. A fuse has a 0.015 Ω resistance and a maximum power dissipation of 2.5 W. What maximum current will the fuse be able to carry?

31. An electric motor delivers 0.45 hp, and its efficiency is 90%. Determine the electric current that it draws when it is connected directly across a 120 V power supply.

32. A light bulb has a resistance of 80 Ω, and it carries a current of 1.2 A for 8.0 h. (a) Determine the power dissipation. (b) If electric energy costs 3.0 ¢/(kW·h), how much does it cost to run the light?

33. In order to brew his hourly coffee, a student uses a resistance heater to heat 1.0 kg of water from an initial temperature of 20°C to the boiling point. The resistance of the heater is 200 Ω, and it carries a 4.0 A average current. If only 80% of the total heat is applied to the water, how long does it take to boil the water?

34. If electric energy costs 5.0 ¢/(kW·h), how much does it cost to run an electric motor for 5.0 h if it has an efficiency of 80% and develops an average output power of 15 hp?

20-6 COMBINATIONS OF RESISTORS

Two or more resistors are often connected by metallic wires of negligible resistance. The total equivalent resistance of the combination depends on their arrangement. There are two basic configurations.

Resistors in Series

Two or more resistors are said to be in series when they are connected end to end such that all the resistors always carry equal electric currents.

Consider three resistors that are connected in series with some power supply so that a total potential difference U_{TOT} exists across the group. Suppose that the resistors have resistances R_1, R_2 and R_3 and that the potential differences between the ends of the individual resistors are U_1, U_2 and U_3, respectively (Fig. 20-10). Since electric charge is

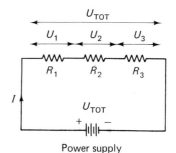

Figure 20-10
Resistances in series.

always conserved and there is only one possible path for the current, *the electric current is the same at each point in the circuit, and each resistance carries the same electric current*. When some charge Q is displaced through the series combination, its total change in electric energy $W_{TOT} = QU_{TOT}$ is equal to the sum of its energy changes as it is displaced through the individual resistances.

$$W_{TOT} = QU_{TOT} = W_1 + W_2 + W_3$$
$$= QU_1 + QU_2 + QU_3$$

or
$$U_{TOT} = U_1 + U_2 + U_3$$

In general, the total potential difference across a series combination of circuit components is equal to the sum of the potential differences across the individual components.

$$U_{TOT} = U_1 + U_2 + U_3 + \cdots = \sum_i U_i \quad (20\text{-}14)$$

Since each resistance carries the same current I, from the definition the total resistance is:

$$R_{TOT} = \frac{U_{TOT}}{I} = \frac{U_1 + U_2 + U_3}{I} = R_1 + R_2 + R_3$$

since $R_1 = U_1/I$, $R_2 = U_2/I$ and $R_3 = U_3/I$. Thus the total resistance of resistances in series is equal to the sum of the individual resistances.

$$R_{TOT} = R_1 + R_2 + R_3 \cdots = \sum R \quad (20\text{-}15)$$

Example 20-12

Resistors that have resistances $R_1 = 6.0\ \Omega$, $R_2 = 8.0\ \Omega$, $R_3 = 10.0\ \Omega$ and $R_4 = 12.0\ \Omega$ are connected in series with a power supply so that a total potential difference $U_{TOT} = 7.20\ V$ exists across the combination. Determine (a) the current I drawn by the resistors, (b) the power P_{TOT} dissipated in the combination, (c) the power P_2 dissipated in the 8.0 Ω resistance, (d) the potential drop U_4 across the 12.0 Ω resistor.

Solution:

(a) $R_{TOT} = \sum R$
$$= 6.0\ \Omega + 8.0\ \Omega + 10.0\ \Omega + 12.0\ \Omega$$
$$= 36.0\ \Omega$$

Therefore, the current

$$I = \frac{U_{TOT}}{R_{TOT}} = \frac{7.20\ V}{36.0\ \Omega} = 0.200\ A$$

(b) $P_{TOT} = I^2 R_{TOT} = (0.200\ A)^2(36.0\ \Omega)$
$$= 1.44\ W$$

(c) $P_2 = I^2 R_2 = (0.200\ A)^2(8.0\ \Omega) = 0.32\ W$

(d) $U_4 = IR_4 = (0.200\ A)(12.0\ \Omega) = 2.40\ V$

Resistors in Parallel

When two or more resistors are connected in parallel, they form different paths for the electric current, but the potential difference between the ends of each resistor is identical. If the potential drops across the parallel resistors were not equal, a potential difference would also exist between their interconnected ends, and some charge would flow from one resistor to the other to neutralize that potential difference.

Consider three resistors that have resistances R_1, R_2 and R_3 that are connected in parallel with some power supply such that an equal potential difference U exists across each resistor (Fig. 20-11). Since electric

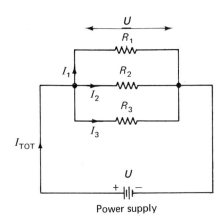

Figure 20-11
Resistances in parallel.

charge is conserved, the total electric current I_{TOT} carried by the combination of resistors in parallel is equal to the sum of the electric currents in the individual resistors. Thus, if I_1, I_2 and I_3 are the currents in R_1, R_2 and R_3, respectively:

$$I_{TOT} = I_1 + I_2 + I_3$$

In general, for any number of resistors in parallel.

$$I_{TOT} = I_1 + I_2 + I_3 + \cdots = \sum I \quad (20\text{-}16)$$

From the definition of resistance, the total equivalent resistance of the parallel combination $R_{TOT} = U/I_{TOT}$; therefore, $I_{TOT} = U/R_{TOT}$. Similarly $R_1 = U/I_1$, $R_2 = U/I_2$ and $R_3 = U/I_3$. Thus

$$I_{TOT} = \frac{U}{R_{TOT}} = I_1 + I_2 + I_3 = \frac{U}{R_1} + \frac{U}{R_2} + \frac{U}{R_3}$$

and $1/R_{TOT} = 1/R_1 + 1/R_2 + 1/R_3$. In general for a parallel combination of resistors, the reciprocal of the total equivalent resistance is equal to the sum of the reciprocals of the individual resistances.

$$\frac{1}{R_{TOT}} = \frac{1}{R_1} + \frac{1}{R_2} + \frac{1}{R_3} + = \sum \frac{1}{R} \quad (20\text{-}17)$$

Example 20-13

Resistors that have resistances $R_1 = 10.0\ \Omega$, $R_2 = 30.0\ \Omega$ and $R_3 = 5.0\ \Omega$ are connected in parallel with a power supply such that an equal potential difference $U = 3.0\ V$ exists between the ends of each resistor. Calculate (a) the total resistance R_{TOT}, (b) the current I_3 carried by the 5.0 Ω resistor, (c) the total power P dissipated in the combination, (d) the power P_2 dissipated in the 30.0 Ω resistor.

Solution:

(a) $\dfrac{1}{R_{TOT}} = \dfrac{1}{R_1} + \dfrac{1}{R_2} + \dfrac{1}{R_3}$

$\qquad = \dfrac{1}{10.0\ \Omega} + \dfrac{1}{30.0\ \Omega} + \dfrac{1}{5.0\ \Omega} = \dfrac{10.0}{30\ \Omega}$

$\qquad R_{TOT} = 3.0\ \Omega$

(b) $\qquad I_3 = \dfrac{U}{R_3} = \dfrac{3.0\ V}{5.0\ \Omega} = 0.60\ A$

(c) $P = I_{TOT}^2 R_{TOT} = \dfrac{U^2}{R_{TOT}} = \dfrac{(3.0\ V)^2}{3.00\ \Omega} = 3.0\ W$

(d) $\qquad P_2 = \dfrac{U^2}{R_2} = \dfrac{(3.0\ V)^2}{30.0\ \Omega} = 0.30\ W$

To determine the total equivalent resistance of a complex combination of resistors, compute the resistance of each group in parallel and each group in series, and replace these groups by their equivalent resistance.

Repeat the process until the entire combination is reduced to a single value for resistance.

Example 20-14

Determine the total equivalent resistance of the complex combinations of five resistors illustrated in Fig. 20-12.

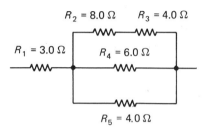

$R_2 = 8.0\ \Omega \qquad R_3 = 4.0\ \Omega$

$R_1 = 3.0\ \Omega \qquad R_4 = 6.0\ \Omega$

$R_5 = 4.0\ \Omega$

Figure 20-12

Solution:

(a) The equivalent resistance R_{23} of the series combination of R_2 and R_3:

$$R_{23} = R_2 + R_3 = 8.0\ \Omega + 4.0\ \Omega = 12.0\ \Omega$$

(b) The total equivalent resistance R_{11} of the parallel combination of resistors R_{23}, R_4 and R_5:

$$\frac{1}{R_{11}} = \frac{1}{R_{23}} + \frac{1}{R_4} + \frac{1}{R_5}$$

$$= \frac{1}{12.0\ \Omega} + \frac{1}{6.0\ \Omega} + \frac{1}{4.0\ \Omega} = \frac{6.0}{12.0\ \Omega}$$

$$R_{11} = 2.0\ \Omega.$$

(c) The equivalent resistance of the series combination involving R_{11} and R_1 is equal to the total resistance

$$R_{TOT} = 2.0\ \Omega + 3.0\ \Omega = 5.0\ \Omega$$

Problems

35. Determine the total resistance of resistors $R_1 = 5.0\ \Omega$, $R_2 = 15.0\ \Omega$ and $R_3 = 20.0\ \Omega$ when they are connected (a) in series, (b) in parallel, (c) with a parallel combination of R_1 and R_2 in series with R_3.

36. Three resistors that have resistances of 7.0 Ω, 8.0 Ω and 12.0 Ω are connected in series with a power supply so that a potential difference of 54.0 V exists across the combination. Calculate (a) the electric current drawn by the resistors, (b) the power dissipated in each resistor, (c) the total power dissipated, (d) the potential drop across each resistor.

37. Four resistors with resistances of 3.0 Ω, 5.0 Ω, 7.5 Ω and 5.0 Ω are connected in parallel with a power supply such that an equal potential difference of 3.0 V exists between the ends of each resistor. Calculate (a) the total resistance, (b) the current carried by each resistor, (c) the total power dissipated, (d) the power dissipated in each resistor.

38. Compute the total equivalent resistance of a parallel combination of 6.0 Ω and 9.0 Ω resistors in series with a 1.5 Ω resistor.

39. Determine the total equivalent resistance of the resistor combination in Fig. 20-13: (a) when $R_1 = 14.0 \, \Omega$, $R_2 = 4.9 \, \Omega$, $R_3 = 3.0 \, \Omega$, $R_4 = 4.0 \, \Omega$, $R_5 = 15.0 \, \Omega$, $R_6 = 8.0 \, \Omega$ and $R_7 = 24.0 \, \Omega$; (b) when $R_1 = 18.0 \, \Omega$, $R_2 = 2.7 \, \Omega$, $R_3 = 5.0 \, \Omega$, $R_4 = 4.0 \, \Omega$, $R_5 = 27 \, \Omega$, $R_6 = 11.0 \, \Omega$ and $R_7 = 3.3 \, \Omega$.

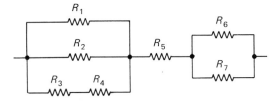

Figure 20-13

REVIEW PROBLEMS

1. Determine the average electric current when $+4.5$ nC and -6.0 nC of charge flow in opposite directions past a fixed point in a conductor is 30 μs.

2. Determine the voltage drop in a 220 Ω resistor when it carries 25 mA.

3. A copper bus bar is 2.5 mm thick, 2.0 cm wide and 75 cm long. Determine (a) its resistance at 20°C and (b) the potential difference between its ends at 20°C if it carries 100 A.

4. Determine the resistivity of a wire that is 30 m long with a diameter of 1.2 mm if its resistance is 0.371 Ω.

5. A wire has a resistivity of 1.2×10^{-5} Ω·m and a diameter of 0.75 mm. What length of wire would be required to make the heating element for a toaster that draws 6.0 A from a 110 V source?

6. An aluminum bar is $\frac{1}{2}$ cm × 2 cm in cross-section, and it is 30 m long. Calculate (a) its resistance at 20°C and (b) its resistivity at 100°C.

7. A resistance thermometer has a resistance of 10 Ω at 0°C, and in a furnace its resistance is 45 Ω. If the temperature coefficient of resistance is 0.003 6/°C, what is the temperature of the furnace?

8. A 15.0 m long copper wire has a diameter of 0.98 mm. (a) What is its resistance at 20°C? (b) What is its resistance at 150°C?

9. What is the power rating of an electric motor that draws 8.0 A from a 50 V source?

10. What length of nichrome wire with a diameter of 0.80 mm has a resistance of 32 Ω at 250°C?

11. What minimum power rating must a 1.0 kΩ resistor have if it is located in a circuit (a) where the current is 12 mA? (b) Where it is subjected to a voltage of 15.0 V?

12. Determine the maximum current that a 0.021 0 Ω fuse can carry if its maximum power dissipation is 4.40 W.

13. If electric energy costs 6.0 ¢/(kW·h), how much does it cost to run an electric motor for 24 h if it has an efficiency of 85% and develops an output power of 6.0 hp?

14. Find the total resistance when three resistors with resistances of 220 Ω, 100 Ω and 470 Ω are connected (a) in series, (b) in parallel, and (c) with a series combination of 220 Ω and 100 Ω in parallel with the 470 Ω.

15. A 22 Ω resistor is connected in series with an unknown resistor and power supply. If the potential drop in the 22 Ω resistor is 30 V and the potential drop in the unknown resistor is 48 V, what is the value of the unknown resistance?

16. Given the circuit in Fig. 20-14 where R_1 = 20 kΩ, R_2 = 30 kΩ, R_3 = 15 kΩ, R_4 = 10 kΩ, R_5 = 25 kΩ and R_6 = 5.0 kΩ,

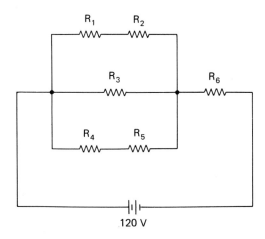

Figure 20-14

find (a) the total equivalent resistance, (b) the current in each resistor, (c) the potential drop in each resistor, (d) the power dissipated in each resistor.

An *electric circuit* is any system of conductors that are connected so that an electric charge can flow in one or more closed paths. Electric circuits often contain many individual components, such as resistors, capacitors, inductors, diodes and transistors, which are interconnected by metallic conductors. Each component usually performs a special function in the total circuit; for example, resistors control the magnitude of the current, and capacitors store electric energy.

When particular types of charge, such as negatively charged electrons, always move in the same direction through a conductor, the flow of charge is called *direct current* or *dc*. If a direct current exists, electric charge is actually displaced from one point to another in the conductor. A *steady state current* exists when the time rate of flow of electric charge is constant. When a conductor carries an *alternating current* (*ac*), all of the moving charges continually change their directions of motion, and they vibrate about fixed positions at their locations in the conductor.

A *circuit diagram* is a schematic representation of a circuit illustrating the various interconnections between the individual components. Each component of the circuit is represented by a special standard symbol, and the interconnections are represented by horizontal and vertical lines:

21

Direct Current (dc) Circuits

Connected junction   Crossing wires connected   Crossing wires not connected

The simplest electric circuit consists of a source of electric energy that is connected to a single load. There are many kinds of load; however, because they all tend to dissipate electric energy, they are often represented by a single resistor. Other, more complex electric circuits are designed to perform special functions, such as measurement, amplification or rectification.

21-1 SOURCES OF ELECTRIC ENERGY

With the possible exception of superconductors, all components of a circuit dissipate electric energy; therefore a source of electric energy is required to maintain an electric current. A device in which other energy forms are reversibly converted into electric energy is called a *seat* (or *source*) *of an electromotive force*. As charges pass through a source of an electromotive force, the source does work to increase the energy of the charges. The source moves positive charges from a point of low potential to a point of higher, more positive potential, and it displaces negative charges in the opposite direction.

The work done per unit charge that passes through the source is called the *electromotive force* or *emf*.* If it does work W to transmit and increase the energy of charge Q, the average emf of the source is:

$$\varepsilon = W/Q \qquad (21\text{-}1)$$

In SI the emf of a source has units of joules per coulomb or volts. A source has an emf of one volt when it does one joule of work to transmit one coulomb of charge.

There are many different sources of emf.

Chemical Sources

Atoms of different materials have different structures. Consequently some materials have a greater affinity (attraction) for electrons than others. In many cases, when atoms unite to form a molecule, one or more electrons

---

*Pronounced ee-em-ef. Note that, in spite of its name, emf is not a force.

may either be transferred from one atom to another or their orbits are displaced so that they are closer to the atom(s) with the greatest electron affinity (Chapter 32). As a result, many molecules, such as water, have a non-symmetric distribution of charge, i.e. their centers of positive and negative charge do not coincide. These molecules are called *polar molecules*.

When some other polar molecule is mixed with water molecules, it may dissociate into two or more ions. For example, hydrochloric acid (HCl) is a polar molecule consisting of a single hydrogen (H) atom that is united with a single chlorine (Cl) atom. The center of negative charge is closer to the chlorine atom because it has a greater electron affinity than the hydrogen. When the acid molecule is placed in water, the negative side of the acid molecule is attracted to the positive side of water molecules, and the positive side of the acid molecule is attracted to the negative side of the water molecules. In many cases these attractions are strong enough to dissociate the acid molecules. As the acid molecule breaks apart, an electron is transferred from the hydrogen atom (leaving it as a positive ion H^+) to the chlorine atom (making it a negative ion Cl^-) (Fig. 2-1). This process is called *electrolytic dissociation* and the solution that contains the ions is called an *electrolyte*.

If a zinc (Zn) plate is dipped into the electrolyte, some of the zinc dissolves. As each zinc atom enters the solution, it ionizes, leaving two electrons on the zinc plate, and then combines chemically with one or more negative ions of the electrolyte (Fig. 21-2).

$$Zn \longrightarrow Zn^{++} + 2e^-$$

then

$$Zn^{++} + 2Cl^- \longrightarrow ZnCl_2$$

The negative charge builds up until the zinc plate attracts the zinc ions back at the same rate that the zinc dissolves. However, in this equilibrium state the zinc plate possesses a

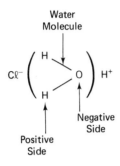

Figure 21-1
Electrolytic dissociation in water.

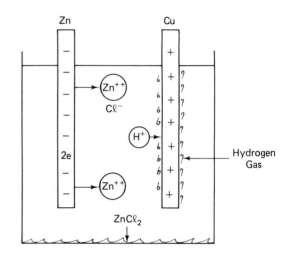

Figure 21-2

net negative charge, and it has a negative electric potential with respect to the electrolyte.

Hydrogen has a greater electron affinity than copper; therefore, if a copper electrode is dipped into the dilute acid electrolyte, some of the hydrogen ions acquire electrons from the copper atoms. Each pair of neutralized hydrogen atoms forms a molecule of hydrogen gas, which bubbles up the electrode to the surface of the electrolyte. As the copper atoms lose electrons to the hydrogen ions, a net positive charge builds up on the copper electrode. Eventually an equilibrium state is reached, in which the copper electrode attracts electrons electrostatically from the hydrogen atoms at the same rate that it loses electrons. At this point, however, the copper electrode has a net positive charge and a positive electric potential with respect to the electrolyte. The positive electrode is called the *anode*, and the negative electrode is called the *cathode*.

A potential difference exists between the negatively charged zinc electrode and the positively charged copper plate. If the electrodes are externally connected by a conductor, electrons flow from the zinc through the conductor to the copper. This process disturbs the equilibrium between both electrodes and the electrolyte so that more zinc ionizes and enters the solution and more hydrogen ions are neutralized at the copper electrode. Therefore

the external flow of electrons continues. This source of an electromotive force is called a *Voltaic cell* (Fig. 21-3).

As the cell discharges, the zinc electrode and the hydrogen ions in the electrolyte are consumed; consequently the Voltaic cell eventually ceases to function. Also, the hydrogen gas tends to accumulate on the copper electrode, shielding it from the electrolyte so that hydrogen ions cannot be neutralized, and the cell stops operating. This is called *polarization*. In order to reduce the effects of polarization, other chemicals called *depolarizers* are often added to chemical cells.

A chemical cell is represented schematically by the symbol:

The positive electrode is designated by the longer line.

There are two fundamental types of cell. *Primary cells* are nonreversible cells that must eventually be discarded because one of the electrodes is consumed, such as the zinc elec-

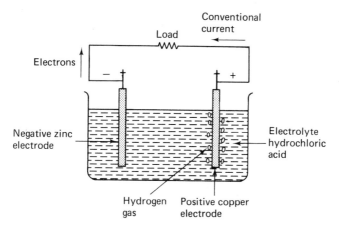

Figure 21-3
A simple Voltaic cell.

trode in the Voltaic cell. *Secondary cells* are reversible cells that may be recharged by reversing the direction of the electric current.

A *Daniell cell* is a relatively reliable primary cell that produces a steady current and

an emf of approximately 1.08 V. It consists of a zinc cathode in a saturated zinc sulphate solution that is separated from a copper anode in a saturated copper sulphate solution by a porous barrier (Fig. 21-4).

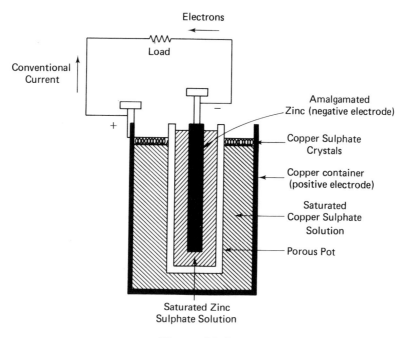

Figure 21-4
The Daniell cell.

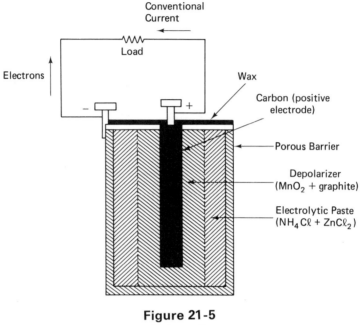

Figure 21-5
A dry cell.

Dry cells are common primary cells that are portable and produce an emf of approximately 1.5 V. A simple dry cell consists of a carbon anode that is completely surrounded by a depolarizer (powdered manganese dioxide and graphite) and separated from a zinc cathode by an electrolytic paste (ammonium chloride and zinc chloride) and a porous barrier (Fig. 21-5).

Recently, dry cells with longer lifetimes have been developed. *Alkaline-manganese cells* are similar to dry cells, but they have higher capacity. They consist of a zinc anode and a manganese dioxide cathode in a potassium hydroxide electrolyte. A *mercury cell* consists of a mercuric oxide (HgO) and graphite anode, a zinc cathode, and a potassium hydroxide (KOH) electrolyte (Fig. 21-6). It develops an almost constant emf of approximately 1.4 V, and it has a relatively high capacity (in ampere-hours).

Lead acid storage cells are secondary cells that produce emf's of approximately 2 V. Groups of these cells are often connected in series to form *storage batteries*. A single storage cell consists of a lead (Pb) electrode and a lead dioxide (PbO_2) electrode in an electrolytic solution of dilute sulphuric acid (H_2SO_4) (Fig. 21-7). When it delivers energy to an externally connected load, the lead electrode is negative, and the lead dioxide is the positive plate. As the storage cell discharges, both electrodes become coated with lead sulphate, and the concentration of the sulphuric acid in the electrolyte is reduced. Eventually both plates become completely coated, and the cell cannot function.

Nickel-cadmium cells are rugged secondary cells that develop a relatively constant emf of approximately 1.2 V. The positive electrode consists of nickel hydroxide with graphite, and the negative electrode is a mixture of cadmium oxide and iron oxide.

To recharge a secondary cell, a source

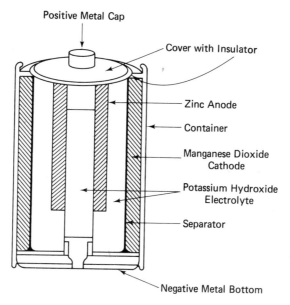

Figure 21-6
Alkaline-manganese cell.

such as a generator is used to reverse the direction of the electric current. This reverses the chemical reactions.

The potential difference produced by most chemical cells varies with their use and they are extremely sensitive to variations in temperature.

A Weston cadmium cell produces an al-most constant emf of 1.018 V; consequently it is often used as a standard source. The negative electrode is an amalgam of approximately 12% cadmium in mercury; a paste of mercurous sulphate (Hg_2SO_4) depolarizer floats on the top of a liquid—mercury positive electrode, and the electrolyte is a saturated solution of cadmium sulphate (Fig. 21-8).

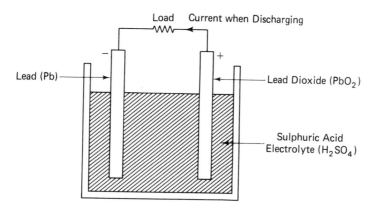

Figure 21-7

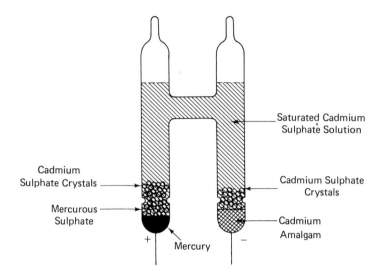

Figure 21-8
Weston standard cell.

Fuel Cells

A *fuel cell* is also a device in which chemical energy is converted directly into electric energy. These devices are improving rapidly, and they will eventually become a common and important source of emf.

A simple fuel cell consists of two hollow porous tubes that are coated with a catalyst and immersed in an electrolyte, such as potassium hydroxide (KOH). Hydrogen gas is passed into one tube (the negative electrode), and oxygen gas is fed into the other (the positive electrode) (Fig. 21-9).

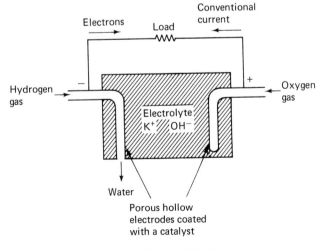

Figure 21-9
A fuel cell.

At the negative electrode the hydrogen gas reacts with the negative hydroxide ion (OH⁻) of the electrolyte to form water and releases electrons (e⁻) in the process.

$$2H_2 + 4OH^- \longrightarrow 4H_2O + 4e^-$$

These electrons flow through the externally connected load to the positive electrode, where they combine with oxygen and water to produce hydroxide ions.

$$4e^- + O_2 + 2H_2O \longrightarrow 4(OH^-)$$

Note that, while hydroxide ions are consumed at the negative electrode, an equal number are created at the positive electrode so that the electrolyte is not changed by the reactions.

Photoelectric Cells

Some materials emit electrons when they are irradiated with light; consequently they can be used to convert radiant energy directly into electric energy. These devices are discussed later.*

Generators

An emf is induced in any conductor that experiences a changing magnetic field. This variation in magnetic field intensity may be produced by moving the conductor through the magnetic field. A *generator* or *dynamo* is a device in which mechanical energy is used to rotate conductors in a magnetic field so that an emf is produced; therefore generators convert mechanical energy into electrical energy (Chapter 23).

Thermoelectricity

All entities tend to exist in the lowest possible energy state. Heat is a form of energy; there-

---

*See photoelectric effect and solar cells.

fore, when one end of a metal rod is maintained at a higher temperature than the other end, some negatively charged electrons flow from the hotter to the cooler area to reduce their energy. A negative charge builds up at the cooler end as the electrons accumulate, and a positive charge due to the deficiency of electrons is produced at the hotter end so that a potential difference is induced in the rod.

A *thermocouple* is a device that is used to measure temperatures. It consists of two dissimilar metals that are connected so that they form two junctions. One junction is maintained at a constant standard temperature, e.g. inserted in ice water, whereas the other junction experiences the temperature to be measured (Fig. 21-10). The induced potential

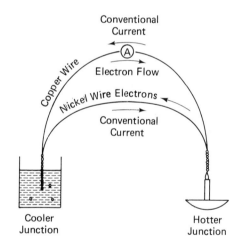

Figure 21-10
A thermocouple.

difference and the current in the device depend on the temperature difference between the junctions. Thermocouples are calibrated so that the unknown temperature can be determined from the current or the induced potential difference.

Piezoelectricity

When crystals of some dielectric materials, such as quartz or Rochelle salt, are mechanically stressed, net charges of opposite polarity may be induced on opposing crystal faces because of the displacement of electrons. Thus an electric field and an emf are produced between diametrically opposite faces in the crystal. This phenomenon is called the *piezoelectric effect*. Also, the same crystal is distorted if it is subjected to an electric field; this is called *electrostriction*.

By applying a rapidly alternating electric field to a piezoelectric crystal, it is possible to produce high frequency mechanical vibrations; therefore piezoelectric crystals can be used as transducers for ultrasonic waves. Piezoelectric materials are also used in some microphones and phonograph pickups to convert mechanical pressures into small electrical impulses.

Problems

1. Determine the emf of a source that does 2.4×10^{-19} J of work as it transmits a single electron.

2. Calculate the power developed by a 9.0 V source when it transmits 5.0×10^{-6} C of charge in each second.

21-2 ELECTROLYSIS

Electrolysis is essentially the reverse of a production of an emf in a chemical cell. Unlike metallic conductors, an electrolyte carries a current by means of the flow of ions. Therefore, if two electrodes in an electrolytic solution are connected to an external source of emf, the electrolyte conducts. The conduction is due to the chemical reactions involving electrons at the electrodes. Electrons are deposited at the positive electrode (anode) by one substance and acquired at the negative electrode (cathode) by another substance. This process is known as *electrolysis*. It is used in *electroplating* and for the *decomposition* of compounds, such as water (with sulphuric acid) into the constituent elements hydrogen and oxygen.

To *electroplate* an object, it is made the cathode, the electroplating material is the anode, and the electrolyte used contains ions of the plating material. The material enters the solution as an ion at the anode and is neutralized and deposited at the cathode (Fig. 21-11). When water (with sulphuric acid) is electrolytically dissociated, hydrogen gas forms at the cathode, and oxygen gas is liberated at the anode.

The basic laws governing electrolysis were discovered by Michael Faraday (1791–1867). He found that:

1. The mass m of an element liberated at either electrode varies directly with the total charge Q passed through the electrolyte.

$$m \propto Q \quad \text{or} \quad m = zQ \quad (21\text{-}2)$$

The constant z is known as the *electrochemical equivalent*.

2. The liberated mass m of an element varies directly with its molar mass A and inversely with the number n of elementary charges on its ions.

$$m \propto \frac{A}{n}$$

Therefore

$$m \propto \frac{QA}{n} \quad \text{or} \quad m = \frac{QA}{Fn} \quad (21\text{-}3)$$

The universal constant $F = 9.648\ 7 \times 10^4$ C is known as a *faraday*.

Example 21-1

What thickness of zinc ($n = 2$, density $\rho = 7\ 150$ kg/m^3) will be deposited on an iron sheet with a total surface area $S = 25.0$ cm^2 in 2.00 h if the sheet is the cathode in a zinc sulphate electrolyte where the average current is 5.00 A?

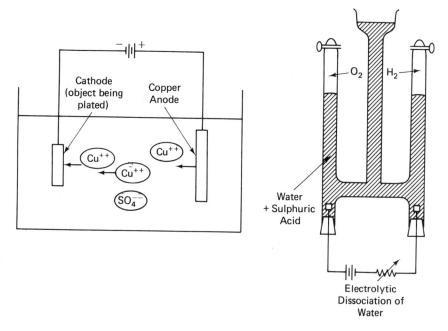

Figure 21-11
Copper plating.

Solution:

The total charge $Q = It = (5.00\ A)(2.00 \times 3\ 600\ s) = 36\ 000$ C. From the periodic table (Appendix 2), since the atomic mass of zinc is 65.38 u, the mass of 1 mol $A = 65.38$ g; thus the mass of zinc deposited

$$m = \frac{QA}{Fn} = \frac{(36\ 000\ C)(65.38\ g)}{(9.65 \times 10^4\ C)(2)} = 12.2\ g$$

Since the density $\rho = \frac{m}{V}$, the volume of zinc deposited is:

$$V = \frac{m}{\rho} = \frac{12.2 \times 10^{-3}\ kg}{7.15 \times 10^3\ kg/m^3}$$
$$= 1.71 \times 10^{-6}\ m^3 = 1.71\ cm^3$$

Thus the thickness is:

$$h = \frac{V}{S} = \frac{1.71\ cm^3}{25.0\ cm^2} = 6.82 \times 10^{-2}\ cm$$

Problems

3. How many grams of silver ($n = 1$) are deposited on a copper goblet at the cathode if there is a 3.0 A current in a silver nitrate electrolyte for 2.0 h?

4. What thickness of copper ($n = 2$ and $\rho = 8\ 890\ kg/m^3$) is deposited on an iron plate with a total surface area of 185 cm² in 6.0 h if it is the cathode in a copper sulphate electrolyte and the average current is 8.0 A?

5. If electricity costs 5.3 ¢/(kW·h) and a 30 V potential difference is maintained between anode and cathode, how much does it cost to electroplate a dish with a surface area of 400 cm² to a thickness of 0.050 mm with silver ($n = 1$ and $\rho = 7\ 150\ kg/m^3$), using a silver nitrate electrolyte.

21-3 INTERNAL RESISTANCE AND BATTERIES

Even though they are expressed in the same units (volts), the emf $\mathcal{E}$ of a source is not the same as the potential difference U between its terminals. Each source has some *internal resistance r*. Consequently a part of the total electric energy that it generates is dissipated internally, and as it discharges, the terminal potential difference U is less than its emf. If the source carries some electric current I, the internal potential drop $= Ir$; therefore from the conservation of energy law:

$$\mathcal{E} = U + Ir \quad \text{or} \quad U = \mathcal{E} - Ir \quad (21\text{-}4)$$

Note that the larger the electric current I in the source, the greater the difference between its emf $\mathcal{E}$ and the terminal potential difference U. If the terminals of the source are not connected, i.e. if they are externally insulated from each other, no current exists, and the terminal potential difference is equal to the emf. This potential difference is called the *open-circuit voltage*.

Consider a simple circuit consisting of a single load of resistance R connected to the terminals of a source that has an internal resistance r and produces an emf $\mathcal{E}$ (Fig. 21-12). If the source does work W to transmit

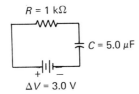

$R = 1 \text{ k}\Omega$

$C = 5.0 \text{ }\mu\text{F}$

$\Delta V = 3.0 \text{ V}$

Figure 21-12

some net charge Q in an elapsed time t, it generates a total average power:

$$\bar{P}_T = \frac{W}{t} = \frac{Q\mathcal{E}}{t} = I\mathcal{E} \quad (21\text{-}5)$$

where $I = Q/t$ is the average current in the circuit. According to the conservation of energy law, when the circuit carries a steady current, it must also completely dissipate this power. Some power (I^2R) is dissipated in the load, and the remainder (I^2r) is dissipated by the internal resistance of the source. Thus

$$\bar{P}_T = I\mathcal{E} = I^2R + I^2r = I^2(R + r)$$

and $\quad \mathcal{E} = I(R + r) \quad \text{or} \quad I = \dfrac{\mathcal{E}}{R + r} \quad (21\text{-}6)$

Note that, as the resistance R of the load decreases, the electric current I increases. If the external terminals of the source are *shorted* (connected by some conductor of negligible resistance), $R = 0$ and the electric current $I = \mathcal{E}/r$. This is called the *short-circuit current*.

Example 21-2

A dc generator has an open circuit voltage of 34.5 V. When it is connected to the terminals of the generator, a 22.0 Ω resistor draws an electric current of 1.40 A. Determine (a) the internal resistance r of the generator, (b) the terminal potential difference of the generator, (c) the short-circuit current.

Solution:

(a) $\quad r = \dfrac{\mathcal{E} - IR}{I} = \dfrac{34.5 \text{ V} - (1.40 \text{ A})(22.0 \text{ }\Omega)}{1.40 \text{ A}}$

$\qquad\qquad = 2.64 \text{ }\Omega$

(b) $\quad U = \mathcal{E} - Ir = IR = (1.40 \text{ A})(22.0 \text{ }\Omega)$

$\qquad\qquad = 30.8 \text{ V}$

(c) When $R = 0$,

$$I = \frac{\mathcal{E}}{r} = \frac{34.5 \text{ V}}{2.64 \text{ }\Omega} = 13.1 \text{ A}$$

Combinations of Cells

Any combination of cells is called a *battery*, and it is represented schematically by the symbol:

$$-\!\vdash\!\mid\!\mid\!\vdash\!+$$

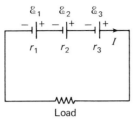

(a) Series aiding.
Total emf $\mathscr{E}_T = \mathscr{E}_1 + \mathscr{E}_2 + \mathscr{E}_3$.
Total internal resistance $r_T = r_1 + r_2 + r_3$.
Each cell carries the same current I.

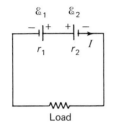

(b) Series opposing.
Total emf $\mathscr{E}_T = \mathscr{E}_1 - \mathscr{E}_2$.
Total internal resistance $r_T = r_1 + r_2$.
Each cell carries the same current I.

Figure 21-13
Cells in series.

Cells in Series. Two or more cells are in series when they are connected so that they provide only one path for an electric current and all the cells carry identical electric currents. When the cells are arranged with the negative electrode of one connected to the positive electrode of the next (Fig. 21-13a), they aid each other; if two like electrodes are connected, the cells oppose each other (Fig. 21-13b).

Each cell does work to change the total energy of the same charge as it passes through the system; therefore the total emf of the series combination is equal to the algebraic sum of the individual emfs.

$$\mathscr{E}_{TOT} = \sum_{\text{algebraic}} \mathscr{E} \quad \text{(in series)} \quad (21\text{-}7)$$

Also, since the cells are in series, the total internal resistance r_T is equal to the sum of the individual internal resistances:

$$r_T = \sum r$$

Cells in Parallel. A group of cells are connected in parallel when they provide different paths for an electric current. When *iden-*

tical cells are connected in parallel with their positive electrodes joined and their negative electrodes joined, their total emf is equal to the emf of a single cell, and all cells always carry identical electric currents (Fig. 21-14). For example, consider the total work W that is done as a total charge Q is transmitted through three identical cells that have equal emfs $\mathscr{E}$. The cells always carry identical cur-

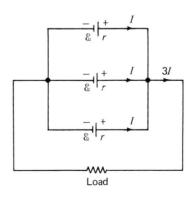

Load

Figure 21-14
Identical cells in parallel.

rents, and they each transmit a fraction $Q/3$ of the total charge Q; therefore the total work done is:

$$W = \frac{Q}{3}\varepsilon + \frac{Q}{3}\varepsilon + \frac{Q}{3}\varepsilon = Q\varepsilon$$

Thus the total equivalent emf $\varepsilon = W/Q$ is equal to the emf of a single cell.

If r_T is the total internal resistance and r is the internal resistance of each cell in parallel, then:

$$\frac{1}{r_T} = \frac{1}{r} + \frac{1}{r} + \cdots = \frac{n}{r} \qquad (21\text{-}8)$$

where n is the number of *identical* cells; therefore $r_T = r/n$.

Recharging a Battery. When a battery is recharged, work is done by some other source to reverse the direction of the electric current. The battery and the source are connected so that the emf ε_B of the battery opposes the emf ε_s of the source (Fig. 21-15). If the inter-

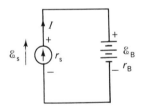

Figure 21-15
Recharging a battery.

nal resistances of the battery and source are r_B and r_s respectively, when the circuit carries a current I, then the total series resistance of the sources $= r_B + r_s$. This resistance dissipates electric energy at a rate $I^2(r_B + r_s)$. Therefore, since the battery opposes the source, the total power delivered to the circuit $P = I\varepsilon_s - I\varepsilon_B = I^2(r_B + r_s)$ and

$$I = \frac{\varepsilon_s - \varepsilon_B}{r_B + r_s} \qquad (21\text{-}9)$$

Example 21-3

Determine the terminal potential difference of a source if it sustains a charging current of 15 A through a battery that has an emf of 12 V and an internal resistance of 0.20 Ω.

Solution:

$$U = \varepsilon_s - Ir_s = \varepsilon_B + Ir_B$$
$$= 12\text{ V} + (15\text{ A})(0.20\text{ Ω}) = 15\text{ V}.$$

The total electric resistance (including all internal resistances) for a complete circuit is:

$$R_t = \frac{\text{net emf of the sources}}{\text{current}} = \frac{\Sigma\varepsilon}{I}$$

Problems

6. Calculate the terminal potential difference of a 9.0 V battery if its internal resistance is 0.25 Ω and it carries a current of 300 mA.

7. Determine the electric current in a source that has an emf of 6.0 V and an internal resistance of 0.50 Ω when its terminal potential difference is 5.8 V.

8. Explain why the headlights of a car become dimmer when the car is started. *Hint:* The starter draws a large electric current.

9. A 440 Ω load draws a steady current of 20 mA when it is connected to the terminals of a battery that has an open-circuit voltage of 9.0 V. Determine (a) the internal resistance of the battery, (b) the terminal potential difference of the battery, and (c) the short-circuit current.

10. A storage battery has an emf of 11.0 V and an internal resistance of 0.50 Ω. If the battery is connected to a 30.0 Ω resistor, determine (a) the current in the resistor and (b) the terminal voltage of the battery.

11. Calculate the charging current when a battery with an emf of 6.0 V and an inter-

nal resistance of $0.25\ \Omega$ is charged by a source that has a terminal potential difference of 8.0 V.

21-4 KIRCHHOFF'S RULES—NETWORK ANALYSIS

Many circuits contain more than one source of emf and several devices that dissipate electric energy. These components are usually interconnected by metallic conductors that cross regularly and are often joined to form a complex system called a *network*. A *mesh* or *loop* is any group of conductors that are connected so that they form a closed path in a network (Fig. 21-16). Points where three or more components are joined are called *principal nodes* or *junctions*. Sources of emf convert other energy forms into electric energy; therefore they are called *active* devices. Components of a network that dissipate electric energy, such as resistors, are called *passive* devices.

In many cases it is not possible to analyze a complex network in terms of a net emf and a total resistance. However two fundamental rules formulated by Gustav Kirchhoff (1824–1887) may be used. These rules are called *Kirchhoff's laws*.

Kirchhoff's first law is essentially a restatement of the conservation of charge. It states that:

> *At any instant, the total electric current arriving at any point in a network is equal to the total electric current leaving that point.*

This law merely implies that, when a charge flows into a point, an equal charge must leave that point. If a net charge accumulates at any point, the electric potential of that point changes.

Kirchhoff's second law is a restatement of the conservation of energy. It states that:

> *The algebraic sum of the emfs is equal to the algebraic sum of the potential drops around any mesh in a network.*

or

> *The algebraic sum of the potential changes around any mesh is equal to zero.*

Sources of emf may do work to move some charge around any closed path in a network, but the energy of that charge is dissipated in various devices. When it returns to the starting point, where the electric potential is unchanged, the charge has the same energy that it started with.

These rules may be used to solve for currents and voltages in relatively complex circuits. The procedure is illustrated in Example 21-4.

Example 21-4

Given the network in Fig. 21-16a, determine (a) the electric current in the resistors R_1, R_2 and R_3, and (b) the potential difference between points A and B.

Solution:

(a) 1. It is usually convenient to redraw the network, indicating the internal resistances separately from the sources. If terminal potential differences are given, this step may be omitted.

    2. In this case there are two independent closed loops;* therefore denote the currents in each as I_1 and I_2, and assume that they are directed as indicated in Fig. 21-16b. If the assumed direction is incorrect, we will obtain a negative value, but this will not affect other values.

    3. By applying Kirchhoff's first law to junction A or B, we see that the current in the resistance R_3 is $I_1 + I_2$.

    4. Indicate higher and lower poten-

---

*We could take the total outside circuit instead of one of these.

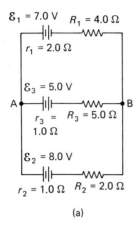

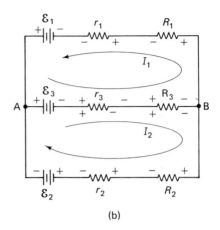

(a)   (b)

Figure 21-16

tial terminals of each component with + and − signs, respectively. Note that conventional currents I_1 and I_2 are directed *from* higher (+) *to* lower (−) potential terminals in resistors and other *passive* components, whereas the polarity of a source of emf must be as indicated on the schematic.

5. According to Kirchhoff's second law the algebraic sum of potential changes in each loop (indicated by the currents) is zero. Positive potential changes correspond to an increase in potential (minus to plus), and negative changes are decreases in potential. Thus starting at point A and in the directions indicated by the currents,* we obtain:

Mesh 1: $-\mathcal{E}_2 - r_3(I_1 + I_2) - R_3(I_1 + I_2)$
$- I_1 R_1 - I_1 r_1 + \mathcal{E}_1 = 0$

Substituting the values and rearranging, we obtain:

$$2.0 \text{ V} = (12.0 \ \Omega)I_1 + (6.0 \ \Omega)I_2 \quad (1)$$

---

*Remember that the current in the central branch is $I_1 + I_2$ and that the effects of the currents in the sources are considered in terms of the potential drops in their internal resistances.

Mesh 2: $-\mathcal{E}_3 - r_3(I_1 + I_2) - R_3(I_1 + I_2)$
$- I_2 R_2 - I_2 r_2 - \mathcal{E}_2 = 0$

Thus $-13 \text{ V} = (6.0 \ \Omega)I_1 + (9.0 \ \Omega)I_2 \quad (2)$

The two simultaneous equations (1) and (2) may now be solved by substitution or determinants to give:

$$I_1 = 1\tfrac{1}{3} \text{ A} \quad \text{and} \quad I_2 = -2\tfrac{1}{3} \text{ A}$$

The negative sign for the current I_2 implies that it is in the opposite direction to that indicated.

(b) Between A and B the potential difference is merely the algebraic sum of the potential changes.

$$U_{AB} = -\mathcal{E}_1 + I_1 r_1 + I_1 R_1$$
$$= -(7.0 \text{ V}) + (1\tfrac{1}{3} \text{ A})(2.0 \ \Omega + 4.0 \ \Omega)$$
$$= 1.0 \text{ V}$$

Note that the same result can also be obtained by considering either of the other branches because they are in parallel.

Problems

12. Determine the electric current in each branch of the network illustrated in Fig. 21-17, when $\mathcal{E}_1 = 1.5$ V, $\mathcal{E}_2 = 6.0$ V, $r_1 = 0.5 \ \Omega$, $r_2 = 1.0 \ \Omega$, $R_1 = 5.0 \ \Omega$, $R_2 = 3.0 \ \Omega$ and $R_3 = 10.0 \ \Omega$.

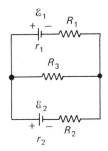

Figure 21-17

13. Determine the current in each resistor and the potential difference between points A and B in the network represented by Fig. 21-18.

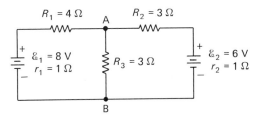

Figure 21-18

14. If the voltages $U_1 = 30$ V and $U_2 = 20$ V, find (a) the current in each resistor, (b) the potential difference between A and B, (c) the power dissipated in each resistor of the network illustrated in Fig. 21-19.

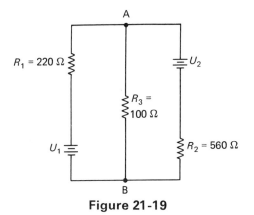

Figure 21-19

21-5 MEASURING INSTRUMENTS

A *D'Arsonval galvanometer* is one of the basic devices that is used in many measuring instruments. This device operates on the principle that a current carrying wire coil experiences a torque when it is placed in a magnetic field.

A wire coil is suspended so that it can deflect about its own axis in a constant magnetic field, and a spiral spring is used to provide an opposing torque. When it carries an electric current, the coil experiences a torque, and it deflects by distorting the spiral spring. The magnitude of the deflection is directly related to the magnitude of the current in the coil, and it is indicated by a pointer, which is attached to the coil (Chapter 23). In circuit diagrams galvanometers are represented by the symbol:

The dc Ammeter

An *ammeter* is an instrument that can be used to measure electric currents. It is represented by the symbol:

Isolated galvanometers are ammeters, but they are usually so sensitive that a very small electric current produces a full scale deflection of the pointer. Ammeters that measure larger currents are constructed by connecting a very low resistance, called a *shunt*, in parallel with a galvanometer. Most of the current is then diverted away from the galvanometer through the shunt (Fig. 21-20).

In order to measure an electric current, an ammeter must be connected in series so that it forms a part of the total current cir-

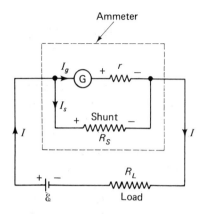

Figure 21-20
A dc ammeter in a simple circuit.

cuit. The ammeter should not affect the characteristics of a circuit by dissipating significant quantities of electric energy; therefore its total resistance must be quite small. The magnitude of the shunt resistance R_s, determines the magnitude of the circuit current I that will give a full-scale deflection of the galvanometer. In the ammeter a part I_s of the total current is diverted through the shunt, and the remainder I_g produces a deflection in the galvanometer. According to Kirchhoff's laws, $I = I_g + I_s$, and, if r is the internal resistance of the galvanometer, $I_g r - I_s R_s = 0$. Thus

$$\frac{R_s}{r} = \frac{I_g}{I_s} = \frac{I_g}{I - I_g} \qquad (21\text{-}10)$$

*Example 21-5*_____

A galvanometer has an internal resistance $r = 27\ \Omega$, and it indicates a full-scale deflection when it carries a current $I_g = 200$ mA. Calculate the magnitude of a shunt resistance that should be connected in parallel with the galvanometer to produce an ammeter with a 2.0 A full-scale reading.

Solution:

From Eq. (21-10),

$$R_s = \frac{r I_g}{I - I_g} = \frac{(27\ \Omega)(0.200\ A)}{(2.0\ A - 0.200\ A)} = 3.0\ \Omega$$

The dc Voltmeter_____

A *voltmeter* is an instrument that is designed to measure potential differences; it is represented by the symbol:

The device consists of a very sensitive galvanometer that is connected in series with a very large resistance called a *multiplier* resistor. To measure the potential difference between two points in a circuit, the voltmeter must be connected in parallel with that circuit element (Fig. 21-21). Note that the potential drop is directly related to the current in the circuit element, and some current I_g is diverted through the voltmeter during the measurement process. However, because the voltmeter has a very high resistance, the current that it draws is usually insignificant.

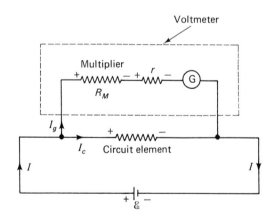

Figure 21-21
A voltmeter which is being used to measure the potential drop in a circuit element.

The potential difference U_F that produces a full-scale deflection of the voltmeter depends on the magnitude of R_M of the multiplier resistance. If the galvanometer indicates a full-scale deflection when it carries some current I_g, then the total potential drop in between the voltmeter terminals is:

$$U_F = I_g(R_M + r) \qquad (21\text{-}11)$$

where r is the internal resistance of the galvanometer. The *sensitivity* of a voltmeter is the inverse of the current that produces a full-scale deflection in the galvanometer (*sensitivity* $= 1/I_g$).

*Example 21-6*_____

A galvanometer has an internal resistance $r = 50\ \Omega$, and it indicates a full-scale deflection when it carries a current $I_g = 100\ \mu A$. (a) Calculate the magnitude of a multiplier resistance that should be connected in series with the galvanometer to produce a voltmeter with a 20 V full-scale reading. (b) Determine the sensitivity of the voltmeter.

Solution:

(a) From Eq. (21-11)

$$R_M = \frac{U_F}{I_g} - r = \left(\frac{20\ \text{V}}{1.00 \times 10^{-4}\ \text{A}}\right) - 50\ \Omega$$
$$= 199\ 950\ \Omega$$

(b) Sensitivity

$$= \frac{1}{I_g} = \frac{1}{1.00 \times 10^{-4}\ \text{A}} = 1.00 \times 10^4\ /\text{A}$$

or $\qquad 10\ \text{k}\Omega/\text{V}$

_____

The Ohmmeter_____

An *ohmmeter* is a convenient instrument that allows us to determine rapidly the approximate values of resistances; it does not give accurate values for very high or very low resistances, but it is easy to use.

The device consists of a galvanometer that is connected in series with a source of emf and an adjustable resistor (Fig. 21-22). When the leads of the ohmmeter are connected, the resistor is adjusted ("zeroed") until the circuit draws a current that produces a full-scale deflection of the galvanometer. If an external resistor is then connected between the ohmmeter leads, the total resistance of

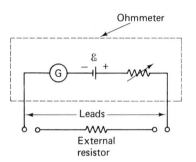

Figure 21-22
An ohmmeter.

the circuit is increased. Consequently less current is drawn, and the galvanometer deflection is reduced. The scale for resistance is therefore in the reverse direction to that for current or voltage, and it is not linear.

Potentiometer_____

In many networks a *voltage divider* circuit is used to produce different values of potential difference from a single source of emf. The source is connected in series with a string of resistors or to a single resistor that may be tapped at different points.

Consider a simple potential divider consisting of two resistors that have resistances $R_1 = 20\ \Omega$ and $R_2 = 30\ \Omega$ connected in series with a source that develops a terminal potential difference $U = 200\ \text{V}$ (Fig. 21-23). The total resistance of the resistors $R_T = R_1 + R_2 = 50\ \Omega$, and the circuit draws a current $I = U/R_T = 4.0\ \text{A}$. Therefore, the potential difference that is produced between the terminals of R_1 is:

$$U_1 = IR_1 = (4.0\ \text{A})(20\ \Omega) = 80\ \text{V}$$

Similarly the potential difference between the terminals of R_2 is:

$$U_2 = IR_2 = 120\ \text{V}$$

Since the current is constant, the ratio of the resistances is the same as the ratio of the voltage drops across their terminals.

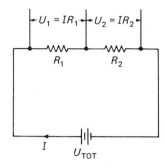

Figure 21-23
A simple voltage divider.

$$R_1/R_2 = U_1/U_2 \qquad (21\text{-}12)$$

Unfortunately, when a load is connected in parallel with one of these resistors, it changes the resistance of the circuit, and the potential difference changes. This effect can be reduced by including a *bleeder* resistor in parallel with the load. The magnitude of this resistor is chosen so that it tends to draw a constant current from the source.

The accuracy of measurements obtained with many instruments is limited because the instrument dissipates electric energy and changes the characteristics of the circuit during the measurement process. Accurate measurements can be obtained by *comparing* one device with another (some standard) in a special type of measuring instrument called a *bridge*.

A *potentiometer* is a type of bridge that can be used to measure accurately the emf of a source by comparing it with a known emf from some standard source, such as a Weston cell. Potentiometers are also used to calibrate ammeters and voltmeters.

A schematic of a potentiometer is illustrated in Fig. 21-24. The operation of the instrument is based on the voltage divider principle. Resistor AB can be tapped at any

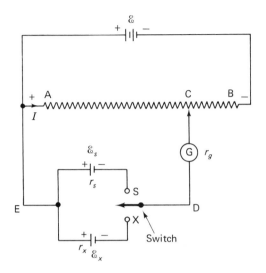

Figure 21-24
A potentiometer schematic.

point C along its length, and it draws a constant current I from the working source ε. By adjusting point C, the potential drop between A and C can be varied, since the resistance of the resistor segment AC usually is directly proportional to its length.

To measure the emf ε_x of a source, the switch is thrown to position X, and the contact point C is adjusted until the galvanometer indicates that there is no current in the source ε_x. At this point the potentiometer is said to be balanced. Using Kirchhoff's second law around the mesh ACDE of the balanced potentiometer, we obtain:

$$-IR_x + 0(r_g + r_x) + \varepsilon_x = 0$$
or $$\varepsilon_x = IR_x \qquad (1)$$

where R_x is the resistance of the resistor segment AC and r_g and r_x are the internal resistances of the galvanometer and the source ε_x, respectively.

When the switch is thrown to position S, the standard source ε_s replaces the source ε_x in the circuit. The contact point C is readjusted until the galvanometer indicates that the potentiometer is balanced. Then

$$\mathcal{E}_s = IR_s \qquad (2)$$

where R_s is the resistance of the different resistor segment AC. Comparing Eqs. (1) and (2), we obtain:

$$\mathcal{E}_x/\mathcal{E}_s = R_x/R_s \quad \text{or} \quad \mathcal{E}_x = \mathcal{E}_s \frac{R_x}{R_s} \quad (21\text{-}13)$$

Therefore, if the emf $\mathcal{E}_s$ of the standard source is known, the emf $\mathcal{E}_x$ of some other source can be determined by balancing the potentiometer and measuring the ratio of the resistances R_x and R_s. This measurement process is very accurate because there is no energy dissipation (or current) in the sources $\mathcal{E}_x$ or $\mathcal{E}_s$ when the potentiometer is balanced.

In practice a variable resistor is often added in series with the galvanometer to protect it from large currents.

The Wheatstone Bridge

A *Wheatstone bridge* is an instrument in which an unknown resistance R_x can be accurately measured by a comparison process with three adjustable precision resistances R_1, R_2 and R_3 that have known values. The typical bridge circuit is illustrated in Fig. 21-25.

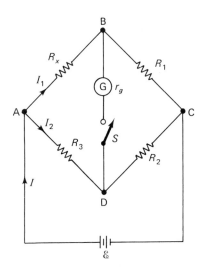

Figure 21-25
A Wheatstone bridge schematic.

To measure the unknown resistance R_x, the switch S is closed, and the bridge is balanced by adjusting one or more of the other three resistances until the galvanometer indicates that there is no current between points B and D. When Kirchhoff's first law is applied to the junctions B and D of the balanced bridge, it is obvious that R_1 carries the same current I_1 as the unknown resistance R_x, and that R_2 carries the same current I_2 as R_3. Using Kirchhoff's second law around the mesh ABD, we see that $-I_1R_x + 0r_g + I_2R_3 = 0$, where r_g is the internal resistance of the galvanometer. Thus

$$I_1R_x = I_2R_3 \quad \text{or} \quad \frac{I_2}{I_1} = \frac{R_x}{R_3} \quad (1)$$

Similarly, applying Kirchhoff's second law to mesh BCD, we obtain:

$$I_1R_1 = I_2R_2 \quad \text{or} \quad \frac{I_2}{I_1} = \frac{R_1}{R_3} \quad (2)$$

Equating (1) and (2), we obtain:

$$\frac{I_2}{I_1} = \frac{R_x}{R_3} = \frac{R_1}{R_2}$$

Therefore, the unknown resistance is:

$$R_x = \frac{R_3 R_1}{R_2} \quad (21\text{-}14)$$

Problems

15. A galvanometer has an internal resistance of 48 Ω, and it indicates a full-scale deflection when it carries a current of 100 mA. Calculate the magnitude of the shunt resistance that should be connected to the galvanometer to make an ammeter with a 2.5 A full-scale reading, and describe how the shunt should be connected.

16. A galvanometer has an internal resistance of 60 Ω, and it indicates a full-scale deflection when it carries a 200 μA current. (a) Calculate the value of a multiplier resistor that must be used to make a voltmeter with a 100 V full-scale reading. (b) Describe how the multiplier resistor is connected. (c) Compute the sensitivity of the voltmeter.

17. A single galvanometer has an internal resistance of 59 Ω, and it indicates a full-scale deflection when it carries a 25 mA current. Describe how to convert it into (a) an ammeter with a 1.5 A full-scale reading, (b) a voltmeter with a 350 V full-scale reading.

18. Describe how a potential difference of 35 V can be obtained from a source that delivers a terminal potential difference of 80 V.

19. When the switch is thrown to position X, the potentiometer in Fig. 21-24 is balanced by a 924.3 unit scale reading of AC; however, if the switch is thrown to position S, the potentiometer is balanced by a 618.1 unit scale reading. Calculate the emf ε_x of the source if the standard emf $\varepsilon_s = 1.018$ V.

20. The Wheatstone bridge in Fig. 21-25 is balanced when $R_1 = 23.1$ Ω, $R_2 = 43.6$ Ω and $R_3 = 56.4$ Ω. Calculate the unknown resistance R_x.

21-6 THERMIONIC VACUUM TUBES

Vacuum tubes still have many important applications, but they are gradually being replaced by solid-state devices, such as transistors and thyristors.

The operation of all vacuum tubes depends on the emission of electrons from a material. Normally an electron maintains some orbit around an atomic nucleus, but the electron may be able to overcome the electrostatic attraction to the nucleus and to break free of the atom if it is given sufficient energy. The extra energy may come from many sources, such as heat, strong electric fields, irradiation or bombardment with other particles.

Electron emission from a heated material is called *thermionic emission*; it was discovered by Thomas Edison in 1889 while he was investigating the properties of light bulbs. The significance of thermionic emission was not realized until 1904 when John Fleming developed the first vacuum tube diode.

The Vacuum Tube Diode

A *vacuum tube diode* consists of a heated electrode called the *cathode* and a metal *plate* or *anode*, which are sealed in an evacuated envelope (Fig. 21-26). Some vacuum tubes have directly heated cathodes that generate heat internally by dissipating electric energy, but the cathode usually is indirectly heated by a separate resistance heater. Most directly heated cathodes are tungsten or thoriated tungsten filaments. A common indirectly heated cathode consists of a hollow nickel cylinder that is coated with barium oxide and strontium oxide; the heater is located inside the cylinder, and the cathode is completely surrounded by a cylindrical plate.

A cathode must be heated to some minimum temperature before the heat energy is sufficient to produce thermionic emission of electrons from its surface. If the plate is maintained at some positive electric potential with respect to the cathode, some of these electrons flow to the plate, and the diode conducts. The plate is not heated, and electrons are not emitted from its surface; therefore, when the anode has a negative potential with respect to the cathode, no electrons flow, and the tube does not conduct. A vacuum tube diode only passes electric current in one direction. When it conducts, the electrons always flow from cathode to anode. The conventional current is in the opposite direction. If it is connected in series with a source that produces an alternating potential difference, the vacuum tube diode only conducts when its plate has a positive electric potential; this is

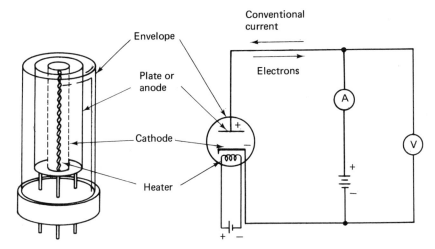

Figure 21-26
The vacuum tube diode.

known as *rectification*, and the diode behaves as a *rectifier*.

The electric current that is transmitted in a vacuum tube diode depends on the potential difference between the cathode and the plate (*the plate potential V*) and the temperature of the cathode.

Suppose that the diode is maintained at a constant positive plate potential. When the cathode is heated, its temperature increases until thermionic emission begins, and the diode conducts. As the temperature of the cathode rises, the rate of electron emission increases until electrons are emitted from the cathode faster than they are attracted to the plate. A net positive charge forms on the cathode as it emits electrons. Consequently the electrons tend to accumulate, forming a cloud of negative charge, called a *space charge*, around the cathode. This space charge repels other electrons back to the cathode, and the current *saturates* (becomes constant).

Consider the operation of the diode when the temperature of the cathode is constant and it emits electrons. Initially the current I increases as the plate potential V increases,

but eventually the rate at which the electrons are attracted to the plate is limited to the rate of thermionic emission from the cathode, and again the current saturates. Before it saturates, the current I is approximately proportional to the positive plate potential to the 3/2 of power.

$$I \approx CV^{3/2}$$

This relationship is called the *Langmuir-Child law*; the proportionality constant C depends on the structure of the tube.

The Triode

In 1906 L DeForest (1873–1961) added a third element called a *grid* between the cathode and plate of a vacuum tube and produced a *triode*. The grid is located close to the cathode, and it is constructed from a spiral or mesh of fine wire so that electrons can pass through it on the way to the plate. The relatively large current in the vacuum tube can be controlled by applying a small electric potential to the grid. If the grid is given a negative electric potential, it tends to repel electrons back to the

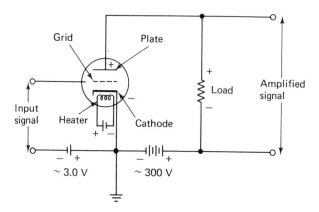

Figure 21-27
An amplifier circuit.

cathode, and the current in the tube is reduced. As the grid potential is made less negative, it allows more electrons to pass through it, and the current increases.

During the normal operation of a triode, the plate has a positive potential, and the grid has a small negative potential or *bias* with respect to the cathode. Small fluctuations in the potential difference between the grid and the cathode produce large variations in the current between the plate and the cathode; this is the fundamental principle of an *amplifier*. An amplifier circuit is illustrated in Fig. 21-27. The small signal (from an antenna, for example) is used to vary the potential difference between the cathode and the grid; this produces similar larger variations in the current between the plate and the cathode.

Triodes have relatively poor high-frequency operating characteristics because of capacitance between the three elements. This capacitance is reduced by including one or two more elements between the cathode and the plate. These devices are called *tetrodes* and *pentodes*, respectively.

In other tubes the envelope is filled with a gas. As the electrons accelerate in the electric field between the cathode and the plate, they may gain sufficient energy to ionize some

of the gas molecules, freeing other electrons. The electrons emitted from the cathode and those freed during the ionization process are then accelerated, and they can ionize other gas molecules. The operating characteristics of gas-filled tubes are usually quite different from their vacuum tube counterparts.

21-7 CATHODE RAYS

Discharge tubes are used to investigate the passage of electricity through gases at various pressures. They consist of a glass envelope with two electrodes. The *cathode* is an electrode that is maintained at a negative electric potential with respect to the positive electrode, which is called the *anode*. A vacuum pump is used to reduce the pressure of the gas in the glass envelope, and the potential difference between the anode and cathode may be varied by adjusting the voltage divider contact (Fig. 21-28).

At atmospheric pressure, air is a relatively good insulator, and electric field strengths of approximately 3×10^6 V/m are required before discharge occurs. However, at low pressures (of the order of 5 mm of

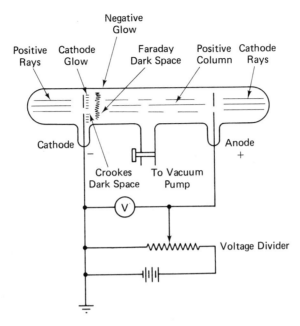

Figure 21-28
Discharge tube.

mercury) gases become relatively good conductors, and electric discharges are produced by much smaller electric fields. The gas in the tube glows during the discharge, and the tube is practically filled with a column of light known as the *positive column*. If the pressure is reduce further, the cathode becomes covered with a soft light known as the *cathode glow*. At the same time there is a dark space (*Crookes' dark space*), a luminous (*negative glow*) column and a second dark space (*Faraday's dark space*).

Different gases glow with different characteristic colors. This glow gradually disappears as the gas pressure is reduced even farther.

When the glass envelope is almost evacuated (the gas pressure is less than 10^{-3} mm of mercury) at the end of the envelope nearest the anode, the glass begins to glow, emitting a fluorescent green light. This glow is produced because of the bombardment of the

glass by rays that emanate from the cathode. These rays, called *cathode rays*, are really high-speed electrons. Cathode rays have the following characteristics:

1. If there are no external influences, they travel in straight lines. This can be demonstrated by placing a solid object in their path; a sharp dark shadow of the object appears in the area where the glass glows.

2. Cathode rays are emitted perpendicular to the surface of the cathode. They can be made to converge or diverge by changing the shape and curvature of the cathode surface (Fig. 21-29).

3. They possess kinetic energy and momentum. When cathode rays strike an object, the object recoils, and its temperature increases as some of the kinetic energy is converted into heat.

4. Cathode rays produce fluorescence in many materials. High-energy cathode rays

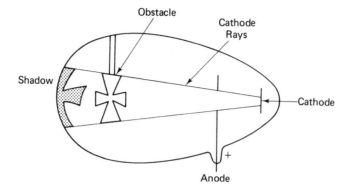

Figure 21-29

produce X-rays when they impinge on metal anodes.

5. Electric and magnetic fields can be used to deflect cathode rays. The direction of the deflection indicates that they possess negative charge. In 1897, J. J. Thomson (1856–1940) measured the magnitude of the deflections in electric and magnetic fields. He determined that all cathode rays (electrons) have the same charge to mass ratio; $e/m = 1.7589 \times 10^{11}$ C/kg. Note that this does not imply that they all have the same charge or mass. However, during the period from 1909 to 1911, Robert Millikan (1868–1953) found that all net charges were integer multiples of 1.6021×10^{-19} C, and he concluded that this value was a single electronic charge e. Thus the mass m of a single electron could be determined:

$$m = \frac{1.6021 \times 10^{-19} \text{ C}}{1.7589 \times 10^{11} \text{ C/kg}} = 9.1085 \times 10^{-31} \text{ kg}$$

Cathode Ray Tubes

Cathode ray tubes are extremely important devices that are used in oscilloscopes and television sets. The construction of a simple cathode ray tube is illustrated in Fig. 21-30.

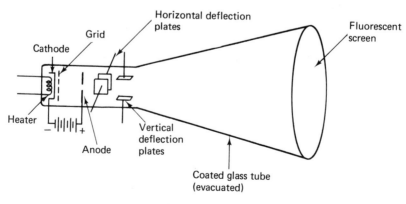

Figure 21-30
A cathode ray tube.

In operation electrons are emitted from the heated cathode, and they are accelerated through the control grid towards the anode. The electrons pass through a hole in the anode and continue into the regions between the pairs of horizontal and vertical deflecting plates. By applying potential differences between each set of plates, the electrons may be deflected horizontally and vertically. The magnitude of these deflections is proportional to the electric fields between the plates. In many cathode ray tubes (television sets), the electron beam is deflected by magnetic fields rather than electrostatic fields, but the result is identical. Once they have passed through the deflecting plates, the electrons continue in almost straight lines to the screen where they produce fluorescence.

REVIEW PROBLEMS

1. Determine the work done by a 9.0 V cell when it transmits a single electron.

2. Calculate the power developed by a 1.5 V cell if it transmits 2.5 μC of charge per second?

3. Calculate the terminal potential difference of a 6.0 V battery if its internal resistance is 0.20 Ω and it carries 500 mA.

4. Determine the internal resistance of a source that has an open circuit voltage of 8.0 V if it develops a 7.8 V terminal potential difference when it delivers 500 mA.

5. A battery has an open circuit voltage of 9.0 V and a 0.40 Ω internal resistance. If the battery is connected to a 15 Ω resistor, (a) what is the current in the circuit? (b) What is the terminal potential difference of the battery?

6. A 220 Ω load draws 50 mA when it is connected to the terminal of a 12 V battery. Determine (a) the internal resistance

of the battery, (b) the terminal potential difference, (c) the short-circuit current.

7. A battery has an open-circuit voltage of 12.0 V. When it is connected to a battery, a 500 Ω load draws a steady 20 mA current. Determine (a) the internal resistance of the battery, (b) the terminal potential difference of the battery, (c) the short-circuit current.

8. (a) Find the current in each branch of the circuit on Fig. 21-31. (b) What is the potential difference between A and B? (c) What power is dissipated in each resistance?

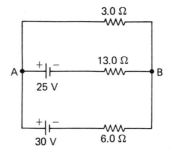

Figure 21-31

9. (a) Use Kirchhoff's laws to calculate the currents in each resistor of the circuit in Fig. 21-32. (b) Find the potential difference between A and B. (c) What power is dissipated in each resistor?

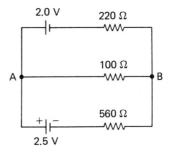

Figure 21-32

10. A galvanometer has a resistance of 250 Ω, and it requires 1.0 mA for a full-scale deflection. How would you convert it into (a) a voltmeter with a 30 V full-scale reading? (b) an ammeter with a 10.0 mA full-scale reading? (c) an ohmmeter?

11. A galvanometer has an internal resistance of 250 Ω, and it indicates a full-scale deflection when it carries a 200 mA current. (a) Describe how you would convert the galvanometer into a dc voltmeter with a 100 V full-scale reading, and calculate the values of any resistors involved. (b) Describe how you would convert the galvanometer into an ammeter with a 5.0 A full-scale reading, and calculate the values of any resistors involved.

12. A Wheatstone bridge is connected as shown in Fig. 21-33. The galvanometer has no current when $R_1 = 200 \, \Omega$, $R_2 =$

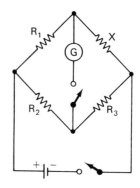

Figure 21-33

800 Ω and $R_3 = 60 \, \Omega$ at room temperature 20°C. (a) What is the resistance of the unknown wire X? (b) If wire X has a temperature coefficient of resistance $\alpha = 3.6 \times 10^{-3}/°C$ at 20°C, what is its resistance at $-50°C$?

Any region that contains a net electric charge has an electric potential, and it can do work by attracting or repelling other charges. This property is utilized in devices called *capacitors* that store electric potential energy for short periods of time in terms of a potential difference and an electric field.

There are many different types of capacitors, and they have many applications. Any system of conductors may be classified as a capacitor, but most capacitors consist of two or more closely spaced conductors separated by an insulator. In operation the conductors normally have equal magnitude but opposite polarity charges. Capacitors are represented schematically by the symbols:

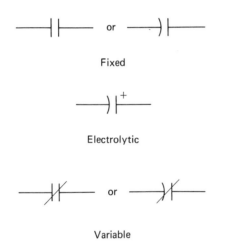

Fixed

Electrolytic

Variable

22

Capacitance and Dielectrics

22-1 CAPACITANCE

The electric potential of any conductor is directly proportional to its net charge, but different conductors which have the same net charge may have completely different electric potentials. The proportionality constant between the net charge Q and the corresponding electric potential V of a conductor is defined as its *capacitance C*.

$$C = \frac{Q}{V} \qquad (22\text{-}1)$$

Therefore *capacitance* is the term that describes the change in electric potential of a conductor when it is given a particular net charge. It is directly related to the size, shape and nature of the conductor.

The SI unit of capacitance is the *farad* (F). A capacitor has a capacitance of 1 F if its potential changes by 1 V when its charge changes by 1 C (1 F = 1 C/V). In practice the farad unit is generally too large, and capacitances are usually expressed in microfarads, nanofarads or even picofarads.

*Example 22-1*_____

Determine the capacitance of a metallic sphere that has a radius $r = 1.8$ cm if it is in a vacuum.

Solution:

For a spherical conductor of radius r and charge Q, the electrical potential is:

$$V = \frac{1}{4\pi\epsilon_0}\left(\frac{Q}{r}\right)$$

Thus

$$C = \frac{Q}{V} = 4\pi\epsilon_0 r$$

$$= 4\pi(8.85 \times 10^{-12} \text{ F/m})(1.8 \times 10^{-2} \text{ m})$$

$$= 2.0 \times 10^{-12} \text{ F} \quad \text{or} \quad 2.0 \text{ pF}.$$

The Parallel Plate Capacitor_____

One of the simplest forms of capacitor consists of two parallel metal plates that are separated by air, an insulator or a vacuum. An external source of electric energy, such as a battery or a power supply, must do work in order to produce equal magnitude but opposite charges $+Q$ and $-Q$ on the metal plates so that a potential difference U exists between them (Fig. 22-1). In effect a charge Q is transferred from one plate to the other. The capacitance C is the ratio of the charge Q on either plate to the corresponding potential difference U between them.

$$C = \frac{Q}{U} \qquad (22\text{-}2)$$

If the charged plates have relatively large surface areas A compared with their distance d apart, the electric field strength due to the equal and opposite charges tends to cancel at distant points exterior to the region between the plates. However between the plates a small positive test charge would be repelled by the positive plate and attracted to the negative plate. Consequently, if distortions of the field at the ends of the plates are ignored, the lines of electric force are parallel to each

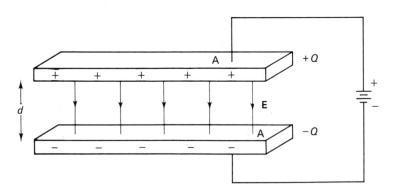

Figure 22-1
A charged parallel plate capacitor.

other and perpendicular to the surface of the plates. The electric field is approximately confined to the region between the plates and is equal to zero elsewhere.

If the plates are separated by a vacuum (with a permittivity ϵ_0), the electric field strength between the plates has a magnitude:

$$E = \frac{Q}{\epsilon_0 A} \qquad (22\text{-}3)$$

Also, from Eq. 19-12 if d is the distance between the plates, the potential gradient is:

$$\frac{U}{s} = \frac{U}{d} = E \qquad (22\text{-}4)$$

Thus the potential difference between the plates is:

$$U = Ed = \frac{Qd}{\epsilon_0 A} \qquad (22\text{-}5)$$

Therefore the capacitance of the parallel plate capacitor is:

$$C = \frac{Q}{U} = \frac{Q}{(Qd/\epsilon_0 A)} = \frac{\epsilon_0 A}{d} \qquad (22\text{-}6)$$

Note that in SI, permittivities can also be expressed in units of farads per meter (F/m).

Example 22-2

A simple capacitor is composed of two identical parallel metal plates that are located a distance $d = 1.20$ mm apart. If the plates each have surface areas $A = 8.00$ cm², determine (a) the capacitance if a vacuum exists between the plates and (b) the charge on each plate when they have a potential difference of 30.0 V.

Solution:

(a) $\quad C = \frac{\epsilon_0 A}{d}$

$\qquad = \dfrac{(8.85 \times 10^{-12} \text{ F/m})(8.00 \times 10^{-4} \text{ m}^2)}{1.2 \times 10^{-3} \text{ m}}$

$\qquad = 5.90 \times 10^{-12}$ F or 5.90 pF

(b) $\quad Q = CU = (5.90 \times 10^{-12}$ F$)(30.0$ V$)$

$\qquad = 1.77 \times 10^{-10}$ C

Problems

1. Determine the net charge on a 20 μF capacitor when it is connected to a 120 V power supply.

2. Determine the capacitance of a metallic sphere if it is in a vacuum and it has a radius of 2.7 cm.

3. Determine the radius of a metallic sphere if it is in a vacuum and it has a capacitance of 8.0 pF.

4. Opposite polarity charges of magnitude 3.6×10^{-3} C produce a potential difference of 720 V in a capacitor. What is the capacitance?

5. A simple parallel plate capacitor has a capacitance of 0.25 μF. If the charge on each plate has a magnitude of 5.0 μC, what is the potential difference between the plates?

6. A simple capacitor is composed of two parallel metal plates that are 1.1 mm apart and have equal surface areas of 8.2 cm². If a vacuum exists between the plates, determine (a) the capacitance and (b) the charge on each plate when they have a potential difference of 50 V.

7. A simple capacitor is composed of two parallel metal plates that have equal surface areas of 12.0 cm². If a vacuum exists between the plates, how far apart must they be in order to give a capacitance of 2.0 pF?

22-2 DIELECTRICS

A *dielectric* is an insulator in which an electric field can be produced without a significant flow of charge.

Michael Faraday (1791–1867) investigated the affects on capacitance when dielectrics such as glass, mica and oil, were

Table 22-1
Typical Properties of Dielectrics

Material	Dielectric Constant κ	Dielectric Strength E_{max} (kV/cm)
Air (1 atm)	1.000 6	29.5
Bakelite	4.4—5.4	158
Glass—Pyrex	4.8	130
Mica	6.9—9.2	1 970
Nylon	3.5	470
Oil	2.1	157
Paper	2.5—4.0	200
Paraffin	2.0—2.5	250
Polyethylene	2.3	470
Polystyrene	2.6	235
Porcelain	6.0—10.5	79
Rubber (hard)	2.8	275
Teflon	2.1	590
Vacuum	1.000 0	
Water	80	

inserted between the conductors of a capacitor. He found that, compared with a vacuum, the dielectrics increased the capacitance by a factor greater than one. This factor is called the *dielectric constant* κ. The dielectric constant depends only on the nature of the dielectric: it is independent of the charge, shape and dimensions of the capacitor. Some typical dielectric constants for different dielectric materials are listed in Table 22-1.

For example, in the case of a simple parallel plate capacitor with plates of area A and separation d, the capacitance with a vacuum between the plates is:

$$C_0 = \frac{\epsilon_0 A}{d}$$

With a dielectric (dielectric constant κ between the plates), the same capacitor has a capacitance:

$$C = \frac{\kappa \epsilon_0 A}{d} = \frac{\epsilon A}{d} \qquad (22\text{-}7)$$

where $\epsilon = \kappa \epsilon_0$ is called the *permittivity of the dielectric*. The dimensionless dielectric constant:

$$\kappa = \epsilon/\epsilon_0 = C/C_0 \qquad (22\text{-}8)$$

Example 22-3

A simple parallel plate capacitor has plates of area $A = 8.0$ cm^2 that are 0.15 cm apart. Determine (a) the capacitance C_0 with a vacuum between the plates and (b) the capacitance C when an insulator, which has a dielectric constant $\kappa = 4.0$, is inserted between the plates.

Solution:

(a) $C_0 = \dfrac{\epsilon_0 A}{d}$

$$= \frac{(8.85 \times 10^{-12} \text{ F/m})(8.0 \times 10^{-4} \text{ m}^2)}{1.5 \times 10^{-3} \text{ m}}$$

$$= 4.7 \times 10^{-12} \text{ F} = 4.7 \text{ pF}$$

(b) $C = \kappa C_0 = (4.0)(4.7 \text{ pF}) = 19 \text{ pF}$

Polarization of the Dielectric

The centers of positive and negative charge of some dielectric molecules, such as water, do not coincide even if there is no electric field. These molecules are called *polar molecules*, and they act like permanent atomic dipoles. A *dipole* consists of two equal magnitude but opposite polarity charges separated by some distance.

Other dielectric molecules become dipoles only in the presence of an external electric field that exerts an electric force on the electrons and protons in the molecules. These charged particles cannot flow through the dielectric because it is an insulator, but the orbits of the electrons around the atomic nuclei are displaced by a small amount. As a result the centers of positive and negative charge no longer coincide in the molecules, and atomic dipoles are produced.

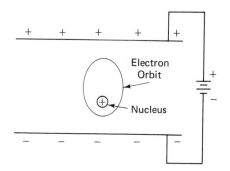

Figure 22-2

Displacement of electron orbits in an electric field are due to the Coulomb forces between the electrons and the charged plates.

This displacement of the charge centers in an electric field is called *polarization* (Fig. 22-2). The dielectric constant is a measure

of the degree of polarization of a dielectric in an electric field.

Consider a dielectric material that is located between the metal plates of a simple parallel plate capacitor. When the capacitor is charged, the atomic dipoles align in the electric field with their centers of negative charge closer to the positive plate of the capacitor (Fig. 22-3). In the interior of the dielectric, the effects of polarization are relatively small because of the proximity of the opposite polarity charge centers. However a layer of negative charge exists on the surface of the dielectric that is adjacent to the positively charged metal plate, and a layer of positive charge exists adjacent to the negative capacitor plate. These surface charges in the dielectric are called *surface polarization charges* Q_p. They produce an electric field $\mathbf{E}_p$ in the interior of the dielectric that opposes

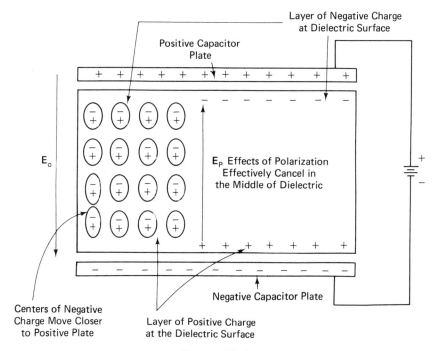

Figure 22-3

Polarization of a dielectric.

the field E_0 due to the charges on the capacitor plates. As a result the net electric field strength E is reduced in the dielectric to:

$$E = (E_0 - E_p) \text{ in the direction of } E_0$$

The dielectric also reduces the potential gradient and the potential difference between the capacitor plates even though they still contain the original net charge. Therefore the capacitance of the capacitor is increased. That is, $U = Ed$ decreases and thus $C = Q/U$ increases. Therefore for the same potential difference, each plate of a capacitor has a greater net charge when a dielectric is present.

Dielectric Strength

A dielectric not only modifies the electric field between capacitor plates but also insulates one plate from the other and prevents an electrical discharge between opposite polarity charges. However, when the electric field strength reaches some critical magnitude in the dielectric, an electric discharge occurs from one plate through the dielectric to the other plate. The *dielectric strength* is the magnitude of the maximum electric field strength that a dielectric can withstand before it loses its insulating properties. That is,

$$\text{Dielectric strength } E_{max} = \frac{U_{max}}{d} \quad (22\text{-}9)$$

Some typical values are listed in Table 22-1.

Example 22-4

The plates of a parallel plate capacitor are separated by sheets of mica that are 0.080 0 mm thick. Determine the maximum potential difference that the capacitor can withstand.

Solution:

From Eq. 22-9 and Table 20-1, we obtain

$$U_{max} = E_{max}d = (1\ 970 \text{ kV/cm})(8.00 \times 10^{-3} \text{ cm})$$
$$= 15.8 \text{ kV}$$

Problems

8. A capacitor has a capacitance of 3.2 μF with no dielectric and a capacitance of 8.6 μF when a particular dielectric is present. Determine (a) the dielectric constant and (b) the permittivity of the dielectric material.

9. A capacitor is constructed from two aluminum discs with diameters of 5.00 cm separated by a sheet of polystyrene 1.20 mm thick. (a) Calculate the capacitance. (b) If the capacitor is charged to a potential difference of 30.0 V, what is the charge on each plate? (c) Determine the maximum potential difference that the capacitor can withstand.

10. A 30 V potential difference is applied to a parallel plate capacitor that has a vacuum between the plates and a capacitance of 25 μF. The capacitor is then isolated; when a dielectric is inserted, the capacitance increases to 75 μF. Determine (a) the charge on each capacitor plate, (b) the dielectric constant of the dielectric material, (c) the permittivity of the dielectric.

11. Determine the maximum potential difference that a 1.6 cm thick sheet of ebonite can withstand if its dielectric strength is 8.0×10^4 kV/m.

22-3 COMBINATIONS OF CAPACITORS

In many electric circuits two or more capacitors are combined to produce some new value of capacitance. There are two common basic configurations.

Parallel

Two or more capacitors are said to be connected in *parallel* when they are arranged, as

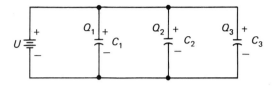

Figure 22-4
Three capacitors in parallel.

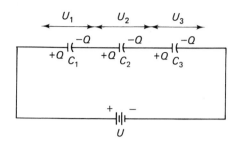

Figure 22-5
Three capacitors in series.

in Fig. 22-4, so that the *pairs of capacitor plates always have equal potential differences.* When the capacitors are charged, all the positively charged plates are interconnected, and all the negatively charged plates are interconnected. If the capacitors had unequal potential differences, a potential difference would also exist between the interconnected plates, and charge would flow until the potential of each interconnected plate was identical.

Consider three capacitors, which have capacitances C_1, C_2 and C_3. When the capacitors are connected in parallel with some source of electric energy that creates a potential difference U, charges of magnitude Q_1, Q_2 and Q_3 appear on the plates of the capacitors as shown in Fig. 22-4. Thus from the definition of capacitance:

$$U = \frac{Q_1}{C_1} = \frac{Q_2}{C_2} = \frac{Q_3}{C_3}$$

The total charge Q_{TOT} displaced is equal to the sum of the charges displaced by the capacitors:

$$Q_{TOT} = Q_1 + Q_2 + Q_3 = UC_1 + UC_2 + UC_3$$
$$= U(C_1 + C_2 + C_3)$$

Thus the total capacitance of the parallel combination is:

$$C_{TOT} = \frac{Q_{TOT}}{U} = C_1 + C_2 + C_3$$

or $\quad C_{TOT} = C_1 + C_2 + \cdots = \sum_i C_i \quad$ (22-10)

Series

Two or more capacitors are said to be in series when they are connected, as in Fig.

22-5, so that *equal magnitude charges always exist on all capacitor plates.* When the capacitors are charged, the positive plate of one capacitor is connected to the negative plate of its neighbor.

Consider the series combination of three capacitors, which have capacitances C_1, C_2 and C_3. When a total potential difference U is applied across the combination, some net charge $+Q$ forms on one plate of C_1. An equal but opposite charge $-Q$ is induced on the other plate of C_1, but, since C_1 and C_2 are connected and charge is conserved, this leaves a charge $+Q$ on one plate of C_2. This charge in turn induces a charge $-Q$ on the other plate of C_2. The same arguments apply to all other plates of capacitors that are connected in series; each plate in each capacitor has the same magnitude of net charge.

If U_1, U_2 and U_3 are the potential differences between the pairs of plates in the capacitors C_1, C_2 and C_3 respectively, then $Q = C_1 U_1 = C_2 U_2 = C_3 U_3$. But the total potential difference is:

$$U = U_1 + U_2 + U_3 = \frac{Q}{C_1} + \frac{Q}{C_2} + \frac{Q}{C_3}$$
$$= Q\left(\frac{1}{C_1} + \frac{1}{C_2} + \frac{1}{C_3}\right)$$

If C_{TOT} is the total capacitance, then:

$$\frac{1}{C_{TOT}} = \frac{U}{Q} = \frac{1}{C_1} + \frac{1}{C_2} + \frac{1}{C_3}$$

In general the reciprocal of the total capacitance of a system of capacitors in series is equal to the sum of the reciprocals of the individual capacitances.

$$\frac{1}{C_{TOT}} = \frac{1}{C_1} + \frac{1}{C_2} + \cdots = \sum_i \frac{1}{C_i} \quad (22\text{-}11)$$

Capacitors are often connected in a complex combination of series and parallel configurations that is equivalent to a single capacitance. In order to determine the equivalent capacitance of the total system, calculate the capacitance of each group in series and each group in parallel, and replace these groups by their equivalent capacitance. Repeat the process until the entire system is reduced to a single value for capacitance.

Example 22-5

Four capacitors have capacitances $C_1 = 2.0$ μF, $C_2 = 1.5$ μF, $C_3 = 3.0$ μF and $C_4 = 6.0$ μF. Determine the total capacitance when they are connected (a) in parallel, (b) in series, (c) with a series combination of C_2 and C_3 in parallel with a series combination of C_1 and C_4 (Fig. 22-6). (d) If a 50 V potential difference is applied across the whole system, determine the charge and potential difference for each capacitor in part (c).

Solution:

(a)
$$C_{TOT} = C_1 + C_2 + C_3 + C_4$$
$$= 2.0 \ \mu F + 1.5 \ \mu F + 3.0 \ \mu F$$
$$+ 6.0 \ \mu F$$
$$= 12.5 \ \mu F$$

(b)
$$\frac{1}{C_{TOT}} = \frac{1}{C_1} + \frac{1}{C_2} + \frac{1}{C_3} + \frac{1}{C_4}$$
$$= \frac{1}{2.0 \ \mu F} + \frac{1}{1.5 \ \mu F} + \frac{1}{3.0 \ \mu F}$$
$$+ \frac{1}{6.0 \ \mu F}$$
$$= \frac{10.0}{6.0 \ \mu F}$$

Thus

$$C_{TOT} = 6.0 \ \mu F / 10.0 = 0.60 \ \mu F.$$

(c) The total capacitance C_{23} due to the series combination of C_2 and C_3 is given by:

$$\frac{1}{C_{23}} = \frac{1}{C_2} + \frac{1}{C_3} = \frac{1}{1.5 \ \mu F} + \frac{1}{3.0 \ \mu F}$$
$$= \frac{2.0 + 1.0}{3.0 \ \mu F} = \frac{3.0}{3.0 \ \mu F}$$

or $C_{23} = 1.0$ μF. The two capacitors C_2 and C_3 can be replaced by a single equivalent capacitor C_{23}. Similarly the total capacitance C_{14} due to the series combination of C_1 and C_4 is given by:

$$\frac{1}{C_{14}} = \frac{1}{C_1} + \frac{1}{C_4} = \frac{1}{2.0 \ \mu F} + \frac{1}{6.0 \ \mu F} = \frac{4.0}{6.0 \ \mu F}$$

or $C_{14} = 1.5$ μF is equivalent to the series combination of C_1 and C_4. Therefore the total capacitance is the sum of the parallel combination of C_{14} and C_{23} (Fig. 22-6b):

$$C_{TOT} = C_{14} + C_{23} = 2.5 \ \mu F$$

(d) The charge on the equivalent capacitor C_{23} is given by:

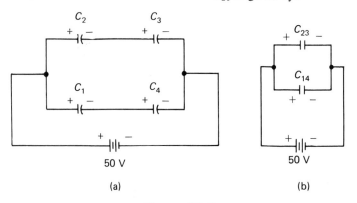

(a)   (b)

Figure 22-6

$$Q_2 = C_{23}U = (1.0 \times 10^{-6} \text{ F})(50 \text{ V})$$
$$= 5.0 \times 10^{-5} \text{ C}$$

Since they are connected in series, the capacitors C_2 and C_3 have this charge. Similarly the charges on C_1 and C_4 are:

$$Q_1 = C_{14}U = (1.5 \times 10^{-6} \text{ F})(50 \text{ V})$$
$$= 7.5 \times 10^{-5} \text{ C}$$

The potential differences between the plates:

$$U_1 = \frac{Q_1}{C_1} = \frac{7.5 \times 10^{-5} \text{ C}}{2.0 \times 10^{-6} \text{ F}} = 37.5 \text{ V},$$

$$U_2 = \frac{Q_2}{C_2} = \frac{50 \text{ } \mu\text{C}}{1.5 \text{ } \mu\text{F}} = 33\tfrac{1}{3} \text{ V}$$

$$U_3 = \frac{Q_2}{C_3} = \frac{50 \text{ } \mu\text{C}}{3.0 \text{ } \mu\text{F}} = 16\tfrac{2}{3} \text{ V},$$

$$U_4 = \frac{Q_1}{C_4} = \frac{75 \text{ } \mu\text{C}}{6.0 \text{ } \mu\text{F}} = 12.5 \text{ V}$$

Note that $U_1 + U_4 = 50 \text{ V}$ and $U_2 + U_3 = 50 \text{ V}$ as a check.

Multiplate Capacitors

Multiplate capacitors are modifications of simple parallel plate capacitors. They consist of two sets of parallel plates with alternate plates connected so that each successive pair of plates is in parallel (Fig. 22-7).

A multiplate capacitor containing N identical parallel plates is equivalent to $(N - 1)$ simple parallel plates capacitors, each of capacitance C in *parallel*; thus the total capacitance is

$$C_{\text{TOT}} = (N - 1)C = (N - 1)\frac{\kappa\epsilon_0 A}{d} \quad (22\text{-}12)$$

where κ is the dielectric constant of the material between the plates, ϵ_0 is the permittivity of free space, A is the cross-sectional area of the plates and d is the distance between successive plates.

Example 22-6

A multiplate capacitor is constructed from 20 parallel sheets of aluminum foil with areas $A = 0.50 \text{ cm}^2$ separated by mica films 0.150 mm thick with a dielectric constant $\kappa = 7.0$. Determine (a) the capacitance and (b) the maximum potential difference that can be applied.

Solution:

(a)
$$C = (N - 1)\frac{\kappa\epsilon_0 A}{d} = (20 - 1)\frac{(7.0)(8.85 \times 10^{-12} \text{ F/m})(5.0 \times 10^{-5} \text{ m}^2)}{1.50 \times 10^{-4} \text{ m}}$$
$$= 3.9 \times 10^{-10} \text{ F} \quad \text{or} \quad 39 \text{ pF}$$

(b)
$$U_{\text{max}} = E_{\text{max}}d = (1\ 970 \text{ kV/cm})(1.50 \times 10^{-2} \text{ cm}) = 29.6 \text{ kV}$$

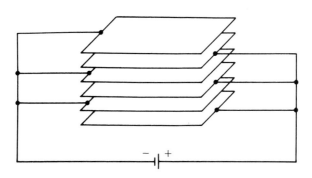

Figure 22-7
A multiplate capacitor.

Problems

12. Determine the total capacitance of two capacitors that have capacitances of 5.0 μF and 15.0 μF (a) if they are connected in parallel, (b) if they are connected in series.

13. Calculate the total capacitance of three capacitors that have capacitances $C_1 = 4.0$ pF, $C_2 = 12.0$ pF and $C_3 = 6.0$ pF (a) if they are connected in parallel, (b) if they are connected in series, (c) if they are connected so that a series combination of C_1 and C_2 is in parallel with C_3.

14. Determine the total capacitance of four capacitors that have capacitances $C_1 = 30$ μF, $C_2 = 50$ μF, $C_3 = 15$ μF and $C_4 = 25$ μF, if they are connected: (a) in parallel; (b) in series; (c) with a series combination of C_1, C_2 and C_3 in parallel with C_4; (d) with a series combination of C_1 and C_2 in parallel with a series combination of C_3 and C_4.

15. Calculate the charge and potential difference for each capacitor in part (c) of Problem 14 if the potential difference across the whole system is 100 V.

16. Calculate the charge and potential difference for each capacitor in part (d) of Problem 14, if the potential difference across the whole system is 30 V.

17. How can a number of 3.0 μF capacitors be arranged to give a total capacitance equal to (a) 12.0 μF? (b) 1.0 μF? (c) 7.0 μF?

18. A multiplate capacitor is constructed from 14 parallel sheets of aluminum foil with areas of 0.75 cm² separated by paper (dielectric constant 3.0) sheets 0.080 mm thick. Determine the capacitance and the maximum potential difference that can be applied.

19. A multiplate capacitor is to be made from 80 parallel sheets of aluminum foil separated by mica (dielectric constant 7.5). If the capacitor must have a capacitance of 2.50 nF and be able to withstand a maximum potential difference of 75.0 kV, determine (a) the minimum thickness of the mica sheets and (b) the corresponding areas of the plates.

22-4 ENERGY STORED IN CAPACITORS

In electric circuits capacitors are usually used to store electric energy for short periods of time. Work must be done in order to produce a net charge, and the energy of a capacitor is stored in terms of the potential energy that exists because of an accumulation of like charges.

The potential difference between the plates of an uncharged capacitor is equal to zero, but some work is required to transfer a small unit of charge, such as that carried by an electron, from one plate A to the other plate B. As the electron is removed, plate A acquires a net positive charge, and work must be done in order to overcome the coulomb attractive force that tends to return the electron to plate A. When the electron is deposited on plate B, a potential difference (and an electric field) exists between the capacitor plates (Fig. 22-8).

A greater amount of work is now required to remove a second electron from plate A to plate B because larger coulomb (electrostatic) forces exist and they tend to return the electrons to plate A. As the charges build up on the capacitor plates, they exert greater coulomb forces; therefore progressively larger amounts of work are required to produce a unit increase in the charges.

The electric potential of each plate in an

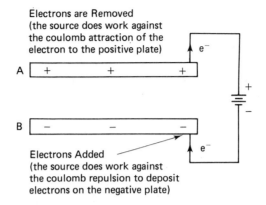

Figure 22-8
Charging a capacitor.

uncharged capacitor is equal to zero. When work is done in order to transfer a charge of magnitude Q in small increments from one capacitor plate to the other, some final potential difference U exists between the plates. Therefore the average potential difference between the plates during the charging process is:

$$\bar{U} = \frac{(0 + U)}{2} = \frac{U}{2}$$

Thus the work done is:

$$W = Q\bar{U} = \frac{QU}{2}$$

Once it has been charged, a capacitor possesses an electric potential energy that is due to the coulomb repulsion between the like charges on each plate. When the plates are interconnected with a conductor, the capacitor returns energy to an electric circuit since the like charges repel each other, and

some charges flow to neutralize the areas of net charge. The energy returned to the circuit is equal to the work done in charging the capacitor; therefore the electric potential energy of the charged capacitor is:

$$E_p = W = \frac{QU}{2} = \frac{Q^2}{2C} = \frac{CU^2}{2} \quad (22\text{-}13)$$

since $\qquad C = Q/U$.

*Example 22-7*_____

A multiplate capacitor has 50 plates, each with an area $A = 10.0 \text{ cm}^2$. If the plates are separated by strips of mica (dielectric constant $\kappa = 8.00$) of thickness $d = 0.75$ mm and the potential difference between successive plates is 35.0 V, determine (a) the stored electric potential energy and (b) the total charge Q on the capacitor.

Solution:

(a) $\qquad C = (N - 1)\frac{\kappa\epsilon_0 A}{d} = \frac{(50 - 1)(8.00)[8.85 \times 10^{-12} \text{ F/m}](10.0 \times 10^{-4} \text{ m}^2)}{7.5 \times 10^{-4} \text{ m}}$

$\qquad\qquad = 4.63 \times 10^{-9}$ F   or   4.63 nF.

Thus $\qquad E_p = \tfrac{1}{2}CU^2 = \tfrac{1}{2}(4.63 \times 10^{-9} \text{ F})(35 \text{ V})^2 = 2.84 \times 10^{-6}$ J

(b) $\qquad Q = CU = (4.63 \times 10^{-9} \text{ F})(35.0 \text{ V}) = 1.62 \times 10^{-7}$ C

*Problems*_____

20. How much electric potential energy is stored in a 6.0 μF capacitor if it has a potential difference of 120 V between its plates?

21. A simple parallel plate capacitor has plates of area $A = 8.0$ cm² that are separated by a neoprene strip (dielectric constant $= 6.7$) of thickness $d = 0.75$ mm. If the potential difference between the plates is 115 V, calculate (a) the stored electric potential energy and (b) the net charge on each capacitor plate.

22. A multiplate capacitor has 16 plates. Each plate has an area of 4.0 cm², and it is separated from its neighbor by a 1.2 mm thick sheet of teflon ($\kappa = 2.0$). If the potential difference between successive plates is 36 V, determine (a) the stored electric potential energy and (b) the charge on the capacitor.

23. A simple parallel plate capacitor has plates of area $A = 5.0$ cm² that are 0.075 cm apart in a vacuum. The capacitor is charged so that the potential difference between the plates is 250 V; the plates are then isolated, and an insulator with a dielectric constant of 3.5 is inserted between them. Determine (a) the charge on the capacitor plates, (b) the final potential difference between the plates, (c) the electric potential energy stored by the capacitor.

24. Three capacitors with capacitances of 200 μF, 300 μF and 600 μF are connected in parallel and are charged to a potential difference of 50 V. What is the total electric potential energy stored by the system?

25. If the three capacitors in Problem 24 are connected in series and charged so that the total potential difference across the combination is 120 V, determine the total

electric potential energy stored in the system.

26. A 250 pF capacitor is connected to a power supply that charges it to a potential difference of 50 V. The power supply is then removed, and the charged capacitor is connected in series with a second uncharged 250 pF capacitor. Determine the electric potential energy stored in the two capacitors.

22-5 TYPES OF CAPACITORS

The properties of a capacitor are mainly due to the dielectric material that is used in its construction; consequently capacitors are usually classified in terms of their dielectric.

Variable air capacitors consist of two sets of parallel metal plates that are separated by air. One set of plates is free to rotate into the spaces between the other set of plates so that the magnitude of the adjacent areas can change, and consequently the capacitance can be varied (Fig. 22-9a).

Paper capacitors are constructed by rolling two long strips of metal foil with two long strips of a paper dielectric that have been treated with wax or mineral oil; the roll is then sealed in a metal container (Fig. 22-9b).

In a *plastic film capacitor* two long strips of metal foil are tightly rolled with two long strips of a plastic film dielectric, and the entire capacitor is encased with some protecting material. Its structure is similar to that of a paper capacitor. Some of the plastics commonly used are *mylar, teflon, kodar* and *polystyrene*. The development of new plastics is resulting in a rapid improvement of plastic film capacitors.

Mica capacitors are constructed by separating two or more parallel sheets of metal foil with thin sheets of a mica dielectric; the entire capacitor is usually enclosed in a metal

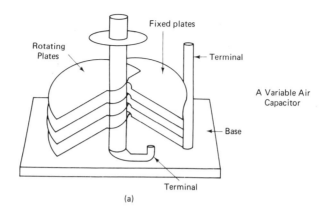

Fixed plates

Rotating Plates

Terminal

A Variable Air Capacitor

Base

Terminal

(a)

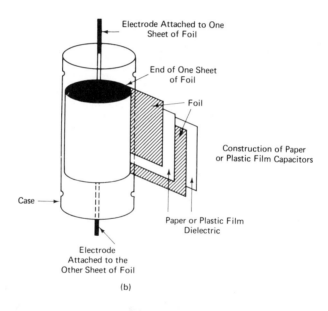

Electrode Attached to One Sheet of Foil

End of One Sheet of Foil

Foil

Construction of Paper or Plastic Film Capacitors

Case

Paper or Plastic Film Dielectric

Electrode Attached to the Other Sheet of Foil

(b)

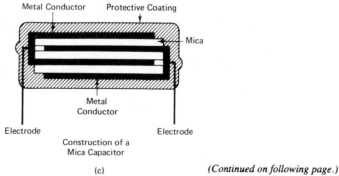

Metal Conductor

Protective Coating

Mica

Metal Conductor

Electrode

Electrode

Construction of a Mica Capacitor

(c)

(Continued on following page.)

Figure 22-9
Types of capacitors.

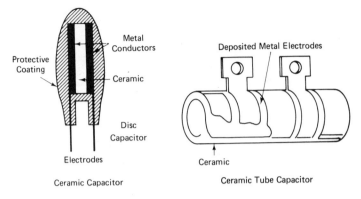

Ceramic Capacitor

Ceramic Tube Capacitor

(d)

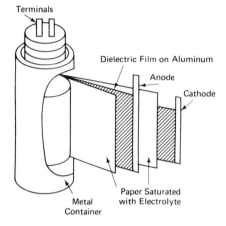

Electrolytic Capacitor

(e)

Figure 22-9 (cont.)
Types of capacitors.

or bakelite case. In *silvered mica capacitors* the metal foils are replaced by thin layers of silver that are deposited onto each side of the mica dielectric (Fig. 22-9c).

Ceramic capacitors consist of two or more metallic conductors that are separated by a ceramic dielectric; the capacitor is usually varnished or encased by some protecting material (Fig. 22-9d).

Some *electrolytic capacitors* are non-polar, but most have a definite polarity so that in operation one plate (the *anode*) is maintained at a positive electric potential with respect to the other plate (the *cathode*). The anode consists of an aluminum foil or a sintered tantalum powder; the dielectric is a thin oxide layer on the surface of the anode, and an electrolyte (Chapter 21) forms the cathode (Fig. 22-9e).

Usually a porous paper is saturated with the electrolyte and then wound into a roll with the coated anode and a metallic cathode terminal. The system is then placed in an aluminum or plastic container.

22-6 RC CIRCUITS

Charging a Capacitor

Consider a series circuit containing a capacitor, a resistor, a switch, and a source of emf (Fig. 22-10a). When the switch is thrown to position X, the source does work to transfer charges from one capacitor plate through the resistor to the other capacitor plate. At first there is a rapid transfer of charges, but the rate of charge flow decreases because of the increasing coulomb forces as the net charge builds up on each capacitor plate. After some elapsed time the rate of charge flow becomes very small, and the transient current virtually ceases.

Suppose that the capacitor was initially uncharged and has a capacitance C. After the switch has been closed for an elapsed time t, the source has transferred some net charge Q between the capacitor plates, and there is an instantaneous current i in the circuit.*
From the definition of capacitance, the instantaneous potential difference between the capacitor plates equals Q/C. Thus applying Kirchhoff's second law, we obtain:

$$U - iR - \frac{Q}{C} = 0 \quad \text{or} \quad i = \frac{U - Q/C}{R} \quad (22\text{-}14)$$

where U is the terminal potential difference developed by the source and R is the resistance of the resistor in ohms. But the instantaneous current i is equal to the time rate of charge flow $\Delta Q/\Delta t$, the slope of the tangent to the net charge versus elapsed time curve at the instant t (Fig. 22-10b). Thus

$$U - \left(\frac{\Delta Q}{\Delta t}\right)R - \frac{Q}{C} = 0$$

By the use of integral calculus, it can be shown that the solution to this equation is:

$$Q = CU(1 - e^{-t/RC}) \quad (22\text{-}15)$$

where $e = 2.718$ is a constant. This exponential function is plotted in Fig. 22-10b. Note that, as the elapsed time becomes very large ($t \rightarrow \infty$), the charge approaches a maximum value $Q_0 = CU$. Also, from Eq. 22-14, the instantaneous current is:

$$i = \frac{U - Q/C}{R} = \frac{U}{R} - \frac{CU}{RC}(1 - e^{-t/RC})$$
$$= \frac{U}{R}e^{-t/RC} \quad (22\text{-}16)$$

Therefore the instantaneous current decreases exponentially from an initial value U/R. Finally the potential difference u_c *across the capacitor* after an elapsed time t is:

$$u_c = \frac{Q}{C} = U(1 - e^{-t/RC}) = U - iR \quad (22\text{-}17)$$

---

*Lower case letters such as i and u will be used to denote time varying currents and voltages; capitals indicate steady values or constants.

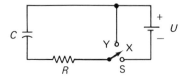

(a) When the switch S is thrown to position X the capacitor charges, and when the switch is at Y, the capacitor discharges.

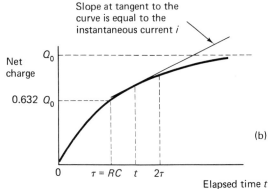

(b) Net charge on each capacitor plate versus elapsed time for the charging circuit.

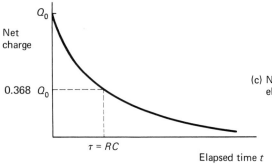

(c) Net charge on each capacitor plate versus elapsed time for the discharging circuit.

Figure 22-10
Charging and discharging a capacitor.

The quantity

$$\tau = RC \qquad (22\text{-}18)$$

is called the *time constant* for the series resistor and capacitor (RC) circuit because it is an indication of how rapidly the capacitor charges. When the elapsed time t is equal to the time constant $\tau = RC$, the charge on each capacitor plate has a magnitude:

$$Q = CU(1 - e^{-1}) = 0.632CU$$

That is, the capacitor is charged to approximately 63% of the maximum value $Q_0 = CU$. In practice a capacitor is usually considered to be fully charged when the elapsed time t is equal to five time constants.

*Example 22-8*_____

An uncharged 20 μF capacitor is connected in series with a 250 Ω resistor and a source that develops a terminal potential difference of 50 V. Determine (a) the time constant of the circuit, (b) the initial current in the circuit, (c) the charge Q on the capacitor, (d) the instantaneous potential difference u_c across the capacitor, (e) the instantaneous current i after 10.0 ms.

Solution:

(a) $\quad \tau = RC = (250\ \Omega)(20 \times 10^{-6}\ \text{F})$
$\quad\quad = 5.0 \times 10^{-3}\ \text{s} = 5.0\ \text{ms}$

(b) When $\quad t = 0, i = \dfrac{U}{R} = \dfrac{50\ \text{V}}{250\ \Omega} = 0.20\ \text{A}$

(c) $Q = CU(1 - e^{-t/\tau})$
$\quad\quad = (20 \times 10^{-6}\ \text{F})(50\ \text{V})(1 - e^{-10\ \text{ms}/5.0\ \text{ms}})$
$\quad\quad = 8.6 \times 10^{-4}\ \text{C}$

(d) $\quad\quad u_c = U(1 - e^{-t/\tau})$
$\quad\quad = (50\ \text{V})(1 - e^{-10\ \text{ms}/5.0\ \text{ms}}) = 43.2\ \text{V}$

(e) $\quad\quad i = \dfrac{U}{R}e^{-t/\tau} = \dfrac{50\ \text{V}}{250\ \Omega}(e^{-10\ \text{ms}/5.0\ \text{ms}})$
$\quad\quad = 2.7 \times 10^{-2}\ \text{A} = 27\ \text{mA}$

Discharging a Capacitor

At the instant when the capacitor has some potential difference U_0, if the switch S in Fig. 22-10a is thrown to position Y, there is a flow of electrons from the negatively charged capacitor plate through the resistor to the other capacitor plate; the conventional current is in the opposite direction. The flow of electrons tends to neutralize the charges on the capacitor plates. Consequently the coulomb forces decrease, and the rate of charge flow decreases.

After an elapsed time t the remaining charge on each capacitor plate is reduced to magnitude Q, and there is an instantaneous current i in the circuit. Applying Kirchhoff's second law, since there is no emf, we obtain:

$$\frac{Q}{C} + iR = 0 \quad \text{or} \quad i = \frac{-Q}{RC} = \frac{\Delta Q}{\Delta t} \quad (22\text{-}19)$$

where C is the capacitance in farads, R is the resistance in ohms and $i = \Delta Q/\Delta t$ is the time rate of charge flow. The negative sign merely

implies that the discharging current is in the opposite direction to the charging current. If the initial charge on the capacitor was Q_0, this equation has the solution:

$$Q = Q_0 e^{-t/RC} = Q_0 e^{-t/\tau} \quad (22\text{-}20)$$

The corresponding graph of charge versus elapsed time is shown in Fig. 22-10c. Therefore the instantaneous current is:

$$i = -\frac{Q}{RC} = -\frac{Q_0 e^{-t/RC}}{RC} = -\frac{U_0}{R}e^{-t/RC} \quad (22\text{-}21)$$

and the instantaneous voltage across the capacitor is:

$$u_c = -iR = U_0 e^{-t/\tau} \quad (22\text{-}22)$$

When the elapsed time $t = \tau = RC$, the charge on the capacitor is:

$$Q = Q_0 e^{-1} = 0.368 Q_0$$

That is, the capacitor has discharged to approximately 37% of its initial charge.

Problems

27. Determine the fraction of the maximum possible charge on a capacitor when it has been charging for an elapsed time equal to five time constants.

28. Determine the fraction of the total initial charge Q_0 that remains on the capacitor when it has been discharging for an elapsed time equal to five time constants.

29. An uncharged 50 μF capacitor is connected in series with a 1.2 kΩ resistor and a source that develops a terminal potential difference of 250 V. (a) Determine the time constant for the circuit and the initial current. (b) Compute the instantaneous current, the charge of the capacitor and the potential difference between the capacitor plates after 0.18 s.

30. A 10.0 kΩ resistor is connected in series with a 250 μF capacitor and a power supply that develops 150 V. (a) How long does it take to charge the capacitor to a

potential difference of 80 V? (b) What is the current in the circuit at that instant?

31. A 50 μF capacitor with an initial charge of 2.0 μC is connected in series with a 500 Ω resistor. Determine (a) the time constant and (b) the current in the circuit after an elapsed time of 0.10 s.

32. A 100 μF capacitor is charged to a 30 V potential difference and then is discharged through a 22 kΩ resistor. (a) How much energy is dissipated if the capacitor is totally discharged? (b) What is the time constant? (c) How long does it take to reduce the potential difference across the capacitor to 10 V?

33. An uncharged 250 μF capacitor is connected in series with a 200 Ω resistor and a source that delivers a terminal potential difference of 30 V. After an elapsed time of 0.20 s the source of emf is instantaneously removed, and the capacitor is allowed to discharge through the resistor. Determine the current in the circuit after a *total* elapsed time of (a) 0.15 s and (b) 0.25 s.

REVIEW PROBLEMS

1. Determine the potential difference across a 25 μF capacitor when it is charged to 40 μC.

2. A simple parallel plate capacitor has two metal plates 0.60 mm apart with equal surface areas of 3.5 cm^2 in air. Calculate (a) the capacitance and (b) the charge on the capacitor when a 25 V potential difference is applied.

3. A capacitor has a 60 μF capacitance with a vacuum between its plates and a 240 μF capacitance when a dielectric is inserted. (a) What is the dielectric constant? (b) Determine the permittivity of the dielectric.

4. Determine the maximum potential difference that a 2.0 cm thick sheet of ebonite can withstand if its dielectric strength is 8.0 $\times$ 10^4 kV/m.

5. Determine the capacitance of a multi-parallel plate capacitor that has 16 plates of area 8.0 cm^2 that are separated by a 0.95 mm thick dielectric with dielectric constant $\kappa = 6.5$.

6. A multiparallel plate capacitor is to be constructed from aluminum foil with a teflon dielectric. If the capacitor is to have 30 plates and a capacitance of 1.5 pF and it must withstand up to 15 kV, find (a) the minimum thickness of the teflon dielectric and (b) the corresponding area of each plate.

7. (a) Determine the total equivalent capacitance of the system of capacitors in Fig. 22-11.

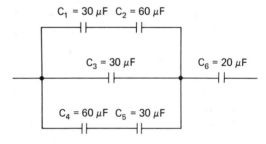

$C_1 = 30\ \mu F$ $C_2 = 60\ \mu F$

$C_3 = 30\ \mu F$ $C_6 = 20\ \mu F$

$C_4 = 60\ \mu F$ $C_5 = 30\ \mu F$

Figure 22-11

(b) If a 300 V potential difference is applied across the whole system, compute the potential difference across *each* capacitor.

8. How much electric potential energy is stored in a 150 nF capacitor when the potential difference between its plates is 30 V?

9. Three capacitors with capacitances 100 μF, 200 μF and 300 μF are connected in parallel and charged to a potential differ-

ence of 30 V. What is the stored electric potential energy?

10. A multiplate capacitor has 20 parallel metal plates each with area of 3.0 cm². The plates are separated from their neighbors by 0.80 mm thick sheets of mica with a dielectric constant of 5.0. If the potential difference between successive plates is 70 V, determine (a) the capacitance, (b) the stored electric potential energy, (c) the charge on the capacitor.

11. An uncharged 20 μF capacitor is connected in series with a 2.2 kΩ resistor and a source that develops a terminal potential difference of 120 V. (a) Determine the time constant for the circuit. (b) Compute the initial current in the circuit. (c) Calculate the charge on the capacitor after 132 ms. (d) Determine the instantaneous current in the circuit after 132 ms. (e) Determine the potential difference between the capacitor plates after 132 ms.

Electricity and magnetism are directly related because both types of phenomena can be attributed to interactions between electric charges. However, whereas electric forces are always exerted between charges whether they are at rest or in motion, magnetic forces only occur between moving charges.

The operation of many devices, such as galvanometers, electric motors, mass spectrometers, particle accelerators and some cathode ray tubes, depends on magnetic forces that are exerted on charges that move through a magnetic field.

Magnetic fields originate from moving electric charges; therefore electric currents are surrounded by magnetic fields. The operation of electromagnets, magnetic relays and switches, inductors and many other devices depends on the magnetic fields of electric currents.

23

Magnetic Fields

23-1 MAGNETIC POLES

The first observation of a magnetic phenomenon is usually attributed to the ancient Greeks. They noticed that pieces of a special kind of rock attracted each other. This type of rock is called *lodestone* or *magnetite*. It is a compound of iron and oxygen (Fe_3O_4) that is found in many parts of the world. A sample of lodestone is a *natural magnet*; it also attracts pieces of some other materials, such as iron, cobalt and nickel, that may not be magnetic.

An iron bar (or needle) becomes magnetized when it is stroked many times in the same direction along its length with one end of a natural magnet. The iron bar becomes an artificial magnet, but it possesses magnetic properties that are similar to those of a natural magnet. If iron filings are sprinkled on a magnet, they tend to cluster around two points, one at each end of the magnet. These points, where the magnetic attractions are the strongest, are called *magnetic poles* (Fig. 23-1).

Figure 23-1
Iron filings cluster around two points called magnetic poles.

Magnets may be made in many different shapes, but they all have similar properties.

A crude *magnetic compass* can be made by freely suspending a bar magnet from a string so that the long axis of the magnet is approximately horizontal. If there are no other magnetic materials nearby, the suspended magnet eventually aligns approximately in a north–south direction. The same magnetic pole of the magnet always indicates a northerly direction; therefore it is called a *north-seeking pole* or a *north pole*. The other magnetic pole is called a *south-seeking pole* or a *south pole*.

When the north pole of one magnet is brought close to the south pole of another, the two magnets attract each other; however, if a north pole of one magnet is moved close to the north pole of the second, the magnets repel each other (Fig. 23-2). *Similar magnetic poles repel, and unlike magnetic poles attract each other.*

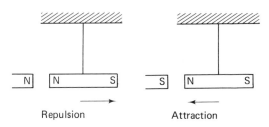

Figure 23-2
Magnetic forces between magnetic poles.

The earth itself behaves like a very large magnet with magnetic poles close to the geographical north and south poles. However

the earth's magnetic pole that is close to the geographic north pole is a south-seeking or south magnetic pole because it attracts the north-seeking poles of compasses and magnets. Similarly the earth's north-seeking magnetic pole is located near the geographic south pole. The earth's magnetic field is due to electric currents in the molten iron core (Fig. 23-3). The magnetic poles of the earth are actually located below the earth's surface. As a magnetic compass approaches the earth's magnetic poles, it tends to point at some angle (called the *angle of dip*) below the horizontal.

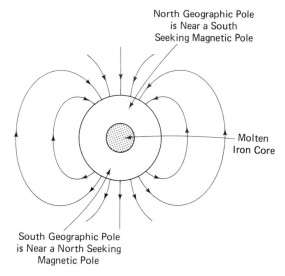

Figure 23-3
The earth's magnetic field is due to currents in the molten iron core.

Even though a magnetic pole was recently isolated, *magnetic poles normally occur in unlike pairs*. If a magnet is cut into two or more pieces, each piece will contain two unlike poles (Fig. 23-4).

A magnetic pole may attract other dissimilar magnetic poles and pieces of some unmagnetized materials. However, magnetic *repulsion* can only occur between like magnetic poles.

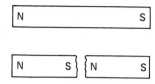

Figure 23-4
When a magnet is cut into two or more pieces, each piece retains two unlike poles.

*Problems*_____

1. How can you tell whether a material is magnetized?

2. Gyroscopes rather than magnetic compasses must be used in the Arctic and Antarctic regions. Explain why.

23-2 MAGNETIC FIELDS

It is convenient to consider magnetic interactions between moving electric charges in terms of magnetic fields. An electric field is produced in a region surrounding an isolated stationary electric charge, but, if the charge is moving, it sets up both electric and magnetic fields. It is useful to introduce the concept of a magnetic field in terms of the force that would be experienced by an isolated north magnet pole. A *magnetic field exists in any region where an isolated magnetic pole experiences a magnetic force. The magnetic field may be represented by lines of magnetic force corresponding to the paths along which an isolated north magnetic pole would move in the field.* Also, at any location *the direction of a magnetic line of force is taken as the direction in which a free, isolated north magnetic pole would move if it were placed at that point.* Magnetic fields represented by magnetic lines of force are often mapped with the aid of a small magnetic compass. At each position in space the direction indicated by the north-seeking pole of the compass corresponds to the direction of the magnetic field (Fig. 23-5). They can also be mapped by sprinkling iron filings on a sheet of paper that covers a magnet or system of magnets.

Magnetic lines of force always form closed loops. There is no source or sink of a mag-

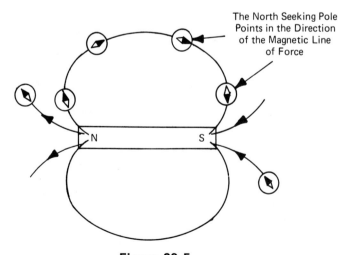

The North Seeking Pole Points in the Direction of the Magnetic Line of Force

Figure 23-5
Mapping a magnetic field with the aid of a small compass.

netic line of force. *Inside* a magnet, the lines of force continue *from* the south pole to the north pole.

Finally, *magnetic lines of force never intersect.* If they did, it would imply that there would be two separate forces on an isolated north pole.

While a small magnetic compass may be used to determine the direction of a magnetic field, it gives little indication of the field's magnitude. By analogy with electric fields, the magnitude of a magnetic field is represented by the number of lines of magnetic force that are drawn through a unit perpendicular area. The field is strong where the concentration of lines of force is large, and it is weak where the lines of force are far apart. The number of lines of force Φ_m is called the *magnetic flux* that passes through the area A; in SI it is expressed in units of webers (Wb). A *weber* is the magnetic flux that, when threading a circuit of one turn, produces an emf of one volt when it is reduced uniformly to zero in one second.

Magnetic Flux Density or Induction

The number of lines of flux Φ_m per unit perpendicular area A in a magnetic field is called the *magnetic flux density B.*

$$B = \frac{\Phi_m}{A_1} \qquad (23\text{-}1)$$

The SI unit of magnetic flux density is the telsa (T): $1\,T = 1\,Wb/m^2$ In practice, the magnitude of the flux density at any location is probed in terms of the magnetic forces on moving electric charges.

Consider the effects of constant electric and magnetic fields on some electric charge Q. When the charge is stationary, it experiences some force $\mathbf{F} = Q\mathbf{E}$ due to the electric field strength $\mathbf{E}$; however, if the charge is moving, it is affected by *both* the electric *and* the magnetic fields. Therefore the difference between the forces experienced by the charge

when it moved as compared with when it was stationary corresponds to the magnetic force.

A magnetic field at any location is completely specified by a vector quantity called the *magnetic induction* or *magnetic flux density* **B**. The direction of the magnetic flux density at any location is taken as the direction of the line of magnetic force, but its magnitude is defined in terms of its effect on moving electric charges.

Consider an electric charge that moves with a velocity **v** at some angle θ with respect to the magnetic field. Experiments show that the magnitude F of the magnetic force experienced by the charge is always directly proportional to the product of the magnitude Q of the charge, its speed v and $\sin\theta$:

$$F \propto Qv \sin\theta$$

The magnitude B of the magnetic flux density is defined as the proportionality constant:

$$F = BQv \sin\theta \quad \text{or} \quad B = \frac{F}{Qv \sin\theta} \qquad (23\text{-}2)$$

The magnetic force **F** is always *perpendicular* to both the velocity **v** of the charge and the magnetic field represented by **B**. The magnetic force on a charge Q is completely specified by the *vector* product:

$$\mathbf{F} = Q(\mathbf{v} \times \mathbf{B}) \qquad (23\text{-}3)$$

The quantity $\mathbf{v} \times \mathbf{B} = vB \sin\theta$ in the direction specified by the right-hand rule, which states: rotate the fingers of the right hand in the natural direction that they curl from the direction of **v** to the direction of **B**; the straightened thumb (perpendicular to **v** and **B**) indicates the direction of the magnetic force **F** (Fig. 23-6).

The SI unit of magnetic flux density, a *telsa* (T), is the magnetic flux density where a charge of 1 C moving at a speed of 1 m/s perpendicular to the magnetic field ($\theta = 90°$) experiences a magnetic force of 1 N.

$$1\,T = \frac{1\,N}{C \cdot m/s} = \frac{1\,N}{A \cdot m} \quad \text{since} \quad \frac{1\,C}{s} = 1\,A$$

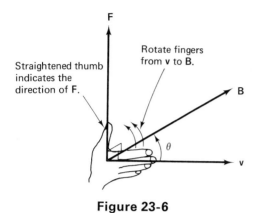

Figure 23-6

Also

$$1\,\text{Wb} = 1\,\text{T}\cdot\text{m}^2$$

*Example 23-1*_____

Calculate the magnetic force on an electron in a television tube if it moves horizontally due west at a speed of 2.5×10^6 m/s through the magnetic deflecting coils where the magnetic flux density of 40 mT is horizontal but due north.

Solution:

The magnitude of the force is:

$F = QvB \sin \theta$
$\quad = (1.6 \times 10^{-19}\,\text{C})(2.5 \times 10^6\,\text{m/s})(0.040\,\text{T}) \sin 90°$
$\quad = 1.6 \times 10^{-14}\,\text{N}$

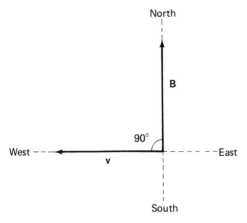

Figure 23-7

The corresponding vectors are illustrated in Fig. 23-7 c. Note that $\mathbf{v} \times \mathbf{B}$ is directed downwards, but *the charge Q on an electron is negative*; therefore it experiences a force in the opposite direction, i.e. upwards. Alternately we could use a similar left-hand rule for negative charges.

_____

The deflection of electric charges in magnetic fields is a three dimensional phenomenon. The force experienced by the charge is perpendicular to the plane defined by its initial velocity $\mathbf{v}$ and the magnetic flux density $\mathbf{B}$. For convenience, when a vector is directed into the page, it is represented by the symbol $\times$ (indicating the "tail" of an arrow); where it is directed out of the page it is represented by $\cdot$ (the "point" of an arrow).

When a charge Q passes simultaneously through electric and magnetic fields, it experiences a net force equal to the vector sum of the electric and magnetic forces:

$$\text{net } \mathbf{F} = Q\mathbf{E} + Q(\mathbf{v} \times \mathbf{B}) \quad \text{(vector sum)} \quad (23\text{-}4)$$

where $\mathbf{E}$ is the electric field strength, $\mathbf{B}$ is the magnetic flux density and $\mathbf{v}$ is the velocity of the charge. This equation is sometimes called the *Lorentz relation*.

*Problems*_____

3. An electron moves horizontally at 6.0×10^6 m/s through a perpendicular magnetic field in the deflection coils of a television. If the magnetic flux density is 3.5 mT, determine (a) the magnitude of the magnetic force on the electron and (b) the acceleration of the electron.

4. When an electron moves at 5.0×10^5 m/s at an angle of 60° with respect to a magnetic field, it experiences a force of 3.2×10^{-16} N. Calculate the magnitude of the flux density.

5. Determine the magnetic force on a proton when it moves due west with a speed of 5.0×10^5 m/s through a magnetic field

where the flux density is due south and has a magnitude of 2.5 mT.

6. An electron is moving horizontally due north with a speed of 6.25×10^6 m/s at a point where the earth's magnetic field has a flux density of 25 μT that is due north but inclined at 30° below the horizontal. (a) Calculate the magnetic force experienced by the electron and (b) compare the magnetic force with the weight of the electron.

23-3 CHARGES IN CIRCULAR MOTION

Consider a particle that has a positive electric charge Q and moves in a vacuum with an instantaneous velocity **v** through a uniform perpendicular magnetic field where the flux density is **B** (Fig. 23-8). The instantaneous

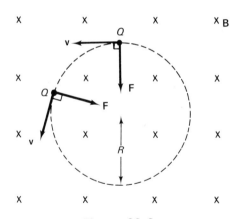

Figure 23-8
Path of a positively charged particle in a uniform magnetic field.

magnetic force **F** experienced by the particle is perpendicular to both **v** and **B**, and it has a magnitude:

$$F = QvB \sin 90° = QvB \qquad (i)$$

This force is always perpendicular to the path of the particle. Therefore it only changes the direction of the motion; it does not affect the particle's speed v. Even when the charged particle is deflected in the uniform magnetic field, its path is always perpendicular to the flux density. Consequently the magnitude of the magnetic force is constant. Under the influence of the magnetic field, the charged particle is deflected into a circular motion at a constant speed v. The centripetal force that is required to maintain the circular orbit is supplied by the magnetic force. Thus, if the particle has a mass m, the centripetal force is:

$$F = \frac{mv^2}{R} = QvB$$

and the radius of the circular path is:

$$R = \frac{mv}{QB} \qquad (23\text{-}5)$$

The time required for one complete orbit (the period T) of the charged particle:

$$T = \frac{2\pi R}{v} = \frac{2\pi m}{QB} \qquad (23\text{-}6)$$

is completely independent of the radius of orbit and the speed of the particle!* Note that, since the magnetic force is perpendicular to the displacement at all points around the circular path, the magnetic force does no work.

To investigate the structure of matter, scientists often initiate reactions in a material by bombarding it with high-speed ions. These high-speed ions may be produced in a variety of different devices called *particle accelerators*.

One of the most common particle accelerators, the *cyclotron*, was invented by Ernest Lawrence (1901–1958) in 1932. This device consists of two flat, hollow metal chambers

*Unless the particle travels close to the speed of light, in which case the mass must be replaced by the relativistic mass $m = m_0/\sqrt{1 - v^2/c^2}$, where m_0 is the rest mass and c is the speed of light.

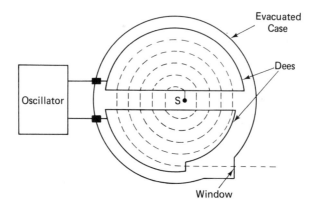

Figure 23-9
Cyclotron.

(called *dees* because of their shape), which are located in an evacuated container (Fig. 23-9). An oscillator is used to produce an alternating electric field in the region between the flat sides of the two dees, and the system is located in a nearly uniform, strong magnetic field between the poles of a strong electromagnet.

In operation the electric fields are used to increase the speed of an ion while the perpendicular magnetic field confines the ion to a series of semicircular orbits of increasing radius. A slow-moving ion is produced by a source S near the center of the system, and it is deflected into a semicircular path by the uniform magnetic field. When the ion reaches the gap between the dees, it is accelerated by the electric field, and its speed increases as it moves from one dee into the other. In the second dee the ion is deflected into a semi-circular arc of greater radius than before because of its increased speed. As the ion arrives at the gap between the dees, the oscil-lator reverses the electric field so that the ion accelerates and increases its speed as it returns to the first dee. This process is repeated many times; the ion increases its speed (and kinetic energy) until it is allowed to exit through a thin window.

The oscillator has a constant frequency that is synchronized to change the polarity of

the electric field after each half orbit of the ion so that the ion always increases its speed as it moves through the gap between the dees. Cyclotrons have been used to accelerate he-lium ions to kinetic energies of up to 40 MeV $(6.4 \times 10^{-12}$ J).

The Synchrotron

There is an upper limit to the speed to which an ion can be accelerated in a cyclotron. As the ion approaches the speed of light, its relativistic mass increases, and the period of its orbit increases. Consequently the synchro-nization between the oscillator and the orbit-ing ion is lost.

In a *synchrotron* the cyclotron technique is used to increase the speed of the orbiting ion, and the oscillator frequency is gradually reduced, i.e. the period increases, to correct for relativistic motion. Synchrotrons have accelerated electrons to kinetic energies of

$$30 \text{ GeV}, (4.8 \times 10^{-9} \text{ J})$$

The Charge to Mass Ratio of Electrons

In 1897, J. J. Thomson used electric and mag-netic forces to determine the ratio of charge

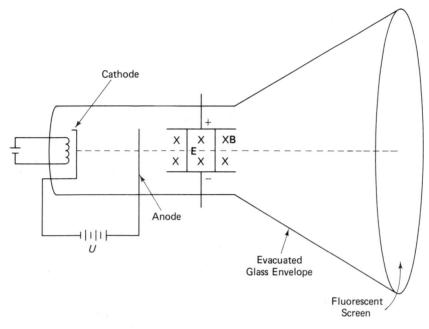

Figure 23-10
Thomson's charge to mass ratio experiment.

e to mass m for electrons. A version of Thomson's apparatus is illustrated in Fig. 23-10. Electrons are emitted thermionically from a heated cathode, and they are accelerated from rest to speed v by the applied potential difference U between the cathode and the anode. According to the work-energy principle, the work done by the accelerating potential difference U is equal to the gain in the kinetic energy of the electron. Thus the work done by the field is:

$$W = eU = \frac{1}{2}mv^2 \quad \text{and} \quad v^2 = \frac{2eU}{m} \quad (1)$$

After it passes through a hole in the anode, an electron continues at a constant speed v into a region where perpendicular electric *and* magnetic fields are present. These fields are adjusted until they exert equal and opposite forces on the electrons, and the electron beam continues undeflected to the screen. In Fig. 23-10 an electron with a negative charge $-e$ and a speed v is deflected upwards by the

electric force $\mathbf{F}_E = -e\mathbf{E}$ and downwards by the magnetic force $\mathbf{F}_M = -e(\mathbf{v} \times \mathbf{B})$; therefore when there is no net deflection:

$$eE = evB$$

or

$$v = E/B$$

where E and B are the magnitudes of the electric field strength and the magnetic flux density, respectively. Thus from (1),

$$v^2 = \frac{E^2}{B^2} = \frac{2eU}{m}$$

or

$$\frac{e}{m} = \frac{E^2}{2UB^2} \quad (23\text{-}7)$$

The magnitude of the electric field strength may be determined from the potential difference U_d and the perpendicular distance d between the parallel deflecting plates, i.e.

$$E = U_d$$

Modern measurements of the charge-to-mass ratio for electrons give a value $e/m = 1.758\ 803 \times 10^{11}$ C/kg.

The Mass Spectrometer

A *mass spectrometer* or *mass spectrograph* is a device in which magnetic deflections are employed to measure accurately the masses of atoms that have a known charge. One type of mass spectrometer, built by Bainbridge, is illustrated in Fig. 23-11. In operation positive

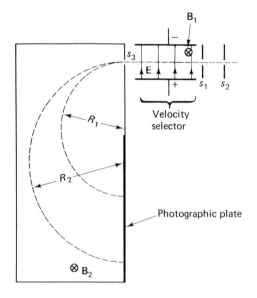

Figure 23-11
A mass spectrometer.

ions are produced by some source, and they are accelerated in an electric field. A narrow beam of ions passes through slits S_1 and S_2 into the *velocity selector*, where they are subjected simultaneously to perpendicular uniform electric and magnetic fields. If **E** is the uniform electric field strength and **B**$_1$ is the uniform magnetic flux density, ions with a charge Q and speed v would be deflected upwards by the electric force $\mathbf{F}_E = Q\mathbf{E}$ and downwards by the magnetic force $\mathbf{F}_M = Q(\mathbf{v} \times \mathbf{B}_1)$. Therefore they experience no net deflection when:

$$F_E = QE = QvB_1 = F_M$$

Thus only ions that have a speed:

$$v = E/B_1 \qquad (23\text{-}8)$$

pass undeflected by the velocity selector through the slit S_3. After the ions pass through S_3, they enter a region where they are deflected into semicircular paths by another uniform perpendicular magnetic field of known magnetic flux density **B**$_2$. If the ion has charge Q, mass m and speed v, the radius of its semicircular path is given by Eq. 23-4.

$$R = \frac{mv}{QB_2}$$

Since all ions have the same charge Q and speed v, the radius of their semicircular path is:

$$R = \left(\frac{v}{QB_2}\right)m = (\text{constant})\, m \qquad (23\text{-}9)$$

Therefore ions of different masses are deflected into semicircular paths of different radius, and they arrive at different positions on the photographic plate. When the exposed photographic plate is developed, a series of dark lines indicates the positions where the different ions converged. This ordered array indicating the relative atomic masses is called a *mass spectrum*. The relative intensities of the lines in a mass spectrum are an indication of the relative numbers of the different atomic masses.

Example 23-2

Positive ions with a charge of 1.6×10^{-19} C enter a velocity selector where the perpendicular magnetic flux density has a magnitude $B_1 = 0.60$ T and the deflecting plates are 1.5 cm apart at a potential difference of 45 kV. The ions that emerge from the velocity selector enter a mass spectrometer where they are deflected into a semicircular path of radius $R = 62.5$ cm by a perpendicular magnetic flux density of magnitude $B_2 = 1.0$ T. Determine (a) the speed and (b) the mass of the ions that emerge from the velocity selector.

Solution:

(a) $\quad v = \dfrac{E}{B} = \dfrac{U_d}{dB} = \dfrac{45 \times 10^3 \text{ V}}{(1.5 \times 10^{-2} \text{ m})(0.60 \text{ T})}$

$\qquad = 5.0 \times 10^6 \text{ m/s}$

(b) $m = \dfrac{RQB_2}{v} = \dfrac{(0.625 \text{ m})(1.6 \times 10^{-19} \text{ C})(1.0 \text{ T})}{5.0 \times 10^6 \text{ m/s}}$

$\qquad = 2.0 \times 10^{-26} \text{ kg}$

From mass spectra J. J. Thomson determined that atoms of the same element do not always have the same mass. Atoms of the same element that have different masses are called *isotopes*.

Magnetohydrodynamic Generator

A *magnetohydrodynamic generator* (MHD) is a device that can be used to convert heat into electricity without using rotating mechanisms. It consists of a combustion chamber where an ionized gas (called a *plasma*) is heated to temperatures in excess of 4 000 K. The plasma is then drawn at high speed into a region containing two parallel metal conductors and a perpendicular magnetic field (Fig. 23-12). As they flow through the magnetic field, positive ions are deflected onto one metal plate while negative ions are deflected onto the other. The ions are neutralized when they give up their charges to the metal plates, thus creating a potential difference between them. When the plates are connected via an external load, there is a current from the anode to the cathode.

These devices must operate at high temperatures, and they are normally physically larger then conventional generators. They will problably find extensive use in the near future.

Problems

7. An electron moves at a speed of 5.0×10^6 m/s in a circular path of radius 3.6×10^{-5} m through a uniform perpendicular magnetic field. Determine the magnitude of the magnetic flux density.

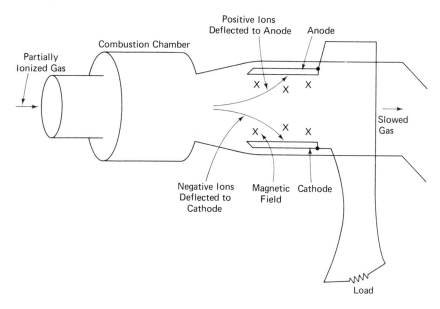

Figure 23-12
A magnetohydrodynamic generator.

8. A proton moves in a circular path of radius 0.30 m through a perpendicular uniform magnetic field where the flux density has a magnitude of 1.2 T. Determine (a) the speed and (b) the kinetic energy of the proton.

9. The parallel electrostatic deflecting plates of a velocity selector are 1.0 cm apart, and the perpendicular magnetic flux density is 0.55 T. Determine the potential difference that must be applied between the plates so that ions with a speed of 6.0×10^6 m/s are not deflected.

10. Positive ions with a charge of 1.6×10^{-19} C enter a velocity selector where the perpendicular magnetic flux density has a magnitude $B_1 = 0.50$ T and the deflecting plates are 1.4 cm apart at a potential difference of 28 kV. The ions that emerge from velocity selector then enter a mass spectrometer where they are deflected into semicircular paths of radii $R_1 = 27.8$ cm, $R_2 = 97.4$ cm and $R_3 = 75.2$ cm by a uniform perpendicular flux density of magnitude $B_2 = 0.60$ T. Determine (a) the speed and (b) the masses of the atoms that emerge from the velocity selector.

23-4 THE MAGNETIC FIELDS OF CURRENTS

Hans Cristian Oerstead first observed that, when a magnetic compass is placed near a current-carrying conductor, the compass needle is deflected from its regular north–south alignment. We now know that all magnetic fields originate from moving charges or an electric current which is a flow of charges in unison.

The Biot-Savart Law

At some point P, the magnitude of the magnetic flux density due to the current I in a small segment Δl of a current-carrying conductor is given by the *Biot-Savart Law:*

$$\Delta B_p = \left(\frac{\mu}{4\pi}\right)\left(\frac{I\,\Delta l \sin \theta}{r^2}\right) \qquad (23\text{-}10)$$

where r is the distance from the segment Δl to the point P and θ is the angle between the conventional electric current I and the direction to P (Fig. 23-13). The factor μ is a con-

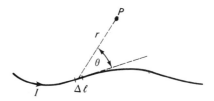

Figure 23-13

stant called the *permeability of the medium;* its value depends on the nature of the medium surrounding the conductor. For free space (vacuum) and approximately for air, its value is:

$$\mu = \mu_0 = 4\pi \times 10^{-7} \text{ T·m/A}$$

Another right-hand rule may be used to determine the direction of the magnetic flux density. "Hold" the conductor in the right hand with the straightened thumb pointing in the direction of the conventional current. The curled fingers indicate the direction of the magnetic flux density surrounding the conductor (Fig. 23-14).

At any point, the total magnetic flux density **B** due to a long current-carrying conduc-

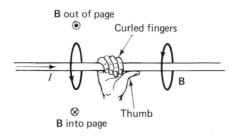

Figure 23-14

tor is equal to the vector sum of the magnetic flux densities $\Delta \mathbf{B}_i$ due to each small segment of that conductor.

$$\mathbf{B} = \sum_{\substack{\text{vector} \\ \text{sum}}} \Delta \mathbf{B}_i$$

These calculations are usually quite complex and usually require a knowledge of integral calculus; therefore we shall only consider the results of the more important currents.

Magnetic Flux Density Near a Long Straight Current

According to the right-hand rule, the magnetic lines of force form concentric circles around a long straight current I. The magnetic flux density at a perpendicular distance r from the current has a magnitude of:

$$B = \frac{\mu I}{2\pi r} \quad \text{(long straight current)} \quad (23\text{-}11)$$

where μ is the permeability of the surrounding medium.

Example 23-3

Determine the magnetic flux density in air at a point 5.0 cm from a long straight wire that carries 1.2 A.

Solution:

$$B = \frac{\mu_0 I}{2\pi r} = \frac{(4\pi \times 10^{-7} \text{ T·m/A})(1.2 \text{ A})}{2\pi(5.0 \times 10^{-2} \text{ m})}$$
$$= 4.8 \times 10^{-6} \text{ T}$$

Note that the earth's magnetic field $\sim 10^{-5}$ T. Therefore wires should either be shielded or kept away from any compasses that are used for navigation.

The Magnetic Flux Density at the Center of a Current Loop

Consider a circular wire loop that has a radius r and carries an electric current I (Fig. 23-15). The right-hand rule indicates that the mag-

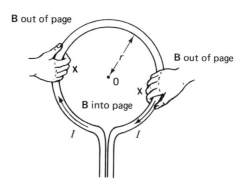

B out of page

B out of page

B into page

Figure 23-15
When the conventional current is clockwise, the magnetic flux density due to each segment of a loop is directed into the page inside the loop and out of the page outside the loop.

netic flux density in the interior of the loop due to each current-carrying segment is in the same direction. The flux density is in the opposite direction at all points exterior to the loop. Note that each segment is the same distance (equal to the radius r) from the center 0 of the loop and that the radius is always perpendicular to the loop, i.e. $\theta = 90°$. Therefore, from the Biot-Savart law, the magnitude of the flux density *at the center* of the current loop is:

$$B_0 = \frac{\mu I}{2r} \quad \text{(center of current loop)} \quad (23\text{-}12)$$

and it is in the direction perpendicular to the plane of the loop. Note that the flux density **B** is not uniform at all points in the interior of the loop.

The direction of the magnetic field due to current-carrying loops, coils or solenoids may also be determined by a different right-hand rule: Align the curled fingers of the right hand in the direction of the conventional electric current; the straightened thumb points in the direction of the magnetic flux density **B** in the interior of the loop (Fig. 23-16).

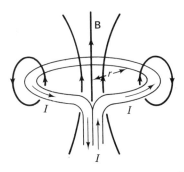

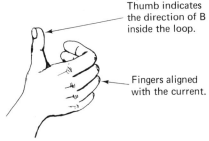

Thumb indicates the direction of B inside the loop.

Fingers aligned with the current.

Figure 23-16
A right-hand rule.

Magnetic Flux Density Along the Axis of a Flat Coil

At the center 0 of a flat coil that is composed of N similar current loops of radii r, the magnetic flux density is merely N times that of a single current loop in a direction that may be determined from a right-hand rule.

$$B_0 = \frac{N\mu I}{2r} \quad \text{(center of flat coil)} \quad (23\text{-}13)$$

At some point P along the axis of the flat coil at a distance s from its center 0, (Fig. 23-17), the magnitude of the flux density is:

$$B = \left(\frac{\mu}{2}\right)\left[\frac{NIr^2}{(r^2 + s^2)^{3/2}}\right] \quad \begin{array}{l}\text{(At a distance } s \text{ along} \\ \text{the axis of the coil)}\end{array}$$

$$(23\text{-}14)$$

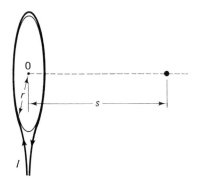

Figure 23-17

where μ is the permeability of the medium, r is the radius of the N turns and I is the electric current in the coil.

Example 23-4

A flat circular coil has a radius of 12.0 cm, and it carries a 5.0 A current. If the coil has 50 turns and it is located in air, determine the magnitude of the magnetic flux density (a) at the center of the coil and (b) at a point on the axis a distance of 10.0 cm from the center of the coil.

Solution:

(a) $\quad B = \frac{N\mu_0 I}{2r}$

$\qquad = \frac{(50 \text{ turns})(4\pi \times 10^{-7} \text{ T}\cdot\text{m/A})(5.0 \text{ A})}{2(0.12 \text{ m})}$

$\qquad = 1.3 \times 10^{-3} \text{ T}$

(b) $\quad B = \frac{\mu_0}{2}\frac{NIr^2}{(r^2 + s^2)^{3/2}} = \frac{(4\pi \times 10^{-7} \text{ T}\cdot\text{m/A})(50 \text{ turns})(5.0 \text{ A})(0.12 \text{ m})^2}{2[(0.12 \text{ m})^2 + (0.10 \text{ m})^2]^{3/2}} = 5.9 \times 10^{-4} \text{ T}$

Magnetic Flux Density in a Solenoid and Toroid

A *solenoid* is a single wire that is wound into a helix (Fig. 23-18). If the length l of a current-carrying solenoid is much greater than

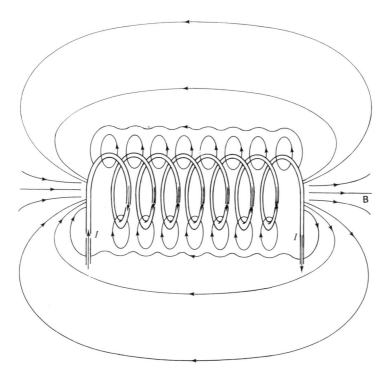

Figure 23-18
Magnetic field of a long solenoid.

its diameter, the magnetic field in its interior (but away from the ends) is very uniform. At a point on the interior axis of the solenoid, the magnetic flux density has a magnitude:

$$B = \frac{N\mu I}{l} = n\mu I \quad \begin{array}{l}\text{(interior axis of}\\\text{a solenoid)}\end{array} \quad (23\text{-}15)$$

where μ is the permeability of the material inside the solenoid, I is the current and $n = N/l$ is the number of turns per unit length of the solenoid. The direction of the flux density may be determined by a right-hand rule.

One type of solenoid called a *toroid* is frequently used in magnetic measurements and in electronic circuits. It consists of a tightly wound coil of wire around a doughnut-shaped material, such as iron or a nonmagnetic material (Fig. 23-19). The magnetic field is very uniform inside the ring or core of material, and there is almost no magnetic field outside of the toroidal core.

Example 23-5

The core of a toroid has a mean radius $r = 15$ cm, and it is wound with 5 000 turns of wire.

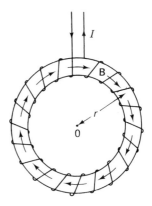

Figure 23-19
A toroid.

When the current in the wire is 3.0 A, the magnetic flux density inside the core has a magnitude of 1.5 T. Determine the permeability of the core.

Solution:

Since the length of the core $l = 2\pi r$, from Eq. 23-15:

$$\mu = \frac{Bl}{NI} = \frac{(1.5 \text{ T})(2\pi \times 0.15 \text{ m})}{(5\,000 \text{ turns})(3.0 \text{ A})}$$
$$= 9.4 \times 10^{-5} \text{ T·m/A}$$

Problems

11. The magnetic flux density has a magnitude of 5.0 μT in air at a point 5.0 cm from a long straight wire. Compute the current in the wire.

12. Determine the magnetic flux density at the center of a circular wire loop that has a diameter of 20 cm if it carries a clockwise 8.0 A current and is located in air.

13. Calculate the current in a circular wire loop of radius 10.0 cm if the magnetic flux density at its center has a magnitude of 35 μT in air.

14. A flat circular coil has a 30.0 cm diameter, and it carries a 6.0 A current. If the coil has 70 turns and is located in air, determine the magnitude of the magnetic flux density (a) at the center of the coil and (b) at a point on the axis a distance of 15.0 cm from the center of the coil.

15. A flat circular coil has a diameter of 20 cm and 50 turns of wire. If the magnetic flux density at the center of the coil has a magnitude of 5.0×10^{-4} T in air, compute the current in the coil.

16. The core of a toroid has a mean diameter of 20 cm and is wound with 4 000 turns of wire. When the current in the wire is 3.0 A, calculate the magnitude of the magnetic flux density in the core if the core is composed of a material that has

a permeability of (a) $4\pi \times 10^{-7}$ T·m/A, (b) 1.5×10^{-5} T·m/A.

17. A flat circular coil has 50 turns of diameter 24 cm, and it carries a 2.5 A current. At some instant an electron is located in a vacuum at a point along the axis but 10 cm from the center of the coil. Determine the magnitude of the magnetic force experienced by the electron if it moves with a speed of 7.0×10^6 m/s perpendicular to the axis of the coil.

18. A flat circular coil with 80 turns of diameter 2.0 cm is used to deflect an electon beam that is perpendicular to the axis of the coil in a vacuum. If the electrons move at speeds of 5.2×10^6 m/s, what current in the coil would produce a magnetic force of 1.4×10^{-15} N on each electron at the instant when it is at a point along the axis 2.5 cm from the coil center?

23-5 MAGNETIC FORCE ON A CURRENT

Consider a small straight current-carrying conductor of length l and cross-sectional area A that is located in a magnetic field where the flux density is **B** (Fig. 23-20). If the conductor

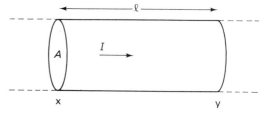

Figure 23-20

is composed of a material that has N free charges per unit volume, it has a total of NAl free charges. The electric current I consists of these charges moving with some average velocity **v** through the conductor. As each free charge moves, it experiences a magnetic force of magnitude:

$$F_i = QvB \sin \theta$$

where Q is the magnitude of the charge and θ is the angle between its velocity **v** and the flux density **B**. Since the conductor contains NAl such moving charges, it experiences a total magnetic force of magnitude:

$$F = NAlF_i = (NAl)(QvB \sin \theta) \qquad (1)$$

However, the total free charge in the segment XY of the conductor is:

$$Q_{\text{TOT}} = NAlQ$$

If this total charge moves through end Y in elapsed time t, the average current is:

$$I = \frac{Q_{\text{TOT}}}{t} = \frac{NAlQ}{t} = NAvQ$$

where $v = l/t$ is the average speed of the charges. Thus in (1)

$$F = NAvQlB \sin \theta$$

or $\qquad\qquad F = IlB \sin \theta \qquad (23\text{-}16)$

The direction of the total force is the same as the direction of the forces on the individual charges; it may be determined by the right-hand rule. For convenience the magnetic force on a current-carrying wire is often expressed as the vector product:

$$\mathbf{F} = l(\mathbf{I} \times \mathbf{B}) \qquad (23\text{-}17)$$

where l is the length of the conductor, **I** is the conventional current (in the direction in which positive charges flow) and **B** is the magnetic flux density. In Fig. 23-21 the magnetic force on the conductor is directed into the page.

*Example 23-6*_____

A 12.0 cm segment of a wire conductor is located in a magnetic field where the flux density has a magnitude of 2.5×10^{-2} T. Calculate the magnitude of the magnetic force on the wire when it carries a 5.0 A conventional current in a direction inclined at 30° with the flux density.

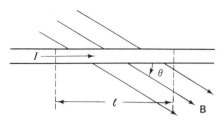

Figure 23-21

A current-carrying conductor in a magnetic field. The resultant magnetic force is directed into the page.

Solution:

$$\begin{aligned} F &= IlB \sin \theta \\ &= (5.0 \text{ A})(0.12 \text{ m})(2.5 \times 10^{-2} \text{ T}) \sin 30° \\ &= 7.5 \times 10^{-3} \text{ N} \end{aligned}$$

Magnetic lines of force surround any current-carrying conductor; therefore in an external magnetic field the total flux density is equal to the vector sum of the flux densities of the external field and the field due to the current. In Fig. 23-22 for example, the external field is from left to right, and the field caused by the current is counterclockwise. Therefore, the total flux density is reduced in

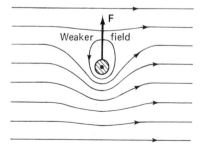

= current out of the page

Figure 23-22

Lines of force around an electric current in a magnetic field. The magnetic force **F** is toward the region where the magnetic flux density is smallest.

the region above the current (where the fields oppose), and it is increased in the region below the current (where the fields are in the same direction). The magnetic force experienced by the conductor is always directed away from the region where the magnetic flux density is largest towards the region where the flux density is smallest.

*Problems*_____

19. Determine the direction of the magnetic force on a horizontal conductor if it carries a conventional current directed towards the north and is placed in a magnetic field where the magnetic flux density is (a) towards the north, (b) towards the east, (c) towards the south, (d) towards the north but 30° below the horizontal.

20. A 4.0 cm wire conductor is located at a point where the magnetic flux density has a magnitude of 0.75 T. Calculate the magnitude of the magnetic force on the conductor when it carries a 100 mA conventional current in a direction inclined at 120° with respect to the flux density.

21. A 2.0 A conventional current is maintained in a southerly direction through a horizontal 6.0 cm wire conductor. Calculate the force due to the earth's magnetic field when the conductor is located where the earth's magnetic flux density is 25 μT towards the north but 30° below the horizontal.

23-6 MAGNETIC FORCES BETWEEN CURRENTS

An electric current sets up a magnetic field, but electric currents also experience magnetic forces when they are located in magnetic fields; therefore electric currents may exert magnetic forces on each other.

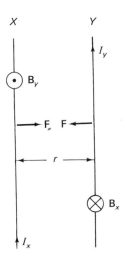

Figure 23-23
Parallel current-carrying conductors a distance *r* apart produce an attractive force.

Consider two parallel, long straight conductors X and Y that are a distance r apart and carry electric currents I_x and I_y, respectively (Fig. 23-23). At wire Y, the magnetic flux density due to the current I_x in conductor X has a magnitude given by Eq. 23-11:

$$B_x = \frac{\mu I_x}{2\pi r}$$

and it is perpendicular to conductor Y. Since $\theta = 90°$, from Eq. 23-16, the magnitude of the magnetic force that is exerted on a segment Δl_y of conductor Y by the magnetic field B_x is:

$$F_y = I_y \, \Delta l_y B_x \sin \theta = I_y \, \Delta l_y B_x$$

Substituting for B_x from Eq. 23-11, we obtain:

$$F_y = I_y \, \Delta l_y \left[\frac{\mu I_x}{2\pi r}\right] = \frac{\mu \, \Delta l_y}{2\pi} \frac{I_x I_y}{r} \quad (23\text{-}18)$$

The force per unit length has a magnitude:

$$\frac{F_y}{\Delta l_y} = \frac{\mu}{2\pi} \frac{I_x I_y}{r} \quad (23\text{-}19)$$

Similarly the magnitude of the magnetic force per unit length exerted on conductor X by the magnetic field due to the current I_y in conductor Y is:

$$\frac{F_x}{\Delta l_x} = \frac{\mu}{2\pi}\frac{I_x I_y}{r}$$

Note that the force per unit length has the same magnitude for each conductor; this is a direct consequence of Newton's third law.

In each case the force has a direction given by the vector product $\mathbf{I} \times \mathbf{B}$. When the currents I_x and I_y are parallel, the forces are attractive; when the currents are antiparallel, the forces are repulsive.

The lines of force surrounding the parallel current-carrying conductors are illustrated in Fig. 23-24. If the currents are parallel, the total magnetic field is weaker in the region between the conductors where the fields due to the currents oppose each other. The conductors experience magnetic forces towards this region and they attract each other.

When the currents are antiparallel, their magnetic fields are parallel, and the total field is strongest in the region between the conductors. The conductors experience magnetic forces away from the region where the field is strongest, and they repel each other.

In SI the base unit of electric current (the ampere) is defined in terms of the magnetic force between two long straight currents. If two parallel conductors are exactly 1.0 m apart in free space and carry equal currents of exactly 1.0 A, each conductor experiences a magnetic force per unit length of 2.0×10^{-7} N/m. That is, substituting the values in Eq. 23-19, we obtain:

$$\begin{aligned}\frac{F}{\Delta l} &= \frac{\mu_0}{2\pi}\frac{I_x I_y}{r} \\ &= \frac{(4\pi \times 10^{-7}\ \text{T·m/A})(1.0\ \text{A})(1.0\ \text{A})}{2\pi(1.0\ \text{m})} \\ &= 2.0 \times 10^{-7}\ \frac{\text{N}}{\text{m}}\end{aligned}$$

Problems_____

22. Two straight parallel wires are 2.0 mm apart in free space, and they carry equal magnitude currents. If each wire experiences a repulsive magnetic force per unit length of 1.6×10^{-5} N/m, calculate the electric currents that they carry.

23. Two straight parallel wires are 1.5 mm

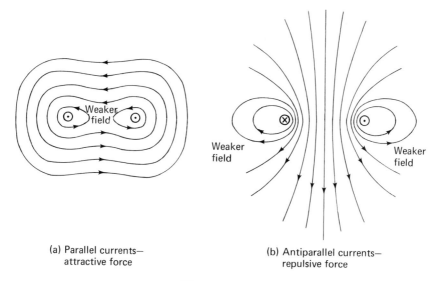

(a) Parallel currents—
attractive force

(b) Antiparallel currents—
repulsive force

Figure 23-24
Lines of magnetic force surrounding parallel
current-carrying conductors.

apart in free space, and they carry parallel currents of 12 A and 5.0 A. Determine the magnetic force on a 2.5 m long section of one wire.

23-7 MAGNETIC TORQUE ON A CURRENT LOOP

We have seen that a current-carrying conductor may experience a magnetic force when it is located in a magnetic field. If the conductor is shaped into a loop or a coil, its opposite sides carry antiparallel currents, and they experience equal and opposite forces that tend to rotate the loop until its plane is perpendicular to the magnetic field.

Consider a rectangular loop PQRS that is constrained so that it may rotate about a central axis parallel to the sides PQ and RS, and it is located in a uniform magnetic field with its axis perpendicular to the flux density **B** (Fig. 23-25). Suppose that the loop carries a current I and its plane is inclined at some angle θ with respect to **B**. The forces on the loop segments PS and QR have the same lines of action because the loop is constrained to

rotate about a perpendicular axis; therefore they produce no torque on the loop. However, the loop segments PQ and RS also carry antiparallel currents, and they are always perpendicular to **B**. Therefore they experience equal and opposite forces of magnitude:

$$F = IxB \sin 90° = IxB$$

where x is the length of each segment. Since these equal and opposite forces do not have the same lines of action; they constitute a couple that gives rise to a net torque about the axis 0 of the loop. The perpendicular distance between the lines of action of the forces $d = y \cos \theta$; thus the total torque has a magnitude:

$$\tau_0 = dF = (y \cos \theta)(IxB) = BxyI \cos \theta$$

where xy is the *area A* of the loop. Thus

$$\tau_0 = BAI \cos \theta \qquad (23\text{-}20)$$

Note that, when the plane of the loop is parallel to the magnetic flux density B, the angle $\theta = 0°$ and the torque has a maximum value equal to BAI; but if $\theta = 90°$, the plane of the loop is perpendicular to **B**, and the torque is zero. The magnetic torque always tends to align the loop perpendicularly to the flux density **B**.

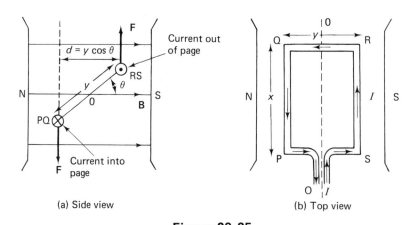

(a) Side view   (b) Top view

Figure 23-25
A current-carrying loop in a uniform magnetic field.

For a coil that is composed of N current loops of any shape, the total torque has a magnitude given by:

$$\tau_0 = BAIN \cos \theta \qquad (23\text{-}21)$$

where A is the area of the coil, I is the current and θ is the angle between the plane of the coil and the flux density $\mathbf{B}$.

Example 23-7

A 25 cm $\times$ 10 cm coil that has 50 turns is mounted in a uniform magnetic field with its plane inclined at an angle $\theta = 60°$ with respect to the flux density. Calculate the magnitude of the magnetic flux density if the coil experiences a torque of 5.0×10^{-3} m·N when it carries a 2.0 A current.

Solution:

$$B = \frac{\tau_0}{AIN \cos \theta}$$

$$= \frac{5.0 \times 10^{-3}\ \text{m·N}}{(0.25\text{m} \times 0.10\ \text{m})(2.0\ \text{A})(50\ \text{turns})(\cos 60°)}$$

$$= 4.0 \times 10^{-3}\ \text{T}$$

The operations of moving-coil meters and electric motors depend on the magnetic torques that are experienced by current-carrying coils in magnetic fields. Electric motors are discussed in Chapter 25.

The D'Arsonval Galvanometer

A *D'Arsonval galvanometer* is a device that is sensitive to electric currents; its basic structure is illustrated in Fig. 23-26. A wire coil is wound lengthwise around a cylindrical soft iron core, and the system is mounted on bearings in a cylindrical cavity between the poles of a permanent magnet. Lines of magnetic force concentrate in the soft iron core, creating a more uniform magnetic field in the air gap between the permanent magnet and the iron core (Fig. 23-27). The magnetic torque on the coil depends on the current that it carries, and the corresponding deflection of

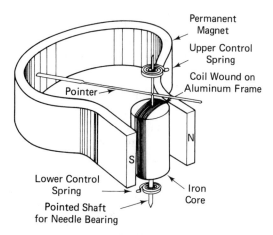

Figure 23-26
A moving-coil galvanometer.

the coil and pointer is controlled by two spiral springs, one at each end of the iron core. When the coil deflects, it distorts the spiral springs, and they develop a countertorque that is proportional to their distortion. Thus the magnitude of the deflection depends on the current in the coil.

Problems

24. A 10 cm $\times$ 50 cm coil that has 20 turns is mounted in a uniform magnetic field with its plane inclined at 45° with the flux

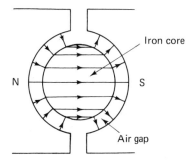

Figure 23-27
Lines of force take the "easiest path" through the soft iron core; as a result, the magnetic field is more uniform in the air gap.

density **B**. If the coil carries a 1.5 A current and the magnitude of **B** is 2.0×10^{-2} T, determine the magnetic torque on the coil.

25. A 15 cm × 7.0 cm coil has 30 turns and is mounted in a uniform magnetic field with its plane inclined at 60° with the flux density. If the magnetic torque on the coil is 1.7×10^{-3} m·N and **B** has a magnitude of 3.0×10^{-3} T, calculate the current in the coil.

REVIEW PROBLEMS

1. An electron moves horizontally due south at 6.0×10^6 m/s through a uniform flux density of 3.0×10^{-5} T that is due north but inclined at 60° below the horizontal. Calculate the magnetic force on the electron.

2. An electron moves horizontally due north with a speed of 5.0×10^6 m/s at a point where the earth's magnetic field has a magnetic flux density of 2.5×10^{-5} T that is due north but inclined at 30° below the horizontal. Calculate the magnetic force experienced by the electron.

3. An alpha particle (charge $+3.2 \times 10^{-19}$ C and mass 6.7×10^{-27} kg) moves in a circular path of radius 1.2 m through a uniform magnetic field where the flux density is 1.3 T. Determine (a) the speed and (b) the kinetic energy of the alpha particle.

4. Nitrogen has two isotopes of masses 14.007 5 u and 15.004 9 u. If the *radius* of the path described by the lighter isotope in a mass spectrograph is exactly 15 cm, find the separation between the traces produced by the two isotopes on the photographic plate.

5. The parallel electrostatic deflecting plates of a velocity selector are 1.5 cm apart, and the perpendicular magnetic flux density is 1.2 T. (a) Determine the potential difference that must be applied between the plates so that ions with a speed of 5.2×10^6 m/s are not deflected. (b) If these ions have a charge of 1.6×10^{-19} C and they pass into the mass spectrometer where they are deflected in circular arcs of radii 6.07 cm and 6.50 cm by a uniform magnetic flux density of 0.50 T, what are their masses?

6. Compute the flux density in air at the center of a circular wire loop of diameter 1.5 cm if it carries 300 mA.

7. A flat 30 turn coil has a diameter of 12 cm. If the current in the coil is 3.6 A, calculate the flux density in air (a) at the center of the coil and (b) at a point on the coil axis 15 cm from the coil.

8. The core of a toroid has a mean diameter of 15 cm, and it has 3 000 turns. When the current is 3.0 A, the flux density in the coil is 0.286 T. Calculate the permeability of the core.

9. The core of a toroid has a mean diameter of 20 cm, and it is wound with 4 000 turns of wire. If the current in the wire is 3.0 A and the core has a permeability of 1.5×10^{-5} T·m/A, calculate the magnitude of the magnetic flux density in the core.

10. A wire conductor is located at a point where the flux density is 0.82 T. If the wire carries 2.5 A in a direction inclined at 42° with the magnetic field, what is the magnitude of the magnetic force per unit length on the wire?

11. Two straight parallel wires are 2.3 mm apart in air, and they carry antiparallel currents of 200 mA and 800 mA. What is the magnetic force per unit length between the wires?

12. A 12 turn rectangular coil with dimen-

sions 4.0 cm $\times$ 9.0 cm is mounted in a uniform magnetic field with its plane inclined at 25° with a uniform magnetic flux density of 2.5 mT. What is the torque on the coil when it carries 5.4 A?

13. Determine the torque on a 50 turn flat coil of dimensions 20 cm $\times$ 12 cm that carries a 12 A current when its plane is inclined at 30° to the flux density of magnitude 1.25 T.

In an atom, one or more negatively charged electrons orbit a positively charged nucleus at high speeds; at the same time the electrons and the nucleus spin about their axes. Some magnetic effects are due to the spinning of the positively charged atomic nucleus, but most magnetic properties of materials originate from the motions of the negatively charged orbital electrons.

Some materials increase the magnetic flux density, and others decrease the flux density at their locations when they are placed in an external magnetic field. This effect is usually very small, but in a few substances, which are called *ferromagnetic materials*, the magnetic flux density is increased significantly. These ferromagnetic materials are used to increase the magnetic flux density in many devices, such as transformers, motors, generators, loudspeakers and electromagnets.

24-1 THE ATOMIC THEORY OF MAGNETISM

An electron has a negative charge $-e$, and its orbit about an atomic nucleus is equivalent to a small current loop. The magnitude of the equivalent current is $I = -e/T$, where T is the period of the electron orbit. This electron current loop sets up a small magnetic field in a direction that may be determined by a *left*-hand rule for *negative* charges; that is, if the fingers of the left hand are curled in the direction of the electron's orbit, the straightened thumb indicates the direction of the magnetic field. In an unmagnetized material the small magnetic fields due to all of the electron current loops have random orientations, and there is no net effect. However, when a material is magnetized, the electron current loops are aligned to some degree, and they produce a net magnetic field inside the material.

24

Magnetic Properties of Materials

Magnetization

Each electron current loop in a material experiences a magnetic torque when it is placed in an external magnetic field. If the plane of an electron orbit is inclined at some angle θ with the flux density $\mathbf{B}$, according to Eq. 23-20 the electron current loop experiences a magnetic torque of magnitude:

$$\tau_0 = BAI \cos\theta = Bm \cos\theta \qquad (24\text{-}1)$$

where A is the area of the electron orbit, I is the equivalent current and $m = IA$ is the magnitude of a vector quantity called the *magnetic moment*. The direction of the magnetic moment $\mathbf{m}$ is taken as the direction of the magnetic field that is set up by the electron current loop (Fig. 24-1).

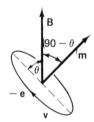

Figure 24-1

The orbit of a negatively charged electron is equivalent to a current loop that sets up a small magnetic field represented by the magnetic moment $\mathbf{m}$.

Materials contain very large numbers of electrons in different orbits around the atomic nuclei, and each electron orbit gives rise to a small magnetic moment. The net magnetic moment (the vector sum) per unit volume is a vector quantity, which is called the *magnetization* $\mathbf{M}$ of the material. In SI, magnetization is expressed in units of amperes per meter (A/m).

Magnetic Permeability

If there is some alignment of their magnetic moments, the electron current loops set up a magnetic field, which may change the net magnetic field in the interior of the material. At any point near a system of moving charges, the magnetic flux density depends on the *magnetic permeability* μ of the medium. The permeability is an indication of the degree that the medium will become magnetized when it is located in an external magnetic field. If a material replaces a vacuum at some point in a magnetic field, the flux density at that location changes by some factor, called the *relative permeability* κ_M of that material. The relative permeability of a substance is a dimensionless quantity that is defined as the ratio of the permeability μ of that substance to the permeability μ_0 of free space.

$$\kappa_M = \mu/\mu_0 \qquad (24\text{-}2)$$

Most materials have very small mag-

Table 24-1

Typical Magnetic Properties of Ferromagnetic Materials at Room Temperature

Ferromagnetic Material	Maximum Relative Permeability	Saturation Flux Density/T
Iron	5 300	2.15
Cobalt	260	1.8
Nickel	600	0.6
Iron—Silicon	7 000	2.0
Permalloy (55% Fe, 45% Ni)	28 000	1.6
Mumetal (5% Cu, 2% Cr, 77% Ni, 16% Fe)	100 000	0.7

netizations, and their relative permeabilities are almost equal to one; but some materials, such as iron, have very large magnetizations, and they have large relative permeabilities (Table 24-1).

Problems

1. Calculate the magnitude of the magnetic moment of an electron if it is in a circular orbit of radius 5.3×10^{-11} m at a speed of 2.2×10^6 m/s.

2. Determine the magnetic torque on an electron current loop if the electron orbit has a period of 1.5×10^{-16} s, radius of 4.37×10^{-11} m and the plane of the orbit is inclined at $30°$ with the flux density of magnitude 0.50 T.

3. Determine the maximum permeability of (a) iron, and (b) mumetal.

4. Calculate the relative permeability of a material that has a permeability of 1.9×10^{-4} T·m/A.

24-2 MAGNETIC INTENSITY OR FIELD STRENGTH

A magnetic field surrounds any system of moving charges, but the magnetic flux density at any nearby point may be modified by the magnetic field due to the magnetization of the medium. Therefore it is convenient to define another vector, called the *magnetic field strength* **H**, that is completely independent of a the medium. If the medium has a permeability μ at some point where the flux density is **B**, the *magnetic field strength* is defined as:

$$\mathbf{H} = \frac{\mathbf{B}}{\mu} \qquad (24\text{-}3)$$

In SI, magnetic field strengths are expressed in units of amperes per meter (A/M).

The magnetization **M** of a material is directly proportional to the field strength **H**

at its location. The dimensionless proportionality constant is called the *magnetic susceptibility* X_m of the material:

$$X_m = \frac{\mathbf{M}}{\mathbf{H}} \qquad (24\text{-}4)$$

At a point where the magnetic field strength is **H** and **M** is the magnetization of the medium, the flux density is:

$$\mathbf{B} = \mu_0(\mathbf{H} + \mathbf{M}) \qquad (24\text{-}5)$$

where μ_0 is the permeability of free space. Comparing Eqs. 24-3 and 24-5, we obtain:

$$\mathbf{B} = \mu\mathbf{H} = \mu_0(\mathbf{H} + \mathbf{M})$$

Thus the relative permeability of the medium is:

$$\kappa_m = \frac{\mu}{\mu_0} = \frac{\mathbf{H} + \mathbf{M}}{\mathbf{H}} = 1 + \frac{\mathbf{M}}{\mathbf{H}} = 1 + X_m$$
$$(24\text{-}6)$$

Example 24-1

The magnetic field strength has a magnitude of 50 A/m at a point in a material where the magnitude of the magnetic flux density is 3.15×10^{-4} T. Determine (a) the permeability, (b) the relative permeability, (c) the magnetization, (d) the magnetic susceptibility of the material medium.

Solution:

(a) $\qquad \mu = \dfrac{B}{H} = \dfrac{3.15 \times 10^{-4}\text{ T}}{50\text{ A/m}}$
$$= 6.3 \times 10^{-6}\text{ T·m/A}$$

(b) $\qquad \kappa_m = \dfrac{\mu}{\mu_0} = \dfrac{6.3 \times 10^{-6}\text{ T·m/A}}{4\pi \times 10^{-7}\text{ T·m/A}} = 5.0$

(c) $\mathbf{B} = \mu_0(\mathbf{H} + \mathbf{M})$ with **B** parallel to **H**. Thus

$$M = \frac{B}{\mu_0} - H$$
$$= \frac{3.15 \times 10^{-4}\text{ T}}{4\pi \times 10^{-7}\text{ T·m/A}} - 50\text{ A/m}$$
$$= 200\text{ A/m}$$

in a direction that is parallel to **B** and **H**.

(d) $\qquad X_m = \dfrac{\mathbf{M}}{\mathbf{H}} = \dfrac{200\text{ A/M}}{50\text{ A/M}} = 4.0$

Note $\kappa_M = 1 + X_m$.

All materials are composed of moving charged particles, and they are affected to some extent when they are located in a magnetic field.

*Problem*_____

5. The magnetic field strength has a magnitude of 22.6 A/m at a point in a material where the magnitude of the magnetic flux density is 5.20×10^{-3} T. Determine (a) the permeability, (b) the relative permeability, (c) the magnetization, (d) the magnetic susceptibility of the material medium.

24-3 DIAMAGNETISM AND PARAMAGNETISM

Diamagnetic materials, such as bismuth, copper, silver, gold and the inert gases, have small negative-valued magnetic susceptibilities, and their relative permeabilities are slightly less than one. When a diamagnetic material is placed in a magnetic field, the flux density **B** is *reduced* at points inside the diamagnetic material.

Diamagnetic materials have no net magnetic moments if there is no external magnetic field; however, when they are placed in a magnetic field, there is a net increase in their magnetic moment antiparallel to the magnetic field strength. The magnetization **M** of this material is negative valued because it opposes the external magnetic field; therefore the magnetic susceptibility $X_m = $ **M/H** is also negative.

Superconductors are strong diamagnetic materials because they tend to expel all magnetic fields from their interior, i.e. **H** = −**M** for a perfectly diamagnetic material.

Some materials, such as platinum, rare earth metals and atoms or ions of iron, cobalt and nickel, have small positive magnetic susceptibilities, and their relative permeabilities are slightly greater than one. These materials

are called *paramagnetic* because the individual atoms possess net magnetic moments that experience magnetic torques and tend to align parallel to any external magnetic field. Thus paramagnetic materials have small positive magnetizations, and they tend to increase the flux density at their location. The net magnetic moments are due to the electron orbits, the spins of the negatively charged electrons and the spin of the positively charged nucleus.

*Problems*_____

6. How would the net dipole moment of a diamagnetic atom orientate itself in a magnetic field? Why?

7. How would the net magnetic moment of a paramagnetic atom tend to orientate itself in a magnetic field? Why?

24-4 FERROMAGNETISM

Crystals of the elements iron, cobalt, nickel, gadolinium and dysprosium, and of many of their compounds have very large magnetic susceptibilities, and their relative permeabilities are much greater than one. These materials are called *ferromagnetic*, and they may be made into permanent magnets.

The existence of large magnetizations in ferromagnetic solids is due to interaction between the electron spins of neighboring atoms. This type of interaction is called *exchange coupling*, and it tends to align the magnetic moments of groups of atoms. A *domain* is a region in a ferromagnetic solid that is occupied by a group of atoms that have parallel magnetic moments; the boundary between two domains is called a *domain wall*. When a ferromagnetic solid is not magnetized, it contains a large number of magnetic domains, and the directions of their magnetic moments have random orientations.

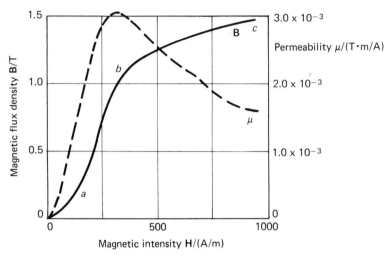

Figure 24-2
A magnetization curve.

Magnetization Curve

Consider the magnetization of a ferromagnetic solid in an external magnetic field (Fig. 24-2). At point 0 the solid is not magnetized, and the net magnetic moments of the domains have random orientations. In the region from O to a, the magnetization **M** and the flux density **B** in the solid increase as the applied field strength **H** increases. The domains that are parallel or nearly parallel to **H** increase in size as the magnetic moments of atoms near the domain walls in neighboring domains rotate until they are also aligned in the magnetic field (Fig. 24-3). Between a and b the domains that have magnetic moments parallel to **H** continue to increase in size; however, if the applied field is strong enough, the magnetic moments of all atoms in some of the other domains may rotate simultaneously until they become aligned parallel to **H**. Finally at point c all of the magnetic moments are aligned parallel to **H**; therefore the solid is fully magnetized and is said to be *saturated*.

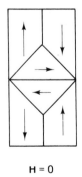

H = 0

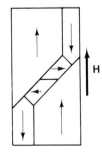

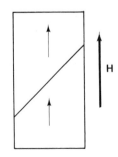

Figure 24-3
Domain growth.

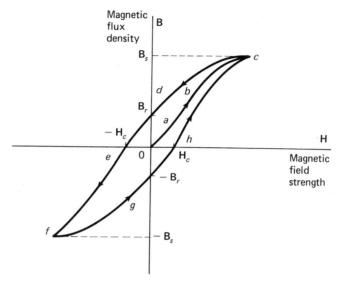

Figure 24-4
A hysteresis loop.

Hysteresis

At normal temperatures, if the applied field **H** is removed from a magnetically saturated solid, the demagnetization curve does not return along its initial path from *c* to O (Fig. 24-4). From *c* to *d*, as the applied field **H** is reduced to zero, the solid retains some magnetization because of the exchange coupling between the electron spins. The magnetic flux density B_r when the applied field strength $H = 0$ is called the *retentivity* (or *remanent induction*); it corresponds to a *permanent magnetization*. To reduce the flux density to zero inside the solid (from *d* to *e*), the magnetic field strength **H** of the applied field must be reversed to magnitude $-H_c$, which is called the *coercive force*. If the value of the reverse magnetic field strength is increased, the solid eventually becomes fully magnetized (at point *f*) in the reverse direction, and it saturates at flux density $-B_s$.

Between *f* and *g*, as the magnitude of the field strength **H** is reduced to zero, the solid retains some magnetization, and the mag-

netic field strength must be increased to a value equal to the coercive force at point *h* to demagnetize the material completely. In the region from *h* to *c*, as the applied field strength increases, the magnetization increases until at *c* the sample is saturated once again.

The curves representing the magnetization and reverse demagnetization of solid do not coincide: The flux density lags the field strength. This is called *hysteresis*, and in Figure 24-4 the loop *cdefghc* is called a *hysteresis loop*. Ferromagnetic solids that have large hysteresis loops are called *hard*, and those with small hysteresis loops are called *soft*.

Note that the relationship between the flux density **B** and the field strength **H** is not linear for ferromagnetic materials. Consequently their permeabilities μ and magnetic susceptibilities X_m are not constant. From the definition of field strength, the permeability is:

$$\mu = \mathbf{B}/\mathbf{H}$$

When a ferromagnetic solid undergoes a

magnetization cycle, the magnetic moments of its atoms continually change their orientations. Consequently some of the magnetic energy is dissipated as heat. The area that is bounded by the corresponding hysteresis loop is directly proportional to the energy that is dissipated in one magnetization cycle. These *hysteresis losses* are often significant, and they may produce serious heating in ac circuits.

Soft iron has a relatively large permeability, but its hysteresis losses are quite small; therefore it is often used in ac circuits as cores for electromagnets and as rotors in instruments and electric motors. Hard iron has a rather large hysteresis loop, but it also has a high retentivity; therefore it is used for permanent magnets.

During a magnetization process the size and shape of a ferromagnetic solid changes as the magnetic moments become aligned; this is called *magnetostriction*. Ultrasonic waves may be produced with high-frequency magnetostriction oscillators.

As a ferromagnetic solid is heated, the thermal vibration of the atoms weakens the exchange coupling between the electron spins. Above some critical temperature (770°C for iron), ferromagnetism is destroyed, and the sample becomes paramagnetic.

When a ferromagnetic solid is placed in a magnetic field, it increases the magnetic flux density at its location as it becomes magnetized. Therefore, since lines of magnetic force always form closed paths, they tend to concentrate in ferromagnetic solids. By analogy with the electrical resistance of conductors, ferromagnetic solids are said to have a low *reluctance* because they constitute an "easy path" for the lines of force.

In order to demagnetize a ferromagnetic material, the reverse magnetic field strength must be decreased each time in a cycle (Fig. 24-5).

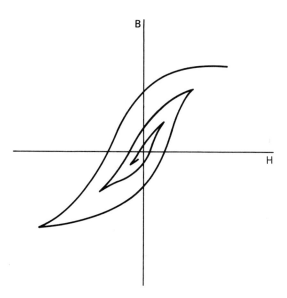

Figure 24-5
Demagnetization of a ferromagnetic material.

*Problems*_____

8. The area of a hysteresis loop has the same dimensions as the product of the flux density **B** and the field strength **H**. Use dimensional algebra to show that this is equivalent to units of energy per unit volume.

9. The magnetic flux at the center of a solenoid is 8.4×10^{-6} Wb in air. If a ferromagnetic core is inserted in the solenoid, the magnetic flux increases to 1.2×10^{-3} Wb. Calculate the relative permeability of the ferromagnetic material.

10. Some ferromagnetic material A has a lower retentivity but a higher coercive force than a second ferromagnetic material B. Which material would you use to make a permanent magnet? Why?

11. Describe how to demagnetize a ferromagnetic material by means of a series of demagnetization cycles.

12. Given the corresponding magnitudes of the flux density at certain magnetic field strengths, plot the magnetization curve and a graph of the permeability versus the field strength. Determine (a) the

approximate saturation flux density, (b) the maximum permeability, (c) the maximum relative permeability, (d) the magnetization when the permeability is at a maximum.

$H/(A/m)$	50	100	150	200	250	300	400	500
$B/(T)$	0.09	0.45	0.94	1.18	1.30	1.37	1.47	1.53

13. Compare the advantages and disadvantages of soft iron versus hard iron cores for electromagnets.

14. A 1 500 turn coil has a cross-sectional area of 1.2 cm² and a length of 24 cm, and it carries a 240 mA current. Calculate the field strength and the flux density at the center of the coil if it has (a) an air core or (b) a ferromagnetic core with a relative permeability of 1 200.

24-5 MAGNETIC DEVICES

The operation of many devices depends on the ability of ferromagnetic solids to control the magnitude and direction of the flux density at points in a magnetic field. A cylindrical soft iron core (or *rotor*) is used to produce a more uniform magnetic field for the wire coil in galvanometers, electric motors and generators. The lines of force take the lowest reluctance path between the poles of the permanent magnet (Fig. 23-18).

In many electric circuits, some components must be shielded from the stray magnetic fields of the electric currents; this is accomplished by placing the components within the cavity of a hollow ferromagnetic cylinder. The lines of force are distorted so that they pass through the high permeability ferromagnetic shield, leaving the cavity relatively free from external magnetic influences (Fig. 24-6).

Electromagnets

An *electromagnet* consists of a solenoid (or coil) that is usually wound around a high permeability ferromagnetic core. Any electric current in the solenoid magnetizes the core. Electromagnets are capable of lifting very heavy loads of ferromagnetic solids, such as scrap iron and steel, and are used to separate ferromagnetic solids from other materials. Dynamic loudspeakers and microphones utilize electromagnets to convert electric impulses into mechanical sound waves and vice versa.

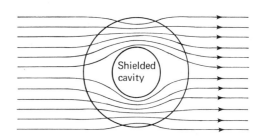

Figure 24-6
Magnetic shielding.

Relays and Switches

A simple magnetic *relay* is a type of switch in which an electromagnet is used to control the current in a circuit (Fig. 24-7). When switch S of the control circuit is closed, the soft iron core of the electromagnet is magnetized in the field that is set up by the current in the solenoid. As the soft iron is attracted to the core of the electromagnet, the armature pivots and closes the contacts, completing the main circuit. If switch S is opened, the core of the electromagnet becomes demagnetized, and the spring breaks the contacts in the main circuit.

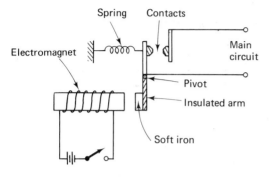

Figure 24-7
A simple relay.

Reed relays and *switches* consist of a pair of overlapping ferromagnetic reeds that are sealed in a glass tube containing an inert gas. In a magnetic field both reeds become magnetized in the same direction, but their overlapping ends become opposite magnetic poles. Consequently they attract each other and make contact (Fig. 24-8). When the external

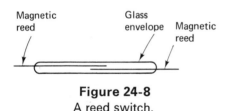

Figure 24-8
A reed switch.

magnetic field is removed, the reeds spring apart, breaking the contact. The external magnetic field may be produced either by permanent magnets or by placing the reed switch at the center of a solenoid. Reed switches and relays are small and light and may be activated by small electric currents.

Moving-Iron Meters

Moving-iron meters utilize the magnetic forces between an electromagnet and a piece of soft iron to deflect a pointer. They may be used to measure both alternating and direct electric currents. In a *repulsion* type of moving-iron meter, two parallel pieces of soft iron are located inside a solenoid. One piece of soft iron is rigidly fixed while the other is attached to a pivoted armature and a pointer (Fig. 24-8). A current in the solenoid mag-

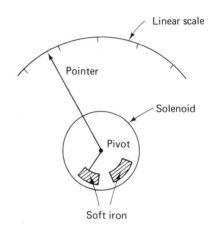

Figure 24-9
A repulsion type moving-iron meter.

netizes both pieces of soft iron in the same direction, and they repel each other. As one piece of soft iron is repelled, the pointer deflects across the scale, the amount of the deflection being directly proportional to the current in the solenoid.

Electrodynamometer

An *electrodynamometer* is an extremely useful device that can be used in the construction of ac voltmeters, ammeters, wattmeters and frequency meters. It consists of a pointer on a moving coil that is free to rotate in the magnetic field of two stationary coils that are connected in series with the movable coil (Fig. 24-10). As the moving coil rotates, deflecting a pointer over the scale, it distorts spiral springs, which develop a counter torque. Therefore the deflection of the pointer

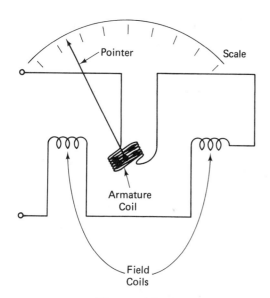

Figure 24-10
Electrodynamometer schematic.

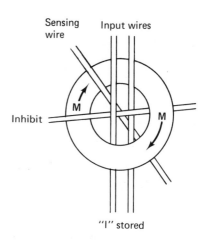

Figure 24-11
Ferrite memory core. If "I" is represented by a clockwise magnetization **M**, then "O" is represented by a counterclockwise magnetization.

depends on the current in the coils. Note that, if the direction of the current reverses, the direction of the field also reverses so that the pointer deflection remains clockwise over the scale; consequently, these devices can be used with alternating currents.

Ferrite Ring Computer Memory

Ferrites are mixtures of iron oxide and a compound of a divalent metal, such as zinc oxide. These materials have very large electric resistivities in addition to large magnetic saturations; they are often used in devices that must operate at high frequencies, e.g. inductors and transformers.

Many computers still store information in terms of magnetic fields in ring-shaped discs of a ferrite material. Each ferrite ring is very small (less than 1/8 cm in diameter), and it is threaded by four wires (Fig. 24-11).

In operation these memory cores are always magnetized in one direction or the other around the ring. The direction of the

magnetization may be reversed by an opposing net magnetic field of two parallel currents in the two input wires simultaneously. Very little energy is required to reverse the magnetization to saturation; however, if only one input wire carries a current, it does not affect the magnetization of the core. If two parallel currents last for approximately 1 μs, they are sufficient to magnetize the core.

The ones and zeros of the binary system are represented by magnetic flux in opposite directions in the ferrite core. For example, if one is represented by a clockwise magnetization, then zero corresponds to a counterclockwise magnetization.

When the magnetization of the ferrite core is reversed, the changing magnetic field induces a current in the sensing wire. To "read" the computer memory, parallel currents in the input wires induce a current in the sensing wire only if there is a change in the direction of the magnetization. Therefore the sensing wire detects the direction in which the core is magnetized.

If the remaining "inhibit" wire carries a

current that is antiparallel to the current in the two input wires, it prevents a charge in the magnetization of the core because its magnetic field opposes the field of the currents in the input wires.

More modern computers use solid state devices to store information. These devices are considered in Chapter 33.

REVIEW PROBLEMS

1. Calculate the magnitude of the magnetic moment of an electron in a circular orbit of radius 2.4×10^{-10} m at a speed of 6.5×10^6 m/s.

2. The flux density in a material has a magnitude of 5.0×10^{-2} T when the field strength is 295 A/m. (a) What is the relative permeability of the material? (b) What is the magnetization? (c) Calculate the magnetic susceptibility.

3. A material has a flux density of 44.1 mT when the field strength is 292 A/m. (a) What is the relative permeability of the material? (b) What is the magnetization? (c) Determine the magnetic susceptibility.

4. A 150 turn toroid on a ferromagnetic core (relative permeability of 120) has a cross-sectional area of 0.80 cm² and an average circumference of 25 cm and carries a 180 mA current, calculate (a) the magnetic flux density at the center of the core; (b) the magnetic field strength, (c) the magnetization of the core.

5. The field strength has a magnitude of 22.6 A/m at a point in a material where the magnitude of the flux density is 5.20×10^{-3} T. Determine (a) the permeability of the material, (b) the relative permeability of the material, (c) the magnetization of the material, (d) the magnetic susceptibility.

6. A 100 turn toroidal coil with an average circumference of 1.5 cm has an iron core with a relative permeability of 2 500. If the coil carries 120 mA current, (a) calculate the magnetic flux density in the core. (b) What is the field strength in the core? (c) Calculate the magnetization of the core. (d) Determine the magnetic susceptibility of the core.

7. A toroid with 1 200 turns is wound on an iron ring with a 4.0 cm² cross-sectional area, 80 cm mean circumference, and a 1 500 relative permeability. If the windings carry 0.12 A, find (a) the magnetic flux density, (b) the magnetic field strength, (c) the magnetic flux, (d) the permeability of iron, (e) the magnetization of the core.

A potential difference produces an electric current in a conductor, and the current gives rise to a magnetic field. Michael Faraday and Joseph Henry discovered independently that the reverse phenomenon also occurs. An emf is induced in any conductor when it moves through a magnetic field or if it experiences a varying magnetic field; this is called *electromagnetic induction*. The operation of several important devices, such as generators, transformers and inductors, depends on the principles of electromagnetic induction.

The normal operation of modern industries and households requires vast quantities of energy. Most of this energy is usually obtained from some distant source, and it must be converted into electrical energy before it can be transmitted economically to the consumer. Some energy sources, such as waterfalls and geysers, produce mechanical energy, and many other sources, such as nuclear and fossil fuels, are used to produce mechanical energy in heat engines. Generators are then used to convert the mechanical energy into electric energy. Transmission losses are reduced by using transformers to increase the potential difference before the electrical energy is transmitted. At the receiver other transformers are used to reduce the potential difference to the required value.

25-1 FARADAY'S LAW AND INDUCED EMF

While Joseph Henry was the first person to observe electromagnetic induction, the phenomenon was discovered independently a few months later by Michael Faraday. Faraday conducted a series of experiments in order to determine the properties of electromagnetic induction.

A simple example of electromagnetic induction is illustrated in Fig. 25-1. When the pole of a magnet is moved toward the coil, the galvanometer indicates the presence of

25

Electro-magnetic Induction

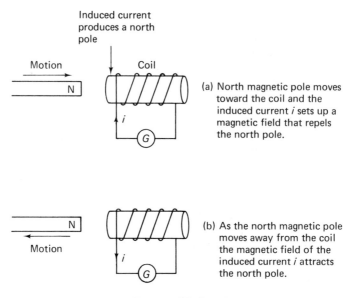

Figure 25-1
Electromagnetic induction.

an induced electric current* in the coil. If the magnetic pole is moved in the opposite direction (away from the coil), the galvanometer's deflection shows that the induced current is in the opposite direction. Electric currents are induced in the coil only when there is a relative motion between the magnetic pole and the coil. The direction of the induced current may be predicted by a simple rule that was suggested by H. Lenz. *Lenz's law* states that:

> *The induced electric current is always in a direction so that it sets up a magnetic field that opposes the original change that caused it.*

For example, as a north magnetic pole moves towards a coil, the direction of the induced current is such that it sets up a magnetic field that repels the north magnetic pole. The motion of the north pole towards the coil causes the induced current, and the magnetic field of the induced current opposes the mo-

---

*Varying currents are represented by i and steady currents by I.

tion of the magnetic pole. This is a direct consequence of the conservation of energy. Work must be done by some external agent in order to move the north magnetic pole towards the coil and to create the electric energy. The induced current could not be in the opposite direction because it would then set up a magnetic field that attracted the north magnetic pole and no external work would be required to move the pole and to create the electric energy. Similarly, when the north magnetic pole is moved away from the coil, it is attracted by the magnetic field of the induced current in the coil.

Michael Faraday investigated the induced current in one coil due to the magnetic field of a current in another adjacent coil (Fig. 25-2). When the switch is closed, a magnetic field builds up while the current in coil 1 quickly increases from zero to some steady value. As the magnetic field of the current in coil 1 increases, the galvanometer indicates that an electric current is induced in coil 2. In accordance with Lenz's law, the direction

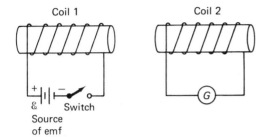

Coil 1   Coil 2

$\mathcal{E}$ Switch
Source
of emf

G

Figure 25-2

of the induced current is such that its magnetic field opposes the changes of the magnetic field due to the changing current in coil 1. When the current in coil 1 is steady, there is no induced current in coil 2. If the switch is then opened, the current in coil 1 decreases rapidly to zero, and a transient current is induced in coil 2. Thus *a current is only induced in the second coil when the current in the first coil changes or when the coils are moved relative to each other.*

Faraday determined that the magnitude of the induced emf $\mathcal{E}$ is equal to the product of the number of turns N in the circuit and the time rate of change of the magnetic flux linking that circuit. Thus, if there is a time rate of change of magnetic flux $\Delta\Phi_m/\Delta t$, the induced emf in a coil of N turns is:

$$\mathcal{E} = -N\frac{\Delta\Phi_m}{\Delta t} \qquad (25\text{-}1)$$

This expression is known as *Faraday's law*. The negative sign is required because of Lenz's law: It indicates that the induced emf opposes the change that produced it.

Example 25-1

A flat circular coil of radius $r = 1.50$ cm has 5 turns, and it is located in a magnetic field with its plane perpendicular to the flux density. Determine the emf induced in the coil at the instant when the flux density changes at a rate of 295 T/s.

Solution:

The area of the coil $A = \pi r^2$, since the flux density **B** is perpendicular to the plane of the coil, the magnetic flux through the coil $\Phi_m = BA$. Thus

$$\mathcal{E} = -N\frac{\Delta\Phi_m}{\Delta t} = -N\frac{\Delta(BA)}{\Delta t} = -NA\frac{\Delta B}{\Delta t}$$

Since A is a constant:

$$\mathcal{E} = -N\pi r^2\left(\frac{\Delta B}{\Delta t}\right)$$
$$= -(5 \text{ turns})\pi(1.50 \times 10^{-2} \text{ m})^2\left(295\frac{T}{s}\right)$$
$$= -1.04 \text{ V}$$

Problems

1. Determine the average emf induced in a 12 turn solenoid if it experiences a time rate of change of magnetic flux of -4.17 Wb/s.

2. Calculate the change in the magnetic flux linking a 25 turn coil if an average emf of -12.0 V is maintained during an elapsed time of 50 μs.

3. A solenoid of radius $r = 2.00$ cm has 60 turns, and it is located in a magnetic field with its axis parallel to the flux density. Determine the emf induced in the solenoid if the flux density changes at a rate of 217 T/s.

4. A flat 20 turn coil with a radius of 5.0 cm is located in a magnetic field with its plane perpendicular to the magnetic flux density. Determine (a) the change in magnetic flux and (b) the change in the flux density that would produce an average emf of 5.0 V in the coil for 200 μs.

25-2 A MOVING CONDUCTOR IN A MAGNETIC FIELD

Consider a wire conductor CD of length l that moves with a velocity **v** through a magnetic field where the flux density is **B**. Suppose

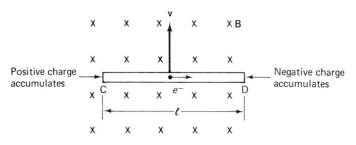

Figure 25-3
Electromagnetic induction in a conductor
that moves through a magnetic field.

that the length of the wire, its velocity and the flux density are mutually perpendicular (Fig. 25-3). The conductor contains some negatively charged electrons that are relatively free to flow. These "free electrons" (of charge $-e$) move with the conductor through the magnetic field; therefore according to Eq. 23-2 they experience a magnetic force of magnitude:

$$F_M = -evB \sin 90° = -evB$$

that deflects them towards end D of the conductor. As the electrons accumulate, a negative charge builds up at end D of the conductor, and a positive charge is formed at the end C because it has a deficiency of electrons. Thus an emf ε and an electric field strength $\mathbf{E}$ are induced in the wire, and they exert electric forces that oppose the magnetic forces on the electrons.

The induced emf increases until an equilibrium state is reached where the electric force $\mathbf{F}_\varepsilon = -e\mathbf{E}$ exactly balances the magnetic force $\mathbf{F}_M$, i.e. $-eE = -evB$, and the flow of charge ceases. Therefore in equilibrium the induced electric field strength has a magnitude:

$$E = vB$$

and the potential difference between the ends of the wire (the induced emf):

$$\varepsilon = El = vBl \qquad (25\text{-}2)$$

In the more general case where the length

l of the wire is perpendicular to the velocity and the magnetic field and the velocity is inclined at some angle θ with respect to the magnetic flux density $\mathbf{B}$, the induced emf is:

$$\varepsilon = vBl \sin \theta \qquad (25\text{-}3)$$

Note that an emf is only induced when the conductor moves through the magnetic field. If the flux density is constant and the conductor is stationary or moves parallel (or antiparallel) to the magnetic flux density, there is no induced emf.

The same induced emf is present even when the ends of the wire are connected externally by some conductor, but in this case the charges do not accumulate because they flow around a circuit (Fig. 25-4). The conventional electric current that is induced in the wire is directed from D towards C (opposite to the electron flow). A right-hand rule may be used to determine the direction of this current: *align the fingers of the right hand in the direction of the velocity* $\mathbf{v}$ *and rotate in the direction that they curl to the direction of the magnetic flux density* $\mathbf{B}$; *the straightened thumb indicates the direction of the conventional current* $\mathbf{I}$ (Fig. 25-4).

The induced current I experiences a magnetic force $\mathbf{F}_M$ in a direction opposite to the velocity.

$$\mathbf{F}_M = l(\mathbf{I} \times \mathbf{B})$$

This magnetic force $\mathbf{F}_M$ opposes the motion

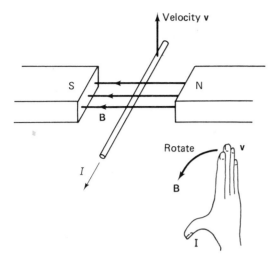

Figure 25-4
Right-hand rule.

of the wire CD through the magnetic field. Consequently an equal and opposite mechanical force (parallel to the velocity) must be applied in order to maintain the current. Thus work must be done by some external agent in order to produce an emf. This is another example of Lenz's law.

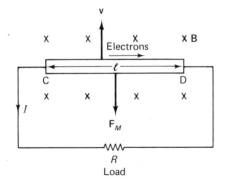

Figure 25-5
Electromagnetic induction showing the magnetic force F_M on the induced current in the conductor CD is in the opposite direction to velocity **v**.

Example 25-2

A wire conductor has a length $l = 8.0$ cm and a resistance $r = 1.0\,\Omega$, and it moves at a speed $v = 1.5$ m/s through a magnetic field where the flux density has a magnitude $B = 0.75$ T. If the length of the wire, its velocity and the flux density are mutually perpendicular and the ends of the wire are connected to an $8.0\,\Omega$ resistor, determine (a) the induced emf ε and (b) the instantaneous current I in the wire.

Solution:

(a) $\varepsilon = vlB \sin\theta$
$= (1.5$ m/s$)(8.0 \times 10^{-2}$ m$)(0.75$ T$) \sin 90°$
$= 0.090$ V

(b) $I = \dfrac{\varepsilon}{R_{TOT}} = \dfrac{\varepsilon}{r + R} = \dfrac{0.090 \text{ V}}{1.0\,\Omega + 8.0\,\Omega}$
$= 0.010$ A or 10 mA

since the wire is in series with the resistor.

Problems

5. A wire conductor has a length of 15 cm, and it moves with a speed of 2.0 m/s through a magnetic field where the magnetic flux density has a magnitude of 0.50 T. Determine the emf that is induced in the wire (a) when its length, its velocity and the flux density are mutually perpendicular and (b) when the length of the wire and the magnetic flux density are mutually perpendicular but the velocity of the wire is inclined at 30° with the magnetic flux density.

6. A wire conductor has a resistance of 0.50 Ω and a length of 15.0 cm, and it moves at a speed of 3.0 m/s through a magnetic field where the flux density has a magnitude of 1.2 T. If the length of the wire, its velocity and the magnetic flux density are mutually perpendicular, determine (a) the induced emf and (b) the electric field strength in the wire. (c) If the ends of the wire are connected to an 8.5 Ω resistor, determine the current in the circuit.

25-3 GENERATORS AND ELECTRIC MOTORS

A *generator* (or *dynamo*) is a device that is employed to convert mechanical energy into electrical energy. It is usually represented schematically by the symbol:

ac Generators

The structure of a simple *ac generator* or *alternator* is illustrated in Fig. 25-6. A closely wound coil is mounted in a magnetic field of a permanent magnet or an electromagnet so that it is free to rotate about an axis that is perpendicular to the magnetic flux density. The coil is usually wound on a laminated soft iron core to increase the magnetic flux.

Circular metal conductors called *slip rings* are attached to the terminals of the coil, and graphite brushes make electrical contact as they slide over the slip rings. In operation the coil is rotated by the application of mechanical energy, and the induced electric current is delivered through the slip rings and brushes to the external load.

Consider a flat coil of wire of N turns that has a cross-sectional area A and that rotates with a constant angular velocity ω in a uniform magnetic field where the flux density is **B**. As the loop rotates, the opposite sides that are parallel to the axis of rotation 0 move in opposite directions through the lines of force.

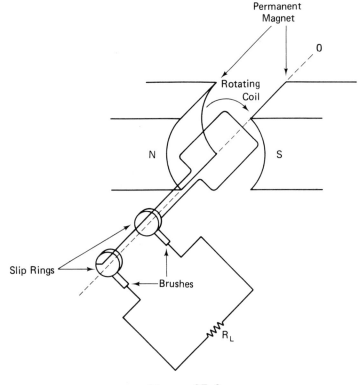

Figure 25-6
An ac generator.

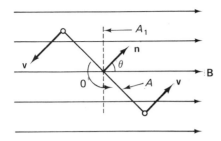

Figure 25-7

The position of the coil after an elapsed time t.

Suppose that the normal **n** (the perpendicular) to the plane of the loop was initially parallel to the lines of force. After an elapsed time t if the normal to the coil is inclined at some angle θ with the magnetic flux density (Fig. 25-7), the component of the area A that is perpendicular to **B** is:

$$A_1 = A \cos \theta$$

Therefore at that instant the magnetic flux through the coil is:

$$\Phi_m = BA_1 = BA \cos \theta$$

and the instantaneous induced emf* is:

$$\mathcal{E} = -N \lim_{\Delta t \to 0} \frac{\Delta \Phi_m}{\Delta t} = NBA\omega \sin (\omega t)$$
$$= NBA\omega \sin \theta \qquad (25\text{-}4)$$

The angle $\theta = \omega t$ is called the *phase angle*.

The induced emf is a sinusoidal function of the angular displacement θ of the coil. It continually reverses direction or *alternates* as the coil rotates (Fig. 25-8). One complete rotation of the coil corresponds to a single cycle of the induced emf. During the cycle the magnitude of the induced emf varies from

---

*The angular displacement $\theta = \omega t$; thus

$$\mathcal{E} = -N \frac{d}{dt} BA \cos (\omega t) = NBA\omega \sin (\omega t)$$
$$= NBA\omega \sin \theta$$

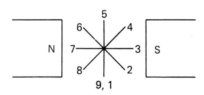

(a) Coil positions

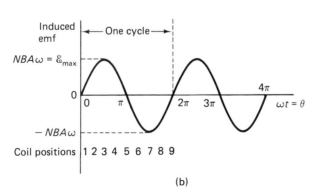

(b)

Figure 25-8
The alternating induced emf.

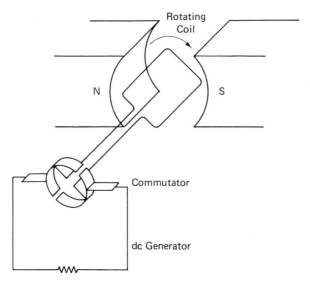

Figure 25-9
A dc generator.

zero (when $\theta = 0°$ and $180°$) to a maximum
value of $NBA\omega$ (when $\theta = 90°$ or $270°$).

dc Generators

A simple ac generator may be converted into
a dc generator by replacing the two slip rings
with a single split ring called a *commutator*
(Fig. 25-9). Each terminal of the coil is connected to a different half of the commutator;
as the coil rotates, the brushes alternately
make contact with each of the halves. At the
instant when the induced emf reverses polarity, the brushes change contacts on the commutator so that the pulsating potential
difference is always in the same direction
across the load (Fig. 25-10). To reduce the
fluctuations (or *ripple*) in the emf, most dc
generators are constructed with a number of
different coils that are wound on the armature
with their planes at different orientations. As
the induced emf in one coil becomes zero,
the emf induced in a perpendicular coil is a
maximum; therefore the net emf is much
smoother.

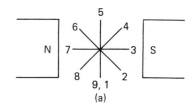

(a)

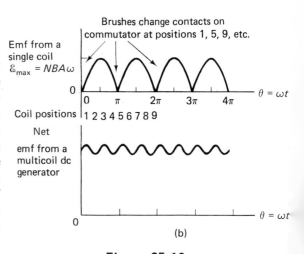

(b)

Figure 25-10
The pulsating emf between the brushes of a
dc generator.

Classification and Structure of Generators

Generators (and electric motors) have two main sections, a rotating part called the *rotor* that turns in a stationary frame or yoke called a *stator*.

dc Generators. In dc generators the armature is always wound on the rotor, and the field coils are located in the stator. The stator consists of an annealed steel frame that supports the laminated steel pole pieces of the field coils and acts as a low reluctance path for the lines of force. Armature windings are located in slots in the cylindrical steel core of the rotor. Brushes make electric contact with a multisegment commutator (Fig. 25-11).

Generators in which permanent magnets are used to produce the magnetic fields are called *magnetos*. In most generators (and electric motors) the magnetic fields are developed with electromagnets, and the device is often classified in terms of the interconnections between the electromagnet windings and the armature coils. If the electromagnet of a dc generator draws its energy from a separate source and is not interconnected with the armature coils, it is called *separately excited*. In a *self-exciting* dc generator some of the current that is induced in the armature coils is utilized in the electromagnet (Fig. 25-12).

The electromagnet windings (or *field coil*) of a *series-wound* dc generator are connected in series with the armature coil(s). If the current drawn by the external load increases, it produces a stronger magnetic field in the electromagnet, and the induced emf increases.

A dc generator is called *shunt-wound* if its electromagnet windings are connected in parallel with the armature coils. In this device only a part of the total induced current is used by the electromagnet. As the current through the external load increases, the current in the electromagnet decreases, and the magnetic field and the induced emf also decrease.

In a *compound-wound* dc generator the armature coils are connected both in series and parallel with the electromagnet windings.

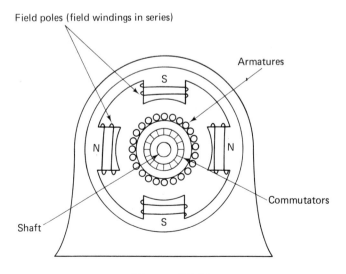

Figure 25-11
A four-pole dc generator.

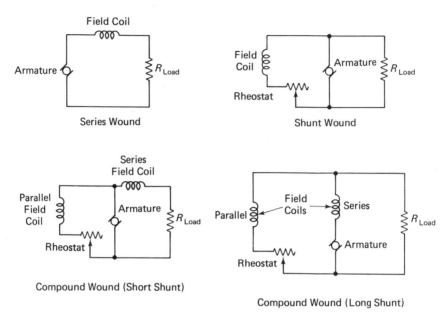

Figure 25-12
Self-excited dc generators (dc motors).

These devices can maintain a relatively steady emf because the effects of the series and shunt windings tend to cancel.

ac Generators. Most modern ac generators (alternators) are of the *revolving field* type in which the armature coils are located in fixed positions around a rotating magnetic field. Direct current is supplied from a separate source through brushes and slip rings to the field coil windings on the rotor so that the magnetic field rotates with the rotor. Two different types of field windings are used on the rotor. *Salient-pole* windings consist of coils (wound around laminated cores) connected in series so that alternate poles have opposite polarities (Fig. 25-13). *High-speed alternators* have coils wound lengthwise in slots around a solid cylindrical rotor. The armature coils are wound on laminated iron cores and are wedged in slots on the inner surface of the stator. Revolving field genera-

tors normally develop larger emfs than rotating coil generators, and they require less insulation because there are only two slip rings and brushes (for the field coils) (Fig. 25-14).

Most typical rotating field generators also have a small dc generator built in to supply the dc current for the field coils. This dc generator is called an *exciter*. The frequency of the ac from these generators depends on the number of poles and the angular velocity of the rotor. Since a pair of opposite polarity poles induce one cycle in the armature coil in each rotation of the rotor, the frequency of the ac output is:

$$f = (p/2)\omega \qquad (25-5)$$

where p is the number of poles and ω is the angular velocity of the rotor in rev/s.

*Example 25-3*_____
At what angular velocity must an 18 pole alternator be driven in order to produce an

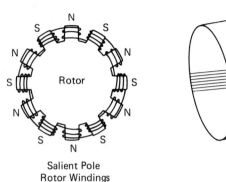

Salient Pole
Rotor Windings

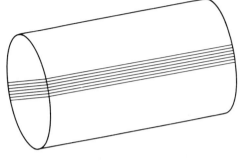

Wound Rotor

Figure 25-13

ac output at a frequency of 60 Hz in the armature coil?

Solution:

$$f = (p/2)\omega$$

Thus

$$\omega = \frac{2f}{p} = \frac{2(60 \text{ Hz})}{18 \text{ poles/rev}} = \frac{120 \text{ rev}}{18 \text{ s}}$$

$$= \frac{120 \text{ rev}}{(18/60) \text{ min}} = 400 \text{ rpm}$$

Revolving field *polyphase alternators* are
frequently used to generate two or more emfs
simultaneously. These devices have two or
more single-phase windings distributed sym-

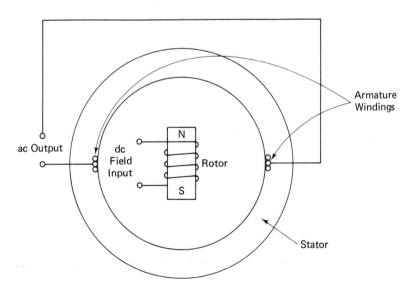

Figure 25-14
A rotating field generator.

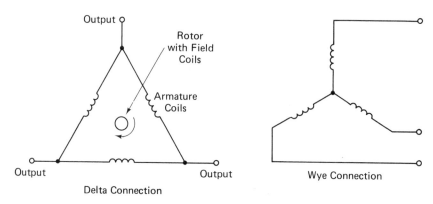

Figure 25-15
A three-phase alternator in which the field
coils are attached to a rotor.

metrically around the stator. The most common type is a *three*-phase system in which three single-phase armature coils are arranged in a *wye* or *delta* configuration so that there is a phase angle of 120° between the outputs (Figs. 25-15 and 25-16).

Electric Motors

Electric motors and generators have similar structures, but they perform opposite functions. In an electric motor electric energy is converted into mechanical energy. A potential difference is applied to the terminals of the coil, and the current-carrying coil experiences a magnetic torque when it is located in the magnetic field (Sect. 23-7). This torque tends to rotate the armature. Commutators are also used in dc motors to reverse the direction of the electric current as the armature rotates so that the magnetic torque is always in the same direction.

In operation a motor acts to some extent like a generator, and a generator acts like a motor.

The Motor Effect in Generators

In a generator, mechanical energy is used to rotate a coil in a magnetic field, and a current

is induced in the coil. However magnetic forces are then exerted on the induced currents, and these forces produce a torque on the coil in exactly the same way that a torque is produced in an electric motor. In accordance with Lenz's law this torque opposes the rotary motion of the coil, and work must be done by an external agent to keep the coil rotating.

Back emf in Motors

Electric energy is converted into mechanical energy in electric motors. A magnetic torque is exerted on a current-carrying coil that is located in a magnetic field; this torque tends to make the coil rotate. However, as in the case of a generator, an emf is then induced in the coil as it rotates in the magnetic field. This emf is called the *back* or *counter emf* because it acts in opposition to the potential difference U that is applied to the terminals of the coil. At some instant when the back emf is $\mathcal{E}_B$, the instantaneous current in the coil is:

$$i = \frac{\text{net emf}}{\text{total resistance}} = \frac{U - \mathcal{E}_B}{R} \quad (25\text{-}6)$$

where R is the resistance of the coil.

Note that from Eq. 25-4 the back emf $\mathcal{E}_B = NBA\, \omega \sin(\omega t)$ increases as the angular

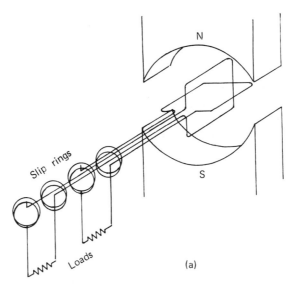

(a)

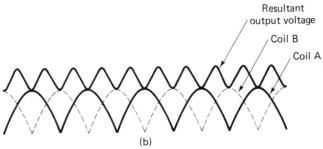

(b)

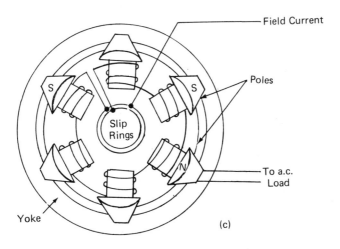

(c)

Figure 25-16

(a) Two-phase ac generator. (b) Output
from a two-phase alternator. (c) Six-pole
ac generator.

velocity ω of the coil increases. Therefore the net emf and current are always the greatest when an electric motor is started, and they decrease as the angular velocity of the armature increases. Since the magnetic torque is directly proportional to the current, the torque also decreases as the angular velocity increases. To reduce the starting current, an adjustable rheostat is often connected in series with the armature coil.

*Example 25-4*_____

A 100 V potential difference is applied between the terminals of a $2.5\,\Omega$ coil in a dc series motor; when the armature rotates at the rated angular velocity of 2 000 rpm, there is a 4.0 A current in the coil. Determine (a) the current in the coil at the instant when the motor is started, (b) the back emf that is developed when the motor reaches its rated angular velocity, (c) the resistance of a rheostat that would limit the starting current to 10 A if it was connected in series with the armature coil.

Solution:

(a) When the motor is started $\omega = 0$ and $\mathcal{E}_B = 0$; thus the instantaneous current is:

$$i = \frac{U}{R} = \frac{100\text{ V}}{2.5\,\Omega} = 40\text{ A}$$

(b) $\mathcal{E}_B = U - iR = 100\text{ V} - (4.0\text{ A})(2.5\,\Omega)$
$= 90\text{ V}.$

(c) If r is the starting resistance of the rheostat, the total series resistance of the armature and rheostat is:

$$R_{TOT} = r + R = r + 2.5\,\Omega$$

Thus

$$R_{TOT} = \frac{\text{applied potential difference}}{\text{electric current}}$$
$$= \frac{100\text{ V}}{10\text{ A}} = 10\,\Omega$$

and

$$r = 10\,\Omega - 2.5\,\Omega = 7.5\,\Omega$$

Classification and Structure of Motors_____

Motors generally have the same structure as generators.

dc Motors. In almost all dc motors the same power supply is used for both the armature and the field coils. To solve problems involving dc motors, it is often useful to replace the armature by its equivalent circuit consisting of a back emf and the armature resistance in series.

*Example 25-5*_____

A dc series motor has a field resistance of $59\,\Omega$, and the electromagnetic windings are connected in series with a $1.0\,\Omega$ armature coil. If the motor is connected to a 115 V power supply and it develops a back emf of 109 V when it runs at the rated speed, compute (a) the field current and the current in the armature coil, (b) the power supplied to the motor, (c) the output power, (d) the efficiency of the motor.

Solution:

(a) Since the field windings and the armature are in series, they carry the same current and the total resistance:

$$R_T = 59\,\Omega + 1.0\,\Omega = 60\,\Omega$$

Thus

$$I = \frac{U - \mathcal{E}_B}{R_T} = \frac{115\text{ V} - 109\text{ V}}{60\,\Omega} = 0.10\text{ A}$$

(b) $P_{in} = IU = (0.10\text{ A})(115\text{ V}) = 11.5\text{ W}$
(c) $P_{out} = I\mathcal{E}_B = (0.10\text{ A})(109\text{ V}) = 10.9\text{ W}$

Note that the power loss to heat $P_{loss} = I^2 R_T$ $= (0.10\text{ A})^2(60\,\Omega) = 0.60\text{ W}$ and $P_{in} = P_{out} + P_{loss}$

(d) $\eta = \dfrac{P_{out}}{P_{in}} = \dfrac{10.9\text{ W}}{11.5\text{ W}} = 0.95$ or 95%

_____

Series-wound motors are also called *universal motors* because they can be operated on either ac or dc. The no-load speed of a universal motor is high, and it decreases as the load increases. Their normal operating speed varies from 3 500 rpm to 10 000 rpm. These devices are frequently used with adjustable resistances to control the speed. They are found in vacuum cleaners, hair dryers, sewing machines and relatively large power tools.

ac Motors. Most ac motors are *induction motors* that utilize revolving magnetic fields (Fig. 25-17). In these devices alternating emf's are applied to the windings of electromagnets around the armature. The magnetic field produced by these time-varying emf's induces electric currents in the armature coils. According to Lenz's law, the magnetic field

of the induced current in the armature opposes the magnetic field of the electromagnets, and the corresponding magnetic repulsion produces a rotation of the armature.

The *stator* of *polyphase induction motors* is made from a laminated steel core with slots in which insulated field coils are located. When these coils are connected to polyphase ac, they produce a rotating magnetic field.

Squirrel-cage motors have a *rotor* containing a laminated iron core with slots in which copper (or aluminum) bars are located. These bars are interconnected by solid copper rings at each end. When a rotating magnetic field is produced in the field coils, currents are induced in the squirrel cage, producing a field in the rotor core. The field poles induced in

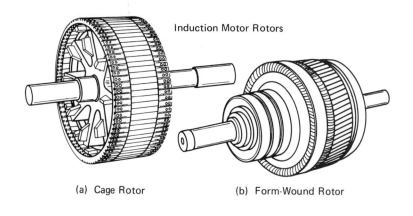

Induction Motor Rotors

(a) Cage Rotor   (b) Form-Wound Rotor

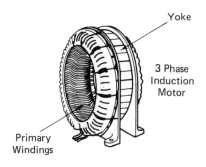

Yoke

3 Phase
Induction
Motor

Primary
Windings

(c) Stator

Figure 25-17
Induction motor rotors.

the rotor are then attracted to the opposite poles in the armature, and the rotor turns as the field is rotated.

Wound-rotor motors have rotor windings of insulated wire coils interconnected so that they have the same number of poles as the stator. In each case the rotor field cannot rotate as fast as the armature field in the stator; otherwise there would be no induction. The angular velocity of the armature field is called the *synchronous speed* of the motor:

$$N_s = \frac{f(60)}{(p/2)} \text{ rpm} \qquad (25\text{-}7)$$

where f is the frequency of the ac and p is the number of poles. The difference between the synchronous speed N_s and the operating speed N is called *slip*; it is usually expressed as a percentage.

$$\text{Percentage slip} = \left(\frac{N_s - N}{N_s}\right) 100\% \quad (25\text{-}8)$$

*Example 25-6*_____

A three-phase, squirrel-cage induction motor with six poles is operated on a 60 Hz mains at 1 140 rpm. Determine the percentage slip.

Solution:

$$N_s = \frac{f(60)}{(p/2)} \text{ rpm} = \frac{(60)(60)}{3} \text{ rpm} = 1\ 200 \text{ rpm}$$

$$\text{percentage slip} = \left(\frac{N_s - N}{N_s}\right) 100\%$$

$$= \left(\frac{1\ 200 \text{ rpm} - 1\ 140 \text{ rpm}}{1\ 200 \text{ rpm}}\right) 100\%$$

$$= 5\%$$

If direct current is supplied to the field coils of the rotor and polyphase ac is applied to the armature coils on the stator, the rotor rotates at the same speed as the armature field. This type of motor is called a *synchronous motor*.

*Problems*_____

7. A 120 turn coil has a length of 20 cm and a width of 10 cm, and it rotates about a perpendicular axis in a uniform magnetic field where the flux density is 1.2 T. Calculate the angular velocity in rpm of the coil if the maximum induced emf is 150 V.

8. A 300 turn coil has a length of 25 cm, a width of 15 cm and a resistance of 2.5 Ω. The coil rotates with a constant angular velocity of 3 000 rpm about an axis that is perpendicular to a uniform magnetic field where the flux density has a magnitude of 0.90 T. If the normal to the plane of the coil was initially parallel to the magnetic flux density, calculate (a) the maximum induced emf, (b) the instantaneous induced emf after an elapsed time of 27 ms, (c) the maximum induced current in the coil, (d) the current in the coil after 27 ms, (e) the average emf that is developed during one complete cycle.

9. A 110 V potential difference is applied between the terminals of a 2.75 Ω coil in a dc series motor; when the armature rotates at the rated angular velocity of 2 500 rpm, there is an 8.0 A current in the coil. Determine (a) the current in the coil at the instant when the motor is started, (b) the back emf that is developed when the motor reaches its rated angular velocity, (c) the resistance of a rheostat that would limit the starting current to 11.0 A.

10. A dc shunt motor has a field resistance of 60 Ω, and the field circuit is connected in parallel with the 2.0 Ω armature coil. If the motor is connected to a 120 V power supply, it develops a 110 V back emf when it runs at its rated speed. Calculate (a) the current in the armature coil, (b) the field current, (c) the total power supplied to the motor, (d) the power output from the motor.

11. A dc shunt generator has a field resistance of 220 Ω in parallel with an armature resistance of 2.0 Ω. If the generator deliv-

ers an output power of 4.4 kW at a potential difference of 110 V, calculate (a) the total current in the external load, (b) the total resistance of the generator, (c) the field current, (d) the induced current in the armature, (e) the induced emf, (f) the efficiency of the generator.

12. A two-pole induction motor is operated on a 60 Hz mains at 3 560 rpm. Determine the percentage slip.

13. A six-pole, squirrel-cage induction motor operating on a 60 Hz mains has a 2% slip. Determine the rotor speed.

25-4 MUTUAL INDUCTION

The emf and current that are induced in one coil when it experiences a change in the magnetic flux of a magnetic field due to current in a second adjacent coil. (Sect. 25-1) are an example of a phenomenon called *mutual induction*.

Consider two circuits that are located so that the magnetic field of the instantaneous current i_1 in circuit 1 (the *primary*) threads circuit 2 (the *secondary*) (Fig. 25-18). An emf and current are induced in the secondary only while it experiences a change in magnetic flux. If there is no relative motion between the circuits, the magnetic flux Φ_1 that links the secondary may only be changed by varying

the current i_1 in the primary. Since the magnetic flux Φ_1 is directly proportional to the instantaneous current i_1 in the primary, the time rate of change of magnetic flux is directly proportional to the time rate of change of current.

$$\frac{\Delta\Phi_1}{\Delta t} \propto \frac{\Delta i_1}{\Delta t}$$

Thus according to Faraday's law, the average induced emf in the secondary is:

$$\bar{\mathcal{E}}_2 = -N_2 \frac{\Delta\Phi_1}{\Delta t} = -M \frac{\Delta i_1}{\Delta t} \qquad (25\text{-}9)$$

where N_2 is the number of turns in the secondary and M is a constant called the *mutual inductance* of the system.

Similarly, the average emf that is induced in circuit 1 by a time rate of change of current $\Delta i_2/\Delta t$ in circuit 2 is given by:

$$\bar{\mathcal{E}}_1 = -N_1 \frac{\Delta\Phi_2}{\Delta t} = -M \frac{\Delta i_2}{\Delta t}$$

where N_1 is the number of turns in circuit 1 and $\Delta\Phi_2$ is the change in the flux linking circuit 1 due to a change Δi_2 of the current in circuit 2. Note that the mutual inductance M is the same in each case. Any two closed adjacent circuits have a mutual inductance. In accordance with Lenz's law the induced current sets up a magnetic field that opposes the change in the magnetic field of the current in the other circuit.

The *mutual inductance* of any pair of adjacent circuits is defined as the ratio of the induced emf in one circuit to the time rate of change of current in the other.

$$M = \frac{-\mathcal{E}_2}{\Delta i_1/\Delta t} = \frac{-\mathcal{E}_1}{\Delta i_2/\Delta t} \qquad (25\text{-}10)$$

Solving for the mutual inductance in Eq. 25-9, we obtain:

$$M = N_2 \frac{(\Delta\Phi_1/\Delta t)}{(\Delta i_1/\Delta t)} = N_2 \frac{\Delta\Phi_1}{\Delta i_1} = \frac{N_2\Phi_1}{i_1} \qquad (25\text{-}11)$$

since the current i_1 in the primary is always directly proportional to the magnetic flux Φ_1 at the secondary. Therefore the mutual inductance always has a positive value, and

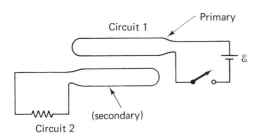

Figure 25-18
Mutual induction.

its magnitude depends on the geometry of the circuits. The term $N_2\Phi_1$ is often called the *flux linkage* in the secondary due to the current i_1 in the primary. Similarly

$$M = \frac{N_1\Phi_2}{i_2}$$

where Φ_2 is the flux at circuit 1 due to the current i_2 in circuit 2. Thus mutual inductance is also equal to the flux linkage in one circuit per unit current in the other.

In SI, the derived unit for mutual inductance is called the *henry* (H). A pair of adjacent circuits is said to have a mutual inductance of one henry if a time rate of change of current equal to one ampere per second in one circuit induces an emf of one volt in the other [1 H = 1 V/(1 A/s) = 1 V·s/A].

Problems _____

14. Calculate the mutual inductance of a pair of adjacent coils if an average emf of 120 V is induced in the secondary by a current in the primary that changes at a rate equal to 20.0 A/s.

15. The mutual inductance of a pair of adjacent circuits is 0.25 H and the secondary has 50 turns. If the current in the primary increases from 1.5 A to 8.1 A in 30 ms, calculate (a) the average induced emf in the secondary and (b) the change in the magnetic flux at the secondary.

16. If the mutual inductance of a pair of circuits is 0.75 H, calculate the flux linkage in the secondary when the current in the primary is 40 A.

25-5 SELF-INDUCTION

While a changing electric current induces emf's in neighboring circuits, it also induces an emf in its own circuit. As the current varies in a single isolated circuit, its magnetic field changes, and there is a corresponding change in the magnetic flux linking that circuit. The changing magnetic flux induces an emf that opposes the original change of the current in the same circuit. This phenomenon is called *self-induction*, and the induced emf is called the *counter emf*.

The induced *counter emf* ε is directly proportional to the time rate of change of current $\Delta i/\Delta t$ in the same circuit:

$$\varepsilon = -L\frac{\Delta i}{\Delta t} = -N\frac{\Delta\Phi}{\Delta t} \qquad (25\text{-}12)$$

where $\Delta\Phi$ is the change in magnetic flux, N is the number of turns and L is the proportionality constant called the *self-inductance* of the circuit. Thus the *self-inductance* of a circuit is defined as the ratio of the induced counter emf to the time rate of change of current in that circuit:

$$L = -\frac{\varepsilon}{\Delta i/\Delta t} \qquad (25\text{-}13)$$

The SI unit of self-inductance is the henry (H). In a circuit that has a self-inductance of one henry, a time rate of change of current equal to one ampere per second induces a counter emf of one volt. Any closed circuit has a self-inductance. By analogy with mutual inductance, the self-inductance of a circuit is:

$$L = \frac{N\Phi}{i} \qquad (25\text{-}14)$$

where N is the number of turns and Φ is the magnetic flux at the circuit due to the instantaneous current i. Thus self-inductance also depends on the geometry of the circuit.

Example 25-7 _____

A back emf of 12 V is induced in a coil in which the current increases at a rate of 48 A/s. Calculate (a) the self-inductance and (b) the flux linkage when the current is 10 mA.

Solution:

(a) $\qquad L = -\frac{\varepsilon}{\Delta i/\Delta t} = \frac{12\text{ V}}{48\text{ A/s}} = 0.25$ H

(b) $\qquad N\Phi = Li = (0.25 \text{ H})(0.010 \text{ A})$
$\qquad\qquad\qquad = 2.5 \times 10^{-3} \text{ Wb}$

Some devices called *inductors* are often included in circuits because of their inductances. These devices are used to store energy for a period of time in terms of their magnetic field, and they return the energy to the circuit as the field collapses. Because the induced emf in an inductor opposes the changing magnetic field that causes it, inductors are used to filter or smooth variations in currents. In circuit schematics inductors are usually represented by the symbols:

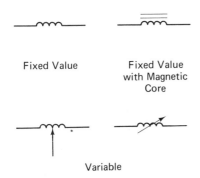

Fixed Value

Fixed Value with Magnetic Core

Variable

Figure 25-19
Inductors.

Most common inductors are *coils* or *chokes* and consist of coils of insulated wire. The type of core and the number of windings depends on the frequency at which the inductor is to operate. An inductor that operates at low frequencies usually consists of many turns of wire on a laminated iron core. Since iron is a conductor of electricity, any change in magnetic flux normally induces small circulating currents in the iron core of the inductor. These small currents are called *eddy currents*. They are undesirable because they generate heat losses in the iron core and set up magnetic fields in opposition to the changing magnetic flux that causes them.

Eddy currents are normally reduced by using laminated iron cores. A number of sheets of iron are coated with an insulator such as varnish. They are then stuck together and arranged so that their flat surfaces are aligned parallel to the magnetic flux density.

Low-frequency inductors normally have large inductances, and they are physically large. The inductance of some low-frequency inductors may be adjusted by moving the iron core (or *slug*) into or out of the coil; this is called *permeability tuning*.

High-frequency inductors normally have fewer turns of wire, and the windings are arranged so that the effects of capacitance between the turns are minimized. These inductors have air cores, powdered iron cores or ferrite cores.

The Self-Inductance of a Coil

Consider a coil of N turns that carries an instantaneous current i and has a length l, which is much greater than its cross-sectional area A. At the center of the coil, the flux density due to the current i is perpendicular to the cross-sectional area A, and it has a magnitude:

$$B = \frac{\mu Ni}{l}$$

where μ is the permeability of the core. Therefore the magnetic flux threading the coil is:

$$\Phi = BA = \left(\frac{\mu Ni}{l}\right)A$$

Substituting this expression in Eq. 25-14, we obtain the self-inductance of the coil:

$$L = \frac{N\Phi}{i} = \frac{N}{i}\left(\frac{\mu NiA}{l}\right) = \frac{\mu N^2 A}{l} \quad (25\text{-}15)$$

Note that, while this expression is accurate for toroids, it is only an approximation for coils and solenoids.

In a *ferrite bead* inductor, a core (or bead) is placed around a wire. The inductance may be increased if the wire is threaded several times through the same bead. These inductors have little capacitance and low dc resistances.

Problems

17. A constant back emf of 90 V is induced in a coil in which the current decreases from 210 mA to 30 mA in 1.5 ms. Calculate (a) the self-inductance of the coil and (b) the flux linkage when the current is 160 mA.

18. How long does it take the current in a 60 mH coil to rise from 100 mA to 450 mA if the back emf is constant at 1.2 V?

19. Calculate the self-inductance of a 750 turn coil that has a cross-sectional area of 1.2 cm², a length of 25 cm and an iron core with a relative permeability of 192.

20. How many turns of wire must be wound on a core with a relative permeability of 3 000 that is 1 cm long and has a diameter of 2.0 mm in order to obtain an inductance of 30 mH?

21. When the current in a 1 400 turn coil changes from 400 mA to 50 mA in 70 ms, it induces a counter emf of 1.5 V. If the length of the coil is 12 cm and its cross-sectional area is 0.75 cm², calculate the relative permeability of its core.

25-6 COMBINATIONS OF INDUCTANCES

Many circuits contain two or more interconnected inductances, and each pair of inductances often interacts by mutual inductance. The result may be quite complex; therefore we shall only consider combinations of noninteracting inductances that are arranged so that there is no mutual inductance.

Noninteracting Inductors in Series

When two or more noninteracting inductors are connected in series, the time rate of

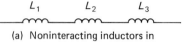

(a) Noninteracting inductors in series. The total self inductance $L_{TOT} = L_1 + L_2 + L_3$.

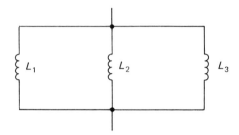

(b) Noninteracting inductors in parallel.
$$\frac{1}{L_{TOT}} = \frac{1}{L_1} + \frac{1}{L_2} + \frac{1}{L_3}.$$

Figure 25-20
Noninteracting inductors.

change of current $\Delta i / \Delta t$ is the same in each inductor. The total induced counter emf $\mathcal{E}_{TOT}$ is equal to the sum of the counter emfs that are developed by self-induction in each inductor. For example, if three noninteracting inductors with self-inductances L_1, L_2 and L_3 are connected in series (Fig. 25-20a) and their self-induced counter emfs are $\mathcal{E}_1$, $\mathcal{E}_2$ and $\mathcal{E}_3$ respectively, then

$$\mathcal{E}_{TOT} = \mathcal{E}_1 + \mathcal{E}_2 + \mathcal{E}_3$$
$$= -L_1 \frac{\Delta i}{\Delta t} - L_2 \frac{\Delta i}{\Delta t} - L_3 \frac{\Delta i}{\Delta t} = -L_{TOT} \frac{\Delta i}{\Delta t}$$

Therefore, the total inductance $L_{TOT} = L_1 + L_2 + L_3$, and in general:

$$L_{TOT} = \sum_i L_i \qquad (25\text{-}16)$$

Noninteracting Inductors in Parallel

If two or more noninteracting inductors are connected in parallel (Fig. 25-20b), the same counter emf $\mathcal{E}$ is induced in each. From the conservation of charge, at any instant the total time rate of change of current $\Delta i_{TOT} / \Delta t$ in the combination is equal to the sum of the time rates of current changes in the individual inductors. In a parallel combination of three inductors with inductances L_1, L_2 and L_3:

$$\frac{\Delta i_{\text{TOT}}}{\Delta t} = \frac{\Delta i_1}{\Delta t} + \frac{\Delta i_2}{\Delta t} + \frac{\Delta i_3}{\Delta t} = \frac{-\mathcal{E}}{L_1} - \frac{-\mathcal{E}}{L_2} - \frac{-\mathcal{E}}{L_3}$$

$$= \frac{-\mathcal{E}}{L_{\text{TOT}}}$$

Thus, the total self-inductance L_{TOT} of the combination may be determined from:

$$\frac{1}{L_{\text{TOT}}} = \frac{1}{L_1} + \frac{1}{L_2} + \frac{1}{L_3}$$

or in general

$$\frac{1}{L_{\text{TOT}}} = \sum_i \frac{1}{L_i} \qquad (25\text{-}17)$$

*Problems*_____

22. Determine the total inductance of three noninteracting inductors with inductances $L_1 = 2.0$ H, $L_2 = 3.0$ H and $L_3 = 5.0$ H when they are connected (a) in series, (b) in parallel, (c) with a series combination of L_1 and L_2 in parallel with L_3.

23. Two noninteracting inductors with inductances of 30 mH and 50 mH are connected in series. Compute the total counter emf in the combination at the instant when the time rate of change of current is 2.0 A/s.

24. Two noninteracting inductors with inductances of 80 mH and 120 mH are connected in parallel. Determine (a) the total time rate of change of current and (b) the rate of change of the current in each inductor when the induced counter emf is 24 V.

25. If you are given four 30 mH inductors, how could you connect two or more of them to give the following total inductances if there are no interactions: (a) 120 mH, (b) 7.5 mH and (c) 45 mH?

25-7 INDUCTORS IN DC SERIES CIRCUITS

Inductors normally have a measurable electric resistance; since they represent only one path for a current, a single inductor is equivalent to an inductance in series with a resistance.

Growth of Current_____

Consider the circuit schematic in Fig. 25-21.

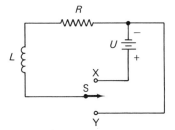

Figure 25-21
A series circuit containing an inductor and a resistor.

When the switch S is thrown to position X, a source of emf (which develops a terminal potential difference of U) is connected in series with an inductor of self-inductance L and the *total* resistance R of the external circuit. As the electric current increases, its changing magnetic field induces a counter emf by self-induction in the inductor. After the switch has been closed for an elapsed time t, if i is the instantaneous current and $\Delta i/\Delta t$ is the instantaneous time rate of change of current in the circuit, from Kirchhoff's second law:

$$U - L\frac{\Delta i}{\Delta t} = iR \qquad (25\text{-}18)$$

This equation must be solved by integral calculus. The solution is:

$$i = \frac{U}{R}(1 - e^{-Rt/L}) \qquad (25\text{-}19)$$

The relationship between the elapsed time t and the current i in the circuit is plotted in Fig. 25-22a.

Solving Eq. 25-18 for the instantaneous time rate of change of current and using Eq. 25-19, we obtain:

$$\frac{\Delta i}{\Delta t} = \frac{U - iR}{L} = \frac{U - U(1 - e^{-t/\tau})}{L} = \frac{Ue^{-t/\tau}}{L}$$
$$(25\text{-}20)$$

This is equal to the slope of the tangent to the current versus elapsed time curve at the instant t (Fig. 25-22a). As the elapsed time t

becomes very large, the current approaches a maximum *steady* value $I_{max} = U/R$, and consequently the induced counter emf approaches zero.

The quantity

$$\tau = L/R \qquad (25\text{-}21)$$

has dimensions of time and is called the *time constant* for the series inductor and resistor (*L-R*) circuit. After the switch has been closed for an elapsed time equal to the time constant, the current in the circuit equals:

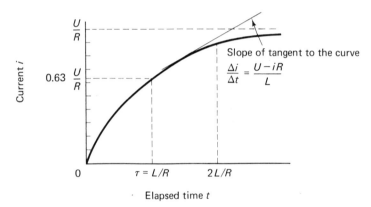

(a) The growth of current
in an *L-R* series circuit

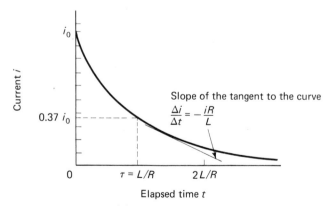

(b) The decay of current
in an *L-R* series circuit

Figure 25-22
Growth and decay of current.

$$i = \frac{U}{R}(1 - e^{-1}) = 0.632\frac{U}{R}$$

that is, the current is approximately 63% of the maximum value.

*Example 25-8*_____

An inductor with a self-inductance $L = 30$ mH and a resistance of 5.0 Ω is connected in series with a resistor that has a resistance of 10 Ω and a source of emf that develops a terminal potential difference $U = 30$ V. Determine (a) the maximum steady current in the circuit, (b) the time constant, (c) the instantaneous current and the potential drop across the inductance after an elapsed time of 4.0 ms.

Solution:

(a) $\quad I_{max} = \dfrac{U}{R_{TOT}} = \dfrac{30 \text{ V}}{5.0 \text{ Ω} + 10 \text{ Ω}} = 2.0 \text{ A}$

(b) $\quad \tau = \dfrac{L}{R} = \dfrac{3.0 \times 10^{-2} \text{ H}}{5.0 \text{ Ω} + 10 \text{ Ω}} = 2.0 \text{ ms}$

(c) The instantaneous current is:

$$i = \frac{U}{R_{TOT}}(1 - e^{-t/\tau})$$
$$= 2.0 \text{ A}(1 - e^{-4.0 \text{ ms}/2.0 \text{ ms}}) = 1.7 \text{ A}$$
$$v_L = -L\frac{\Delta i}{\Delta t} = -Ue^{-t/\tau} = -(30 \text{ V})e^{-2}$$
$$= -4.06 \text{ V}$$

The Decay of Current_____

If the source of emf is quickly removed from the circuit of Fig. 25-21 by throwing the switch S from position X to position Y, the current immediately begins to decrease from some initial value i_0. In the inductor the changing magnetic field of the decreasing current induces a counter emf that opposes the reduction of the current. After some elapsed time t, if i is the instantaneous current and $\Delta i/\Delta t$ is the instantaneous time rate of change of current, from Kirchhoff's second law:

$$-L\frac{\Delta i}{\Delta t} = iR \qquad (25\text{-}22)$$

This equation has the solution:

$$i = i_0 e^{-Rt/L} = i_0 e^{-t/\tau} \qquad (25\text{-}23)$$

This relationship between the current i and the elapsed time t is plotted in Fig. 25-22b. When the elapsed time t is equal to the time constant $\tau = L/R$ of the circuit, the current has reduced to approximately 37% of its initial value i_0.

*Problems*_____

26. An inductor with a self-inductance of 50 mH and a resistance of 3.0 Ω is connected in series with a 7.0 Ω resistor and a source of emf that delivers a terminal potential difference of 60 V. Calculate (a) the maximum current in the circuit, (b) the time constant, (c) the instantaneous current, the potential drop across the inductance and time rate of change of current after an elapsed time of 15 ms.

27. If the source of emf is shorted from the circuit of Problem 25-26 when the current is 5.0 A, calculate the instantaneous current and time rate of change of current after an additional elapsed time of 10 ms.

28. A 250 turn solenoid has a cross-sectional area of 0.500 cm², a length of 10.0 cm and an iron core with a relative permeability of 153. The solenoid has a resistance of 2.50 Ω, and it is connected in series with an 9.50 Ω resistor and a source of emf that develops a terminal potential difference of 36.0 V. Calculate (a) the self-inductance of the solenoid, (b) the maximum current in the circuit, (c) the time constant, (d) the instantaneous current and time rate of change of current after an elapsed time of 2.00 ms.

25-8 MAGNETIC ENERGY

An induced counter emf always opposes any change of current in an inductor. Work must be done by some external agent in order to produce a current by displacing charge in opposition to the counter emf and resistance in an inductive circuit. If the current in the circuit increases, the magnetic field around the inductor increases as the work done by the external agent in displacing charge against a counter emf is *reversibly* converted into magnetic energy. This energy is "stored" in the magnetic field of the current in the inductor. In addition, some of the work that is done is usually irreversibly converted into heat because of the resistance of the circuit.

When the external agent is shorted from the circuit, the magnetic field around the inductor collapses, and the self-induced counter emf opposes the change. The energy that was stored in the magnetic field then does an equivalent amount of work by opposing the decrease in electric current.

Consider an inductor with a self-inductance L that is connected to a source of emf. After an elapsed time t, if the instantaneous current i is increasing at a rate $\Delta i/\Delta t$, the self-induced counter emf in the inductor is:

$$\mathcal{E} = -L\frac{\Delta i}{\Delta t}$$

Therefore the work done by the source in displacing a very small charge ΔQ through the inductor against the counter emf:

$$\Delta W = -\mathcal{E}\Delta Q = L\frac{\Delta i}{\Delta t}\Delta Q = L\frac{\Delta Q}{\Delta t}\Delta i = Li\,\Delta i$$

where Δi is the small increment in the current and the instantaneous current $i = \Delta Q/\Delta t$. In increasing the current from zero to some final value i, the total work done by the source against the counter emfs is:

$$W = \Sigma\,\Delta W = \Sigma\,Li\,\Delta i = \tfrac{1}{2}Li^2 = E^* \quad (25\text{-}24)$$

This is equivalent to the energy E that is stored in the magnetic field. Note that this

---

$$^*W = \int_0^i Li\,di = \tfrac{1}{2}Li^2$$

is only a part of the total work done by the source, some of the applied energy is usually dissipated as heat in the resistances.

Example 25-9

Calculate the energy that is stored in the magnetic field of a 150 mA current in a 40 mH inductor.

Solution:

$$E = \tfrac{1}{2}Li^2 = \tfrac{1}{2}(4.0 \times 10^{-2}\text{ H})(0.15\text{ A})^2$$
$$= 4.5 \times 10^{-4}\text{ J}.$$

Problems

29. Calculate the energy stored in the magnetic field of a 2.0 A current in a 1.5 H inductor.

30. An inductor has a self-inductance of 2.0 H, and its instantaneous current of 3.00 A is increasing at a rate of 50.0 A/s. Calculate (a) the energy stored in the magnetic field of the current in the inductor and (b) the rate at which this energy is being stored. *Hint: P = &i.*

31. A 300 turn solenoid has a cross-sectional area of 0.50 cm², a length of 15 cm and an iron core with a relative permeability of 156. Determine the magnetic energy that is stored in the magnetic field of a 1.5 A current in the solenoid.

32. A 1.0 H inductor with a resistance of 50 Ω is connected in series with a 150 Ω resistor, a switch and a battery with a terminal potential difference of 12 V. (a) Calculate the time constant of the circuit. (b) Determine the current in the circuit when the switch has been closed for an elapsed time equal to the time constant. (c) What is the potential energy stored in the magnetic field around the inductor when the switch has been closed for 5.0 ms?

25-9 THE TRANSFORMER

A *transformer* is a device in which the principles of induction are employed to increase or decrease the amplitude of a fluctuating emf. It is represented schematically by the symbol:

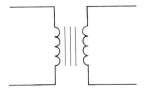

The device consists of two coils, the primary and the secondary, that are wound on a common laminated iron core (Fig. 25-23).

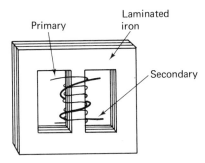

Figure 25-23
A concentric coil transformer.

In operation power is applied to the primary coil and is removed at the secondary coil. As the current in the primary varies, the magnetic flux in the iron core changes. If the resistance of the primary is negligible, the self-induced emf $\mathcal{E}_p$ in the primary that is due to the changing magnetic flux is equal and opposite to the applied potential difference. Thus

$$\mathcal{E}_p = -N_p \frac{\Delta\Phi}{\Delta t} \qquad (i)$$

where N_p is the number of turns in the pri-

mary and $\Delta\Phi/\Delta t$ is the time rate of change of magnetic flux due to the changing current i_p in the primary.

In an ideal transformer there is no flux leakage, and the same flux links both the primary and the secondary coils. Therefore, the emf $\mathcal{E}_s$ that is induced in the secondary by the changing magnetic flux in the iron core is:

$$\mathcal{E}_s = -N_s \frac{\Delta\Phi}{\Delta t} \qquad (ii)$$

where N_s is the number of turns in the secondary. Dividing (ii) by (i) we obtain

$$\frac{\mathcal{E}_s}{\mathcal{E}_p} = \frac{N_s}{N_p} \qquad (25\text{-}25)$$

In a *step-up transformer* the secondary has more turns than the primary, and the output emf $\mathcal{E}_s$ at the secondary is greater than the input emf $\mathcal{E}_p$ at the primary. If the primary has more turns than the secondary, the output emf $\mathcal{E}_s$ is less than the input emf $\mathcal{E}_p$, and the device is called a *step-down transformer*.

Most transformers have relatively high efficiencies (in excess of 97%); in addition to flux leakage and the resistance of the windings, there are losses due to hysteresis of the iron core and eddy currents. To minimize losses, the coils may be concentric windings of high conductivity wire on a laminated iron core (Fig. 25-23). Hysteresis losses may be reduced by the use of materials with small hysteresis loops, but generally it is more convenient to use high permeability iron to ensure large magnetic fluxes. In a perfect transformer, which has a 100% efficiency, the output power P_{out} is equal to the input power P_{in}, thus

$$P_{\text{out}} = \mathcal{E}_s i_s = P_{\text{in}} = \mathcal{E}_p i_p$$

where i_p and i_s are the instantaneous currents in the primary and secondary, respectively. Therefore

$$\frac{\mathcal{E}_s}{\mathcal{E}_p} = \frac{I_p}{I_s} = \frac{N_s}{N_p} \qquad (25\text{-}26)$$

When a transformer steps up the voltage, it also reduces the current and vice versa.

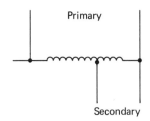

Figure 25-24
Autotransformer.

In an *autotransformer* the same coil is used as the primary and the secondary (Fig. 25-24). These devices are economical and efficient, but they have a limited transformation ratio.

Transformers are used to step up or step down voltages and currents in the transmission of ac power. Step-down transformers are used to obtain low voltages from a high-voltage line or to produce high currents.

*Example 25-10*_____

An ideal step-down transformer produces a 230 A current at the secondary for use in an arc welder. If the transformer draws 1.15 A at 230 V from the 60 Hz mains, determine (a) the ratio of turns in the transformer and (b) the voltage at the secondary.

Solution:

(a) $$\frac{N_s}{N_p} = \frac{I_p}{I_s} = \frac{1.15 \text{ A}}{230 \text{ A}} = \frac{1}{200}$$

(b) $$\mathcal{E}_s = \mathcal{E}_p \frac{I_p}{I_s} = (230 \text{ V})\frac{(1.15 \text{ A})}{(230 \text{ A})} = 1.15 \text{ V}$$

The Induction Coil_____

An *induction coil* (Fig. 25-25) is a type of transformer that is used to produce high voltages. The primary coil (a few turns of heavy insulated wire) and the secondary coil (many turns of thin wire covered with silk) are both wound on a laminated soft iron core. Current enters the primary coil through screw B and magnetizes the iron core, which attracts the soft iron hammer H. As the hammer moves towards the core, the platinum contact A breaks, and the primary current ceases. The iron core is then demagnetized, and hammer H springs back, closing the primary circuit once more. Very high voltages are produced

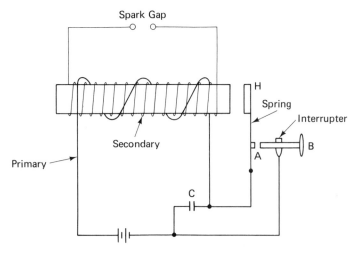

Figure 25-25
Induction coil.

in the secondary at each make or break of the contact. The capacitor C is used to reduce arcing at contact A.

Induction coils are used to produce the high voltages required for the spark plugs in an automobile ignition system (Fig. 25-26). However in this case a rotating shaft drives a cam that makes and breaks the contacts. One end of the secondary is grounded, and the other is connected to a rotating arm, which is a part of the *distributor*. As the arm moves, it makes contact with different metal connectors, each of which are attached to one end of a different spark plug.

*Problems*_____

33. A 15 MW ac generator delivers power at 200 kV to an ideal step-up transformer that has a 500 turn primary coil and a 6 000 turn secondary coil. Determine (a) the current and (b) the emf induced in the secondary.

34. An ideal transformer is used in an arc welder to produce a 250 A current at 2.0 V at the secondary. If the primary is connected to a 220 V, 60 Hz mains, determine (a) the ratio of turns in the transformer and (b) the current in the primary.

35. An ideal step-down transformer delivers 30 V to a 3.0 Ω load at the secondary. If the primary current is 5.0 A, determine (a) the primary voltage and (b) the secondary current.

36. An ideal transformer with a 600 turn primary and a 30 turn secondary is connected to a 110 V, 60 Hz mains. What voltage is delivered at the secondary?

REVIEW PROBLEMS

1. Determine the average emf that is induced in a 25 turn coil if it experiences a steady change in magnetic flux from 4.0×10^{-3} Wb to 3.6×10^{-2} Wb in 4.0 ms.

2. Compute the change in magnetic flux through a 30 turn coil if an average emf of 45 V is maintained for 12 μs.

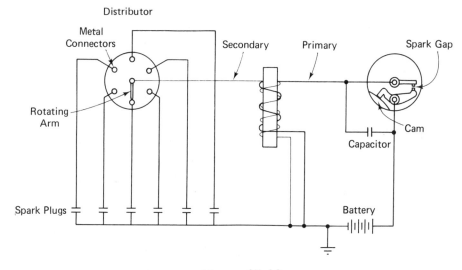

Figure 25-26
Automobile ignition system.

3. A 50 turn solenoid with a radius of 1.8 cm is located in a magnetic field with its plane perpendicular to the magnetic flux density. Calculate the average emf induced in the solenoid when the magnetic flux density increases steadily from 2.5 mT to 0.18 T in 200 μs.

4. A wire conductor has a length of 2.0 cm, and it moves at 5.0 m/s through a uniform magnetic field where the flux density is 0.78 T. If the velocity of the wire makes an angle of 42° with the magnetic flux density and the wire has a 0.50 Ω resistance, (a) calculate the induced emf. (b) If the ends of the wire are connected to a 4.5 Ω resistor, calculate the current in the circuit.

5. A coil with 250 turns and a diameter of 10 cm has its axis parallel to a magnetic field of 5.0×10^{-2} T. If the field is reduced steadily to 3.0×10^{-2} T in 200 μs, what is the average induced emf.

6. A dc series motor operates at 115 V, and the coils have a 3.00 Ω resistance. As the armature rotates at the rated angular velocity, there is an 8.0 A current in the armature coil. Calculate (a) the starting current in the coil; (b) the back emf developed at the rated angular velocity, (c) the resistance of a rheostat that would limit the starting current to 11.5 A.

7. A dc series motor has a field resistance of 49 Ω, and the electromagnet windings are connected in series with a 1.0 Ω armature coil. If the motor is connected to a 120 V power supply and it develops a back emf of 110 V when it runs at the rated speed, find (a) the field current and the current in the armature coil, (b) the power supplied to the motor, (c) the output power, (d) the efficiency of the motor.

8. A dc shunt motor has a field resistance of 80 Ω, and the field circuit is connected in parallel with a 2.0 Ω armature coil. If the

motor is connected to a 110 V power supply, it develops a 100 V back emf when it runs at the rated speed. Calculate: (a) the current in the armature coil, (b) the field current, (c) the total power supplied to the motor, (d) the output power from the motor.

9. A dc shunt motor has a field resistance of 110 Ω, and the field circuit is parallel with a 2.0 Ω armature coil. If the motor is connected to a 115 V power supply, it develops 105 V back emf at the rated speed. Calculate (a) the current in the armature, (b) the field current, (c) the power supplied to the motor, (d) the output power, (e) the efficiency of the motor.

10. Calculate the mutual inductance between a pair of adjacent coils if an average emf of 65 V is induced in one when the current in the other changes at a rate of 40 A/s.

11. The mutual inductance of a pair of adjacent coils is 80 mH, and the secondary has 30 turns. If the current in the primary changes from 300 mA to 800 mA in 5.0 ms, calculate: (a) the average induced emf in the secondary, (b) the change in the magnetic flux at the secondary, (c) the flux linkage in the secondary when current in the primary is 500 mA.

12. An average 50 V back emf is induced in an isolated circuit in which the current decreases from 300 mA to 50 mA in 300 μs. Calculate (a) the self-inductance of the circuit and (b) the flux linkage when the current is 150 mA.

13. A 2.0 H coil carries a current of 0.50 A. How much energy is stored in it?

14. Determine the total inductance when three noninteracting inductors with inductances $L_1 = 30$ mH, $L_2 = 60$ mH and $L_3 = 80$ mH when they are connected (a) in series, (b) in parallel, (c) with a series combination of L_2 and L_3 in

parallel with L_1. (d) What is the total emf induced in the series combination at the instant when the current changes at a rate of 5.0 A/s?

15. A 20 Ω coil is 20 cm long and 2.0 cm in diameter, and it has 500 turns. If the coil is connected in series with an 80 Ω resistor and a source that develops a terminal potential difference of 30 V, determine (a) the self-inductance of the coil, (b) the time constant for the circuit, (c) the initial current in the circuit, (d) the current in the circuit after 14.9 μs.

16. An 80 mH inductor with a 6.0 Ω resistance is connected in series with a 14 Ω resistor and a power supply. At the instant when the current in the circuit is 800 mA, the power supply is removed by shorting it out of the circuit. Calculate the instantaneous current and the voltage drop across the inductor after an additional 8.0 ms.

17. A 10 mH inductor with a resistance of 15 Ω is connected in series with a 45 Ω resistance and a 20 V supply. Determine (a) the time constant of the circuit, (b) the current in the circuit after an elapsed time equal to the time constant, (c) the potential energy stored by the inductor at that time.

18. A 250 turn solenoid has a cross-sectional area of 0.500 cm², a length of 10.0 cm and an ion core with a relative permeability of 153. The solenoid has a resistance of 2.5 Ω, and it is connected in series with a 9.50 Ω resistor and a source of emf that develops a terminal potential difference of 36.0 V. Calculate (a) the self-inductance of the solenoid, (b) the maximum current in the circuit, (c) the time constant, (d) the instantaneous current after 2.00 ms, (e) the energy stored in the magnetic field after 2.00 ms.

19. An ideal step-down transformer is used to produce 6.0 V for a door bell. If the primary is connected to a 120 V, 60 Hz mains, determine the ratio of turns in the transformer.

20. An ideal step-down transformer produces a 240 A current at its secondary for use in an arc welder. If the transformer draws 1.5 A at 220 V from the 60 Hz mains, determine (a) the ratio of turns in the transformer and (b) the voltage at the secondary.

21. A six-pole, squirrel-cage induction motor is operated on a 50 Hz mains at 980 rpm. Calculate the percentage slip.

22. A four-pole induction motor operated on a 60 Hz mains has a 2.5% slip. Determine the rotor speed.

23. A 25 turn rectangular coil with dimensions 5.00 cm × 8.00 cm is mounted in air in a uniform magnetic field with its plane inclined at 25° with a uniform flux density of 20 mT. What is the torque on the coil when it carries 2.8 A?

24. A 150 turn solenoid with a radius of 1.2 cm is located in a magnetic field with its plane inclined at 20° with the magnetic induction. Calculate the emf in the solenoid when the magnetic flux density decreases at a rate of 750 T/s.

25. A dc shunt-wound generator has an 80 Ω field coil and a 12 Ω armature coil. If a 220 Ω load is connected to the generator, it draws 5.0 A. Calculate (a) the potential difference produced across the armature coil, (b) the current in the field coil, (c) the output power of the generator.

26. A dc compound-wound motor has a series field resistance of 12.0 Ω and a shunt resistance of 48 Ω. If the armature resistance is 4.0 Ω and the motor draws 3.0 A when connected to a 120 V power supply, find (a) the back emf and (b) the efficiency.

27. A 20 Ω coil with an iron core of relative permeability 1 200 is 12 cm long and 0.60 cm in diameter and has 300 turns. If the coil is connected in series with an 80 Ω resistor and a source that develops a terminal potential difference of 30 V, determine (a) the self-inductance of the coil, (b) the time constant for the circuit, (c) the initial current in the circuit, (d) the current in the circuit after 960 μs, (e) the potential difference across the inductor after 960 μs, (f) the energy stored in the magnetic field when the current is 250 mA.

An *alternating current* (ac) periodically reverses its direction as the free charges vibrate to and fro; there is no net transfer of charge around the circuit. Most of our electricity is generated as ac because the time-varying voltage may be stepped up and down in transformers to reduce transmission losses. A large number of devices utilize ac directly, but other devices require dc; therefore special circuits are frequently used to rectify the alternating current. Many of the fundamental concepts of dc circuits are valid for ac circuits, but there are also several distinct differences. In this chapter some of these differences are discussed.

Light, X-rays, microwaves, radiowaves and television waves are all forms of electromagnetic waves. These electromagnetic waves all have the same speed in a vacuum, and with the exception of phenomena produced by differences in their wavelengths and frequencies, they all have similar properties. Many types of electromagnetic waves arc used as carrier waves in our communications systems. Some electromagnetic waves have the ability to penetrate solid materials and are used to probe the structure of matter and to detect and locate defects.

26-1 EFFECTIVE CURRENT AND VOLTAGE

Most alternating currents are produced in alternators. The instantaneous emf that is induced in an alternator coil as it rotates in a uniform magnetic field is given by the sinusoidal function:

$$e = \mathcal{E}_p \sin \theta \qquad (26\text{-}1)$$

where $\mathcal{E}_p$ is the maximum or peak value of the induced emf and θ is the phase angle, corresponding to the angle between the normal to the coil and the direction of the magnetic field. One complete rotation of the alternator coil produces one *cycle* of the induced emf (Fig. 26-1). If the normal to the plane of the

26

Alternating Current and Electro- magnetic Waves

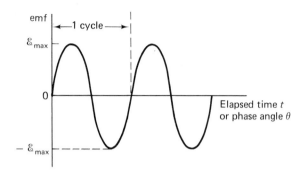

Figure 26-1
Alternating emf.

coil is initially parallel to the magnetic flux density, during one cycle the induced emf changes from zero to a peak value $\mathcal{E}_P$, back through zero to a peak value $\mathcal{E}_P$ in the opposite direction, and finally returns to zero. The time required for one complete cycle is called the *period T*, and the number of cycles per second is the *frequency f* of the alternating emf. In North America the "mains" frequency is usually 60 Hz. If the alternator coil has an average angular velocity ω, in an elapsed time t it rotates through some angle $\theta = \omega t = 2\pi ft$. Thus the instantaneous emf is:

$$e = \mathcal{E}_P \sin \theta = \mathcal{E}_P \sin (2\pi ft) \qquad (26\text{-}2)$$

When an alternating emf is applied to the terminals of a metallic resistor, since the total circuit resistance R is constant, the instantaneous current is:

$$i = \frac{e}{R} = \left(\frac{\mathcal{E}_P}{R}\right) \sin \theta = I_P \sin \theta = I_P \sin (2\pi ft)$$
$$(26\text{-}3)$$

where $I_P = \mathcal{E}_P/R$ is the maximum or *peak value* of the alternating current. Note that the current in a *pure* resistor has the same phase angle θ as the applied voltage; therefore the current is said to be *in phase* with the voltage (Fig. 26-2).

There is no *net* flow of charge in an alternating current; therefore the average current in one complete cycle is zero. However electric power is still dissipated as heat when an alternating current is applied to a resistor. At any instant the power that is being dissipated in a metallic resistor that has a constant resistance R and that carries an instantaneous current $i = I_P \sin (2\pi ft)$ is:

$$P = i^2 R = I_P^2 R \sin^2 (2\pi ft)$$

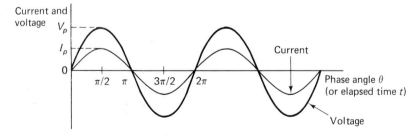

Figure 26-2
A pure resistive circuit. The current is in phase with the voltage.

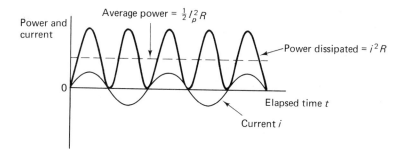

Figure 26-3
The average power dissipated by an alternating current in a pure resistor.

During one complete cycle (Fig. 26-3) the average value of $\sin^2(2\pi ft) = \frac{1}{2}$; thus the average power dissipated is:

$$\bar{P} = \frac{I_P^2 R}{2} \qquad (26\text{-}4)$$

The *effective* (or *rms**) *value* I of the alternating current is equivalent to the steady direct current that produces the same average power dissipation in the resistor:

$$\bar{P} = I^2 R = \frac{I_P^2 R}{2}$$

Thus
$$I = \frac{I_P}{\sqrt{2}} = 0.707 I_P \qquad (26\text{-}5)$$

Similarly, if $v = V_P \sin(2\pi ft)$ is the instantaneous voltage across the resistor, the instantaneous power dissipated $P = v^2/R$, and during one complete cycle the average power dissipated in the resistor is:

$$P = \frac{V_P^2}{2R}$$

where V_P is the peak voltage. The *effective value* V of the alternating voltage is the steady voltage that produces the same power dissipation. Thus

$$V = \frac{V_P}{\sqrt{2}} = 0.707 V_P \qquad (26\text{-}6)$$

Alternating currents and voltages are nor-

mally measured and specified in terms of their effective values.

Example 26-1

When a metallic resistance heater is connected to a standard 120 V, 60 Hz mains, it dissipates an average power of 180 W. Calculate (a) the peak voltage, (b) the effective current in the heater, (c) the peak current in the heater.

Solution:

(a) $\qquad V_P = \dfrac{V}{0.707} = \dfrac{120\ \text{V}}{0.707} = 170\ \text{V}$

(b) The average power $\bar{P} = IV$; thus

$$I = \frac{\bar{P}}{V} = \frac{180\ \text{W}}{120\ \text{V}} = 1.50\ \text{A}$$

(c) $\qquad I_P = \dfrac{I}{0.707} = \dfrac{1.5\ \text{A}}{0.707} = 2.12\ \text{A}$

Problems

1. Calculate the peak value of an 18.0 A alternating current.

2. Calculate the peak voltage from a European 220 V, 50 Hz mains.

3. If the peak voltage of a 1.5 kHz audio oscillator is 15 V, determine the instantaneous voltage when the phase angle is (a) 30°, (b) 80°, (c) 150°, (d) 290°.

---

*Root mean square.

4. If an oscillator develops a peak voltage of 25 V, what effective current would it deliver to a 220 Ω load?

5. A circuit draws 5.00 A from a 120 V, 60 Hz mains. Calculate (a) the peak current in the circuit, (b) the average power dissipated in the circuit, (c) the instantaneous current and voltage 2.00 ms after they pass the zero value.

6. The peak voltage from a 500 Hz oscillator is 12.0 V. Calculate (a) the effective voltage, (b) the instantaneous voltage when the phase angle is 120°, (c) the instantaneous voltage 2.30 ms after it passes through zero.

7. What is the maximum effective alternating potential difference that can be applied between two electrodes if they are separated by a 1.5 cm thick bakelite insulator that has a dielectric strength of 158 kV/cm.

8. Calculate the potential difference that would be indicated on an ac voltmeter when it is connected in parallel with a 120 Ω resistor that dissipates an average power of 60 W.

9. A 300 Ω heater filament is connected to a 120 V, 60 Hz mains. How many kilowatt hours of electric energy does it dissipate in 10 min?

26-2 ALTERNATING CURRENT IN A PURE INDUCTIVE CIRCUIT

In pure resistive circuits the alternating current is always in phase with the voltage. However, alternating currents vary with time, therefore most ac circuits also have an inductance. This inductance produces a phase difference between the simultaneous values of the current and voltage.

Phase Relationship

Consider a simple circuit consisting of a source that produces an alternating potential difference $v_L = V_P \sin(2\pi f t)$ with a frequency f between the terminals of an inductor (Fig. 26-4a). If the inductor has a self-inductance L and no resistance or capacitance, at the instant when the alternating current is changing at a rate $\Delta i / \Delta t$, the induced emf in the inductor is:

$$e_L = -L \frac{\Delta i}{\Delta t}$$

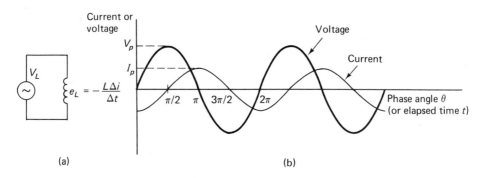

(a)   (b)

Figure 26-4
A pure inductive circuit. Current *lags* the voltage by ¼ cycle or a phase angle of $\pi/2$.

The time rate of change of current $\Delta i/\Delta t$ corresponds to the slope of the current versus time graph (Fig. 26-4b); this slope is zero when the current is maximum, and it is a maximum when the current is zero. Therefore the induced emf is not in phase with the current. Applying Kirchhoff's second law around the circuit, we obtain:

$$v_L - L\frac{\Delta i}{\Delta t} = 0$$

therefore $\quad \dfrac{\Delta i}{\Delta t} = \dfrac{v_L}{L} = \dfrac{V_P}{L}\sin{(2\pi ft)}\quad$ (26-7)

Through the use of integral calculus, it can be shown that the instantaneous current in the inductor is:

$$i = -\frac{V_P}{2\pi fL}\cos{(2\pi ft)} = \frac{V_P}{2\pi fL}\sin\left(2\pi ft - \frac{\pi}{2}\right)$$
(26-8)

and the peak current is:

$$I_P = \frac{V_P}{2\pi fL} \qquad (26\text{-}9)$$

Thus

$$i = I_P\cos{(2\pi ft)} = I_P\sin{(2\pi ft - \pi/2)}\quad(26\text{-}10)$$

In a pure inductive circuit the instantaneous current i lags the voltage by a phase angle of $\pi/2$ *radians (90°).* This relationship is illustrated in Fig. 26-4b; the peak current occurs $\frac{1}{4}$ cycle *after* the peak voltage; thus the current is at a maximum when the voltage is zero, and the voltage is at a maximum when the current is zero.

In practice inductors always have some electric resistance, and the current lags the voltage by some phase angle ϕ less than $\pi/2$ radians. The general phase relationship between the instantaneous values of voltage and the current is $v = V_P\sin{(2\pi ft)}$ and $i = I_P\sin{(2\pi ft - \phi)}$.

Inductive Reactance

Even if a circuit could have an inductance and no electric resistance, the induced emf would still oppose *changes* in the current; thus

inductances oppose *alternating* currents. By analogy with dc resistance, in Eq. 26-9 the opposition of an inductance to an alternating current is described by the term $2\pi fL$; this is called the *inductive reactance* X_L. The *inductive reactance* of an inductance L is defined as the ratio of the effective voltage V to the effective current I. It also equals the ratio of the peak voltage V_P to the peak current I_P. Thus

$$X_L = 2\pi fL = \frac{V_P}{I_P} = \frac{V}{I}\qquad(26\text{-}11)$$

Note that the inductive reactance is independent of the instantaneous voltage and current; its magnitude depends only on the inductance L and the frequency f. In SI, inductive reactances are expressed in units of ohms.

Example 26-2

Determine (a) the inductive reactance and (b) the effective current when a 30.0 mH inductor with a negligible resistance is connected to a 12.0 V, 1.30 kHz oscillator.

Solution:

(a) $\quad X_L = 2\pi fL$
$\qquad = 2\pi(1.30 \times 10^3\text{ Hz})(30.0 \times 10^{-3}\text{ H})$
$\qquad = 245\ \Omega$

(b) $\quad I = \dfrac{V}{X_L} = \dfrac{12.0\text{ V}}{245\ \Omega} = 0.0490\text{ A}\quad\text{or}\quad 49.0\text{ mA}$

Problems

10. Calculate the inductive reactance when a 2.0 H choke is connected to a 220 V, 50 Hz source.

11. A pure inductive circuit draws a 200 μA current from a 100 V, 500 kHz source. What is the inductance of the circuit?

12. An inductor has an inductive reactance of 500 Ω in a circuit when the ac frequency is 1.0 kHz. What is its inductive reactance when the ac frequency is 50 kHz?

13. At what frequency will a 120 μH inductor have an inductive reactance of 240 Ω?

14. What effective and peak currents will a 0.50 H choke draw from a 110 V, 60 Hz mains if its resistance and capacitance are negligible?

26-3 ALTERNATING CURRENTS IN PURE CAPACITIVE CIRCUITS

In normal operation there is no direct internal transfer of charge between the plates of a capacitor because they are separated by an insulator. A steady current cannot be maintained in a circuit containing a capacitor because there is no conduction between the capacitor plates: The capacitor acts as an open circuit. However capacitors do not prevent time-varying currents in a circuit because the time-varying net charges on the capacitor plates produce time-varying electric fields in the region between the plates. Even though there is no transfer of charge, these time-varying fields are equivalent to an electric current called a *displacement current*.

Phase Relationship

When some alternating potential difference $v_c = V_P \sin(2\pi f t)$ with a frequency f is applied to a capacitor, the net charge on the capacitor plates changes with time (Fig. 26-5). If C is the capacitance of the capacitor and there is no resistance or inductance, at the instant when the potential difference is v_c, the charge on the capacitor $q = Cv_c$.

Since the capacitance C is constant, the instantaneous current in the circuit is:

$$i = \frac{\Delta q}{\Delta t} = C \frac{\Delta v_c}{\Delta t} \qquad (26\text{-}12)$$

The time rate of change of the potential difference $\Delta v_c/\Delta t$ is the slope of the voltage–time graph (Fig. 26-5b). Therefore the current is zero when the voltage is a maximum, and the current is at a maximum when the voltage is zero. Thus the current is not in phase with the applied voltage. Through the use of differential calculus, it can be shown that the instantaneous current

$$\begin{aligned} i &= 2\pi f C V_P \cos(2\pi f t) \\ &= 2\pi f C V_P \sin(2\pi f t + \pi/2) \quad (26\text{-}13) \end{aligned}$$

Therefore, the peak current is:

$$I_P = (2\pi f C) V_P \qquad (26\text{-}14)$$

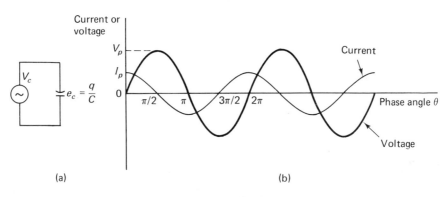

(a)   (b)

Figure 26-5
A pure capacitive circuit. Current *leads* the voltage by ¼ cycle or a phase angle of $\pi/2$.

and

$$i = I_P \cos(2\pi ft) = I_P \sin(2\pi ft + \pi/2) \quad (26\text{-}15)$$

In a pure capacitive circuit the instantaneous current i leads the applied voltage by a phase angle ϕ of $\pi/2$ radians. The peak current occurs $\frac{1}{4}$ cycle before the peak voltage.

Capacitive Reactance

Electrostatic forces between electric charges tend to oppose increases in the net charges on the plates of a capacitor. Therefore capacitors oppose electric currents. This opposition of a capacitance to an alternating current is called the *capacitive reactance X_c.* The capacitive reactance X_c of a capacitance C is defined as the ratio of the effective voltage V to the effective current I; it also equals the ratio of the peak voltage V_P to the peak current I_P. Thus, from Eq. 26-14:

$$X_c = \frac{V}{I} = \frac{V_P}{I_P} = \frac{1}{2\pi fC} \quad (26\text{-}16)$$

In SI, capacitive reactance is expressed in units of *ohms*.

Example 26-3

Determine the capacitive reactance of a 200 pF capacitor in a circuit where the ac frequency is 1.5 MHz.

Solution:

$$X_c = \frac{1}{2\pi fC} = \frac{1}{2\pi(1.5 \times 10^6 \text{ Hz})(2.0 \times 10^{-10} \text{ F})}$$
$$= 530 \ \Omega$$

Problems

15. Calculate the capacitive reactance of a 1.5 μF capacitor when it is connected to a 110 V, 60 Hz source.

16. (a) What is the capacitive reactance of a pure capacitive circuit that draws 50 mA from a 110 V, 60 Hz source? (b) Calculate the capacitance of the circuit.

17. At what frequency will a 20 μF capacitor have a reactance of 220 Ω?

18. What value capacitor will have a reactance of 60 Ω at 1.2 kHz?

19. What effective and peak currents are there in a 15 nF pure capacitive circuit that is connected to a 120 V, 60 Hz mains?

26-4 IMPEDANCE OF SERIES CIRCUITS

Consider a circuit composed of a resistance R, capacitance C and inductance L in series with a source of alternating voltage (Fig. 26-6). The collective opposition of the resist-

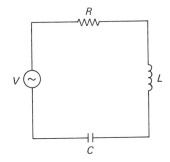

Figure 26-6
A series *LCR* circuit.

ance, capacitance and inductance to an alternating current is called the *impedance Z.* It is defined as the ratio of the applied effective voltage V to the effective current I in the circuit:

$$Z = \frac{V}{I} \quad (26\text{-}17)$$

In SI the unit for impedance is the *ohm*.

According to Kirchhoff's laws, at any instant the emf of the source must equal the sum of the instantaneous potential drops around the circuit, and, excluding the capaci-

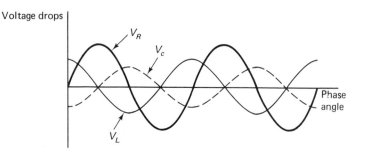

Figure 26-7

Phase relationships between the potential drops across a resistance, capacitance and inductance in series.

tor, the current i must be the same at each point in the circuit. However, while the current is in phase with the voltage in a resistance, in a capacitance the voltage *lags* the current by a phase angle of $\pi/2$, and in an inductance the voltage *leads* the current by a phase angle of $\pi/2$ (Fig. 26-7). Since we are normally only concerned with the effective values of voltage, these differences in phase angles must be considered. For simplicity, the effective potential drops in a series ac circuit may be represented by vectors. The effective potential drop V_R across the resistance R is usually drawn in the positive x-direction, and phase angles are measured with respect to this direction. Thus, since the potential drop V_L across the inductance L

leads V_R by a phase angle of $\pi/2$, V_L is drawn in the positive y-direction. Finally, the potential drop V_c across the capacitance C is drawn in the negative y-direction because it *lags* V_R by a phase angle of $\pi/2$ (Fig. 26-8). The total effective potential drop is represented by the *vector sum* of the potential drops in the circuit components.

$$\mathbf{V} = \mathbf{V_R} + \mathbf{V_L} + \mathbf{V_c}$$

The *phase angle* ϕ between the current and the voltage in the circuit corresponds to the angle between $\mathbf{V}$ and $\mathbf{V_R}$. When V_L is greater than V_c, the phase angle is positive, and the voltage leads the current; when V_L is less than V_c, the phase angle is negative, and the voltage lags the current.

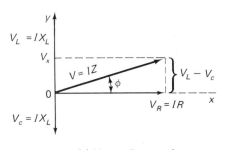

(a) Vector diagram of effective voltages

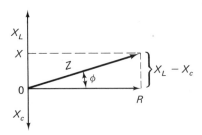

(b) Vector diagram to find the impedance

Figure 26-8

Suppose that X_L and X_c are the inductive and capacitive reactances, respectively, and I is the effective current in the ac series circuit. Then $V_R = IR$, $V_L = IX_L$ and $V_c = IX_c$.

Since V_L and V_c are 180° out of phase, the total effective *reactance* potential drop $V_x = V_L - V_c = I(X_L - X_c)$. Consequently, the total reactance of the series circuit $X = X_L - X_c$.

Using the Pythagorean theorem, we obtain the magnitude of the total effective potential drop in the ac series circuit:

$$V = \sqrt{V_R^2 + (V_L - V_c)^2}$$
$$= I\sqrt{R^2 + (X_L - X_c)^2} \quad (26\text{-}18)$$

Therefore, the *impedance* equals:

$$Z = \frac{V}{I} = \sqrt{R^2 + (X_L - X_c)^2}$$
$$= \sqrt{R^2 + \left(2\pi fL - \frac{1}{2\pi fC}\right)^2} \quad (26\text{-}19)$$

The phase angle ϕ may be determined by:

$$\tan \phi = \frac{V_L - V_c}{V_c} = \frac{X_L - X_c}{R} \quad (26\text{-}20)$$

Note that, since the effective current I is constant, we may use vectors to determine the total impedance Z of the ac series circuit (Fig. 26-8b).

Example 26-4

Calculate (a) the impedance Z, (b) the phase angle ϕ, (c) the effective current when a 120 V, 60 Hz source is connected in an ac series circuit that has a total resistance $R = 1.20 \text{ k}\Omega$, a total inductance $L = 300 \text{ mH}$ and a total capacitance $C = 2.00 \ \mu\text{F}$.

Solution:

(a) $X_L = 2\pi fL = 2\pi(60 \text{ Hz})(300 \times 10^{-3} \text{ H})$
$= 113 \ \Omega$

$X_c = \dfrac{1}{2\pi fC} = \dfrac{1}{2\pi(60 \text{ Hz})(2.00 \times 10^{-6} \text{ F})}$
$= 1.33 \times 10^3 \ \Omega$

Thus

Z
$= \sqrt{R^2 + (X_L - X_c)^2}$
$= \sqrt{(1.20 \times 10^3 \ \Omega)^2 + (113 \ \Omega - 1.33 \times 10^3 \ \Omega)^2}$
$= 1.71 \times 10^3 \ \Omega$

(b) $\tan \phi = \dfrac{X_L - X_c}{R}$

$= \dfrac{(113 \ \Omega - 1.33 \times 10^3 \ \Omega)}{1.20 \times 10^3 \ \Omega}$

$= -1.014$

and $\phi = -45.4°$

The voltage *lags* the current by a 45.4° phase angle.

(c) $I = \dfrac{V}{Z} = \dfrac{120 \text{ V}}{1.71 \times 10^3 \ \Omega} = 7.02 \times 10^{-2} \text{ A}$

or 70.2 mA

Problems

20. An ac circuit consists of a resistor and a pure inductor in series with an ac source. If the effective potential drops across the resistor and inductor are 30 V and 40 V, respectively, determine (a) the voltage of the source and (b) the phase angle.

21. An ac circuit consists of a resistor and a capacitor in series with an ac source. If the effective potential drops across the resistor and capacitor are 120 V and 50 V, respectively, calculate (a) the voltage of the source and (b) the phase angle.

22. A 200 mH coil with a resistance of 22 Ω is connected in series with a 100 Ω resistor and a 30 V, 500 Hz ac source. Calculate (a) the impedance, (b) the phase angle, (c) the effective current in the circuit.

23. A 300 mH inductor draws 120 mA when it is connected to the terminals of a 60 V, 250 Hz ac source. Determine (a) the reactance and (b) the resistance of the inductor.

24. Calculate (a) the impedance, (b) the phase angle, (c) the effective current when a

220 pF capacitor and a 470 kΩ resistor are connected in series with a 250 V, 1.5 kHz ac source.

25. Determine (a) the impedance, (b) the phase angle, (c) the effective current when a 220 V, 50 Hz source is connected in an ac series circuit that has a resistance of 220 Ω, an inductance of 2.0 H and a capacitance of 50 μF.

26. A 500 nF capacitor, a 300 Ω resistor and a 150 mH inductor with a resistance of 60 Ω are connected in series with a 500 Hz ac source. If the circuit draws 250 mA, calculate (a) the impedance, (b) the voltage of the source, (c) the phase angle.

26-5 POWER DISSIPATION IN AC CIRCUITS

Capacitors and inductors store electric energy in terms of electric and magnetic fields, respectively. These devices eventually return all of the stored energy to the circuit; therefore no electric power is dissipated because of the capacitance or inductance of a circuit. However electric power is dissipated as heat in the resistances of the circuit. Therefore, if I is the effective current and V_R is the effective potential drop in the total resistance R of the circuit, the average power dissipation is:

$$\bar{P} = IV_R = I^2 R$$

But in Fig. 26-8a it can be seen that:

$$V_R = V \cos \phi \qquad (26\text{-}21)$$

where V is the total effective potential drop in the circuit and ϕ is the difference in phase between the effective current and the effective voltage. Thus

$$\bar{P} = IV_R = IV \cos \phi \qquad (26\text{-}22)$$

The product of the effective current and the effective voltage is known as the *apparent power*. In order to distinguish it from the actual power that is dissipated in the circuit,

the apparent power is usually expressed in units of volts amperes. The term $\cos \phi$ is called the *power factor* of the ac circuit; it is equal to the ratio of the true power dissipated to the apparent power:

$$\cos \phi = \frac{\bar{P}}{IV} = \frac{V_R}{V} = \frac{IR}{IZ} = \frac{R}{Z}$$
$$= \frac{R}{\sqrt{R^2 + (X_L - X_c)^2}} \qquad (26\text{-}23)$$

where X_L and X_c are the inductive and capacitive reactances, respectively, and Z is the impedance of the circuit. Power factors are frequently expressed as a percentage.

At any instant the electric power that is being consumed in an ac circuit is equal to the product of the instantaneous values of the current and voltage. Consequently the power varies during each cycle of the alternating current. If the current and voltage are not in phase, the power often assumes negative values as the capacitance and inductance return stored energy to the circuit. When the total reactance $X = (X_L - X_c)$ is large, the power factor $\cos \phi$ is small, and the circuit dissipates only a fraction of the apparent power IV that it draws from the source. As a result, the circuit draws a large current from the source, and the power losses $I^2 R$ in the transmission lines are large. In many cases capacitors are added in parallel to the load to increase the power factor and to reduce transmission losses.

Example 26-5

A 1.20 kW ac motor has a power factor of 80% when it is connected to a 120 V, 60 Hz mains. (a) What effective current does the motor draw? (b) What effective current would it draw if its power factor was 100%?

Solution:

(a) $I = \dfrac{\bar{P}}{V \cos \phi} = \dfrac{(1.20 \times 10^3 \text{ W})}{(120 \text{ V})(0.80)} = 12.5 \text{ A}$

(b) $I = \dfrac{\bar{P}}{V \cos \phi} = \dfrac{(1.20 \times 10^3 \text{ W})}{(120 \text{ V})(1.00)} = 10.0 \text{ A}$

Problems

27. A 5.0 kW ac motor has a power factor of 65% when it is connected to a 200 V, 50 Hz mains. What effective current does the motor draw?

28. An ac circuit has an inductance of 500 mH, a capacitance of 35 μF and a resistance of 80 Ω in series with a 120 V, 60 Hz ac source. Calculate (a) the power factor, (b) the effective current, (c) the average power dissipated in the circuit.

29. An ac circuit has an inductance of 300 mH, a capacitance of 50 μF and a resistance of 220 Ω in series with a 220 V, 50 Hz ac source. Calculate (a) the inductive and capacitive reactances, (b) the impedance, (c) the power factor, (d) the effective current, (e) the average power dissipated in the circuit, (f) the effective potential drop across each component.

26-6 RESONANCE

From Eq. 26-19 it can be seen that the impedance Z of an ac series circuit depends on the frequency f of the applied alternating voltage V. The impedance of the circuit is at a minimum when the inductive reactance X_L is equal to the capacitive reactance X_c:

$$Z = \sqrt{R^2 + (X_L - X_c)^2} = R \quad \text{(when } X_L = X_c\text{)}$$

the pure resistance, and the circuit draws a *maximum effective current* $I = V/R$ from the source. This condition is called *electric resonance*. When resonance occurs:

$$2\pi f_0 L = \frac{1}{2\pi f_0 C}$$

Therefore the frequency of the applied voltage at resonance equals

$$f_0 = \frac{1}{2\pi\sqrt{LC}} \qquad (26\text{-}24)$$

This is known as the *resonant frequency* of the

circuit. Note that in an ac series circuit at resonance, the total reactance $X = X_L - X_c = 0$. Consequently the current is always in phase with the applied voltage, and the power factor $\cos \phi = 1$.

At resonance the effective potential drops in the capacitance and inductance of an ac series circuit are equal in magnitude but 180° out of phase. The capacitance and inductance of the circuit store equal energies but during different half cycles of the alternating current. As the energy stored in the electric field of the capacitance increases, the energy stored in the magnetic field of the inductance decreases and vice versa. Some energy is always dissipated in the resistance of the circuit; therefore, if the magnitude of the effective current is to be maintained, an ac source must continually supply energy at the resonant frequency of the circuit. Without the ac source, the effective current decreases with time, and the oscillation is said to be damped. Circuits that have specific ac frequencies are called *oscillators*.

Resonant circuits have important applications in communications systems. They are used to produce electric oscillations and to detect modulated carrier waves at specific frequencies. These special circuits normally contain adjustable capacitors or inductors so that they may be tuned to different frequencies. For example, in order to receive a specific modulated carrier wave, the antenna circuit of a radio receiver is tuned to resonance at the frequency of the carrier wave. The carrier wave produces electric oscillations in the tuned circuit; these oscillations are then amplified and demodulated. When the frequency of the carrier wave is exactly the same as the resonant frequency, there is a maximum current I_{max} in the tuned circuit. However the antenna circuit also detects carrier waves at frequencies close to the resonant frequency, but the circuit impedance is larger for these frequencies, and they produce smaller currents.

Example 26-6

Determine the resonant frequency of an ac series antenna circuit that has a resistance $R = 20\ \Omega$, an inductance $L = 280\ \mu H$ and a capacitance $C = 70.0$ pF.

Solution:

$$f_0 = \frac{1}{2\pi\sqrt{LC}}$$

$$= \frac{1}{2\pi\sqrt{(2.80 \times 10^{-4}\ H)(7.00 \times 10^{-11}\ F)}}$$

$$= 1.14 \times 10^6\ Hz \quad \text{or} \quad 1.14\ MHz$$

Problems

30. Determine the resonant frequency of an ac series antenna circuit that has a 7.5 Ω resistance, an inductance of 27 μH and a capacitance of 300 pF.

31. What inductance must be placed in series with a 20 μF capacitor to produce a circuit with a resonant frequency of 3.0 kHz?

32. What capacitance must be connected in series with a 500 μH coil to produce a circuit with a resonant frequency of 1.5 MHz?

26-7 AC POWER TRANSMISSION

At present there are four methods that are used to produce most commercial electric power:

1. *Internal combustion engines*, such as diesels, in which fuel is burned and the rotary output is used to turn the rotor of a generator. This type of power plant is used at isolated locations or in regions that have vast quantities of natural gas or fuel oil.

2. *Fossil fuels*, such as coal, oil or natural gas, are used to heat water, producing high-pressure steam in a boiler. The steam then flows past the blades of a turbine into a low-pressure area, and the rotation of the turbine is transferred to a generator.

3. In *hydroelectric power plants* a dam is used to create a large reservoir of water. As the water is allowed to fall down a penstock,

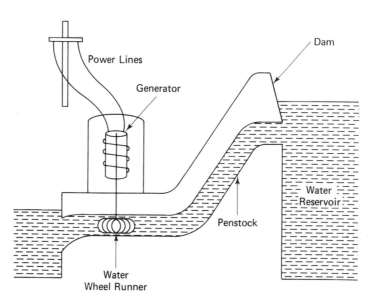

Figure 26-9

its potential energy is converted into kinetic energy, and at the bottom the fast-moving water turns the blades of a turbine. This rotary motion is then transferred to a generator (Fig. 26-9).

4. A *nuclear reactor* can be used to generate a considerable amount of heat, which superheats steam to high pressures. The steam then passes through the blades of a turbine, the shaft of which is connected directly to the armature of a generator (Chapter 34).

Once the electric energy has been produced, it is normally transmitted from one place to another in a conductor called a *transmission line* (Fig. 26-10). However, there are a number of ways in which energy is lost during the transmission process:

Line Resistance Losses. With the possible exception of superconductors, all materials have a measurable electric resistance. Consequently some electric energy is dissipated as heat during the transmission process. If R is the resistance of the transmission line and i is the instantaneous current that it carries, the rate at which electric energy is dissipated as heat is:

$$P_{LOSS} = i^2 R \qquad (26\text{-}25)$$

and the potential drop in the transmission line is:

$$U = iR \qquad (26\text{-}26)$$

These lines' resistive losses are usually reduced by using a step-up transformer to decrease the current before the energy is transferred to the transmission line. At the receiver the current may be increased once more in a step-down transformer (Fig. 26-10).

Transformer Losses. While transformers are normally quite efficient ($\sim 97\%$), there are losses due to hysteresis of the core, flux leakage, eddy currents in the core and resistance of the windings.

Leakage Losses. There is an upper limit to the transmission voltage because high voltage ionizes the air surrounding the transmission lines and there is some leakage of current. These losses increase in a damp climate.

Power Factor. If the power factor of the ac circuit is not unity, the circuit draws a large current from the source, and there are large power losses ($I^2 R$) in the transmission lines.

*Example 26-7*_____

A 6.0 MW generator delivers power at 2.0 kV to an ideal transformer that steps up the voltage to 150 kV before the energy is transmitted along a 125 Ω transmission line. At the receiver an ideal transformer steps down the

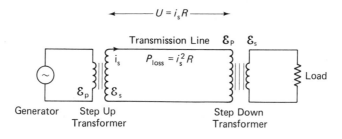

Figure 26-10
The transmission of electric energy from the source to the load.

voltage before the power is delivered at 11.6 kV to the load. If the power factor is unity and there are no leakage losses, calculate (a) the power lost in the transmission line, (b) the voltage at the primary of the step-down transformer, (c) the power delivered to the load, (d) the ratio of turns in the transformers, (e) the total current in the load.

Solution:

(a) The output power from the step-up transformer is equal to the power delivered by the generator P_{gen}; therefore the current at the secondary is:

$$i_s = \frac{P_{gen}}{\mathcal{E}_s} = \frac{6.0 \times 10^6 \text{ W}}{150 \times 10^3 \text{ V}} = 40 \text{ A}$$

Thus the power loss is:

$$P_{loss} = i_s^2 R = (40 \text{ A})^2(125 \text{ }\Omega)$$
$$= 2.0 \times 10^5 \text{ W} = 200 \text{ kW}$$

(b) In the line

$$U = i_s R = (40 \text{ A})(125 \text{ }\Omega) = 5.0 \times 10^3 \text{ V}$$

Thus at the step-down transformer:

$$\mathcal{E}_p = 150 \text{ kV} - 5 \text{ kV} = 145 \text{ kV}$$

(c) $P_{load} = P_{gen} - P_{loss} = 6.0 \text{ MW} - 0.20 \text{ MW}$
$$= 5.8 \text{ MW}$$

(d) Step up

$$\frac{N_s}{N_p} = \frac{\mathcal{E}_s}{\mathcal{E}_p} = \frac{150 \text{ kV}}{2.0 \text{ kV}} = 75$$

Step down

$$\frac{N_s}{N_p} = \frac{\mathcal{E}_s}{\mathcal{E}_p} = \frac{11.6 \text{ kV}}{145 \text{ kV}} = \frac{2.0}{25}$$

(e) $i_{load} = \frac{P_{load}}{\mathcal{E}_s} = \frac{5.8 \times 10^6 \text{ W}}{11.6 \times 10^3 \text{ V}} = 500 \text{ A}$

At some locations rectifiers are used to produce direct current after the voltage has been stepped up; the electric energy is then transmitted as high-voltage dc. Superconducting transmission lines are also being used with considerable success both in Europe and North America.

Problems

In each of the following problems assume a power factor of unity and no leakage losses:

33. A 50 kW generator delivers power at 2.0 kV to a 40 Ω transmission line. Calculate the power lost in transmission (a) if the generator delivers the power directly to the line, (b) if an ideal transformer (100% efficient) steps up the potential difference to 25 kV before the power is supplied to the line. (c) Determine the ratio of the number of turns in the coils of the step-up transformer.

34. A power transmission system consists of a 300 kW generator that delivers power at 1.50 kV to a 300 turn primary coil of an ideal step-up transformer. The 6 000 turn secondary coil of the transformer delivers power to a 60.0 Ω transmission line. At the receiver, the energy is passed into an ideal step-down transformer, and the secondary coil produces a 15.0 kV potential difference across a load. If there are no leakage losses, calculate (a) the voltage in the transmission line, (b) the power loss in the line, (c) the power delivered to the receiver, (d) the current in the load, (e) the ratio of the turns in the step-down transformer.

35. A 750 kW generator delivers power to a 500 Ω transmission line. Calculate the power loss in the line (a) if the generator supplies power at 50 kV directly to the line, (b) if a transformer steps up the voltage to 250 kV before the power is delivered to the line. (c) What is the ratio of the turns in the step-up transformer?

36. Power from a generator is delivered to a transmission line after the voltage has been stepped up in an ideal transformer. At the receiver 50 kV is delivered to the primary coil of an ideal step-down trans-

former in which the ratio of turns is 5 to 1 and a 400 kW load draws power from the secondary. If the potential drop and power loss in the transmission line are 1.0 kV and 6.5 kW, respectively, calculate (a) the potential difference across the load, (b) the current in the load, (c) the minimum power of the generator, (d) the voltage at the secondary of the step-up transformer.

26-8 ELECTROMAGNETIC WAVES

By the middle of the nineteenth century, there had been a number of important observations concerning the interrelationships between electricity and magnetism. Electric fields were known to exist in the neighborhood of electric charges; these fields exerted forces on other charges, and they could produce electric currents. Oersted had discovered that, in addition to electric fields, magnetic fields are set up around moving charges. Finally, according to Faraday's law of induction, a changing magnetic field induces an emf and consequently an electric field in a conductor.

James Clerk Maxwell mathematically analyzed these phenomena by proposing the existence of a symmetrical relationship between electric and magnetic fields. He assumed that *a changing electric field produces a magnetic field, and a changing magnetic field produces an electric field*. With this assumption Maxwell developed eight equations* to describe the interrelationships, and he showed that electromagnetic energy could propagate as *electromagnetic waves* consisting of fluctuating electric and magnetic fields. Maxwell also determined that electromagnetic waves travel at the speed of light, and they

---

*These were eventually simplified to four equations that are called *Maxwell's equations*.

do not require a material medium. He correctly concluded that light itself propagates as an electromagnetic wave, and he predicted that *electromagnetic waves would be produced by accelerating electric charges*.

Eventually, in 1887 Heinrich Hertz verified Maxwell's theory when he generated and detected electromagnetic waves with oscillatory circuits.

The Nature of Electromagnetic Waves

Consider the time-varying electric and magnetic fields that are produced in the region around a dipole antenna consisting of two straight conductors that are connected to an oscillator (Fig. 26-11). The oscillator continually changes the charge distribution and consequently the electric and magnetic fields as it accelerates electrons up and down the antenna. When the opposite polarity net charges on the antenna sections increase, the electric field increases; the magnetic field is proportional to the current in the antenna.

At the instant when the current is at a maximum, there are no net charges on the antenna sections; therefore the magnetic field is at a maximum when the electric field is at a minimum. The current and magnetic field decrease as the net charges increase. Consequently the electric field is at a maximum when the magnetic field is at a minimum. The oscillator continually reverses the polarity of the fields. If the oscillator has a relatively high frequency, the electric and magnetic fields change rapidly, but it takes a finite time for these changes to be transmitted to distant points. As a result some of the distant parts of the time-varying fields become detached, and they propagate (as electromagnetic waves) away from the antenna.

Close to the antenna the electric and magnetic fields are 90° out of phase, but these

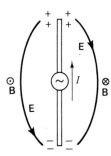

(a) Current I and magnetic induction **B** decrease as the net charges and electric field intensity **E** increase.

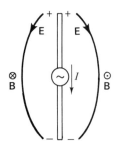

(b) I and **B** increase as net charges and **E** decrease.

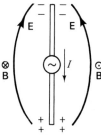

(c) I and **B** decrease as net charges and **E** increase.

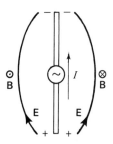

(d) I and **B** increase as net charges and **E** decrease.

Figure 26-11
Fluctuating electric and magnetic fields around a dipole antenna.

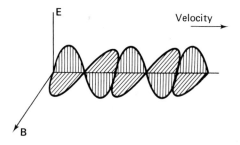

Figure 26-12
Electromagnetic wave. The time-varying electric and magnetic fields and the direction of propagation are mutually perpendicular.

fields are in phase at points that are distant from the antenna. Electromagnetic waves are transverse waves; the electric and magnetic fields and the direction of propagation are mutually perpendicular (Fig. 26-12). Maxwell also showed that in a material that has a permittivity ϵ and a permeability μ, the speed of electromagnetic waves is:

$$c = \frac{1}{\sqrt{\epsilon \mu}} \qquad (26\text{-}27)$$

Thus in free space:

$$c = \frac{1}{\sqrt{\epsilon_0 \mu_0}}$$

$$= \frac{1}{\sqrt{(8.85 \times 10^{-12} \text{ F/m})(4\pi \times 10^{-7} \text{ T·m/A})}}$$

$$= 3.00 \times 10^8 \text{ m/s}$$

which is equal to the speed of light.

The power that is radiated as an electromagnetic wave is proportional to the fourth power of the frequency, but significant power is only radiated when the dimensions of the antenna are comparable to the wavelength of the electromagnetic wave. Electromagnetic waves possess both energy and momentum. The time-varying electric field of an electromagnetic wave can be used to induce an alternating current in a receiver antenna.

The Electromagnetic Spectrum

The different types of electromagnetic waves have many similar properties, and they all originate at accelerating charges. However, if their frequencies are significantly different, electromagnetic waves exhibit many dissimilar characteristics; therefore they are usually classified in terms of their frequency (or wavelength). An organized array of the wavelengths of electromagnetic radiation is called a *spectrum*; the *electromagnetic spectrum* is given in Table 26-1. Light occupies only a very small region of the electromagnetic spectrum.

26-9 THE RELATIVISTIC DOPPLER EFFECT

When there is relative motion between an observer and a source of electromagnetic waves, the frequency of the radiation that is received by the observer is different from that emitted by the source. This difference in frequency is due to a Doppler effect, but it is not the same as the Doppler effect for sound because of the effects of relativity.

According to Einstein's theory of relativity, the speed of light that is measured by an observer is independent of the relative mo-

Table 26-1
The Electromagnetic Spectrum

Frequency (Hz)		Wavelength (m)
10^{25}		
	cosmic rays	10^{-16}
10^{23}		
		10^{-14}
10^{21}		
	gamma rays	10^{-12}
10^{19}		
	X-rays	10^{-10}
10^{17}		
	ultra-violet	10^{-8}
10^{15}		
	visible light blue	10^{-6}
	red	
10^{13}	infrared	
		10^{-4}
10^{11}		
	microwaves	10^{-2}
10^{9}		
	radar	
	U.H.F.	1
	V.H.F. F.M. radio	
	television	
10^{7}	shortwave radio	
	A.M. radio	10^{2}
10^{5}	medium wave radio	
		10^{4}
10^{3}	long wave radio	
		10^{6}
10		

tion between that observer and the source of light. Even if the observer moves at a very high speed towards the light source, he would still measure the speed of light as 3.00×10^8 m/s. Also, it is not possible to determine whether the source or the observer is moving, only their relative motion can be computed.

When a source emits electromagnetic

radiation with a frequency f, the frequency received by an observer is:

$$F = f\sqrt{\frac{1 + v/c}{1 - v/c}} \qquad (26\text{-}28)$$

where v is the *relative* speed of *approach* between the source and the observer. If the source and observer are moving apart, the signs in Eq. 26-28 are interchanged, and v is the relative speed of separation.

In many cases the frequency change $F - f = \Delta f$ is much less than the transmitted frequency f, and to a first approximation Eq. 26-28 reduces to:

$$\frac{\Delta f}{f} = \frac{v}{c} \qquad (26\text{-}29)$$

When the source approaches the observer, the observed frequency F is greater than the transmitted frequency f; but if the source and observer are moving apart, the observed frequency F is less than the transmitted frequency.

Example 26-8

Two aircraft are flying directly towards each other with a relative speed of approach $v = 500$ m/s. If the pilot of the first aircraft transmits a VHF signal at exactly 150 MHz, at what frequency does the pilot of the second aircraft receive the message?

Solution:

$$\Delta f = \frac{fv}{c} = \frac{(150 \times 10^6 \text{ Hz})(500 \text{ m/s})}{(3.0 \times 10^8 \text{ m/s})} = 250 \text{ Hz}$$

Since the aircraft are approaching each other:

$$F = f + \Delta f = 150 \text{ MHz} + 0.000\,25 \text{ MHz}$$
$$= 150.000\,25 \text{ MHz}$$

The relativistic Doppler effect has been used to determine the speeds of stars relative to the earth. Stars appear to be moving away from the earth; consequently we receive their light at a frequency that is lower than the transmitted frequency. This is often called the *red shift* because the frequency shift is towards the red end of the visible spectrum. The speeds of satellites, aircraft and cars may also be determined from the Doppler frequency changes in their transmissions.

Problems

37. As a space ship approaches the earth, an astronaut transmits a 500 MHz UHF signal, and an observer at the earth's surface receives the signal at a frequency that is 10.5 kHz greater than the transmitted frequency. What is the relative speed of the spaceship with respect to the earth?

38. A helium–neon gas laser emits monochromatic light with a frequency of 4.74×10^{14} Hz. What is the relative speed of the laser with respect to a receiver if the receiver detects a 20 Hz frequency shift in the laser light?

REVIEW PROBLEMS

1. If a 100 W bulb is connected to a 120 V, 60 Hz mains, calculate (a) the peak voltage, (b) the effective current, (c) the peak current in the filament.

2. A motor of a power saw draws 8.0 A from a 120 V, 60 Hz mains. Determine (a) the peak current in the circuit, (b) the average power dissipated by the motor, (c) the instantaneous current and voltage 2.5 ms after they pass through the zero value.

3. If the peak voltage of a 1.5 kHz audio oscillator is 18.0 V, calculate (a) the effective voltage and (b) the instantaneous voltage 4.7 ms after it passes through zero.

4. What is the maximum effective alternating potential difference that can be applied to a capacitor if the capacitor

plates are separated by a 1.2 mm sheet of mica that has a dielectric strength of 1970 kV/cm?

5. A 15 Ω heater filament is connected to a 120 V, 60 Hz mains. How many kilowatt hours of energy does it dissipate in 3.0 h? (b) If electricity costs 6¢ per kilowatt hour, how much does it cost to run the heater for 3.0 h?

6. Calculate the inductive reactance when an 80 mH choke is connected to a 30 V, 60 Hz source.

7. A pure inductive circuit draws 30.0 mA from a 5.0 V, 250 Hz source. What is the inductance of the circuit?

8. An inductor has an inductive reactance of 392 Ω at 1.5 kHz. What is its inductive reactance at 700 Hz?

9. What is the capacitive reactance of a pure capacitive circuit that has a capacitance of 200 μF when it is connected to a 220 V, 50 Hz source?

10. Determine (a) the capacitive reactance and (b) the capacitance of a pure capacitive circuit that draws 300 mA from a 120 V, 60 Hz mains.

11. If a capacitor has a capacitive reactance of 620 Ω at 300 Hz, what is its capacitive reactance at 1.2 kHz?

12. An ac circuit consists of a resistor and a pure inductor in series with an ac source. If the effective potential drops across the resistor and inductor are 15 V and 26 V, respectively, determine (a) the voltage of the source and (b) the phase angle.

13. An 80 mH choke with a 12 Ω resistance is connected in series with a 47 Ω resistor and a 15 V, 60 Hz ac source. Calculate (a) the impedance, (b) the phase angle, (c) the effective current in the circuit.

14. A 30 mH inductor draws 20 mA when it is connected to the terminals of a 12 V,

120 Hz ac source. Determine (a) the impedance, (b) the inductive reactance, (c) the resistance of the inductor.

15. Determine (a) the impedance, (b) the phase angle, (c) the effective current when a 300 pF capacitor and 560 kΩ resistor are connected in series with a 30 V, 1.2 kHz ac source.

16. A 600 nF capacitor, a 400 Ω resistor and a 180 mH inductor with a 20 Ω resistance are connected in series with a 700 Hz source. If the circuit draws 200 mA, calculate (a) the reactance, (b) the impedance, (c) the phase angle, (d) the voltage of the source.

17. A 750 W ac motor has a power factor of 72% when it is connected to a 120 V, 60 Hz mains. (a) What effective current does it draw? (b) What effective current would it draw if the power factor was 100%?

18. An ac circuit has an inductance of 300 mH, a capacitance of 50 μF and a 300 Ω resistance in series with a 120 V, 60 Hz source. Calculate (a) the impedance of the circuit, (b) the phase angle, (c) the effective current in the circuit, (d) the power factor, (e) the average power dissipated in the circuit.

19. Determine the resonant frequency of an ac series circuit that has a 120 Ω resistance, an inductance of 30 μH and a capacitance of 200 pF.

20. What inductance must be connected in series with a 200 pF capacitor in order to produce a circuit with a resonant frequency of 25 kHz?

21. Two satellites travel directly towards each other with a relative speed of approach of 3.2 × 10⁴ km/h. If one satellite transmits a VHF signal at exactly 150 MHz, at what frequency does the second satellite receive the signal?

22. A 36 MW generator delivers power at

15.0 kV to the primary of an ideal step-up transformer that has a 500 turn primary coil and a 6 000 turn secondary. The secondary is connected to a 150 Ω transmission line, and at the other end the energy passes to the primary of an ideal step-down transformer that has a 5 000 turn primary and a 250 turn secondary. If there are no leakage losses, calculate (a) the power loss in the transmission line, (b) the power delivered to a load at the receiver, (c) the current delivered to the load at the secondary of the step-down transformer.

Light is the form of radiant energy that has the capacity to stimulate the sensation of vision. The eye is only sensitive to a part of the total radiant energy that is emitted by matter. However, in order to be visible, light from an object must enter the eye. This light may be produced by the object or merely reflected from it.

Bodies of matter that emit light as a part of their radiant energy are called *luminous bodies*. The sun and other stars are natural sources of light. Many artificial light sources, such as incandescent lamps, have been developed.

An object does not have to be a luminous body to be visible: most objects become visible because they reflect light. If light from a luminous body is incident on a surface, some of the light is reflected, and that surface is said to be *illuminated*.

The branch of physics called *optics* is concerned with light and the phenomena and effects related to light. The science that is concerned with the measurement of light is called *photometry*.

Light is essential in many aspects of our lives because we frequently rely on vision to detect and locate objects. Many measurement processes utilize light to transmit information about the relative size and shape of objects.

The improper illumination of a working area increases eye strain, brings on fatigue and reduces the general efficiency of workers; therefore the choice of a suitable lighting system is of great importance. Adequate illumination of a working surface is essential, but excess illumination should also be avoided because it causes glare and shadows.

27
Light and Illumination

27-1 THE NATURE OF LIGHT

Light can be transformed into other energy forms, such as heat. Energy may be transferred from one place to another either by the motion of a particle (kinetic energy) or by the propagation of a vibratory disturbance known as a wave. Therefore it is reasonable to assume that light must be wave-like or particle-like.

Some scientists, such as Isaac Newton, believed the corpuscular theory which held that light was a particle and travelled in straight lines. However, others such as Christian Huygens developed a wave theory for light.

At the beginning of the nineteenth century, Young and Fresnel demonstrated that light exhibited the wave-like characteristics of diffraction and interference; these characteristics cannot be explained in terms of the corpuscular theory. In order to explain the wave theory, Young assumed the existence of an invisible, all-pervading medium called an *ether*, through which the light waves were supposed to travel. The existence of an ether was generally accepted until the latter part of the nineteenth century when a series of special experiments proved that it did not exist.

In 1864 James Maxwell predicted mathematically that electromagnetic radiation propagated as a transverse wave that required no material medium. The wavelengths of visible electromagnetic radiations are between approximately 380 nm and 760 nm; each wavelength corresponds to a different color of light. Since it propagates as a transverse wave, light may be polarized (Chapter 30).

By the twentieth century most scientists believed the wave theory for light. The corpuscular theory had been abandoned because it could not explain many experimental observations. However a series of new experimental observations involving interactions between light and matter led to the development of *quantum theory* because they could not be explained in terms of wave theory (Chapter 31).

Quantum theory is based on the principle that the exchange of energy between electromagnetic radiation and matter is always in discrete amounts of energy called *quanta*. Each quantum of light has some *particle-like*

characteristics, and it is known as a *photon*. Wave theory predicts a continuous range of energies for electromagnetic radiation.

The apparent contradiction concerning the wave-like and particle-like characteristics of light is merely due to our inability to compare light with a single macroscopic model. Light possesses both particle-like and wave-like characteristics. Wave theory may be used to describe the propagation of light, but particle-like characteristics explain the interactions between light and matter; this is known as the *duality concept*.

Problem

1. List several phenomena that may be explained in terms of (a) the wave-like and (b) the particle-like characteristics of light.

27-2 RAYS AND SHADOWS

Light exhibits many wave-like characteristics, but it is often convenient to consider its motion in terms of straight lines called *rays*, which represent the path of the light energy. The term ray is also used to describe the path of other forms of electromagnetic energy, such as X-rays, gamma rays and cosmic rays.

Geometrical optics is founded on the assumption that light travels in straight lines represented by rays. Even though light really propagates as a wave, the concepts of geometrical optics may be used to describe many optical phenomena, such as reflection, refraction and shadows.

When it is irradiated with light, an object may transmit, reflect and absorb some of the light energy. *Transparent* materials transmit light without significantly scattering the light energy; objects may be seen with very little distortion through transparent materials. Materials that transmit and scatter light energy are called *translucent*; objects cannot be clearly seen through translucent materials.

Opaque materials do not transmit light, they block the passage of light energy.

Shadows are produced when the path of light is blocked by an opaque object; they are a direct consequence of the straight-line propagation of light energy. If an opaque object blocks light from a small (point) light source, it casts a sharp shadow on a screen (Fig. 27-1a). The region on the screen where all of the light rays from the source are excluded is called the *umbra*.

Consider the shadow that is formed when an opaque object is placed between an extended light source *AB* and a screen (Fig. 27-1b). In this case there are two types of shadow regions: the *umbra* where all of the rays are excluded, and the *penumbra* where some of the light rays are excluded, but not from every part of the source. For example, light from point *C* of the source can illuminate all points of the screen except points between *S* and *R*. The illumination of the penumbra regions

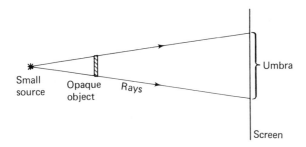

(a) The shadow cast by an opaque object when it blocks the light from a point source

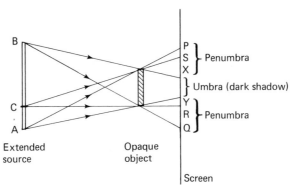

(b) The shadow cast by an opaque object when it blocks light from an extended source

Figure 27-1
Shadows that are cast when an opaque object is placed in the path of light from a source.

increases with the distance from the umbra region.

A *solar eclipse* is produced when the moon moves between the earth and the sun, and its shadow is cast on the earth's surface. The solar eclipse is said to be *total* in the umbra regions and *partial* in the penumbra regions of the moon's shadow. During a *lunar eclipse* the earth is located between the moon and the sun, and its shadow covers at least a part of the moon's surface.

Problems

2. Are shadows always sharp? Explain.

3. List several undesirable effects of shadows in working areas.

27-3 LIGHT SOURCES

Light energy constitutes only a fraction of the total electromagnetic radiation that is emitted from the sun. This electromagnetic radiation is the result of a continuous series of nuclear reactions, during which some of the sun's mass is converted directly into energy in the form of electromagnetic radiation (Chapter 34).

Many kinds of artificial light sources have also been produced. Most of these sources generate light by one of the following processes.

Incandescence

Incandescent lamps consist of a spiral wire filament, usually made of tungsten, that is located in a glass bulb. To reduce the evaporation of the filament at high temperatures, the bulb is either evacuated, or it contains an inert gas, such as argon or nitrogen, under low pressure. The mounting base also serves as the electrical connection between the bulb socket and the filament.

The filament has a high electric resistance; therefore, if it carries an electric current, it is maintained at a high temperature, and it dissipates electric energy as heat and light. In normal operation, at a temperature of approximately 3 000°C, the filament emits light that is very similar to sunlight, but the efficiency of the incandescent lamp is quite low because most of the energy is dissipated as heat.

Gas Discharge

A simple gas discharge tube consists of a glass tube containing a low-pressure gas and two electrodes. When an electric current is maintained through the low-pressure gas, the gas atoms become excited into high-energy states. The excited gas atoms return to their low-energy states by emitting light with discrete wavelengths. Most gas discharge tubes emit light that is rather different from sunlight; therefore they only have limited use in general lighting systems. However gas discharge tubes are used for special types of lighting, such as neon signs and sunlamps.

Fluorescence

A *fluorescent lamp* consists of a long glass tube that contains mercury vapor and some argon gas under low pressure, and the inner surface of the tube is coated with a special chemical phosphor. When there is an electric discharge through the mercury vapor, some of the mercury atoms are excited into a high-energy state, and they emit ultraviolet radiation as they return to their lower energy states. The ultraviolet radiation impinges on the chemical phosphor, and some of the phosphor atoms are excited into higher energy states, where they emit visible light as they reduce their energy.

Fluorescent lamps have a relatively high efficiency, and they have long lifetimes. The nature of the visible light that is emitted from a fluorescent lamp depends on the type of phosphor coating. Different types of fluorescent lamps are often used in factories, offices, schools and houses.

*Problem*_____

4. Describe how the three types of artificial sources affect the color of an object.

27-4 CHARACTERISTICS OF LIGHT SOURCES

All known sources emit significant amounts of other forms of electromagnetic radiation, such as heat, simultaneously with the light. The *radiant flux* (or *radiant power*) P of a source is defined as the time rate at which it radiates all forms of electromagnetic energy. Since it has dimensions of power, the SI unit for radiant flux is the *watt*.

Initially, when a lamp is "turned-on" (supplied with electrical power), its tempera-

ture increases rapidly, but it soon reaches an equilibrium state by dissipating energy as heat and light at the same rate that it receives the electric energy. At this point the temperature of the filament remains almost constant, and the input electric power is approximately equal to the radiant flux.

Normally we are interested in the visible electromagnetic energy with wavelengths approximately between 380 nm and 760 nm, and this usually constitutes only a fraction of the total energy that is radiated by a source. The response of the human eye is not equal even for visible light of different wavelengths. If the light is relatively strong, the eye is most sensitive to yellow–green light with a wavelength of 555 nm. Relatively low-power radiation at a wavelength of 555 nm produces a visual sensation that appears to be equally as strong as higher power radiations at other wavelengths. The relative response of a human eye to light of various wavelengths is illustrated in the *eye sensitivity curve* (or *spectral sensitivity curve*) (Fig. 27-2). This curve

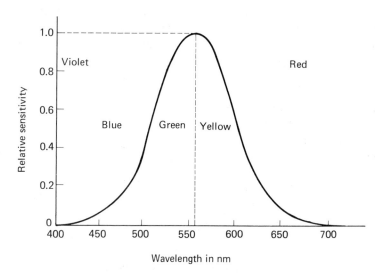

Figure 27-2
The eye sensitivity curve.

indicates the reciprocals of the relative powers of radiation at the different wavelengths that are required to stimulate equal sensations of brightness in the human eye.

Example 27-1

Calculate the power P that must be radiated by a laser at a wavelength of 500 nm in order to produce the same brightness as a power $P_M = 6.6$ mW at 555 nm.

Solution:

From Fig. 27-2, the relative sensitivity of the eye at 500 nm: $P_M/P = 0.33$. Thus,

$$P = \frac{P_m}{0.33} = \frac{6.6 \text{ mW}}{0.33} = 20 \text{ mW}$$

The *luminous flux* Φ is the time rate of flow of the electromagnetic energy that has the ability to stimulate the sensation of vision. It is normally expressed in units of *lumens* (abbreviation lm). Each color of light has a different wavelength, and it stimulates the sensation of vision by an amount that is proportional to the product of its power and the corresponding relative sensitivity of the eye. In many cases, visible light consists of electromagnetic radiations of many different wavelengths, and the luminous flux is proportional to the sum of the visual sensations that are produced by the individual colors.

Light sources are rated in terms of their ability to stimulate vision. Since the sensitivity of the eye is involved, it is convenient to describe light in terms of special units.

By analogy with two-dimensional angles, the region surrounding any point in space may be specified in terms of three-dimensional *solid angles*, which are expressed in supplementary base units called *steradians* (symbol sr). The total surface *area $4\pi r^2$* of a sphere is said to subtend a solid angle of 4π *steradians* at the center of that sphere. One *steradian* is defined as the solid angle subtended at the center of a sphere by a surface area equal to

the square of the radius r. Since r^2 subtends a solid angle of 1 sr, some surface area A_1 subtends a solid angle:

$$\omega = \frac{A_1}{r^2} \text{ steradians} \qquad (27\text{-}1)$$

where A_1 and r^2 have the same units and A_1 is perpendicular to r (Fig. 27-3). Note that the

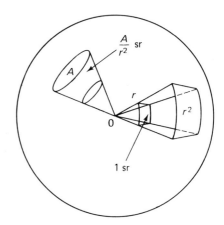

Figure 27-3
Solid angles in steradians subtended at the center of a sphere by various surface areas.

solid angle is independent of the distance from the source.

The *luminous intensity* I in a particular direction from a small source is defined as the luminous flux Φ per unit solid angle ω subtended at the source.

$$I = \frac{\Phi}{\omega} \qquad (27\text{-}2)$$

If Φ is the luminous flux through a perpendicular area A_1 at a distance r from the source, the luminous intensity at that location is:

$$I = \frac{\Phi}{\omega} = \frac{\Phi r^2}{A_1} \qquad (27\text{-}3)$$

Luminous intensity is expressed in terms of a base SI unit called the *candela* (cd). One candela is approximately equal to one international candle. The standard one candela is defined as the luminous intensity from one sixtieth of a square centimeter of a blackbody radiator that is maintained at a temperature of freezing platinum (2 046 K) and a pressure of 101 325 Pa. For convenience, carbon filament lamps are also used as standard sources. The luminous intensity in a particular direction is called the *candle power* (cp) of the source.

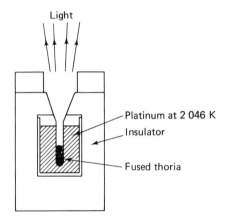

Figure 27-4
The standard light source that is used to define a candela.

Luminous flux is expressed in derived units called *lumens* (lm). If a point light source radiates with a luminous intensity of one candela uniformly in all directions, the luminous flux through each unit solid angle is defined as one *lumen*.

Most sources do not radiate light uniformly in all directions, and their luminous intensity depends on their orientation. In many cases reflectors, lenses, diffusers and shields are used to control the luminous intensity in the different directions from the source. These variations in the luminous intensity are represented by a *candle power distribution curve*, which is a polar graph of the luminous intensities at various orientations in a plane around the source or complete light fixture (Fig. 27-5).

Considering all directions in three dimensions, the average luminous intensity of a source is called its *mean spherical luminous intensity* $\bar{I}$. Since the total solid angle around a source $\omega_T = 4\pi$ sr, from Eq. 27-2 the *total luminous flux* of the source is:

$$\Phi_T = \bar{I}\omega_T = 4\pi\bar{I} \qquad (27\text{-}4)$$

Example 27-2

A small light source has a mean spherical luminous intensity of 31 cd, and a reflector concentrates all of the light onto a perpendicular area $A_1 = 3.75$ m² at a distance of 5.00 m from the source. Calculate (a) the total luminous flux from the source and (b) the luminous intensity of the light when viewed from the illuminated area.

Solution:

(a) $\quad \Phi_T = 4\pi\bar{I} = (4\pi \text{ sr})(31 \text{ cd}) = 390 \text{ lm}$

(b) $\quad I = \dfrac{\Phi_T}{\omega} = \dfrac{\Phi_T r^2}{A_1} = \dfrac{(390 \text{ lm})(5.00 \text{ m})^2}{3.75 \text{ m}^2}$

$\qquad = 2\ 600 \text{ cd}$

Artificial light sources are often rated in terms of their ability to convert other energy forms into visible light energy. The *luminous efficacy** K of a light source is defined as the ratio of its useful output power, i.e. the total luminous flux Φ_T, to the total input power P. In the case of electric lights, the input power is equal to the radiant flux P; therefore the luminous efficacy equals:

$$K = \frac{\Phi_T}{P} \qquad (27\text{-}5)$$

Luminous efficacies are *not* expressed as a percentage, rather they have units of lumens per watt.

Example 27-3

Calculate the luminous efficacy of a small 200 W lamp if its mean spherical luminous intensity is 294 cd.

---

*This used to be called *luminous efficiency*.

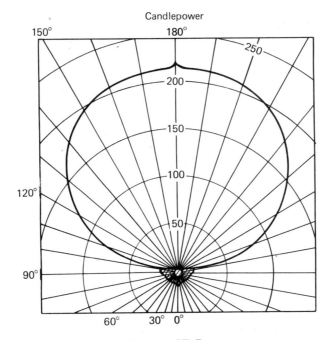

Figure 27-5
Candle power distribution curve for a light fixture composed of a shade and an incandescent lamp.

Solution:

$$K = \frac{\Phi_T}{P} = \frac{4\pi\bar{I}}{P} = \frac{(4\pi \text{ sr})(294 \text{ cd})}{200 \text{ W}} = 18.5 \frac{\text{lm}}{\text{W}}$$

The luminous efficacies of most artificial light sources are quite small because most of the input power is converted into heat. Fluorescent lamps have higher luminous efficacies than incandescent lamps because they emit less heat.

Problems

5. Determine the solid angle that is subtended at the center of a sphere of radius 3.0 ft by a surface area of 18 in².

6. Calculate the solid angle that is subtended at the center of a sphere of radius 200 cm by a surface area of 80 cm².

7. Calculate the luminous intensity of light where a luminous flux of 200 lm passes through a 0.80 sr solid angle from the source.

8. Determine the candle power of a small light source in a direction where it emits a luminous flux of 500 lm in a solid angle of 2.0 sr.

9. A small light source has a mean spherical luminous intensity of 100 cd, and a reflector concentrates all of the light onto a perpendicular area of 15 m² at a distance of 3.0 m from the source. Compute (a) the total luminous flux from the source and (b) the intensity of the light when it is viewed from the illuminated area.

10. A reflector concentrates all of the light from a small 40 cp source into a solid

Table 27-1

Typical Ratings of Some Electric Light Sources

Sources	Total Luminous Flux/(lm)	Luminous Efficacy (lm/W)
40 W tungsten filament	470	11.8
60 W tungsten filament	830	13.8
100 W tungsten filament	1 630	16.3
150 W tungsten filament	2 700	18.0
100 W clear mercury lamp	3 600	36
40 W fluorescent 48 in long	2 800	70
60 W fluorescent 48 in long	4 000	67
100 W fluorescent 48 in long	5 500	55

angle of 0.50 sr. Calculate the luminous intensity of the light beam.

11. A reflector concentrates all of the light from a small source into a solid angle of 0.20 sr. The luminous flux in the direction of the light beam is 2 000 lm. Calculate the candle power of the source.

12. Determine the luminous efficacy of a 56 W, slim-line fluorescent lamp that is 72 in long if its mean spherical luminous intensity is 330 cd.

13. A reflector concentrates all of the light from a small 100 W source onto a perpendicular surface of area 4.5 ft² at a distance of 1.5 ft from the source. If the luminous intensity in the direction of the light beam is 900 cd, calculate the luminous efficacy of the source.

27-5 ILLUMINATION OF A SURFACE

A surface is said to be illuminated when it is irradiated with visible light from any number of sources. The *illuminance E* of a surface is defined as the *total* incident luminous flux Φ per unit surface area A.

$$E = \frac{\Phi}{A} \qquad (27-6)$$

The SI unit for illuminance is called the *lux* (lx), which is the illuminance when a surface receives one lumen of luminous flux per square meter (1 lx = 1 lm/m²). In the British Engineering System the unit of illuminance is called a *footcandle* (fc), which is equivalent to one lumen flux per square foot (1 fc = 1 lm/ft²).

Many different light sources may contribute to the illumination of the same surface. The total illuminance of a surface A is equal to the *sum* of the illuminances due to the luminous flux that it receives from each light source:

$$E_{\text{TOT}} = E_1 + E_2 + \cdots = \Sigma E = \frac{1}{A} \Sigma \Phi \quad (27-7)$$

*Example 27-4*_____

Calculate the total illuminance of a 2.0 m² surface if it receives 20 lm from one light source and 80 lm from a second light source.

Solution:

$$E_{\text{TOT}} = \frac{1}{A} \Sigma \Phi = \frac{1}{2.0 \text{ m}^2}(20 \text{ lm} + 80 \text{ lm})$$
$$= 50 \text{ lx}$$

Consider the illumination of a flat surface area A that is located at a distance r from a single point source S of light (Fig. 27-6). At the center of the area A, if the normal **n** to the surface is inclined at an angle θ with respect to the direction of the source S, the component of A that is perpendicular to the direction of S is:

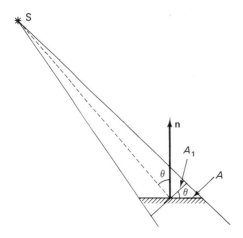

Figure 27-6
The illumination of a surface.

$$A_1 = A \cos \theta$$

Therefore, the area A subtends a solid angle at the source:

$$\omega = \frac{A_1}{r^2} = \frac{A \cos \theta}{r^2} \qquad (27\text{-}8)$$

When the luminous intensity of the source is I, from Eq. 27-2 the luminous flux through a solid angle ω is:

$$\Phi = I\omega$$

Therefore, the luminous flux that is incident on the surface area A is:

$$\Phi = I\omega = \frac{IA \cos \theta}{r^2} \qquad (27\text{-}9)$$

The corresponding illuminance of that surface is:

$$E = \frac{\Phi}{A} = \frac{(I \cos \theta)}{r^2} \qquad (27\text{-}10)$$

Example 27-5

A small lamp S has a luminous intensity of 250 cd, and it is located 120 cm above a horizontal work bench. Calculate the illuminance of the work bench at a point P that is located 50.0 cm to one side of the position 0 directly below the lamp (Fig. 27-7).

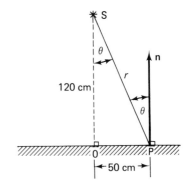

Figure 27-7
The illumination of a work bench.

Solution:

The distance of P from the source is:

$$r = \sqrt{OP^2 + SO^2} = \sqrt{(0.50 \text{ m})^2 + (1.20 \text{ m})^2}$$
$$= 1.30 \text{ m}$$

If θ is the angle between the normal n to the surface at P and the direction from P to the source S, then

$$\cos \theta = SO/SP = 1.20 \text{ m}/1.30 \text{ m}$$

Thus

$$E = \frac{I \cos \theta}{r^2} = \frac{(250 \text{ cd})(1.20 \text{ m}/1.30 \text{ m})}{(1.30 \text{ m})^2} = 137 \text{ lx}$$

For the special case when the illuminated area is perpendicular to the direction from the source, $\theta = 0$ and the illuminance $E = I/r^2$. If there is no absorption or scattering of the light from a point source, the luminous intensity I is constant in a particular direction; therefore, the illuminance is:

$$E \propto 1/r^2$$

and

$$\frac{E_1}{E_2} = \frac{r_2^2}{r_1^2} = \left(\frac{r_2}{r_1}\right)^2 \qquad (27\text{-}11)$$

where E_1 and E_2 are the illuminances at distances r_1 and r_2, respectively. This is another example of an inverse square law.

*Example 27-6*_____

A point on a surface has an illuminance $E_1 = 121$ fc when it is irradiated with light from a small lamp at a distance $r_1 = 4.0$ ft. How far from the same point should the lamp be located in order to reduce the illuminance to 100 fc?

Solution:

$$r_2 = r_1 \sqrt{\frac{E_1}{E_2}} = (4.0 \text{ ft}) \sqrt{\frac{121 \text{ fc}}{100 \text{ fc}}} = 4.4 \text{ ft}$$

_____

Lighting systems are usually composed of a number of different light fixtures, and any point in the room may receive diffused and reflected light from its surroundings as well as direct light from each source. In practice the illuminance at some location in a room is often very difficult to calculate because of the complexity of the sources and the reflecting surfaces, but illuminances are easily measured with photoelectric devices, such as footcandle meters. The *zonal cavity method** is now used for illumination design in North America.

*Problems*_____

14. Calculate the total illuminance of a 5.0 ft² surface if it receives 30 lm from one light source and 20 lm from a second light source.

15. Calculate the total luminous flux that is incident on a surface with an area of 4.0 m² if it has a uniform illuminance of 80 lx.

16. A point source has a luminous intensity of 200 cd, and it is located 8.0 ft above a horizontal working area. Calculate the illuminance of the working area (a) at a point directly below the source and (b) at a point directly below the source and (b) at a point directly below the source.

_____

*Specified and adopted in 1964 by the Illuminating Engineering Society.

Table 27-2
Typical Recommended Illumination Levels

Area or Task	Illuminance	
	Fc	Lx
Chemical laboratory	50	540
Electrical testing	100	1 100
Machine shops	100	1 100
Medium bench and machine work	100	1 100
Regular office work	90	1 000
Store windows	180	2 000
Outside building construction	10	107
Reading and writing	75	800

at a point 6 ft to one side of the point that is directly below the source.

17. Three 100 cd point sources are located in a line 4.0 m apart, and each is suspended 3.0 m above a work bench. Calculate the total illuminance of the work bench (a) at a point directly below the center source and (b) at a point directly below one of the end sources. Ignore reflections.

18. A point on a surface has an illuminance of 180 lx when it is irradiated with light from a small lamp at a distance of 5.0 m. Calculate the illuminance at the same point if the lamp is moved 4.0 m farther away.

_____

27-6 LUMINANCE

A surface of area A may emit light if it is part of a luminous body, or it may reflect the light from some source. In each case the area A has a *brightness*, and the light from it may illuminate some other surface.

In practice only a few light sources have dimensions small enough so that they may be considered to be point sources. Most sources emit light from extended surface areas. Also, because of their luminous intensities, light sources may have different appearances to an observer; some sources appear to be *brighter* than others. Illuminance is independent of the nature of the illuminated surface, but two different surfaces that have the same illuminance may also have different brightnesses because they reflect different amounts of the incident light.

Consider a surface area A that either emits or reflects light with a luminous intensity I in a direction that is inclined at some angle θ with respect to the normal **n** to that surface (Fig. 27-8). In the direction θ, the

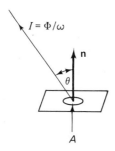

Figure 27-8
Luminance of an extended surface.

luminance L of the area A is the ratio of the luminous intensity I to the component of the area A perpendicular to the direction specified by θ:

$$L = \frac{I}{A_{\text{perp}}} = \frac{I}{A \cos \theta} = \frac{\Phi/\omega}{A \cos \theta} \quad (27\text{-}12)$$

where Φ is the luminous flux through the solid angle ω in the direction θ and $A_{\text{perp}} = A \cos \theta$ is the component of A that is perpendicular to that direction. Luminance is expressed in candela per square meter (cd/m^2) in SI and in *footlamberts* (fL) in the British Engineering System.

$$1 \text{ fL} = \frac{1}{\pi} \text{ cd/ft}^2 \quad \text{or} \quad 1 \text{ cd/ft}^2 = \pi \text{ fL}$$

Any surface will normally reflect a fraction of the total incident light, and it will transmit or absorb the rest. The *reflectance* ρ of a surface is defined as the ratio of the reflected light to the incident light.

$$\rho = \frac{\text{reflected light}}{\text{incident light}} \quad (27\text{-}13)$$

Reflectance usually depends on the nature of the surface, shiny surfaces have high reflectances whereas dull surfaces have low reflectances. Similarly the *transmittance* τ of a surface is defined as the ratio of the transmitted light to the incident light.

$$\tau = \frac{\text{transmitted light}}{\text{incident light}} \quad (27\text{-}14)$$

Since excessive luminance causes glare, strong light sources are usually shielded with shades or diffusing globes, and glossy surfaces are avoided.

Problems

19. Determine the luminance of an area of 8.0 in^2 in a direction that is inclined at an angle of 60° with the normal to the surface if the corresponding luminous intensity is 50 cd.

20. A frosted glass cover is used to diffuse the light from a fluorescent lamp. Each square meter of the glass surface receives 80 lm, and it transmits 72 lm while reflecting 6 lm and absorbing the remaining luminous flux. Calculate the reflectance and the transmittance of the glass cover.

27-7 PHOTOMETERS

We are able to see in candlelight as well as in bright sunlight because our eyes adjust to the degree of illumination. However, because of these adjustments, the eye is not capable of directly comparing luminous intensities of sources, but it can detect small differences in the brightness of adjacent surfaces.

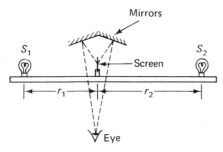

Figure 27-9
A Bunsen grease spot photometer.

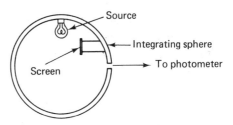

Figure 27-10
The integrating sphere.

Devices that are used to compare and measure the luminous intensities of light sources are called *photometers*. There are many types of photometers: One of the simplest is the *Bunsen grease spot photometer*, illustrated in Fig. 27-9. Two sources are arranged so that they illuminate opposite sides of an opaque white paper screen that contains a translucent grease spot. A pair of mirrors enables an observer to view both sides of the screen simultaneously. To compare the luminous intensities I_1 and I_2 of two sources S_1 and S_2, respectively, the screen is moved until it is illuminated equally on both sides. At this point the grease spot almost disappears; since the screen is perpendicular to the direction from the sources, from Eq. 27-10:

$$E_1 = E_2 \quad \text{or} \quad I_1/r_1^2 = I_2/r_2^2 \quad (27\text{-}15)$$

where r_1 and r_2 are the distances of the screen from the sources S_1 and S_2, respectively.

In a *flicker photometer* two illuminated surfaces are viewed alternately at a frequency that enables the eye to detect a flicker when the surfaces have unequal illuminances. The sources are moved relative to the surfaces until the flicker in the brightness disappears, and then the distances from the screen to the sources are measured. The luminous intensities of sources of different colors are often compared with flicker photometers.

Photometers are also used in conjunction with a device called an *integrating sphere* (Fig. 27-10) to measure the mean spherical luminous intensity of a source. The source is placed in a large hollow sphere that is coated with a uniform layer of a white diffusing surface, such as zinc oxide. Light from the source is reflected many times from the interior wall of the integrating sphere. If the inner surface of the integrating sphere is perfectly diffusing, all points around the wall of the sphere are equally illuminated by the reflected light. Some light is transmitted through a small window in the sphere to a photometer where it is compared with some standard. The luminous intensity of the transmitted light corresponds to the mean spherical luminous intensity of the source in the integrating sphere.

Problems

21. A standard 90 cd source and an incandescent lamp illuminate opposite sides of a screen in a Bunsen photometer. When the illuminance is matched, the standard is 1.5 ft from the screen, and the incandescent lamp is 2.0 ft from the screen. Calculate the luminous intensity of the incandescent lamp.

22. The light from an integrating sphere illuminates one side of a photometer screen, and a 75 cd source illuminates the other side of the screen. When the illuminances are equal, the standard is 100 cm from the screen, and the integrating sphere is 120

cm from the screen. Calculate (a) the mean spherical luminous intensity and (b) the total luminous flux of the source in the integrating sphere.

REVIEW PROBLEMS

1. Determine the solid angle subtended at the center of a sphere of radius 2.5 m by a surface area of 2.5 cm².

2. Determine the luminous intensity of a light that radiates a luminous flux of 250 lm through a 0.85 sr solid angle.

3. Determine the candlepower of a small light source in a direction where it emits a luminous flux of 350 lm in a solid angle of 1.7 sr.

4. A small light source has a mean spherical luminous intensity of 24.0 cd, and a reflector concentrates all of the light into a solid angle of 1.3 sr. Calculate (a) the total luminous flux of the source and (b) the intensity of the light in the illuminated area.

5. The bulb of a search light has a mean spherical luminous intensity of 75 cd, but a reflector and lens concentrate the light onto an area of 25.0 ft in diameter 900 ft from the source. Calculate the luminous intensity of the light in the direction of the light beam.

6. What is the luminous intensity of a spotlight with a 35 cd bulb if the beam is concentrated on a 500 ft² area of a vertical wall at a distance of 50 ft?

7. A 100 W, 110 V tungsten lamp has a luminous efficacy of 10 lm/W. (a) What is its mean spherical luminous intensity? (b) At what distance from the lamp is the maximum illuminance 5.0 fc?

8. A 500 W, 110 V tungsten lamp has a luminous efficacy of 12 lm/W. (a) Find the mean spherical luminous intensity. (b) If 60 % of the total light emitted by the lamp illuminates an area of 30 m², what is the illuminance?

9. (a) Use the data from Table 27-1 to find the mean spherical luminous intensity of a 100 W tungsten lamp. (b) If 70 % of the emitted light illuminates a 50 ft² area, what is the illuminance?

10. (a) Use the data from Table 27-1 to find the mean spherical luminous intensity of a 150 W tungsten lamp. (b) At what distance from the lamp is the maximum illuminance 12 fc?

11. Calculate the total illuminance of a 3.0 m² surface if it receives 35 lm from one light source and 25 lm from a second light source.

12. Determine the total luminous flux incident on a 6.0 ft² area if it has a uniform illuminance of 45 fc.

13. A point light source has a luminous intensity of 180 cd, and it is located 3.0 m above a horizontal working surface. Calculate the illuminance of the working surface at a point (a) directly below the source and (b) 4.0 m to one side of the point that is directly below the source.

14. Two 750 cd light sources are 10 ft apart and 8.0 ft above a horizontal working surface. Determine the illuminance at a point (a) directly below one of the sources and (b) on the working surface midway between the sources.

15. Determine the illuminance at the edge of a circular table of radius 2.0 m if a 500 cd source is suspended 3.0 m above its center.

16. A lighting system has two lamps 25 ft apart, each 12 ft above a work plane. If

the lamps have luminous intensities of 750 cd, what is the illuminance at the point on the work plane that is 9.0 ft from the point below one lamp and 16 ft from the point below the other?

17. A small lamp is 2.0 m above a table. To what distance must it be lowered to increase the illuminance by a factor of 3?

18. Determine the luminance of an area of 12.0 in² in a direction that is inclined at 37° with the normal to the surface if the luminous intensity is 28 cd.

19. A small lamp 10 ft away has an intensity that is 16 times greater than that of a second identical lamp. How far away is the second lamp?

In this chapter we consider the properties of light and optical instruments by assuming that light travels in straight lines represented by rays. Even though this assumption is not strictly correct, it enables us to describe many optical phenomena. We only consider light, but many of the principles of geometrical optics may also be applied to other forms of electromagnetic radiations.

Objects do not have to be luminous bodies in order to be visible, most objects are visible because they reflect light. Objects that have irregular surfaces reflect light in many directions simultaneously; this type of reflection is called *diffuse* (Fig. 28-1). The reflection

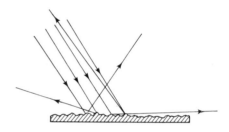

Figure 28-1
Diffuse reflection—the reflected light is scattered in all directions.

is perfectly diffuse if the reflecting surface has an equal luminance (or brightness) in all directions.

A few objects, such as mirrors, have smooth shiny surfaces, and there is a symmetrical relationship between the incident and the reflected light; this type of reflection is called *specular* or *regular* (Fig. 28-2). Mirrors and reflectors are used extensively to control the direction and concentration of light and other forms of electromagnetic energy.

28-1 THE LAWS OF REFLECTION

Consider a beam of light that is incident on a flat reflecting surface; the direction of the energy flow is represented by rays. The *angle*

28

Reflection and Mirrors

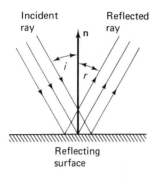

Figure 28-2
Specular reflection.

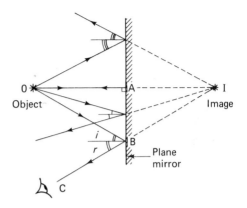

Figure 28-3
A ray diagram showing the location of an image I that is formed by a plane mirror.

of incidence i corresponds to the angle between the incident ray and the normal to the reflecting surface. Similarly the *angle of reflection r* is the angle between the reflected ray and the normal to the reflecting surface. The laws of specular reflection were discussed in Chapter 17; these laws state:

1. The angle of incidence is equal to the angle of reflection.

2. The incident ray *i*, the reflected ray *r* and the normal **n** to the reflecting surface all lie in the same plane.

The Plane Mirror

Consider a point source of light 0 that is located in front of a *plane* (flat) *mirror*. Some light rays from the source are incident on the surface of the mirror where they undergo specular reflection. To an observer, all of the reflected light appears to emanate from some point I called the *image*, which is located *behind* the surface of the mirror.

The exact location of the image I may be determined from a *ray diagram* (Fig. 28-3). Any number of rays may be drawn from the object to the surface of the mirror, and the laws of reflection are used to determine the direction of the reflected rays. For simplicity we only need to consider two light rays

(OA and OB, for example) and their reflections AO and BC. In each case the angle of incidence is equal to the angle of reflection. Note that the light ray OA is incident perpendicular to the surface of the mirror; therefore it is reflected back along its own path. The extensions of all reflected rays meet at a common point I behind the surface of the mirror. This point corresponds to the location of the image.

Triangles OAB and IAB are similar, and they have a common side AB; therefore the distance OA from the object O to the mirror is equal to the distance AI of the image I behind the surface of the mirror. The image I is called a *virtual image*; light rays do not actually pass through a virtual image, and it cannot be projected onto a screen. If light rays pass through an image, it can be projected onto a screen, and it is called a *real image*.

Extended objects are composed of a large number of points, and a plane mirror forms an image of each point. The extended object may be a luminous body or any other object that reflects some light in the direction of the mirror. In each case the plane mirror forms an upright virtual image of the entire extended object (Fig. 28-4). From the ray diagram it

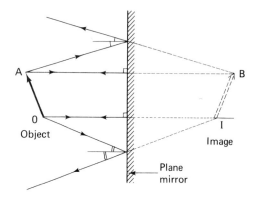

Figure 28-4
A ray diagram showing the location of an extended image I that is formed by a plane mirror.

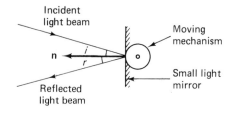

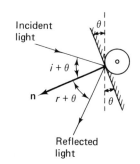

Figure 28-5
The optical lever.

can be seen that the image and the object are symmetrically located with respect to the mirror, and the image is the same size as the object. The image is also *laterally inverted*; the left side of the object becomes the right side of the image and vice versa.

The Optical Lever

Measurements are often made with some instruments, such as galvanometers, by noting the deflection of a pointer over a scale. The accuracy of these instruments is limited to some extent by the length of the pointer; long pointers are too heavy for the moving mechanism, but short pointers require compressed scales, which are difficult to read. In many cases the principles of specular reflection are used to increase the sensitivity and accuracy of these measuring instruments. A small, light plane mirror is attached to the rotating mechanism, and it reflects a narrow beam of light that acts as a pointer (Fig. 28-5). Therefore the pointer can be made any length without affecting the moving mechanism. When the moving mechanism and mirror are deflected through an angle θ, both the angle of incidence i and the angle of reflection r

increase by the same angle θ. Consequently the angular difference between the incident and reflected rays changes by 2θ. The increased length and the doubled angular deflection of the pointer increase the sensitivity and accuracy of the instrument.

The Sextant

A sextant is a device that is used to measure the angle between a heavenly body (star, moon, etc.) and the horizon. Navigators and surveyors make accurate measurements of these angles so that they can determine positions and directions on the earth's surface. A surveyor's *theodolite* can be used to measure both horizontal and vertical angles, but it requires a firm foundation, and it cannot be used at sea or in an aircraft.

A marine sextant (Fig. 28-6a) consists of a *horizon glass* that has one-half clear and one-half silvered, and an *index mirror* that can be rotated by moving an *index arm* over a graduated scale. To operate the device, an observer views the horizon through a telescopic (or sighting tube) eyepiece and the clear half of the horizon glass and adjusts the index bar until the image of the heavenly body coincides with the horizon. Since the altitude of the heavenly body is twice the angle through which the index mirror is rotated (optical lever principle), the graduated scale has angles marked at twice their actual value.

In airborn sextants (Fig. 28-6b) an air bubble in a liquid is used to indicate the horizontal. The angle between the mirrors is adjusted (by means of a knob) until the images of the heavenly body and the bubble coincide. Special averaging devices are normally used to obtain an average altitude over a fixed time interval, and this value is then read from a scale.

*Problems*_____

1. Discuss the laws of specular reflection in terms of Huygens' principle.

2. Name three instruments that could employ optical levers.

3. A 6.0 ft man stands in front of a vertical plane mirror. What must be the minimum length of the mirror if the man is to see his complete image?

4. A narrow beam of light is reflected from a small mirror that is attached to a galvanometer coil, and it makes a light spot on a scale 50 cm from the mirror. If the galvanometer coil rotates 3°, how far will the light spot move over the scale?

28-2 SPHERICAL MIRRORS

Some optical instruments utilize curved mirrors to concentrate or disperse light energy. In many cases the reflecting surfaces of these curved mirrors are shaped as if they are a portion of a spherical surface, and they are

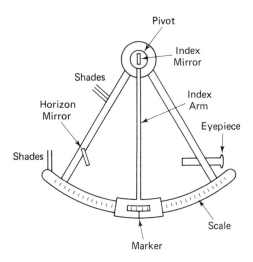

(a) Marine Sextant

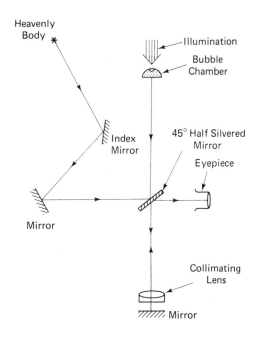

(b) Optical System of an Airborn Bubble Sextant

Figure 28-6

called *spherical mirrors*. There are two types of spherical mirrors: If the inner spherical surface acts as the reflector the mirror is called *concave*; if the outer spherical surface acts as the reflector, the mirror is called *convex*.

Nomenclature

The reflecting surface of a spherical mirror is a portion of an imaginary sphere. The center of this imaginary sphere is called the *center of curvature C* of the mirror, and the *radius of curvature R* of the mirror is the radius of the sphere (Fig. 28-7). A spherical mirror usually has a circular boundary, and its diameter *MN* is called the *aperture*; the aperture of a spherical mirror is usually much smaller than its radius of curvature. The center of the mirror is called its *vertex V*, and the straight line that passes through the vertex *V* and the center of curvature *C* is called the *principal axis PA*.

The radius of curvature *R* is a perpendicular to any point on the reflecting surface of a spherical mirror. Consequently the normal to the reflecting surface is directed towards the center of curvature of a concave mirror and away from the center of curvature of a convex mirror. At the reflecting surface of a spherical mirror, the angle of incidence *i* of any ray is equal to its angle of reflection *r*.

Rays that are incident parallel to the principal axis all pass approximately through a single point on the principal axis after reflection from the surface of a *concave* spherical mirror; this point is called the *principal focus F* of the mirror.* Similarly, any rays that are parallel to the principal axis of a *convex* mirror *diverge* after reflection approximately *as if they came from a* single point called the *principal focus F*. The principal focus of a

convex mirror is located on the principal axis behind the surface of the mirror (Fig. 28-7). The distance between the principal focus *F* and the vertex *V* is called the *focal length f* of the mirror.

Consider a ray AB that is parallel to the principal axis of a spherical mirror. After reflection from the surface of the mirror, the ray passes through the principal focus *F* of a concave mirror, or it diverges as if it came from the principal focus of a convex mirror (Fig. 28-7). Since the angle of incidence *i* is equal to the angle of reflection *r*, the angles BCV = *i* = CBF; thus triangle CBF is an isosceles triangle, and its sides CF and FB are equal. If the ray AB is close to the principal axis, then FB $\approx$ FV = *f* (the focal length); therefore, the radius of curvature is:

$$R = CV = CF + FV$$
$$= FB + FV \approx 2FV = 2f$$

Thus

$$R \approx 2f \qquad (28\text{-}1)$$

The principal focus *F* is approximately midway between the center of curvature C and the vertex V of the spherical mirrors.

Ray Diagrams

Consider an object OP that is located in front of a spherical mirror. A spherical mirror usually forms an image IJ that is quite different from the image that would be formed in a plane mirror. For example, the *object distance p* between the object O and the vertex *V* of the mirror may not be equal to the *image distance q* between the image I and the vertex *V*.

Ray diagrams are scale drawings that may be used to locate the images. If two or more rays are drawn from the same point on the object, the intersection of the reflected rays or the extensions of the reflected rays correspond to the location of the image. The laws of reflection may be applied to any number of rays that originate from the object and are

---

*Parallel rays that are inclined at some angle with the principal axis are focused to a point that is *not* on the principal axis.

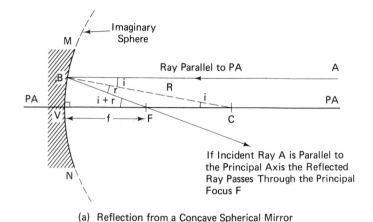

(a) Reflection from a <u>Concave</u> Spherical Mirror

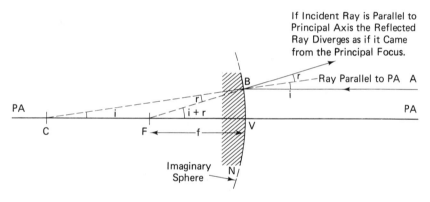

(b) Reflection from a <u>Convex</u> Spherical Mirror

Figure 28-7

reflected from the surface of the mirror, but usually only two rays are required to complete a ray diagram. Solid lines represent the actual paths of light rays whereas dotted lines are used to indicate the directions from which the rays *appear* to originate. For convenience any two of the following rays are normally used in ray diagrams (Fig. 28-8).

1. When a ray PQ that is parallel to the principal axis is reflected, it passes through the principal focus *F* of a concave mirror (or it diverges as if it came from the principal focus of a convex mirror).

2. If a ray PN passes through the center of curvature of a concave mirror (or is directed towards the center of curvature of a convex mirror), it is reflected back along its own path.

3. Any ray PT that passes through the principal focus of a concave mirror (or is directed towards the principal focus of a convex mirror) is reflected parallel to the principal axis (along TS).

The image location J of the object point P corresponds to the intersection of the reflected rays (or to the extensions of those reflected rays behind the mirror). Note that for *real* images the reverse is true; if the object is placed at point J, the concave mirror forms the image at P. Any such pair of points (for example, J and P in Fig. 28-8a) are called *conjugate foci*; when the object is located at one of these points, its image is formed at the other.

Images

The perpendicular line from point J to the principal axis of the mirror constitutes the complete image IJ of the extended object OP. Images may be (a) real or virtual, (b) upright or inverted and (c) their size may be different from that of the object.

When the object distance *p* from a *concave spherical mirror* is greater than the focal length *f*, the image is always inverted, and it is real because light rays actually pass through it (Figs. 28-8 and 28-9). If an object is placed in front of a *convex spherical mirror* or between the principal focus and the vertex of a *concave spherical mirror*, the image is upright, and it is *virtual* because it is located behind the surface of the mirror where the light rays cannot penetrate.

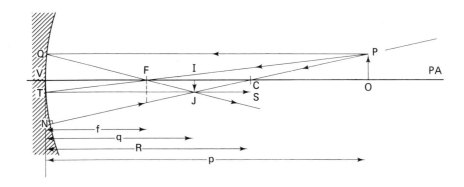

(a) Concave Mirror with Object Distance p
Greater than the Radius of Curvature R.
Image is Real, Diminished, and Inverted

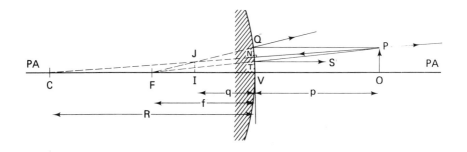

(b) Concave Mirror. Image is Always
Virtual, Upright, and Diminished

Figure 28-8
Ray diagram location of images.

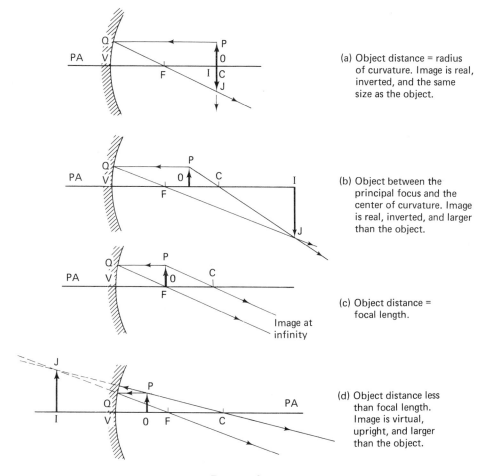

(a) Object distance = radius of curvature. Image is real, inverted, and the same size as the object.

(b) Object between the principal focus and the center of curvature. Image is real, inverted, and larger than the object.

(c) Object distance = focal length.

(d) Object distance less than focal length. Image is virtual, upright, and larger than the object.

Figure 28-9
Images formed in a concave spherical mirror.

The Mirror Equation

When an object OP is placed in front of a spherical mirror, the location of its image IJ may be determined from a ray diagram (Fig. 28-10a). Triangles PVO and JVI are similar because they have equal angles; therefore

$$\frac{IJ}{OP} = \frac{IV}{OV} = \frac{q}{p} \qquad (28\text{-}2)$$

where $p = OV$ is the object distance and $q = IV$ is the image distance. Triangles PCO and JCI are also similar; thus

$$\frac{IJ}{OP} = \frac{IC}{OC} = \frac{R - q}{p - R} \qquad (28\text{-}3)$$

Equating Eqs. 28-2 and 28-3 and rearranging terms, we obtain:

$$\frac{1}{p} + \frac{1}{q} = \frac{2}{R} \qquad (28\text{-}4)$$

If object OP is located close to the principal axis and is small compared with the radius of curvature, the ray PQ is close to the principal axis, and the focal length $f \approx R/2$. Thus

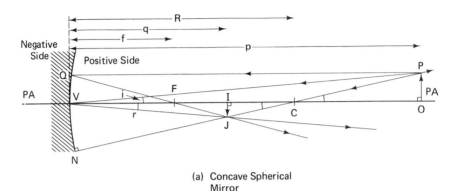

(a) Concave Spherical
Mirror

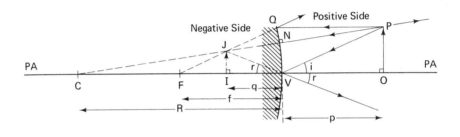

(b) Convex Spherical
Mirror

Figure 28-10

$$\frac{1}{p} + \frac{1}{q} = \frac{2}{R} \approx \frac{1}{f} \qquad (28\text{-}5)$$

This expression is called the *mirror equation*. It is valid for both concave and convex spherical mirrors if the following sign convention is used.

All quantities describing the positions of items that are located in front of the reflecting (shiny) side of the mirror are positive valued, whereas quantities describing positions behind the reflecting surface are negative valued.

For example, the center of curvature C and the principal focus F are located behind the reflecting surface of a convex mirror, thus the radius of curvature R and the focal length f are both negative. A concave mirror has its center of curvature and principal focus in front of the reflecting surface; therefore its

radius of curvature R and focal length f are both positive. Similarly real objects and images are located in front of the reflecting surface, and the corresponding object distance p and image distance q are positive. Virtual objects and images are located behind the reflecting surfaces, and both their object distance p and image distance q are negative.

Linear Magnification

The ratio of the length (height or width) of an image H_i to the corresponding length H_0 of the object is called the *linear magnification* M; it is positive valued when the image is erect and negative valued when the image is inverted. In Fig. 28-10a, the linear magnification is negative because the image is inverted; thus from the similar triangles PVO and JVI:

$$M = \frac{\text{image height}}{\text{object height}} = \frac{H_i}{H_0} \quad (28\text{-}6)$$

$$= -\frac{IJ}{OP} = -\frac{q}{p}$$

since the image distance q and object distance p are both positive.

Before the mirror equation is used to solve a problem, it is usually useful to sketch the system.

*Example 28-1*_____

An object of height $H_0 = 2.0$ cm is placed 20 cm in front of a convex spherical mirror that has a focal length of 30 cm. Find the position and size of the image.

Solution:

Figure 28-8b indicates that the image should be virtual, upright and diminished. We are given $p = +20$ cm, and, since the mirror is convex, $f = -30$ cm. Thus from Eq. 28-5:

$$\frac{1}{q} = \frac{1}{f} - \frac{1}{p} = \frac{1}{(-30 \text{ cm})} - \frac{1}{20 \text{ cm}}$$

$$= \frac{-2 - 3}{60 \text{ cm}} = \frac{-5}{60 \text{ cm}}$$

or $\quad q = \frac{-60 \text{ cm}}{5} = -12$ cm

The negative sign indicates that a virtual image is formed behind the reflecting surface. Suppose that H_i is the image height; the linear magnification $M = H_i/H_0 = -q/p$. Thus

$$H_i = -H_0 \frac{q}{p} = \frac{-(2.0 \text{ cm})(-12 \text{ cm})}{(20 \text{ cm})}$$

$$= +1.2 \text{ cm}$$

The positive sign implies that the image is upright.

*Example 28-2*_____

When an object is placed 10 cm in front of a spherical mirror, an erect vertical image that is 1.5 times as large as the object is formed. Determine the focal length of the mirror.

Solution:

$$q = -Mp = -(1.5)(10 \text{ cm}) = -15 \text{ cm}$$

The negative sign implies that the image is virtual. Thus

$$\frac{1}{f} = \frac{1}{p} + \frac{1}{q} = \frac{1}{10 \text{ cm}} + \frac{1}{(-15 \text{ cm})}$$

$$= \frac{3 - 2}{30 \text{ cm}} = \frac{1}{30 \text{ cm}}$$

or $f = +30$ cm. The mirror is concave because the focal length is positive valued. Fig. 28-9d illustrates the system.

*Problems*_____

5. An object 2.5 cm high is placed 5.0 cm in front of a concave spherical mirror that has a radius of curvature of 20 cm. (a) Determine the position and size of the image. (b) Describe the image.

6. Determine the position and size of an image that is formed when a 2.0 cm high object is located 30 cm in front of a convex spherical mirror that has a radius of curvature of 15 cm.

7. If an object is located 18 cm in front of a spherical mirror, a virtual image that is 0.60 times as large as the object is formed. Determine the focal length of the mirror, and sketch the ray diagram.

8. A concave spherical mirror is used as a shaving mirror because it produces virtual upright images when the object distance is less than the focal length. Determine the magnification if a man's face is 20 cm from a shaving mirror that has a focal length of 40 cm.

9. A reflecting telescope utilizes a concave spherical mirror with a focal length of 16 m to form a real image of the moon. If the moon is 3.84×10^8 m from the mirror and has a diameter of 3.49×10^6 m, calculate the diameter of the image.

10. A convex spherical mirror forms a virtual image that is 0.80 times the size of the object. If the image is 8.0 in behind the mirror, determine (a) the position of the object and (b) the radius of curvature of the mirror. (c) Draw the ray diagram.

11. A concave spherical mirror forms a real image that is 1.25 times the size of the object. If the focal length of the mirror is 50 cm, determine (a) the location of the object and (b) the location of the image. (c) Draw the ray diagram and describe the image.

12. A concave spherical mirror has a radius of curvature of 80 cm. Determine the position and size of the corresponding image when a 3.0 cm high object is located the following distances from the reflecting surface of the mirror: (a) 100 cm, (b) 80 cm, (c) 60 cm, (d) 40 cm, (e) 20 cm.

13. A technician uses a concave mirror with a radius of curvature of 5.00 cm in order to check visually a circuit that is located in an awkward position. Determine the location of the image and the magnification when a circuit element is 1.50 cm from the mirror.

14. A concave spherical mirror with a radius of curvature of 5.0 ft is used to project an image onto a screen that is 15 ft from the mirror. Where must the object be located? Draw the ray diagram.

28-3 SPHERICAL ABERRATION AND PARABOLIC REFLECTORS

Rays that are parallel to the principal axis of a concave spherical mirror pass approximately through the same point (the principal focus) only if the aperture of the mirror is much smaller than its radius of curvature. If the aperture of the mirror is not relatively

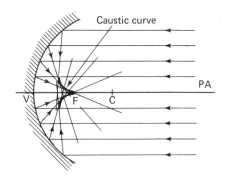

Figure 28-11
Spherical aberration.

small, the parallel rays that are farthest from the principal axis are reflected through points on the principal axis that are closer to the vertex of the mirror (Fig. 28-11). This is called *spherical aberration*; it results in blurred images. The envelope of the reflected rays in the neighborhood of the principal focus is called the *caustic curve*.

Spherical aberration is eliminated if a *parabolic mirror* (Fig. 28-12) is used instead of a spherical mirror when a large aperture is required. Parallel rays always pass through the same point after they have been reflected

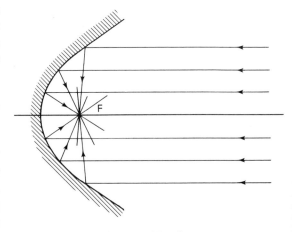

Figure 28-12
A parabolic mirror.

from a parabolic mirror. Parabolic reflectors are used instead of spherical reflectors in large reflecting telescopes (Fig. 28-13).

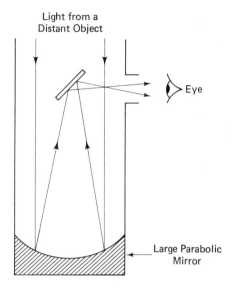

Light from a
Distant Object

Eye

Large Parabolic
Mirror

Figure 28-13
A reflecting telescope.

When a light source is placed at the principal focus of a parabolic mirror, parallel light rays emerge after reflection from the mirror's surface. This principle is used in automobile headlights and in searchlights.

REVIEW PROBLEMS

1. What kind of mirror would be used in a searchlight? Explain why.

2. Describe the advantages and disadvantages of spherical mirrors and parabolic mirrors.

3. Sketch the image(s) formed when an object is located in front of two mirrors that are perpendicular to each other.

4. An object 5.0 in high is placed 15 in in front of a concave spherical mirror that has a radius of curvature of 10 in. Calculate the position and size of the image.

5. An object 3.0 cm high is placed 8.0 cm in front of a concave spherical mirror of small aperture that has a focal length of 12 cm. Calculate (a) the position, (b) magnification, (c) height of the image.

6. An object 5.0 cm high is placed 15 cm in front of a concave spherical mirror that has a radius of curvature of 10 cm. (a) Sketch the system. (b) Calculate the position and size of the image. (c) Is the image real or virtual? Why?

7. An object 1.0 cm high is placed 18 cm in front of a concave spherical mirror that has a focal length of 6.0 cm. (a) What is the position and size of the image? (b) Sketch the system.

8. When an object is placed 12 cm in front of a spherical mirror, a real image 0.75 times as large as the object is formed. (a) Determine the focal length of the mirror. (b) Sketch the system.

9. A 2.0 cm high object is located 12 cm in front of a convex spherical mirror that has a radius of curvature of 20 cm. Find the position and size of the image.

10. When a 2.5 cm object is located 20 cm in front of a spherical mirror, a virtual image that is 0.6 times as large as the object is formed. Calculate (a) the focal length of the mirror and (b) the position and size of the image.

Transparent materials reflect some light from their surfaces, but they also transmit a fraction of any incident light energy. The speed of light changes as it passes from one transparent medium into another, and this causes the light rays to bend unless they are incident perpendicularly to the interface between the two media; this bending of light rays is called *refraction*. The nature of the two media and the orientation of the interface between the media determine the degree to which the light rays are refracted. A *lens* is a piece of transparent material, such as glass, that is constructed with one or two curved regular surfaces that refract light.

29-1 REFRACTION

The speed of light depends on the nature of the medium through which it travels. Light travels at its greatest speed of 3.00×10^8 m/s or 1.86×10^5 mi/s in a vacuum; its speed is reduced when it enters a material medium. When light passes at an oblique angle from one material into another, its speed changes; this causes the paths of the light rays to bend. This phenomenon is called *refraction*. Light rays are bent towards the normal (perpendicular) to the boundary as they pass from an optically less dense material into an optically denser material where the speed of light is reduced (Fig. 29-1). If light passes from an optically dense material into an optically less dense material, its speed increases, and the light rays are bent away from the normal to the boundary. The *angle of incidence i* is the angle between the incident ray and the normal to the boundary, and the *angle of refraction r* is the angle between the normal and the refracted ray.

Note that the path of a refracted light ray is reversible. That is, if the direction of the refracted light ray BC is reversed, it retraces its original path, and it is refracted (at the boundary) along the path of the original incident ray.

29

Refraction and Lenses

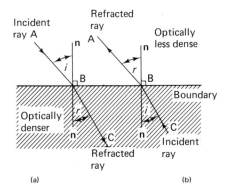

(a) (b)

Figure 29-1
Refraction of light. (a) Light rays are bent towards the normal as they pass from optically less dense into optically denser materials. (b) Light rays are bent away from the normal when they pass from optically denser into optically less dense materials.

Consider a plane light wave that passes at some oblique angle of incidence α_1 from one medium 1, where the speed of light is v_1 into another medium 2, where the speed of light is v_2 (Fig. 29-2). At the instant when point A of the wavefront AC is incident at the boundary, point C must still travel some

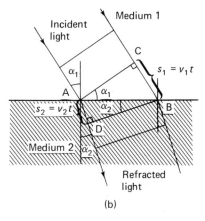

(b)

Figure 29-2
Refraction is caused by the differences in the speed of light of different materials.

distance s_1 in the medium 1 before it too reaches the boundary at point B. Therefore point C of the wavefront remains in medium 1 for an extra elapsed time $t = s_1/v_1$, during which point A is able to travel some distance $s_2 = v_2\,t$ to point D in medium 2. Since their wavespeeds are different, the wavefronts of the incident and refracted waves are not parallel.

Rays are always perpendicular to the wavefronts; therefore angle CAB is equal to the angle of incidence α_1, angle ABD is equal to the angle of refraction α_2, and triangles ABC and ABD are right angled. Thus

$$\sin \alpha_1 = \frac{CB}{AB} = \frac{v_1 t_1}{AB}$$

and

$$\sin \alpha_2 = \frac{AD}{AB} = \frac{v_2 t}{AB}$$

Therefore

$$\frac{\sin \alpha_1}{\sin \alpha_2} = \frac{v_1}{v_2} = \text{constant } n_{12} \qquad (29\text{-}1)$$

since the speeds v_1 and v_2 in the two media are constant. The constant n_{12} is called the *relative index of refraction* between the two media.

The *laws of refraction* at a boundary between two media are:

1. The ratio of the sine of the incident angle α_1 to the sine of the angle of refraction α_2 is a constant that is equal to the ratio of the speeds of light in the two media.

2. The incident ray, the refracted ray and the normal to the boundary at the point of incidence lie in the same plane.

The ratio of the speed of light c in a vacuum to the speed of light v in a material medium is defined as the *index of refraction n* (or *optical density*) of that material:

$$n = c/v \qquad (29\text{-}2)$$

Note that, since light travels fastest in a vacuum, the index of refraction of a material medium is always greater than one. Some typical values for the indices of refraction of a few materials are listed in Table 29-1. These values depend on the wavelength of the incident light and the temperature of the system.

If n_1 and n_2 are the indices of refraction of media 1 and 2, the corresponding speeds of light in these media are $v_1 = c/n_1$ and $v_2 = c/n_2$, respectively. Therefore Eq. 29-1 may be rewritten as:

$$\frac{\sin \alpha_1}{\sin \alpha_2} = \frac{v_1}{v_2} = \frac{n_2}{n_1} = n_{12} \qquad (29\text{-}3)$$

This relationship is called *Snell's law*. It was discovered by W. Snell in the seventeenth century.

*Example 29-1*_____

A narrow beam of light in air impinges on a flint glass plate at an angle of incidence $\alpha_1 = 52°$. Calculate (a) the angle of refraction α_2 and (b) the speed of light v_2 in flint glass.

Solution:

(a) From Snell's law and Table 29-1:

$$\sin \alpha_2 = \frac{n_1 \sin \alpha_1}{n_2} = \frac{(1.00)(\sin 52°)}{1.58}$$
$$= 0.500$$

Thus $\alpha_2 = 30°$.

(b) $$v_2 = \frac{c}{n_2} = \frac{3.00 \times 10^8 \text{ m/s}}{1.58}$$
$$= 1.90 \times 10^8 \text{ m/s}$$

When light passes at an oblique angle through several media, it is refracted at each boundary between the pairs of different media. The laws of refraction may be applied at each boundary. According to Snell's law, for multiple refractions at the boundaries between three parallel layers of different transparent media:

$$n_1 \sin \alpha_2 = n_2 \sin \alpha_2 = n_3 \sin \alpha_3$$

where n_1, n_2 and n_3 are the indices of refraction of the media and α_1, α_2 and α_3 are the corresponding angles between the light rays and the normals to the boundaries between the media.

Real and Apparent Depth or Thickness_____

When an observer in air views objects that are located either in or behind a transparent medium that is optically denser than air, the object appears to be closer to the observer

Table 29-1

Typical Indices of Refraction for Light with a Wavelength of 589.3 nm at 20°C

Substance	Index of Refraction	Substance	Index of Refraction
Vacuum	1.000 00	*Solids*	
Air	1.000 29	Crown glass	1.50
		Flint glass	1.58
Liquids		Dense crown glass	1.60
Benzene	1.50	Dense flint glass	1.72
Ethyl alcohol	1.36	Canada balsam	1.53
Water	1.33	Diamond	2.42
		Perspex	1.50
		Quartz (fused)	1.46

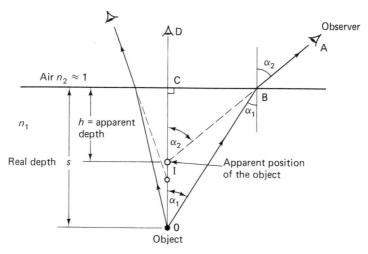

Figure 29-3
Real and apparent depth of an object.

than its actual distance. This effect is due to refraction of the light rays at the boundary between the two media. Light rays from the object are refracted away from the normal as they pass into the air. Consequently they appear to originate from points that are closer to the boundary (Fig. 29-3). The difference between the actual distance s and the apparent distance h from the boundary depends on the angle from which the object is viewed by the observer; the larger the angle of refraction α_2, the closer the object appears to be to the surface.

If the observer is located in air (which has an index of refraction $n_2 \approx 1$) and views the object from point D in a direction that is perpendicular to the boundary, the ratio of the real to apparent depth is approximately equal to the index of refraction n_1 of the optically denser medium:

$$\frac{s}{h} = \frac{n_1}{n_2} \approx n_1 \qquad (29\text{-}4)$$

*Example 29-2*_____

If an object is 5.00 cm from an observer, how much closer does it appear to be when it is

viewed in air from a perpendicular direction than through a plate of dense flint glass that is 2.00 cm thick?

Solution:

The only difference is the apparent thickness of the glass. Thus from Table 29-1 and Eq. 29-4:

$$h = \frac{s}{n_1} = \frac{2.00 \text{ cm}}{1.72} = 1.16 \text{ cm}$$

Thus the object appears closer by a distance $s - h = 0.84$ cm.

Atmospheric Refraction_____

The speed of light is greater in warm air than in cool air; therefore the cool air has a slightly larger index of refraction than the warm air. In summer cool air is frequently located above warmer air near the earth's surface. Consequently, when light passes obliquely from the cooler into the warmer air, it is refracted towards the earth's surface. As a result an observer often sees inverted *mirages* of distant objects as if they are below the horizon.

In winter these atmospheric conditions are often reversed; warm air may be located above a layer of cooler air near the earth's surface. Light rays that pass obliquely from the cooler into the warmer air are refracted upwards, away from the earth's surface. Consequently an observer may see an upright mirage *looming* above the horizon.

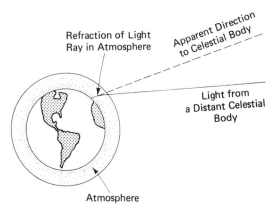

Refraction of Light Ray in Atmosphere

Apparent Direction to Celestial Body

Light from a Distant Celestial Body

Atmosphere

Figure 29-4
Refraction of light by the earth's atmosphere.

Light from distant stars and other celestial bodies reaches the earth after passing through the near vacuum of space; if this light penetrates the earth's atmosphere at an oblique angle, it is refracted downwards (Fig. 29-4). For this reason the *angle of elevation* between the earth's horizon and the direction to a celestial body usually appears to be greater than the actual elevation. When a navigator or a surveyor determines his location by measuring the elevation of celestial bodies with a sextant, he must apply a correction for the refraction of light in the earth's atmosphere.

Total Internal Reflection

When light passes from one transparent material into another, some of the light energy is reflected at the boundary between the two materials. The fraction of the total incident light energy that is reflected depends on the nature of the two transparent materials and the angle of incidence of the light at the boundary between them. As the angle of incidence is increased, a greater fraction of the incident light is reflected. For example, at large angles of incidence a plate of glass behaves like a mirror.

Consider a beam of light that passes from one transparent material, which has an index of refraction n_1, into another transparent material, which has a smaller index of refraction n_2. At small angles of incidence some light is reflected, and some is transmitted after it is refracted *away* from the normal at the boundary. As the angle of incidence increases, a greater fraction of the incident light is reflected, and the angle of refraction of the transmitted light approaches a maximum value of 90° (Fig. 29-5).

When the angle of refraction is 90°, the transmitted light emerges parallel to the boundary; the corresponding angle of incidence is called the *critical angle* α_c. Thus according to Snell's law:

$$\frac{\sin \alpha_c}{\sin 90°} = \sin \alpha_c = \frac{n_2}{n_1} \qquad (29\text{-}5)$$

If the angle of incidence exceeds the critical angle, all of the incident light is reflected at the boundary between the two transparent materials. Note that this total internal reflection occurs from a single surface.

Example 29-3

Determine the critical angle between crown glass and air.

Solution:

$$\sin \alpha_c = \frac{n_2}{n_1} = \frac{1.00}{1.50} = 0.667.$$

Thus $\alpha_c = 41.8°$

The index of refraction of a transparent material depends on the amount of impurities

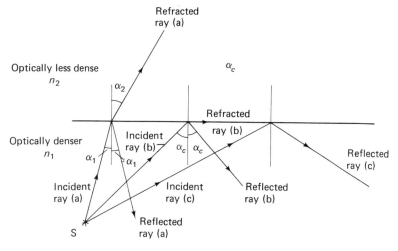

Figure 29-5
Internal reflection. (a) When $\alpha_1 < \alpha_c$ some light is transmitted, and some is reflected. (b) When $\alpha_1 = \alpha_c$, the refracted light is parallel to the boundary and the rest is reflected. (c) When $\alpha_1 > \alpha_c$ all incident light is reflected at the boundary.

that it contains; therefore the impurity concentration may be determined by measuring the index of refraction. In liquids the index of refraction may be accurately and conveniently determined by measuring the critical angle.

The recent development of special fibers has vastly improved our ability to control the path of light energy. These fibers are relatively flexible, and they can be bent into different shapes. Light is fed into one end of the fiber, and it undergoes multiple internal reflections as it is transmitted through the fiber (Fig. 29-6).

Special fibers of high purity glass about as thick as a human hair are now being used in lightwave communication systems as guides for modulated light. These fibers are low cost and operate at high speeds with an attenuation of less than 1 dB/km. Because of the high frequency of light, a single fiber can transmit with a high data concentration replacing many conventional cables. In practice many

of these fibers may be bundled together in a cable that increases the overall strength in addition to the amount of information that can be transmitted.

These systems are very suitable for the transmission of digital (on/off) information. Therefore, the electrical impulses to be sent must first be encoded into a digital signal and they must be decoded at the receiver.

A typical optical communications system has three components (Fig. 29-6a):

1. A *transmitter* consisting of a solid state laser (germanium-aluminum-arsenide) or a light emitting diode* (LED) which converts the electrical energy into light. Each may be modulated with a multiplexed signal. The laser produces a well-collimated beam of light with a narrower range of wavelengths than an LED. As a result it produces lower *chromatic dispersions*

---

*See Chapter 33.

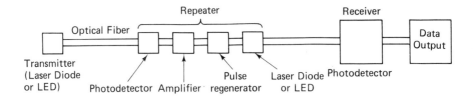

(a) Optical Communication System

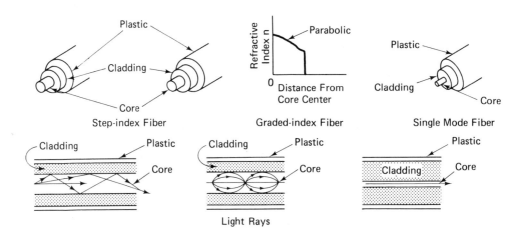

(b) Types of Optical Fibers

Figure 29-6
Optical communications system.

(the spreading of the pulse due to differences in speeds of the different wavelengths) than LED's in optical fibers.

2. The *receiver* is a photodetector that converts the light energy back to electrical energy. Repeaters, which are back-to-back transmitters and receivers, are required if the transmitting distance is greater than about 5 km.

3. An *optical fiber* is the transmitting medium; the transmitter and receiver are coupled directly to a low-loss optical fiber in which the light travels. Each fiber has three distinct regions: an *inner core*, which transmits the light; a *cladding layer*, with a lower refractive index, around the core which reflects the light; and an outer *plastic coating*, which protects the fiber from scratches and abrasions (Fig. 29-6b).

There are three types of optical fibers that are currently in use:

1. *Step-index fibers* are those in which the refractive index changes abruptly at the boundary between the core and the cladding. In these fibers the light can travel different paths and therefore any pulse spreads out while it is being transmitted; this is called *modal dispersion*.

2. *Graded-index fibers* are those in which the refractive index decreases parabolically

with the distance from the core center. This reduces the modal dispersion and increases the bandwidth because the light rays tend to travel at the same speed.

3. *Single-mode fibers* have very large bandwidths because their core size is small and the difference in the indices of refraction between the core and the cladding is small, so that only one mode is allowed to propagate. However, these devices are more difficult to manufacture, splice and align.

Fiber optics also has many other important applications. For example, in medicine, one end of a very fine fiber can be guided into the interior of a living organ, and the physician can then observe the internal operation of that organ. Scientists have already used fibers to observe the interior of a beating human heart.

*Problems*_____

1. A narrow beam of light impinges on water from air at an angle of incidence of 30°. Determine (a) the angle of refraction and (b) the speed of light in water.

2. (a) Determine the frequency of light that has a wavelength of 589.3 nm in a vacuum. (b) What is the wavelength of this light in flint glass?

3. A narrow beam of light is incident from air onto the surface of a transparent material. If the angle of incidence is 48° and the angle of refraction is 32°, calculate (a) the index of refraction of the material and (b) the speed of light in that material.

4. Determine the wavelength of a light in a liquid that has an index of refraction of 1.42 if its wavelength in a vacuum is 550 nm.

5. A plate of crown glass is completely immersed in water, and a beam of light passes from air through the water to the glass plate. If the angle of incidence of the light at the boundary between the air

and water is 38°, determine the angle of refraction in the glass plate.

6. A completely submerged diver shines a narrow beam of light towards the surface of the water. If the angle of incidence of the light beam is 25°, determine the angle of refraction in air.

7. A navigator measures the elevation of a star with a sextant. If the elevation is 20.00° and the index of refraction for air is 1.000 3, determine the true elevation of the star. *Hint:* Elevation is measured with respect to the horizon.

8. A container of water appears to be 1.2 ft deep when it is viewed from a perpendicular direction. What is its real depth?

9. A microscope is focused from above onto a flaw in a glass block. If the flaw appears to be 0.90 cm below the surface of the block and its actual distance from the surface is 1.40 cm, determine the index of refraction for the glass.

10. Determine the critical angle between (a) diamond and air and (b) crown glass and water.

11. A swimmer is 6.0 ft below the surface of water. Determine the radius of the circle at the surface of the water through which he is able to see objects that are located in air.

29-2 PRISMS

A total internal reflection from a polished glass surface with air is a very efficient process: There are only minute losses in light energy. The critical angle from glass to air is less than 45°; therefore, when a light ray in the glass is incident at an angle of 45° at the glass–air boundary, it undergoes total internal reflection. Many optical instruments contain glass prisms with polished sides and

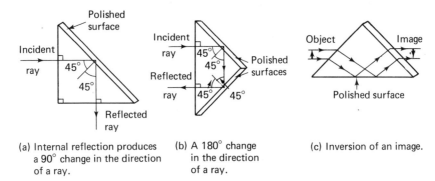

(a) Internal reflection produces a 90° change in the direction of a ray.

(b) A 180° change in the direction of a ray.

(c) Inversion of an image.

Figure 29-7
A 45°, 45°, 90° prism may be used to change the direction of light rays or to invert images by total internal reflection.

angles of 45°, 45° and 90°; these prisms efficiently change the direction of light or they invert images (Fig. 29-7).

Consider a ray of light that passes from a medium with an index of refraction n_1 through an optically denser prism that has an index of refraction n_2 and an angle A between the refracting surfaces (Fig. 29-8). The ray PQ is refracted towards the normal to the first surface as it enters the optically denser prism; then it is refracted away from the normal at the second surface as it reenters the less dense medium. The angular difference direction of the initial ray PQ and the direction of the ray RS that emerges from the prism is called the *angular deviation δ*.

The smallest value of angular deviation is called the *angle of minimum deviation δ_m*; it occurs when the path of the ray through the prism is symmetrical. Thus, in the special case of minimum angular deviation, $\alpha_1 = \alpha_4$ and $\alpha_2 = \alpha_3$.

29-3 LENSES

The laws of refraction also apply to light that is refracted at curved boundaries between two transparent media. Transparent objects that have two regular surfaces that alter the shape of the wavefront of light are called *lenses*. Either one or both of the regular surfaces of a lens are curved, and the lens is named in terms of the shapes of these surfaces (Fig. 29-9).

Most lenses have spherical surfaces, and they are made of materials, such as glass, that are optically denser than air. Light travels slower in glass than in air; therefore, when a light wave passes from air into a glass lens, the part of the wavefront that passes through the thicker portion of the lens spends more time in the lens, and it lags behind other parts of the wavefront (Fig. 29-10). As a result the

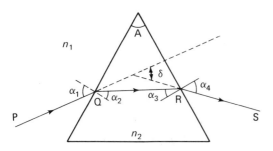

Figure 29-8
Double refraction in a prism.

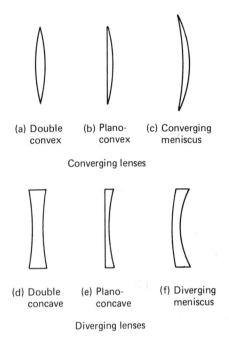

(a) Double
convex

(b) Plano-
convex

(c) Converging
meniscus

Converging lenses

(d) Double
concave

(e) Plano-
concave

(f) Diverging
meniscus

Diverging lenses

Figure 29-9
Lenses.

wavefront is distorted by the lens, the amount of distortion depends on the index of refraction and the shape of the lens.

A *converging lens* is usually thicker at its center*; its curved surfaces tend to refract incident parallel light rays through a common point. A *diverging lens* is usually thinner at its center*; its curved surfaces tend to refract incident parallel light rays as if they originated at a common point.

Nomenclature

Refracting surfaces of *spherical lenses* are portions of imaginary spheres. The centers of these spheres are called the *centers of curvature* C_1 and C_2 of the lens, and the radii R_1

---

*Unless the lens has a lower index of refraction than the surrounding medium, in which case the reverse is true, i.e. converging lenses are thinner at their centers and diverging lenses are thicker at their centers.

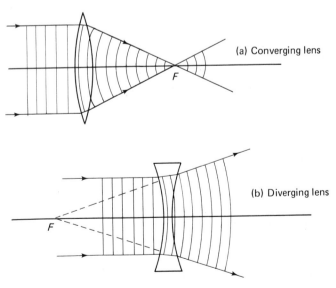

(a) Converging lens

F

(b) Diverging lens

F

Figure 29-10
Lenses change the shape of the wavefront.

and R_2 of these spheres are called the *radii of curvature*. The diameter of a lens is called the *aperture*, and it is usually much smaller than the radii of curvature. The amount of light that passes through a lens depends on its aperture; lenses with large apertures form bright images. After they pass through a particular point called the *optical center* O of the lens, all rays emerge from the lens in a direction that is parallel to their incident direction. In thin lenses the optical center is located approximately at the center of the lens. The *principal axis* PA of the lens is a straight line that passes through the centers of curvature and the optical center. The distance between the object and the optical center is called the *object distance p*, and the distance between the image and the optical center is called the *image distance q*. Rays that are initially parallel to the principal axis are refracted through (or diverge as if they came from) a common point called the *principal focus F* after they emerge from the lens. The distance between the optical center and the principal focus is called the *focal length f*. Note that there is a principal focus on each side of a lens, and these foci are equidistant from the optical center. The focus that is located on the same side of the lens as the incident light is called the *first* (or *virtual*) *principal focus* F_1. No light is refracted through the first principal focus. The other focus F_2 is called the *second* (or *real*) *principal focus* and is located on the side of the lens where the refracted light emerges.

The Lensmaker's Equation

The focal length f of a thin lens depends on the radii of curvature R_1 and R_2 of its surfaces and on the difference between the indices of refraction n_2 and n_1 of the lens and the surrounding medium, respectively. For lenses that have radii of curvature much greater than their thickness:

$$\frac{1}{f} \approx (n_2 - n_1)\left(\frac{1}{R_1} - \frac{1}{R_2}\right) \quad \text{(thin lenses)}$$

$$(29\text{-}6)$$

This relationship is known as the *lensmaker's equation*. It is valid for both converging and diverging thin lenses if the following sign convention is applied.

Sign Convention

The radius of curvature of a single refracting surface is positive valued when the corresponding center of curvature is located on the side of the lens where the refracted light emerges. If the center of curvature is on the same side of the lens as the incident light, the radius of curvature is negative valued. Thus, when the light is incident on a convex surface, the radius of curvature is positive valued; if the light is incident on a concave surface, the radius of curvature is negative valued (Fig. 29-11).

This sign convention produces positive values for the focal lengths of converging lenses and negative values for the focal lengths of diverging lenses. The radius of curvature of a plane (flat) surface is infinite.

Example 29-4

If a double convex crown glass lens has radii of curvature of 30 cm and 20 cm, what is its focal length in air?

Solution:

The center of curvature of the first surface lies on the side where the refracted light emerges. Therefore $R_1 = +30$ cm, but the center of curvature of the second surface is on the same side as the incident light; thus, $R_2 = -20$ cm. See Fig. 29-12. From Eq. 29-6 and Table 29-1:

$$\frac{1}{f} = (n_2 - n_1)\left(\frac{1}{R_1} - \frac{1}{R_2}\right)$$

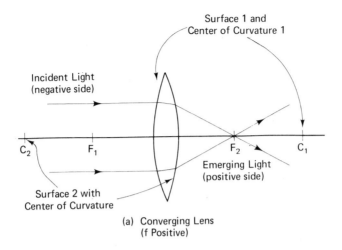

Figure 29-11
Sign convention.

$$= (1.50 - 1.00)\left(\frac{1}{30 \text{ cm}} - \frac{1}{-20 \text{ cm}}\right)$$

$$= 0.50\left(\frac{2+3}{60 \text{ cm}}\right) = \frac{2.5}{60 \text{ cm}}$$

Therefore $f = 24$ cm.

Ray Diagrams

The nature and location of an image that is formed by a thin lens may be determined with a ray diagram. Consider two or more rays that are drawn from the same point on an object. These rays are refracted as they pass

through a lens, and the intersection of the refracted rays or the extension of the refracted rays corresponds to the image location. For convenience any two of the following rays are normally used (Fig. 29-12).

1. When a ray PQ that is parallel to the principal axis is incident on a thin lens, it is refracted through the second principal focus F_2 of a converging lens, or it diverges as if it originated at the first principal focus F_1 of a diverging lens.

2. Any ray PO that passes through the optical center O of any lens is not deviated.

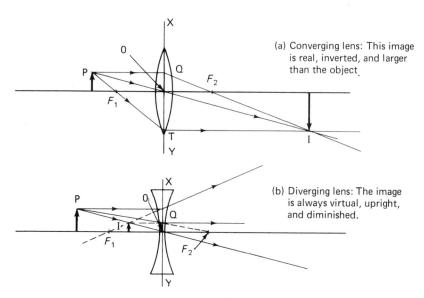

(a) Converging lens: This image is real, inverted, and larger than the object.

(b) Diverging lens: The image is always virtual, upright, and diminished.

Figure 29-12
Ray diagrams.

3. If a ray PT passes through the first principal focus F_1 of a converging lens or if it is directed towards the second principal focus F_2 of a diverging lens, it is refracted by the lens, and it emerges parallel to the principal axis.

Refraction occurs at both surfaces of a lens; however, for convenience in ray diagrams, the direction changes are assumed to occur at the plane XY that is perpendicular to the principal axis and contains the optical center O. Light rays actually pass through lenses; therefore *real images* are formed on the side of the lens where the refracted light emerges, and they can be projected onto a screen. *Virtual images* are located on the same side of the lens as the incident light. These images cannot be projected onto a screen because no light actually passes through them, but the refracted light rays appear to originate there. Light rays actually emanate from *real objects* that are located on the same side of the lens as the incident light. However virtual objects may be produced in systems containing more than one lens; these objects are located on the side of a lens where the refracted light emerges, and therefore no light actually originates at *virtual* objects.

The images that are formed by diverging lenses are always virtual, upright and diminished. Some of the images that are formed by converging lenses are illustrated in Fig. 29-13. Converging lenses may also be used to produce parallel rays of light by placing a light source at the first principal focus.

The Lens Equation

Consider the image IJ that is formed when an object PN is placed in front of a thin converging lens (Fig. 29-14). Triangles PNO and JIO are similar because they have equal angles. Therefore

$$\frac{IJ}{PN} = \frac{IO}{NO} = \frac{q}{p}$$

Since the ray PQ is parallel to the principal axis, the object height PN = QO. Also, triangles QOF_2 and JIF_2 are similar. Thus

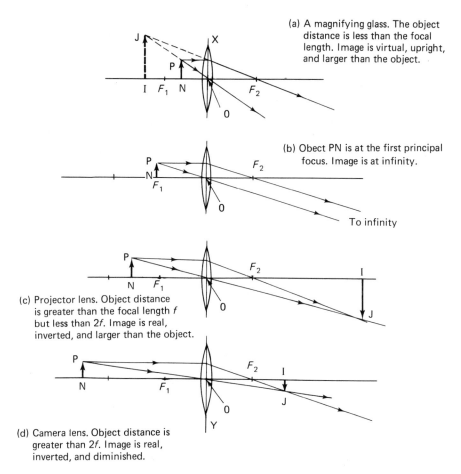

(a) A magnifying glass. The object distance is less than the focal length. Image is virtual, upright, and larger than the object.

(b) Obect PN is at the first principal focus. Image is at infinity.

To infinity

(c) Projector lens. Object distance is greater than the focal length f but less than $2f$. Image is real, inverted, and larger than the object.

(d) Camera lens. Object distance is greater than $2f$. Image is real, inverted, and diminished.

Figure 29-13
Images formed by converging lenses.

$$\frac{IJ}{PN} = \frac{IJ}{QO} = \frac{IF_2}{OF_2} = \frac{q-f}{f} = \frac{q}{f} - 1$$

Equating these expressions and rearranging the terms, we obtain:

$$\frac{1}{p} + \frac{1}{q} = \frac{1}{f} \qquad (29\text{-}7)$$

This is called the *lens equation*, and it is valid for both converging and diverging thin lenses if the following sign conventions are used:

1. The focal lengths f of converging lenses are positive valued, and the focal lengths f of diverging lenses are negative valued.

2. Object distances p and image distances q of real objects and images are positive valued, and those of virtual objects and images are negative valued.

Linear Magnification

As in the case of mirrors, the ratio of the length (height or width) of the image to the corresponding length of the object is called the *linear magnification M*.

$$M = \frac{\text{image length}}{\text{object length}} = \frac{H_i}{H_0} = -\frac{q}{p} \qquad (29\text{-}8)$$

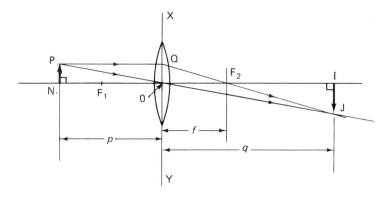

Figure 29-14

Linear magnificantion is positive valued when the image is erect and negative valued when the image is inverted.

The Power of a Lens

The *power P* of a lens in units of *diopters* is defined as the reciprocal of its focal length f in meters.

$$P \text{ (diopters)} = 1/f \text{ (meters)} \qquad (29\text{-}9)$$

Lenses with short focal lengths have larger powers than those with long focal lengths because they produce larger distortions of the wavefronts.

Converging lenses have positive-valued powers because their focal lengths are positive, but diverging lenses have negative-valued focal lengths and powers.

Before the thin lens equation is used, it is advisable to sketch the ray diagram of the system and to apply the sign conventions to express the known quantities in terms of their symbols.

Example 29-5

Determine the position and height H_i of the image that is formed when an object of height $H_0 = 3.0$ cm is placed 20 cm in front of a

Summary of the Sign Conventions for Thin Lenses

Positive Valued	Negative Valued
1. Focal lengths f and powers P of converging lenses.	Focal lengths f and powers P of diverging lenses.
2. Object distances p and image distances q of real objects and images.	Object distances p and image distances q of virtual objects and images.
3. Radii of curvature when the centers of curvature are on the side of the lens where the refracted light emerges.	Radii of curvature when the centers of curvature are on the same side of the lens as the incident light.
4. Linear magnification M and image length H_i when the image is erect.	Linear magnification M and image length H_i when the image is inverted.

thin diverging lens that has a focal length of 5.0 cm.

Solution:

The ray diagram is similar to Fig. 29-12b. We are given $p = 20$ cm, and, since it is a diverging lens, $f = -5.0$ cm. Thus

$$\frac{1}{q} = \frac{1}{f} - \frac{1}{p} = \frac{1}{(-5.0 \text{ cm})} - \frac{1}{20 \text{ cm}}$$

$$= \frac{-4 - 1}{20 \text{ cm}} = \frac{-5}{20 \text{ cm}}$$

or $\qquad q = \frac{-20 \text{ cm}}{5} = -4.0$ cm

The negative sign implies that the image is virtual and that it is located on the same side of the lens as the object. From Eq. 29-8 the linear magnification $M = H_i/H_0 = -q/p$; thus the image height is:

$$H_i = \frac{-H_0 q}{p} = \frac{-(3.0 \text{ cm})(-4.0 \text{ cm})}{(20 \text{ cm})} = 0.60 \text{ cm}$$

*Example 29-6*_____

The lens of a movie projector forms inverted images 7.00 m long on a screen that is 30.0 m away. If the film size is 35 mm, calculate (a) the focal length and (b) the power of the lens.

Solution:

(a) $H_0 = 35$ mm, and, since the images are real and inverted, $q = +30$ m and $H_i = -7.0$ m. Thus

$$M = \frac{H_i}{H_0} = \frac{-7.0 \text{ m}}{3.5 \times 10^{-2} \text{ m}} = -200 = \frac{-q}{p}$$

or $\qquad p = q/200$

Therefore

$$\frac{1}{f} = \frac{1}{p} + \frac{1}{q} = \frac{200}{q} + \frac{1}{q} = \frac{201}{q}$$

or $\qquad f = \frac{q}{201} = \frac{30.0 \text{ m}}{201} = 0.149$ m

(b) $\qquad P = \frac{1}{f} = \frac{201}{30 \text{ m}} = 6.7$ diopters

_____

*Problems*_____

12. Determine the focal length in air of a double convex flint glass lens that has radii of curvature of 8.0 in and 12.0 in.

13. Calculate the focal length in air of a plano-concave dense crown glass lens if the radius of curvature of the concave surface is 15 cm.

14. Compute the radius of curvature of the convex surface of a plano-convex lens that is made from crown glass if its focal length in air is 14 cm.

15. Determine the focal length and power of a double concave crown glass lens in water if its radii of curvature are 6.0 cm and 9.0 cm.

16. Compute the radius of curvature of the spherical surface of a plano-convex lens if it has a power of 1.5 diopters in air and it is made from dense flint glass.

17. Determine the position and height of the image that is formed when a 1.5 cm object is placed 12 cm in front of a converging lens that has a focal length of 9.0 cm.

18. Calculate the position and size of the image that is formed when a 20 mm object is placed 100 mm in front of a diverging lens that has a focal length of 70 mm.

19. When a 3.0 in object is placed 12 in in front of a lens, an inverted real image 27 in high is formed. Determine the focal length of the lens.

20. A crown glass double convex lens has radii of curvature of 12.0 cm and 15.0 cm. Determine the position and size of the image that is formed when a 2.0 cm object is placed 8.0 cm in front of the lens in air. Describe the image.

21. A diverging meniscus is made from dense crown glass, and it has radii of curvature of 15 cm and 10 cm. Determine the position and size of the image when a 3.0 cm object is located 25 cm in front of the lens in air.

22. An upright virtual image 120 mm high is

formed when an object that is 80 mm high is placed 60 mm from a lens. Determine (a) the location of the image and (b) the focal length of the lens.

23. A converging lens with a focal length of 9.0 in is used to form an image that is three times as large as the object. Determine how far from the lens should the object be (a) if the image is real and (b) if the image is virtual.

24. A camera that has a lens with a focal length of 75 cm is located in a satellite. If the camera takes photographs from an altitude of 150 km, how large is the image of a 100 m rocket on the film?

29-4 SIMPLE OPTICAL INSTRUMENTS

The human eye is a very important optical instrument because it enables us to see, but it does have limitations. We frequently rely on other optical instruments to improve the visual abilities of the eye.

The Human Eye

The structure of a human eye is illustrated in Fig. 29-15. Light enters the eye through a transparent membrane called the *cornea*; it then passes through a transparent liquid (the *aqueous humor*) to a converging lens composed of a number of transparent flexible fibers that are held in place by the *ciliary muscles*. These ciliary muscles also adjust the shape and focal length of the lens; this adjustment is called *accommodation*. The size of the *pupil* or aperture of the lens is controlled by a colored diaphragm called the *iris*.

After the light has been refracted by the lens, it passes through a jelly-like matter called the *vitreous humor*, and, if the eye functions correctly, sharp inverted images are formed on the *retina*. The retina is the inner surface at the rear of the eye and is composed of a layer of nerve tissue containing millions of nerve endings called *rods* and *cones*. Stimulations of the rods and cones are transmitted through the optic nerve to the brain where they are analyzed and the sensation of vision is produced. Note that, even though an inverted image is formed on the retina, the brain

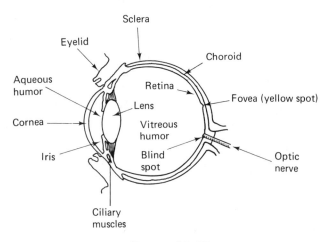

Figure 29-15
The human eye.

interprets it as an erect image. Cones are sensitive to color, and they are used for daylight vision. The concentration of cones is greatest at the *fovea* (or *yellow spot*) at the center of the retina. The eye is most sensitive to details of images that are formed in this region. In dim blue light rods are more sensitive than cones; therefore we use rods for night vision. Rods are also very sensitive to movements. No rods or cones are located at the *blind spot* where the optic nerve enters the eye; therefore this area is insensitive to light.

There are three common defects of the human eye:

1. *Myopia or nearsightedness.* If the ciliary muscles are not strong enough to increase the focal length of the lens or if the eyeball is elongated, images of distant objects are focused to points in front of the retina. Only images of close objects are focused on the retina; blurred images are formed of distant objects. An eyeglass with a suitable diverging lens is used to correct for this defect.

2. *Hyperopia or farsightedness.* When the eyeball is too short or the focal length of the lens is too long, images are formed behind the retina. Images of distant objects are in focus, but blurred images are formed of close objects. It is corrected by using eyeglasses with suitable converging lenses.

3. *Astigmatism.* Ocular astigmatism is frequently due to irregularities in the curvature of the cornea. As a result of these irregularities, the focal length in one plane may be quite different from that in another plane, and some images are sharp while others are blurred. For example, the image of vertical lines may be sharp while images of horizontal lines are blurred or not visible! Eyeglasses with cylindrical lenses are used to correct for this condition.

Example 29-7

A nearsighted man has a greatest distance of distinct vision of 2.50 m. What power eyeglass lens should he use in order to see clearly distant objects?

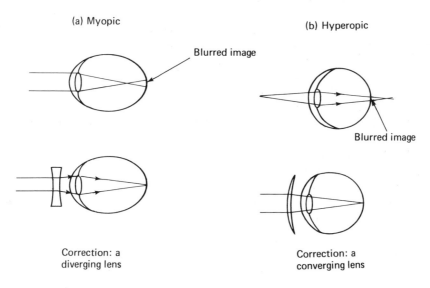

(a) Myopic

Blurred image

Correction: a diverging lens

(b) Hyperopic

Blurred image

Correction: a converging lens

Figure 29-16
Defects of the human eye.

Solution:

The lens must form virtual images at 2.50 m from the eye. Thus $p = \infty$, $q = -2.50$ m and the power of the lens:

$$P = \frac{1}{f(\text{m})} = \frac{1}{p} + \frac{1}{q} = 0 + \frac{1}{-2.50 \text{ m}}$$

$$= -0.40 \text{ diopters}$$

Example 29-8

A farsighted man can only see objects distinctly when they are at least 125 cm away. What power of eyeglass lens should he use in order to distinguish clearly objects at a distance of 25 cm?

Solution:

The lens must form virtual images 125 cm from the eye. Thus $p = 0.25$ m, $q = -1.25$ m and the power of the lens:

$$P = \frac{1}{f(\text{m})} = \frac{1}{q} + \frac{1}{p} = \frac{1}{-1.25 \text{ m}} + \frac{1}{0.25 \text{ m}}$$

$$= \frac{4}{1.25 \text{ m}} = 3.2 \text{ diopters}$$

Cameras

In inexpensive cameras a single converging lens is used to produce small inverted real images of distant objects; these images are recorded on light-sensitive film. Light can only enter the camera through the lens. The duration of the exposure is controlled by a *shutter*, which opens and allows light to enter the camera for a predetermined *exposure time*. An adjustable *diaphragm* effectively changes the aperture of the lens, and it controls the amount of light that is allowed to enter the camera. The *aperture* or *f-number* of a camera lens is normally specified as a fraction of the focal length. For example, a setting of $f/16$ implies that the diameter of the aperture is one-sixteenth of the focal length of the camera lens. The illuminance E of the film is directly proportional to the exposure time t and the square of the aperture d of the lens and

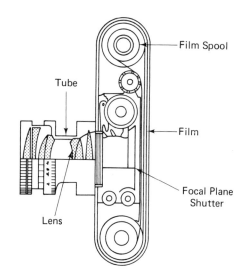

Figure 29-17
A simple camera.

inversely proportional to the focal length f of the lens:

$$E \propto \frac{d^2 t}{f} \qquad (29\text{-}10)$$

Example 29-9

If an exposure meter indicates that the correct f-value of a camera lens is $d_1 = f/5.6$ for an exposure time $t_1 = 1/125$ s, what f-value should be used for an exposure time $t_2 = 1/500$ s?

Solution:

Since the focal length is constant $d_1^2 t_1 = d_2^2 t_2$ or

$$d_2 = d_1\sqrt{t_1/t_2} = \frac{f}{5.6}\sqrt{\frac{(1/125) \text{ s}}{(1/500) \text{ s}}} = \frac{2f}{5.6} = \frac{f}{2.8}$$

In many cameras the lens may be moved so that the image distance from the lens to the film may be changed to produce sharp images of objects that are different distances from the lens. Once the position of the lens has been set, sharp images are only formed of objects that are a particular distance from the lens. If an object is located at some other

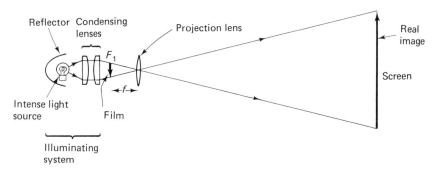

Figure 29-18
A projector.

distance, the lens focuses its image to a point that is not on the film, and the film records a blurred circular image of each point of the object. These blurred circular images are called *circles of confusion*. The farther the object is from the point that is focused to a sharp image on the film, the larger the circles of confusion are. If the diameter of a circle of confusion is sufficiently small, it appears to be sharp to the human eye. Objects that are within a certain range, which is called the *depth of field*, are focused to relatively sharp images on the film. The diameter of the circles of confusion may be reduced and the depth of field increased by reducing the aperture of the lens. However this also reduces the illumination of the film; therefore the exposure time must be increased.

Expensive cameras usually contain a number of lenses to correct for spherical and chromatic aberrations.

Projectors

An *illuminating system* of a projector consists of an intense light source, a reflector and a pair of *condensing lenses* that direct the light onto a film. The film is located slightly farther than the focal length from a converging *projection lens* that forms real enlarged and inverted images of the objects on the film (Fig. 29-18). Normally the film is inverted so

that the observer sees an upright image on the screen.

The Magnifier

A normal human eye sees objects most distinctly when they are 25 cm away. If an object is farther than 25 cm, the eye is unable to distinguish fine details; if the object is closer than 25 cm, blurred images are formed on the retina.

Virtual erect and enlarged images are formed by a single converging lens when the object distance is less than the focal length. Therefore a single converging lens acts as a *magnifier* (or magnifying glass); it enables us to see clearly objects that are closer to the eye than 25 cm. The object O is located closer to the magnifier than the focal length; when the eye is close to the magnifier, it sees the virtual enlarged image most clearly when the image distance is 25 cm (Fig. 29-19). Thus, since $q = -25$ cm, from the thin lens equation:

$$p = \frac{qf}{q-f} = \frac{(-25 \text{ cm})f}{-25 \text{ cm} - f} = \frac{(25 \text{ cm})f}{25 \text{ cm} + f}$$

and the linear magnification is:

$$M = -\frac{q}{p} = -\frac{(-25 \text{ cm})}{(25 \text{ cm})f}(25 \text{ cm} + f)$$

$$= \frac{25 \text{ cm}}{f} + 1 \qquad (29\text{-}11)$$

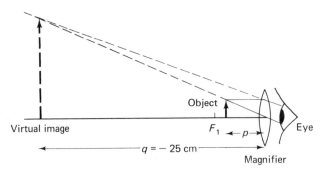

Figure 29-19
A magnifier.

Example 29-10

A converging lens with a focal length of 3.50 cm acts as a magnifier. If the lens forms a virtual image 25.0 cm from the eye, determine (a) the object distance and (b) the linear magnification.

Solution:

(a) $f = +3.50$ cm   and $q = -25.0$ cm.

Thus from the thin lens equation:

$$p = \frac{qf}{q - f} = \frac{(-25.0 \text{ cm})(3.50 \text{ cm})}{(-25.0 \text{ cm}) - (3.50 \text{ cm})}$$
$$= 3.07 \text{ cm}$$

(b) $M = -\dfrac{q}{p} = -\dfrac{(-25.0 \text{ cm})}{3.07 \text{ cm}} = 8.14*$

Problems

25. An exposure meter indicates that the correct *f*-value of a camera lens is *f*/8 for an exposure time of 1/60 s. What *f*-value should be used for an exposure time of 1/15 s?

26. A camera lens has an aperture of 30.0 mm. If its focal length is 20.0 mm, what is its *f*-value?

---

*Or from $M = \dfrac{25 \text{ cm}}{f} + 1.$

27. A nearsighted man has a greatest distance of distinct vision of 1.5 m. What power of eyeglass lens should he use in order to see clearly distant objects?

28. A farsighted man has a least distance of distinct vision of 2.5 m. Calculate the focal length and power of the eyeglasses that he requires to restore normal vision.

29. A magnifier with a focal length of 2.5 cm forms a virtual image 25 cm from the eye. Determine (a) the object distance and (b) the linear magnification.

29-5 COMBINATIONS OF LENSES

Consider a number of thin lenses that are placed in contact. If the lenses have powers $P_1, P_2, P_3, \ldots$ and their focal lengths are $f_1, f_2, f_3, \ldots$, the total power of the lens combination is:

$$P_T = P_1 + P_2 + P_3 + \cdots \quad (29\text{-}12)$$

Thus the focal length f_T of the lens combination is given by:

$$\frac{1}{f_T} = \frac{1}{f_1} + \frac{1}{f_2} + \frac{1}{f_3} + \cdots \quad (29\text{-}13)$$
(thin lenses in contact)

Many optical instruments contain combinations of two or more lenses that are not in contact, but each lens of the system refracts the incident light. The image that is formed by the first lens (the closest lens to the object) becomes the object for the second lens, and the image formed by the second lens becomes the object for the third lens, etc. Note that, if the first lens forms an image behind the front surface of the second lens, that image becomes a *virtual object* (and has a negative value for its *object* distance) for the second lens.

In order to locate the final image that is formed after the light has passed through a lens combination, it is important to proceed in stages:

1. Ignore the effects of the other lenses in the system, and draw the ray diagram or use the lens equation to find the position of the image I_1 that is formed by the first lens.

2. The image I_1 becomes the object for the second lens. Determine the distance of I_1 from the optical center of the second lens; this distance becomes the object distance p_2 (which is negative for virtual objects) for the second lens.

3. Use a ray diagram or the lens equation to find the location of the image I_2 formed by the second lens.

4. Repeat the process for all lenses in the system.

5. Each lens of the system magnifies the image that is formed by the preceding lens; therefore the total magnification M_T of the system is equal to the product of the individual magnifications produced by the individual lenses:

$$M_T = M_1 M_2 M_3 \cdots \qquad (29\text{-}14)$$

*Example 29-11*_____

A converging lens A with a focal length $f_A = 6.00$ cm and a diverging lens B with a focal length $f_B = -9.00$ cm are located 4.00 cm apart (Fig. 29-20). If an object 4.00 cm high is located a distance $p_A = 30.0$ cm in front of

the converging lens A, find (a) the distance q_B of the final image from lens B, (b) the total magnification, (c) the image size.

Solution:

(a) From the thin lens equation, for A:

$$q_A = \frac{p_A f_A}{p_A - f_A} = \frac{(30.00 \text{ cm})(6.00 \text{ cm})}{(30.0 \text{ cm}) - (6.00 \text{ cm})}$$
$$= 7.50 \text{ cm}$$

This image becomes the (virtual) object for lens B an object distance $p_B = -3.50$ cm, i.e. behind lens B. Thus

$$q_B = \frac{p_B f_B}{p_B - f_B} = \frac{(-3.50 \text{ cm})(-9.00 \text{ cm})}{(-3.50 \text{ cm}) - (-9.00 \text{ cm})}$$
$$= 5.73 \text{ cm}$$

(b) $$M_T = M_A M_B = \left(-\frac{q_A}{p_A}\right)\left(-\frac{q_B}{p_B}\right)$$
$$= \left(\frac{-7.50 \text{ cm}}{30.0 \text{ cm}}\right)\left(\frac{-5.73 \text{ cm}}{-3.50 \text{ cm}}\right) = -0.409$$

(c) $$H_i = H_0 M_T = (4.00 \text{ cm})(-0.409)$$
$$= -1.64 \text{ cm}$$

That is, an inverted image 1.64 cm high.

_____

The Compound Microscope_____

A *compound microscope* is a device that is used to produce very large magnifications. The basic instrument consists of two converging lenses (or corrected converging lens systems) that are arranged on a common principal axis (Fig. 29-21). The lens that is closest to the object is called the *objective*; it has a very short focal length, and it forms a real inverted and enlarged image $I_1 J_1$ of any object PQ that is located slightly beyond its first principal focus F_1^o. A simple magnifier with a moderately short focal length acts as the *eyepiece* (or *ocular*) of the microscope. The eyepiece is arranged so that it forms a virtual enlarged image $I_2 J_2$ of the real image $I_1 J_1$; this final image $I_2 J_2$ is 25 cm (the distance of most distinct vision) from the eyepiece.

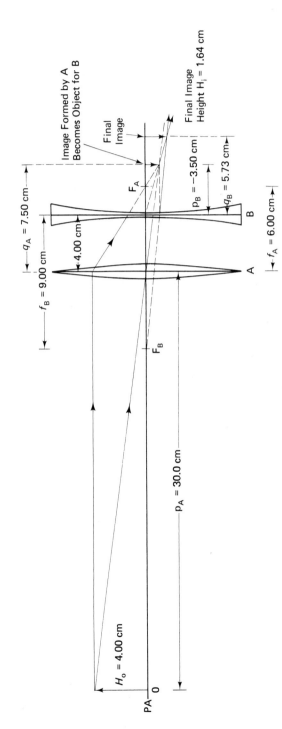

Figure 29-20

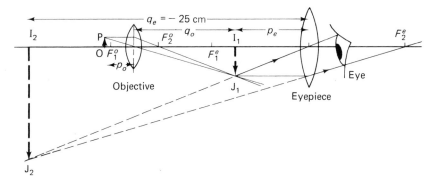

Figure 29-21
A compound microscope.

Example 29-12

The objective and eyepiece of a compound microscope are 20.00 cm apart, and they have focal lengths $f_0 = 7.5$ mm and $f_e = 50$ mm, respectively. If the final virtual image I_2J_2 is formed 25 cm from the eyepiece, determine (a) the distance p_0 of the object from the objective and (b) the total magnification M_T.

Solution:

(a) From the thin lens equation, the distance p_e of the real image I_1J_1 from the eyepiece is:

$$p_e = \frac{q_e f_e}{q_e - f_e} = \frac{(-25 \text{ cm})(5.0 \text{ cm})}{-25 \text{ cm} - 5.0 \text{ cm}}$$
$$= 4.17 \text{ cm}$$

Thus the distance of the real image I_1J_1 from the objective $q_0 = 20.00 \text{ cm} - 4.17 \text{ cm} = 15.83$ cm. From the thin lens equation, the distance of the object from the objective is:

$$p_0 = \frac{q_0 f_0}{q_0 - f_0} = \frac{(15.83 \text{ cm})(0.75 \text{ cm})}{15.83 \text{ cm} - 0.75 \text{ cm}}$$
$$= 0.79 \text{ cm}$$

(b) $\quad M_T = M_0 M_e = \left(\frac{-q_0}{p_0}\right)\left(\frac{-q_e}{p_e}\right)$

$$= \left(\frac{-15.83 \text{ cm}}{0.79 \text{ cm}}\right)\left(\frac{+25 \text{ cm}}{4.17 \text{ cm}}\right) = -120$$

Microscopes normally have a *turret nose* with three compound objectives with different powers. These objectives are moved into alignment with the eyepiece by rotating the turret.

Telescopes

A basic *refracting astronomical telescope* also consists of two converging lenses that are arranged on a common principal axis with the distance between the lenses approximately equal to the sum of their focal lengths (Fig. 29-22). The *objective* has a large aperture and a long focal length f_0; it produces a real inverted and diminished image I_1J_1 of a distant object. The eyepiece is a simple magnifier with a moderately short focal length f_e; it forms a virtual enlarged image I_2J_2 of the real image I_1J_1.

Telescopes are normally used to observe distant objects, and they form final virtual images that are smaller than the object and inverted. The *angular magnification* that is produced by a telescope is defined as the ratio of the angle β that is subtended at the eye by the image to the angle α that would be subtended at the naked eye by the object. It is equivalent to the negative ratio of the objective focal length f_0 to the eyepiece focal length f_e:

$$M = \frac{\beta}{\alpha} = -\frac{f_0}{f_e} \qquad (29\text{-}15)$$

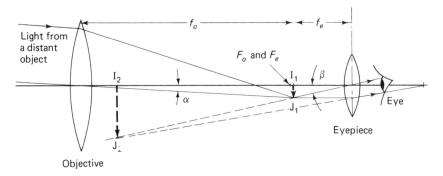

Figure 29-22
A refracting astronomical telescope.

Binoculars are also a form of astronomical telescope in which light is internally reflected in two 45° prisms. These prisms shorten the length of the telescope and rectify the final image (Fig. 29-23).

A *Galilean telescope* (or *opera glass*) also produces an erect image. The objective is a converging lens, but a diverging lens with a short (negative) focal length is used as the eyepiece. The distance between the objective and the eyepiece is approximately equal to the difference in the magnitudes of their focal lengths. Consequently Galilean telescopes are shorter than astronomical telescopes, but they have smaller fields of view.

Terrestrial telescopes consist of three converging lenses. The third lens is inserted between the objective and the eyepiece to rectify the image. However the addition of the extra lens increases the length of the device, and it reduces the intensity of the final image.

A surveyor's *theodolite* essentially consists of a telescope mounted on graduated plates. These instruments may be used to measure accurately horizontal and vertical angles.

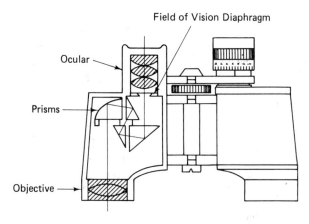

Figure 29-23
Prism binocular.

*Problems*_____

30. A thin diverging lens with a focal length of 20 in and a thin converging lens with a focal length of 10 in are placed in contact. Determine the position and size of the image that is formed by the combination when a 0.25 in object is located 15 in in front of the combination.

31. An object is located 8.0 mm from the objective of a compound microscope and the final image is formed 25 cm from the eyepiece. (a) What is the distance between the objective lens and the eyepiece if their focal lengths are 7.5 mm and 30 mm respectively? (b) Calculate the linear magnification.

32. An astronomical telescope has an objective lens with a focal length of 200 cm and it produces an angular magnification of −150 for distant objects. Calculate (a) the focal length of the eyepiece and (b) the distance between the objective and the eyepiece.

33. A Galilean telescope has a converging objective lens with a focal length of 200 mm, and a diverging lens with a focal length of 50 mm as the eyepiece. Calculate (a) the angular magnification and (b) the distance between the lenses when the telescope is focused on distant objects.

29-6 LENS DEFECTS

Spherical lenses are relatively easy to manufacture, but they are subject to a number of defects: spherical aberration, distortion and chromatic aberration.

Spherical Aberration_____

Incident light rays that are parallel to the principal axis of a spherical lens only pass through a common focus if the aperture of the lens is small. If the aperture is not very small, the parallel rays that are incident nearer to the perimeter of the lens are refracted through larger angles, and they are brought to a focus at points that are closer to the lens than the principal focus (Fig. 29-24). This is called *spherical aberration*.

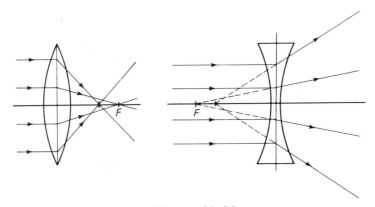

Figure 29-24
Spherical aberration. Incident rays near the perimeter of the lens are brought to focus at points closer to the lens than the principal focus.

To minimize spherical aberrations, lenses are often designed so that the incident light is deviated by equal amounts at each refracting surface. A diaphragm may be used to limit the effective aperture of the lens, but it also reduces the amount of light that is transmitted through the lens.

Distortion

When a diaphragm is used to control spherical aberration by limiting the aperture of a lens, it also produces a variation in the linear magnification at different points on the image; this is called *distortion*. If the diaphragm is located between the object and the lens, the object distance is increased for points of the object that are farthest from the principal axis. As a result the linear magnification of these points is smaller than the linear magnification of object points that are located nearer to the principal axis, and the lens forms distorted *barrel-shaped* images of square objects. This type of distortion is often called *barrel distortion.*

(a) Object   (b) Barrel distortion   (c) Pin-cushion distortion

Figure 29-25
Distortions.

When a diaphragm is positioned between the lens and the image, it effectively increases the image distances for points that are farthest from the principal axis. In this case the linear magnification is the least for points on the principal axis, and the lens forms distorted *pin-cushion* images of square objects; this is called a *pin-cushion distortion.*

Chromatic Aberration

The index of refraction of a material depends on the wavelength of the incident light. A material has a larger index of refraction for blue light than for red light; therefore a lens refracts different colors of incident light through different angles; this is called *chromatic aberration*. The lens forms a separate image for each color of light that emanates from the object, and the total image becomes blurred.

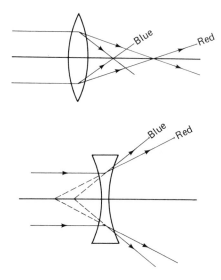

Figure 29-26
Chromatic aberration.

Special lenses that have been corrected for chromatic aberrations are called *achromatic lenses*. These lenses are really two lenses in contact, a converging lens and a diverging lens, but the two component lenses are made from different materials, such as crown glass and flint glass. The achromatic lens is designed so that the color dispersion in the converging lens is equal and opposite to the color dispersion in the diverging lens, but one of these component lenses has a greater power than the other. Consequently an achromatic

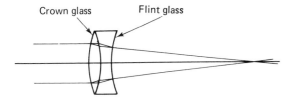

Crown glass   Flint glass

Figure 29-27
An achromatic lens.

lens has a net power, but it produces no color dispersion.

REVIEW PROBLEMS

1. (a) Determine the frequency of laser light that has a wavelength of 633 nm in a vacuum. (b) What is the wavelength of this light in water?

2. A container of water appears to be 60 cm deep when it is viewed from above. What is its real depth?

3. Determine the critical angle between (a) air and dense crown glass and (b) benzene and water.

4. Determine the critical angle between water and air at 20°C.

5. A container is filled to a depth of 2.0 ft with water and a 1.5 ft layer of ethyl alcohol on top of the water. What is the total apparent depth of the liquid?

6. A ray of light enters a sample of glass at an incident angle of 30°. If the index of refraction for the glass is 1.5, (a) what is the speed of light in the glass? (b) What is the angle of refraction?

7. An object 5.0 cm high is placed 30 cm in front of a converging lens of focal length 15 cm. What is the position and size of the image? Is it real or virtual?

8. A lens has a focal length of 20 cm. What is its power in diopters?

9. An object 5.0 in high is placed 25 in in front of a thin lens of focal length -10 in. (a) What kind of lens is used? (b) Sketch the system. (c) Use the lens equation to determine the position and size of the image. (d) Is the image real or virtual?

10. At what distance from a converging lens of focal length 5.6 cm must an object be placed in order that a real image be formed twice the size of the object?

11. An exposure meter indicates that the correct f-value of a camera lens is $f/16$ for a $\frac{1}{125}$ s exposure time. What f-value should be used for an exposure time of $\frac{1}{1\,000}$ s?

12. A nearsighted man has a greatest distance of distinct vision of 2.50 m. What power eyeglasses should he use in order to see distant objects clearly?

13. A farsighted man has a least distance of distinct vision of 3.0 m. What power eyeglasses does he require in order to restore normal vision?

14. A magnifier with a focal length of 1.8 cm forms a virtual image 25 cm from the eye. Determine (a) the object distance and (b) the linear magnification.

15. An object 3.0 in high is placed 30 in in front of a diverging lens of focal length 20 in and a converging lens of focal length 15 in that are in contact. (a) Determine the position and size of the image. (b) Sketch the system.

16. Determine the angular magnification of a refracting astronomical telescope that has an objective with a focal length of 180 cm and an eyepiece with a focal length of 3.0 cm.

17. An object 12 cm high is placed 20 cm in front of a converging lens of focal length 15 cm and a diverging lens with a focal

length of 20 cm is placed 80 cm behind the converging lens. Determine the position and size of the image formed by the lens system.

18. An object 2.0 cm high is placed 12 cm in front of a diverging lens of focal length 6.0 cm. If a converging lens of focal length 4.0 cm is placed 2.0 cm behind the diverging lens, calculate (a) the position of the final image and (b) the magnification of the whole system. (c) Determine the size of the final image.

19. The objective and eyepiece of a com-pound microscope are 18 cm apart, and they have focal lengths of 6.0 mm and 3.0 cm, respectively. If it forms a final virtual image 25 cm from the eyepiece, find the distance of the object from the objective and the total magnification.

20. The objective and eyepiece of a com-pound microscope have focal lengths of 6.0 mm and 4.0 cm, respectively. If an object is 6.2 mm from the objective, how far apart must the lenses be in order to produce a final virtual image 25 cm from the eyepiece?

The study of the wave-like nature of light and the interactions between light and matter is called *physical optics*. In this chapter we only consider the phenomena that are due to the wave-like nature of light. Interactions between light and matter are discussed in Chapter 31.

Electromagnetic radiation exhibits the wave-like characteristics of interference and diffraction; these properties are frequently used to determine the wavelength of the electromagnetic waves or to make precision measurements. Colored lights, paints and dyes are often used to improve our visual environment or to distinguish one item from another. The phenomenon of color is attributed to the different visual sensations that are produced by the different wavelengths of visible light.

30-1 COLOR

In 1666 Isaac Newton passed a narrow beam of "white" sunlight through a single glass prism so that the refracted light formed a continual elongated band of different colors. He was able to reconstruct the white light by passing the band of colored light through an inverted prism. From this therefore he concluded that white light is composed of a mixture of colored lights. The refractive separation of white light into its component colors is called *dispersion;* it occurs because the speeds of the different colors of light are not equal in the prism. All colors of light have equal speeds in a vacuum, but in any other transparent medium the shorter wavelengths travel slower than the longer wavelengths. Consequently the medium has a larger index of refraction for the shorter wavelengths. Violet light has a shorter wavelength than red light for example; therefore a glass prism has a larger index of refraction for the violet light, and it produces a greater angular deviation for the violet light (Fig. 30-1).

30

Wave
Optics

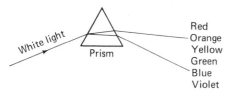

Figure 30-1
Formation of a spectrum.

Color Classification. A color may be described in terms of three variables: hue, lightness and saturation. *Hue* is the dominant color, such as red, orange, yellow, green, blue, violet, purple and the intermediate colors. *Lightness* is a comparison of the quantities of light that are reflected by different surfaces when they are irradiated with the same white light. *Saturation* is a measure of the degree of concentration of wavelengths that form the color. Thus a single wavelength light has a high saturation, but white light has a low saturation because it is composed of many wavelengths of light.

Addition of Colored Lights. When colored lights are mixed, they may produce a light that has a different color. Yellow light may be produced from a combination of red and green lights. This is called *color addition* because the resulting colored light contains all the wavelengths of the constituent lights (Fig. 30-2a). In practice, white light or any color can be reproduced by adding various intensities of only three colors of light; these are called *primary colors*. The most effective primary colors are red, green and blue. White light is also produced when specific intensities of two colors (blue and yellow, red and cyan or green and magenta) are added. If two colors can produce white light, they are known as *complementary colors*.

Color Subtraction. When a colored object is irradiated with white light, it selectively absorbs many of the incident wavelengths (Fig. 30-2b); if the object is opaque, we see the color that is produced by the reflected light. For example, an opaque object is blue if it reflects only blue light and absorbs red, yellow and green light. However colored objects do not usually reflect a narrow band of wavelengths; rather they absorb the complementary color of their reflected light. Therefore a blue object absorbs the wavelengths that would produce the complementary color yellow, and it reflects the wavelengths that produce blue.

Pigments, such as paints and dyes, also produce color by a process of selective absorption of specific wavelenghts from the incident

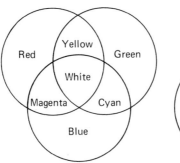

(a) Addition of colored lights

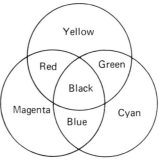

(b) Color subtraction in pigments and color filters

Figure 30-2

light. When different pigments are mixed, each color selectively absorbs specific wavelengths of any incident light; the resulting color is due to the remaining reflected wavelengths. For example, a mixture of yellow and blue paints produces green paint because the yellow paint absorbs violet and blue light while the blue paint absorbs red and yellow light; only green light is reflected.

Transparent color filters also produce color by a process of selective absorption. Some wavelengths of incident light are selectively absorbed in the filter. The transmitted wavelengths produce the color.

*Problems*_____

1. Describe how the colors are produced (a) in a color television set and (b) in a color slide.

2. Explain what happens when a green pigment is viewed under (a) white light, (b) blue light, (c) magenta light.

3. Why do objects have different colors when they are viewed in sunlight and under fluorescent lights?

30-2 YOUNG'S DOUBLE SLIT

The wave-like nature of light was known before Maxwell developed the theory of electromagnetic waves. Huygen's principle (1678) had been used to explain successfully the reflection and refraction of light, but the wave-like nature was not confirmed until 1801 when Thomas Young produced interference with light. In his original experiment Young oberved interference between the two spherical waves that were produced when he allowed sunlight to pass through two small pinholes.

An improved version of Young's experimental arrangement is illustrated in Fig. 30-3. *Monochromatic light* (light that has a single constant wavelength) is incident on a single narrow slit S. According to Huygen's principle, the slit S becomes a source of a new light wave that illuminates the two slits S_1 and S_2. Finally slits S_1 and S_2 become *coherent* sources of new light waves (their phase difference is constant) that interfere to produce a series of bright and dark lines (or *fringes*)

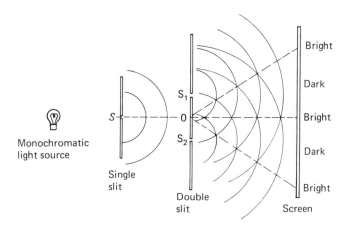

Figure 30-3
Young's experiment. The circular arcs represent successive crests of the light waves.

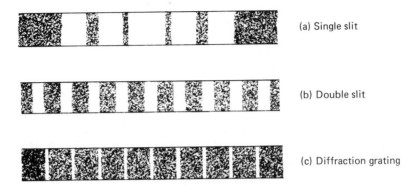

(a) Single slit

(b) Double slit

(c) Diffraction grating

Figure 30-4
Interference fringes.

on the screen (Fig. 30-4). When the light waves from S_1 and S_2 arrive in phase, they *interfere constructively* to produce the *bright* bands on the screen. The *dark bands* represent the regions where the two waves *interfere destructively*.

Consider two parallel slits S_1 and S_2 that are an equal distance L from a flat screen (Fig. 30-5). If monochromatic light waves leave S_1 and S_2 in phase, they also arrive in phase and interfere constructively to produce a bright fringe at P_0, which is equidistant from S_1 and S_2. Bright fringers also occur at

other points P where the path difference $S_2P - S_1P = S_2Q$ is an integer multiple of the wavelenght:

$$S_2O = N\lambda \quad \text{(bright fringes)} \qquad (30\text{-}1)$$

where N is an integer or zero and λ is the wavelength of the monochromatic light. Similarly dark fringes are located at points where the waves arrive 180° out of phase, corresponding to a path difference of an odd integer multiple of half the wavelength:

$$S_2Q = (2N - 1)\frac{\lambda}{2} \qquad (30\text{-}2)$$

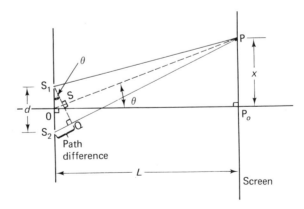

Figure 30-5
The path difference for the light that passes through the two slits.

If O is midway between the slits, the right triangles S_1OS and POP_0 are similar since angles S_1OP and OPP_0 are equal (alternate angles); therefore angles OS_1Q and POP_0 are equal. Since $S_1S_2 = d$, the separation of the slits is very small, angle S_1QS_2 is approximately a right angle, and triangles S_1S_2Q and POP_0 are similar. Thus

$$\frac{S_2Q}{S_1S_2} = \frac{PP_0}{OP} \quad \text{or} \quad \frac{S_2Q}{d} = \frac{x}{OP}$$

where x is the distance of P from the central bright fringe at P_0. However the angle θ is usually very small, and the length $OP \approx L$, Therefore

$$\frac{S_2Q}{d} \approx \frac{x}{L} \quad \text{or} \quad x = \frac{(S_2Q)L}{d} \quad (30\text{-}3)$$

Substituting for the path differences from Eqs. 30-1 and 30-2, we see that the distances of the fringes from the central bright fringe P_0 are given by:

$$x = \frac{N\lambda L}{d} \quad \text{(bright fringes)} \quad (30\text{-}4)$$

$$x = \frac{(2N-1)\lambda L}{2d} \quad \text{(dark fringes)} \quad (30\text{-}5)$$

The *order* of the fringes is designated by the integer N. For example, the zero-, first-, and second-order fringes correspond to values $N = 0$, 1 and 2 respectively. Note that equally spaced fringes are symmetrically located on both sides of the central bright fringe. The wavelength of the light may be determined from the distance between fringes.

*Example 30-1*_____

Monochromatic light from a helium–neon gas laser is incident on two narrow slits that are 0.800 mm apart and 2.00 m from a screen. If the average distance between the zero- $(N = 0)$ and second-order $(N = 2)$ bright fringes on the screen is 3.16 mm, calculate the wavelength and frequency of the laser light.

Solution:

From Eq. 30-4:

$$\lambda = \frac{xd}{NL} = \frac{(3.16 \times 10^{-3} \text{ m})(8.00 \times 10^{-4} \text{ m})}{2(2.00 \text{ m})}$$
$$= 6.32 \times 10^{-7} \text{ m}$$

and

$$f = \frac{c}{\lambda} = \frac{3.00 \times 10^8 \text{ m/s}}{6.32 \times 10^{-7} \text{ m}} = 4.75 \times 10^{14} \text{ Hz}$$

*Problems*_____

4. Compare the interference pattern produced by coherent sources with that produced by incoherent sources.

5. Monochromatic light is incident on two narrow slits that are 0.500 mm apart and 1.00 m from a screen. If the average distance between the zero- and first-order bright fringe is 1.00 mm, determine (a) the wavelength and frequency of the light and (b) the average distance of the second dark fringe from the zero-order bright fringe.

6. Blue monochromatic light with a wavelength of 4.50×10^{-7} m is incident on two narrow slits that are 0.900 mm apart and 120 cm from a screen. Determine the average distances of (a) the first-order bright fringe, (b) the third-order bright fringe, (c) the second dark fringe from the zero-order bright fringe.

30-3 THE DIFFRACTION GRATING

A *transmission diffraction grating* consists of a large number of parallel, equally spaced grooves (usually several thousand per centimeter) that are cut in a flat glass plate; the transparent spaces between the opaque grooves are equivalent to narrow slits. In a *reflection diffraction grating* the grooves are cut in a flat metal plate; fringe patterns are produced by the interference of the light that is reflected from the regions between the grooves. These high-quality gratings are often used as molds to make inexpensive replica gratings from plastic.

Diffraction gratings are normally used to accurately determine wavelengths because they produce interference fringes that are much narrower and more intense than those obtained with a double slit.

When a parallel beam of monochromatic light is incident perpendicular to the surface of a transmission diffraction grating, each slit becomes a coherent source of secondary waves that are initially in phase (Fig. 30-6).

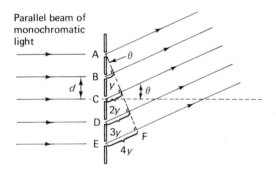

Figure 30-6
The diffraction grating.

In a direction specified by an angle θ, the secondary waves must travel different distances to the wavefront AF. The path difference between the waves from *successive* slits is:

$$y = d \sin \theta$$

where d is the distance between the centers of adjacent slits. The secondary waves only interfere constructively to produce a bright fringe if they are in phase at the wavefront, i.e., if the path difference y of the waves from adjacent slits is zero or an integer multiple of the wavelength. Thus bright lines representing interference maxima occur symmetrically on

both sides of the central maximum in specific directions θ_N that satisfy the relationship:

$$N\lambda = d \sin \theta_N \qquad (30\text{-}6)$$

where λ is the wavelength, d is the distance between the centers of adjacent slits and N is zero or an integer. This relationship is called the *grating equation*. The *order* of the spectrum is specified by the value of N.

Example 30-2

When monochromatic light is incident on a diffraction grating that has 8 000 lines/cm, the average angular deviation of the first-order image is $21°$;. Calculate (a) the wavelength of the light and (b) the angular deviation of the second-order image.

Solution:

(a) $\quad d = \dfrac{1}{8\,000}$ cm

$\qquad = 1.25 \times 10^{-4}$ cm $= 1.25 \times 10^{-6}$ m

Thus $\quad \lambda = \dfrac{d \sin \theta_N}{N}$

$\qquad = \dfrac{(1.25 \times 10^{-6}\text{ m})(\sin 21°)}{1}$

$\qquad = 4.48 \times 10^{-7}$ m

(b) $\quad \sin \theta_2 = \dfrac{2\lambda}{d} = \dfrac{2(4.48 \times 10^{-7}\text{ m})}{1.25 \times 10^{-6}\text{ m}}$

$\qquad = 0.717$, thus $\theta_2 = 45.8°$

Since the angular dispersion depends on the wavelength of the incident light, diffraction gratings and prisms may be used to form *spectra* by separating the light into an ordered array of its component wavelengths.

The Spectrometer

A *spectrometer* is a device that is used to observe and analyze the spectra of light sources. It consists of a *collimator*, a prism or a diffraction grating and a telescope (Fig. 30-7). Light enters the collimator through an adjustable narrow slit S that is located at the

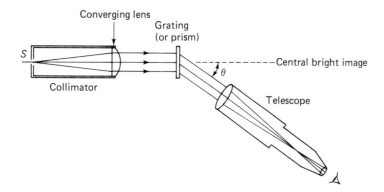

Figure 30-7
A spectrometer.

principal focus of the converging lens. The parallel rays of light that emerge from the collimator are incident on the prism or diffraction grating, where the various component wavelengths undergo different angular dispersions. The colored images of the slit are viewed through a telescope that can move in a circular arc about the central prism or grating to measure the angular deviations θ.

In a *spectrograph* the telescope is replaced by a camera to obtain photographs of spectra.

Problems

7. Monochromatic light is incident perpendicular to the surface of a diffraction grating that has 10 000 lines/cm. If the angular deviation of the second-order image is 52.0°, calculate (a) the wavelength of the light and (b) the angular deviation of the first-order image.

8. Monochromatic green light with a wavelength of 500 nm is incident perpendicular to the surface of a diffraction grating that has 6 000 lines/cm. Calculate the angular deviation of the first-, second- and third-order images.

9. Yellow monochromatic light with a wavelength of 577 nm is incident perpendicular

to the surface of diffraction grating. If the angular deviation of the second-order image is 35°, calculate the linear density of the lines on the grating.

10. Red monochromatic light with a wavelength of 630 nm is incident perpendicular to the surface of a diffraction grating that has 12 700 lines/inch. Calculate the angular deviation of the first- and second-order images.

30-4 SINGLE SLIT DIFFRACTION

Bright and dark interference fringes are also formed when light is passed through a single narrow slit or aperture. These fringes are produced by the interference of light from different points *in the same slit*.

Consider the interference fringes that are formed when a parallel beam of monochromatic light is incident on a perpendicular narrow single slit (Fig. 30-8). The location of the bright and dark fringes may be predicted in terms of the path difference y of the light from the extremities A and B of the slit.

When the path difference $y = 0$, the light

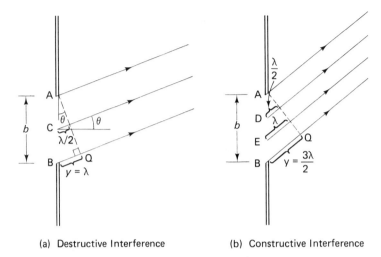

(a) Destructive Interference   (b) Constructive Interference

Figure 30-8

The formation of single slit interference
fringes.

from all points of the slit is in phase, and the
central bright fringe is formed. In the direction where the path difference $y = \lambda/2$, the
light from A and B is 180° out of phase, but
the light from all other points in the slit is
not cancelled. Consequently this light also
forms part of the central bright fringe.

The first dark fringe occurs in some direction θ in which the path difference y is equal
to the wavelength λ of the incident light. In
this direction the light from A is 180° out of
phase with the light from the center C of the
slit because the corresponding path difference
$y = \lambda/2$. Similarly the light from any other
point between A and C is 180° out of phase
with the light from the corresponding point
between C and B; therefore there is complete
destructive interference. In general, dark
fringes are produced in all directions θ where
the path difference y between the light from
the extremities A and B is an integer multiple
of the wavelength.

$$y = N\lambda$$

where N is a positive integer. Therefore in the
right-angled triangle ABQ:

$$b \sin \theta = y = N\lambda \quad \text{(dark fringes)} \quad (30\text{-}7)$$

where b is the width of the slit.

In order to determine the type of fringe
in the direction θ where the path difference
y is $3(\lambda/2)$, it is convenient to separate the slit
into three equal sections AD, DE and EB
(Fig. 30-8a). There is complete destructive
interference between the light from sections
AD and DE, but the remaining light from EB
produces a bright fringe. Bright fringes occur
when the path difference is:

$$y = (2N + 1)(\lambda/2) \quad \text{or} \quad y = 0$$

where N is a positive integer. Thus

$$b \sin \theta = (2N + 1)\lambda/2 \quad \text{or} \quad 0 \quad (30\text{-}8)$$
$$\text{(bright fringes)}$$

Note that, as the width of the slit is decreased,
the angular separation of the fringes increases.

The ability of an optical instrument to
form separate distinguishable images of
objects that have small angular separations is

called its *resolving power*. Light is diffracted as it passes through the aperture of a lens; therefore each image that is formed by the lens is surrounded by a series of interference fringes. These fringes limit the resolving powers of instruments such as telescopes and microscopes. For example, if two objects have small angular separations at the objective lens, a telescope may form two images that have overlapping central bright fringes. These images cannot be distinguished from each other, and the objects are *not resolved*. Two objects are said to be *just resolved* by an optical instrument when the central bright maximum of one image coincides with the first dark minimum of the second image. In this case the separate images can be distinguished.

The minimum angular separation $\theta_{\min}$ at which two objects can be just resolved is directly proportional to the wavelength λ of the light and inversely proportional to the aperture D of the objective lens:

$$\theta_{\min} \propto \lambda/D \qquad (30\text{-}9)$$

For a circular aperture such as that found in telescopes and microscopes, the minimum angular resolution (in radians) is:

$$\theta_1 = 1.220 \frac{\lambda}{D} \qquad (30\text{-}10)$$

where D is the diameter of the aperture that limits the light (normally the objective). In these calculations white light (daylight) is usually assumed to have an effective wavelength of 560 nm.

If the aperture D of a telescope or microscope is increased, its resolving power improves because the minimum angular separation between just resolved objects is reduced. Similarly the shorter the wavelength of the light, the greater the resolving power of a microscope.

Example 30-3

What is the minimum aperture of the objective of a telescope that produces an angular resolution of 5.0″ of arc?

Solution:

$$\theta = 5.0'' = \left(\frac{5.0}{3\,600}\right)^{\circ} = \left(\frac{5.0}{3\,600}\right)\left(\frac{\pi}{180}\,\text{rad}\right)$$
$$= 2.42 \times 10^{-5}\,\text{rad}$$

Thus

$$D = \frac{1.22\lambda}{\theta} = \frac{(1.22)(5.6 \times 10^{-5}\,\text{cm})}{2.42 \times 10^{-5}\,\text{rad}} = 2.8\,\text{cm}$$

Problems

11. An interference pattern is formed on a screen when a parallel beam of monochromatic light with a wavelength of 500 nm is incident on a 0.25 nm single slit. If the slit is 50 cm from the screen, calculate the width of the central bright band.

12. Determine the minimum angle of resolution in seconds of arc of a small telescope on a theodolite if its diameter is 1.5 cm.

30-5 THIN FILMS

Fringes are also produced by the interference of the light that is reflected from the two surfaces of a transparent thin film. This effect gives rise to the colors of soap bubbles and oil films when they are viewed under white light.

Consider a parallel beam of monochromatic light that is incident from a medium 1, which has an index of refraction n_1, onto a thin transparent film, which has a thickness d and an index of refraction n_2 (Fig. 30-9). The light undergoes multiple partial reflections from both surfaces of the film. An incident ray AB is partially reflected (BC) and partially refracted (BD) at the top surface of

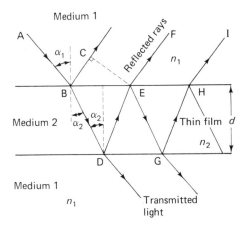

Figure 30-9
Multiple reflections from the surfaces of a thin film.

surfaces of the air film. The interference of this reflected light produces a fringe pattern. As light is reflected from the denser medium (the glass) at the bottom of the air wedge, it undergoes a 180° phase shift. Therefore there is a dark fringe where the two glass plates touch.

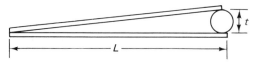

Figure 30-10
An air wedge.

the film. The refracted ray BD is then partially reflected by the bottom surface of the film, and the reflected ray DE is again partially transmitted (EF) and partially reflected (EG) at the top surface. As this process continues, the intensity of the *internally* reflected light decreases rapidly. The phase differences between the reflected rays BC, EF, HI, ... arise because they travel different paths to the wavefront and rays undergo a 180° phase shift when they are reflected from the surface of a denser medium. No phase change occurs when light is reflected from the less dense medium.

The Air Wedge

The principles of thin film interference are often used to obtain precise measurements of small dimensions, such as the diameter of a human hair or a sheet of paper. An *air wedge* is formed by locating the small object between the ends of two optically flat glass plates; the other ends of the plates are held in contact (Fig. 30-10). When the air wedge is irradiated with monochromatic light, some of the light is reflected at both the top and the bottom

If monochromatic light with a wavelength λ is incident perpendicular to the surface of the air wedge, bright fringes occur where the thickness of the air film:

$$t = (2N - 1)\frac{\lambda}{4} \quad \text{(bright fringes)} \quad (30\text{-}11)$$

The integer N represents the number of bright fringes between the dark fringe where the glass plates touch and the various points along the air wedge. For example, the first bright fringe (corresponding to $N = 1$) is produced when the wave that is reflected from the bottom surface travels a total distance of $\lambda/2$ in the air film. This wave experiences a 180° phase shift as it is reflected from the bottom surface; consequently it emerges in phase with the light that is reflected by the upper surface of the air film.

Dark fringes occur when the wave that is reflected from the bottom surface travels an integer number of wavelengths in the air film. At these locations, the thickness of the air film:

$$t = (N - 1)\frac{\lambda}{2} \quad \text{(dark fringes)} \quad (30\text{-}12)$$

Example 30-4

An air wedge is made by placing a fine wire between the ends of a pair of flat glass plates and clamping the other ends of the plates together. Twenty bright fringes can be seen between the clamped ends and the wire when the air wedge is viewed from above under monochromatic laser light. If the wavelength of the laser light is 633 nm, what is the diameter of the wire?

Solution:

$$t = (2N - 1)(\lambda/4) = \frac{[2(20) - 1](6.33 \times 10^{-7} \text{ m})}{4}$$

$$= 6.17 \times 10^{-6} \text{ m}$$

Interference fringes are also used to compare the dimensions of objects and to detect wear in the working components of machinery and deviations in flat surfaces. Two objects are located some distance apart between two optically flat glass plates, and the system is viewed under monochromatic light. No fringes are observed if the objects have the same thickness, but fringes are produced if one object is smaller than the other. The difference in size can be determined from the number of fringes.

Newton's Rings. An air wedge may also be formed between a plano-convex lens with a long focal length and a flat glass surface (Fig. 30-11). This system produces circular fringes, which are centered at the point of contact between the lens and the flat surface. These circular fringe patterns are called *Newton's rings*. They are used to determine the radius of curvature of the lens and to detect deformities in its curved surface.

Optical Coatings. In many optical instruments the lenses and prisms are coated with a thin transparent film of a material such as calcium or magnesium fluoride; these films tend to reduce reflections. At normal incidence there is destructive interference between

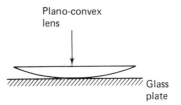

Plano-convex lens

Glass plate

Figure 30-11
Newton's rings.

the light that is reflected from the two surfaces of the film when the thickness of the coating is:

$$t = \frac{\lambda}{4n} \qquad (30\text{-}13)$$

where n is the index of refraction of the film. Note that the thickness t depends on the wavelength λ. Normally for white light we choose the thickness that produces destructive interference for a wavelength at the center of the visible spectrum.

The Michelson Interferometer

Interference fringes may be used to measure distances or changes in length very accurately by means of an *interferometer* (Fig. 30-12) developed by Michelson in 1881. The device consists of a mirror A that is coated so that it only reflects half of the incident light, a glass plate C and two perpendicular mirrors B and D. When light from a source is incident on mirror A, it is separated into two components. One part X is reflected to B, which in turn reflects the light back (along its original path) to A where a part is transmitted to the eye. The second component Y is transmitted through A to mirror D, which reflects it back to A and then to the eye. The glass plate is used to compensate for the extra distance that component X travels through a glass medium. When the optical paths AB and AD are equal, the two components arrive in phase and interfere constructively at the eye. However, if one distance is a quarter wavelength different from

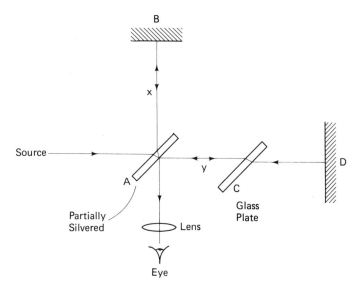

Figure 30-12
The Michelson interferometer.

the other, they produce destructive interference at the eye. Therefore, if one of the distances varies, a series of fringes moves past a reference, by counting the fringes the change in length can be accurately determined.

Problems

13. An air wedge is made by placing a sheet of paper between the ends of a pair of optically flat glass plates and clamping the other ends of the plates together. When the air wedge is viewed from above under monochromatic light with a wavelength of 436 nm, 19 bright fringes can be seen between the clamped ends and the sheet of paper. Calculate the thickness of the paper.

14. An air wedge is formed by placing a fiber between the ends of two flat glass plates and clamping the other ends of the plates together. When it is viewed from above under monochromatic blue light, the average distance between successive fringes is 0.18 cm. If the wavelength of the light is 460 nm and the plates are 5.0 cm long, determine (a) the diameter of the fiber and (b) the thickness of the air film at the fifth bright fringe.

15. Determine the thickness of an optical coating of a material that has an index of refraction of 1.36 if it is to reflect a minimum amount of light with a wavelength of 520 nm.

16. Describe three uses of a Michelson interferometer.

30-6 POLARIZATION

The characteristic properties of interference and diffraction clearly demonstrate the wavelike nature of electromagnetic radiation, but they do not indicate whether these waves are longitudinal or transverse. However, according to Maxwell's theory, electromagnetic waves must be transverse because the variations in their electric and magnetic fields are perpendicular to their velocity; this is verified by polarization.

The velocity and the variations of the electric and magnetic fields of an electromagnetic wave are always mutually perpendicular. At the wavefront the orientations of the field vibrations are restricted to the plane that is perpendicular to the direction in which the wave propagates. In most cases the variations of the electric and magnetic fields have completely random orientations in that plane; these electromagnetic waves are *unpolarized*. The orientations of the field vibrations of a *polarized* electromagnetic wave are not random: They form specific and consistent patterns.

By convention, the direction of the variations in the electric field is used to specify the type of polarization (Fig. 30-13). For

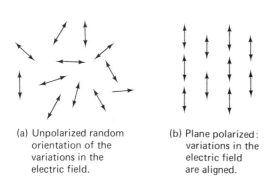

(a) Unpolarized random orientation of the variations in the electric field.

(b) Plane polarized: variations in the electric field are aligned.

Figure 30-13

example, in electromagnetic waves that are generated by accelerating charges in a dipole antenna, the variations in the electric field are always parallel to the antenna. These electromagnetic waves are called *plane* (or *linearly*) *polarized* because the electric field variations are always in a single plane that is defined by the antenna and the direction of propagation. Maximum reception occurs when the receiver antenna is aligned so that it is parallel to the variations in the electric field of the incident electromagnetic wave.

Electromagnetic waves may also have *elliptic* or *circular polarizations* in which the terminal point of the electric field strength vector traces an ellipse or a circle about the direction of propagation.

The light that is emitted by most luminescent bodies is unpolarized, but devices called *polarizers* may be used to obtain polarized light. There are several polarization processes:

1. *Reflection.* When an unpolarized beam of light is incident on the polished surface of a transparent material, such as glass, some light is reflected and some is refracted. At a particular angle of incidence α called the *polarizing angle* ($\sim 57°$ for glass), the reflected light is partially plane polarized.

2. *Scattering.* Scattering is more pronounced for the shorter wavelengths of light, and light that has been scattered by small particles is plane polarized. During the daytime the sky appears to be blue because we mainly see light that has been scattered in the atmosphere. At sunset the direct sunlight must pass through a thicker layer of air and dust before it reaches the earth; the sunset appears to be red because we see the light that has not been scattered.

3. *Double Refraction.* Crystals of most materials, such as calcite, quartz and ice, are anisotropic; their physical properties are not the same in all directions. When an unpolarized light ray is incident on one of these crystals, the refracted light separates in two. These two refracted rays are polarized perpendicular to each other. This is known as *double refraction.* One of the refracted rays is called the *ordinary ray*, and it always obeys Snell's laws. Normally the other refracted ray does not obey Snell's laws and is called the *extraordinary ray*.

4. *Selective Absorption.* Light rays undergo double refraction in other materials, such

as tourmaline, but these crystals absorb one component of the polarized light. This property of selective absorption is called *dichroism*. If unpolarized light is passed into a *dichroic crystal*, the light that emerges is polarized in one particular plane, and all of the other light is absorbed. Tourmaline crystals are colored; therefore they have limited application in polarizing devices.

In 1932 E. H. Land invented a new polarizing material called *polaroid*. If consists of thin dichroic crystals of iodosulphate that are embedded in a thin sheet of cellulose. Improved polaroid sheets are composed of aligned polymeric iodine molecules that are embedded in a polyvinyl alcohol film. The film is usually sandwiched between two plastic or glass plates for protection. During the manufacturing process, the film is stretched to align the organic chain molecules before it is treated with the iodine. *H-polaroid* is made by impregnating the stretched film with iodine, and the iodine atoms align along the parallel stretched chain molecules. The more stable *K-polaroid* is made by dehydrating the stretched film after it has been immersed in an iodine solution.

The naked human eye cannot distinguish between polarized and unpolarized light; therefore we normally use a polarizing device to analyze polarization. For example, when unpolarized light is incident on a polaroid polarizer, only light that is polarized in one particular plane is transmitted A second sheet of polaroid may be used as an *analyzer* to investigate the nature of the polarization. When the analyzer and polarizer are aligned to transmit light that is polarized in the same plane, the intensity of the transmitted light has some maximum value I_m (Fig. 30-14). If

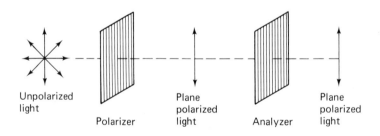

(a) Maximum intensity of polarized light is transmitted when the polarizer and analyzer are aligned.

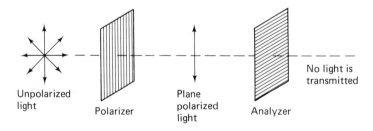

(b) No light is transmitted when the analyzer is aligned perpendicular to the polarizer.

Figure 30-14

the analyzer is now rotated through some angle θ with respect to the polarizer, the intensity of the transmitted light decreases to:

$$I = I_m \cos^2 \theta \qquad (30\text{-}14)$$

This is called *Malus' law*. No light is transmitted when the analyzer is aligned perpendicular to the polarizer ($\theta = 90°$).

Reflected light is partially polarized in a plane that is parallel to the reflecting surface. Polaroid sheets are used in some sunglasses, camera filters and windows to reduce glare by absorbing reflected sunlight.

Photoelasticity

Some transparent noncrystalline materials, such as glass and perspex, are normally isotropic, but they become doubly refracting when they are strained. This phenomenon is known as *photoelasticity;* it is used to detect strains in transparent devices and to analyze the strains in models of structures and machinery. A polarizer and an analyzer are usually arranged for extinction, i.e. so that they do not transmit light. The strained material is located between them (Fig. 30-14b). Plane polarized light from the polarizer is doubly refracted in the strained regions, and, since their speeds are different, the two refracted rays are not necessarily in phase when they emerge from the transparent material. The analyzer only transmits the components of the ordinary and extraordinary rays that are polarized in the same plane. The two plane-polarized rays produce an interference pattern. Note that there is no interference between light rays that have polarizations perpendicular to each other. A series of bright interference fringes can be seen if the strained material is viewed through the analyzer. The regions of greatest strain have the largest number of fringes.

Optical Rotation

When some substances, such as a sugar solution transmit plane polarized light, they rotate the plane of the polarization. The amount of rotation depends on the nature and thickness of the substance. The angle of rotation is measured by placing the substance between a polarizer and an analyzer; the device is called a *polarimeter*. In other special devices called *saccharimeters*, the principles of optical rotation are used to measure the concentration of sugar solutions. The angle of rotation that is produced in a fixed thickness of solution is proportional to the sugar content.

Liquid Crystal Displays (LCD)

1. A *dynamic-scattering LCD* consists of a *nematic liquid crystal* (rod-shaped molecules) between two glass plates that have a very thin coating of a metallic electrode on their inner surfaces. The front plate has a transparent electrode containing the message, e.g. a seven-segment display, and the rear plate has a continuous electrode that may be either silvered for a reflective type of display or transparent for a transmissive type of display (which requires a light source behind it). Normally the molecules of the nematic liquid are aligned and transmit light; however, when an electric field is applied to a section of the crystal, that section becomes opaque and scatters light. Thus numbers, letters, etc. can be produced by energizing different segments of the electrodes on the front face.

2. *Field-effect LCD's* are more popular than dynamic-scattering LCD's, and they consume less power. They consist of a helical-shaped nematic liquid crystal between two glass plates containing transparent metallic electrodes. The system is located between two polarizing filters (Fig. 30-15).

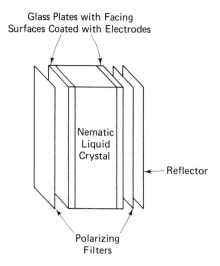

Glass Plates with Facing
Surfaces Coated with Electrodes

Nematic
Liquid
Crystal

Reflector

Polarizing
Filters

Figure 30-15
A field-effect LCD.

Normally, as polarized light passes through the liquid crystal, it rotates the plane of polarization by 90°, but the second polarizing filter is aligned to transmit the light. However, when the electrodes are energized, the liquid crystal untwists, and it no longer changes the plane of polarization; therefore the second polarizing filter does not transmit the light.

LCD's are now quite common. They are reliable, visible even in bright light and can be made quite large and with custom characters. Their low power consumption makes them ideal for battery-operated instruments and wrist watches. However they are relatively slow and usually will not operate below −10° C.

*Problems*_____

17. A maximum light intensity I_m passes through a polaroid when it is aligned in a vertical polarizing direction; the intensity I of the transmitted light is reduced when the polaroid is rotated. Determine the ratio I/I_m when the polaroid is rotated so that its polarizing direction is (a) 30°, (b) 50°, (c) 75° with respect to the vertical.

18. Could the properties of polarization be used to modulate laser light? Explain.

19. List three devices that employ a LCD.

REVIEW PROBLEMS

1. Monochromatic light is incident on two narrow slits 0.40 mm apart that are 120 cm from a screen. If the incident light frequency is 633 nm, determine the average distance between (a) the zero- and second-order bright fringes and (b) the zero-order bright fringe and the second-order dark fringe.

2. A Young's double slit is illuminated with monochromatic light from a single source of wavelength 600 nm. If the slit separation is 1.5×10^{-3} cm and they are 1.8 m from a screen, what is the separation between the third- and zero-order bright fringes?

3. A diffraction grating with 7 500 lines/cm is illuminated with monochromatic light of wavelength 500 nm. (a) At what angle does the second-order bright spot appear? (b) If the grating is 3.0 m from the screen, how far is it from the zero- to the second-order bright spot?

4. Monochromatic light is incident perpendicular to the surface of a diffraction grating containing 13 000 lines/in. If the angular deviation of the first-order image is 16.7°, calculate (a) the wavelength of the light and (b) the angular deviation of the second-order image.

5. A single slit 2.0×10^{-5} m wide is illuminated by light with a wavelength of 400 nm. If the slit is 5.0 m from the screen, (a) how far from the central bright spot does the second-order dark fringe appear? (b) At what distance from the central bright spot does the second-order bright fringe appear?

6. An air wedge is made by placing an aluminum foil between the ends of a pair of flat glass plates and clamping the other ends of the plates together. If 120 dark fringes are counted between the clamped ends and the foil when it is illuminated with monochromatic light of wavelength 500 nm, what is the thickness of the foil?

7. What thickness optical coating of a material with an index of refraction of 1.25 reflects a minimum amount of light of wavelength 580 nm?

8. Describe how you could use interference to measure large distances very accurately.

By the beginning of the twentieth century scientists had made many observations, such as atomic spectra and blackbody radiation, that could not be explained satisfactorily by the laws of classical physics. In their attempts to describe physical observations involving very small particles, such as electrons and photons, scientists formed macroscopic (large scale) models of the system, and they applied the laws of classical mechanics to their models. However they often found that, in spite of their imagination and inventiveness, they could not describe or predict the scientific observations from their macroscopic models!

There followed a most exciting period of scientific discovery. Quantum physics was conceived, the atom was probed, and its structure was successfully described. These discoveries have led to an enormous number of technological advances, ranging from nuclear power to the more recent developments of the transistor, integrated circuits and the laser.

31

Quantum Physics and the Atom

31-1 SPECTRA

All substances radiate electromagnetic waves when they are heated. These electromagnetic waves may be analyzed by means of a *spectrometer*, a device that utilizes a prism or diffraction grating in order to organize the radiation into an ordered array of wavelengths called a *spectrum*. There are five basic types of spectra.

Continuous Emission Spectra and Blackbody Radiation. The spectrum of the radiation emitted by a hot solid consists of a continuous band of wavelengths; the wavelength of the maximum intensity depends on the temperature, but the spectrum of all hot solids is similar. In 1900, Max Planck produced a satisfactory explanation of blackbody radiation by considering the atoms of the blackbody as harmonic oscillators that continually emit and absorb energy, but he had to make two assumptions:

1. The atomic oscillators can only have discrete energies or states given by:

$$E = nhf \qquad (31\text{-}1)$$

where f is the oscillator's natural frequency, $h = 6.6262 \times 10^{-34}$ J·s is called *Planck's constant* and n is a positive integer called a *quantum number*.

2. Atomic oscillators emit radiation in discrete energy packages called *quanta*. This energy is produced when the atomic oscillator changes from one energy state $n_2 hf$ to another $n_1 hf$, emitting quanta with energies corresponding to integer multiples of hf. Thus for quanta of emitted energy, $\Delta E = n_1 hf - n_2 hf = nhf$, where $n = n_1 - n_2$ is an integer.

Line Emission Spectra. When gas or vapor atoms (under low pressure) are excited, the system glows. The spectrum of the emitted light consists of a series of different-colored lines of various intensities separated by dark areas. Each chemical element has its own unique line emission spectrum; the chemical composition of a vaporized material may be determined by a comparison of its spectrum with known spectra.

Band Spectra. The spectrum of radiation emitted from molecules (composed of two or more atoms) consists of a series of regularly spaced lines in the form of bands, rather than the single lines that characterize the emission spectra of single atoms.

Continuous (Band) Absorption Spectra. When light from a hot solid (which produces a continuous emission spectrum) is passed through a liquid or solid, sections of the spectrum are absorbed. The resulting spectrum has large dark areas corresponding to the band of wavelengths that are absorbed by the liquid or solid. For example, a red glass filter only transmits red light, absorbing all other wavelengths.

Line Absorption Spectra. If light that gives a continuous emission spectrum is passed through a gas or vapor, certain discrete wavelengths are absorbed. The resulting spectrum consists of a series of dark lines (corresponding to the wavelengths absorbed) on a continuous background. Bright lines of the line emission spectrum and dark lines of

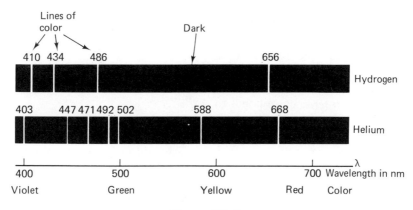

Figure 31-1
Line spectra.

the line absorption spectrum from the same vapor have exactly corresponding wavelengths.

The emission spectrum of the sun shows a series of dark lines on a continuous background; these lines are called *Fraunhofer lines*. The sun itself emits a continuous spectrum, but the Fraunhofer lines are produced as a result of the absorption of certain wavelengths by a layer of cooler gases surrounding the sun.

The Hydrogen Spectrum. Atomic hydrogen has the simplest spectrum, consisting of a series of regularly spaced lines. In 1884 Balmer produced a mathematical relationship that could be used to determine the wavelengths of *some* of the lines in the hydrogen spectrum. He found that the wavelengths of *some* lines could be represented by the equation:

$$\lambda = \frac{3.645 \times 10^{-7} n^2}{n^2 - 4} \text{ meters} \qquad (31\text{-}2)$$

where n may assume integer values greater than 2.

Many other attempts to predict the remaining lines followed. In general all of these predictions can be summarized by the single equation:

$$\frac{1}{\lambda} = R\left(\frac{1}{m^2} - \frac{1}{n^2}\right) \qquad (31\text{-}3)$$

where $R = 1.097\ 373\ 1 \times 10^7/\text{m}$ and is called the *Rydberg constant*, m and n are integers with n greater than m. The *Balmer series* corresponds to $m = 2$ and $n = 3, 4, 5 \ldots$; the *Lyman series* corresponds to $m = 1$, $n = 2, 3, 4 \ldots$; the *Paschen series* corresponds to $m = 3$, $n = 4, 5, 6 \ldots$ etc.

*Problems*_____

1. Determine the wavelengths of the first three spectra lines of atomic hydrogen (a) in the Balmer series, (b) in the Paschen series, (c) in the Lyman series.

2. What kind of spectra is produced by the light from (a) tungsten filament incandescent lamps? (b) Fluorescent lamps? (c) Carbon arc lamps? (d) Sodium vapor lamps?

31-2 THE PHOTOELECTRIC EFFECT

By the end of the nineteenth century, scientists had observed that some surfaces emitted electrons when they were irradiated with light. This phenomenon is called the *photoelectric effect*, and the emitted electrons are called *photoelectrons*.

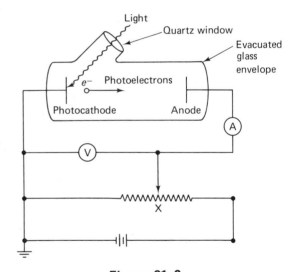

Figure 31-2

Apparatus to measure the photoelectric effect.

Consider the apparatus in Fig. 31-2. Light passes through a quartz window and irradiates the photocathode, causing the emission of photoelectrons, which are accelerated (by the accelerating potential V) towards the anode. A quartz window is used because quartz transmits ultraviolet light as well as visible light. It is found that:

1. If the frequency f of the incident light and the accelerating potential V are kept constant, the number of photoelectrons emitted by the photocathode (the photoelectric current) is directly proportional to the intensity I of the incident light. This result is expected because the intensity of the incident light is directly proportional to its time rate of energy flow, and light with a higher rate of energy flow is able to free more photoelectrons per unit time.

2. Photoelectrons are ejected from a particular photocathode material only if the incident light has a frequency greater than some limiting value called the threshold frequency f_0. If the incident light has a frequency less than the threshold frequency of the photocathode material, no photoelectrons are emitted regardless of the incident light intensity! This threshold frequency is a characteristic of the *photocathode material*, and it varies for different substances. The existence of a threshold frequency cannot be explained in terms of the classical wave theory for light.

Einstein's Theory. Max Planck had proposed the existence of small energy packets, or quanta, in the neighborhood of an emitter of radiation. In 1905 Albert Einstein modified this theory by suggesting that, instead of propagating as a wave, light energy is confined to small energy packets called *photons* that retain their identity. A single photon has an energy $E = hf$, where h is Planck's constant and f is the frequency of the light.

The photoelectrons are bound to the atoms in the photocathode and a minimum energy, which is called the *photoelectric work function* or *threshold energy* ϕ, is required to free them. Photoelectric work functions usually have values between 1.6×10^{-19} J and 10^{-18} J. If the incident light energy were wave-

like, it would continually impart energy to the photocathode, and photoelectrons would eventually be freed, regardless of the incident light frequency. However, since light energy exists in quanta, it exhibits particle-like characteristics; in order to free a photoelectron from the photocathode, the individual photons must have energy equal to or greater than the photoelectric work function ϕ. Thus the incident light must have a frequency greater than or equal to the threshold frequency f_0 where $\phi = hf_0$.

3. In accordance with the conservation of energy, when a photoelectron is emitted, it has a maximum kinetic energy E_{kmax} equal to the difference between the incident photon energy $E = hf$ and the photoelectric work function $\phi = hf_0$ of the surface. Thus

$$E_{kmax} = hf - \phi = h(f - f_0) \quad (31\text{-}4)$$

If the incident light frequency f is constant, *the maximum kinetic energy of the photoelectrons that are ejected from a particular photocathode material is independent of the incident light intensity.* A more intense light of the same frequency merely produces more photoelectrons but with the same kinetic energies.

The maximum kinetic energy of the ejected photoelectrons may be determined by reversing the polarity of the accelerating potential V so that it retards the photoelectrons. The retarding potential that is just sufficient to stop the most energetic photoelectrons is called the *stopping potential* V_s. From the definition of electric potential, the work done in stopping the photoelectrons is:

$$W = E_{kmax} = V_s e$$

where e is the charge. Therefore

$$V_s e = E_{kmax} = hf - \phi = h(f - f_0) \quad (31\text{-}5)$$

If c is the speed and λ is the wavelength

of the light, then $c = f\lambda$; thus, Eq. 30-9 may be written as:

$$V_s e = E_{kmax} = \frac{hc}{\lambda} - \phi = hc\left(\frac{1}{\lambda} - \frac{1}{\lambda_0}\right) \quad (31\text{-}6)$$

where $\lambda_0 = c/f_0$ is called the *threshold wavelength* of the *photocathode*. Photoelectrons are not ejected by the photocathode when the irradiating light has a wavelength *greater* than the threshold wavelength.

Energies of particles such as electrons are often expressed in terms of electron volts (eV), which is a special energy unit permitted for use with SI. One *electron volt* is the energy acquired by an electron as it accelerates in a vacuum through a potential difference of one volt (1 eV $\times$ 1.602 $\times 10^{-19}$ J). Thus the maximum kinetic energy in electron volts is numerically equal to the stopping potential.

Equation 31-5 is the equation of a straight line with a slope equal to Planck's constant h. The intercept on the frequency axis corresponds to the threshold frequency f_0, and the intercept of the extended line with the maximum kinetic energy axis is equal to the work function ϕ of the photocathode material (Fig. 31-3).

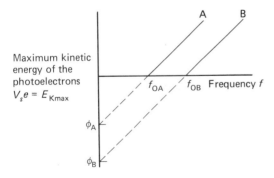

Figure 31-3

Relationship between the maximum kinetic energy of photoelectrons and the incident light frequency for two different photocathode materials A and B.

4. Most photoelectrons originate from points within 10^{-6} m of the photocathode surface; therefore the photoelectric effect depends on the nature of that surface. *Photoelectric yield* or efficacy of a surface is defined as the photoelectric current (amperes) per unit luminous flux (lumens) of the irradiation.

Example 31-1

When monochromatic light with a wavelength of 430 nm irradiates a photocathode, a minimum stopping potential of -0.65 V is required to stop the photoelectric current. Determine (a) the photoelectric work function of the photocathode and (b) the threshold wavelength.

Solution:

(a) $c = 3.0 \times 10^8$ m/s, and for an electron $m = 9.1 \times 10^{-31}$ kg and $e = -1.6 \times 10^{-19}$C. Thus from Eq. 31-6:

$$\phi = \frac{hc}{\lambda} - V_s e$$

$$= \frac{(6.63 \times 10^{-34} \text{ J·s})(3.0 \times 10^8 \text{ m/s})}{4.3 \times 10^{-7} \text{ m}}$$

$$- (-0.65 \text{ V})(-1.6 \times 10^{-19} \text{ C})$$

$$= 3.6 \times 10^{-19} \text{ J}$$

(b) $$\phi = \frac{hc}{\lambda_0},$$

thus

$$\lambda_0 = \frac{hc}{\phi}$$

$$= \frac{(6.63 \times 10^{-34} \text{ J·s})(3.0 \times 10^8 \text{ m/s})}{3.6 \times 10^{-19} \text{ J}}$$

$$= 5.5 \times 10^{-7} \text{ m}$$

Photoelectric cells are constructed by coating the cathode of a small vacuum tube with a thin layer of photoelectric material. Photoelectric currents are directly proportional to the intensity of the irradiating light; therefore photoelectric cells are often used to detect variations in light intensity.

Problems_____

3. Determine the number of photons emitted per second by a laser that operates at 2.5 mW and emits monochromatic light with a wavelength of 632.8 nm.

4. If the photoelectric work function of a material is 3.53 eV, determine (a) the threshold frequency and (b) the threshold wavelength.

5. If the threshold wavelength of a surface is 350 nm, determine (a) the photoelectric work function and (b) the maximum kinetic energy of the ejected photoelectrons when the surface is irradiated with monochromatic light of wavelength 250 nm.

6. When the monochromatic light with a frequency of 8.00×10^{14} Hz irradiates a photocathode, photoelectrons are ejected with a maximum speed of 6.25×10^5 m/s. Determine (a) the threshold frequency of the surface, (b) the photoelectric work function of the surface, (c) the threshold wavelength of the surface, (d) the stopping potential for the photoelectrons.

7. When a potassium photocathode is irradiated with monochromatic light, a minimum stopping potential of -1.2 V is required to bring all the photoelectrons to rest. Determine the wavelength of the irradiating light if the photoelectric work function for potassium is 2.24 eV.

8. A particular photocathode was irradiated, and the stopping potentials were determined for a series of different monochromatic lights:

Plot a graph of stopping potential versus the inverse wavelength of the irradiating light, and from this graph determine (a) Planck's constant, (b) the threshold frequency and the threshold wavelength, (c) the work function of the photocathode.

31-3 THE RUTHERFORD ATOM

All substances are composed of one or more pure basic substances that cannot be chemically decomposed; these pure substances are called *elements*. While there are in excess of one hundred known elements (some of which no longer exist naturally), they rarely occur in the pure state but are usually mixed or chemically combined with other elements. In a mixture of two or more elements, each element retains its own chemical and physical properties. For example, air is mainly a mixture of the elements oxygen and nitrogen although other gases are also present.

A chemical combination of two or more elements results in a substance called a *compound* that behaves quite differently from its constituent elements. For example, table salt is a chemical combination of the very reactive metal sodium and the poisonous gas chlorine, yet table salt is a very important, edible compound.

The smallest particles of an element that can combine chemically are called *atoms*. They are not visible even under the most powerful microscope. When two or more atoms combine chemically, they form *molecules* that are the smallest particles of compounds.

Wavelength of irradiating light/nm	320	400	480	540
Stopping potential/V	-1.58	-0.81	-0.20	-0.002

Since 1951, by international agreement the relative masses of atoms have been compared with the most abundant neutral isotope of carbon-12, which is taken as exactly twelve *unified atomic mass units* (symbol u). Thus 1 u is 1/12 the mass of a neutral atom of carbon 12.

In the late nineteenth century a vast amount of evidence had been accumulated to prove the existence of atoms, but the structure of the atom itself was not known. In 1911 Ernest Rutherford proposed a model in order to explain new experimental observations.

Rutherford's graduate students Hans Geiger and Ernest Marsden had investigated the effects of bombarding a thin metal foil with fast alpha particles. Alpha particles are positively charged helium atoms which have lost all of their electrons; they are emitted from some radioactive substances, such as radium (Chapter 34). Geiger and Marsden formed the fast alpha particles into a single unidirectional stream by passing them through a lead collimator. The stream of alpha particles were scattered when they impinged on thin metal foils, and the scattered particles were detected by a movable fluorescent zinc sulphide screen (Fig. 31-4). In order to minimize the absorption of alpha particles, the whole system was located in an evacuated chamber. When the experiment was performed, it was found that most alpha particles experienced little deflection, but some were substantially deflected, and a few were even deflected through more than 90°!

Rutherford performed a series of scattering calculations in order to find a model that would correctly explain the results of this experiment. He concluded that the atom must consist of a small, very dense, positively charged nucleus surrounded by a "cloud" of small, light, negatively charged orbiting electrons. The total atom is usually electrically neutral, and most of its volume is composed of empty space. Electrons must be in continual motion about the nucleus; otherwise they would be attracted towards the nucleus because of the electrostatic force between unlike charges. The electrostatic force merely provides the centripetal force that is necessary to keep the electrons orbiting the nucleus.

This result is in direct conflict with the classical theory of electromagnetism, which predicts that accelerating charges, such as the orbiting electrons, emit electromagnetic radiation. According to the classical theory, as the electron loses energy, it slows down and gradually spirals into the nucleus. This does not occur.

Rutherford's atomic model successfully explains the results of the scattering experi-

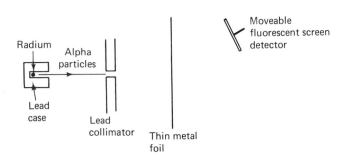

Figure 31-4
Geiger and Marsden apparatus.

ments, but it does not explain why the orbiting electrons do not emit electromagnetic radiation.

31-4 THE BOHR ATOM

Many scientists attempted to explain the line spectra of the elements, but they were only partially successful until 1913 when Niels Bohr developed a theory that satisfactorily explained the line emission spectrum of gaseous hydrogen. It should be noted that, while the Bohr atomic model is useful and instructive, the theory has been superseded by wave mechanics.

Bohr made use of the basic Rutherford model by assuming that the electrons were in circular orbits about the small, dense, positively charged nucleus, but he also made important two postulates.

Postulate 1. An electron can only exist in orbits where its angular momentum b is an integer multiple of $h/2\pi$.

If the electron has a mass m and a tangential speed v as it orbits the nucleus at a distance r, its angular momentum is:

$$b = mvr = \frac{nh}{2\pi} \qquad (31\text{-}7)$$

where n is an integer called the *principal quantum number* and h is Planck's constant. If this condition is not met, the electron cannot exist in the orbit, and it radiates energy (losing angular momentum) until it reaches a stable state.

Consider a single electron that is in a circular orbit of radius r about an atomic nucleus containing Z protons. If the orbit is stable, Bohr's first postulate implies that the tangential speed of the orbital electron* is:

---

*The nucleus actually moves, but the effect of its motion is not large.

$$v = \frac{nh}{2\pi mr} \qquad (31\text{-}8)$$

and its total mechanical energy is:

$$E_n = \frac{-Z^2 e^4 m}{8\epsilon_0^2 h^2}\left(\frac{1}{n^2}\right) \qquad (31\text{-}9)$$

where ϵ_0 is the permittivity of free space, e is the charge and m is the mass of an electron. Note that the orbital energy is a function of the principal quantum number n, which may only have positive integer values. The different energies corresponding to different values of n are called *energy levels*. The lowest possible energy level (which occurs when $n = 1$) is called the *ground state*, and the higher energy levels are called *excited states*. Systems normally attain the lowest energy configuration when they are stable; therefore the electrons usually exist in their lowest possible energy states in an atom.

In a hydrogen atom a single electron orbits a nucleus consisting of a single proton. Thus $Z = 1$ and

$$E_n = \frac{-e^4 m}{8\epsilon_0^2 h^2}\left(\frac{1}{n^2}\right) =$$

$$-\frac{(1.6 \times 10^{-19}\text{C})^4(9.1 \times 10^{-31}\text{ kg})}{8(8.85 \times 10^{-12}\text{ F/m})^2(6.63 \times 10^{-34}\text{ J} \cdot \text{s})^2}\left(\frac{1}{n^2}\right)$$

or

$$E_n = -\frac{2.17 \times 10^{-18}\text{ J}}{n^2} = -\frac{13.6\text{ eV}}{n^2} \qquad (31\text{-}10)$$

By substituting various integer values for n in this equation, we can determine the values of the stable energy levels of the electron in a hydrogen atom. These values are often arrayed in terms of an *energy level diagram* (Fig. 31-5).

Postulate 2. When it is in a stable orbit (conforming to the first postulate), an electron can orbit the nucleus indefinitely without radiating energy. If the electron changes from one stable orbit to any other, it loses or gains energy in discrete amounts (quanta) equal to the energy difference between the two energy states.

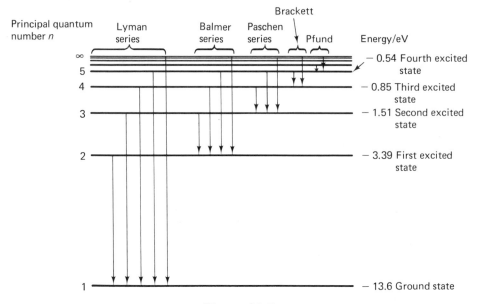

Figure 31-5
Energy level diagram for a hydrogen atom.

As an electron changes its energy from some higher initial energy state E_i to some lower final energy state E_f, it emits a photon with a precisely defined frequency f and wavelength λ according to:

$$hf = \frac{hc}{\lambda} = E_i - E_f \qquad (31\text{-}11)$$

Thus

$$f = \frac{E_i - E_f}{h} = \frac{Z^2 e^4 m}{8\epsilon_0^2 h^3}\left(\frac{1}{n_f^2} - \frac{1}{n_i^2}\right)$$

$$= Z^2 cR\left(\frac{1}{n_f^2} - \frac{1}{n_i^2}\right) \qquad (31\text{-}12)$$

where $R = \dfrac{e^4 m}{8\epsilon_0^2 h^3 c} = 1.097 \times 10^7/\text{m}$ is the Rydberg constant.

Also $\qquad \lambda = \dfrac{c}{f} = \dfrac{ch}{E_i - E_f} \qquad (31\text{-}13)$

Note that, in moving from some higher energy state to a lower energy state, the electron may pass through several intermediate states, emitting the corresponding photon for each transition that it makes. Transitions between electron energy levels are statistical processes. The frequencies and wavelengths of the photons that are emitted when the electron changes its energy level correspond to the experimental line spectra obtained for hydrogen. Even though the Bohr model successfully predicts the wavelengths for the hydrogen spectrum, it is not accurate for other atoms.

Example 31-2

Determine the wavelengths of the photons that are emitted from hydrogen atoms when electrons move from the second excited state ($n = 3$) to the ground state ($n = 1$).

Solution:

There are actually two possibilities, giving rise to photons with three different wavelengths:

1. The electron may make a direct transition between the second excited state and the ground state. From Eq. 31-10, the energies of these states are $E_3 = -\dfrac{13.6\text{ eV}}{(3)^2} = -1.51$ eV and $E_1 = -13.6\text{ eV}$. Thus in Eq. 31-13:

$$\lambda_{31} = \frac{ch}{E_3 - E_1} = \frac{(3.00 \times 10^8 \text{ m/s})(6.63 \times 10^{-34} \text{ J·s})}{[(-1.51 \text{ eV}) - (-13.6 \text{ eV})](1.6 \times 10^{-19} \text{ J/eV})} = 1.03 \times 10^{-7} \text{ m}$$

2. The electron may move to an intermediate state (the first excited state with an energy $E_2 = -\dfrac{13.6 \text{ eV}}{(2)^2} = -3.4 \text{ eV}$, emitting a photon with a wavelength:

$$\lambda_{32} = \frac{ch}{E_3 - E_2} = \frac{(3.00 \times 10^8 \text{ m/s})(6.63 \times 10^{-34} \text{ J·s})}{[(-1.51 \text{ eV}) - (-3.4 \text{ eV})](1.6 \times 10^{-19} \text{ J/eV})} = 6.6 \times 10^{-7} \text{ m}$$

It will then move to the ground state, emitting a second photon with a wavelength:

$$\lambda_{21} = \frac{ch}{E_2 - E_1} = \frac{(3.00 \times 10^8 \text{ m/s})(6.63 \times 10^{-34} \text{ J·s})}{[(-3.4 \text{ eV}) - (-13.6 \text{ eV})](1.6 \times 10^{-19} \text{ J/eV})} = 1.2 \times 10^{-7} \text{ m}$$

An electron may move into an excited state when it is subjected to an external energy source, such as heat, light or electric fields, or even by interaction with some other particle. The energy that is absorbed by the electron as it makes the transition between energy levels is exactly equal to the energy difference between its final and initial states. Electrons often return to their ground state by eventually emitting a photon. In a hydrogen atom for example, when an electron in the ground state absorbs $(13.6 - 3.39)$ eV of energy, it is excited into the first excited state. It may then return to the ground state by emitting a photon with exactly the same energy.

If it absorbs energy in excess of the negative ground state energy, an electron is completely removed from the atom, and the atom is said to be *ionized*. The minimum energy required to remove an electron from an atom is called the *ionization energy*, and it is numerically equal to the energy of the ground state. Thus for hydrogen the ionization energy is 13.6 eV. Any energy in excess of the ionization energy is absorbed by the electron as kinetic energy.

Luminescence and Phosphorescence. When a material absorbs energy, some electrons are excited into higher energy states. These electrons return to their ground state by the spontaneous emission of one or more photons. If the time delay between energy absorption and photon emission is less than 10^{-8} s, the material is called *fluorescent*. If the time delay is greater than 10^{-8} s, it is called *phosphorescent*.

Many semiconductor materials, such as zinc sulphide and zinc oxide, act as phosphors. They are utilized in fluorescent screens, such as the coating of a television screen, and in fluorescent lamps.

Problems

9. Determine the frequencies and wavelengths of the photons that are emitted from a hydrogen atom when its electron makes the following transitions between energy states: (a) from the third excited state to the ground state, (b) from the fourth excited state to the ground state, (c) from the fifth excited state to the ground state, (d) from the fourth excited state to the first excited state, (e) from the fifth excited state to the first excited state, (f) from the fifth excited state to the second excited state.

10. A sample of gaseous hydrogen in the ground state is subjected to external energy of 12.6 eV. Determine the wavelengths of the photons emitted by the gas.

11. A free electron moves with a speed of 3.6×10^6 m/s when it joins with a stationary proton to form a hydrogen atom

in its ground state. If a single photon was emitted, determine its wavelength.

12. Compute the energy of a photon if it has a wavelength of 555 nm.

13. Determine the minimum frequency of an incident photon that can ionize a hydrogen atom that originally was in the third excited state.

31-5 QUANTUM NUMBERS

The theory proposed by Bohr successfully predicts the spectral lines of hydrogen isotopes and hydrogen-like atoms containing a single orbital electron, but it does not explain the relative intensities of these lines. Bohr's theory is also inadequate for more complex atoms containing two or more orbital electrons because substantial interactions occur, not only between the nucleus and each orbital electron, but also between the orbital electrons themselves. Finally, when a material is subjected to magnetic or electric fields, its spectral lines are finely split into a number of lines.

In his theory Bohr assumed that the orbiting electron performed pure circular motion about the nucleus. The *principal quantum number n* (which takes positive integer values) quantizes the relative separation between the electron and the nucleus. If it possesses a certain energy, the angular momentum of an object is at a maximum when it performs pure circular motion. Any deviation from circular motion reduces its angular momentum. The angular momentum of an orbital electron is also quantized and is described in terms of another quantum number called the *orbital quantum number l*. For an orbit that is characterized by the principal quantum number n, it is found that the orbital quantum number l can only have positive integer values of $[l = 0, 1, 2, \ldots (n - 1)]$.

Each orbital electron also has a magnetic

moment, which is antiparallel to its orbital angular momentum **b**. The orientation of the angular momentum vector of the orbiting electron with respect to an applied external magnetic field strength **H** is also quantized and is characterized by a *magnetic quantum number m*, which can only have *integer* values (including zero) from $-l$ to $+l$.

In 1925 Goudsmit and Uhlenbeck observed that further fine splitting of the energy levels could be explained in terms of a spinning electron. They proposed that the electron spins about its own axis as it orbits the nucleus. Therefore the spinning electron possesses angular momentum and, because of its negative charge, also has a magnetic moment. This electron spin is quantized and has a value of $\frac{1}{2}(h/2\pi)$. The spin magnetic moment aligns itself either parallel or antiparallel to an external magnetic field intensity. Therefore the *spin quantum number s* can only have values of $+\frac{1}{2}$ or $-\frac{1}{2}$.

The Pauli Exclusion Principle

In all atoms, as the electrons orbit the nucleus, their energy state is characterized by four quantum numbers, but how are the electrons distributed among the available states? After analysis of many experimental observations, W. Pauli in 1925 generalized the results in terms of an exclusion principle. The *Pauli exclusion principle* states that:

No two electrons of the same atom can have identical values for all four quantum numbers: At least one quantum number must be different.

Therefore in the ground state the electrons occupy the lowest possible energy levels subject to the Pauli exclusion principle. For example, lithium has three orbital electrons and in the ground state these electrons have quantum numbers: $n = 1, l = 0, m = 0, s = -\frac{1}{2}$ (the lowest energy state); $n = 1, l = 0, m = 0, s = +\frac{1}{2}$; and $n = 2, l = 0, m = 0, s = -\frac{1}{2}$.

The Pauli exclusion principle also has important consequences when atoms unite to form molecules or solids because it then applies to the whole system. This results in "energy bands," which are discussed in Chapter 32.

According to the Bohr model an atomic nucleus is surrounded by a number of electrons, each of which has a definite orbit and energy. The various orbits are characterized by the principal quantum number n. This concept is statistically modified in quantum mechanics because the electron orbits are not precisely located; however the most probable orbits may be determined.

In an atom, an *electron shell* is defined as the group of all possible electronic orbits with the same principal quantum number. The Pauli exclusion principle limits the number of electrons in each shell. Therefore in a particular shell all electrons have the same principal quantum number n, but they differ in at least one of the other three quantum numbers. The maximum number of electrons that may occupy an electron shell that is characterized by the principal quantum number n is given by $2n^2$. Electron shells are labelled K, L, M, ... corresponding to principal quantum numbers of 1, 2, 3, ..., respectively.

Electron shells are divided into *subshells*. In a particular subshell all electrons have the same orbital quantum number l as well as the same principal quantum number n. Subshells are labelled s, p, d and f, corresponding to orbital quantum numbers of 0, 1, 2 and 3, respectively. For example, an electron with quantum numbers $n = 2$ and $l = 1$ is located in electron shell L and subshell p.

In atoms containing three or more orbital electrons, the *outer shell electrons*, which have the largest principal quantum number, are located farthest from the nucleus. These outer shell electrons are shielded from the nucleus by the full or closed inner shells of electrons, and less energy is required to remove them from the atom.

Problems

14. List all the possible orbital, magnetic and spin quantum numbers of an electron if its principal quantum number (a) $n = 2$ and (b) $n = 3$.

15. In which shells and subshells are the electrons of an atom if their quantum numbers are: (a) $n = 1$, $l = 0$, (b) $n = 3$, $l = 2$, (c) $n = 4$, $l = 1$, (d) $n = 4$, $l = 0$.

16. Determine the maximum number of orbital electrons that may occupy the following electronic shells: (a) K shell, (b) L shell, (c) M shell.

31-6 THE PERIODIC TABLE

One of the most significant discoveries in science was made independently by Meyer and Mendeleév in 1869. They found that, if the elements were listed in order of increasing atomic mass, elements with similar chemical and physical properties occur at regular intervals; this is known as the *periodic* law. They constructed a *periodic table* by listing the elements in order of increasing atomic mass, locating elements with similar properties in vertical columns (see Appendix 2).

The relative atomic mass of an element may vary because different isotopes of that element contain different numbers of neutrons. In a *modern periodic table* the elements (characterized by the number of protons in their nucleus) are arranged in order of increasing atomic number Z. Elements with similar properties are listed in vertical columns called groups. The elements in group I (with the exception of hydrogen H) are quite active metallic substances and are easily ionized; they are called *alkali metals*. Group II elements, called the *alkaline earths*, are not as active. Under normal conditions the elements of group VII, the *halides*, are chemically very active gases. The elements of group VIII, the *inert gases*, are chemically very inactive; they were not discovered until after Meyer and Mendeleév had proposed the original periodic table. Rows of elements beginning at group I (an alkali metal) and ending at group VIII (an inert gas) are called *periods*.

The periodic table may be analyzed in terms of the occupation of the electron shells and subshells and the chemical activities of the elements. In a stable configuration the total energy of the orbital electrons is at a minimum; therefore electrons tend to occupy the shells and subshells corresponding to the lowest energy states. Electrons that are located in incomplete outer shells are mainly responsible for chemical activity; these electrons are called *valence electrons*. Elements of the same group in the periodic table are physically and chemically similar because they have the same number of valence electrons. Inert gases are very stable and inactive because their outer shell is complete, and they have no valence electrons. The neutral inert gas helium (He) has two orbital electrons; therefore the first electron shell (the K shell) may contain a maximum of two electrons with different spin quantum numbers.

Lithium (Li) has three orbital electrons: Two complete the K shell and the third is located farther from the nucleus in the s subshell ($l = 0$) of the L shell ($n = 2$). This outer shell, or valence electron, is easier to remove from the atom, and as a result lithium is a relatively reactive element.

Beryllium has four orbital electrons: Two complete the K shell, and two with opposite spins complete the s subshell of the L shell. Beryllium has an extra nuclear proton, and the coulomb attraction between the nucleus and the outer shell electrons is therefore greater than it is for lithium. Consequently more energy is required to ionize beryllium, and it is not as active.

Boron has a complete K shell and a complete s subshell of the L shell. The fifth electron is located in the p subshell ($l = 1$) of the L shell. The outer shell electrons are even more tightly bound because of the even larger nuclear charge.

Fluorine has a complete K shell and requires only one more electron in order to complete its outer L shell. The valence electrons are tightly bound because of the relatively large nuclear charge. Fluorine is quite chemically active because it readily accepts an electron from another atom in order to complete its L shell and to reach the stable electron configuration of neon.

The periodic table is not only useful because it groups elements with similar physical and chemical properties, but it also indicates the electron configurations of the elements.

Problems

17. Determine how many valence electrons there are in neutral atoms of (a) silicon, (b) aluminum, (c) xenon, (d) chlorine, (e) boron, (f) phosphorus.

18. Why do copper, silver and gold have similar electronic properties?

19. Why do silicon and germanium have similar electronic properties? Describe some properties of these elements.

31-7 MATTER WAVES

Interference and diffraction experiments have proved conclusively that light possesses wavelike characteristics, but quantum theory indicates that light also behaves like a particle. These "particles of light," called photons, possess both momentum and a discrete energy.

Louis de Broglie in 1925 extended this duality concept to all matter. He proposed that all matter possesses both particle and wave-like characteristics; these matter waves are now called *de Broglie waves*. De Broglie assumed that the relationship between the particle-like characteristic (the momentum p)

and the wave-like characteristic (the wavelength λ) of matter waves was the same as that for photons:

$$p = \frac{h}{\lambda} \qquad (31\text{-}14)$$

where h = Planck's constant. In a vacuum all electromagnetic waves travel with the same speed (the speed of light), but de Broglie waves move with the speed of the corresponding particle. Matter waves do not disperse like other waves, rather they continue in a specific direction. Thus, if a particle of mass m moves with a speed v, its momentum $\mathbf{p} = m\mathbf{v}$, and its de Broglie wavelength is:

$$\lambda = \frac{h}{p} = \frac{h}{mv} \qquad (31\text{-}15)$$

*Example 31-3*_____

Determine the de Broglie wavelength of an electron that moves with a speed of 3.0×10^6 m/s.

$$\lambda = \frac{h}{mv}$$

$$= \frac{6.63 \times 10^{-34} \text{ J} \cdot \text{s}}{(9.1 \times 10^{-31} \text{ kg})(3.0 \times 10^6 \text{ m/s})}$$

$$= 2.4 \times 10^{-10} \text{ m}$$

If a particle moves at a speed v comparable to the speed of light $c = 3.0 \times 10^8$ m/s, Eq. 31-15 must be modified to account for relativity by replacing the *rest mass* m_0 with relativistic mass:

$$m = \frac{m_0}{\sqrt{1 - v^2/c^2}}$$

De Broglie's assumptions were verified by C. Davisson and L. Germer when they diffracted a beam of electrons from the atomic planes in a crystal. Their calculations of the electron's wavelength corresponded to the de Broglie wavelength. Other matter waves, such as protons and neutrons, may also be diffracted from the atomic planes of a crystal.

Wave Theory_____

Heisenberg and Schrödinger in 1926 independently developed different approaches to quantum mechanics by assuming the existence of matter waves. According to Heisenberg, the wave particle duality of matter leads to a certain imprecision in scientific measurements. Schematically matter waves may be represented as a particle that is located somewhere within a de Broglie wave (Fig. 31-6). According to de Broglie's hypothesis (Eq. 31-15), if the de Broglie wavelength is large, the momentum is small and vice versa.

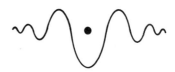

Figure 31-6
Schematic representation of a matter wave.

Some measuring process is required in order to determine the position or momentum of an object. For example, this book must be illuminated with light before it can be seen. Since light possesses momentum, as the photons continually strike the book, it recoils slightly, but the book is much more massive than the incident photons; therefore the recoil is extremely small.

When photons or some other particles are used to determine the position of a small particle, such as an electron, the recoil may be quite substantial. The measurement process disturbs the particle, and its new position and momentum are still unknown.

Quantum mechanics is founded on the concepts of statistics and most probable conditions rather than precise measurements and predictions. A limit for the accuracy of experimental observations was proposed by Heisenberg in 1927. *Heisenberg's uncertainty principle* specifies that:

$$(\Delta x)(\Delta p) \geq h/2\pi \qquad (31\text{-}16)$$

where Δx and Δp are the uncertainties in position and momentum, respectively, and h is Planck's constant.

The wave theory for the hydrogen atom assumes that the electron is a matter wave and produces a standing wave as it orbits the nucleus. An electron is assumed to be in a stable orbit without radiating energy only when the length of its orbital path is an integer multiple of de Broglie wavelengths. Thus, if an electron has mass m, speed v and de Broglie wavelength λ, it is in a stable circular orbit of radius r when the circumference of the orbit is:

$$2\pi r = n\lambda = \frac{nh}{mv} \qquad (31\text{-}17)$$

where n is a positive integer and h is Planck's constant (Fig. 31-7).

If the electron is not in a stable orbit, the standing wave interferes destructively, and the electron loses energy until a stable orbit is achieved. Note that Eq. 31-17 corresponds to Bohr's first postulate, Eq. 31-7.

*Problems*_____

20. Determine the de Broglie wavelength of an electron that has a speed of 5.0×10^6 m/s.

21. Determine the de Broglie wavelength of an electron that has a kinetic energy of 4 000 eV.

22. It has been shown that electrons do not exist in an atomic nucleus. Determine the uncertainty that would exist in the momentum of an electron if its uncertainty in position is 2.5×10^{-15} m, the diameter of a nucleus.

31-8 LASERS

Most sources, such as fluorescent lights, tungsten lamps and the sun, produce light that is incoherent (not in phase) and not monochromatic. This type of light cannot be used as a carrier wave for the transmission of information because its amplitude and total wave form vary.

The theory of the laser (Light Amplification by Stimulated Emission of Radiation) was developed about 1958, and the first successful laser was constructed in 1960. Laser light is coherent, and in most cases its wavelengths are sharply defined. Although they are relatively new inventions, lasers already have many uses in technology, and their possible future applications are enormous.

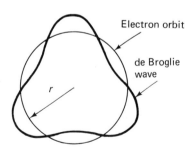

(a) Stable electron orbit showing a circumference equal to 3 de Broglie wavelengths

(b) Destructive interference of the de Broglie wave

Figure 31-7

The Laser Principle

Consider a simple material that has an energy level diagram consisting of two energy states, a lower energy state and an excited state. If the material is irradiated with electromagnetic energy (from an intense light, for example), the electrons may be excited into the higher energy state; this process is called *pumping*. Eventually each excited electron randomly returns to its lower energy state by emitting a photon with a discrete wavelength and energy corresponding to the energy difference between the two energy states. This random process is known as *spontaneous emission*; most sources emit light spontaneously (Fig. 31-8). The light waves produced by spontaneous emission have a random phase and are said to be *incoherent*. Note that the excitation energy must be greater than or at least equal to the energy of the emitted photon.

When it absorbs a photon, an atom is raised to an excited state. As it returns to the lower state, the atom spontaneously emits another photon with a discrete energy and wavelength λ. However, if another photon with the same wavelength λ passes in the vicinity of an atom while it is in the excited state, the excited atom is stimulated to emit its photon in phase with the incident photon and in the same direction. This is called *stimulated emission*, and the emitted light (which is in phase) is called *coherent* light (Fig. 31-9).

Consider a group of like atoms. When some are excited (by a pumping process) into a higher state, a few return to their lower state by spontaneous emission. In the process some of the spontaneously emitted photons stimulate other excited atoms, causing them to emit their photons in phase. It is therefore desirable to have more atoms in the excited state than in the lower state in order to

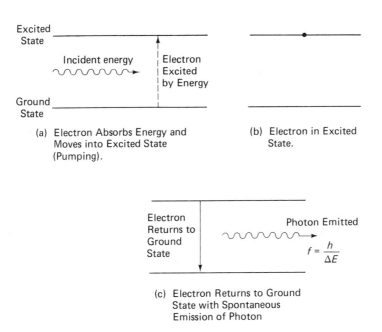

(a) Electron Absorbs Energy and Moves into Excited State (Pumping).

(b) Electron in Excited State.

(c) Electron Returns to Ground State with Spontaneous Emission of Photon

Figure 31-8
Spontaneous emission.

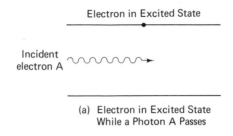

Electron in Excited State

Incident
electron A

(a) Electron in Excited State
While a Photon A Passes

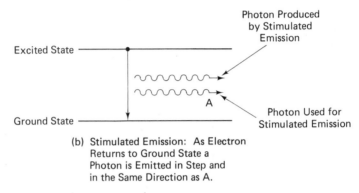

Photon Produced
by Stimulated
Emission

Excited State

Ground State

A

Photon Used for
Stimulated Emission

(b) Stimulated Emission: As Electron
Returns to Ground State a
Photon is Emitted in Step and
in the Same Direction as A.

Figure 31-9
Stimulated emission.

increase the possibility of stimulated emission. This is called a *population inversion.* Note that, even when a population inversion exists, large numbers of photons are still emitted spontaneously.

The intensity and directional properties of the coherent laser light are enhanced by placing a reflecting surface at each end of the system (Fig. 31-8). These reflecting surfaces are parallel so that any light that is incident perpendicular to one surface is reflected continually back and forth between the two surfaces. Any light that does not move continually between the two reflecting surfaces is lost; therefore only light that moves in one direction is allowed to increase in intensity. After a very short time this unidirectional beam of coherent light becomes very intense, and the probability of stimulating other excited atoms to emit their photons

with the same phase and in the same direction increases.

Some lasers may be triggered so that they release all of their intense coherent light at once. Others, called *pulsed lasers*, emit their light at regular intervals. In *continuous lasers* one reflecting surface is replaced by a partially reflecting surface, and some light is allowed to pass through; however most is reflected internally in order to stimulate continually emission from the excited atoms.

Lasers may be constructed from solids, liquids and gases. One of the most common lasers utilizes the gases helium and neon. It consists of a glass capillary tube that contains a plasma of approximately 85% helium and 15% neon gas. Brewster mirrors are located at each end of the capillary tube, and one of these mirrors is partially reflective (Fig. 31-10). Helium and neon have an excited

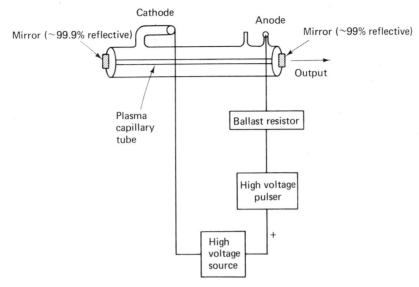

Figure 31-10
Schematic of a helium-neon gas laser.

state with approximately the same energy. When a high voltage ($\sim 2\,000$ V) is applied, the electrical energy pumps the helium atoms into an excited state. The helium atoms may then transfer their energy to the neon atoms, creating a population inversion of excited neon atoms, and photons are emitted as the neon atoms return to their lower energy state. Helium-neon gas lasers usually emit coherent light with a wavelength of 632.8 nm. All other wavelengths are suppressed.

Other lasers have been constructed from different materials, such as argon ions, helium with cadmium, krypton, oxygen with nitrogen, organic liquids, ruby rods, YIG (yttrium garnate), and silicon crystals. It is now believed that a laser can be constructed from any material that will fluoresce. These lasers vary considerably in size and shape, and they produce different wavelengths and output powers.

There are a vast number of laser applications in technology. Just a few areas and uses include:

1. *Medicine.* Laser is used as a sterile surgical cutting instrument. It is also used to weld the retina of the eye and has possible use as an artificial eye for the blind.

2. *Surveying.* Laser light is used to determine straight lines and to measure distances very accurately by interference techniques.

3. *Alignment.* The directional properties of laser light make it useful in aligning objects, such as aircraft wings, communications systems, tunnel-digging equipment and photographic reproduction.*

4. *Cutting and boring instruments.* High output lasers, such as a carbon dioxide-nitrogen laser, are quite capable of burning through metals and even firebrick. They are used for cutting and trimming dies, paper, clothes and

---

*Such as in the production of the masks used in the fabrication of integrated circuits.

microelectronic devices and in some cases as a scribe in the separation of integrated circuits.

5. *Communications.* Because of its coherence and wavelength, a single laser beam has the ability to act as an extraordinary carrier wave that (when multiplexed) is capable of carrying thousands of signals (such as telephone calls) simultaneously. However, because light is affected by weather conditions, optical fibers are frequently used to guide the light from the transmitter to the receiver.

6. *Nuclear fusion.* At present the possibility of using a high-energy laser beam to initiate a controlled nuclear fusion process is being investigated.

7. *Printing.* The production of a relief printing plate called a laser plate involves multiple applications of a laser beam. This eliminates the requirement of a photographic copying system.

8. *Weapons.* The high power output of a laser may eventually be utilized as a weapon. Laser beams are already utilized in tracking devices.

9. *Research.* Lasers are used to excite atoms in materials, to sense rotations and to measure earth strains. They even have a possible future in a laser engine that may propel space vehicles.

10. *Testing and measurements.* Very precise measurements can be made with laser light; small deviations in a surface can be measured by interference techniques. It is possible to measure heights of buildings, people and even plants accurately from an aircraft! Laser beams are also used to detect and measure air pollution.

11. *Readout systems.* Scientists at Bell Laboratories have developed a type of liquid crystal that can be used to display information "readout" from electronic instruments and computers. This liquid crystal consists of a thin film of an organic liquid mixture* between two glass plates. The inner surfaces of the glass plates are coated with indium-tin oxide and silane.

A focused infrared laser beam traces the message on the liquid crystal. The laser light is converted into heat by the indium-tin oxide layer, and the heat produces a phase change in the organic liquid, making it opaque. The liquid crystal is illuminated by a light source, and the opaque areas appear as a visible message. This type of readout system may retain the message for a long time without any rewriting process. Messages are cleared by applying a variable voltage of approximately 40 V at a frequency of 1 500 Hz to the indium-tin oxide coating.

12. *Holography.* One of the most important applications of laser light is the production of holograms (types of three-dimensional pictures) and the reconstruction of images from holograms. A hologram is produced by separating a single laser beam into two parts. One part illuminates the object being photographed while the second part acts as a reference beam. The interference between the reference beam and the reflections from the object produce an interference pattern in the emulsion of the photographic plate (Fig. 31-11a). To reconstruct the image, the photographic plate is illuminated by the diverging beam of a laser (Fig. 31-11b).

Holograms themselves have many applications, a few of which are:

1. *Motion pictures.* Three-dimensional motion pictures have already been produced using holograms.

---

*90% p-methoxy benzylidene-p-n-butylanaline and 10% cholesteryl nonanoate.

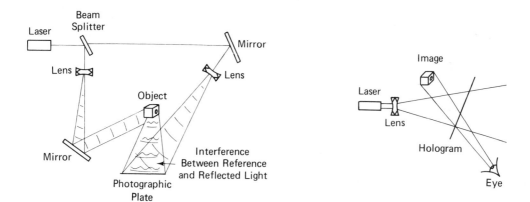

Figure 31-11
(a) Production of a hologram. (b) Reproduction of the image.

2. *Maps.* Holographic relief maps that can be seen with the aid of a special viewer are now available.

3. *Video discs and video tapes.*

4. *Computer memory.* Holograms have a promising future in data storage systems. An array of very small holograms, each formed by exposing a data mask, is stored on a single photographic plate. The digit "one" is represented by a hole and "zero" by the absence of a hole in the mask.

Two steps are required to retrieve a particular piece of information. Initially an optical deflector is used to direct a laser beam on to the desired hologram on the photographic plate. Then the image of the original data mask falls on an array of photo detectors that

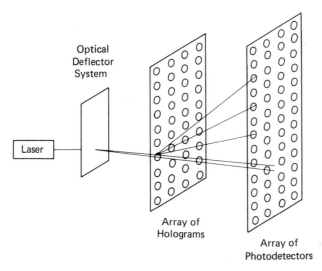

Figure 31-12

are selectively interrogated, converting data into binary logic.

These systems are fast, relatively inexpensive and have very high storage densities.

REVIEW PROBLEMS

1. When monochromatic light with a frequency of 9.2×10^{14} Hz irradiates a photocathode, photoelectrons are ejected with a maximum speed of 7.0×10^5 m/s. Determine (a) the threshold frequency of the surface, (b) the photoelectric work function, (c) the threshold wavelength, (d) the stopping potential.

2. When monochromatic light with a wavelength of 360 nm irradiates a photocathode, photoelectrons are ejected with a maximum speed of 5.4×10^5 m/s. Determine (a) the threshold frequency of the photocathode, (b) the photoelectric work function, (c) the threshold wavelength, (d) the stopping potential.

3. Lithium has a photoelectric work function of 2.42 eV. If a lithium photocathode is irradiated with monochromatic light of wavelength 420 nm, calculate (a) the threshold frequency, (b) the maximum kinetic energy of the photoelectrons, (c) the stopping potential.

4. When a lithium photocathode with a photoelectric work function of 2.42 eV is irradiated with monochromatic light, a minimum stopping potential of -0.85 V is required to stop the photoelectric current. Determine the wavelength of the incident light.

5. Determine the number of photons that are emitted by a 1.5 mW laser in 5.0 s if it emits monochromatic light with a wavelength of 632 nm.

6. Determine the number of photons that are emitted by a 3.0 mW laser in 3.0 s if it emits monochromatic light with a wavelength of 633 nm.

7. Determine the number of photons that are emitted by a 2.5 mW laser in 8 s if it emits monochromatic light with a wavelength of 633 nm.

8. The radius of a second Bohr orbit is 2.12×10^{-10} m for the hydrogen atom. Determine the speed (tangential) of the electron in its circular orbit.

9. Determine the wavelength of the photon that is emitted when an electron changes its Bohr orbit from the second excited state to the first excited state in hydrogen.

10. A free electron moving at 6.0×10^6 m/s joins a stationary proton to form a hydrogen atom in its ground state. If a single photon is emitted, what is its wavelength?

11. List all the possible orbital, magnetic and spin quantum numbers of an electron if its principal quantum number is 4.

12. What is the de Broglie wavelength of an electron that is moving at 2.5×10^6 m/s?

13. Determine the de Broglie wavelength of a proton moving at 3.0×10^5 m/s?

14. Calculate the de Broglie wavelength of an orbital electron in a hydrogen atom if it is (a) in the ground state and (b) in the third excited state.

15. If the uncertainty in the momentum of an electron is 4.0×10^{-19} kg·m/s, what is the approximate uncertainty in its position?

Solids are rigid substances that tend to retain their shape and volume. Two or more atoms often reduce their total energy by "uniting" to form a molecule by a process called *bonding*. The molecule is stable if its total energy is less than the sum of the energies originally possessed by the isolated atoms. Similarly, under certain conditions of temperature and pressure, large numbers of atoms or molecules may reduce their total energy by bonding together to form a stable solid. Solids are often classified according to their atomic structure. In *amorphous* solids the positions of the atoms are irregular, but *crystalline* solids are characterized by a definite regularity in the atomic or molecular positions. The three-dimensional array of atoms, ions or molecules in a crystalline solid is called a *crystal lattice*. It is the nature of this crystal lattice and type of bonding that determines many of the physical properties of the solid. An understanding of crystal structures is often essential for the development and use of materials.

32-1 BONDING

Valence electrons are mainly responsible for many of the physical and chemical properties of atoms. It is often convenient to consider individual atoms in terms of an outer shell of valence electrons and an inner positively charged *ion core* that contains the atomic nucleus and the complete inner shell electrons (Fig. 32-1).

Atoms and molecules attract each other to form solids under certain conditions when the solid state corresponds to the minimum energy configuration. The attractive force is always electrostatic in nature, and the type of bonding is classified according to the dominant type of electrostatic attraction involved. In a stable solid the attractive forces are balanced by the electrostatic repulsions between the positively charged nuclei. Different types of bonds have different strengths; however the

32

The Solid State

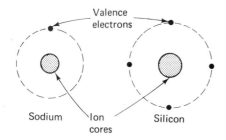

Figure 32-1
Representation of atoms in terms of the outer shell valence electrons and the positive ion cores.

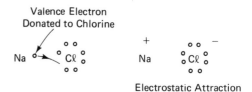

Electrostatic Attraction

Figure 32-2
The ionic bond.

differences between the types of bond are not sharply defined. Some of the more important types of bonds are described below.

Ionic (Electrovalent) Bond

Elements from the left side of the periodic table, such as the alkali metals, have relatively low ionization potentials. Therefore they are called *electropositive* because comparatively little energy is required to remove their valence electron(s) and to leave them as positive ions. However, with the loss of their valence electrons, these elements now have closed outer electron shells and chemically are relatively stable.

Some elements, such as the halides, are strongly *electronegative* because they require only one more electron in order to form the stable electron configuration of an inert gas with a complete outer electron shell. These elements readily accept electrons, becoming negative ions and releasing energy in the process. Their *electron affinity*, which is the energy required to remove the extra electron from the negative ion, is quite high.

In an ionic bond (Fig. 32-2) an electropositive element, such as sodium (Na), is ionized, and its valence electron is accepted by an electronegative element, such as chlorine (Cl). This leaves both elements with complete outer electron shells but with opposite charges. Consequently the two ions attract each other electrostatically. The total energy of the molecule is much less than that of the individual neutral atoms; therefore ionic bonds are quite strong. Also, an ionic molecule is a *polar* molecule because its centers of negative and positive charge do not coincide.

Ionic crystals are usually transparent or colored, relatively hard materials with relatively high melting points. They are electrical insulators in the solid state, but they may form conducting solutions when dissolved in a solvent.

Covalent Bond

A covalent bond consists of a pair of electrons that are shared between two atoms; each atom contributes one of its valence electrons to the bond (Fig. 32-3). The shared electrons orbit both atoms, but they are most frequently located in the region between the two atoms.

Covalent crystals of silicon (Si) and germanium (Ge) are extremely important semiconductor materials and are used in the construction of transistors, integrated circuits and other solid-state devices. Characteristically, covalent crystals (diamond for example) are very hard, transparent materials with very high melting points. They are formed between group IV elements, groups III and V, or groups II and VI of the periodic table.

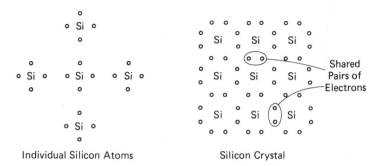

Figure 32-3
Covalent bond.

The Metallic Bond

The metallic bond is rather complex. It usually occurs in crystals of metallic elements, such as sodium, copper, silver and gold, where the valence electrons are loosely bound. When a group of metal atoms are brought close together, the valence electrons are not bound to any particular metal ion core but are free to move. These electrons are therefore called *free electrons*. The metallic bond is a result of the electrostatic attraction between the fixed metal ion cores and the free electrons (Fig. 32-4). Since free electrons are always close to several metallic ion cores in the crystal simultaneously, the metallic bond reduces the energy of the free electrons.

Metallic crystals are generally opaque, shiny and have moderately high melting points. They are also strong, ductile and good conductors of heat and electricity.

32-2 CRYSTAL STRUCTURE

Most materials that are used in engineering are crystalline in nature and are characterized by a regular three-dimensional array of atoms, ions or molecules called a *crystal lattice*. In the study of crystals, which is called *crystallography*, it is usually assumed that the lattice points (occupied by the atoms, ions or molecules) are symmetrical. A *crystal plane* is a geometric plane formed by the lattice points. Parallel crystal planes are usually regularly spaced.

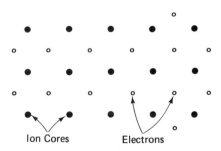

Each Electron is Attracted Towards Several Ion Cores Simultaneously and Ion Cores Attracted Towards Each Electron

Figure 32-4
Metallic bond.

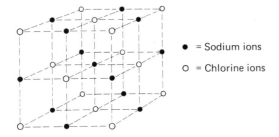

Figure 32-5
Sodium chloride crystal showing lattice planes.

● = Sodium ions

○ = Chlorine ions

Many elements and compounds may form more than one type of crystal lattice, depending on external conditions. The properties of a crystal depend substantially on the nature of its crystal lattice. For example, diamond and graphite are both composed of carbon atoms, but their lattice structure and therefore their properties are quite different.

Materials may exist in a crystalline form either as a single crystal or as a polycrystal. A *single crystal* has the same crystal lattice orientation at all interior points, and each crystal plane is continuous throughout the crystal. *Polycrystals* are composed of a number of similar interior crystals called *grains*; however the lattice orientation of each grain is different from that of its neighbors and they may also differ in size and shape.

Single crystals are more expensive, but they are used extensively in research and the construction of electronic devices, such as transistors and integrated circuits. The vast majority of commercially used crystalline materials are polycrystals.

Crystal Strength

In spite of their symmetry, crystals usually possess small defects that have a marked effect on their physical properties. Line imperfections, called *dislocations*, often exist in the crystal planes (Fig. 32-6). When a crystal is stressed, these dislocations tend to move; consequently their presence weakens the strength of the crystal. In order to increase the strength of a crystal, the dislocations must be removed or constrained so that they cannot move.

The grains of a polycrystal have different lattice orientations, and they are separated by *grain boundaries*. Each dislocation is constrained within single grains, and usually it cannot move from one grain to another. The physical strength of a polycrystal depends on the size and orientation of its grains. Stronger polycrystals have large numbers of small grains. The number of grains may be controlled to some extent during the solidification process either by seeding the "melt" (molten material) with small particles (the grains tend to form on these seeds) or by rapid cooling and solidification. Metal polycrystals also harden if they are subjected to large plastic deformations (hammered or rolled). This process increases the size of the grain bounda-

(a) Edge dislocation

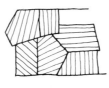

(b) Crystal grains

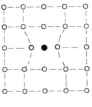

(c) Impurity

Figure 32-6

ries, thus restricting the motion of dislocations.

Impurities also affect the physical properties of a material. They are often deliberately added to a pure substance in order to change its characteristics. *Alloys* are formed when a metal is solidified from a melt with some other metal or nonmetal. Although they have the characteristics of metals, alloys are usually quite different physically from their individual components. They are often stronger because the motion of dislocation is usually stopped at an impurity, and they may possess many other desirable qualities, such as a resistance to corrosion. For example, steel is an alloy composed mainly of iron and carbon; the presence of a small amount of carbon (from 0.15% to 1.5%) substantially increases the strength and hardness of the iron. Other impurities, such as vanadium, nickel and cobalt, are often added to change the corrosive properties, melting point and hardness of steel.

Impurities, such as arsenic, phosphorus, indium and boron, are added to single crystals of silicon or germanium by a process called *doping* in order to change their electrical characteristics. These "doped" crystals are used extensively in construction of semiconductor electronic devices.

Problems_____

1. List three common alloys.

2. Why are impurities often added to materials?

3. Describe the difference between a single crystal and a polycrystal.

32-3 X-RAYS

In 1895, while he was conducting a series of experiments involving electrical discharges through gases, Wilhelm Röntgen covered a discharge tube with black cardboard. By accident he noticed fluorescence on a nearby screen of barium platinocyanide, and he ultimately determined that this fluorescence was caused by some "agents" that managed to pass through the cardboard from the glass walls of the discharge tube. Röntgen could not explain the nature of these agents; therefore he called them *X-rays*.

X-rays have a wide variety of applications in modern technology. They have the ability to penetrate lighter chemical elements, and they may be detected on a fluorescent screen or photographic plate. Their ability to penetrate skin and flesh is used to advantage in medicine where pictures of bones and vital organs are essential in the diagnosis of fractures and many diseases and disorders. Crystal structure is investigated by the diffraction of X-rays from the planes of atoms, which constitute the crystal. X-ray spectra are also used in the analysis of materials.

Today X-rays are produced in a *Coolidge tube* by bombarding a solid heavy metal anode target, such as copper and nickel, with high-energy electrons (cathode rays) (Fig. 32-7). The interaction of these electrons with the target atoms produces two different types of X-rays.

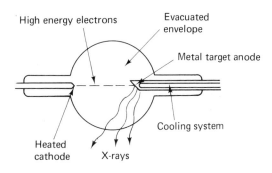

Figure 32-7
A Coolidge tube.

Continuous X-rays

When the elections gradually interact with the metal atoms of the anode, the kinetic energy of most fast electrons is converted into heat; therefore the anode nust be continually cooled. Some of the fast electrons are rapidly decelerated as they interact with the metal atoms of the anode, and in this process part of their kinetic energy may be converted into electromagnetic radiation in the form of *continuous* or *Bremsstrahlung X-rays*. If, in a single interaction, a fast electron loses all of its kinetic energy, a maximum energy (minimum wavelength) X-ray is produced. Thus, if V is the accelerating potential between the cathode and anode, the maximum energy of the X-rays is.

$$E_{max} = hf_{max} = \frac{hc}{\lambda_{min}} = eV \qquad (32\text{-}1)$$

where h = Planck's constant, f_{max} = maximum frequency, λ_{min} = minimum wavelength of the X-ray, c = speed of light and e = electron charge.

Example 32-1

Determine the minimum wavelength of X-rays that are emitted from a Coolidge tube when the accelerating potential is 30.0 kV.

Solution:

$$\lambda_{min} = \frac{hc}{eV} = \frac{(6.63 \times 10^{-34} \text{ J·s})(3.00 \times 10^8 \text{ m/s})}{(1.60 \times 10^{-19} \text{ C})(30.0 \times 10^3 \text{ V})}$$
$$= 4.14 \times 10^{-11} \text{ m}$$

Bremsstrahlung X-rays may have any wavelength greater than λ_{min} if only part of the kinetic energy possessed by a fast electron is converted into an X-ray (Fig. 32-8).

Characteristic X-rays

If the accelerating potential is high enough, the energetic electrons may have sufficient kinetic energy to *ionize* the atoms *of the anode* by removing an *inner* shell electron. An electron in an outer higher energy shell of the same anode atom may then reduce its energy by filling the vacancy in the inner shell. The energy lost by this orbital electron as it changes shells is a characteristic of the discrete electron energy levels of the anode atoms (Fig. 32-9). Therefore this energy, which is converted into electromagnetic radiation in the form of an X-ray, has a discrete energy *characteristic of the anode material*. If the orbital electron reduces its energy from E_2 (L-shell) to E_1 (K-shell) for example,

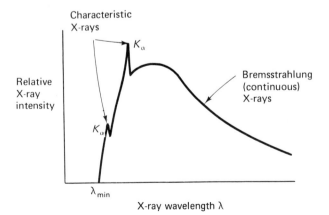

Figure 32-8
A typical X-ray spectrum.

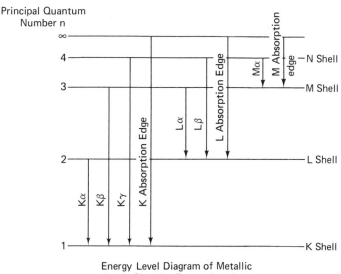

Energy Level Diagram of Metallic
Anode Showing Energy Level
Transitions for Corresponding
Characteristic X-rays.

Figure 32-9
Energy level diagram of metallic anode
showing energy level transitions for corresponding characteristic X-rays.

the characteristic X-ray, which is called the K_α X-ray has an energy E and frequency f corresponding to:

$$E = E_2 - E_1 = hf \qquad (32\text{-}2)$$

where h = Planck's constant.

X-ray Diffraction

Max Von Laue in 1912 suggested that regularly spaced atomic planes of a crystal could act as a diffraction grating for short wavelength X-rays. Thus, if d is the distance between successive parallel crystal planes, according to the diffraction grating equation, X-rays of wavelength λ are diffracted through some angle θ such that:

$$N\lambda = d \sin \theta \qquad (32\text{-}3)$$

where N is the order of the spectrum.

The experiment was first performed by W. Friedrich and P. Kipping, they allowed some X-rays to pass through a small hole in a lead shield and to form a narrow beam. This narrow beam of X-rays was diffracted by the atomic planes of a crystal and detected with a photographic plate (Fig. 32-10). The exposure to the diffracted X-rays produced a regular array of spots on the photographic plate. This type of array is now called a *Laue pattern*. Each spot in a Laue pattern is produced when X-rays are diffracted from a set of parallel crystal planes and interfere constructively at the photographic plate.

Later in the same year, W. Bragg developed an X-ray spectrometer (Fig. 32-11). A crystal is mounted so that it can be rotated at the center of curvature of the circular path taken by some detector. X-rays pass through a narrow slit in a lead shield and form a wide pencil beam that is then scattered from the various atomic planes in the crystal.

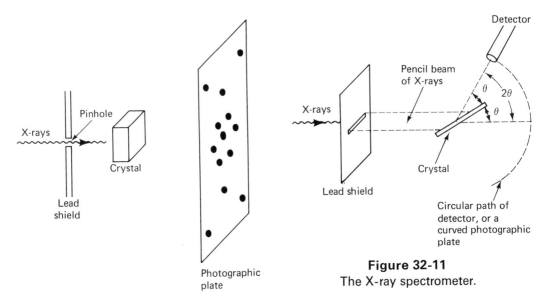

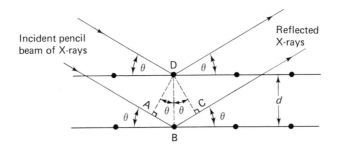

Figure 32-11
The X-ray spectrometer.

Figure 32-10
Von Laue diffraction of X-rays.

Consider a crystal with a set of parallel atomic planes regularly spaced a distance d apart; each atom scatters any incident X-rays. If X-rays of wavelength λ are incident on the crystal at some angle θ with respect to the atomic planes (Fig. 32-12), the scattered X-rays interfere constructively when:

$$2d \sin \theta = N\lambda \qquad (32\text{-}4)$$

where N is an integer. This equation is known as *Bragg's rule*. Note that, in order to detect the point of constructive interference, the detector must be positioned along the line that makes an angle 2θ at the crystal with respect to the direction of the incident X-rays. Remember that, since the maximum value of $\sin \theta = 1$, Bragg's rule only applies if $\lambda \leqq 2d$; this is why short wavelength radiation is required.

X-rays of the same wavelength produce consistent interference patterns from each

Figure 32-12
Constructive interference by the reflection of X-rays from adjacent atomic planes in a crystal.

set of parallel atomic planes in the same crystal. Therefore, if the wavelength of the X-rays is known, the spacing d between adjacent parallel atomic planes may be determined. In addition, if the spacing between the atomic planes is known, the wavelength of the incident X-rays may be determined.

In order to perform scattering experiments, monochromatic (single wavelength) X-rays are required. The Bremsstrahlung and unwanted characteristic X-rays are absorbed by metal filters (of nickel for example).

Example 32-2

A beam of characteristic X-rays with a wavelength of 68.9 pm is incident on a sodium chloride (table salt) crystal in an X-ray spectrometer. If the detector is moved (from the zero to the first-order constructive interference points) through an arc that subtends an angle $2\theta = 12.58°$ at the crystal, determine the distance d between adjacent parallel atomic planes.

Solution:

$$d = \frac{N\lambda}{2\sin\theta} = \frac{(1)(6.89 \times 10^{-11})}{2(\sin 6.29°)}$$
$$= 3.14 \times 10^{-10} \text{ m} = 0.314 \text{ nm}$$

Problems

4. Determine the minimum wavelength of X-rays that are emitted from a Coolidge tube when the accelerating potential is 50 kV.

5. What accelerating potential is required to produce X-rays with a minimum wavelength of 30.0 pm from a Coolidge tube?

6. Characteristic X-rays with a wavelength of 0.155 nm are incident on a crystal in an X-ray spectrometer, and the detector must be moved through an arc of 14.4° between the zero-and first-order constructive interference. Determine (a) the distance between the crystal planes and (b) the angle for the second-order maximum.

7. A monochromatic beam of X-rays is incident on a crystal in an X-ray spectrometer. If the distance between the adjacent parallel atomic planes in the crystal is 0.28 nm and the angle between the zero- and first-order interference is 11.2°, determine the wavelength of the X-ray.

32-4 ENERGY BANDS

Atoms that contain many orbital electrons have quite complex energy level diagrams because of the interactions between these electrons. The energy level diagrams of molecules are even more complicated because the rotation and vibrational energies of the atoms in a molecule are also quantitized. These energies produce the flutings in the band spectra of molecules.

Crystals are composed of large numbers of atoms that are arranged in a regular three-dimensional array. Consider the formation of a crystalline solid when N atoms bond together ($N \sim 10^{20}$). As the atoms move closer to each other, the orbits of the higher energy outer shell electrons, the valence electrons, begin to overlap (Fig. 32-13a). The Pauli exclusion principle then applies not only to the electrons in the individual atoms, but also to the electrons with overlapping orbits. As a result the energy level corresponding to the outer electron orbit splits into $\sim N$ very closely spaced levels.* As the spacing between the atoms is reduced, other inner energy orbits may also begin to overlap and their energy levels also separate into $\sim N$ very closely spaced levels (Fig. 32-14). Each group of split energy levels is called an *energy band*. In the ground state, the valence electrons of all the atoms occupy positions in the *valence band,*

---

*s subshells into $2N$ levels, p subshells into $6N$ levels, etc.

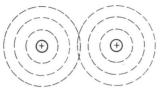 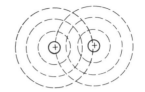

(a) Schematic representation of the overlapping of outer electron orbits

(b) Schematic representation of the overlapping of inner shell electron orbits when the atoms are close

Figure 32-13
Overlapping of electron orbits.

and inner shell core electrons are located below the valence band in full energy bands with lower energies. The regions between the energy bands are called *energy gaps*. Electrons cannot exist in a stable crystal if their energy is located in an energy gap. Other energy bands corresponding to available energy states above the valence band are called *conduction bands*. These correspond to the excited states of atoms. Normally only the conduction band immediately above the

valence band is considered. When an electron has such an energy that it is located in the conduction band, it is able to move freely through the crystal and may act as a charge carrier for electric current.

Energy band diagrams are graphical representations of the available energy states in a crystalline solid. These diagrams are often used in order to analyze the electric properties of solids since the location of the conduction band and the number of electrons

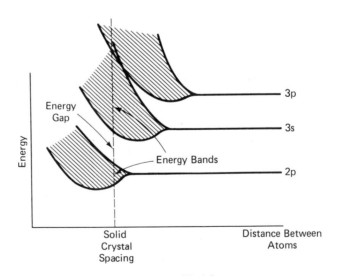

Figure 32-14
Outer electron orbits overlap and their energy levels split to form energy bands.

in the valence band usually characterize the electrical properties of the various crystalline solids. Electric current will only pass through a crystal when at least one of its energy bands is partially full or partially empty. There is no electric current in a completely full or a completely empty band. Energy must first be· supplied in order to remove an electron from a full band into an empty band.

Conductors—Metals

Good conductors of electricity possess electrons that are not tightly bound to any particular atom, they are free to move through the crystal. There are two common types of energy band diagram:

1. In some metals, such as the alkali metals sodium and potassium, only a *single* valence electron exists in the outer orbit of each atom (an *s* orbit). A crystal of *N* atoms may contain 2 *N* electrons in each *s* band since each *s* subshell may contain two electrons with opposite spins. Therefore the valence band is only partially filled, and electrons in the valence band have higher energy states immediately accessible to them (Fig. 32-15a). Very

little energy is required to move a valence electron from one atom (corresponding to one energy state) to some other atom (corresponding to some other energy state in the same energy band).

2. In other metals, such as magnesium, the valence band is full, but it overlaps a higher energy band (Fig. 32-15b). An electron may move into a higher energy state from one atom to a neighbor without the input of much energy.

Semiconductors

Semiconductor materials, such as silicon and germanium, are formed when atoms bond covalently, where each outer (valence) electron is shared between two atoms. In order to produce an electric current, a valence electron must be removed from the bond (which corresponds to the valence band) and located at some other point in the crystal where it is free to move (equivalent to being located in the conduction band). This requires the input of additional energy.

At low temperatures the energy band diagram of a pure semiconductor consists of a valence band that is full and a conduction

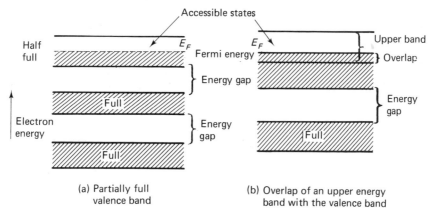

(a) Partially full valence band

(b) Overlap of an upper energy band with the valence band

Figure 32-15
Energy band diagrams of metals.

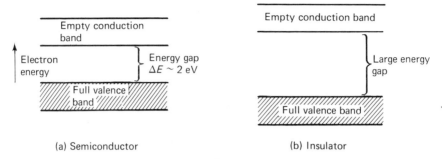

Figure 32-16
Energy band diagrams.

band that is completely empty of electrons. The energy required to remove an electron from a covalent bond to some other point corresponds to the magnitude of the energy gap ΔE between the top of the valence band and the bottom of the conduction band (Fig. 32-16a). In most semiconductors this energy gap is approximately 2 eV or less; therefore an energy input of approximately 2 eV is required before there is an electric current.

Insulators

In insulators such as diamond, the electrons are very tightly bound to the individual atoms, and large energies are required to free them. This energy corresponds to the large energy gap between the full valence band and the empty conduction band (Fig. 32-16b).

Problems

8. Describe the difference between conductors, semiconductors and insulators.

9. What is the relationship between energy level diagrams for atoms and energy band diagrams for crystals?

32-5 ELECTRON STATISTICS

The motion of electrons is responsible for all current in a crystal. In some cases, however, the current may be an indirect consequence of the electron motion, and analysis of the Hall effect shows that the current is due to the motion of positive charges.

In order to conduct electricity, electrons must be freed from their individual atoms so that they may move through the crystal. Therefore, in the case of semiconductors and insulators, the electrons must be excited (by the input of energy) from the valence band (corresponding to the bonded electron) into the conduction band.

A symmetry exists between a completely full valence band and a completely empty conduction band. When an electron moves from a full valence band into an empty conduction band, it leaves a vacancy at the bond that corresponds to a space in the valence band. An electron from some other bond in the crystal may move into this vacancy, but in doing so it leaves a vacancy elsewhere. As the electron moves in one direction in order to fill a vacancy, the vacancy, which is called a *hole*, "moves" in the opposite direction (Fig. 32-17). When an electron is removed from an atom, it leaves the atom with a net positive charge e. Therefore, as the hole "moves," it has the same effect as the motion of a positive charge, and it may contribute to the electric current. A *hole* is an imaginary positively charged particle that occupies empty energy bands analogous to the occupation of full energy bands by electrons. The

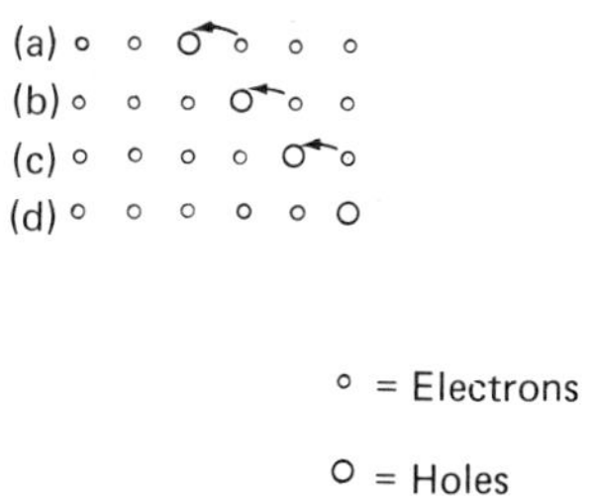

(a) ○ ○ O⌒○ ○ ○
(b) ○ ○ ○ O⌒○ ○
(c) ○ ○ ○ ○ O⌒○
(d) ○ ○ ○ ○ ○ O

○ = Electrons

O = Holes

Figure 32-17

Schematic representation of relative electron and hole motion. The holes "move" to the right as the electrons move to the left.

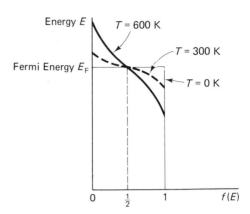

Probability That an Electron Occupies a State With an Energy E

Figure 32-18

The Fermi function at constant temperatures.

energy of a hole is therefore measured in the opposite direction to that for electrons. An increase in electron energy upwards on the energy band diagram is equivalent to an increase in hole energy downwards. In many cases electric current is the result of both electron and hole flow in opposite directions; the total current is the sum of both components.

The magnitude of an electric current in a crystal is directly related to the number of free electrons in the conduction band and free holes in the valence band. These free particles, which are responsible for electric current, are often called *charge carriers*. It can be shown from statistics that the probability that an electron occupies a particular state with an energy E is given by:

$$f(E) = \frac{1}{e^{(E-E_F)/kT} + 1} \quad \text{(Fermi function)} \quad (32\text{-}5)$$

where T is the absolute temperature, k is Boltzmann's constant and E_F is a constant for the material; it is called the *Fermi energy* and Eq. 32-5 is called the *Fermi function*.

Consider the graphical representations of the Fermi function at different constant temperatures (Fig. 32-18). Since heat is a form of energy, as the temperature increases, some electrons may occupy higher energy states. The Fermi energy E_F separates the occupied and unoccupied states at absolute zero. At any other temperature, the probability that

an electron has an energy equal to the Fermi energy is always $\frac{1}{2}$. Thus, when $E = E_F$ in Eq. 32-5, $f(E) = \frac{1}{2}$. However it should be noted that in pure semiconductors and insulators the Fermi energy is located almost exactly at the center of the energy gap between the conduction and valance bands. Consequently electrons cannot exist at the Fermi energy in in these materials (Fig. 32-19b). The Fermi energy in metals is the highest occupied energy level at absolute zero, and it is located somewhere in an energy band.

In metals as the temperature increases, the thermal vibrations of the atoms cause an increase in electrical resistance. The opposite is true for semiconductors and insulators. Since heat is a form of energy, as the temperature of a semiconductor increases, the energy of the electrons also increases. If sufficient heat energy is available, some valence electrons may be freed from the bonds, and they can then act as charge carriers. In the energy band diagram, this is equivalent to moving an electron from the valence band to the conduction band and simultaneously moving a hole from the conduction band to the valence band (Fig. 32-19). As the temperature of a semiconductor increases, more heat

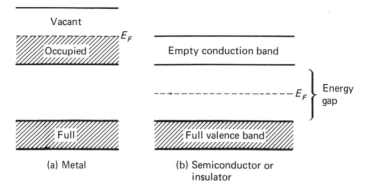

Figure 32-19
Location of the Fermi energy level E_F at absolute zero.

energy is available to produce more free charge carriers; therefore the electrical resistance decreases. Other energy forms, such as light, electric and magnetic fields and γ-rays, may also increase the number of charge carriers and reduce the electrical resistance of semiconductors. This variation of electrical resistance is utilized in some semiconductor devices, such as *thermistors*, in order to detect and measure extremely small temperature changes. Other semiconductor devices are used to detect γ-rays, light, X-rays and other forms of radiation.

If the Fermi function(Fig. 32-18) is superimposed on the energy band diagram for a semiconductor (Fig. 32-19b), the energy band diagram (Fig. 32-20) indicates the probability of finding electrons at certain energy levels in

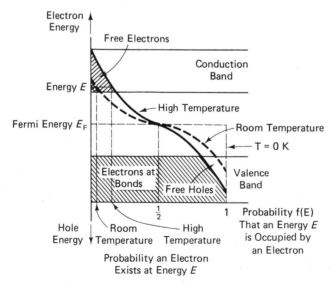

Figure 32-20
The Fermi function at various temperatures.

the energy bands. The Fermi function is symmetrical with respect to the Fermi energy. At absolute zero all electrons are located in the valence band of a semiconductor; as the temperature increases, the probability $f(E)$ of finding an electron in the conduction band increases. In pure (*intrinsic*) semiconductors, as each electron is excited into the conduction band, it leaves a hole in the valence band. Therefore the probability of finding an electron in the conduction band is the same as the probability of finding a hole in the valence band.

Free electrons and free holes may reunite, emitting a photon in the process. This recombination may occur directly, but more often it is enhanced when a free electron or free hole is trapped at an impurity or a surface boundary. The emitted photon has an energy equal to the total energy lost by the electron and hole.

Problems

10. Prove that the probability of an electron having an energy equal to the Fermi energy is always $\frac{1}{2}$.

11. Show that at $T = 0\,K$ the probability that energies less than the Fermi energy are occupied is 100% and that energies greater than the Fermi energy are not occupied by electrons.

12. Why does the electric resistance of carbon composition resistors decrease as the temperature increases?

32-6 SEMICONDUCTORS

The most commonly used semiconductor materials are silicon and germanium, which are elements from group IV of the periodic table. They are formed into almost perfect single crystals, and almost all impurities are removed by a zone refining process. Other group IV elements, such as selenium and tellurium, have a limited use; however they are not as abundant, and their electrical characteristics are not as stable.

Some semiconductor devices are constructed from covalent compounds of two different elements. An element from group III of the periodic table is covalently bonded with a group V element or a group II element with a group VI element so that each atom in the crystal has unlike nearest neighbors.

Intrinsic Semiconductors

Consider a pure (*intrinsic*) semiconductor compound of a single crystal, tetravalent (4 valence electrons) element from group IV of the periodic table. Each interior atom is bonded covalently to its four nearest neighbors, and in the ground state it is surrounded by eight valence electrons, four of its own and one from each of its four nearest neighbors (Fig. 32-21). Each pair of valence electrons constitutes a covalent bond. In order to create a free electron and free hole, an electron must be removed from a bond and located elsewhere in the crystal. This requires an energy input ΔE that is equivalent to the energy gap between the valence and conduction bands. If more energy is applied, then more electrons are able to make the transition from the valence to the conduction band, leaving more free holes in the valence band. Note that in intrinsic semiconductors the number of free electrons is equal to the number of free holes.

Semiconductor single crystals are often *doped* by the addition of a controlled amount of an impurity, such as arsenic, phosphorous, indium, boron, aluminum or gallium. Doping changes the electrical characteristics of a semiconductor.

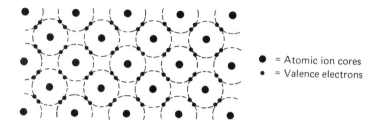

Figure 32-21
Each atom has four nearest neighbors and is
surrounded by eight valence electrons.

● = Atomic ion cores
• = Valence electrons

n-type Extrinsic Semiconductors

Consider the doping of a tetravalent single crystal semiconductor by the addition of an impurity element from group V of the periodic table, such as arsenic or phosphorous, which has five valence electrons. The impurity atoms occupy some of the lattice sites normally occupied by tetravalent atoms. Even though the crystal is still electrically neutral, only four valence electrons are required for the chemical bonds. The fifth valence electron of the impurity atom is only loosely bound to the atomic ion core. Consequently only a relatively small amount of energy is is required in order to make a free charge carrier by removing the "extra" electron from the impurity atom (Fig. 32-22a). Since the impurity atoms are responsible for the extra electrons, they are called *donors*, and this type of impurity semiconductor is called *n-type extrinsic* because the extra negatively charged electrons act as the charge carriers. Note that in an *n*-type extrinsic semiconductor electrons become charge carriers, but no corresponding holes are produced; therefore the number of

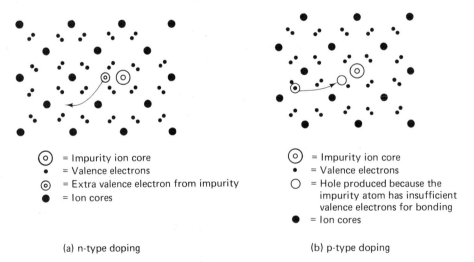

⊚ = Impurity ion core
• = Valence electrons
◉ = Extra valence electron from impurity
● = Ion cores

(a) n-type doping

⊚ = Impurity ion core
• = Valence electrons
○ = Hole produced because the
  impurity atom has insufficient
  valence electrons for bonding
● = Ion cores

(b) p-type doping

Figure 32-22
Doping.

free electrons is not equal to the number of free holes. The most numerous type of charge carrier is known as the *majority carrier*. Electrons are the majority carriers in *n*-type extrinsic semiconductors; the holes are called *minority carriers*.

In the energy band diagram of an *n*-type extrinsic semiconductor (Fig. 32-23b), a *donor energy level*, which is due to the impurity atoms, is situated just below the conduction band. Only a small amount of energy ΔE_d (which is of the order of 0.01 eV) is required to transfer an electron from this donor level into the conduction band where it may act as a free charge carrier. This energy is equivalent to the energy that is required to remove the "extra" electron from the impurity atom.

p-type Extrinsic Semiconductors

Another type of semiconductor is produced by doping a tetravalent single crystal semiconductor with an impurity, such as boron, aluminum or gallium (an element from group III of the periodic table), which only has *three* valence electrons. The impurity atoms occupy the regular lattice sites, but they do not have sufficient valence electrons to complete all the chemical bonds. Even though the material is electrically neutral, a hole exists at one of the bonds around each impurity atom (Fig. 32-22b). When a small amount of energy ΔE_A (also ~ 0.01 eV) is applied, this hole may be filled by an electron moving from a bond at a neighboring tetravalent atom; as a result the hole is transferred to that tetravalent atom. In this process the electron merely changes its location at bond sites, and its energy is not much higher. However the hole at the tetravalent atom is free to act as a free charge carrier. The semiconductor is called *p-type* (positive) *extrinsic*. This type of impurity is called an *acceptor* because it accepts electrons from other atoms. Free holes are produced with no corresponding free electrons; therefore the number of free holes is greater than the number of free electrons. Free holes are the majority carriers while free electrons are the minority carriers.

In the corresponding energy band diagram of a *p*-type extrinsic semiconductor (Fig. 32-23c), an *acceptor energy level*, due to the hole at the bond of the impurity atom, is situated just above the valence band. Only a small amount of energy ΔE_A is required to remove an electron from the valence band to the acceptor energy level, creating a free hole in the valence band. This energy is equivalent to the energy that is required to move an

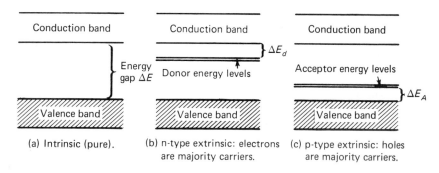

(a) Intrinsic (pure).

(b) n-type extrinsic: electrons are majority carriers.

(c) p-type extrinsic: holes are majority carriers.

Figure 32-23
Energy band diagrams of different semiconductors.

electron from a neighboring bond into the hole at the impurity.

If both free electrons and free holes act as charge carriers, the total conductivity of a semiconductor is:

$$\sigma = \sigma_n + \sigma_p \qquad (32\text{-}6)$$

where σ_n and σ_p are the conductivities due to the free electrons and free holes, respectively.

Temperature Effects

The Fermi energy level of an intrinsic semiconductor is located approximately at the center of the energy gap between the valence band and the conduction band. When donor (n-type) impurities are added, electrons may occupy positions in donor energy levels just below the conduction band; thus the Fermi energy level (corresponding to a 50% probability of occupancy) must be located closer to the conduction band. As the concentration of the impurity atoms increases, the probability of occupancy of the higher energy states increases, and the Fermi level moves closer to the conduction band. In some cases the impurity concentration is extremely high, and the Fermi energy level is actually located in the conduction band. Electronic devices, such as tunnel diodes, are constructed from such very heavily doped materials (Chapter 33).

At absolute zero all valence electrons are located in the full valence band and the full donor energy levels since this configuration corresponds to the lowest energy states available. As the temperature increases, electrons are transferred from the donor level into the conduction band, and the semiconductor behaves extrinsically. Normally at room temperatures the thermal energy may only be sufficient to remove the "extra" valence electrons from the impurity atoms (Fig. 32-24). At higher temperatures sufficient heat energy becomes available to transfer electrons not only from the donor levels into the conduction

band but also directly from the valence band to the conduction band. The heat energy is now sufficient to remove valence electrons from the chemical bonds. Since the number of donor electrons (due to the impurities) is much less than the number of electrons in the valence band, at high temperatures most free electrons enter the conduction band directly from the valence band, and the semiconductor behaves intrinsically. The temperature at which the impurity semiconductor behaves approximately intrinsically is called the *intrinsic temperature*.

Below the intrinsic temperature most conduction electrons in the conduction band come from the donor energy levels. Above the intrinsic temperature most free electrons come from the valence band, and the probability that an electron occupies an energy in the valence band is reduced.

In a p-type semiconductor the acceptor impurities are responsible for acceptor energy levels close to the valence band. Electrons may be excited from the valence band into the acceptor level, creating free holes in the valence band. Electrons from neighboring bonds move to fill holes at the impurity atom, creating holes at their original bond. Therefore at low temperatures the valence electrons from the individual atoms are located in the valence band (or the acceptor energy levels). Consequently the Fermi energy level is closer to the valence band.

The higher the acceptor impurity concentration is, the closer the Fermi energy level is to the valence band. If the acceptor impurity concentration is very high, the Fermi energy level may be located in the valence band.

At absolute zero all valence electrons are located in the valence band. As the temperature increases, valence electrons may move into the empty acceptor levels, creating free holes in the valence band. At high temperatures all acceptor energy states are occupied, but sufficient heat energy enables electrons to

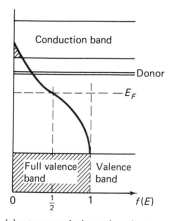

(a) n-type extrinsic semiconductor:
The Fermi energy level E_F is
closer to the conduction band.

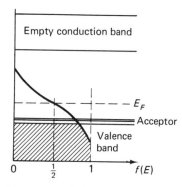

(b) p-type extrinsic semiconductor:
The Fermi energy level E_F is
closer to the valence band.

Figure 32-24
The Fermi energy level and Fermi function
at room temperatures.

transfer directly from the valence band into the conduction band. This type of semiconductor also behaves intrinsically.

Below the intrinsic temperature most free holes are due to electrons moving from the valence band into the acceptor energy levels. Above the intrinsic temperature electrons move from the valence band directly into the conduction band, and there is a higher probability that conduction electrons occupy the higher energy states.

*Problems*_____

13. Discuss the structures of extrinsic semiconductors.

14. Describe the effects of temperature on the Fermi energy level.

_____

32-7 THE HALL EFFECT

Both electrons and holes may act as charge carriers, but measurements of electrical conductivity are not sufficient to determine the polarity of the majority carrier. The polarity of the majority carrier may be determined by the *Hall effect.*

Consider a rectangular bar of a material in which electrons of charge $-e$ act as the majority carriers. When a uniform electric field strength $\mathbf{E}_x$ is imposed along the positive x-axis of the bar (Fig. 32-25), the free electrons experience an average force:

$$\bar{\mathbf{F}}_x = -e\mathbf{E}_x$$

and they tend to move in the negative x-direction with some average drift velocity $-\mathbf{v}_x$.

If a uniform magnetic flux density $\mathbf{B}_y$ is now applied in the positive y-direction perpendicular to the electric field $\mathbf{E}_x$, it produces a magnetic force on the drifting electrons according to:

$$\mathbf{F} = Q(\mathbf{v} \times \mathbf{B}) = -e(-\mathbf{v}_x \times \mathbf{B}_y)$$
$$= ev_x B_y \quad \text{(in the positive } z \text{ direction)}$$

The electrons are deflected towards the front face (in the positive z-direction) of the bar, where a negative charge accumulates. Consequently, from the conservation of charge, a positive charge (caused by the deficiency of

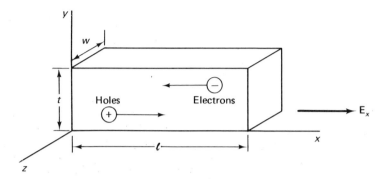

Figure 32-25
Negative electrons drift in the opposite
direction to the electric field $\mathbf{E}_x$.

electrons) is produced on the rear face, and a potential difference U_z and electric field strength $\mathbf{E}_z$ build up between the front and rear faces (Fig. 32-25). The electric field strength $\mathbf{E}_z$ tends to deflect the electrons towards the rear of the bar (Fig. 32-26). As the charge accumulates, the electric field $\mathbf{E}_z$ increases, in magnitude until it exerts a force $\mathbf{F} = -e\mathbf{E}_z$, which exactly balances the magnetic force on the electrons. Then the net force $\Sigma F = e v_x B_y - e E_z = 0$. When the magnitude of the electric field strength is:

$$E_z = v_x B_y \qquad (32\text{-}7)$$

no more electrons are deflected. This electric field strength E_z, which is produced by the accumulation of charge, is called the *Hall field*, and the corresponding potential difference U_H is called the *Hall voltage*. The magnitude and polarity of the Hall voltage U_H may be determined directly with a voltmeter connected between the front and rear faces of the bar.

If free holes of charge $+e$ are the majority carriers, a *positive charge* accumulates on the front face, and a negative charge builds up on the rear face. Note however that the polarity of the Hall field and Hall voltage is opposite

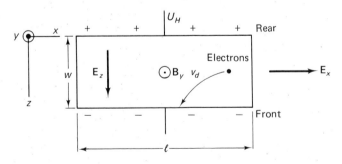

Figure 32-26
Top view of the bar showing the deflection
of electrons by the magnetic field $\mathbf{B}_y$ and
the formation of the electric field $\mathbf{E}_z$.

to that produced when electrons acted as the majority carriers. Consequently the Hall effect may be used to determine the polarity of the majority carrier.

Note that the Hall field and the Hall voltage are directly proportional to the magnetic flux density. Therefore the Hall effect is often used to measure magnetic fields or currents (since a current produces a magnetic field).

*Problem*_____

15. Describe with the aid of diagrams the production of a Hall voltage when holes are majority carriers.

REVIEW PROBLEMS

1. Determine the minimum wavelength of the X-rays emitted from a Coolidge tube when the accelerating potential is 80.0 kV.

2. What minimum potential must be applied to a Coolidge tube with a copper anode in order to produce all characteristic X-rays for copper if K-absorption edge is 0.138 nm?

3. Determine the minimum potential that must be applied to a Coolidge tube with a tungsten anode in order to produce the L_α line that has a wavelength of 0.148 nm.

4. A monochromatic beam of X-rays is incident on a crystal in an X-ray spectrometer. If the distance between adjacent parallel atomic planes in the crystal is 0.42 nm and the angle between the zero-and first-order interference is 10.6°, what is the wavelength of the X-rays?

5. Monochromatic K_α X-rays from a tungsten anode of a Coolidge tube with a wavelength of 21.0 pm are incident on a potassium chloride crystal in an X-ray spectrometer. If the lattice spacing (between the crystal planes) is 0.314 nm, determine the angles (a) between the zero- and first-order interference and (b) between the zero- and second-order interference.

6. Discuss the formation of energy bands in solids. Include a description of donor and acceptor levels.

7. Describe how the Hall effect can be used to measure electric currents (in current probes) and magnetic fields.

33

The Physics of Solid-State Electronic Devices

The development of solid-state devices has revolutionized the electronics industry. Vacuum tubes and cumbersome circuitry have essentially been replaced by solid-state devices, such as transistors, integrated circuits and thin or thick film devices that are smaller, less expensive, more reliable and usually more efficient.

The exploration of space and the development of earth satellites have increased the importance of reducing the size and weight of electronic circuits. Also, even though electricity flows quite rapidly, in computers the time delay of the signal in the interconnections between electronic components is an important consideration; if the interconnections are reduced in size, a computer can perform operations at a faster speed.

Microelectronics involves the miniaturization of regular electronic circuits. A complete electronic circuit, an operational amplifier for example, which contains large numbers of individual interconnected components, such as diodes, resistors, capacitors and transistors, may be formed on a very small single substrate; the complete miniaturized circuit is called an *integrated circuit* (IC).

Integrated circuits are small, light, rugged and reliable. They require less power and lower voltages than the equivalent macroscopic circuits. Consequently they operate at lower temperatures, and individual components may be close together without exceeding the operating temperature limit. Relatively little stray capacitance and short time delays are produced because of the short interconnections among the individual components in the IC. Maintenance is simplified because, if a component of the IC fails, the complete IC is usually replaced. Mass production techniques of planar technology have reduced the cost of many IC's so that they are almost as inexpensive as a single transistor. Eventually most conventional circuits will be replaced by IC's.

Integrated circuits and thin films may be composed of large numbers of individual solid-state devices, yet their dimensions are quite small. The use of microcircuit devices is growing rapidly, and is producing remarkable changes in many industries in addition to the electronics industry.

33-1 FABRICATION OF SINGLE CRYSTAL SEMICONDUCTORS

The basic component of most semiconductor devices is a very pure single crystal of a tetravalent element, such as silicon or germanium. The majority of semiconductor devices are made of silicon, which is one of the most abundant materials at the earth's surface, occuring mainly as silica (SiO_2). Germanium

is normally found in metallic ores of zinc, lead and copper.

The element is extracted from the ore by a series of chemical processes that produce polycrystalline solids about 98 % pure. Special techniques are then used to produce single crystals with impurity levels of less than one part in 10^{10}. There are two common techniques:

Zone Refining. First the impure material is purified by a *zone-refining process* (Fig. 33-1a). An ingot of polycrystal semiconductor material is placed on a quartz boat in a quartz container, and the container is filled with an inert gas. A set of radio frequency induction

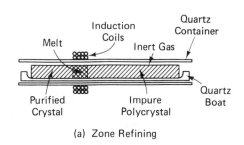

(a) Zone Refining

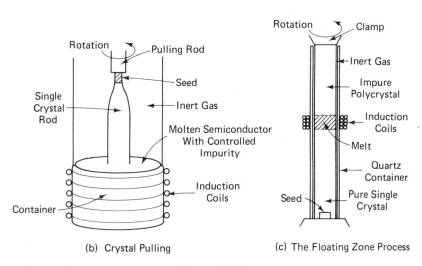

(b) Crystal Pulling   (c) The Floating Zone Process

Figure 33-1
Formation of a single crystal semiconductor.

coils moves slowly from one end of the crystal to the other, melting a small zone of the poly-crystal as it moves. Impurities are swept along at the melt to the end of the crystal, where they are removed.

The single crystal is formed by a *crystal-pulling process* (Fig. 33-1b). A sample of the pure polycrystal is placed in a quartz crucible where it is again melted by a set of induction coils in an inert gas atmosphere. Any *desired* impurities are then added to the molten material. The bottom of a single crystal seed is allowed to make contact with the molten material; it is then slowly raised and rotated. The molten semiconductor material gradually solidifies on the bottom of the seed, forming a single crystal rod with the same lattice structure.

Floating Zone. A *floating-zone process* is very similar to zone refining; it not only forms a single crystal but also purifies the material simultaneously. A polycrystal silicon or germanium rod is clamped vertically in a quartz container, and a single crystal seed is located at the base of the rod. Induction coils melt the material in an inert gas atmosphere. The rod is slowly rotated while the induction coils move upwards, melting a small zone of material as it proceeds (Fig. 33-1c). As the material solidifies, it assumes the lattice structure of the seed, and at the same time impurities are swept along with the melt to the top of the rod, where they are removed.

A single crystal rod is usually between 2.5 cm and 10 cm in diameter and about 30 cm long. It is sliced into wafers about 0.5 mm thick by a high speed diamond saw, and the wafers are ground to smooth surfaces on both sides. Finally one side of the wafer is highly polished. Each wafer forms the basic component or *substrate* of a large number of individual semiconductor devices or inte-grated circuits; each device occupies only a very small area on the total wafer. These devices are separated when the wafer is cut into very small "chips."

33-2 *p-n* JUNCTION DIODES

The majority of active semiconductor devices contain one or more *p-n* junctions, which are produced when a *p*-type semiconductor is joined to an *n*-type semiconductor. Four common methods are used to produce junction diodes with different characteristics.

1. Diffusion. Particles tend to flow naturally from areas of higher to areas of lower concentration. This process is called *diffusion* (Fig. 33-2a). It is similar to the action of a gas as it tends to distribute itself evenly throughout the volume of any container. The diffusion rate depends on the concentration gradient (the change in concentration with distance), the temperature and the nature of the substance present.

At present the majority of *p-n* junction diodes are made by a diffusion process. A wafer of *n*-type single crystal substrate is heated to ~1 000° C in the presence of a *p*-type impurity, either in vapor form or painted on the substrate surface. The *p*-type impurity diffuses into the *n*-type substrate to form a *p-n* junction. The device is then cut and mounted.

2. Alloy Process. In an alloy process (Fig. 33-2b), a pellet of one type of material is heated on a wafer of the other type in an inert gas atmosphere. The materials actually melt at the interface, and a junction is formed when the system is cooled. These devices have larger junction capacitance and current ratings.

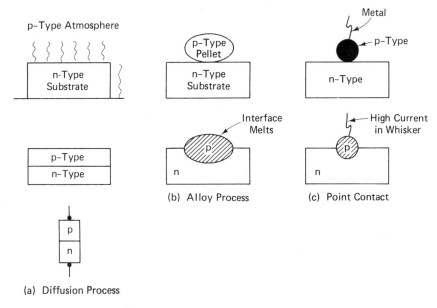

Figure 33-2
Fabrication of p-n junction diodes.

3. Crystal Pulling. During a crystal-pulling process *p-n* junctions can be formed by alternating the type of molten semiconductor material, e.g. from *n* to *p*, as the seed and rod are pulled up and rotated. The rod is then cut into a large number of junction diodes. This type of diode has a high current and power rating.

4. Point Contact. A small quantity of *p*-type material is placed on the end of a fine metal wire (whisker) and then is touched with a *n*-type substrate. When a high current is passed through the whisker, the *p*-type pellet melts into the *n*-type substrate (Fig. 33-2c). These diodes are often used at microwave frequencies.

Operation of a Junction Diode

Before a junction is formed, the Fermi energy level of each semiconductor is different. It is closer to the conduction band in the *n*-type material and closer to the valence band in the *p*-type material of the semiconductor. After a junction has formed, the Fermi energy level (which has an occupation probability of 50%) must be the same in both components. *Differences in the Fermi energy levels give rise to electrical potential differences.*

This "equating" of Fermi energy levels on both sides of the junction arises because of the diffusion of free electrons and free holes. Note that in the diagrams the signs • and ∘ merely indicate an excess of *free* electrons and holes respectively; *they do not represent an area of net change.*

The *p*-region has a high concentration of free holes whereas the *n*-region has a high concentration of free electrons. Free electrons. diffuse from the region of high concentration (the *n*-type side) to the areas of low concentration (the *p*-side of the junction), where they recombine with holes, i.e. they move into a vacancy at an atomic bond. Similarly holes diffuse from the *p*-side to the *n*-side of the junction, where they recombine with electrons.

Most recombination processes occur in a narrow region in the neighborhood of the junction. This region is called the *depletion region* because it has fewer free charge carriers. The thickness of the depletion region depends on the impurity concentrations; usually it is $\sim 10^{-7}$ m. As electrons diffuse from the *n*-side to the *p*-side of the junction, they leave a hole in the donor level on the *n*-side in the depletion region. Similarly, as holes diffuse from the *p*-side to the *n*-side, they leave electrons in the acceptor levels on the *p*-side in the depletion region (Fig. 33-3). The flow of these charges across the junction produces areas of net charge in the depletion region, giving rise to a potential difference called the *barrier voltage V_0*, which tends to oppose the diffusion current. The potential energy eV_0 of the barrier corresponds to the total energy shift in the Fermi energy levels of the components as the junction is formed (Fig. 33-4). Typ-

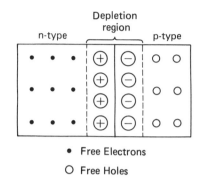

Figure 33-3

n-type has stronger doping and narrower depletion region.

ically this barrier voltage is about 0.3 V for germanium and 0.65 V for silicon diodes.

The width of the depletion region depends on the concentrations of the impurities and the magnitude of any applied potential difference (*bias*). It is narrower on the side where the doping is strongest. For example, if the *n*-type material is more strongly doped than the *p*-type, the electrons in the diffusion cur-

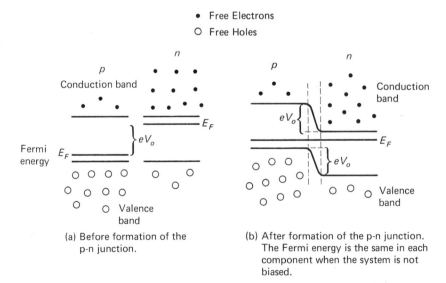

(a) Before formation of the p-n junction.

(b) After formation of the p-n junction. The Fermi energy is the same in each component when the system is not biased.

Figure 33-4

rent must penetrate deeper into the *p*-type material before they are all able to recombine with holes at the bond sites because there are relatively fewer holes. On the other hand, the holes in the diffusion current find a higher concentration of free electrons on the *n*-side, so they recombine nearer the junction.

Electrons on the *p*-side but near the junction may reduce their energy by flowing over the potential barrier into a lower energy state on the *n*-side. Similarly some holes may flow from the *n*-side to the *p*-side of the junction. This produces the *saturation current* I_s (Fig. 33-5). However the concentration of electrons on the *p*-side and holes on the *n*-side of the junction is relatively low; therefore the saturation current is small and approximately constant. The magnitude of the potential hill has almost no affect on the saturation current because the charges always tend to reduce their energy regardless of the amount. In equilibrium the saturation current is equal in magnitude to the diffusion current, and there

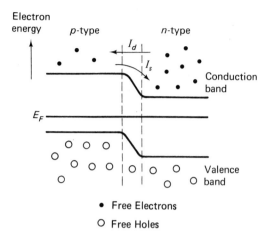

Figure 33-5
In equilibrium, the diffusion currents I_d are equally balanced by the saturation currents I_s.

is no *net* current when there is no external bias.

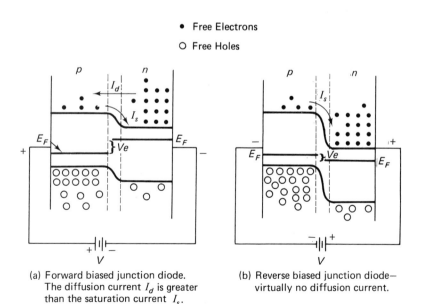

(a) Forward biased junction diode. The diffusion current I_d is greater than the saturation current I_s.

(b) Reverse biased junction diode— virtually no diffusion current.

Figure 33-6

Forward Bias

When a junction diode is forward biased, the p-side is maintained at a positive potential with respect to the n-side (Fig. 33-6a). More electrons are *injected* into the n-region, and more holes are injected into the p-region. This produces an increase in the concentrations of electrons on the n-side and holes on the p-side of the junction. Consequently the *diffusion current* I_d increases. In the energy band dia-grams a forward bias raises the energy bands of the n-type material with respect to the p-type. Electrons on the n-side and holes on the p-side then have greater energies, and they can surmount the potential barrier. The diffusion current increases as the biasing potential V increases.

The forward bias decreases the width of the depletion region because free electrons on the n-side and free holes on the p-side experience electric forces that drive them towards the junction (Fig. 33-7a).

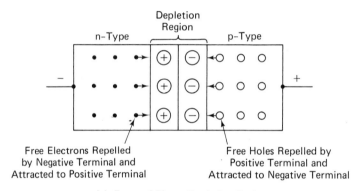

(a) Forward Bias — Depletion Region Narrows

Free Electrons Repelled by Negative Terminal and Attracted to Positive Terminal

Free Holes Repelled by Positive Terminal and Attracted to Negative Terminal

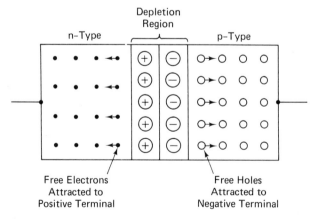

Free Electrons Attracted to Positive Terminal

Free Holes Attracted to Negative Terminal

(b) Depletion Region Widens as Reverse Bias Increases

Figure 33-7

Reverse Bias

A *p-n* junction diode may be reverse biased by making the *p*-side negative with respect to the *n*-side (Fig. 33-6b). The concentrations of electrons on the *n*-side and holes on the *p*-side is reduced; therefore the diffusion current is reduced. In the energy band diagram this is equivalent to a reduction in the energies of the electrons in the *n*-side and holes in the *p*-side so that they cannot surmount the potential barrier. The diffusion current ceases, and only the approximately constant saturation current remains.

An increase in reverse bias increases the width of the depletion region because free electrons on the *n*-side are attracted towards the positive terminal while free holes on the *p*-side move towards the negative terminal (see Fig. 33-7b).

A single *p-n* junction may be used as a rectifier since it favors current in only one direction, provided the breakdown voltage is not exceeded (Sect. 33-3).

Varactor Diodes

Recombination of electrons and holes reduces the number of charge carriers in the depletion region. However in the depletion region two areas of opposite charge are separated by a very small distance at the junction. Consequently it has a capacitance.

When the reverse bias of a junction diode is increased, the width of its depletion region increases, and the depletion capacitance decreases since it varies as an inverse function of the charge separation (for example, the parallel plate capacitor). This type of voltage-sensitive capacitor is called a *varactor diode* or *varicap*; it is used in tuning oscillator and amplifier circuits at high frequencies.

At microwave frequencies these devices are very efficient generators of harmonics when they are subjected to high voltages; therefore they are frequently used as *frequency multipliers*. They are also used as *parametric amplifiers*, operating on the principle that less energy is required to charge a capacitor when the capacitance is low; the device is charged, and its capacitance is increased before it discharges.

The Solar Cell

A solar cell is a *p-n* junction device that converts light energy directly into electrical energy. It usually consists of a thin layer ($\sim 10^{-3}$ in thick) of heavily doped *p*-type silicon on a heavily doped *n*-type silicon wafer. Incident light is able to pass through the thin layer of *p*-type silicon to the junction region,

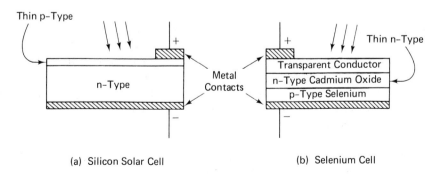

(a) Silicon Solar Cell   (b) Selenium Cell

Figure 33-8
Photovoltaic devices.

where it creates many free electron–hole pairs. A strong reverse-biased electric field in the depletion region causes the free electrons that are produced in the p-side to flow to the n-type material and the negative output terminal while the free holes move to the p-side and the positive output terminal. Therefore a potential difference is created in the device when it is irradiated with light.

Modern solar cells 1 in in diameter can produce approximately 0.6 V and 150 mA in sunlight. They are used to power some electronic equipment and can act as light sensors or receivers for modulated light.

33-3 BREAKDOWN (ZENER) DIODES

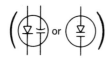

Consider a p-n junction diode that is subjected to a gradually increasing *reverse* bias. When the reverse potential reaches some value called the *breakdown voltage* V_B, there is a very sharp increase in the reverse current (Fig. 33-9). This phenomenon is called *breakdown;* it is not destructive as long as the heat produced is rapidly dissipated. There are two different causes of breakdown.

Avalanche Effect

A strong reverse bias accelerates and increases the kinetic energy of the free electrons and holes that constitute the saturation current. These "energetic particles" interact with valence electrons (at the bonds) in the depletion region. In this process some valence electrons may be excited into the conduction band (i.e. freed from their bond) and holes into the valence band, where they form a part

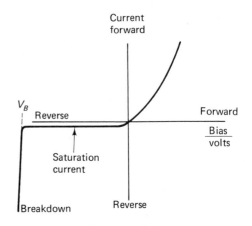

Figure 33-9
Current–voltage characteristics of a p-n junction diode.

of the reverse current. Each freed electron and hole is accelerated in the strong reverse electric field produced by the reverse bias, and in turn they may free other valence electrons by direct interaction. As a result the number of free electrons and holes increases, and the reverse current builds up very rapidly. Since the avalanche effect occurs in the depletion region, it is stronger when this region is wide.

Zener Effect

The transition from n-type to p-type material takes place in the depletion region of a junction diode. Consequently, when a reverse potential is applied, there is a relatively small potential drop in the bulk semiconductor material, but *most of the potential drop occurs in the depletion region*. If the two regions of the diode are heavily doped, strong electric fields are produced across the narrow depletion region. These fields are capable of removing electrons from the chemical bonds in the depletion region, creating additional free electrons and holes.

In the energy band diagram of a junction

diode, the large reverse bias substantially increases the energy of the electrons on the *p*-side of the junction with respect to those on the *n*-side. Electrons in the valence band on the *p*-side may move directly through the depletion region into a similar energy state in the conduction band on the *n*-side (Fig. 33-10). This process is called *tunneling*, and can only be understood in terms of quantum statistics. Note that free holes "move" simultaneously in the opposite direction.

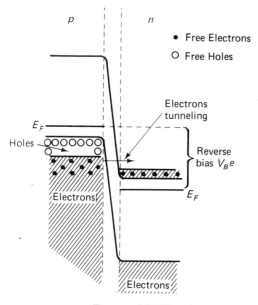

Figure 33-10
Zener tunneling. Electrons on the *p*-side may reduce their energy by moving directly into the conduction band on the *n*-side.

The term *Zener diode* is normally used to describe all semiconductor diodes that are specifically constructed for use in the breakdown region. The sharp increase in current at the breakdown voltage makes the Zener diode an excellent voltage regulator.

33-4 TUNNEL DIODES

In a *tunnel diode* a heavily doped *n*-type semiconductor is separated from a heavily doped *p*-type semiconductor by a very thin depletion region. The Fermi energy level is located in the conduction band of the *n*-type material and in the valence band of the *p*-type material (Fig. 33-11).

When a small forward bias is applied, electrons may tunnel directly from the bottom of the conduction band on the *n*-side, through the thin depletion region and into a similar energy state at the top of the valence band on the *p*-side. Holes tunnel in the opposite direction. If the forward bias is increased, eventually the electrons in the conduction band on the *n*-side have energies that lie in the energy gap on the *p*-side, and no similar energy states are available for a tunneling process; therefore tunneling ceases. As a result the current actually decreases as the voltage is increased until normal *p-n* junction diode operation dominates, and the diffusion current increases as the voltage increases (Fig. 33-12).

A reverse bias again produces tunneling but in the opposite direction as electrons move from the valence band on the *p*-side into similar energies in the conduction band on the *n*-side. As the reverse bias increases, more energy states become available for tunneling, and the reverse current increases. Therefore, in contrast to normal junction diodes, the reverse current of a tunnel diode is not a constant saturation current (Fig. 33-12).

Tunnel diodes can operate at very high frequencies but they have limited use in amplifier and switching circuits because their operation is restricted to low voltages.

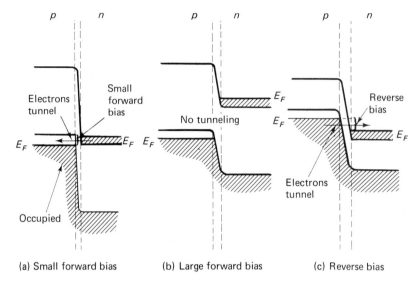

(a) Small forward bias   (b) Large forward bias   (c) Reverse bias

Figure 33-11
The tunnel diode.

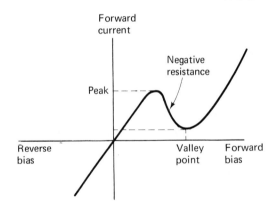

Figure 33-12
Current voltage characteristics of a tunnel diode.

33-5 LIGHT-EMITTING DIODES (LED)

$$\left(\downarrow\!\!\!\!\not\triangledown \right)$$

When a *p-n* junction diode is forward biased, recombination of free electrons and holes occurs in the depletion region, and a photon of energy is emitted spontaneously during the process. Most junction diodes merely dissipate the energy in the form of heat; however in a few semiconductor materials, such as gallium arsenide, gallium phosphide or gallium arsenide with phosphide doping (Ga-As-P), the emitted photon has a wavelength in the visible region. In these devices the photon is produced when an electron moves from the conduction band directly into the valence band, i.e. it recombines directly with a hole. However, by the addition of impurities, it is possible to create donor and acceptor levels at certain locations in the energy gap. An electron may then move from the conduction band into one of these levels (i.e. move to the impurity atom) and then produce a visible photon when it continues to the valence band. Therefore the wavelength emitted by an LED depends to some extent on the impurity concentrations. The intensity of the emitted light is controlled by the potential difference used to forward bias the junction.

The physical structures of typical LED's are illustrated in Fig. 33-13. To reduce losses

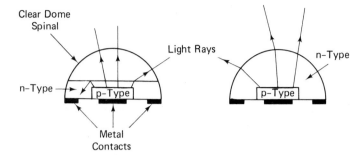

Figure 33-13
An LED.

caused by total internal reflection, the *n*-type material is often dome-shaped or encased in a clear dome-shaped material called a spinal.

At present colors from green through infrared have been produced. Light-emitting diodes are used as small light sources. Often many LED's form an array that is used as a digital readout for electronic instruments. The LED is able to convert electric energy into light very efficiently.

A *laser diode* (Fig. 33-14) consists of an LED with highly polished sides that act as mirrors so that the light is reflected back and forth parallel to the junction. The laser action is produced because recombination is stimulated by photons.

Laser diodes convert electric energy into light relatively efficiently. They are extremely fast and are now being used to generate light which then acts as a carrier wave in some communicatons systems. One laser diode can produce high-frequency light that can be modulated and multiplexed to carry an enormous quantity of information. The light is normally transmitted through an optical fiber to minimize losses.

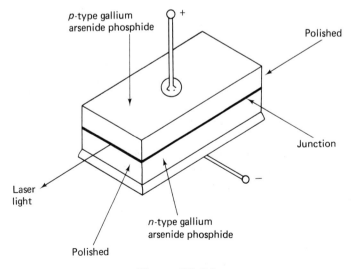

Figure 33-14
The laser diode.

33-6 METAL-SEMICONDUCTOR JUNCTIONS

(—▷|— or —|▷—)

Schottky Barrier Diode. Consider the formation of a junction between a metal and an n-type semiconductor. The Fermi energy levels of the materials are quite different; however, when the junction is formed, electrons must flow from the semiconductor to the metal in order to equalize the Fermi energy level E_F (Fig. 33-15). This produces a negative space charge on the metal side of the junction region that repels other electrons back to the n-type semiconductor. Consequently, if more electrons are to move from the semiconductor to the metal, they must first have sufficient energy to overcome the coulomb potential barrier produced by the space charge.

The metal-semiconductor junction may have rectifying characteristics if the semiconductor is not heavily doped. This type of diode is called a *Schottky diode*. In this case the depletion region (in which the coulomb barrier is effective) is too wide to allow a tunneling process, and electrons may only flow from the semiconductor when they have sufficient energy to surmount the potential barrier.

When the junction of a Schottky diode is forward biased, the metal is maintained at a positive potential with respect to the semiconductor (Fig. 33-15c). Electrons in the conduction band of the semiconductor then require only a small additional energy in order to surmount the coulomb barrier and to flow into the metal. Therefore in the forward-biased Schottky diode the conventional (positive current) is from the metal to the semiconductor.

If the junction of the Schottky diode is reverse biased, the electrons in the conduction band of the semiconductor cannot surmount the greater coulomb barrier. Few electrons in the metal have sufficient energy to flow to the semiconductor unless an extremely large reverse bias is applied and breakdown occurs.

The current–voltage characteristics of a Schottky diode resemble the characteristics of a p-n junction diode, but the Schottky diode can operate at much higher frequencies. Consequently they are used in many computer and switching circuits. Since free electrons are more mobile than free holes, Schottky diodes are normally constructed of an n-type semiconductor and a metal rather than a p-type semiconductor with metal.

Ohmic Contact

The operation of semiconductor devices depends on *ohmic contacts* between the semiconductor and a metallic conductor so that it can be connected in a circuit, i.e. it must obey Ohm's law—the resistance must be constant. However a junction between a metal and a semiconductor acts as a diode! Two common techniques are used to produce ohmic contacts. Ohmic contacts usually result when the semiconductor is damaged by abrasion as metallic leads are soldered or welded to the semiconductor. Also, a semiconductor behaves somewhat like a metal when it is heavily doped or if the Fermi energy level in the metal is higher (lower) than that in the n-type (p-type) semiconductor. For example, in a junction between a metal and a heavily doped n-type semiconductor, the depletion region is quite narrow, and the Fermi energy level of the semiconductor is located in the conduction band. Electrons can then tunnel through the depletion region in either direction; since the current is directly proportional to the applied potential, the junction is said to be *ohmic*.

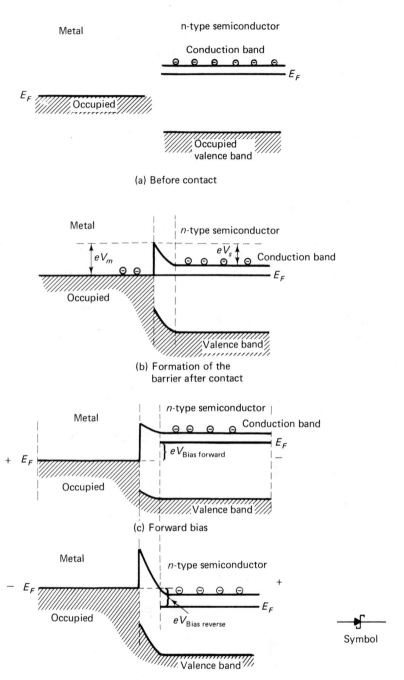

Figure 33-15
The Schottky diode.

Problem

1. Describe the operation of a Schottky diode consisting of a metal and a *p*-type semiconductor.

33-7 THE JUNCTION TRANSISTOR

Transistors may be used as amplifying devices in order to develop voltage, current or power gains in circuits. The first semiconductor amplifier was discovered by J. Bardeen and W. Brittain in 1947, and the junction transistor was developed in 1949 by Schockley.

There are two basic types of junction transistor. An *n-p-n* junction transistor consists of a strongly doped *n*-type region called the *emitter* that is separated from a weakly doped *n*-type region called the *collector* by a thin lightly doped *p*-type region called the *base*. Similarly a *p-n-p* junction transistor is composed of a strongly doped *p*-type emitter that is separated from a weakly doped *p*-type collector by a thin lightly doped *n*-type base (Fig. 33-16). These devices are called *bipolar junction transistors* (BJT) because these both types of charge carriers are involved in their operation.

The fabrication of BJT's is similar to that of junction diodes. Epitaxal transistors and planar technology are discussed in Section 33-11, in each type an *n*-type silicon substrate could be used.

Except for the sign of the majority carrier and the polarity of the biasing, the operation of both types of junction transistor is identical; therefore we shall only consider the operation of the *n-p-n* type. Both types of

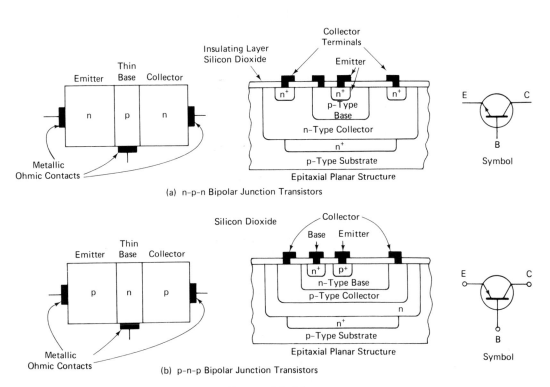

(a) n-p-n Bipolar Junction Transistors

(b) p-n-p Bipolar Junction Transistors

Figure 33-16

transistor perform the same functions in circuits; however, because the two types are available, the versatility of the junction transistor is increased. It should be noted that, since electrons are more mobile than holes, *n-p-n* transistors are favored for circuits where a fast rise time (turns on quickly) is required, and *p-n-p* transistors are used for a fast fall time (turns off quickly).

Consider the energy band diagram of an *n-p-n* bipolar transistor (Fig. 33-17). When the transistor is not biased, the Fermi energy level is the same in each section. However in

normal operation the emitter–base junction (1) is weakly forward biased (~ 0.2 V in germanium and ~ 0.6 V in silicon), and the base–collector junction (2) has a large reverse bias (~ 50 V). Most of the potential drop occurs in the depletion regions around the junctions.

The forward bias of junction (1) increases the concentration of free electrons in the emitter, and their diffusion rate into the base increases. Since the base is only weakly doped *p*-type and is very thin, most of these free electrons pass into the collector region with-

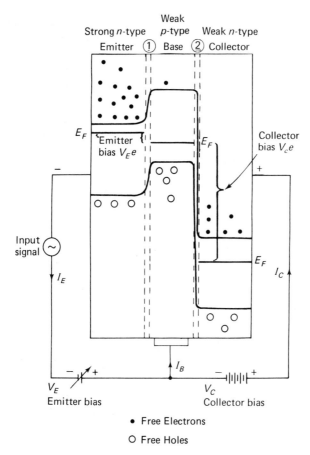

Figure 33-17

Energy band diagram of a forward biased *n-p-n* junction transistor.

out recombining with a hole in the base. Once in the collector, the electrons lose energy by interactions, and because of the large reverse bias, they cannot surmount the potential barrier in order to diffuse back into the base.

Note that from the conservation of charge, the emitter current I_E is equal to the sum of the collector current I_c and the base current I_B:

$$I_E = I_c + I_B$$

Since most of the electrons that diffuse from the emitter reach the collector and only a few combine with holes in the base region, the base current I_B is quite small and $I_c \simeq I_E$. Typically the ratio is:

$$\frac{I_c}{I_E} \sim 0.98$$

In most cases* a resistor is placed in series with the collector circuit, and the output is taken across the resistor and base–collector bias supply. Thus small variations in the input signal produce fluctuations in both I_E and I_c; this in turn produces relatively large fluctuations in potential difference (IR drop) at the output. Therefore bipolar transistors are frequently used in amplifier circuits.

In a forward-biased p-n-p transistor the emitter is made positive and the collector negative with respect to the base when positive holes are injected into the heavily doped p-type emitter.

These bipolar transistors are small, versatile, rugged and reliable. They require no warm-up time or filament heater; therefore they are rapidly replacing vacuum tubes in electronic circuitry. However junction transistors are temperature dependent, and they cannot operate at high frequencies or deliver high output powers although many improvements are being achieved.

Phototransistors

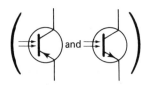

Phototransistors are sensitive to variations in light intensities and are frequently used in detector circuits. Their structure is similar to that of a bipolar transistor except that there may be no base terminal and that light is focused through a lens onto the collector base junction. This produces free electrons and holes and current amplification because the collector current is enhanced.

A *Schottky transistor* consists of a normal junction transistor with a *Schottky diode* connected between the collector and the base. The Schottky diode prevents the transistor from saturating. These devices work at very high speeds and are used in TTL† circuits.

Problem

2. Describe the operation of a *p-n-p* junction transistor.

33-8 FIELD EFFECT TRANSISTOR (FET)

In a junction transistor the emitter and collector currents are controlled by variations in the base current. A field effect transistor (FET) is a unipolar device, i.e. only one type of charge carrier is involved, in which the current is controlled by variations in a potential difference. There are two general categories of FET.

---

*Except common collector circuits.

†Transistor transistor logic.

The Junction Field Effect Transistor (JFET or FET)_____

The JFET was peoposed by W. Shockley in 1952, but it was very difficult to manufacture until 1962 when planar technology was developed. There are two basic types of JFET. An *n-channel JFET* consists of a bar of *n*-type material that has heavily doped *p*-regions called *gates* on each side, leaving a *channel* of *n*-type material between them (Fig. 33-18). Ohmic contacts, called the *source* and *drain*, are made at the ends of the bar. Similarly a *p-channel JFET* consists of a bar of *p*-type material which has heavily doped *n*-type gates.

Except for the sign of the majority carrier and the polarity of the biasing, the operation of both types of JFET is identical. Therefore we shall only consider the *n*-channel JFET.

Electrons enter the *n*-type material through the source. They flow along the *n*-type channel and leave the JFET at the drain.

The junctions between the *n*-type bar and the heavily doped *p*-type gates are reverse biased. As the reverse bias increases, the magnitude of the depletion region must also increase in order to support the increased potential difference. However, because the gates are heavily doped, the change in the area of the depletion region occurs mainly in the

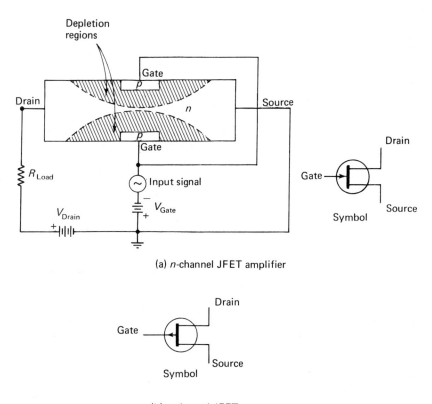

(a) *n*-channel JFET amplifier

(b) *p*-channel JFET

Figure 33-18
The junction field effect transistor.

n-type bar. Consequently, as the reverse bias increases, the depletion regions reduce the cross-sectional area of the channel through which the electrons must flow. Since the electrical resistance is inversely proportional to the cross-sectional area, an increase in the reverse bias effectively increases the resistance of the JFET. Thus the reverse bias between the gates and the source effectively controls the flow of electrons through the channel from the source to the drain.

Note that, as the potential difference between the source and the drain V_{drain} increases, the flow of electrons through the channel and the current I increase. At some reverse bias called the *pinch-off voltage*, the depletion regions may make contact and are said to "pinch-off" the channel (Fig. 33-19). At this point the resistance R of the channel is very large, but the current I due to the flow of electrons from the source to the drain is not zero; if it were the potential drop IR in the channel would also be zero. When the pinch-off condition is reached, the current I saturates; it is only limited by the bulk resistance of the semiconductor material.

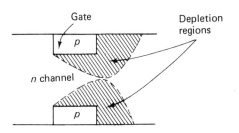

Figure 33-19
The pinch-off condition.

When a very large reverse bias is applied to the gate, the junction breaks down, and there is a very large current. These devices are somewhat analogous to vacuum tube triodes; the gate behaves like the grid. Since the charge carriers do not cross a junction in normal operation, they are often less noisy than bipolar transistors.

Metal Oxide Semiconductor FET (MOSFET) or Insulated Gate FET (IGFET)

Metal oxide semiconductor field effect transistors (MOSFET's) are unipolar devices in which the charge carriers flow through a channel. There are two types.

An n-*channel* MOSFET in the *enhancement mode* consists of a *p*-type substrate containing two heavily doped *n*-type regions, which act as source and drain. The gate consists of a metal layer on top of thin insulating layer of silicon dioxide between the source and drain (Fig. 33-20).

If a positive electric potential is applied to the metal gate, electrons are attracted to the area just below the silicon dioxide insulator in the *p*-type substrate, and a depletion region is formed. The area of the depletion region increases as the magnitude of the positive gate potential increases. When the gate potential is large and positive, the surface of the *p*-type substrate beneath the silicon dioxide insulator has an excess of electrons. Consequently a layer of *n*-type material, called an *inversion layer*, is induced *at the top of the p-type substrate*. The inversion layer is essentially an *n*-type channel through which electrons may flow from the heavily doped *n*-type source to the heavily doped *n*-type drain. The resistance depends on the width of the channel and is controlled by the gate potential. As the gate potential is reduced, the width of the channel decreases, and its resistance increases until the gate potential cannot sustain the inversion layer near the drain. This corresponds to the pinch-off condition, and at this point the current saturates.

In the *depletion mode* a permanent *n*-type channel is constructed between the source and

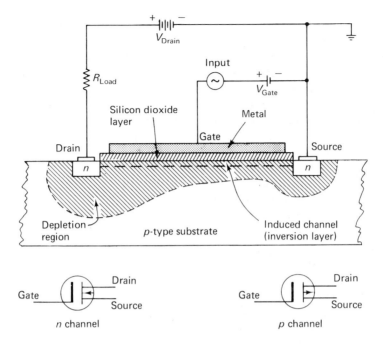

(a) n = Channel MOSFET Amplifier

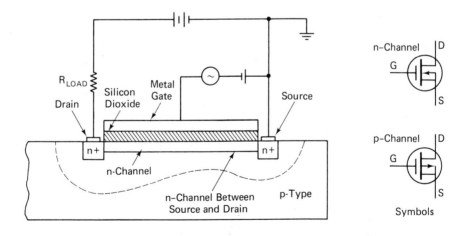

(b) n–Channel MOSFET Depletion Mode.

Figure 33-20
An *n*-channel MOSFET amplifier—enhancement mode.

the drain, and they conduct from source to drain without a gate bias. This mode is also different because the gate is given a *negative* electric potential, which repels electrons from the area below the silicon dioxide, decreasing the channel and increasing the resistance between the source and drain. However these devices can operate in either the enhancement or depletion modes.

Similarly a p-*channel* MOSFET consists of an *n*-type substrate containing heavily doped *p*-type source and drain. As before, the metal gate is isolated from the substrate by a silicon dioxide insulating layer. The operation of these devices is identical to that of the *n*-channel MOSFET's with opposite polarity bias and majority carrier.

In general MOSFET's have very high input impedances, $\sim 10^{12}\,\Omega$, and low noise characteristics. They are frequently used in amplifier, computer-switching and signal-chopping circuits. The rapid development in planar technology and integrated circuits has led to the development of CMOS (complementary MOS circuits) containing *p*-MOS and *n*-MOS devices and requiring very low power for logic circuits.

MOSFET's are also used as components in *charge-coupled devices* (CCD's) which have many applications, including memory circuits. These CCD's consist of a number of MOSFET's on a common substrate, each operating in a capacitive action to store or manipulate charges in the inversion layer

(Fig. 33-21). In the enhancement mode, the electrons in the inversion layer follow the higher potential V_{high} as it is transferred from one gate to the next. These charges are usually detected by measuring either the incremental capacitance, or surface potential, or by collecting them in a reverse-biased junction near the output.

Problems

3. Describe the operation of a *p*-channel JFET.

4. Describe the operations and structures of *p*-channel MOSFET's in both the enhancement and depletion modes.

33-9 UNIJUNCTION TRANSISTOR (UJT)

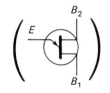

A *unijunction transistor* (UJT) is a three-terminal device with a single junction. It consists of a bar of *n*-type silicon with a *p*-type *emitter* embedded at the approximate center, forming a *p-n* junction (Fig. 33-22).

The equivalent circuit for the device consists of a diode and two resistors (R_{B1} and R_{B2}), representing the resistance of the semiconductor between the emitter and the bases. If the emitter is reverse biased, there is only

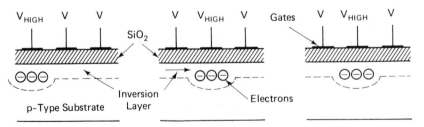

Figure 33-21
Charge coupled device.

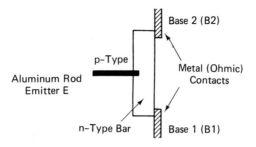

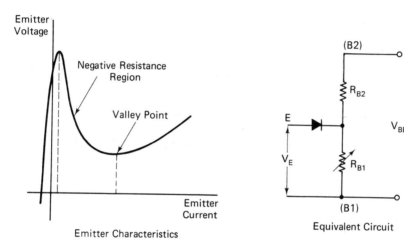

Figure 33-22
The UJT.

a small current, which gradually decreases as the emitter bias increases. When the emitter diode becomes forward biased, it injects holes into base 1 (moving towards the negative terminal of V_{BB}), and electrons move in the opposite direction. Therefore the resistance R_{B1} decreases as the emitter current increases because V_{BB} compensates by supplying more electrons to R_{B1} reducing its potential drop. This produces an area of negative resistance until saturation is reached.

These devices are frequently used in relaxation oscillators and control circuits. They can also be constructed from a *p*-type bar and an *n*-type emitter.

33-10 SILICON CONTROLLED RECTIFIER (SCR)

The silicon controlled rectifier (SCR) is a member of a group of *p-n-p-n* junction devices called *thyristors*. Other thyristors include the bidirectional triode thyristor (TRIAC) and the bidirectional trigger diode (DIAC).

An SCR is a switch or control device that consists of three *p-n* junctions. The outer *n*-type region is called the *cathode*, the outer *p*-type region is called the *anode*, and a thin central *p*-type region acts as the *gate* (Fig. 33-23).

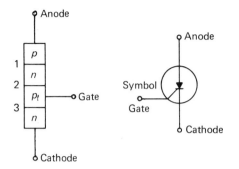

Figure 33-23
An SCR.

The SCR is forward biased when the anode is maintained at a positive potential with respect to the cathode; however the gate bias and the previous state of the SCR control the current. Even though the SCR is forward biased, there are two possible states (Fig. 33-24).

Forward Blocking—Off State

There is almost no current from anode to cathode when junction 3 between the gate and the cathode is reverse biased and the anode potential increases from a negative to a positive value. The gate is open or at a negative potential, and most of the potential difference between the anode and cathode corresponds to a reverse bias at the central *p-n* junction 2, effectively blocking the current. The maximum forward bias that can be applied without the SCR conducting is called the *forward-blocking voltage*. In the forward-blocking state the SCR has a high impedance.

Forward Conducting—On State

When junction 3 between the gate and cathode is forward biased (the gate has a positive potential), electrons diffuse from the cathode

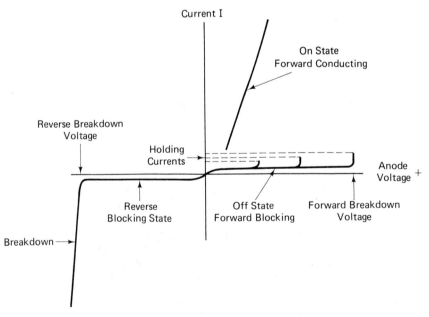

Figure 33-24
Current-voltage characteristics of an SCR.

through the thin gate to the thin central *n*-type region where they are collected. As the negative charge builds up in the central *n*-type region, the forward bias of junction 1 increases, and there is an increase in the diffusion of holes from the anode through the thin central *n*-type region to the gate, where they are collected. Thus the positive charge in the *p*-type gate increases, the forward bias of junction 3 increases, and even more electrons diffuse from the cathode. The diffusion currents of electrons and holes both depend on each other, and the total current increase is regenerative. The voltage required to produce conduction in the device is called the *forward breakover voltage.*

A steady state is reached very quickly (~ 1 μs) when the central junction 2 becomes forward biased because of the accumulation of charge. The current through the SCR depends almost entirely on the external resistance; the internal resistance of the SCR is very small.

The regenerative process makes it rather difficult to switch the device off merely by reversing the bias of the gate. However, when the anode to cathode current is reduced to below a particular value called the *holding current*, electrons and holes recombine faster than they are injected into the system, and the regeneration process ceases.

When the anode and cathode are reverse biased, the anode potential is negative with respect to the cathode potential. In this case junctions 1 and 3 are both reverse biased, and there is very little current.

Light-activated silicon controlled rectifiers (LASCR) consist of an SCR with a transparent window; these devices may be controlled by light or a gate bias. When light energy impinges on a LASCR, free electrons and holes are created. These carriers then start the regenerative process that turns the device on.

The fabrication of monolithic IC's is described below.

SCR's are used as a link between analog and digital electronics. A low power gate input can be used to produce very high power amplifications. These devices are available with high current and voltage ratings, and they operate efficiently with low power losses. They are used as solid-state relays to control and switch alternating currents (i.e. in motor controls, light dimming, and switches). SCR's are also found in regulated power supplies, protective circuits and many other areas of electronics.

TRIAC

Two SCR's connected in inverse parallel on the same semiconductor chip constitute a Triac. These devices can conduct in both directions and they act effectively as an ac switch.

33-11 MICROELECTRONICS-INTEGRATED CIRCUITS

There are three basic types of integrated circuit: *monolithic, thin film* and *thick film* (Fig. 33-25). Monolithic IC's are constructed on a single substrate of a single crystal semiconductor, usually silicon. Thin and thick film IC's are formed on the surface of an insulating material, such as glass or a ceramic. *Hybrid* IC's contain more than a single substrate. The term *hybrid* also applies to combinations of monolithic, thin film and thick film IC's, sometimes even with discrete components.

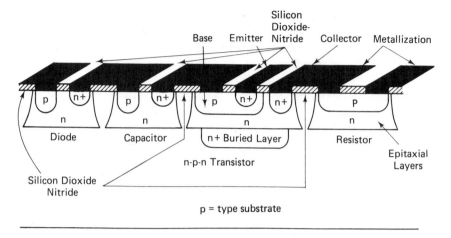

Figure 33-25
Typical structure of a simple IC.

Integrated circuits are also classified according to their function. *Digital* or *logic* IC's are used as switches; they are either "on" or "off." In computers these "on" and "off" states correspond to 0 and 1 respectively. Other ICs are called *linear* or *analog;* this classification includes operational amplifiers. All non digital IC's are called linear or analog IC's even if their operation is non-linear.

The actual cost of IC's has decreased in the last few years because of improvements in fabrication processes. Dust control and the development of new techniques, particularly those involving computer simulations, and new optical systems have resulted in large scale (LSI) and even very large scale (VLSI) integration. A single IC about 5 mm square may now contain thousands of individual circuit elements.

The Mask

1. Once a new circuit has been conceived, a computer is used to design, lay out and even test the circuit by simulation. The precise positions and sizes of the circuit ele-

ments are carefully checked on a large scale diagram constructed by the computer.*

2. A photographic pattern is produced for *each layer* in the integrated circuit by exposing a photographic plate to a computer-controlled light beam.* Each pattern indicates aspects such as the positions and relative sizes of the areas to be doped. These are checked for errors, and then each pattern is reproduced at precise positions on a single transparency to form a *mask* (Fig. 33-26). This can be done photographically using lasers for precision spacing or by a computer-driven electron beam.

The Substrate

A lightly doped (e.g. *p*-type), single crystal silicon wafer between 2.5 cm and 10 cm in diameter and about 0.3 mm thick serves as the base or *substrate* for a large number of IC's as described in Section 33-1 (Fig. 33-27). During

---

*In a few locations these diagrams are still drawn by hand.

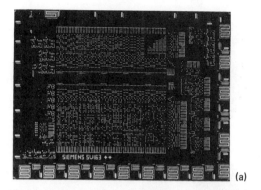

(a)

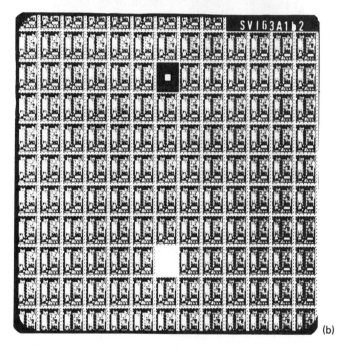

(b)

Figure 33-26
(a) Reticle used for master mask fabrication;
(b) A master mask. (Courtesy of Siemens
AG, Munich)

some stages of fabrication, large numbers of these wafers are processed simultaneously; this mass production results in lower costs. These wafers are ground and polished to a mirror finish.

Before the formation of an epitaxal layer, specific areas (below transistor sites, for example) in the substrate are strongly doped (e.g. $n+$-type), producing a low-resistivity *buried layer* to reduce series resistance for lateral currents. This is accomplished by means of the photolithographic process out-

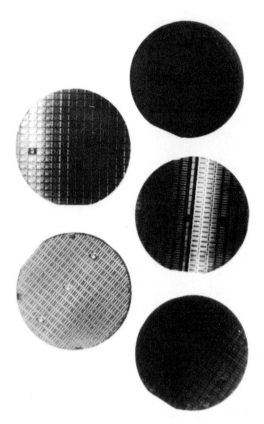

Figure 33-27
An IC at various stages of fabrication on a wafer.

lined later. The wafer is then heated to $\sim 1\,200°$ C in an atmosphere of silicon tetrachloride ($SiCl_4$) and steam (or hydrogen) that contains a suitable impurity (e.g. n-type phosphorus). The silicon tetrachloride decomposes, and the silicon (with the impurity) forms a thin layer ($\sim 20\ \mu$m) on top of the wafer. This layer of doped silicon is called the *epitaxal layer*. It has the same crystal structure as the silicon wafer.

A relatively new technique that is now being used is to deposit a silicon epitaxal layer on a sapphire wafer. These *silicon on sapphire* devices are very stable.

The wafer now undergoes a series of

photolithographic processes; one for each layer in the IC and one for the metal contacts.

Photolithographic Process

1. Oxidation. A number of wafers (~ 100 or more) are placed in a quartz boat that is inserted into a quartz tube in a furnace. The surface of the wafer is oxidized either by heating it to $\sim 1\,100°$ C in an atomsphere of steam or by heating it in an atmosphere of nitrogen and tetraethyl orthosilicate gases that react to form SiO_2, which then deposits on the surface. The thickness of the oxide layer is checked by a visual comparison of its color with an oxide—nitride color chart.

2. Photoresist. The SiO_2 surface is coated with a material called *photoresist* that is sensitive to ultraviolet light and, when polymerized, is resistant to a developer and acid-etching solutions. There are two basic types: *negative photoresist* (more common), which polymerizes when exposed to ultraviolet light; and *positive photoresist*, which is initially a polymer but becomes soluble after it has been exposed to ultraviolet light. A small amount of photoresist in a solvent is applied to the wafer, which is spun to give uniform thickness. The solvent is then evaporated, and the polymer hardens when heated.

3. Mask and Expose. The photoresist layer is covered with a mask and then irradiated with ultraviolet light, which polymerizes the exposed negative photoresist in areas behind the transparent sections of the mask. There are two common practices. The mask may be placed in contact with the photoresist, and the space between the surfaces may even be evacuated to produce a hard contact; this method produces sharp images in the photoresist, but it is hard on the masks. In the second process the mask may be located slightly apart from the coated surface; this technique requires very sophisticated optics to produce the sharp images required, but the lifetime of the mask is greatly extended. In both techniques the alignment of the mask is critical.

4. Develop and Etch. The mask is removed and the areas of *unexposed* negative photoresist are dissolved in a solvent (developer), leaving specific areas of the SiO_2 layer without a photoresist coating. These areas are then etched away when the wafer is dipped in hydrofluoric acid. The hard photoresist is dissolved in a solvent, and the wafer is washed

with ultrapure water. This leaves a "window" to the epitaxal layer at specific locations that are not covered by SiO_2. In some places ionized freon gas is now being used to "*dry etch*" the devices.

5. Diffusion. This process normally occurs in two stages: *predeposition*, in which a high concentration of dopant is deposited on the surface, and *diffusion*, which drives it into the surface. There are three basic techniques for predeposition. In one, the wafers are placed in a quartz boat in a diffusion furnace and surrounded by gases containing the

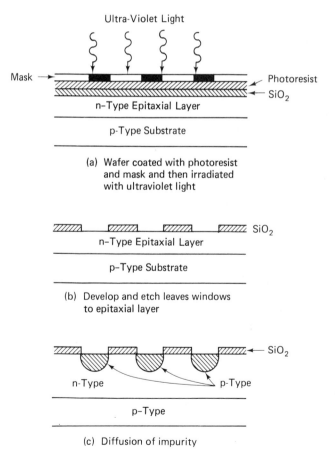

(a) Wafer coated with photoresist and mask and then irradiated with ultraviolet light

(b) Develop and etch leaves windows to epitaxial layer

(c) Diffusion of impurity

Figure 33-28
The photolithographic process.

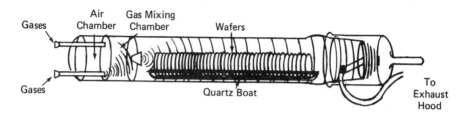

Figure 33-29
Diffusion furnace.

impurity, e.g. boron trichloride or phosphine with nitrogen, oxygen and argon. In a second process solid wafers of the impurity are alternated with the wafers in the quartz boat. In the third technique a liquid impurity is applied to the wafer, which is then spun to distribute the impurity before the wafer is placed in the furnace.

The wafer is then heated in steam and oxygen in the diffusion furnace (Fig. 33-29) to drive the dopant into the exposed areas in the epitaxal layer and to also "grow" an uneven SiO_2 layer on the surface for the next sequence.

This entire process beginning with oxidation is then repeated, using the next mask for the sequence of layers in the IC. An IC usually undergoes five of these diffusion processes in addition to the buried layer. The electrical isolation of the components from the silicon substrate is accomplished with either a high-resistance, reverse-biased *p-n* junction or a layer of SiO_2. This prevents undesirable connections between components.

In some cases *ion implantation* techniques are now used instead of diffusion to drive a dopant into the epitaxal layer. The dopant is ionized, accelerated in a high-potential difference ($\sim 10^5$ V) and directed onto the silicon wafer that has some areas selectively protected by the masking technique. These ions penetrate to a depth that depends on their kinetic energy, and their concentration may be accurately measured in terms of their

charge. However, as they strike the wafer, they damage the crystal lattice, and it must be repaired by an annealing process in a furnace.

Once all layers of the devices in the IC have been completed, a layer of SiO_2 or silicon nitride is deposited over the surface to protect it. A masking and diffusion technique is used to give some areas a strong n+ doping for ohmic contacts, and another oxide or nitride* layer is deposited on the surface.

6. Metallization. A masking technique is now used to form windows at locations where electrical contacts are required. A metal, normally aluminum, is evaporated (in a planetary) over the entire surface of the wafer, making contacts with the diffused areas through these contact windows. The same photoresist procedure is then used to form a protective polymer pattern on the metal, and the unprotected metal is etched away, leaving only the interconnections. A solvent is used to remove the remaining photoresist.

7. Scribe and Mount. Finally each IC on the wafer is tested, and the wafer is scribed and broken into individual IC's, which are mounted and packaged.

Active circuit elements, such as diodes,

---

*The nitride is etched with phosphoric acid, which attacks photoresist; therefore it too is coated with SiO_2.

transistors and SCR's, and all passive elements (resistors, capacitors) except inductors are easily formed by the planar process. Different resistors are obtained by varying the length and cross-sectional area of a segment of doped semiconductor through which the current passes. Capacitors are usually formed from reverse-biased *p-n* junctions. However, because of the complex etching processes, the tolerances of most individual resistors is usually 10% although the ratio of resistors in the same IC may be controlled to better than 1% (Fig. 33-30).

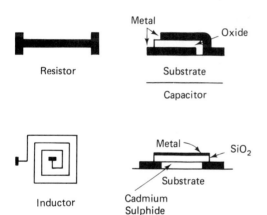

Figure 33-30
Film circuit elements.

Thin and Thick Film IC

Thin and thick film IC's are not as common as monolithic IC's. They are composed of thin layers or conductors, dielectrics and resistors that are deposited on a glass or ceramic substrate.

Thin films ($\sim 1 \ \mu$m thick) are normally made by vacuum evaporation, cathode-sputtering or vapor gas-plating techniques that deposit a thin layer of metal on the substrate. Masking and etching processes similar to those used in the fabrication of monolithic IC's produce the required film

patterns. *Thick films* are formed by forcing conductive and resistive materials through special screens so that they deposit in the required patterns on the substrate. The screens are made by coating a nylon or wire mesh with a photoresist that is covered with a mask and irradiated with ultraviolet light. Unexposed photoresist is then removed with a solvent, leaving windows through which materials can pass when they are deposited onto the substrate.

Precision resistors (1% tolerance) are formed in resistive films that are etched to the correct dimensions. Capacitors consist of a three-layer thin film "sandwich" of two metal plates and a metallic oxide (the dielectric). Inductors consist of a spiral of a metallic conductor on the substrate. Thick film transistors resemble MOSFET's, and they have similar characteristics.

Problems

5. Explain why a crystal wafer must undergo an annealing process after ion implantation.
6. Ion implantation is frequently used to "custom" manufacture devices. Explain why.

33-12 MAGNETIC BUBBLE MEMORIES

Magnetic bubble memories are relatively new devices that have very high storage densities. These devices are reliable, inexpensive, and non-volatile (retain data even after the power is cut off), and yet they are very fast and use very little power.

The device consists of a non-magnetic substrate of a ferrite or synthetic garnet* upon which a thin magnetic garnet epitaxal

---

*Such as gadolinium-gallium garnet, which is known as G^3.

layer is grown. Patterns of permalloy metal are then diffused onto the surface using a two-step photolithographic process. Many different types of patterns have been developed, some of which are TI-bar, Y-bar, contiguous disc, and chevrons. A pair of perpendicular coils, which generate a rotating magnetic field, surrounds the system, and the combination is sandwiched between permanent magnets (Fig. 33-31).

The permanent magnets create the bias field that shrinks the random magnetic domains in the magnetic material into small (~ 30 μm in diameter) cylindrical-shaped magnetic domains called *magnetic bubbles*. The presence of a bubble corresponds to a "one" and its absence is a "zero" in binary code.

To "write" information, a hairpin-shaped conductor loop is energized with a current

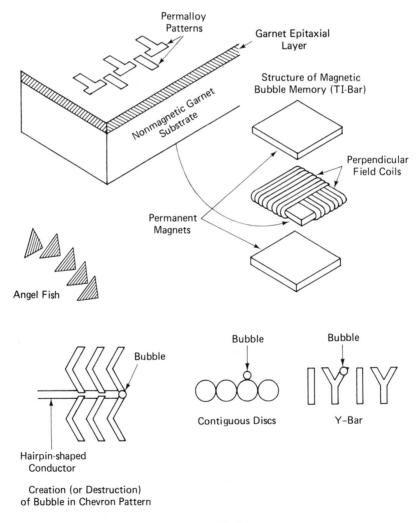

Figure 33-31
Structure of magnetic bubble memory.

pulse *reducing* the bias field so that a bubble is created inside the end of the loop.

Bubbles are moved from one location to another through the magnetic film (similar to the movement in a shift register) as the permalloy patterns assume different magnetizations in the rotating field of the perpendicular coils. Each rotation of the magnetic field moves the bubbles from one permalloy island to the next. The presence of a magnetic bubble may be detected (to read the memory) by the Hall voltage that it induces, or by the rotation of the plane of polarization of light that it produces, or by connecting the permalloy pattern to a bridge circuit which becomes unbalanced when a bubble passes.

A bubble is removed (to erase the memory) by *increasing* the bias field with the application of a current pulse in another conductor similar in shape to the "write" loop, or by moving it to a permalloy guard around the patterns.

REVIEW PROBLEMS

1. Describe the different fabrication processes for p-n junction diodes and where each type is used.

2. Why is the application of the tunnel diode limited?

3. What is meant by the terms n-MOS, p-MOS and CMOS?

4. Explain why the SCR is a very important device.

5. What is an ohmic contact and how is it formed?

6. What are the differences between thin film, thick film, and monolithic IC's?

We know that an atom consists of one or more electrons in orbit around a small, very dense, positively-charged nucleus, but the actual structure of the nucleus is still an active area of research. Some atomic nuclei are unstable and emit radiation as they spontaneously disintegrate. This property of the nucleus is called *radioactivity*. The radiation from a radioactive nucleus can be extremely harmful to living things, but in controlled doses it also has many useful applications. For example, nuclear radiation is used to kill cancer cells, to sterilize insect pests and to destroy harmful bacteria in foods.

During many nuclear reactions small amounts of mass are converted into enormous quantities of energy. Nuclear power is a possible solution to the world's energy crisis; however at present we are only able to control chain reactions in a nuclear fission process, and there are problems involving thermal pollution and the disposal of radioactive wastes. Possibly the most promising energy source of the future is a nuclear fusion process because the fuel is abundant and the process is relatively clean. At present, this type of nuclear reaction is the subject of considerable research.

34

The Atomic Nucleus and Nuclear Energy

34-1 NATURAL RADIOACTIVITY

Natural radioactivity was discovered by A. H. Becquerel in 1896 while he was investigating the possibility of a link between X-rays and fluorescence. Becquerel was studying uranium salts, which become fluorescent in sunlight, when he noticed that uranium emitted some form of radiation even if it had not been exposed to light or other irradiations. This radiation could penetrate black paper. The radiation depended only on the concentration of the uranium. It is not affected by external conditions, such as temperature and pressure variations, or by the chemical structure or physical state of the uranium compounds.

Soon after Becquerel's discovery of natural radioactivity from uranium, Pierre and Marie Curie isolated two new radioactive elements from pitchblende by a fractional crystalization process. These elements are called radium and polonium, and their radiations are much more intense than the radiation from uranium. Many other radioactive materials have since been discovered, and we now know that radioactivity originates at the nucleus of an atom.

There are three different kinds of nuclear radiation. Most radioactive substances emit all three kinds of radiation simultaneously.

Alpha Particles. An *alpha* (α) *particle* is the nucleus of a helium atom. It has a net positive charge, and it is deflected when it passes through a magnetic field (Fig. 34-1).

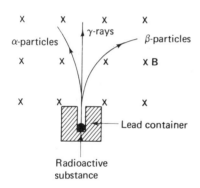

Figure 34-1
The paths of the three types of nuclear radiations in a perpendicular magnetic field.

These particles are the most massive and the least penetrating form of nuclear radiation. Their range in air at atmospheric pressure is normally less than ten centimeters. Most α particles cannot penetrate a single sheet of paper.

Beta Particles. Most *beta* (β) *particles* are electrons that are emitted from an unstable nucleus. They are more penetrating than

α particles and are also deflected when they pass through a magnetic field. A few unstable nuclei emit positively charged particles called *positrons*, which have the same mass and magnitude of charge as electrons. The term *beta emission* is used to describe electron or positron emission from an unstable nucleus; in order to distinguish between these particles, positrons are normally represented by the symbols β^+ or $_1e^0$.

Gamma Rays. Radioactive nuclei also emit *gamma* (γ) *rays* that are a form of high-energy (very short wavelength) electromagnetic radiation. Gamma rays are not deflected in a magnetic field and are the most penetrating kind of nuclear radiation. Their energy E depends on their frequency f and wavelength λ:

$$E = hf = \frac{hc}{\lambda} \qquad (34\text{-}1)$$

where h is Planck's constant and c is the speed of light.

Radiation Detection

Both α and β particles have charge; therefore, when they pass through a gas, they may ionize some of the gas molecules. Ionization also occurs if an orbital electron in a gas molecule absorbs sufficient energy from a γ-ray. These ionizations may be used to detect nuclear radiations (in ionization chambers and Geiger-Müller counters) or to track the path of the radiation (in cloud chambers).

A *cloud chamber* (Fig. 34-2) consists of a piston that fits snugly in a cylinder of gas saturated with water vapor. When the piston is withdrawn, the gas cools and becomes supersaturated. Visible water droplets then form on any ions in the chamber, indicating the tracks of the particles.

Spark chambers consist of a group of parallel plates with a potential difference across them. When an ionizing particle passes,

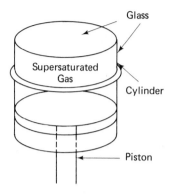

Figure 34-2
A cloud chamber.

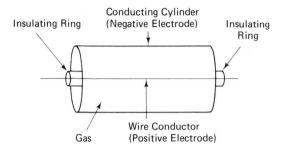

Figure 34-3
Proportion counter.

it leaves a track of sparks in the gaps between the plates.

In *gas detectors*, such as *ionization chambers*, *proportion counters* and *Geiger-Müller counters*, the radiation ionizes gas molecules that are then attracted to electrodes. As the electrons and ions reach these electrodes, they give up their charges, and this charge is measured with an ammeter.

In ionization chambers the electrodes are maintained at a relatively low potential difference, and the current must be amplified to produce a measurable signal. The electrodes in a *proportion counter* are maintained at a higher potential difference so that the elec-

trons produced by ionization gain sufficient kinetic energy to ionize other molecules as they move towards the anode (Fig. 34-3). In a Geiger-Müller operation a very high potential difference is maintained between the electrodes. As a result a single ionization event leads to electric breakdown of the gas, and a spark discharge occurs.

In a *photomultiplier tube* (Fig. 34-4), incident radiation produces a photon by fluorescence (scintillation) that strikes a photocathode, freeing an electron by the photoelectric effect. This photoelectron is then accelerated through large potential differences to a series of specially shaped electrodes. Each time the electron strikes an electrode, it frees other electrons that also

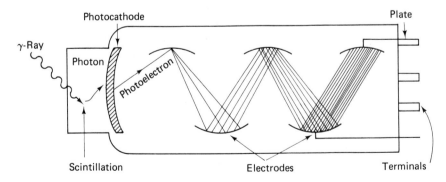

Figure 34-4
Photomultiplier tube.

accelerate to the next electrode. As a result an enormous current amplification is achieved by this cascading effect.

Semiconductor crystals are also used to detect radiations. In these devices the radiation excites electrons into the conduction band, leaving holes in the valence band and producing more free charge carriers that move to electrodes.

Radiation also discharges electroscopes. This principle is used in a *pocket dosimeter.*

Special tags containing a photographic emulsion are usually worn by people who are continually exposed to nuclear radiation. These tags are checked regularly to measure the exposure to nuclear radiation in terms of its effect on the photographic emulsion.

Nuclear radiation also produces fluorescence in some materials. The scintillations are used to detect and measure the radiation.

Problem

1. Determine the frequency and energy of a γ-ray if its wavelength is 1.5 pm.

34-2 NUCLEAR STRUCTURE

Individual chemical elements are characterized by their atomic number Z, which signifies the number of protons in their atomic nuclei. Atoms of different elements always have different atomic numbers, but atoms of the same chemical element always have equal atomic numbers. For example, lithium always has three nuclear protons and an atomic number $Z = 3$. (See the periodic table in the Appendices.) Each nuclear proton has a charge equal to one electronic charge unit $e = 1.6021 \times 10^{-19}$ C and a mass $m_p = 1.6726 \times 10^{-27}$ kg.

Other particles called *neutrons* are also located in the nuclei of most atoms. These particles have no net charges, and their mass $m_n = 1.6748 \times 10^{-27}$ kg is almost the same as the mass of a single proton. The masses of individual protons and neutrons are also approximately equal to one unified atomic mass unit $1 \text{ u} = 1.660\ 5 \times 10^{-27}$ kg. Since they constitute a part of an atomic nucleus, protons and neutrons are collectively called *nucleons.*

Even though atoms of the same chemical element always have equal atomic numbers, their nuclei may contain different numbers of neutrons; therefore they may have different atomic masses. *Isotopes* are atoms of the same chemical element that have different atomic masses because their nuclei have different numbers of neutrons. The *mass number A* of an isotope is equal to the total number of nucleons (protons and neutrons) in its nucleus. Therefore, if N is the number of neutrons in an isotope of element E that has an atomic number Z, the mass number is:

$$A = Z + N \qquad (34\text{-}2)$$

This atom is represented symbolically by $_Z E^A$.

When an atom is characterized in terms of its nuclear structure, i.e. the number of neutrons and protons in its nucleus, it is called a *nuclide.* Isotopes are nuclides that have the same atomic number Z. Nuclides that have the same mass numbers but different atomic numbers are called *isobars.*

Example 34-1

Determine the number of neutrons in the nuclei of the oxygen isotopes (a) $_8 O^{16}$ and (b) $_8 O^{17}$.

Solution:

(a) $\qquad N = A - Z = 16 - 8 = 8.$

(b) $\qquad N = A - Z = 17 - 8 = 9.$

All nucleons have radii $r_0 \approx 1.2 \times 10^{-15}$ m. Since the volume of a nucleus is approximately proportional to the number of nucleons (the mass number A) that it contains, the radius of a nucleus is:

$$R \sim r_0 \sqrt[3]{A} \sim (1.2 \times 10^{-15} \text{ m})\sqrt[3]{A} \qquad (34\text{-}3)$$

The densities of nuclei are extremely large $\sim 2 \times 10^{17}$ kg/m³; at the earth's surface one cubic foot of nuclear material would weigh $\sim 1.5 \times 10^{16}$ lb!

Problems

2. Determine the number of neutrons in the following nuclides (a) $_1H^3$, (b) $_{82}Pb^{207}$, (c) $_{90}Th^{232}$, (d) $_{92}U^{238}$, (e) $_{94}Pu^{239}$.

3. Write the symbolic representation of (a) a bismuth atom that has 126 neutrons in the nucleus, (b) a carbon atom that has 8 neutrons in the nucleus, and (c) a cobalt atom that has 32 neutrons in the nucleus.

4. Determine the approximate nuclear radii of (a) $_{92}U^{238}$, (b) $_8O^{16}$, (c) $_3Li^7$, (d) $_2He^4$.

34-3 BINDING ENERGY

The masses of nuclides may be determined very accurately with a mass spectrograph. From the analysis of these nuclide masses, it is found that the atomic mass of a stable nuclide is less than the sum of the masses of its "free" constituent particles. The mass that is "lost" as the protons, neutrons and electrons combine to form a stable nuclide called the *mass defect* (Δm). It is converted directly into energy:

$$\text{mass defect} = \text{sum of masses of}$$
$$\text{constituent particles} \quad (34\text{-}4)$$
$$- \text{ mass of nuclide}$$

In his theory of special relativity, Albert Einstein determined that there is an equivalence between mass and energy. In fact a mass m is equivalent to an energy:

$$E = mc^2 \quad (34\text{-}5)$$

where c is the speed of light.

Example 34-2

Determine the energy equivalent of a 1.000 u mass.

Solution:

$$1.000 \text{ u} = 1.6605 \times 10^{-27} \text{ kg}.$$

Thus $E = mc^2$

$$= (1.6605 \times 10^{-27} \text{ kg})(2.998 \times 10^8 \text{ m/s})^2$$
$$= 1.492 \times 10^{-10} \text{ J}$$

or $\dfrac{1.492 \times 10^{-10} \text{ J}}{1.602\ 1 \times 10^{-19} \text{ J/eV}} = 9.313 \times 10^8 \text{ eV}$

$$= 931.3 \text{ MeV}$$

Thus 1 u is equivalent to 931 MeV. This is an important result!

A stable nuclide represents a low energy state for its constituent particles. The energy equivalent of the mass defect is called the *binding energy* of the nuclide. It represents the minimum energy that is required to break the nuclide into its constituent particles:

$$\text{Binding energy } E = (\Delta m)c^2 \quad (34\text{-}6)$$

Example 34-3

A helium nuclide $_2He^3$ has a mass of 3.016 030 u, and it is composed of two hydrogen $_1H^1$ atoms and a neutron $_0n^1$. If each isolated $_1H^1$ atom has a mass of 1.007 825 u and an isolated neutron has a mass of 1.008 665 u, find (a) the mass defect and (b) the binding energy.

Solution:

(a) In terms of their masses $\Delta m = [2(_1H^1) + _0n^1] - _2He^3$ or $\Delta m = [2(1.007\ 825 \text{ u}) + 1.008\ 665 \text{ u}] - 3.016\ 030 \text{ u} = 0.008\ 285 \text{ u}$

(b) $E = (\Delta m)c^2 = (0.008\ 285 \text{ u})(931 \text{ MeV/u})$
$$= 7.71 \text{ MeV}$$

That is, a minimum energy of 7.71 MeV is required to break the $_2He^3$ nuclide into two hydrogen $_1H^1$ atoms and a single neutron.

The average binding energy per nucleon is an indication of the relative stability of a nuclide. Nuclides that have mass numbers between approximately 40 and 100 are the most stable because they have the largest

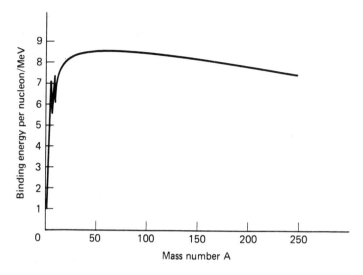

Figure 34-5
The average binding energy per nucleon versus the mass number of the corresponding nucleus.

binding energies per nucleon (Fig. 34-5). The lightest nuclides have the lowest binding energies per nucleon. Consequently a considerable amount of energy is released during *nuclear fusion* when two light nuclides combine to form a more stable heavier nuclide. Similarly energy is released during a *nuclear fission* process when one of the heaviest nuclides, such as $_{92}U^{238}$, breaks into two or more lighter nuclides.

Problems

5. Determine the average binding energy per nucleon in the nuclide $_2He^4$ if its mass is 4.002 603 u, and it is composed of two $_1H^1$ nuclides and two neutrons.

6. Determine the average binding energy per nucleon between the 92 $_1H^1$ nuclides and the 143 neutrons in the nuclide $_{92}U^{235}$ if its mass is 235.043 915 u.

7. Calculate the average binding energy per nucleon between the 17 $_1H^1$ nuclides and the 18 neutrons in the nuclide $_{17}Cl^{35}$ if its mass is 34.968 851 u.

34-4 NUCLEAR DECAY

There are 284 stable nuclides distributed among 83 elements. All but twenty of these elements have two or more stable isotopes. In the nuclei of all stable nuclides except $_1H^1$ or $_2He^3$, the number of neutrons is greater than or equal to the number of protons (Fig. 34-6). The lighter stable nuclides have approximately equal numbers of neutrons and protons, but the heavier stable nuclides have more neutrons than protons.

With the exception of hydrogen, all elements contain two or more positively charged protons in their atomic nuclei, and there are relatively strong Coulomb repulsive forces between these protons. However the existence of many stable nuclides indicates that very strong attractive forces must also be present

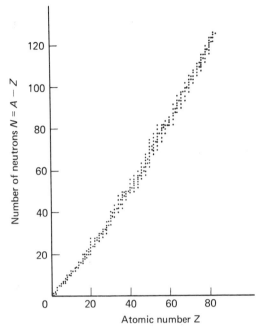

Figure 34-6
The relationship between the number of neutrons and the number of protons in stable nuclides.

in order to bind the nucleons together. These nuclear attractive forces are stronger than Coulomb forces, but they only act over short distances ($\sim 3 \times 10^{-15}$ m). In large atomic nuclei a single nucleon is only attracted to other nucleons that are within this range, but the Coulomb repulsive forces between the protons act over distances that are much greater than nuclear dimensions. As the atomic number Z of the stable nuclides increases, the ratio of the number of neutrons to the number of protons in the nuclei increases. The larger total nuclear attractive force due to the increase in this ratio is required to balance the stronger Coulomb repulsion between the nuclear protons. If the Coulomb repulsive forces are larger than the nuclear attractive forces, the nuclide is unstable, and it disintegrates.

When an atomic nucleus disintegrates by

emitting an α or β particle, its atomic number Z changes. Consequently the atom is actually transformed from one element into another. This process is called *transmutation*. The original unstable nuclide is called the *parent*, and the nuclide that remains after the nuclear disintegration is called the *daughter*. In many cases the daughter is also radioactive and decays farther.

Nuclear reactions are often described in terms of simple balanced equations based on the conservation of the total mass number and total charge (atomic number) of the particles that are involved. In each case the *algebraic sum* of the subscripts (the atomic numbers and charges) is the same on both sides of the equation. The daughter element is determined from the atomic number of the nucleus after it has emitted the α or β particle. The sum of the mass numbers is also the same on both sides of the equation.

Alpha Decay. An α particle is the nucleus of the helium isotope $_2\text{He}^4$ that consists of two protons and two neutrons. Therefore, if an unstable nucleus emits an α particle, its atomic number decreases by two, and its mass number decreases by four. For example, when the unstable uranium nuclide $_{92}\text{U}^{235}$ decays by emitting an α particle, the atomic number of the remaining nuclide is reduced by 2 to 90, and its mass number decreases by 4 to 231. From the periodic table the daughter element is thorium (Th). This nuclear reaction is described by the balanced equation:

$$_{92}\text{U}^{235} \longrightarrow {}_2\text{He}^4 + {}_{90}\text{Th}^{231}$$

Beta Decay. A β particle is a fast electron ($_{-1}e^0$), which has a negative charge and a mass that is much less than 1 u. Therefore, if a nucleus emits a β particle, its atomic number must increase by one, but its mass number remains constant. For example, when a β particle is emitted from the unstable lead nuclide $_{82}\text{Pb}^{210}$, the atomic number increases by 1 to 83, which corresponds to the element bismuth.

$$_{82}Pb^{210} \longrightarrow {}_{83}Bi^{210} + {}_{-1}e^0$$

This type of nuclear decay occurs when there are too many neutrons in the nucleus. Electrons cannot exist in an atomic nucleus; they are created and immediately ejected from the nucleus, along with another particle called an *antineutrino* ($\bar{\nu}$), when a neutron $_0n^1$ is transformed into a proton $_1p^1$.

$$_0n^1 \longrightarrow {}_1p^1 + {}_{-1}e^0 + \bar{\nu}$$

A few unstable nuclides emit positrons ($_1e^0$) in order to increase their neutron to proton ratio. These positrons and neutrinos are ejected when a proton is transformed into a neutron:

$$_1p^1 \longrightarrow {}_0n^1 + {}_1e^0 + \nu$$

Gamma Decay. After an unstable nucleus has disintegrated, it may be left in an excited state, and it reduces its energy by emitting a γ-ray without changing its atomic and mass numbers. Nuclides that have the same atomic number and same mass number but exist in different excited states are called *nuclear isomers*.

Radioactive Series

Radioactive nuclides often disintegrate to form radioactive daughters that decay farther. All *naturally occurring* radioactive nuclides are members of one of three groups of nuclides that are involved in a series of parent-daughter transmutations; each group of nuclides is called a *radioactive series*. Radioactive members of a radioactive series always decay to another nuclide in the same series. All natural radioactive series finally end at a stable lead isotope.

Problems

8. What kind of particle is emitted when (a) the nuclide $_{91}Pa^{234}$ decays to $_{92}U^{234}$, (b)

the nuclide $_6C^{14}$ decays to $_7N^{14}$, (c) the nuclide $_{84}Po^{214}$ decays to $_{82}Pb^{210}$, (d) the nuclide $_8O^{15}$ decays to $_7N^{15}$.

9. Complete the following nuclear equations:
   (a) $_{92}U^{238} \longrightarrow {}_2He^4 + \underline{\quad} + \gamma$
   (b) $_{88}Ra^{228} \longrightarrow {}_{-1}e^0 + \underline{\quad} + \gamma$
   (c) $\underline{\quad} \longrightarrow {}_2He^4 + {}_{88}Ra^{226} + \gamma$
   (d) $_3Li^8 \longrightarrow \underline{\quad} + {}_4Be^8 + \gamma$
   (e) $_{10}Ne^{19} \longrightarrow {}_{+1}e^0 + \underline{\quad} + \gamma$
   (f) $_{88}Ra^{223} \longrightarrow \underline{\quad} + {}_{86}Em^{219} + \gamma$

34-5 NUCLEAR ACTIVITY

The rate at which nuclear disintegrations occur in a material is called its *activity*. Nuclear activity depends on the nature of the material and the number of unstable nuclei that it contains. The SI unit of nuclear activity is called a *becquerel* (Bq); one becquerel is equal to one disintegration per second.*

Radiation doses are expressed in terms of the energy of the ionizing radiation per unit mass of the absorbing material. In SI the unit is called a *gray* (Gy), which is the absorbed dose of 1 J/kg.†

Consider a material in which one kind of unstable parent isotope decays to a stable daughter. Nuclear disintegration is a random process that can only be analyzed in terms of statistics. We cannot predict when a particular nucleus will disintegrate or even the order in which the nuclei will decay. At any instant the activity of a material is directly proportional to the number N of unstable nuclei that it contains. The rate at which the number of unstable nuclei changes as they disintegrate is:

---

*This unit replaces the curie (Ci), which equals 37 GBq, i.e. 3.7×10^{10}/s.

†The *röntgen* (R) is a special unit for X-ray and γ-rays that will be replaced by the gray, 1 R = 2.58×10^{-4} C of ion production per kilogram.

$$\frac{\Delta N}{\Delta t} \propto -N \quad \text{or} \quad \frac{\Delta N}{\Delta t} = \lambda N \quad (34\text{-}7)$$

A negative sign is required because the number of unstable nuclei is decreasing. The proportionality constant λ is known as the *disintegration constant* for the isotope; it is independent of the chemical or physical state of the isotope. From Eq. 34-7 it can be shown that, if the material originally contained N_0 unstable nuclei, after some elapsed time t the number of unstable nuclei that remain is:

$$N = N_0 e^{-\lambda t} \quad (34\text{-}8)$$

This "exponential" function is plotted in Fig. 34-7. The nuclear activity of the sample after the elapsed time t is:

$$A = \lambda N = \lambda N_0 e^{-\lambda t} = A_0 e^{-\lambda t} \quad (34\text{-}9)$$

where A_0 is the initial activity.

The *half life* $T_{1/2}$ of a radioactive isotope is the time required for one half of the original number of its nuclei to disintegrate. Rearranging Eq. 34-8 with $t = T_{1/2}$ and $N = N_0/2$, we obtain:

$$e^{\lambda T_{1/2}} = \frac{N_0}{N} = \frac{N_0}{N_0/2} = 2$$

Therefore, the half life of an isotope that has a disintegration constant λ is:

$$T_{1/2} = \frac{\ln 2}{\lambda} = \frac{0.693}{\lambda} \quad (34\text{-}10)$$

Example 34-4

A substance initially contains 5.0 mg of the unstable thorium isotope $_{90}\text{Th}^{230}$, which has a half life of 8.0×10^4 a.* Determine (a) the disintegration constant λ, (b) the initial activity due to the thorium, (c) the activity of the thorium after 2.4×10^5 a.

Solution:

(a) $\lambda = \dfrac{0.693}{T_{1/2}} = \dfrac{0.693}{8.0 \times 10^4 \text{ a}} = 8.66 \times 10^{-6}/\text{a}$

or $\lambda = \dfrac{8.66 \times 10^{-6}/\text{a}}{(365 \text{ d/a})(24 \text{ h/d})(3600 \text{ s/h})}$

$= 2.75 \times 10^{-13}/\text{s}$

(b) Since 1 mol (230 g) contains 6.023×10^{23} atoms, in 5.0 mg the initial number of unstable $_{90}\text{Th}^{230}$ atoms is:

$$N_0 = \frac{5.0 \times 10^{-3} \text{ g}}{230 \text{ g}}(6.023 \times 10^{23} \text{ atoms})$$

$$= 1.31 \times 10^{19} \text{ atoms}$$

Thus the initial nuclear activity is:

$$A_0 = \lambda N_0 = (2.75 \times 10^{-13}/\text{s})$$
$$\times (1.31 \times 10^{19} \text{ atoms})$$
$$= 3.60 \times 10^6 \text{ Bq}$$

*The SI symbol for years is a.

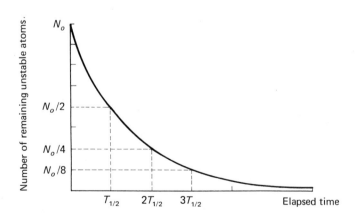

Figure 34-7
Nuclear decay curve.

(c) $A = A_0 e^{-\lambda t}$

$\quad = (3.60 \times 10^6 \text{ Bq}) e^{-(8.66 \times 10^{-6}/\text{a})(2.4 \times 10^5 \text{ a})}$

$\quad = 4.5 \times 10^5 \text{ Bq}$

Note that the elapsed time is equal to three times the half life; thus the activity has been reduced by a factor of $(\frac{1}{2})^3$.

Problems

10. Determine the half life of the radium isotope $_{88}\text{Ra}^{226}$ if its disintegration constant is $1.36 \times 10^{-11}/\text{s}$.

11. If the unstable lead isotope $_{82}\text{Pb}^{212}$ has a half life of 10.6 h, what is its disintegration constant?

12. What fraction of a sample of the unstable bismuth isotope $_{83}\text{Bi}^{210}$ remains after an elapsed time of 25 d its half life is 50 d?

13. A sample of the unstable thallium isotope $_{81}\text{Tl}^{206}$ has an initial activity of 703 kBq and an activity of 159 kBq after 9.0 min. Determine (a) the disintegration constant and (b) the half life of the thallium.

14. A substance initially contains 2.00 mg of the unstable uranium isotope $_{92}\text{U}^{234}$ that has a half life of 2.48×10^5 a. Determine (a) the disintegration constant, (b) the initial activity of the uranium $_{92}\text{U}^{234}$ in the sample, (c) the activity due to that uranium after 7.44×10^5 a.

34-6 ARTIFICIAL TRANSMUTATION

After the discovery of natural radioactivity scientists began to investigate the possibility of producing nuclear disintegration by bombarding stable nuclides with other particles.

Bombardment with Alpha Particles

The first artificial disintegration was observed by Rutherford in 1919 when he bombarded

nitrogen gas with high energy α particles and produced oxygen and high-energy protons. In this reaction, when an α particle overcomes the Coulomb repulsion, it penetrates and combines with the nucleus of a nitrogen atom to form an unstable isotope of fluorine. The unstable fluorine disintegrates by emitting a high-energy proton, leaving a stable isotope of oxygen. This artificially induced transmutation is represented by the equation:

$$_7\text{N}^{14} + {}_2\text{He}^4 \longrightarrow [{}_9\text{F}^{18}] \longrightarrow {}_8\text{O}^{17} + {}_1\text{H}^1$$

Many other relatively light elements undergo similar transmutations when they are bombarded with high-energy α particles.

Cockcroft-Walton Experiment. Between 1930 and 1932, J. D. Cockcroft and E. T. Walton developed a device in which protons could be accelerated by constant high voltages. When they bombarded a lithium target with the protons from their accelerator, Cockcroft and Walton observed that α particles were produced and a considerable amount of energy was released. During this reaction the proton is captured by the lithium nucleus to form an unstable isotope of beryllium that disintegrates into two α particles.

$$_3\text{Li}^7 + {}_1\text{H}^1 \longrightarrow [{}_4\text{Be}^8] \longrightarrow {}_2\text{He}^4 + {}_2\text{He}^4$$

Einstein's mass–energy relationship was quantitatively proved by careful measurements of the energies and masses that are involved in this type of nuclear reaction.

The Neutron

The existence of a particle that has no net charge and a mass approximately equal to 1.0 u was suggested by Rutherford in 1920, but this particle (the *neutron*) was not discovered until 1932.

Bothe and Becker (1930) discovered that a very penetrating form of radiation was produced when beryllium or boron was bombarded with α particles; they assumed that this radiation was high-energy γ-rays. Then in 1932 T. Curie and Joliot observed that the penetrating radiation ejected high-energy protons from substances that contained hydrogen, such as paraffin; they also believed that the radiation was very high-energy γ rays. Finally J. Chadwick in 1932 repeated these experiments. He also passed the penetrative radiation through a cloud chamber containing nitrogen gas and observed the tracks of nitrogen ions when they interacted with the penetrating radiation. Chadwick then completed a series of calculations involving energy and mass conservation and concluded that the penetrating rays were particles that had a mass of approximately 1.0 u and no charge. He called these particles *neutrons* ($_0n^1$). The production of neutrons when α particles bombard beryllium is represented by the balanced equation:

$$_4Be^9 + {_2}He^4 \longrightarrow [_6C^{13}] \longrightarrow {_6}C^{12} + {_0}n^1$$

Neutrons have no net charge; therefore they are not deflected by electric or magnetic fields. Since they do not experience electric forces, neutrons are only stopped by direct collisions with other particles. Many nuclear transmutations may be induced by bombarding materials with neutrons. Slowly moving neutrons, called *thermal neutrons*, are used to induce fission in nuclear reactors.

Photodisintegration

Transmutation may also be induced in some nuclides if they are bombarded with high-energy photons; this process is known as *photodisintegration*. The energy of the incident photon must be greater than or equal to the binding energy of the target nuclide.

Problem

15. Complete the following equations that describe artificial transmutations:
    (a) $_5B^{10} + {_2}He^4 \longrightarrow$ ____ $+ {_1}H^1$
    (b) $_9F^{19} + {_1}H^1 \longrightarrow$ ____ $+ {_2}H^4$
    (c) $_{13}Al^{27} +$ ____ $\longrightarrow {_{14}}Si^{30} + {_1}H^1$
    (d) ____ $+ {_1}H^1 \longrightarrow {_{12}}Mg^{24} + {_2}He^4$
    (e) $_5B^{10} + {_0}n^1 \longrightarrow$ ____ $+ {_3}Li^7$
    (f) $_{13}Al^{27} +$ ____ $\longrightarrow {_{11}}Na^{24} + {_2}He^4$
    (g) $_{15}P^{31} + \gamma \longrightarrow {_{15}}P^{30} +$ ____

34-7 Q-VALUE

Mass is converted into energy when particles combine to form a stable nuclide or when an unstable nuclide disintegrates. The reverse is also true: Energy is converted into mass when a stable nuclide is broken apart or when an unstable nuclide is created. In each case Einstein's relationship $E = mc^2$ may be used to determine the equivalence between the mass and energy.

The energy that is evolved or absorbed during a nuclear reaction is known as the *disintegration energy* or *Q-value*. When mass is converted into energy, the reaction is called *exothermic* or *exoergic*, and the *Q*-value is positive; if energy is converted into mass, the reaction is called *endothermic* or *endoergic*, and the *Q*-value is negative. In any nuclear reaction, the *Q*-value is the *energy equivalent* of:

$$Q = \text{total original mass} - \text{total final mass}$$

$$(34\text{-}11)$$

Example 34-5

Determine the *Q*-value of the following nuclear reaction

$$_9F^{19} + {_1}H^1 \longrightarrow {_8}O^{16} + {_2}He^4$$

if the nuclide masses are $_9F^{19} \equiv 18.998\ 405$ u, $_1H^1 \equiv 1.007\ 825$ u, $_8O^{16} \equiv 15.994\ 915$ u and $_2He^4 \equiv 4.002\ 603$ u.

Solution:

No β particles are involved; therefore the masses of the electrons may be ignored because the number of electrons remains constant. Thus, in terms of mass units, the *Q* value is:

$$Q = (_9F^{19} + _1H^1) - (_8O^{16} + _2He^4)$$
$$= (18.998\ 405\ u + 1.007\ 825\ u)$$
$$- (15.994\ 915\ u + 4.002\ 603\ u)$$
$$= 0.008\ 712\ u \quad or \quad (0.008\ 712)(931\ MeV)$$
$$= 8.11\ MeV$$

Problems

16. Determine the Q-value of the reaction:

$$_{12}Mg^{25} \longrightarrow _{11}Na^{24} + _1H^1$$

if the nuclide masses are $_{11}Na^{24} \equiv 23.990\ 961\ u$, $_{12}Mg^{25} \equiv 24.985\ 838\ u$ and $_1H^1 \equiv 1.007\ 825\ u$.

17. Determine the Q-value of the reaction:

$$_{13}Al^{27} + _0n^1 \longrightarrow _{11}Na^{24} + _2He^4$$

if the nuclide masses are $_{13}Al^{27} \equiv 26.981\ 539\ u$, $_0n^1 \equiv 1.008\ 665\ u$, $_{11}Na^{24} \equiv 23.990\ 961\ u$ and $_2He^4 \equiv 4.002\ 603\ u$.

34-8 CHAIN REACTIONS

Enrico Fermi and his collaborators in 1934 bombarded several elements with neutrons. When the target element was uranium, they observed a number of radioactivities; they believed that these activities were due to *transuranic* elements, i.e. elements with atomic numbers greater than 92. A series of experiments followed, and in 1939 Hahn and Strassmann determined that the radioactivities were due to an unstable barium nuclide rather than a transuranic element. Meitner and Frisch (1939) concluded that the uranium nucleus captured the neutron, and then it split into two fragment nuclei with approximately equal masses. They called this process *nuclear fission.*

It is the isotope $_{92}U^{235}$ that undergoes nuclear fission when uranium is bombarded with slow neutrons. A $_{92}U^{235}$ nucleus captures a thermal neutron, forming the isotope $_{92}U^{236}$, which releases an integer number of neutrons (an average of ~ 2.4) as it divides

into two nearly equal fragments $_{z_1}X^{A_1}$ and $_{z_2}Y^{A_2}$. Many combinations of fission fragments $_{z_1}X^{A_1}$ and $_{z_2}Y^{A_2}$ are possible, and they are usually unstable because they have an excess of neutrons. Consequently they decay by emitting a number of β particles. The fission fragments have larger binding energies per nucleon than the heavier uranium nuclide (see Fig. 34-5). Consequently mass is converted into energy (~ 200 MeV) in nuclear fission. A general nuclear fission process of a $_{92}U^{235}$ nuclide is represented by the equation:

$$_{92}U^{235} + _0n^1 \longrightarrow _{92}U^{236}$$
$$\longrightarrow _{z_1}X^{A_1} + _{z_2}Y^{A_2} + \sim 2.4_0n^1 + \sim 200\ MeV$$

where Z_1 and Z_2 are the atomic numbers, and A_1 and A_2 are the mass numbers of the fragment elements X and Y.

Generally two or three neutrons are released in a single nuclear fission, and each of these neutrons can initiate nuclear fission in other $_{92}U^{235}$ atoms, releasing even more neutrons to continue the fission processes. If these reactions are not controlled, there is a rapid growth in the rate at which the fission processes occur, and the energy that is released also increases. This is called a *chain reaction.*

In order to maintain a chain reaction in a mass of pure fissionable material, such as $_{92}U^{235}$, most of the neutrons that are released in the fission processes must be captured by the nuclei of the $_{92}U^{235}$ atoms. The probability that a neutron will be captured depends on the dimensions (mass) of the fissionable material. If the dimensions are too small, most of the neutrons pass through the fissionable material without being captured, and the chain reaction cannot occur. A mass of fissionable material is said to be of *critical size* if it is just sufficient to sustain a chain reaction. This occurs when an average of only *one* of the neutrons that are produced in *each* fission event is captured and initiates fission in another atom; the rest of the neutrons escape or are captured by other materials.

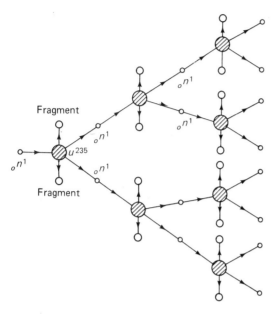

Figure 34-8
A chain reaction.

34-9 NUCLEAR POWER PLANTS

A *nuclear reactor* (or *atomic pile*) is a device in which a self-sustaining chain reaction can be controlled. It is the main component of a nuclear power plant. The first successful reactor was constructed at the University of Chicago in 1942.

There are four basic components in a reactor: a fissionable fuel, a moderator, liquid coolant and control rods.

Reactor Fuels. In most reactors the uranium isotope $_{92}U^{235}$ or the plutonium isotope $_{92}Pu^{239}$ is utilized as the fuel. The transuranic element plutonium does not occur naturally; it is manufactured by bombarding a sample of $_{92}U^{238}$ with neutrons. When a $_{92}U^{238}$ nucleus captures a neutron, it forms the very unstable isotope $_{92}U^{239}$ that decays to the unstable neptunium isotope $_{93}Np^{239}$ by emitting a β particle; the neptunium

decays to $_{94}Pu^{239}$ by emitting another β particle:

$$_{92}U^{238} + {}_0n^1 \longrightarrow {}_{92}U^{239} \longrightarrow {}_{93}Np^{239} + {}_{-1}e^0$$
then $\quad {}_{93}Np^{239} \longrightarrow {}_{94}Pu^{239} + {}_{-1}e^0$

Pellets of fuel are placed in aluminum sheaths to form *fuel bundles* that are located in the *fuel tubes* of the reactor so that they are surrounded by the coolant.

Moderators. The neutrons that are released during nuclear fission have relatively high speeds and are not easily captured by $_{92}U^{235}$ or $_{94}Pu^{239}$ nuclei. In order to reduce the speed of the neutrons and to increase the probability of their capture, the fuel is surrounded by a moderator, such as graphite, ordinary water or heavy water. The fast neutrons reduce their kinetic energy when they make elastic collisions with the carbon or hydrogen nuclei in the moderator. These slow neutrons are called *thermal neutrons*.

Liquid Coolant. The heat that is generated by the nuclear fission processes is removed from the reactor by a liquid coolant, such as ordinary water, heavy water, liquid sodium or an organic liquid. In many reactors this liquid acts as both the moderator and the coolant.

Control Rods. A chain reaction in the reactor is controlled by means of boron or cadmium control rods, which are very effective absorbers of neutrons. When these control rods are inserted into the reactor's core, they retard the chain reaction by capturing some of the neutrons.

The operation of a typical nuclear power plant is illustrated in Fig. 34-9. A controlled amount of heat is generated by nuclear fission in the reactor. This heat is transferred by the coolant to the heat exchanger, where it produces steam. As the steam flows to the lower pressure in the condenser, it turns the blades of the turbine. In the condenser the steam is

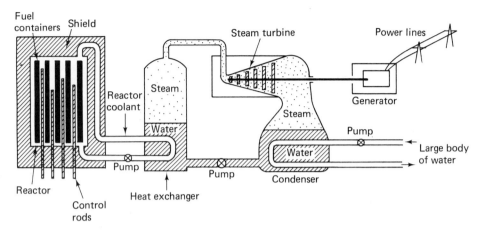

Figure 34-9
A schematic diagram of a nuclear power
plant.

condensed when it makes thermal contact
with water from a lake or river, and it is
returned to the heat exchanger. The shaft of
the turbine is connected directly to the armature of a generator. Electricity is produced as
the steam drives the blades of the turbine.

Nuclear fuels are inexpensive and very
compact, and nuclear power plants are
sources of inexpensive electrical energy. However they produce radioactive wastes, and
they raise the temperature of the body of
water into which waste heat is discharged.
There is also the possibility of a nuclear accident that would release radioactive particles
into the environment.

34-10 NUCLEAR FUSION

Under certain conditions light nuclei may
reduce their energy by combining to form a
heavier nucleus; this process is known as *nuclear fusion*. Before they are able to unite,
the light nuclei must travel at high speeds
(they must have large energies) in order to
overcome the Coulomb repulsion between

their like charges. There is a *net release* of
energy during a nuclear fusion process only
when a dense mass of light nuclei is maintained at a very high temperature. The nuclei
acquire high speeds because of their thermal
energy; consequently the energy that is
released is often called *thermonuclear energy*.

Enormous quantities of energy are produced in stars as part of their mass is converted into energy by continual nuclear fusion
processes. These reactions are sustained by
the extremely high temperatures in the inte-
riors of stars. In our sun a large amount of
energy is released as protons unite to form
helium nuclei. This is a rather slow reaction;
though $\sim 4 \times 10^{14}$ kg of its mass is converted
into energy in a single day, this amount is
much less than the total mass of the sun.
Different nuclear fusion processes involving
other elements occur in other stars.

Controlled thermonuclear reactions are
a promising future source of energy, but at
present we are not able to maintain the light
nuclei at the high temperatures that are
required to sustain nuclear fusion. The hydrogen isotope called *deuteron* $_1H^2$ will probably be involved in controlled thermonuclear

reactions because it is relatively abundant and the fusion of two deuteron nuclei occurs very rapidly. There are two equally probable reactions:

$$_1H^2 + {}_1H^2 \longrightarrow {}_2He^3 + {}_0n^1 + 3.27 \text{ MeV}$$

and

$$_1H^2 + {}_1H^2 \longrightarrow {}_1H^3 + {}_1H^1 + 4.03 \text{ MeV}$$

These reactions produce far fewer radioactive wastes than a nuclear fission process* and there is no possibility of a "melt down" of the reactor walls. Approximately 0.015 6% of hydrogen exists as the isotope deuteron, and it is both easy and inexpensive to separate deuteron from other hydrogen isotopes.

The intense heat required to initiate controlled thermonuclear fusion will probably be generated by a laser beam or intense beams of electrons or ions. This is still a very active area of research.

The deuteron will probably be contained by magnetic fields in a device called a *magnetic bottle* (Fig. 34-10). Conventional recep-

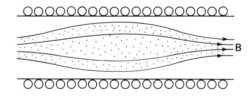

Figure 34-10
A plasma contained in a magnetic bottle.

tacles cannot be used because of the high temperatures. A group of gaseous ions, which is called a *plasma*, is produced by heating a mass of deuteron until it is completely ionized. When the plasma is surrounded by a magnetic field, the ions spiral away from the regions where the magnetic flux density is the greatest. The shape and volume of the plasma

---

*$_1H^3$ has a half life of about 12.4 a.

may be controlled by changing the magnetic fields. As the magnetic flux density increases, the volume of the plasma decreases, and its temperature consequently rises.

REVIEW PROBLEMS

1. A γ-ray has a wavelength 6.0×10^{-13} m. What is its energy equivalence in MeV?

2. Calculate how many neutrons are there in (a) $_{13}Al^{27}$, (b) $_4Be^6$, (c) $_{11}Na^{20}$, (d) $_{92}U^{235}$, (e) $_{82}Pb^{212}$.

3. Write the symbolic representation of (a) a tin atom that has 80 neutrons in the nucleus, (b) a thorium atom that has 142 neutrons in the nucleus, (c) a bismuth atom that has 126 neutrons in the nucleus.

4. What are the approximate nuclear radii of (a) $_4Be^6$ and (b) $_{10}Ne^{20}$?

5. Determine the average binding energy per nucleon between the 6 $_1H^1$ nuclides and 6 neutrons in the nuclide $_6C^{12}$ if its mass is 12.000 000 u.

6. Determine the average binding energy per nucleon between 11 $_1H^1$ nuclides and 12 neutrons in the nuclide $_{11}Na^{23}$ if its mass is 22.987 710 u.

7. Determine the average binding energy per nucleon between 90 $_1H^1$ nuclides and 142 neutrons in the nuclide $_{90}Th^{232}$ if its mass is 232.038 123 u.

8. What kind of particle is emitted when (a) $_{93}Np^{237}$ decays to $_{91}Pa^{233}$, (b) $_{82}Pb^{212}$ decays to $_{83}Bi^{212}$, (c) $_4Be^6$ decays to $_3Li^6$?

9. Complete the following equations:
   (a) $_{13}Al^{24} \longrightarrow {}_1e^0 + \gamma + \underline{\hspace{1cm}}$
   (b) $_{93}Np^{237} \longrightarrow {}_{91}Pa^{233} + \gamma + \underline{\hspace{1cm}}$

10. Determine the decay constant for the cadmium isotope $_{48}Cd^{115}$ if its half life is 53 h.

11. Compute the half life of the indium isotope $_{49}\text{In}^{114}$ if its decay constant is $9.6 \times 10^{-3}/\text{s}$.

12. If the half life of the arsenic isotope $_{33}\text{As}^{76}$ is 26.6 h, what percentage of it remains undecayed after (a) 53.2 h and (b) 106.4 h?

13. A substance initially contains 2.0 mg of the unstable barium isotope $_{56}\text{Ba}^{137}$ that has a half life of 2.6 min. Determine (a) the disintegration constant, (b) the initial activity due to the barium, (c) the activity after 20 min.

14. A substance contains a quantity of the unstable carbon isotope $_6\text{C}^{14}$ that has a half life of 5.7×10^3 a. If the activity of the substance is 167 kBq and it is entirely due to the unstable carbon, determine (a) the disintegration constant of the carbon and (b) the mass of the unstable carbon in the substance.

15. Determine the Q-value in MeV of the following reaction:

$$_1\text{H}^2 + {_4}\text{Be}^9 \longrightarrow {_5}\text{B}^{10} + {_0}\text{n}^1$$

if the nuclide masses are $_1\text{H}^2 = 2.014\ 103$ u, $_4\text{Be}^9 = 9.012\ 186$ u, $_5\text{B}^{10} = 10.012\ 939$ u and $_0\text{n}^1 = 1.008\ 665$ u.

16. If the equation for a positron emission is:

$$_4\text{Be}^6 \longrightarrow {_3}\text{Li}^6 + \beta^+$$

and the masses of the nuclides are: $_4\text{Be}^6 + 6.019\ 780$ u, $_3\text{Li}^6 = 6.015\ 126$ u and $\beta = 0.000\ 549$ u, determine the Q-value of the process. *Hint:* The orbital electrons must be considered:

$$Q = ({_4}\text{Be}^6 - 4\beta^-) - [({_3}\text{Li}^6 - 3\beta^-) + \beta]$$

17. (a) Write a balanced equation for a process during which a $_{92}\text{U}^{235}$ nuclide absorbs a single neutron and then undergoes nuclear fission, producing the nuclides $_{38}\text{St}^{93}$ and $_{54}\text{Xe}^{139}$ with some neutrons. (b) Calculate the energy released during this process if the nuclide masses are $_{92}\text{U}^{235} = 235.043\ 915$ u, $_{38}\text{St}^{93} = 92.914\ 709$ u, $_0\text{n}^1 = 1.008\ 665$ u and $_{54}\text{Xe}^{139} = 138.917\ 840$ u.

Appendices

Appendix 1

CONVERSION FACTORS

Length

1 m = 3.281 ft = 39.37 in
1 km = 0.621 4 mi = 3 281 ft

1 ft = 0.3048 m
1 in = 2.54 cm
1 mi = 1.609 km

Area

1 ha = 10^4 m² = 2.47 acres
1 m² = 10.76 ft² = 1 550 in²
1 cm² = 0.155 in²
1 km² = 0.386 1 mi²

1 ft² = 0.092 90 m²
1 in² = 6.452 cm²
1 acre = 0.405 ha
1 circular mil = 506.7 μm²
1 mi² = 2.59 km²

Volume

1 m³ = 35.31 ft³ = 6.102×10^4 in³
1 L = 10^3 cm³ = 0.264 US Gallon
    = 0.220 Imperial Gallon

1 in³ = 16.39 cm³
1 US gallon = 231 in³ = 3.785 L
1 Imperial gallon = 4.546 L
1 ft³ = 0.028 32 m³

Time

1 min = 60 s
1 d = 86.4 ks

1 h = 60 min = 3 600 s
1 a = 31.5 Ms

Speed or Velocity

1 m/s = 3.60 km/h = 2.25 mi/h
    = 3.281 ft/s

88 ft/s = 60 mi/h
1 ft/s = 0.3048 m/s
1 ft/s² = 0.3048 m/s²

Mass

1 kg = 0.0685 slug
1 u = $1.660 53 \times 10^{-27}$ kg (931 MeV)
1 t = 1 Mg

1 slug = 14.6 kg
1 ton = 2 000 lb
1 grain = $\frac{1}{7\ 000}$ lb = 64.8 mg

Force (weight)

1 N = 0.2248 lb   1 lb = 4.445 N
At the earth's surface 1 kg mass weighs approximately 2.2 lb

Pressure

1 atm = 101.325 kPa = 760 mm mercury = 14.69 lb/in²

Work & Energy

1 kW·h = 3.6 MJ = 1.34 hp·h   1 BTU = 1.06 kJ = 778 ft·lb
1 J = 0.738 ft·lb = 2.78 × 10⁻⁷ kW·h   1 ft·lb = 1.36 J = 1.29 × 10⁻³ BTU
1 eV = 1.6 × 10⁻¹⁹ J   1 hp·h = 2.68 MJ

Power

1 kW = 1.34 hp

$$1 \text{ hp} = 550 \frac{\text{ft} \cdot \text{lb}}{\text{s}} = 746 \text{ W}$$
$$= 2.54 \times 10^3 \text{ BTU/h}$$
$$1 \text{ BTU/h} = 3.93 \times 10^{-4} \text{ hp}$$
$$= 2.93 \times 10^{-4} \text{ kW}$$

Light

1 foot lambert = 3.43 cd/m²   1 fc = 10.76 lx
1 lx = 0.092 9 fc

PHYSICAL CONSTANTS

Constant	Symbol	
Absolute zero	0 K or 0 R	$-273.16°C$ or $-459.7°F$
Acceleration due to gravity at earth's surface	g	9.81 m/s^2 or 32.2 ft/s^2
Avogadro's number	N_0	$6.022\ 5 \times 10^{23}$/mol
Boltzmann constant	k	1.380×10^{-23} J/K
Coulomb constant (free space)	$k = \dfrac{1}{4\pi\epsilon_0}$	9.0×10^9 N·m^2/C^2
Earth mass	M_E	5.975×10^{24} kg or 4.093×10^{23} slug
mean radius	R_E	6.371×10^6 m or 2.09×10^7 ft $= 3.96 \times 10^3$ mi
Electron rest mass	m_e	9.109×10^{-31} kg
Elementary (electronic) charge	e	$1.602\ 1 \times 10^{-19}$ C
Faraday constant	F	$9.648\ 67 \times 10^4$ C
Gas (universal) constant	R	$8.314\ 3$ J/(K·mol)
Gravitational constant	G	6.673×10^{-11} N·m^2/kg^2 or $3.44 \times 10^{-8} \dfrac{\text{lb·ft}^2}{\text{slug}^2}$
Mechanical equivalent of heat	J	$4\ 186$ J/kcal 778 ft·lb/BTU
Moon mass	M_m	7.35×10^{22} kg or 5.02×10^{21} slug
mean radius	R_m	1.739×10^6 m or 5.71×10^6 ft
Neutron rest mass	m_n	$1.674\ 8 \times 10^{-27}$ kg or $1.008\ 665$ u
Permittivity of free space	ϵ_0	8.85×10^{-12} C^2/(N·m^2) or 8.85×10^{-12} F/m
Permeability of free space	μ_0	$4\pi \times 10^{-7}$ Wb/(A·m) or $4\pi \times 10^{-7}$ H/m
Planck's constant	h	$6.626\ 2 \times 10^{-34}$ J·s
Proton rest mass	m_p	$1.672\ 6 \times 10^{-27}$ kg or $1.007\ 342$ u
Rydberg constant	R	$1.097\ 373\ 1 \times 10^7$/m
Sun mass	M_s	1.987×10^{30} kg
mean radius	R_s	6.965×10^8 m
Speed of light in vacuum	c	$2.997\ 925$ m/s or 1.86×10^5 mi/s
Standard atmospheric pressure	p_0	101.325 kPa 760 mm mercury 14.7 lb/in^2
Stefan-Boltzmann constant	σ	$5.669\ 6 \times 10^{-8}$ W/(m^2·K^4)
1 Unified atomic mass unit	u	$1.660\ 53 \times 10^{-27}$ kg or 931 MeV

Appendix 2
PERIODIC TABLE

Legend . . . Atomic number
Element
Atomic mass
u

Period	IA	IIA	IIIA	IVA	VA	VIA	VIIA	VIIIA			IB	IIB	IIIB	IVB	VB	VIB	VIIB	VIIIB
1	1 H 1.0080																	2 He 4.0026
2	3 Li 6.939	4 Be 9.012											5 B 10.81	6 C 12.011	7 N 14.09	8 O 15.999	9 F 18.998	10 Ne 20.18
3	11 Na 22.99	12 Mg 24.31											13 Al 26.98	14 Si 28.09	15 P 30.98	16 S 32.06	17 Cl 35.46	18 Ar 39.95
4	19 K 39.10	20 Ca 40.08	21 Sc 44.96	22 Ti 47.90	23 V 50.95	24 Cr 52.01	25 Mn 54.94	26 Fe 55.85	27 Co 58.93	28 Ni 58.71	29 Cu 63.54	30 Zn 65.38	31 Ga 69.72	32 Ge 72.59	33 As 74.91	34 Se 78.96	35 Br 79.92	36 Kr 83.80
5	37 Rb 85.48	38 Sr 87.63	39 Y 88.92	40 Zr 91.22	41 Nb 92.91	42 Mo 95.95	43 Tc	44 Ru 101.1	45 Rh 102.91	46 Pd 106.4	47 Ag 107.88	48 Cd 112.41	49 In 114.82	50 Sn 118.70	51 Sb 121.76	52 Te 127.61	53 I 126.91	54 Xe 131.30
6	55 Cs 132.91	56 Ba 137.36	57 La* 138.92	72 Hf 178.50	73 Ta 180.95	74 W 183.86	75 Re 186.22	76 Os 190.2	77 Ir 192.2	78 Pt 195.09	79 Au 197.0	80 Hg 200.61	81 Tl 204.39	82 Pb 207.21	83 Bi 209.00	84 Po (209)	85 At (210)	86 Rn (222)
7	87 Fr (223)	88 Ra 226.05	89 Ac** (227)	104 (260)	105 (260)													

*Lanthenum series

58 Ce 140.13	59 Pr 140.92	60 Nd 144.27	61 Pm	62 Sm 150.35	63 Eu 152.0	64 Gd 157.76	65 Tb 158.93	66 Dy 162.51	67 Ho 164.94	68 Er 167.27	69 Tm 168.94	70 Yb 173.04	71 Lu 174.98

**Actinum series

90 Th 232.04	91 Pa 231.04	92 U 238.07	93 Np 237.05	94 Pu (244)	95 Am (243)	96 Cm (247)	97 Bk (247)	98 Cf (251)	99 Es (254)	100 Fm (257)	101 Md (258)	102 No (259)	103 Lw (260)

Period or principal quantum number

CHAPTER 1

2: 2.4×10^4 m²

3: 288 m³

4: 35.6 m³

5: (a) 0.36 m

    (b) 9.3×10^5 m

    (c) 2.6×10^{-4} s

    (d) 1.3×10^{-10} s

    (e) 1.2×10^{11} m

    (f) 8.6×10^{-2} s

    (g) 3×10^{-4} A

    (h) 3.2×10^4 kg

    (i) 8.0×10^{-3} m³

    (j) 7.6×10^{-4} m²

    (k) 9.4×10^{-7} m³

6: (a) 4.6 km

    (b) 2 μs

    (c) 83 GHz

    (d) 0.52 μs or 520 ns

    (e) 12 pF

    (f) 3.2 Mg

    (g) 42 MN/C

    (h) 0.86 MV/m or 860 kV/m

    (i) 7.8 Mg/m³

7: (a) 3×10^6

    (b) 5×10^{-5}

    (c) 4.2×10^{-7}

8: (a) 8×10^{-4}

    (b) 0.5

    (c) 4×10^5

9: 6 800 kg/m³

10: 25 m/s

11: 72 km/h

12: (a) 6.5 km/L

    (b) 15.4 L/100 km

13: 1 069 m

14: 3 281 ft

15: (a) 1.609 km

    (b) 2.4 slug

    (c) 5.33 N

    (d) 9.1 m/s

    (e) 82 ft/s

    (f) 2.3×10^3 kg/m³

    (g) 11.1 m²

    (h) 42 ft³

    (i) 2.1 lb/ft²

    (j) 7 450 W

Answers to Problems

16: (a) 17 N or 3.8 lb
(b) 544 m or 1784 ft
(c) 34 m/s or 113 ft/s
(d) 1.12×10^3 ft^2 or 104 m^2
(e) 1.2 mi^3 or 5.0 km^3
(f) 2 900 kg/m^3 or 5.6 slug/ft^3

17: kg

18: $1/(A \cdot s)$

19: s/A

20: V

21: J

22: (a) correct
(b) correct
(c) incorrect
(d) correct
(e) incorrect

23: $F = k\,Q\,v\,B$

24: $P = k\,I^2\,R$

25: $f = k/\sqrt{L\,C}$

Review Problems

1: (a) 96 m^3
(b) 9.6×10^4 L

2: (a) 4 500 m^3
(b) 4.5×10^6 kg

3: (a) 506 km
(b) 500 km

4: (a) 3 600 cm 3.6×10^7 μm
   3.6×10^{-2} km
(b) 0.188 L 1.88×10^{-4} m^3
   1.88×10^2 cm^3
(c) 1.2×10^8 mg   0.12 Mg
   1.2×10^{11} μg
(d) 3.0×10^3 mm^2 3.0×10^{-3} m^2
(e) 90 cm^3 9×10^{-5} m^3

5: (a) 3.6×10^{-5} m
(b) 2.4×10^{-7} s
(c) 8×10^{-4} A
(d) 1.8×10^7 kg
(e) 2.8×10^{-5} m^3

6: (a) 57 ns
(b) 12 Mg
(c) 340 μm or 0.34 mm
(d) 72 mL
(e) 90 GHz
(f) 0.125 nF or 125 pF

7: 288

8: 16 m/s

9: 64.8 km/h

10: (a) 2 400 kg/m^3
(b) 3.6 W/m^2
(c) 5 GV/m
(d) 26 J/kg
(e) 36 A/m^2
(f) 15 kA/m^2

11: 25 000

12: (a) 1.6×10^5 cm
(b) 175 kg
(c) 15.3 slug/ft^3

13: J

14: C

15: (a) Yes
(b) Yes
(c) No

16: $E = k\,L\,I^2$

17: $P = k\,F\,v$

18: $P = k\,m/Q\,B$

CHAPTER 2

1: (a) 75.4 ft   4.7 ft
(b) 0, 3.0 ft from 12 to 6

2: 150 km east

3: 576 km/h

4: 2.7×10^8 s $= 7.5 \times 10^4$ h

5: 40 km/h north

6: 2.64 km

7: 75 km

8: 880 ft

9: 0.17 m/s^2

10: 93 m/s down

11: 74 N, 21° from 45 N

12: 162 N, 12.5° from 110 N

13: 12.6 km/h, 108°

14: 99 m 60° E of N

15: 12.2 lb 55°

16: 379 km/h 133 T

17: 379 km/h 42° W of N

18: (a) 170 km N
(b) 255 km
(c) 51 km/h
(d) 34 km/h N

19: (a) (255 km/h E, −125 km/h N)
    (b) 257 km/h, 29° S of E
20: (30.3 N horizontal, 17.5 N vertical)
21: (a) 21.2 km/h E, 21.2 km/h N
    (b) 7.5 km/h E, 18.5 km/h N
    (c) −5.1 km/h E, 14.1 km/h N
    (d) −22.7 km/h E, −10.6 km/h N
    (e) 6.2 km/h E, −16.9 km/h N
22: 25 lb, 43 lb
23: (a) 68 N, 25 N
    (b) 20 N, 20 N
    (c) 17.5 N 30.3 N
    (d) 40 N 75 N
24: (a) 37.7 N, 26.4 N
    (b) 4.0 N, 37.8 N
    (c) −45.1 N, 31.6 N
    (d) 21.6 N, −59.2 N
25: 209 N, 46°
26: 4.9 N, 128°
27: 41 N, 64°
28: (a) 43 km, 35.5° N of E
    (b) 43 km, 35.5° S of W
29: 97 km/h 35° E of N
30: (a) 17.1 km 31° S of W
    (b) 30.6 km 12° W of N
    (c) 16.4 km 26° N of W
31: 87 N 29° W of N
32: 95 N·m
33: (a) 46 N·m
    (b) 39 N·m
34: 2.5×10^3 T·m/s west

Review Problems

 2: 60 km/h
 3: 18.6 mi
 4: 3.47 km
 5: 3 m/s²
 6: 120 km/s
 7: 13 N 23°
 8: 302 N
 9: (a) 270 km
    (b) 170 km 37° S of W
    (c) 60 km/h
    (d) 37.8 km/h 37° S of E
10: 75 N, 130 N

11: (a) (53.5 N, 59.5 N)
     (−44.5 N, 46.0 N)
     (−67.7 N, −24.6 N)
     (29.3 N, −41.8 N)
    (b) 48.9 N, 127°
    (c) 195 N, 359°
12: (a) (60 km, 104 km)
     (34.2 km, −94 km)
     (−113 km, −65 km)
     (0 km, 80 km)
    (b) 31 km 36° W of N
    (c) 230 km, 22° W of N
13: 47.7 km/h 57° N of E
14: 3011 ft·lb
15: 259 N·m upwards

CHAPTER 3

 1: 576 m
 2: (b) 50 cm
    (c) 2.5×10^{11} m/s²
 3: (b) 228 m
 4: (a) 6.0 s
    (b) 0 m/s²
    (c) 3.0 s
    (d) ~ 100 m/s² W
    (e) 0 to 6 s
    (f) 6 s to 8 s
    (g) 3 s to 6 s
 5: (b) 200 m W
    (c) 25 m/s W
 6: (a) 2.5 ft/s²
    (b) 585 ft
 7: (b) 5.0×10^9 m/s²
    (c) 100 m
 8: (b) 40 m
    (c) −10 m/s
    (d) 0 m/s²
10: (a) 3.67 ft/s²
    (b) 1 610 ft
11: (a) 0.84 ft/s²
    (b) 513 ft
12: (a) 9.03 ft/s²
    (b) 763 ft

13: (a) 10^6 m/s
   (b) 1.0 m
14: (a) 2.7×10^{11} m/s^2
   (b) 30 cm
15: (a) 130 m/s
   (b) 500m
16: (a) 44.4 m
   (b) -5.6 m/s^2
17: 12 min
18: (a) 82 m/s
   (b) 0.93 s
19: (a) 8.0 m/s^2
   (b) 144 m
20: (a) 10 s
   (b) 250 ft
   (c) 50 ft/s
21: (a) 10.7 s
   (b) 268 ft
22: (a) 5.6 s
   (b) 179 ft/s
23: 207 ft or 63.5 m
24: (a) 13 s
   (b) 141 m
25: 114 m
26: (a) 50.6 m
   (b) 32.5 m/s
27: (a) 127 ft
   (b) 5.6 s
   (c) 90 ft/s
28: (a) 236 m/s
   (b) 2 670 m
29: (a) 5 630 ft
   (b) 37.5 s
30: 798 ft/s
31: (a) 7.91 s
   (b) 6.59 mi
32: 1.96 nm
33: (a) 3.00 s
   (b) 180 m
   (c) 66.8 m/s, 26.1°
34: (a) 123 m
   (b) 10 s
   (c) 849 m
35: 1.02 km
36: 60 s, 6.9×10^4 ft

37: 827 ft
38: 52 km or 980 m

Review Problems

1: (b) 325 m
   (c) -2.5 m/s^2, 4.0 m/s^2
2: (b) 15 s
   (c) 375 m
3: (a) 3.6×10^{14} m/s
   (b) 16.7 ns
4: (a) 5.0 m/s^2
   (b) 360 m
5: (a) 12 m/s
   (b) 48 m
6: (a) 300 m/s^2
   (b) 20 s
7: 7.23 s
8: (a) 0.20 s
   (b) 292 m/s
9: (a) 54.6 m/s up
   (b) 152 m
   (c) 63 m/s down
10: 85 cm
11: (a) 123 m
   (b) 8.00 s
   (c) 1 320 ft
12: (a) 469 ft
   (b) 13.2 s
   (c) 1 320 ft
13: (a) 144 ft
   (b) 5.0 s
   (c) 945 ft

CHAPTER 4

2: 60 N
3: 128 ft/s^2
4: 2.8 m/s^2
5: 30 kg
6: 60 N
7: 9.0 m/s^2
8: 1 m/s^2
9: (a) 48 lb
   (b) 50 ft/s^2

10: 70 m/s²
11: 2.7×10^{-22} N
12: 29.4 N
13: 0.37 slug
14: 2.0 kg
15: (a) 30 lb
(b) 32.8 lb
16: (a) 14 800 N
(b) 4 800 N
(c) 0 N
17: 8.0 ft/s
18: 0.20 m/s²
19: 3.5×10^7 N
20: (a) 504 m
(b) 18.3 s
(c) 54 000 N
21: 1.5×10^6 N
22: (a) 8.0 m/s²
(b) 64 m/s
(c) 256 m
23: (a) 4.2 m/s²
(b) 2.8 N
24: 2.08×10^3 N
25: (a) 156 m, 22.0 m/s
(b) −60.0 m, −14 m/s
26: (a) 8.0 m·s²
(b) 96 m/s
(c) 576 m
27: (a) 135 m
(b) 30 s
28: 1 290 lb
29: 5 400 N
30: (a) −16 ft/s²
(b) 82.5 ft
(c) 3.25 s
31: (a) 20 m/s²
(b) 2.4×10^5 N
32: (a) 5.2 m/s²
(b) 2.4 s

Review Problems
1: 0.80 m/s²
2: 9.0 m/s²
3: (a) 5.6 slug
(b) 30.2 lb

4: (a) 735 N
(b) 647 N
(c) 287 N
(d) 1 950 N
(e) 280 N
(f) 863 N
5: 9.0×10^{10} m/s²
6: 37.5 m/s²
7: 29 700 N
8: (a) 1 940 ft
(b) 22 s
(c) 10 000 lb
9: (a) 24 500 N
(b) 24 900 N
10: (a) 8.0 ft/s²
(b) 5.63 lb
11: (a) 1.5 m/s²
(b) 22.5 m/s
(c) 169 m

CHAPTER 5

1: 5 500 kg/m³
2: 1.99×10^{30} kg
3: 6.3×10^{-3} ft³
4: 1 080 lb/ft³ not gold
1 204 lb/ft³ gold
5: 8 800 kg/m³
6: 0.13 m³
7: 192 N
8: 2.55 in
9: 2 450 N/m, 4.8 cm
10: 1.28
11: 10.6 in²
12: 764 N
13: 0.307 m²
14: 17 000 N
15: 8.63×10^{-4} m²
16: 2.9×10^7 lb/in²
17: (a) 90 kPa
(b) 1.0×10^7 Pa
(c) 1.1×10^{-4}
(d) 100.011 m

18: (a) 2.0×10^5 Pa
(b) 5.0×10^6 Pa
(c) 2.5×10^{-5}
(d) 100.002 5 m
(e) Yes
19: 30.000 586 ft
20: 6.8×10^6 lb/in²
21: -510 cm³
22: 0.678 in³
23: 1.5 MN
24: 3.0×10^5 lb

Review Problems
1: 8.5×10^3 kg/m³
2: 4.1×10^{-2} ft³
3: 12.5 lb/in²
4: 56 kN
5: 8.3×10^3 kg/m³
6: (a) 452 N/m
(b) 6.5 cm
7: 1.75×10^7 Pa
8: 6.28×10^4 lb
9: 1.53 m²
10: (a) 7.2×10^7 N
(b) 1.14×10^7 N
11: 6.9×10^3 lb
12: 16.6×10^{10} Pa
13: 2.86×10^{-2} cm³
14: (a) 7.0×10^6 N
(b) 9.5×10^5 N
15: (a) 9.999 565 ft
(b) 37.2
(c) 7.4×10^6 lb

CHAPTER 6

1: 54 N   68.2°
2: 11.5 lb,   10 lb
3: 707 lb,   500 lb
4: 420 lb
5: 313 lb
6: $A = 112$ lb,   153 from **A**
7: 49 N
8: 11.5°
9: 50 lb,   86.6 lh

10: 433 N, 250 N
11: 142 N
12: 142 N,   145 N
13: 18.3 N,   22.4 N
14: $A = 190$ N, $B = 100$ N
$C = 220$ N, $D = 269$ N
15: 40 lb·ft
16: 75 N·m
17: 167 N
18: 75 N·m
19: (a) 880 lb
(b) 6 160 lb·in
20: (a) 2 670 N
(b) 640 N·m
21: 28 000 N
22: (a) (2.79 cm,   3.86 cm)
(b) (2.81 cm,   3.90 cm)
23: (a) (4.5 in,   5.07 in)
(b) (4.46 in,   5.05 in)
24: 9.25 in
25: 6.09 in
26: 1.66 cm along 3 cm axis
27: 7.29 in
28: (a) 9.45 in from lead
(b) 10.2 in
29: 608 lb·ft
30: 526 lb·ft
31: 38 900 N·m
32: 2.75×10^4 N·m
33: 1065 lb,   435 lb
34: 735 N,   1225 N
35: 7.6 m from A
36: 2.24 m from front
37: 980 N
38: 7690 lb, 8810 lb
39: 1070 lb,   1880 lb
40: 1044 N,   1190 N
41: $R = 1000$ N, $\mathbf{N} = 2200$ N at 63°
42: $R = 1266$ N $\mathbf{N} = 3022$ N 65°
43: $T = 134$ lb $\mathbf{R} = 134$ lb 26.6°
44: $T = 804$ lb $\mathbf{R} = 721$ lb 4.8°
45: $T = 644$ $\mathbf{R} = 769$ lb at 33.1°
46: EF = 1 560 lb tens.
FD = 3 460 lb comp.
FG = 5 770 lb comp.
ED = 0 lb

47: AB = DE = 8 000 lb comp.
AG = EF = 6 930 lb tens.
BC = DC = 7 000 lb comp.
BG = DF = 1 730 lb comp.
GF = 5 200 lb tens.
GC = FC = 1 730 lb tens.

Review Problems

1: 85 N,   90 N
2: 125 N,   43N
3: A = 1 960 N,   B = 2770 N
C = 3 260 N,   D = 4 150 N
4: 104 lb at 35.2°
5: (a) 53 kg,   520 N
(b) 300 N
6: R = 138 N θ = 49.1°
7: 1 070 lb,   1 880 lb
8: 3.39 cm,   3.18 cm
9: (a) 5.89 in
(b) 6.00 in
10: 45 lb
11: 9.2 kN,   10.4 kN
12: (a) 41.7 lb
(b) 25 lb
(c) 25.2 lb at 6.35°
13: (a) 2 140 N
(b) 2 430 N at 40.5°
14: AB = 58.0 kN
AC = 4.70 kN
FG = 6.5 kN
FH = 125 kN
EG = 131 kN

CHAPTER 7

1: 100 lb
2: 30 lb
3: 39.2 N
4: 64.7 N
5: (a) 30.1°
(b) 19.3°
(c) 21.8°
(d) 31°
(e) 26.6°

6: (a) 20.7 N
(b) 44.4 N
(c) 0.47
7: 23.5 lb
8: 344 N
9: 4.2 m/s²
10: (a) 170 ft
(b) 3.9 s
11: (a) 233 ft
(b) 5.3 s
12: (a) 3.32 m/s²
(b) 1.20 s
(c) 3.99 m/s
13: 3.7 m/s
14: (a) 2.0 s
(b) 50 ft
15: (a) 4.52 m/s²
(b) 4.22 N

Review Problems

1: 294 N
2: 26.3 lb
3: 28.8°
4: 1.84 m/s²   35.6 N
5: 9.2 ft/s²
6: 12.1 m
7: 87.8 N

CHAPTER 8

1: 1200 ft·lb
2: (a) 70.7 ft·lb
(b) 70.7 ft·lb
3: 196 J
4: (a) 750 lb
(b) 15 000 ft·lb
5: 9.68×10^5 ft·lb
6: 1.64×10^{-17} J
7: (a) 3530 J
(b) 58.8 W
8: (a) 150 ft·lb
(b) 0.055 hp
9: 31.9 hp
10: 1440 hp
11: 12.2 kW
12: 0.27 hp

13: 81.6 L
14: 23.3 hp
15: 2.96×10^3 hp
or 3.31 MW
16: 3.68 kW
17: 71.4 bhp
18: 672 ft·lb
19: 225 J
20: 1470 J
21: (a) 5000 ft·lb
(b) -2500 ft·lb
22: (a) 204 lb
(b) 26.5 hp
23: 12¢
24: 27¢
25: 56.4 mJ
26: 400 ft·lb
27: 3.13 ft·lb
28: 4.15×10^4 ft·lb
29: 48.5 m/s
30: 126 ft/s
31: (a) 7.0 m/s
(b) 5.05 m/s
32: (a) 38.9 N
(b) 0.242
33: (a) 1.5 lb
(b) 0.22
34: 17.7 lb
35: 4.62×10^4 lb

Review Problems
1: (a) 1.37×10^4 J
(b) 0.31 hp
2: (a) 1.84×10^5
(b) 0.343 hp
3: 0.98 hp
4: 51.2 hp
5: 326 hp
6: 46.5¢
7: 1.92×10^{-17} J
8: 0.44 ft·lb
9: 4.91×10^5 J
10: (a) 0.24 ft·lb
(b) 300 lb/ft
11: (a) 200 lb
(b) 4640 ft·lb

(c) 0.14 hp
(d) 2400 ft·lb
(e) 2240 ft·lb

CHAPTER 9

1: 5.5×10^{-24} kg·m/s
2: 8 800 slug·ft/s east
3: (a) 280 slug·ft/s down
(b) 9 350 lb
4: 146 lb
5: 2.67 m/s
6: 0.18 m/s
7: -3.11 ft/s
8: (a) 1.95×10^4 N
(b) 19.0 m/s^2
9: 1.5×10^5 m/s, 4.0×10^5 m/s
10: -11.1 ft/s, 23.9 ft/s
11: (a) -1.7×10^6 m/s
4.5×10^6 m/s
(b) -1.08×10^6 m/s
4.66×10^6 m/s
(c) -7.8×10^6 m/s
3.2×10^6 m/s
12: (a) 2.96 m/s, 4.16 m/s
(b) -3.15 m/s, 1.65 m/s
13: 29 cm
14: (a) 0.75 in
(b) 0.39 ft·lb
15: 24 mi/h W

Review Problems
1: 2.3 lb·s
2: 300 N·s
3: (a) 0.29 lb·s
(b) -1.56 ft/s
4: 2.7 N·s
5: (a) 2.00 lb·s
(b) 200 lb
6: 824 m/s
7: -6×10^5 m/s
4×10^5 m/s
8: 53.6 ft/s, 43° N of E
9: 6.34×10^6 lb

CHAPTER 10

1: 1.5×10^{-16} s
2: 3.0×10^4 m/s
3: (a) 9.0×10^{22} m/s^2
  (b) 2.2×10^6 m/s
4: (a) 12.2 ft/s
  (b) 130 ft/s^2
5: 48 ft/s
6: 47 ft/s
7: 3.32 m/s
8: (a) 0.016
  (b) 0.16 m/s^2
9: 19.1°
10: 24.7 m/s
11: 14.3 m/s
12: (a) 1.18×10^{11} m
  (b) 88 d
  (c) 47.8 km/s
    97.2 km/s
13: 2.18×10^{-7} lb
14: (a) 3.53×10^{22} N
  (b) 2.97×10^4 m/s
  (c) 3.17×10^7 s
15: 1.62 m/s^2 or 5.31 ft/s^2
16: 7.74 km/s
17: (a) 853 m/s
  (b) 13.8 h
18: 6.25×10^9 J
19: 3.17×10^8 ft·lb
20: 7.68×10^{22} J
21: 2.38 km/s

Review Problems

1: 4.1×10^{-51} N
2: 2.8×10^{-13} N
3: (a) 6.5×10^{-7} s
  (b) 6.9×10^{13} m/s^2
  (c) 1.2×10^{-13} N
4: 665 m
5: 0.0338 m/s^2
6: (a) 6.26 m/s
  (b) 560 N
7: (a) 17 300 mi/h
  (b) 1.51 h
8: 1.98 h

9: (a) 0.269
  (b) 15°
10: (a) 4328 d
  (b) 313 km/s
11: -4.4×10^9 J
12: (a) 3.16×10^{10} J
  (b) 7.25 km/s

CHAPTER 11

1: (a) 1.22 rad
  (b) 573°
  (c) 12 rev
  (d) 8.3 rad
3: (a) 122 rad/s
  (b) 30.5 m/s
4: (a) 0.772 m/s^2
  (b) 39.7 rad/s
  (c) 2.2 rad/s^2
  (d) 56.9 rev
5: (a) 6.28 rad/s^2
  (b) 200 rev
6: (a) 2.5 s
  (b) 31.8 rev
7: (a) 49.7 rad/s^2
  (b) 413 rev
8: (a) 307 rad/s
  (b) 1108 rev
9: 2.16×10^{-6} kg·m^2
10: 0.347 slug·ft^2
11: 1.84×10^{-3} lb·ft
12: 1.5 rad/s^2
13: 0.112 N·m
14: (a) 10.6 cm
  (b) 9.46 cm
15: 1.27 lb·ft
16: 31.2 rad/s^2
17: (a) 17.8 rad/s^2
  (b) 8.91 ft/s^2
18: $7\,M\,R^2/5$
19: 0.0217 slug·ft^2
20: 68.54 ft·lb
21: 200 J
22: 3.74 m/s,   3.43 m/s
23: 5410 N

24: 2443 W

25: 623 N·m, 17.8°

26: 6.70 kg·m²/s

27: 1.06 × 10⁻³⁴ kg·m²/s

28: 3.27 slug·ft²/s

Review Problems

1: (a) 59.6 rad/s
   (b) 16.2 rad/s²

2: (a) −6.28 rad/s²
   (b) 312.5 rev

3: 1216 rpm

4: 8.14 × 10⁻⁵ slug·ft²

5: 7.88 × 10⁻³ N·m

6: 40.9 N·m

7: 48.8 N·m

8: 295 lb

9: 8,38 N·m

10: 704 kW

11: 2.4 kW

12: 3.35 × 10³ ft/s

13: (a) 144 kg·m²
   (b) 1.8 N·m

14: (a) 14.1 N·m
   (b) 5.89 N·m

15: 157°

16: 591 lb

17: 8.86 lb·ft

18: 0.056 kg·m²

19: 1.88 kg·m²/s

20: (a) 1.6 m/s²
   (b) 22.9 N, 24.5 N
   (c) 19.6 J
   (d) 8.5 rad/s

21: 2.96 J

22: 22.6 ft/s

23: 350 lb·ft 0.12°

24: 2.18 slug·ft²/s

CHAPTER 12

1: 43%

2: 62.5 mW

3: 65%

4: 80.3 hp

5: 0.808 hp

6: (a) 335 N
   (b) 2.92
   (c) 100%

7: (a) 2.0
   (b) 2.92
   (c) 68%
   (d) 877 ft·lb
   (e) 277 ft·lb

8: 2.21 lb

9: (a) 60%
   (b) 636 N
   (c) 4410 J
   (d) 2960 J

10: (a) 208 N
   (b) 5.67

11: (a) 30 lb
   (b) 3.3

12: 12

13: (a) 2, 25 N
   (b) 4, 12.5 N

14: 333 rpm

15: (a) 30
   (b) 400 rpm

16: 59%

Review Problems

1: 80%

2: 1.7 kW

3: 34%

4: 88%

5: 47.3 kW

6: (a) 372 N
   (b) 3 720 J
   (c) 186 W
   (d) 2 450 J
   (e) 1.32
   (f) 2.00
   (g) 65.9%

7: 8.8

8: 78 N

9: (a) 50 lb
   (b) 9.0

10: (a) 276 N
   (b) 2.67

11: 17

12: (a) 6
 (b) 5.76
 (c) 4.59×10^4 J
13: (a) 5
 (b) 600 rpm
14: 36%

CHAPTER 13

1: 13 000 lb
3: (a) 312 lb/ft²
 (b) 4250 lb/ft²
 (c) 749 lb/ft²
4: 600 lb/in²
5: (a) 885 kPa
 (b) 100 N
6: (a) 121 kPa
 (b) 116 kPa
 (c) 8.4×10^5 N
 (d) 5.6×10^5 N
7: (a) 16.1 lb/in²
 (b) 15.3 lb/in²
8: 113 kPa
9: 15.9 lb/in²
10: 14.1 lb/in²
11: (a) 36.75 N
 (b) 5000 times
 (c) 8.2 s
12: 42.1 lb
13: (a) 116 lb
 (b) 39 times
 (c) 5.45 hp
14: (a) 206 N
 (b) 250 m
 (c) 10.3 s
15: (a) 9.0 cm
 (b) 212 N
16: 13.4 in
17: (a) 4.36 mN
 (b) 11 mN
 (c) 70.1 mN
18: 1.6×10^{-2} N/m
19: 3.6 mm
20: 2.4 m³
21: 8.33 ft²

22: 220 s
23: 48 m/s
24: 13.5 m/s
 119 kPa
25: (a) 0.72 ft/s
 35.4 lb/in²
 (b) 0.06 ft
 0.01 ft
26: (a) 7.2 m/s
 1.43 MPa
 (b) 2.64 m
27: 1.08×10^{-2} m³/s
28: 4.15 ft
29: 0.43 ft³/s
30: 3.5 ft/s
31: 220 m/s

Review Problems
 1: 3000 N
 2: 4.02×10^8 N
 3: 10.3 m, 33.9 ft
 4: (a) 123 lb/in²
  (b) 24 lb
 5: (a) 16.1 lb/in²
  (b) 13.9 lb/in²
 6: 19 700 lb
 7: (a) 109 N
  (b) 4 500 times
  (c) 4.9 kW
 8: 3.44 cm
 9: 4.18
10: 1.1 mm
11: 0.10 m²
12: 18 lb/in²
13: 6.18 m/s

CHAPTER 14

 1: (a) 26.7°C
  (b) −12°C
  (c) −29°C
  (d) −93°C
  (e) 182°C
  (f) 1649°C
 2: (a) 41°F

(b) 181°F
(c) 464°F
(d) −98°F
(e) −54°F
(f) 3 812°F
3: −40°C
4: 11.4°F, −11.4°C
5: (a) 305 K
(b) 235 K
(c) 635 K
(d) 282 K
(e) 215 K
6: (a) 25°C
(b) 823°C
(c) −249°C
7: 1.51×10^7 J
8: 9.73×10^5 ft·lb
9: 1.45×10^5 J
10: 8.1×10^5 J
11: 8.48×10^4 BTU
12: 2.34×10^5 J
13: 72 BTU
14: 45.7°C
15: 10.3°C
16: 371 J/(kg·°C)
17: 1.45×10^5 J
18: 35.6°C
19: 0.129°C
20: 194°C
21: 1.0×10^7 J
22: 2.26×10^7 J
23: 3.06×10^6 J
24: 2590 BTU
25: 12.5°C
26: 2.29×10^6 J/kg
27: 1.84×10^6 J
28: 3.92×10^3 BTU
29: 100°C
30: 208 lb
31: 840 BTU
32: 50.2 kW
33: 2.93 g
34: 6.52×10^8 J
35: 16.8°C
36: 12.7 hp

37: 0.0132 m³
38: (a) 6.0 kW
(b) 2.3×10^{-4} m³/s
39: (a) 31%
(b) 4.5 g/m³
40: 25%, 30 gr/lb, 35°F
63½°F, 62 gr/lb, 54°F
79°F, 69°F, 64°F
68½°F, 43%, 78 gr/lb
66°F, 28 gr/lb, 34°F
64°F, 72%, 61°F
41: 878 BTU/min
42: 1.20×10^3 BTU/min
4.89 lb/h

Review Problems

1: 3.84×10^5 J
2: 7.04×10^7 J
3: 91°F
4: 1.09×10^5 J
5: 295 g
6: 18.4°C
7: 0.398 BTU/(lb·°F)
8: 62°F
9: 0.172 lb
10: 100°C
11: 29.1°C
12: 2.2×10^8 J
13: 39.6 g
14: (a) 42%
(b) 102 g
15: (a) 50%, 58°F, 72 gr/lb,
13.8 ft³, 30 BTU/lb
(b) 58°F, 47°F, 48 gr/lb
13.6 ft³, 25.3 BTU/lb
(c) 63°F, 52%, 61°F, 13.6 ft³
289 BTU/lb
(d) 95°F, 18%
44 gr/lb, 14.2 ft³, 30 BTU/lb
(e) 70°F, 37%, 60°F, 78 gr/lb,
14.1 ft³
16: (a) 2.6 gal
(b) 4.19 ton
(c) 72%

CHAPTER 15

1: $3.05 \times 10^{-5}/°C$
2: 19.982 m
3: 0.0913 in
4: 0.007 m too low
5: 0.034%
6: 0.024 cm
7: 138°C
8: $1/\{1 + \alpha_0(100°C)\}$
9: (a) 2.4×10^{-4}
　　(b) 1.92×10^5N
10: 4.15×10^5 lb
11: 2.6×10^{-3} ft
12: 29.3 mL
13: 13 512.7 kg/m³
14: 921 kg/m³
15: 0.668 cm³ gasoline overflows
16: 1.78×10^{24}
17: 3.28×10^{23}
18: 6.36 ft³
19: 22.3 L
20: 1.5 atm
21: 0.19
22: 115%
23: 4.3 L
24: 4.33 m
25: (a) 293 K
　　(b) 243 K
　　(c) 2283 K
　　(d) 273 K
　　(e) 295 K
　　(f) 133 K
26: 541 lb/in²
27: 24.6 lb/in²
28: 21.8 atm
29: 9.03 lb/in²
30: 3390 cm³
31: $(16^2/3)\%$
32: 1.2×10^{23}
33: 983 mm
34: 172 K
35: 433 m/s
36: 6.17×10^{-21} J
37: 1.2×10^5 BTU

38: 2.66×10^8 J
39: 0.452 W
40: 2.49 lb
41: (a) −6.62°F
　　(b) 380 BTU/h
42: 0.144 W/(m·K)
43: 649 W
44: (a) 2620 BTU/h
　　(b) 2.86 h·ft²·°F/BTU
45: (a) 2940 BTU/h
　　(b) 2.04 h·ft²·°F/BTU
46: 1.07×10^8 J
47: (a) 0.33 W/°C
　　(b) 200°C
　　(c) 3°C/W
　　(d) 50 W
48: 158 kW
49: 101 W
50: 1610 K

Review Problems

1: $1.1 \times 10^{-5}/°F$
2: 300.227 m
3: 350.492 ft
4: 2201.901 ft
5: 138°C
6: (a) 3.0×10^{-4}
　　(b) 9.0×10^4 lb
7: 1.000 125 m
8: 32.7 cm³
9: 1.14×10^{24}
10: 63.7 m³
11: 5.13×10^5 Pa
12: 1205°C
13: 606°C
14: 605 m/s
15: 6.9×10^{-21} J
16: 1.33×10^8 BTU
17: 4.01×10^6 BTU
18: (a) 6.9 kW
　　(b) 9.16 g
19: 2.59×10^8 J
20: (a) 1.25 W/°C
　　(b) 100°C

(c) 0.80°C/W
(d) 75 W
21: 1310 BTU
22: 438 W
23: 1720 K
24: 7.84 kW

CHAPTER 16

1: 3150 J
2: 1.08×10^5 ft·lb
3: (a) 0
(b) 3900 J
(c) 3900 J
4: -1.8×10^4 J
5: (a) 2.8×10^5 ft·lb
(b) 3.5×10^5 ft·lb
6: (a) 7140 J
(b) 5500 J
(c) 1640 J
7: (a) 2.91×10^4 J
(b) 4.07×10^4 J
8: 2800 J
9: 8.0×10^6 J
10: 7480 J
11: (a) 611 kPa
(b) 414 K
(c) -2130 J
12: (a) 0.075 atm
(b) 98.5 K
(c) 2450 J
13: 75%
14: 9000 J
15: 89%
16: 53%
17: 1332 K
18: (a) 5.45
(b) 5.5 J
19: (a) 3.0
(b) 1120 J
20: 3.93 hp
21: 20.4 ton

Review Problems
1: 1925 ft·lb
2: 1500 J
3: 709 J
4: -1600 J
5: (a) 0 J
(b) 13.3 kJ
(c) 13.3 kJ
6: (a) 31.9 kJ
(b) 24.3 kJ
(c) 7.6 kJ
7: 46.8 kJ
8: (a) 1.27 MPa
(b) -2.7°C
(c) -4220 J
9: 57%
10: 2.45×10^5 J
11: 87%
12: (a) 2.43
(b) 35 J
13: (a) 3.68
(b) 1.14×10^5 J

CHAPTER 17

1: 0.05 s
2: 98 N/m
3: 2.0 lb
4: 4.2 cm/s
5: 2.1 cm/s 175 m/s^2
6: 1.93 m/s
-3100 m/s^2
7: (a) 7.9 N/m
(b) 12.4 cm
8: (a) 0.72 s
(b) 3.65 ft/s
-32 ft/s^2
9: 2.5 s
10: 22.4 ft/s^2
11: 3.0 s
12: (a) -58 m/s^2
0.132 m/s
(b) -29 m/s^2
0.114 m/s

13: (a) 24.5 N/m
(b) 1.11 Hz
(c) -1.46 m/s^2
(d) 0.21 m/s
(e) -0.49 m/s^2
0.20 m/s

14: (a) -2.0×10^4 ft/s
1.6 ft/s
(b) -1.3×10^4 ft/s^2
1.17 ft/s

15: 0.66 m

16: (a) 1.67×10^{-15} s
(b) 6.00×10^{14} Hz

17: 1.0×10^{10} Hz

18: 2.5×10^{17}

19: 0.62 mm

20: (a) 1.9 mm
(b) 1.9 mm

21: 60° or 1.05 rad

22: 2.4° or 0.042 rad

23: (a) s = (3.0 cm)
$$\sin\left\{\frac{2\pi}{3.0 \text{ cm}}\left(x - 5000\,\frac{\text{cm}}{\text{s}}\right)\right\}$$
(b) 0.026 m

24: (a) 200 cm
(b) 7500 cm/s
(c) 37.5 Hz
(d) 27 ms
(e) 1.77 cm

25: 3.2×10^{-4} W/m^2

26: 9.9 W/m^2

27: 1.59×10^{-2} W/m^2

32: 143 km

33: 2.90m
103 MHz

Review Problems

1: 2.4 Hz

2: (a) 3.2 Hz
(b) -8.1 m/s^2
(c) 0.40 m/s

3: 4.43 s

4: 3.06 times

5: (a) 8.95×10^{-28} J
(b) 18.7 cm/s

(c) -1.59×10^5 m/s^2

6: (a) 2.22×10^5 m/s^2
(b) 2.36 m/s
(c) -1.33×10^5 m/s^2
1.88 m/s

7: (a) 2.84×10^{-30} J
(b) 6.32×10^{-15} N/m
(c) 3.77 cm/s
(d) -3.16×10^4 m/s^2
(e) 2.81 cm/s

8: 25 m

9: (a) 2.13 kHz
(b) 469 μs

10: 500 nm

11: (a) s = (2 mm)
$$\sin\left\{\frac{2\pi}{50\text{m}}\left(x - 300\,\frac{\text{m}}{\text{s}}\right)t\right\}$$
(b) -1.2 mm

12: (a) 5.0 m
(b) 30 m/s
(c) 6.0 Hz
(d) 3.8 cm

13: 4.42 mW/m^2

14: (a) 29.2 mW/m^2
(b) 82.6 W
(c) 17.5 mW/m^2

15: 24.9 mW·m^2

17: 160 m
1.88 MHz

CHAPTER 18

1: 61 cm

2: 1.07

3: 158 m/s

4: 253 m/s

5: (a) 198 m/s
(b) 396 m/s

6: 1367 m/s

7: 1480 m/s
3.7 cm

8: 5100 m/s
10.2 cm

9: (a) 327 m/s
   (b) 109 Hz
   (c) 218 Hz
   (d) 436 Hz
10: (a) 146 ft/s
   (b) 7.3 Hz
   (c) 29.2 Hz
   (d) 58.4 Hz
11: 1.08 N
12: 16 cm
13: $n\, v/2\, L$
14: (a) 1125 Hz
      3375 Hz
      7875 Hz
   (b) 2250 Hz
      6750 Hz
      15 750 Hz
15: (a) 1155 ft/s
   (b) 319 m/s
16: 983 m/s
17: 0.96
18: 5 Hz
19: 702 Hz or 698 Hz
20: (a) 1680 Hz
   (b) 1355 Hz
21: (a) 251 Hz
   (b) 252 Hz
22: 1410 Hz
23: 4.34 m/s
24: (a) 7.3 nW/m²
      2.6 nW/m²
   (b) 2.8:1
25: 9:1
26: 60.8 dB
27: (a) 10 μW/m²
   (b) 250 μW/m²
28: 44.8 dB
29: −11.6 dB
30: 78 dB
31: 3.16:1
32: 44 dB
33: 15.8 W
34: 0.70
35: 2.44 s
36: 20 phon

Review Problems
1: (a) 65.3 m/s
   (b) 40.8 Hz
   (c) 122 Hz
   (d) 163 Hz
2: (a) 578 Hz
   (b) 413
3: 5200 m/s
4: (a) 2500 Hz
   (b) 17.5 kHz
5: 4125 m/s
6: 11 Hz
7: 6.77 m/s away
8: $v = c\,(F - f)/(F + f)$
9: (a) 4.97 nW/m²
      1.41 nW/m²
   (b) 3.52:1
10: 76.2 dB
11: 251 μW/m²
12: 64 dB
13: 70 dB
14: 105 dB
15: 14.8 dB
16: −11.6 dB
17: 35.6 dB
18: (a) 1470 s
   (b) 0.51 s
19: (a) 40 phon
   (b) 10 phon

CHAPTER 19

3: −2.03 μC
4: 3.1×10^{13}
5: 3.4×10^{19}
6: 26.7 μA
9: 19.2 mN
10: 1/9
11: 36 F
12: 45 mN,   143°
13: 39 nC
14: (a) 1.6×10^{-26} N
   (b) 1.3×10^{-62} N
   (c) 1.2×10^{36}

15: -4.8×10^{-15} N

16: (a) 1.2×10^7 m/s

    (b) 3.4 ns

17: 10.8×10^6 N/C

18: 4.2 cm

19: 1/9

20: (a) 4.5×10^7 N/C

    $37°$

    (b) 7.2×10^{-12} N

    $217°$

23: 4.8×10^{-17} J

24: 4.8×10^{-15} J

25: 450 V

26: -18 kV

27: (a) 33 kV

    (b) 8.82 cm,   21.4 cm

28: (a) 50.9 kV

    (b) 1.2 kV

29: (a) 12 kV

    (b) 1.0 m from Q_1

30: (a) 4.5 kN/C towards

    negative charge

    (b) -450 V

31: 2.53 μJ

33: 10.4 cm

34: 8.89 kV/m

35: (a) 1.28×10^{-15} N

    (b) 1.4×10^{15} m/s^2

    (c) 3.8 ns, 5.3×10^6 m/s

Review Problems

1: 1.4×10^{21}

2: 96 mN

3: 4.0×10^{-16} N

4: (a) 5.6×10^{-16} N

    (b) 7.44×10^6 m/s

    (c) 12.1 ns

5: (a) 7.8×10^7 N/C

    $59.7°$

    (b) 1.25×10^{-11} N

6: 12 kV

7: (a) 237 kV

    (b) 1.52×10^6 V/m to Q_2

    (c) 2.43×10^{-13} N from Q_2

8: 480 mJ

9: (a) 1.92×10^{-18} J

    (b) 2.05×10^6 m/s

10: 1.1 mJ

11: 8.0×10^{-19} J

    1.33×10^6 m/s

12: (a) 8.9 kV/m

    (b) -1.42×10^{-15} N

    (c) 5.12×10^{-17} J

13: 100 kV/m

14: (a) 15 kV/m

    (b) 2.4×10^{-15} N

    (c) 1.03×10^7 m/s

15: -4.0×10^{-13} N

16: (a) 6.4×10^{-15} N

    (b) 1.67×10^7 m/s

17: (a) 450 kV

    (b) 0.90 J

CHAPTER 20

1: 4.0 mA

2: 40 mA

3: (a) 20 mA

    (b) 40 mA

4: 9.17 Ω

5: 17.7 Ω

6: 114 V

7: 300 mA

8: 13.5 mV

9: 50 Ω

10: (a) 43.4 Ω

    (b) 52.1 V

11: (a) 40 Ω

    (b) 400 mA

12: 2.51×10^{-8} $\Omega \cdot$m

13: 8.5 Ω

14: (a) 26.4 Ω

    (b) 26.6 Ω

15: 148 m

16: 0.54 mm

17: (a) 4.2×10^{-8} $\Omega \cdot$m

    (b) 1.2×10^{-8} $\Omega \cdot$m

18: 276°C

19: (a) 524 Ω
     (b) 577 Ω
20: 4.5 Ω
21: 0.325 Ω
22: 22.8 m
23: 1020°C
24: 55 V
25: (a) 8.1 W
     (b) 2.2 W
26: 15 V
27: 27 mW
28: 500 mA
29: 100 A
30: 13 A
31: 3.1 A
32: (a) 115 W
     (b) 2.8¢
33: 131 s
34: $3.50
35: (a) 40 Ω
     (b) 3.2 Ω
     (c) 23.75 Ω
36: (a) 2.0 A
     (b) 28 W, 32 W
        48 W
     (c) 108 W
     (d) 14 V, 16V
        24 V
37: (a) 1.15 Ω
     (b) I_1 = 1.0 A
        I_2 = I_4 = 600 mA
        I_3 = 400 mA
     (c) 7.8 W
     (d) P_1 = 3 W
        P_2 = P_4 = 1.8 W
        P_3 = 1.2 W
38: 5.1 Ω
39: (a) 23.4 Ω
     (b) 31.4 Ω

Review Problems
1: 350 μA
2: 5.5 V
3: (a) 255 μΩ
     (b) 25.5 mV
4: 1.4 × 10⁻⁸ Ω·m

5: 67.5 cm
6: (a) 7.8 mΩ
     (b) 3.4 × 10⁻⁸ Ω·m
7: 972°C
8: 0.34 Ω
9: 400 W
10: 13.5 m
11: (a) 144 mW
     (b) 255 mW
12: 14.5 A
13: $7.58
14: (a) 790 Ω
     (b) 60 Ω
     (c) 190 Ω
15: 35.2 V
16: (a) 13.7 kΩ
     (b) I_1 = I_2 = 1.52 mA, I_3 = 5.08 mA
        I_4 = I_5 = 2.18 mA
     (c) U_1 = 30.4 V, U_2 = 45.6 V, U_3 =
        76 V
        U_4 = 21.7 V, U_5 = 54.3 V, U_6 =
        44 V
     (d) P_1 = 46.2 mW, P_2 = 69.3 mW, P_3
        = 387 mW
        P_4 = 47.5 mW, P_5 = 119 mW, P_6
        = 385 mW

CHAPTER 21

1: 1.5 V
2: 45 μW
3: 24.2 g
4: 0.35 mm
5: 0.57¢
6: 8.925 V
7: 400 mA
9: (a) 10 Ω
     (b) 8.8 V
     (c) 900 mA
10: (a) 361 mA
     (b) 10.8 V
11: 8.0 A
12: 1/3 A, 2/3 A, 1/3 A
13: (a) 810 mA, 510 mA, 132 A
     (b) 3.96 V

14: (a) 88 mA, 17 mA, 105 mA
    (b) 10.5 V
    (c) 1.72 W, 162 mW, 1.10 W
15: 2.0 Ω
16: (a) 499 940 Ω
    (b) Series with G
    (c) 5.0 kΩ/V
17: (a) 1.0 Ω shunt
    (b) 13 941 Ω multiplier
18: 35 Ω: 45 Ω voltage divider
19: 1.522 V
20: 29.9 Ω

Review Problems
1: 1.44×10^{-18} J
2: 3.75 μW
3: 5.90 V
4: 400 mΩ
5: (a) 584 mA
    (b) 8.77 V
6: (a) 20 Ω
    (b) 11 V
    (c) 600 mA
7: (a) 100 Ω
    (b) 10 V
    (c) 120 mA
8: (a) 4.0 A, 1.0 A, 3.0 A
    (b) 12 V
    (c) 48 W, 54 W, 13 W
9: (a) 5.32 mA
       8.30 mA
       2.98 mA
    (b) 0.83 V
    (c) 6.23 mW
       6.89 mW
       4.97 mV
10: (a) 29 750 Ω multiplier
    (b) 27.8 Ω shunt
11: (a) 250 Ω multiplier
    (b) 10.4 Ω shunt
12: (a) 15 Ω
    (b) 11.22 Ω

CHAPTER 22

1: 2.4 mC
2: 3.0 pF

3: 7.2 cm
4: 5.0 μF
5: 20 V
6: (a) 6.6 pF
    (b) 330 pC
7: 5.3 mm
8: (a) 2.7
    (b) 2.4×10^{-11} F/m
9: (a) 37.7 pF
    (b) 1.13 nC
    (c) 28.2 kV
10: (a) 750 nC
    (b) 3.0
    (c) 2.7×10^{-11} F/m
11: 1280 kV
12: (a) 20.0 μF
    (b) 3.75 μF
13: (a) 22.0 pF
    (b) 2.0 pF
    (c) 9.0 pF
14: (a) 120 μF
    (b) 6.25 μF
    (c) 33 1/3 μF
    (d) 28.1 μF
15: $Q_1 = Q_2 = Q_3 = 833$ μC
    $Q_4 = 2.5$ mC
    $U_1 = 27.8$ V
    $U_2 = 16.7$ V
    $U_3 = 55.5$ V
    $U_4 = 100$ V
16: $Q_1 = Q_2 = 560$ μC
    $Q_3 = Q_4 = 280$ μC
    $U_1 = U_3 = 18.75$ V
    $U_2 = U_4 = 11.25$ V
17: (a) 4 in parallel
    (b) 3 in series
    (c) 3 parallel + 2 series
18: 324 pF
    1.6 kV
19: (a) 0.381 mm
    (b) 1.82 cm²
20: 43 mJ
21: (a) 4.18×10^{-7} J
    (b) 7.27 nC
22: (a) 6.0×10^{-8} J
    (b) 3.3 nC

23: (a) 1.5 nC
　　(b) 71 V
　　(c) 54 nJ
24: 1.4 J
25: 720 mJ
26: 160 nJ
27: 99.3%
28: $6.74 \times 10^{-3} Q_0$
29: (a) 60 ms
　　(b) 208 mA
　　(c) 238 V
30: (a) 1.91 s
　　(b) 7.0 mA
31: (a) 25 ms
　　(b) 1.46 μA
32: (a) 45 mJ
　　(b) 2.2 s
　　(c) 2.42 s
33: (a) 7.33 mA
　　(b) 992 μA

Review Problems

1: 1.6 V
2: (a) 5.16 pF
　　(b) 129 pC
3: (a) 4.0
　　(b) 3.54×10^{-11} F/m
4: 1600 kV
5: 727 pF
6: (a) 254 μm
　　(b) 0.71 cm²
7: (a) 15.6 μF
　　(b) $U_1 = U_5 = 44 \ 1/3$ V
　　　$U_3 = 66 \ 2/3$ V
　　　$U_1 = U_4 = 22 \ 1/3$ V
　　　$U_6 = 233.4$ V
8: 67.5 μJ
9: 270 mJ
10: (a) 315 pF
　　(b) 7.72×10^{-7} J
　　(c) 22 nC
11: (a) 44 ms
　　(b) 54.5 mA
　　(c) 2.28 mC
　　(d) 2.72 mA
　　(e) 114 V

CHAPTER 23

3: 3.36×10^{-15} N
4: 4.62 mT
5: 2.0×10^{-16} N up
6: 1.25×10^{-17} N east
7: 0.79 T
8: (a) 3.4×10^7 m/s
　　(b) 9.9×10^{-13} J
9: 33 kV
10: (a) 4.0×10^6 m/s
　　(b) 6.67×10^{-27} kg
　　　2.34×10^{-26} kg
　　　1.80×10^{-26} kg
11: 1.25 A
12: 50 μT
13: 5.6 A
14: (a) 1.8 mT
　　(b) 620 μT
15: 1.6 A
16: (a) 24 mT
　　(b) 286 mT
17: 3.3×10^{-16} N
18: 6.53 A
19: (a) none
　　(b) into earth
　　(c) none
　　(d) towards west
20: 2.6 mN
21: 1.5 μN east
22: 400 mA
23: 20 mN attractive
24: 21.2 mN
25: 3.6 A

Review Problems

1: 2.5×10^{-17} N west
2: 1.0×10^{-17} east
3: (a) 7.45×10^7 m/s
　　(b) 1.86×10^{-11} J
4: 2.1361 cm
5: (a) 93.6 kV
　　(b) 9.34×10^{-28} kg
　　　1.0×10^{-27} kg
6: 2.51×10^{-5} T

7: (a) 1.13 mT
(b) 5.79×10^{-5} T
8: 1.5×10^{-5} T·m/A
9: 290 mT
10: 1.37 N/m
11: 13.9 μN repulsive
12: 5.29×10^{-4} N·m
13: 15.6 N·m

(c) 2.0×10^6 A/m
(d) 2 499
7: (a) 339 mT
(b) 180 A/m
(c) 1.36×10^{-4} Wb
(d) 1.88×10^{-3} T·m/A
(e) 2.7×10^5 A/m

CHAPTER 24

1: 9.3×10^{-24} A·m²
2: 2.8×10^{-24} N·m
3: (a) 6.660×10^{-3} T·m/A
(b) 0.125 66 T·m/A
4: 151
5: (a) 2.30×10^{-4} T·m/A
(b) 183
(c) 4.12×10^3 A/m
(d) 182
9: 143
12: (a) 1.55 T
(b) 6.3×10^{-3} T·m/A
(c) 5013
(d) 7.2×10^5 A/m
14: (a) 1.9 mT
(b) 2.3 T

Review Problems
1: 1.25×10^{-22} A·m²
2: (a) 135
(b) 39 494 A/m
(c) 134
3: (a) 120
(b) 34 708 A/m
(c) 119
4: (a) 16.3 mT
(b) 108 A/m
(c) 12 900 A/m
5: (a) 2.3×10^{-4} T·mA
(b) 183
(c) 4 115 A/m
(d) 182
6: (a) 2.51 T
(b) 800 A/m

CHAPTER 25

1: 50 V
2: 2.4×10^{-5} Wb
3: 16.3 V
4: (a) -5.0×10^{-5} Wb
(b) -6.4 mT
5: 150 mV
6: (a) 540 mV
(b) 3.6 V/m
(c) 60 mA
7: 497 rpm
8: (a) 3180 V
(b) 2.57 kV
(c) 1.27 kA
(d) 1.03 kA
(e) 0 V
9: (a) 40.0 A
(b) 88 V
(c) 7.25 Ω
10: (a) 5.0 A
(b) 2.0 A
(c) 840 W
(d) 550 W
11: (a) 40.0 A
(b) (220/111)Ω
(c) 500 mA
(d) 40.5 A
(e) 191 V
(f) 57%
12: 1.1%
13: 1176 rpm
14: 6.0 H
15: (a) -55 V
(b) 3.3×10^{-2} Wb
16: 30 Wb

17: (a) 750 mH

(b) 120 mWb

18: 17.5 ms

19: 65 mH

20: 159

21: 195

22: (a) 10.0 H

(b) (30/31) H

(c) 2.5 H

23: 0.16 V

24: (a) -500 A/s

(b) -300 A/s

-200 A/s

25: (a) 4 in series

(b) 4 in parallel

(c) 1 series with

2 parallel

26: (a) 6.0 A

(b) 5.0 ms

(c) 5.7 A

-2.99 V

60 A/s

27: 680 mA

-135 A/s

28: (a) 6.00 mH

(b) 3.00 A

(c) 500 μs

(d) 2.95 A

110 A/s

29: 3.0 J

30: (a) 9.0 J

(b) 300 W

31: 6.6 mJ

32: (a) 5.0 ms

(b) 38 mA

(c) 720 μJ

33: (a) 6.25 A

(b) 2.4 MW

34: (a) 1:11

(b) 2.27 A

35: (a) 60 V

(b) 10 A

36: 5.5 V

Review Problems

1: 200 V

2: -18 μWb

3: -45.2 V

4: (a) 52.2 mV

(b) 10.4 mA

5: 196 V

6: (a) 38.3 A

(b) 91 V

(c) 7.0 Ω

7: (a) 200 mA

(b) 24 W

(c) 22 W

(d) 91.7%

8: (a) 5.0 A

(b) 1.375 A

(c) 701.25 W

(d) 500 W

9: (a) 5.0 A

(b) 1.05 A

(c) 695 W

(d) 525 W

(e) 76%

10: 1.63 H

11: (a) -8.0 V

(b) 1.33 mWb

(c) 40 mWb

12: (a) 60 mH

(b) 9.0 mWb

13: 250 mJ

14: (a) 170 mH

(b) 16 mH

(c) 24.7 mH

(d) -0.85 V

15: (a) 492 μH

(b) 4.92 μs

(c) 0 A

(d) 285 mA

16: (a) 108 mA

(b) 2.17 V

17: (a) 166 μs

(b) 211 mA

(c) 1.78 mJ

18: (a) 6.0 mH

(b) 3.0 A

(c) 500 μs

(d) 2.95 A

(e) 26 mJ

19: 1:20

20: (a) 1:160

(b) 1.375 V

21: 2%

22: 1755 rpm

23: 5.08 N·m

24: −47.8 V

25: (a) 1325 V

(b) 13.75 A

(c) 5.50 kW

26: (a) 79 V

(b) 27.4 %

27: (a) 32 mH

(b) 320 μs

(c) 0 A

(d) 285 mA

(e) 7.2 V

(f) 1.0 mJ

CHAPTER 26

1: 25.5 A

2: 311 V

3: (a) 7.5 V

(b) 14.8 V

(c) 7.5 V

(d) −14.1 V

4: 80.3 mA

5: (a) 7.07 A(b) 600 W

(c) 4.84 A

116 V

6: (a) 8.48 V

(b) 10.4 V

(c) 9.7 V

7: 168 kW

8: 84.9 V

9: 8 000 kW·h

10: 628 Ω

11: 159 mH

12: 25 kΩ

13: 318 kHz

14: 584 mA

825 mA

15: 1.77 kΩ

16: (a) 2.2 kΩ

(b) 1.2 μF

17: 36.2 Hz

18: 2.2 μF

19: 622 μA

880 μA

20: (a) 50 V

(b) 53.1°

21: (a) 130 V

(b) −22.6°

22: (a) 640 Ω

(b) 79°

(c) 47 mA

23: (a) 471 Ω

(b) 168 Ω

24: (a) 673 kΩ

(b) −45.7°

(c) 370 μA

25: (a) 606 Ω

(b) 68.7°

(c) 363 mA

26: (a) 396 Ω

(b) 99 V

(c) −24.8°

27: 35 A

28: (a) 58%

(b) 870 mA

(c) 60.5 W

29: (a) 63.7

(b) 222

(c) 99.1%

(d) 991 mA

(e) 216 W

(f) 93.4 V

63.1 V

218 V

30: 1.77 MHz

31: 140 μH

32: 22.5 pF

33: (a) 25 kW

(b) 160 W

(c) 12.5:1

34: (a) 30.0 kV

(b) 6.00 kW

(c) 294 kW

(d) 19.6 A

(e) 2:1

35: (a) 11.3 MW

(b) 4.5 kW

(c) 5:1

36: (a) 10 kV

(b) 40 A

(c) 406.5 kW

(d) 51 kV

37: 6.3 km/s

38: 12.7 μm/s

Review Problems

1: (a) 170 V

(b) 830 mA

(c) 1.18 A

2: (a) 11.3 A

(b) 960 W

(c) 9.15 A

(d) 137 V

3: (a) 12.7 V

(b) 5.6 V

4: 167 kV

5: (a) 2.88 kW·h

(b) 17.3$^{\text{¢}}$

6: 30.2 Ω

7: 106 mH

8: 183 Ω

9: 16 Ω

10: (a) 400 Ω

(b) 6.6 μF

11: 155 Ω

12: (a) 30 V

(b) 60°

13: (a) 66.3 Ω

(b) 27°

(c) 226 mA

14: (a) 600 Ω

(b) 22.6 Ω

(c) 599.6 Ω

15: (a) 713 Ω

(b) −38.3°

(c) 42.1 μA

16: (a) 792 Ω

379 Ω

(b) 589 Ω

(c) 44.5°

(d) 118 V

17: (a) 8.68 A

(b) 6.25 A

18: (a) 306 Ω

(b) 11.3°

(c) 392 mA

(d) 98.1%

(e) 46.1 W

19: 2.05 MHz

20: 203 mH

21: 150.004 44 MHz

22: (a) 6.0 MW

(b) 30 MW

(c) 4.0 kA

CHAPTER 27

5: 0.014 sr

6: 2.0×10^{-3} sr

7: 250 cd

8: 250 cp

9: (a) 1260 lm

(b) 754 cd

10: 1005 cd

11: 10 000 cd

12: 74 lm/W

13: 18 1m/W

14: 10 fc

15: 320 lm

16: (a) 3.13 fc

(b) 1.6 fc

17: (a) 15.9 lx

(b) 14.0 lx

18: 55.6 lx

19: 5650 fL

20: 7.5%

90%

21: 160 cd

22: (a) 108 cd

(b) 1360 lm

Review Problems

1: 4.0×10^{-5} sr

2: 294 cd

3: 206 cp

4: (a) 302 lm

(b) 232 cd

5: 1.56×10^{6} cd

6: 2200 cd

7: (a) 79.6 cd

(b) 4.0 ft

8: (a) 477 cd
(b) 120 lx
9: (a) 130 cd
(b) 22.9 fc
10: (a) 215 cd
(b) 4.23 ft
11: 20 lx
12: 270 lm
13: (a) 20 lx
(b) 4.32 lx
14: (a) 14.6 fc
(b) 14.3 fc
15: 32 lx
16: 3.78 fc
17: 1.15 m
18: 1320 fL
19: 40 ft

CHAPTER 28

3: 3.0 ft
4: 5.24 cm
5: (a) −10 cm
5 cm
(b) virtual
enlarged
upright
6: −6.0 cm
0.40 cm
7: −27 cm
8: 2
9: 14.5 cm
10: (a) 10 in
(b) −80 in
11: (a) 90 cm
(b) 112.5 cm
(c) real
inverted
enlarged
12: (a) 66 2/3 cm, −2.0 cm
(b) 80 cm, −3.0 cm
(c) 120 cm, −6.0 cm
(d) ∞, −∞
(e) −40 cm, 6.0 cm

13: −3.75 cm
2.5
14: 3.0 ft

Review Problems
4: 7.5 in, −2.5 in
5: (a) −24 cm
(b) +3, 9.0 in
6: (b) 7.5 cm
−2.5 cm
(c) Real
7: 9 cm
−0.5 cm
8: 5.14 cm
9: −5.45 cm
0.91 cm
10: (a) −30 cm
(b) −12 cm
1.5 cm

CHAPTER 29

1: (a) 22.1°
(b) 2.26×10^8 m/s
2: (a) 5.09×10^{14} Hz
(b) 373 nm
3: (a) 1.40
(b) 2.14×10^8 m/s
4: 387 nm
5: 24.2°
6: 34.2°
7: 19° 57′ 10″
8: 1.6 ft
9: 1.56
10: (a) 24.4°
(b) 62.5°
11: 6.84 ft
12: 8.3 in
13: −25 cm
14: −7.0 cm
15: −21.2 cm
−4.72 diopters
−48 cm
16: −48 cm
17: 36 cm, −4.5 cm

18: −41 mm

8.24 mm

19: 10.8 in

20: −20 cm

−5.0 cm

21: −16.7 cm

2.0 cm

22: (a) −90 mm

(b) 180 mm

23: (a) 12.0 in

(b) 6.0 in

24: -5.0×10^{-4} m

25: $f/16$

26: 1.5

27: −2/3 diopter

28: 28 cm, 3.6 diopters

29: (a) 2.27 cm

(b) 11.0

30: −60 in

1.0 in

31: (a) 147 mm

(b) −140

32: (a) 4/3 cm

(b) 201 1/3 cm

33: (a) 4.0

(b) 150 mm

Review Problems

1: (a) 4.74×10^{14} Hz

(b) 476 nm

2: 80 cm

3: (a) 38.7°

(b) 62.5°

4: 48.8°

5: 2.61 ft

6: (a) 2.0×10^8 m/s

(b) 19.5°

7: 30 cm, 5cm, real

8: 5.0 diopters

9: (a) diverging

(c) -7.1 in, $\dfrac{10}{7}$ in

(d) virtual

10: 8.4 cm

11: $f/5.6$

12: −0.40 diopter

13: 3.67 diopters

14: (a) 1.68 cm

(b) 14.9

15: −60 in, 6 in

16: −60

17: −10 cm from 2nd, −18 cm

18: (a) 12 cm from 2nd

(b) 2/3

19: (a) 0.62 cm

(b) −35.9

20: 22 cm

CHAPTER 30

5: (a) 500 nm

6.00×10^{14} Hz

(b) 1.5 mm

6: (a) 0.60 mm

(b) 1.80 mm

(c) 0.90 mm

7: (a) 394 nm

(b) 23.2°

8: 17.5°, 36.9°, 64.2°

9: 4970/cm

10: 18.4°, 39.1°

11: 2.0 mm

12: 9.4 in.

13: 4.03 μm

14: (a) 6.3 μm

(b) 1.0 μm

15: 95.6 nm

17: (a) 0.75

(b) 0.41

(c) 0.067

Review Problems

1: (a) 3.8 mm

(b) 2.85 mm

2: 21.6 cm

3: (a) 48.6°, (b) 340 cm

4: (a) 560 nm, (b) 35°

5: (a) 20 cm, (b) 25 cm

6: 29.8 μm

7: 116 nm

CHAPTER 31

1: (a) 656 nm
    486 nm
    434 nm
  (b) 1.88 μm
    1.28 μm
    1.09 μm
  (c) 122 nm
    103 nm
    97 nm

3: 7.95×10^{15}

4: (a) 8.52×10^{-4} Hz
  (b) 352 nm

5: (a) 5.68×10^{-19} J
  (b) 2.27×10^{-19} J

6: (a) 5.32×10^{14} Hz
  (b) 3.53×10^{-19} J
  (c) 564 nm
  (d) 1.11 V

7: 361 nm

8: (a) 6.63×10^{-34} J$\cdot$s
  (b) 5.50×10^{14} Hz
    546 nm
  (c) 2.3 eV

9: (a) 97.2 nm
  (b) 95.0 nm
  (c) 93.7 nm
  (d) 434 nm
  (e) 410 nm
  (f) 1.09 μm

10: 103 nm
    122 nm
    661 nm

11: 24.6 nm

12: 3.58×10^{-19} J

13: 2.05×10^{14} Hz

14: (a) $l = 0, 1$;
    $m = -1, 0, 1$
    $s = \pm\frac{1}{2}, -\frac{1}{2}$
  (b) $l = 0, 1, 2$
    $m = -2, -1, 0, 1, 2$
    $s = \frac{1}{2}$ or $-\frac{1}{2}$

15: (a) K, s
  (b) M, d
  (c) N, p
  (d) N, s

16: (a) 2
  (b) 8
  (c) 18

17: (a) 4
  (b) 5
  (c) 8
  (d) 7
  (e) 3
  (f) 5

20: 1.46×10^{-10} m

21: 1.94×10^{-11} m

22: 2.65×10^{-14} kg$\cdot$m/s

Review Problems

1: (a) 5.84×10^{14} Hz
  (b) 2.42 eV
  (c) 514 nm
  (d) 1.39 V

2: (a) 6.33×10^{14} Hz
  (b) 2.63 eV
  (c) 474 nm
  (d) 0.83 V

3: (a) 5.84×10^{14} Hz
  (b) 8.64×10^{-20} J
  (c) 0.54 V

4: 380 nm

5: 2.4×10^{16}

6: 2.86×10^{16}

7: 6.36×10^{16}

8: 1.09×10^6 m/s

9: 658 nm

10: 10.7 nm

11: $l = 0, 1, 2, 3$
    $m = -3, -2, -1, 0, 1, 1, 2, 3$
    $s = +\frac{1}{2}, -\frac{1}{2}$

12: 2.91×10^{-10} m

13: 1.32×10^{-12} m

14: (a) 3.33×10^{-10} m
  (b) 1.0×10^{-9} m

15: 2.6×10^{-16} m

CHAPTER 32

4: 24.8 pm

5: 41.4 kV

6: (a) 618 pm
(b) 14.5°
7: 109 pm

Review Problems
1: 15.5 pm
2: 9.01 kV
3: 8.4 kV
4: 77.6 pm
5: 1.9°
3.84°

CHAPTER 34

1: 2.0×10^{20} Hz
1.33×10^{-13} J
2: (a) 2
(b) 125
(c) 142
(d) 146
(e) 145
3: (a) $_{83}Bi^{209}$
(b) $_{6}C^{14}$
(c) $_{27}Co^{59}$
4: (a) 7.44×10^{-15} m
(b) 3.02×10^{-15} m
(c) 2.30×10^{-15} m
(d) 1.90×10^{-15} m
5: 7.07 MeV
6: 7.59 MeV
7: 8.52 MeV
8: (a) β^-
β^-
α
β^+
9: (a) $_{90}Th^{234}$
(b) $_{89}Ac^{228}$
(c) $_{90}Th^{230}$
(d) $_{-1}e^0$ or β^-
(e) $_{9}F^{19}$
(f) $_{2}He^4$ or α
10: 1620 a
11: 1.82×10^{-5}/s
12: 1/32
13: (a) 2.75×10^{-3} s
(b) 4.2 min

14: (a) 8.86×10^{-14}/s
(b) 4.56×10^5 Bq
(c) 5.7×10^4 Bq
15: (a) $_{6}C^{13}$
(b) $_{8}O^{16}$
(c) $_{2}He^4$
(d) $_{13}Al^{27}$
(e) $_{2}He^4$
(f) $_{0}n^1$
(g) $_{0}n^1$
16: -12.1 MeV
17: -3.13 MeV

Review Problems
1: 2.07 MeV
2: (a) 14
(b) 2
(c) 9
(d) 143
(e) 130
3: (a) $_{50}Sn^{130}$
(b) $_{90}Th^{232}$
(c) $_{83}Bi^{209}$
4: (a) 2.18×10^{-15} m
(b) 3.26×10^{-15} m
5: 7.68 MeV
6: 8.19 MeV
7: 7.61 MeV
8: (a) α
(b) β^-
(c) β^+
9: (a) $_{12}Mg^{24}$
(b) α
10: 3.63×10^{-6}/s
11: 72 s
12: (a) 25%
(b) 6.25%
13: (a) 4.4×10^{-3}/s
(b) 3.9×10^{16} Bq
(c) 1.9×10^{14} Bq
14: (a) 3.86×10^{-12}/s
(b) 1.0 μg
15: 4.36 MeV
16: 3.31 MeV
17: (a) $_{92}U^{235} + _{0}n^1 \rightarrow _{38}St^{93} + _{54}Xe^{139} + 4_{0}n^1$
(b) 173 MeV

(Tables are in *italics*; for topics with multiple page references, the most important reference is in **boldface**.)

Index

Work function, 535
Worm gear, 158

X-rays, 557
   Bremsstrahlung, 558
   characteristic, 558
   diffraction, 559
   spectra, 558

Year, 7
Young's double slit, 517
Young's modulus, *61*, **62**, 275

Zener diode, 582
Zener effect, 582
Zonal cavity, 469
Zone refining, 575